Infrared and Raman Spectra of Inorganic and Coordination Compounds

5C

Infrared and Raman Spectra of Inorganic and Coordination Compounds

FOURTH EDITION

KAZUO NAKAMOTO

Wehr Professor of Chemistry
Marquette University

A Wiley-Interscience Publication
JOHN WILEY & SONS
New York • Chichester • Brisbane • Toronto • Singapore

Library of Congress Cataloging in Publication Data:

Nakamoto, Kazuo, 1922–
Infrared and Raman spectra of inorganic and coordination compounds.

"A Wiley-Interscience publication."
Bibliography: p.
Includes index.
1. Infrared spectroscopy. 2. Raman spectroscopy.
I. Title.

QD96.I5N33 1986 543'.08583 86-1345
ISBN 0-471-01066-9

Printed in the United States of America

10 9 8 7 6 5 4 3 2

Preface

The first edition of this book entitled *Infrared Spectra of Inorganic and Coordination Compounds* was published in 1963. Since then, the book was revised in 1970 (second edition) and in 1978 (third edition) to keep up with ever-increasing new literature. The third edition has now become outdated due to rapid progress in inorganic vibrational spectroscopy. The preparation of the fourth edition was begun in the spring of 1982, and it was completed during the summer of 1985.

As I emphasized in the prefaces of the previous editions, "this book is intended to describe fundamental theories of vibrational spectroscopy in a condensed form (Part I) and to illustrate their applications to inorganic (Part II), coordination (Part III), and organometallic compounds (Part IV) using typical examples." In the fourth edition, all of these parts have been updated by adding new references while omitting those shown to be erroneous by later sudies. Part V (Bioinorganic Compounds) has been added to give a glimpse of the current status of bioinorganic vibrational spectroscopy. As in the past editions, I have tried to give a broad and balanced coverage of the field. It was clearly impossible, however, to include all significant work in the limited space available. I only hope that the review articles and reference books which are abundantly quoted throughout this work will compensate for any unbalanced presentation on my part.

I wish to express my sincere thanks to all who helped me in preparing this volume. I am particularly indebted to Professors D. P. Strommen (Carthage College), J. R. Kincaid (Marquette University), D. F. Shriver (Northwestern University), T. Kitagawa (Institute for Molecular Science), T. G. Spiro (Princeton University), T. M. Loehr (Oregon Graduate Center), E. Maslowsky (Loras College), and Dr. R. Czernuszewicz (Princeton University). Thanks are also due to Drs. N. Blom and T. Isobe and Messrs. L. Proniewicz and

A. Bruha of Marquette University, who proofread the reference sections, and to the publishers and colleagues who gave me permission to reproduce some figures in this book. Finally, I would like to thank the National Science Foundation, which provided continuing support for my research during the preparation of this edition.

KAZUO NAKAMOTO

Milwaukee, Wisconsin
May 1986

Contents

Part II. Inorganic Compounds

Part III. Coordination Compounds

Abbreviations

IR, infrared; R, Raman; RR, resonance Raman; *p*, polarized; *dp*, depolarized; *ip*, inverse polarization.

ν, stretching; δ, in-plane bending or deformation; ρ_w, wagging; ρ_r, rocking; ρ_t, twisting; π, out-of-plane bending. Subscripts *a*, *s*, and *d* denote antisymmetric, symmetric, and degenerate modes, respectively. Approximate normal modes of vibration corresponding to these vibrations are given in Figs. III-2 and III-19.

GVF, generalized valence force field; UBF, Urey–Bradley force field.

M, metal; L, ligand; X, halogen; R, alkyl group or cyclopentadienyl (Cp) or other ring compound.

g, gas; l, liquid; s, solid; m or mat, matrix; sol'n or sl, solution.

Me, methyl; Et, ethyl; Bu, butyl; OAc, acetate ion. Abbreviations of the ligands are given when they appear in the text.

In the tables of observed frequencies given in Parts II–IV, values in parentheses are calculated or estimated unless otherwise stated.

Theory of Normal Vibration

Part I

I-1. ORIGIN OF MOLECULAR SPECTRA

As a first approximation, it is possible to separate the energy of a molecule into three additive components associated with (1) the rotation of the molecule as a whole, (2) the vibrations of the constituent atoms, and (3) the motion of the electrons in the molecule.* The translational energy of the molecule may be ignored in this discussion. The basis for this separation lies in the fact that the velocity of electrons is much greater than the vibrational velocity of nuclei, which is again much greater than the velocity of molecular rotation. If a molecule is placed in an electromagnetic field (e.g., light), a transfer of energy from the field to the molecule will occur only when Bohr's frequency condition is satisfied:

$$\Delta E = h\nu \tag{1.1}$$

where ΔE is the difference in energy between two quantized states, h is Planck's constant, and ν is the frequency of the light.† If

$$\Delta E = E'' - E' \tag{1.2}$$

where E'' is a quantized state of higher energy than E', the molecule *absorbs* radiation when it is excited from E' to E'' and *emits* radiation of the same frequency as given by Eq. 1.1 when it reverts from E'' to E'.

Because rotational levels are relatively close to each other, transitions between these levels occur at low frequencies (long wavelengths). In fact, pure rotational spectra appear in the range between 1 cm^{-1} (10^4 μm) and 10^2 cm^{-1} (10^2 μm). The separation of vibrational energy levels is greater, and the transitions occur at higher frequencies (shorter wavelengths) than do the rotational transitions. As a result, pure vibrational spectra are observed in the range between 10^2 cm^{-1} (10^2 μm) and 10^4 cm^{-1} (1 μm). Finally, electronic

---

* Hereafter the word *molecule* may also represent an *ion.*

† The frequency, ν, is converted to the wave number, $\tilde{\nu}$, or the wavelength, λ_ω, through the relation

$$\nu = c\tilde{\nu} = \frac{c}{\lambda_\omega}$$

where c is the velocity of light. For theoretical discussion, ν and $\tilde{\nu}$ are more convenient than λ_ω, since they are proportional to the energy of radiation. More explicit relations between these three units are given in the following table for the region in which vibrational spectra occur.

Frequency (s^{-1})	Wave Number (cm^{-1})	Wavelength (μm)
3×10^{14}	10^4	1
3×10^{13}	10^3	10
3×10^{12}	10^2	10^2

Although the dimensions of ν and $\tilde{\nu}$ differ from one another, it is conventional to use them interchangeably. For example, a phrase such as "a frequency shift of 25 cm^{-1}" is often employed. All the spectral data in this book are given in terms of $\tilde{\nu}$ (cm^{-1}).

energy levels are usually far apart, and electronic spectra are observed in the range between 10^4 cm^{-1} (1 μm) and 10^6 cm^{-1} (10^{-2} μm). Thus pure rotational, vibrational, and electronic spectra are usually observed in the microwave and far-infrared, the infrared, and the visible and ultraviolet regions, respectively. This division into three regions, however, is to some extent arbitrary, for pure rotational spectra may appear in the near-infrared region ($1.5 \sim 0.5 \times 10^4$ cm^{-1}) if transitions to higher excited states are involved, and pure electronic transitions may appear in the near-infrared region if the levels are closely spaced.

Figure I-1 illustrates transitions of the three types mentioned for a diatomic molecule. As the figure shows, rotational intervals tend to increase as the rotational quantum number J increases, whereas vibrational intervals tend to

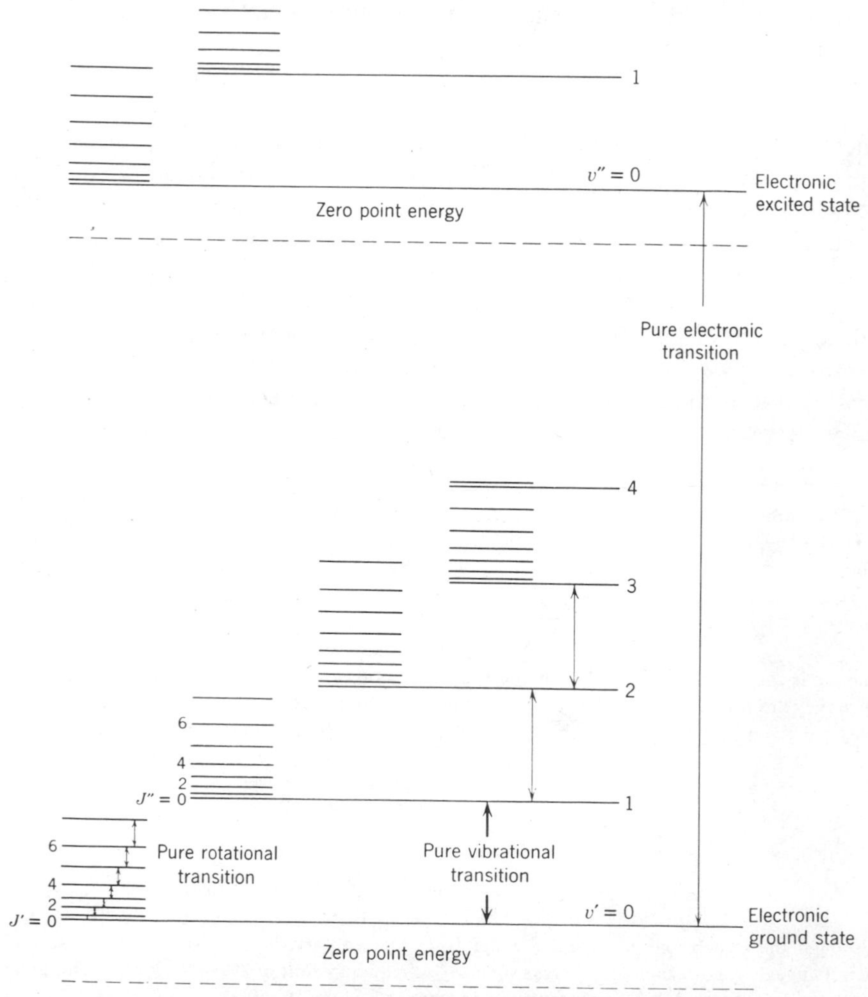

Fig. I-1. Energy levels of a diatomic molecule. (The actual spacings of electronic levels are much larger, and those of rotational levels much smaller, than those shown in the figure.)

decrease as the vibrational quantum number v increases. The dotted line below each electronic level indicates the zero point energy that exists even at a temperature of absolute zero as a result of nuclear vibration. It should be emphasized that not all transitions between these levels are possible. To see whether the transition is *allowed* or *forbidden*, the relevant selection rule must be examined. This, in turn, is determined by the symmetry of the molecule. As will be seen later, vibrational problems like those mentioned above can be solved for polyatomic molecules in an elegant manner by the use of group theory.

Since this book is concerned only with vibrational spectra, no description of electronic and rotational spectra is given. Although vibrational spectra are observed experimentally as infrared or Raman spectra, the physical origins of these two types of spectra are different. Infrared spectra originate in transitions between two vibrational levels of the molecule in the electronic ground state and are usually observed as *absorption spectra* in the infrared region. On the other hand, Raman spectra originate in the electronic polarization caused by ultraviolet or visible light. If a molecule is irradiated by monochromatic light of frequency ν, then, because of electronic polarization induced in the molecule by this incident light, light of frequency ν (*Rayleigh scattering*) as well as of $\nu \pm \nu_i$ (*Raman scattering*) is emitted (ν_i represents a vibrational frequency). Thus the vibrational frequencies are observed as Raman shifts from the incident frequency ν in the ultraviolet or visible region.

Although Raman scattering is much weaker than Rayleigh scattering (by a factor of 10^{-3} to 10^{-4}), it is possible to observe the former by using a strong exciting source. In the past, the mercury lines at 435.8 nm (22,938 cm^{-1}) and 404.7 nm (24,705 cm^{-1}) from a low-pressure mercury arc were used to observe Raman scattering. These lines, however, may be absorbed by a number of compounds which have an electronic absorption band in this region. Recently, several lasers which provide strong monochromatic beams at other frequencies have been developed. Typical laser lines are as follows: Kr^+ (647.1 nm, 15,454 cm^{-1}, red), He–Ne (632.8 nm, 15,803 cm^{-1}, red), Ar^+ (514.5 nm, 19,436 cm^{-1}, green), and Ar^+ (488.0 nm, 20,492 cm^{-1}, blue). With these and other lines, it is possible to measure Raman spectra outside the electronic absorption band. In the case of resonance Raman spectroscopy (Sec. I-21), the exciting frequency is chosen so as to fall inside the electronic band. The degree of resonance enhancement changes as a function of the exciting frequency and reaches a maximum when the frequency approximately coincides with that of the electronic absorption maximum. Recently developed tunable dye-lasers provide a valuable means of studying resonance Raman spectra.

The origin of Raman spectra can be explained by an elementary classical theory. Consider a light wave of frequency ν with an electric field strength E. Since E fluctuates at frequency ν, we can write

$$E = E_0 \cos 2\pi\nu t \tag{1.3}$$

where E_0 is the amplitude and t the time. If a diatomic molecule is irradiated by this light, the dipole moment P given by

$$P = \alpha E = \alpha E_0 \cos 2\pi\nu t \tag{1.4}$$

is induced. Here α is a proportionality constant and is called the *polarizability.* If the molecule is vibrating with frequency ν_1, the nuclear displacement q is written as

$$q = q_0 \cos 2\pi\nu_1 t \tag{1.5}$$

where q_0 is the vibrational amplitude. For small amplitudes of vibration, α is a linear function of q. Thus we can write

$$\alpha = \alpha_0 + \left(\frac{\partial\alpha}{\partial q}\right)_0 q \tag{1.6}$$

Here α_0 is the polarizability at the equilibrium position, and $(\partial\alpha/\partial q)_0$ is the rate of change of α with respect to the change in q, evaluated at the equilibrium position. If we combine Eqs. 1.4–1.6, we have

$$\begin{aligned} P &= \alpha E_0 \cos 2\pi\nu t \\ &= \alpha_0 E_0 \cos 2\pi\nu t + \left(\frac{\partial\alpha}{\partial q}\right)_0 q_0 E_0 \cos 2\pi\nu t \cos 2\pi\nu_1 t \\ &= \alpha_0 E_0 \cos 2\pi\nu t \\ &\quad + \frac{1}{2}\left(\frac{\partial\alpha}{\partial q}\right)_0 q_0 E_0 \{\cos [2\pi(\nu+\nu_1)t] + \cos [2\pi(\nu-\nu_1)t]\} \end{aligned} \tag{1.7}$$

According to classical theory, the first term describes an oscillating dipole which radiates light of frequency ν (Rayleigh scattering). The second term gives the Raman scattering of frequencies $\nu+\nu_1$ (*anti-Stokes*) and $\nu-\nu_1$ (*Stokes*). If $(\partial\alpha/\partial q)_0$ is zero, the second term vanishes. Thus the vibration is not Raman active unless the polarizability changes during the vibration.

Figure I-2 illustrates the mechanisms of *normal* and *resonance Raman* (*RR*) *scattering.* In the former, the energy of the exciting line falls far below that required to excite the first electronic transition. In the latter, the energy of the exciting line coincides with that of an electronic transition.* If the photon is absorbed and then emitted during the process, it is called *resonance fluorescence* (*RF*). Although the conceptual difference between resonance Raman scattering and resonance fluorescence is subtle, there are several experimental differences which can be used to distinguish between these two phenomena. For example, in RF spectra all lines are depolarized, whereas in RR spectra some are polarized and others are depolarized. Additionally, RR bands tend to be broad and weak compared with RF bands.[59,60]

---

* If the exciting line is close to but not inside an electronic absorption band, the process is called "preresonance Raman scattering."

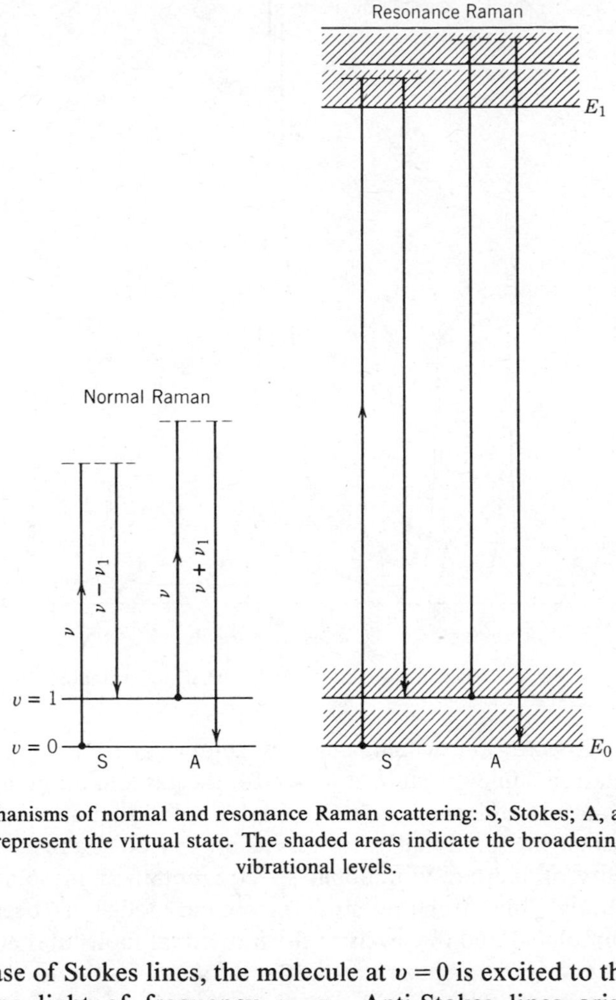

Fig. I-2. Mechanisms of normal and resonance Raman scattering: S, Stokes; A, anti-Stokes. The dashed lines represent the virtual state. The shaded areas indicate the broadening of rotational–vibrational levels.

In the case of Stokes lines, the molecule at $\upsilon = 0$ is excited to the $\upsilon = 1$ state by scattering light of frequency $\nu - \nu_1$. Anti-Stokes lines arise when the molecule initially in the $\upsilon = 1$ state scatters radiation of frequency $\nu + \nu_1$ and reverts to the $\upsilon = 0$ state. Since the population of molecules is larger at $\upsilon = 0$ than at $\upsilon = 1$ (*Maxwell–Boltzmann distribution law*), the Stokes lines are always stronger than the anti-Stokes lines. Thus it is customary to measure Stokes lines in Raman spectroscopy. Figure I-3 illustrates the Raman spectrum (below 500 cm^{-1}) of CCl_4 excited by the blue line (488.0 nm) of an argon-ion laser.

It is to be expected from Fig. I-1 that electronic spectra are very complicated because they are accompanied by vibrational as well as rotational fine structure. The rotational fine structure in the electronic spectrum can be observed if a molecule is simple and the spectrum is measured in the gaseous state under high resolution. The vibrational fine structure of the electronic spectrum is easier to observe than the rotational fine structure, and can provide structural and bonding information about molecules in electronic excited states.

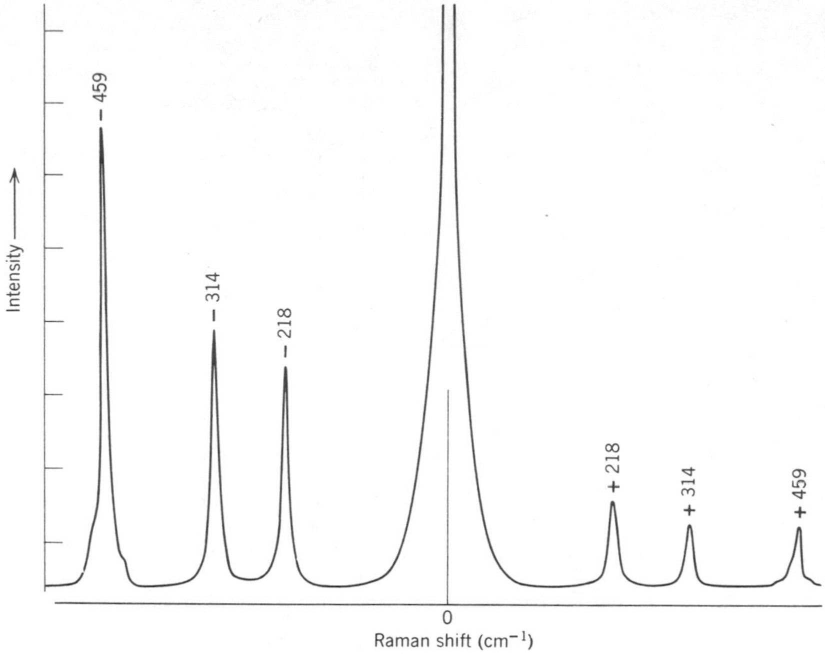

Fig. I-3. Raman spectrum of CCl_4 (488.0 nm excitation).

Vibrational spectra are accompanied by rotational transitions. Figure I-4 shows the rotational fine structure observed for the gaseous ammonia molecule. In most polyatomic molecules, however, such a rotational fine structure is not observed because the rotational levels are closely spaced as a result of relatively large moments of inertia. Vibrational spectra obtained in solution do not exhibit rotational fine structure, since molecular collisions occur before a rotation is completed and the levels of the individual molecules are perturbed differently. Since Raman spectra are often obtained in liquid state, they do not exhibit rotational fine structure.

According to the selection rule for the harmonic oscillator, any transitions corresponding to $\Delta \upsilon = \pm 1$ are allowed (Sec. I-2). Under ordinary conditions, however, only the *fundamentals* that originate in the transition from $\upsilon = 0$ to $\upsilon = 1$ in the electronic ground state can be observed because of the Maxwell–Boltzmann distribution law. In addition to the selection rule for the harmonic oscillator, another restriction results from the symmetry of the molecule (Sec. I-9). Thus the number of allowed transitions in polyatomic molecules is greatly reduced. The *overtones and combination bands** of these fundamentals are forbidden by the selection rule of the harmonic oscillator. However, they are weakly observed in the spectrum because of the anharmonicity of the vibration

* Overtones represent multiples of some fundamental, whereas combination bands arise from the sum or difference of two or more fundamentals.

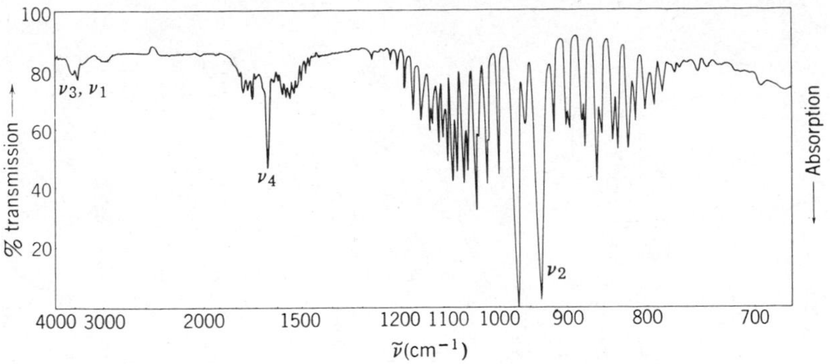

Fig. I-4. Rotational fine structure of gaseous NH_3.

(Sec. I-2). Since they are less important than the fundamentals, they will be discussed only when necessary.

I-2. VIBRATION OF A DIATOMIC MOLECULE

Through quantum mechanical considerations,[4,5] the vibration of two nuclei in a diatomic molecule can be reduced to the motion of a single particle of mass μ, whose displacement q from its equilibrium position is equal to the change of the internuclear distance. The mass μ is called the *reduced mass* and is represented by

$$\frac{1}{\mu}=\frac{1}{m_1}+\frac{1}{m_2} \tag{2.1}$$

where m_1 and m_2 are the masses of the two nuclei. The kinetic energy is then

$$T=\frac{1}{2}\mu\dot{q}^2=\frac{1}{2\mu}p^2 \tag{2.2}$$

where p is the conjugate momentum $\mu\dot{q}$. If a simple parabolic potential function such as that shown in Fig. I-5 is assumed, the system represents a *harmonic oscillator*, and the potential energy is simply given by

$$V=\tfrac{1}{2}Kq^2 \tag{2.3}$$

Here K is the force constant for the vibration. Then the Schrödinger wave equation becomes

$$\frac{d^2\psi}{dq^2}+\frac{8\pi^2\mu}{h^2}\left(E-\frac{1}{2}Kq^2\right)\psi=0 \tag{2.4}$$

If this equation is solved with the condition that ψ must be single valued, finite, and continuous, the eigenvalues are

$$E_v=h\nu(v+\tfrac{1}{2})=hc\tilde{\nu}(v+\tfrac{1}{2}) \tag{2.5}$$

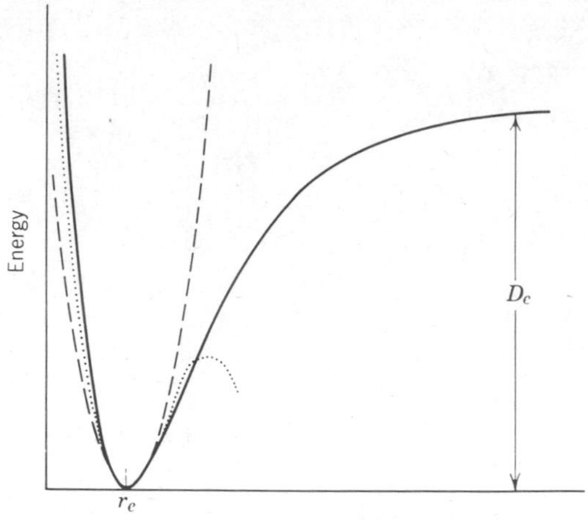

Fig. I-5. Potential curve for a diatomic molecule: actual potential (solid line), parabola (dashed line), and cubic parabola (dotted line).

with the frequency of vibration

$$\nu = \frac{1}{2\pi}\sqrt{\frac{K}{\mu}} \qquad \text{or} \qquad \tilde{\nu} = \frac{1}{2\pi c}\sqrt{\frac{K}{\mu}} \tag{2.6}$$

Here υ is the vibrational quantum number, and it can have the values 0, 1, 2, 3, ...

The corresponding eigenfunctions are

$$\psi_\upsilon = \frac{(\alpha/\pi)^{1/4}}{\sqrt{2^\upsilon \upsilon!}} e^{-\alpha q^2/2} H_\upsilon(\sqrt{\alpha}\, q) \tag{2.7}$$

where $\alpha = 2\pi\sqrt{\mu K}/h = 4\pi^2 \mu\nu/h$, and $H_\upsilon(\sqrt{\alpha}\, q)$ is a Hermite polynomial of the υth degree. Thus the eigenvalues and the corresponding eigenfunctions are

$$\begin{aligned} E_0 &= \tfrac{1}{2} h\nu, & \psi_0 &= (\alpha/\pi)^{1/4} e^{-\alpha q^2/2} \\ E_1 &= \tfrac{3}{2} h\nu, & \psi_1 &= (\alpha/\pi)^{1/4} 2^{1/2} q\, e^{-\alpha q^2/2} \\ &\vdots & &\vdots \end{aligned} \tag{2.8}$$

As Fig. I-5 shows, actual potential curves can be approximated more exactly by adding a cubic term:[2]

$$V = \tfrac{1}{2} K q^2 - G q^3 \qquad (K \gg G) \tag{2.9}$$

Then the eigenvalues are

$$E_\upsilon = hc\omega_e(\upsilon + \tfrac{1}{2}) - hcx_e\omega_e(\upsilon + \tfrac{1}{2})^2 + \cdots \tag{2.10}$$

where ω_e is the wave number corrected for *anharmonicity*, and $x_e\omega_e$ indicates the magnitude of anharmonicity. Table II-1*a* of Part II lists ω_e and $x_e\omega_e$ for a number of diatomic molecules. Equation 2.10 shows that the energy levels of the anharmonic oscillator are not equidistant, and the separation decreases slowly as υ increases. This anharmonicity is responsible for the appearance of overtones and combination vibrations, which are forbidden in the harmonic oscillator. Since the anharmonicity correction has not been made for most polyatomic molecules, in large part because of the complexity of the calculation, the frequencies given in Parts II to V are not corrected for anharmonicity (except those given in Table II-1*a*).

According to Eq. 2.6, the frequency of the vibration in a diatomic molecule is proportional to the square root of K/μ. If K is approximately the same for a series of diatomic molecules, the frequency is inversely proportional to the square root of μ. This point is illustrated by the series H_2, HD, and D_2 shown in Table II-1*a*. If μ is approximately the same for a series of diatomic molecules, the frequency is proportional to the square root of K. This point is illustrated by the series HF, HCl, HBr, and HI. These simple rules, obtained for a diatomic molecule, are helpful in understanding the vibrational spectra of polyatomic molecules.

Figure I-6 indicates the relation between the force constant K, calculated from Eq. 2.6, and the dissociation energy in a series of hydrogen halides. Evidently, the bond becomes stronger as the force constant becomes larger. It should be noted, however, that a general theoretical relation between these two quantities is difficult to derive even for a diatomic molecule. The force constant is a measure of the curvature of the potential well near the equilibrium

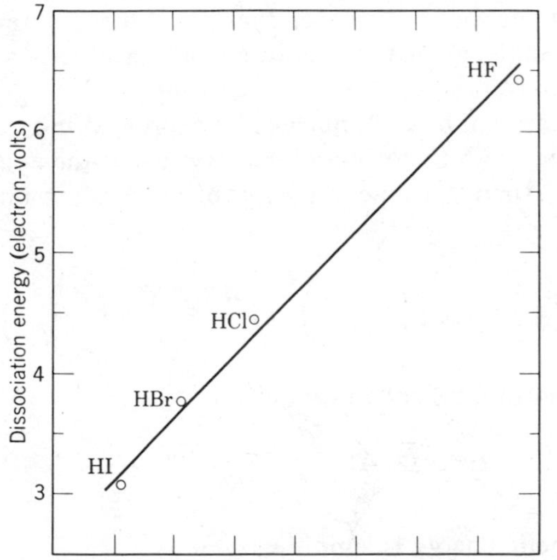

Fig. I-6. Relation between force constant and dissociation energy in hydrogen halides.

position:

$$K = \left(\frac{d^2V}{dq^2}\right)_{q\to 0} \tag{2.11}$$

whereas the dissociation energy D_e is given by the depth of the potential energy curve (Fig. I-5). Thus a large force constant means sharp curvature of the potential well near the bottom but does not necessarily indicate a deep potential well. Usually, however, a larger force constant is an indication of a stronger bond in a series of molecules belonging to the same type (Fig. I-6).

In the case of small molecules, attempts have been made to calculate the force constants by quantum mechanics. The principle of the method is to express the total electronic energy of a molecule as a function of nuclear displacements near the equilibrium position and to calculate its second derivatives, $\partial^2 V/\partial q_i^2$, and so on for each displacement coordinate q_i. Thus far, *ab initio* calculations of force constants have been made for molecules such as HF, H_2O, and NH_3.[61] The force constants thus obtained are in good agreement with those calculated from the analysis of vibrational spectra. Recent progress in computer technology has made it possible to extend this approach to more complex molecules.[61a]

I-3. NORMAL COORDINATES AND NORMAL VIBRATIONS

In diatomic molecules, the vibration of the nuclei occurs only along the line connecting two nuclei. In polyatomic molecules, however, the situation is much more complicated because all the nuclei perform their own harmonic oscillations. It can be shown, however, that any of these extremely complicated vibrations of the molecule may be represented as a superposition of a number of *normal vibrations.*

Let the displacement of each nucleus be expressed in terms of rectangular coordinate systems with the origin of each system at the equilibrium position of each nucleus. Then the kinetic energy of an N-atom molecule would be expressed as

$$T = \frac{1}{2}\sum_N m_N\left[\left(\frac{d\,\Delta x_N}{dt}\right)^2 + \left(\frac{d\,\Delta y_N}{dt}\right)^2 + \left(\frac{d\,\Delta z_N}{dt}\right)^2\right] \tag{3.1}$$

If generalized coordinates such as

$$q_1 = \sqrt{m_1}\,\Delta x_1, \qquad q_2 = \sqrt{m_1}\,\Delta y_1, \qquad q_3 = \sqrt{m_1}\,\Delta z_1, \qquad q_4 = \sqrt{m_2}\,\Delta x_2, \ldots \tag{3.2}$$

are used, the kinetic energy is simply written as

$$T = \frac{1}{2}\sum_i^{3N} \dot{q}_i^2 \tag{3.3}$$

The potential energy of the system is a complex function of all the coordinates involved. For small values of the displacements, it may be expanded in a Taylor's series as

$$V(q_1, q_2, \ldots, q_{3N}) = V_0 + \sum_{i}^{3N} \left(\frac{\partial V}{\partial q_i}\right)_0 q_i + \frac{1}{2} \sum_{i,j}^{3N} \left(\frac{\partial^2 V}{\partial q_i \, \partial q_j}\right)_0 q_i q_j + \cdots \qquad (3.4)$$

where the derivatives are evaluated at $q_i = 0$, the equilibrium position. The constant term V_0 can be taken as zero if the potential energy at $q_i = 0$ is taken as a standard. The $(\partial V/\partial q_i)_0$ terms also become zero, since V must be a minimum at $q_i = 0$. Thus V may be represented by

$$V = \frac{1}{2} \sum_{i,j}^{3N} \left(\frac{\partial^2 V}{\partial q_i \, \partial q_j}\right)_0 q_i q_j = \frac{1}{2} \sum_{i,j}^{3N} b_{ij} q_i q_j \qquad (3.5)$$

neglecting higher-order terms.

If the potential energy given by Eq. 3.5 did not include any cross products such as $q_i q_j$, the problem could be solved directly by using Newton's equation:

$$\frac{d}{dt}\left(\frac{\partial T}{\partial \dot{q}_i}\right) + \frac{\partial V}{\partial q_i} = 0, \qquad i = 1, 2, \ldots, 3N \qquad (3.6)$$

From Eqs. 3.3 and 3.5, Eq. 3.6 is written as

$$\ddot{q}_i + \sum_j b_{ij} q_j = 0, \qquad j = 1, 2, \ldots, 3N \qquad (3.7)$$

If $b_{ij} = 0$ for $i \neq j$, Eq. 3.7 becomes

$$\ddot{q}_i + b_{ii} q_i = 0 \qquad (3.8)$$

and the solution is given by

$$q_i = q_i^0 \sin(\sqrt{b_{ii}}\, t + \delta_i) \qquad (3.9)$$

where q_i^0 and δ_i are the amplitude and the phase constant, respectively.

Since, in general, this simplification is not applicable, the coordinates q_i must be transformed into a set of new coordinates Q_i through the relations

$$\begin{aligned} q_1 &= \sum_i B_{1i} Q_i \\ q_2 &= \sum_i B_{2i} Q_i \\ &\vdots \\ q_k &= \sum_i B_{ki} Q_i \end{aligned} \qquad (3.10)$$

The Q_i are called *normal coordinates* for the system. By appropriate choice of the coefficients B_{ki}, both the potential and the kinetic energies can be written as

$$T = \frac{1}{2} \sum_i \dot{Q}_i^2 \tag{3.11}$$

$$V = \frac{1}{2} \sum_i \lambda_i Q_i^2 \tag{3.12}$$

without any cross products.

If Eqs. 3.11 and 3.12 are combined with Newton's equation (3.6), there results

$$\ddot{Q}_i + \lambda_i Q_i = 0 \tag{3.13}$$

The solution of this equation is given by

$$Q_i = Q_i^0 \sin(\sqrt{\lambda_i}\, t + \delta_i) \tag{3.14}$$

and the frequency is

$$\nu_i = \frac{1}{2\pi} \sqrt{\lambda_i} \tag{3.15}$$

Such a vibration is called a *normal vibration.*

For the general N-atom molecule, it is obvious that the number of the normal vibrations is only $3N-6$, since six coordinates are required to describe the translational and rotational motion of the molecule as a whole. Linear molecules have $3N-5$ normal vibrations, as no rotational freedom exists around the molecular axis. Thus the general form of the molecular vibration is a superposition of the $3N-6$ (or $3N-5$) normal vibrations given by Eq. 3.14.

The physical meaning of the normal vibration may be demonstrated in the following way. As shown in Eq. 3.10, the original displacement coordinate is related to the normal coordinate by

$$q_k = \sum_i B_{ki} Q_i \tag{3.10}$$

Since all the normal vibrations are independent of each other, consideration may be limited to a special case in which only one normal vibration, subscripted by 1, is excited (i.e., $Q_1^0 \neq 0$, $Q_2^0 = Q_3^0 = \cdots = 0$). Then it follows from Eqs. 3.10 and 3.14 that

$$\begin{aligned} q_k &= B_{k1} Q_1 = B_{k1} Q_1^0 \sin(\sqrt{\lambda_1}\, t + \delta_1) \\ &= A_{k1} \sin(\sqrt{\lambda_1}\, t + \delta_1) \end{aligned} \tag{3.16}$$

This relation holds for all k. Thus it is seen that the excitation of one normal vibration of the system causes vibrations, given by Eq. 3.16, of all the nuclei in the system. In other words, in the normal vibration, all the nuclei move with the same frequency and in phase.

This is true for any other normal vibration. Thus Eq. 3.16 may be written in the more general form

$$q_k = A_k \sin(\sqrt{\lambda}\, t + \delta) \tag{3.17}$$

If Eq. 3.17 is combined with Eq. 3.7, there results

$$-\lambda A_k + \sum_j b_{kj} A_j = 0 \tag{3.18}$$

This is a system of first-order simultaneous equations with respect to A. In order for all the A's to be nonzero,

$$\begin{vmatrix} b_{11}-\lambda & b_{12} & b_{13} & \cdots \\ b_{21} & b_{22}-\lambda & b_{23} & \cdots \\ b_{31} & b_{32} & b_{33}-\lambda & \cdots \\ \vdots & \vdots & \vdots & \end{vmatrix} = 0 \tag{3.19}$$

The order of this secular equation is equal to $3N$. Suppose that one root, λ_1, is found for Eq. 3.19. If it is inserted in Eq. 3.18, A_{k1}, $A_{k2}, \ldots$ are obtained for all the nuclei. The same is true for the other roots of Eq. 3.19. Thus the most general solution may be written as a superposition of all the normal vibrations:

$$q_k = \sum_l B_{kl} Q_l^0 \sin(\sqrt{\lambda_l}\, t + \delta_l) \tag{3.20}$$

The general discussion developed above may be understood more easily if we apply it to a simple molecule such as CO_2, which is constrained to move in only one direction. If the mass and the displacement of each atom are defined as follows:

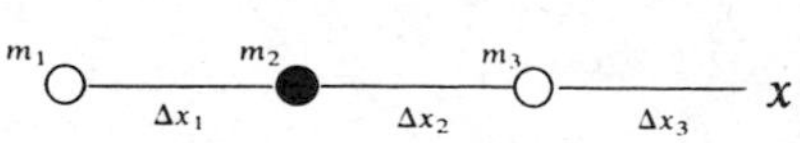

the potential energy is given by

$$V = \tfrac{1}{2} k[(\Delta x_1 - \Delta x_2)^2 + (\Delta x_2 - \Delta x_3)^2] \tag{3.21}$$

Considering that $m_1 = m_3$, we find that the kinetic energy is written as

$$T = \tfrac{1}{2} m_1 (\Delta \dot{x}_1^2 + \Delta \dot{x}_3^2) + \tfrac{1}{2} m_2 \Delta \dot{x}_2^2 \tag{3.22}$$

Using the generalized coordinates defined by Eq. 3.2, we may rewrite these energies as

$$2V = k\left[\left(\frac{q_1}{\sqrt{m_1}} - \frac{q_2}{\sqrt{m_2}}\right)^2 + \left(\frac{q_2}{\sqrt{m_2}} - \frac{q_3}{\sqrt{m_1}}\right)^2\right] \tag{3.23}$$

$$2T = \sum \dot{q}_i^2 \tag{3.24}$$

From comparison of Eq. 3.23 with Eq. 3.5, we obtain

$$b_{11} = \frac{k}{m_1}, \qquad b_{22} = \frac{2k}{m_2}$$

$$b_{12} = b_{21} = -\frac{k}{\sqrt{m_1 m_2}}, \qquad b_{23} = b_{32} = -\frac{k}{\sqrt{m_1 m_2}}$$

$$b_{13} = b_{31} = 0, \qquad b_{33} = \frac{k}{m_1}$$

If these terms are inserted in Eq. 3.19, we obtain the following result:

$$\begin{vmatrix} \dfrac{k}{m_1}-\lambda & -\dfrac{k}{\sqrt{m_1 m_2}} & 0 \\ -\dfrac{k}{\sqrt{m_1 m_2}} & \dfrac{2k}{m_2}-\lambda & -\dfrac{k}{\sqrt{m_1 m_2}} \\ 0 & -\dfrac{k}{\sqrt{m_1 m_2}} & \dfrac{k}{m_1}-\lambda \end{vmatrix} = 0 \qquad (3.25)$$

By solving this secular equation, we obtain three roots:

$$\lambda_1 = \frac{k}{m_1}, \qquad \lambda_2 = k\mu, \qquad \lambda_3 = 0$$

where

$$\mu = \frac{2m_1 + m_2}{m_1 m_2}$$

Equation 3.18 gives the following three equations:

$$-\lambda A_1 + b_{11}A_1 + b_{12}A_2 + b_{13}A_3 = 0$$
$$-\lambda A_2 + b_{21}A_1 + b_{22}A_2 + b_{23}A_3 = 0$$
$$-\lambda A_3 + b_{31}A_1 + b_{32}A_2 + b_{33}A_3 = 0$$

Using Eq. 3.17, we rewrite these as

$$(b_{11} - \lambda)q_1 + b_{12}q_2 + b_{13}q_3 = 0$$
$$b_{21}q_1 + (b_{22} - \lambda)q_2 + b_{23}q_3 = 0$$
$$b_{31}q_1 + b_{32}q_2 + (b_{33} - \lambda)q_3 = 0$$

If $\lambda_1 = k/m_1$ is inserted in the above simultaneous equations, we obtain

$$q_1 = -q_3, \qquad q_2 = 0$$

Similar calculations give

$$q_1 = q_3, \qquad q_2 = -2\sqrt{\frac{m_1}{m_2}}\, q_1 \qquad \text{for } \lambda_2 = k\mu$$

$$q_1 = q_3, \qquad q_2 = \sqrt{\frac{m_2}{m_1}}\, q_1 \qquad \text{for } \lambda_3 = 0$$

The relative displacements are depicted in the following figure:

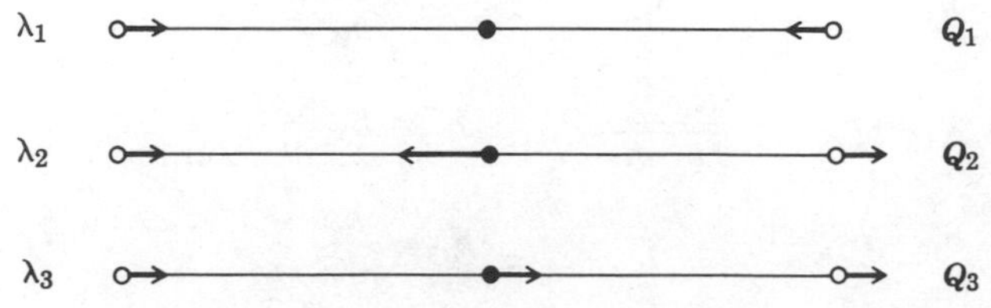

It is easy to see that λ_3 corresponds to the translational mode ($\Delta x_1 = \Delta x_2 = \Delta x_3$). The inclusion of λ_3 could be avoided if we consider the restriction that the center of gravity does not move; $m_1(\Delta x_1 + \Delta x_3) + m_2 \Delta x_2 = 0$.

The relationships between the generalized coordinates and the normal coordinates are given by Eq. 3.10. In the present case, we have

$$q_1 = B_{11}Q_1 + B_{12}Q_2 + B_{13}Q_3$$

$$q_2 = B_{21}Q_1 + B_{22}Q_2 + B_{23}Q_3$$

$$q_3 = B_{31}Q_1 + B_{32}Q_2 + B_{33}Q_3$$

In the normal vibration whose normal coordinate is Q_1, $B_{11}:B_{21}:B_{31}$ gives the ratio of the displacements. From the previous calculation, it is obvious that $B_{11}:B_{21}:B_{31} = 1:0:-1$. Similarly, $B_{12}:B_{22}:B_{32} = 1:-2\sqrt{m_1/m_2}:1$ gives the ratio of the displacements in the normal vibration whose normal coordinate is Q_2. Thus the mode of a normal vibration can be drawn if the normal coordinate is translated into a set of rectangular coordinates, as is shown above.

So far, we have discussed only the vibrations whose displacements occur along the molecular axis. There are, however, two other normal vibrations in which the displacements occur in the direction perpendicular to the molecular axis. They are not treated here, since the calculation is not simple. It is clear that the method described above will become more complicated as a molecule becomes larger. In this respect, the **GF** matrix method described in Sec. I-11 is important in the vibrational analysis of complex molecules.

By using the normal coordinates, the Schrödinger wave equation for the system can be written as

$$\sum_i \frac{\partial^2 \psi_n}{\partial Q_i^2} + \frac{8\pi^2}{h^2}\left(E - \frac{1}{2}\sum_i \lambda_i Q_i^2\right)\psi_n = 0 \tag{3.26}$$

Since the normal coordinates are independent of each other, it is possible to write

$$\psi_n = \psi_1(Q_1)\psi_2(Q_2)\cdots \tag{3.27}$$

and solve the simpler one-dimensional problem.

If Eq. 3.27 is substituted in Eq. 3.26, there results

$$\frac{d^2\psi_i}{dQ_i^2} + \frac{8\pi^2}{h^2}\left(E_i - \frac{1}{2}\lambda_i Q_i^2\right)\psi_i = 0 \tag{3.28}$$

where

$$E = E_1 + E_2 + \cdots$$

with

$$E_i = h\nu_i(v_i + \tfrac{1}{2})$$

$$\nu_i = \frac{1}{2\pi}\sqrt{\lambda_i} \tag{3.29}$$

I-4. SYMMETRY ELEMENTS AND POINT GROUPS[12–17].

As noted before, polyatomic molecules have $3N-6$ or, if linear, $3N-5$ normal vibrations. For any given molecule, however, only vibrations that are permitted by the selection rule for that molecule appear in the infrared and Raman spectra. Since the selection rule is determined by the symmetry of the molecule, this must first be studied.

The spatial geometrical arrangement of the nuclei constituting the molecule determines its symmetry. If a coordinate transformation (a reflection or a rotation or a combination of both) produces a configuration of the nuclei indistinguishable from the original one, this transformation is called a *symmetry operation*, and the molecule is said to have a corresponding *symmetry element*. Molecules may have the following symmetry elements.

(1) Identity *I*

This symmetry element is possessed by every molecule no matter how unsymmetrical it is, the corresponding operation being to leave the molecule unchanged. The inclusion of this element is necessitated by mathematical reasons which will be discussed in Sec. I-6.

(2) A Plane of Symmetry, σ

If reflection of a molecule with respect to some plane produces a configuration indistinguishable from the original one, the plane is called a plane of symmetry.

(3) A Center of Symmetry, *i*

If reflection at the center, that is, inversion, produces a configuration indistinguishable from the original one, the center is called a center of symmetry. This operation changes the signs of all the coordinates involved, $x_i \rightarrow -x_i$, $y_i \rightarrow -y_i$, $z_i \rightarrow -z_i$.

(4) A *p*-Fold Axis of Symmetry, C_p

If rotation through an angle $360°/p$ about an axis produces a configuration indistinguishable from the original one, the axis is called a *p*-fold axis of symmetry, C_p. For example, a twofold axis, C_2, implies that a rotation of 180° about the axis reproduces the original configuration. A molecule may have a two-, three-, four-, five-, or sixfold, or higher axis. A linear molecule has an infinite-fold (denoted by ∞-fold) axis of symmetry, C_∞, since a rotation of $360°/\infty$, that is, an infinitely small angle, transforms the molecule into one indistinguishable from the original.

(5) A *p*-Fold Rotation–Reflection Axis, S_p

If rotation by $360°/p$ about the axis, followed by reflection at a plane perpendicular to the axis, produces a configuration indistinguishable from the original one, the axis is called a *p*-fold rotation–reflection axis. A molecule may have a two-, three-, four, five, or sixfold, or higher, rotation–reflection axis. A symmetrical linear molecule has an S_∞ axis. It is easily seen that the presence of S_p always means the presence of C_p as well as σ when p is odd.

A molecule may have more than one of these symmetry elements. Combination of more and more of these elements produces systems of higher and higher symmetry. Not all combinations of symmetry elements, however, are possible. For example, it is highly improbable that a molecule will have a C_3 and C_4 axis in the same direction because this requires the existence of a 12-fold axis in the molecule. It should also be noted that the presence of some symmetry elements often implies the presence of other elements. For example, if a molecule has two σ-planes at right angles to each other, the line of intersection of these two planes must be a C_2 axis. A possible combination of symmetry operations whose axes intersect at a point is called a *point group.**

Theoretically, an infinite number of point groups exist, since there is no restriction on the order (p) of rotation axes which may exist in an isolated molecule. Practically, however, there are few molecules and ions that possess a rotation axis higher than C_6. Thus most of the compounds discussed in this book belong to the following point groups:

1. $\mathbf{C}_p$. Molecules having only a C_p and no other elements of symmetry: $\mathbf{C}_1$, $\mathbf{C}_2$, $\mathbf{C}_3$, and so on.
2. $\mathbf{C}_{ph}$. Molecules having a C_p and a σ_h perpendicular to it: $\mathbf{C}_{1h} \equiv \mathbf{C}_s$, $\mathbf{C}_{2h}$, $\mathbf{C}_{3h}$, and so on.
3. $\mathbf{C}_{pv}$. Molecules having a C_p and $p\sigma_v$ through it: $\mathbf{C}_{1v} \equiv \mathbf{C}_s$, $\mathbf{C}_{2v}$, $\mathbf{C}_{3v}$, $\mathbf{C}_{4v}, \ldots, \mathbf{C}_{\infty v}$.
4. $\mathbf{D}_p$. Molecules having a C_p and pC_2 perpendicular to the C_p and at equal angles to one another: $\mathbf{D}_1 \equiv \mathbf{C}_2$, $\mathbf{D}_2 \equiv \mathbf{V}$, $\mathbf{D}_3$, $\mathbf{D}_4$, and so on.
5. $\mathbf{D}_{ph}$. Molecules having a C_p, $p\sigma_v$ through it at angles of $360°/2p$ to one another, and a σ_h perpendicular to the C_p: $\mathbf{D}_{1h} \equiv \mathbf{C}_{2v}$, $\mathbf{D}_{2h} \equiv \mathbf{V}_h$, $\mathbf{D}_{3h}$, $\mathbf{D}_{4h}$, $\mathbf{D}_{5h}$, $\mathbf{D}_{6h}, \ldots, \mathbf{D}_{\infty h}$.
6. $\mathbf{D}_{pd}$. Molecules having a C_p, pC_2 perpendicular to it, and $p\sigma_d$ which go through the C_p and bisect the angles between two successive C_2 axes: $\mathbf{D}_{2d} \equiv \mathbf{V}_d$, $\mathbf{D}_{3d}$, $\mathbf{D}_{4d}$, $\mathbf{D}_{5d}$, and so on.
7. $\mathbf{S}_p$. Molecules having only a S_p (p even). For p odd, S_p is equivalent to $C_p \times \sigma_h$, for which other notations such as $\mathbf{C}_{3h}$ are used: $\mathbf{S}_2 \equiv \mathbf{C}_i$, $\mathbf{S}_4$, $\mathbf{S}_6$, and so on.
8. $\mathbf{T}_d$. Molecules having three mutually perpendicular C_2 axes, four C_3 axes, and a σ_d through each pair of C_3 axes: regular tetrahedral molecules.
9. $\mathbf{O}_h$. Molecules having three mutually perpendicular C_4 axes, four C_3 axes, and a center of symmetry, i: regular octahedral and cubic molecules.

* In this respect, point groups differ from space groups, which involve translations and rotations about nonintersecting axes (see Sec. I-23).

Complete listings of the symmetry elements present in each point group are given in the character tables in Appendix I. From the symmetry point of view, molecules belonging to the $\mathbf{C}_1$, $\mathbf{C}_2$, $\mathbf{C}_3$, $\mathbf{D}_2 \equiv \mathbf{V}$, and $\mathbf{D}_3$ groups are optically active since they lack an S_p axis.

I-5. SYMMETRY OF NORMAL VIBRATIONS AND SELECTION RULES

Figure I-7 indicates the normal modes of vibration in CO_2 and H_2O molecules. In each normal vibration, the individual nuclei carry out a simple harmonic motion in the direction indicated by the arrow, and all the nuclei have the same frequency of oscillation (i.e., the frequency of the normal vibration) and are moving in the same phase. Furthermore, the relative lengths of the arrows indicate the relative velocities and the amplitudes for each nucleus.* The ν_2 vibrations in CO_2 are worth comment, since they differ from the others in that two vibrations (ν_{2a} and ν_{2b}) have exactly the same frequency. Apparently, there are an infinite number of normal vibrations of this type, which differ only in their directions perpendicular to the molecular axis. Any of them, however, can be resolved into two vibrations such as ν_{2a} and ν_{2b},

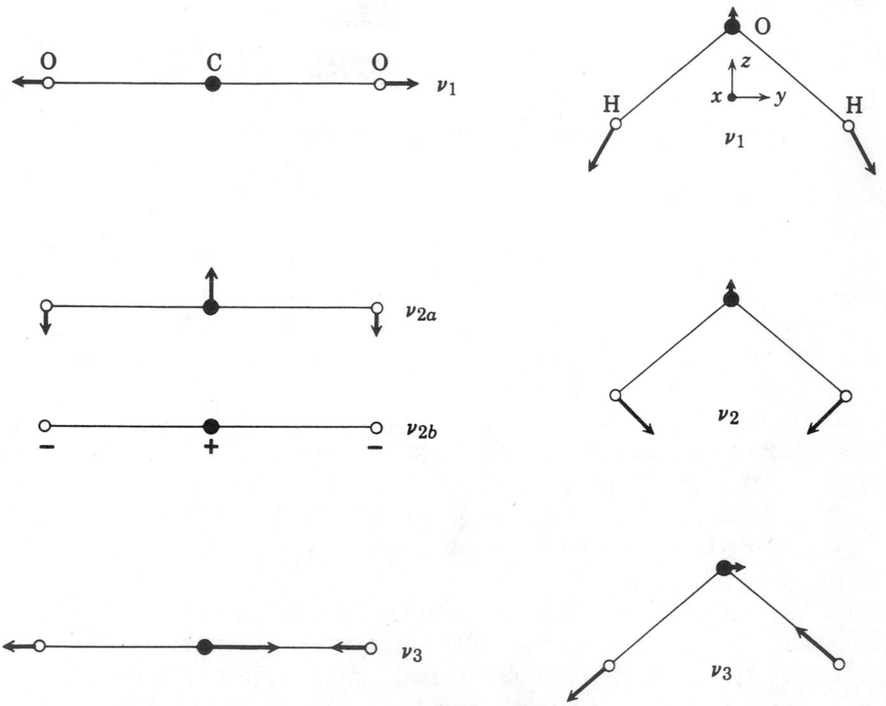

Fig. I-7. Normal modes of vibration in CO_2 and H_2O molecules. (+ and − denote vibrations going upward and downward, respectively, in the direction perpendicular to the paper plane.)

---

* In this respect, all the normal modes of vibration shown in this book are only approximate.

which are perpendicular to each other. In this respect, the ν_2 vibrations in CO_2 are called *doubly degenerate vibrations.* Doubly degenerate vibrations occur only when a molecule has an axis higher than twofold. *Triply degenerate vibrations* also occur in molecules having more than one C_3 axis.

To determine the symmetry of a normal vibration, it is necessary to begin by considering the kinetic and potential energies of the system. These were discussed in Sec. I-3.

$$T=\frac{1}{2}\sum_i \dot{Q}_i^2 \tag{5.1}$$

$$V=\frac{1}{2}\sum_i \lambda_i Q_i^2 \tag{5.2}$$

Consider a case in which a molecule performs only one normal vibration, Q_i. Then $T=\frac{1}{2}\dot{Q}_i^2$ and $V=\frac{1}{2}\lambda_i Q_i^2$. These energies must be invariant when a symmetry operation, R, changes Q_i to RQ_i. Thus

$$T=\tfrac{1}{2}\dot{Q}_i^2=\tfrac{1}{2}(R\dot{Q}_i)^2$$
$$V=\tfrac{1}{2}\lambda_i Q_i^2=\tfrac{1}{2}\lambda_i(RQ_i)^2$$

For these relations to hold, it is necessary that

$$(RQ_i)^2=Q_i^2 \qquad \text{or} \qquad RQ_i=\pm Q_i \tag{5.3}$$

Thus the normal coordinate must change either into itself or into its negative. If $Q_i=RQ_i$, the vibration is said to be *symmetric.* If $Q_i=-RQ_i$, it is said to be *antisymmetric.*

If the vibration is doubly degenerate, we have

$$T=\tfrac{1}{2}\dot{Q}_{ia}^2+\tfrac{1}{2}\dot{Q}_{ib}^2$$
$$V=\tfrac{1}{2}\lambda_i(Q_{ia})^2+\tfrac{1}{2}\lambda_i(Q_{ib})^2$$

In this case, a relation such as

$$(RQ_{ia})^2+(RQ_{ib})^2=Q_{ia}^2+Q_{ib}^2 \tag{5.4}$$

must hold. As will be shown later, such a relationship is expressed more conveniently by using a matrix form:

$$R\begin{bmatrix} Q_{ia} \\ Q_{ib} \end{bmatrix}=\begin{bmatrix} A & B \\ C & D \end{bmatrix}\begin{bmatrix} Q_{ia} \\ Q_{ib} \end{bmatrix}$$

where the values of A, B, C, and D depend on the symmetry operation, R. In any case, the normal vibration must be either symmetric or antisymmetric or degenerate for each symmetry operation.

The symmetry properties of the normal vibrations of the H_2O molecule shown in Fig. I-7 are classified as indicated in Table I-1. Here, +1 and −1 denote symmetric and antisymmetric, respectively. In the ν_1 and ν_2 vibrations,

TABLE I-1

C_{2v}	I	$C_2(z)$	$\sigma_v(xz)$[a]	$\sigma_v(yz)$[a]
Q_1, Q_2	+1	+1	+1	+1
Q_3	+1	−1	−1	+1

[a] σ_v = vertical plane of symmetry.

all the symmetry properties are preserved during the vibration. Therefore they are *symmetric vibrations* and are called, in particular, *totally symmetric vibrations.* In the ν_3 vibration, however, symmetry elements such as C_2 and $\sigma_v(xz)$ are lost. Thus it is called a *nonsymmetric vibration.* If a molecule has a number of symmetry elements the normal vibrations are classified as various species according to the number and the kind of symmetry elements preserved during the vibration.

To determine the activity of the vibrations in the infrared and Raman spectra, the selection rule must be applied to each normal vibration. From a quantum mechanical point of view, *a vibration is active in the infrared spectrum if the dipole moment of the molecule is changed during the vibration, and is active in the Raman spectrum if the polarizability of the molecule is changed during the vibration.* As stated in Sec. I-1, the induced dipole moment P is related to the strength of the electric field E by the relation

$$P = \alpha E^* \tag{5.5}$$

where α is called the *polarizability.* If we resolve P, α, and E in the x, y, and z directions, simple relationships such as

$$P_x = \alpha_x E_x, \qquad P_y = \alpha_y E_y, \qquad \text{and} \qquad P_z = \alpha_z E_z \tag{5.6}$$

do not hold, since the direction of polarization does not coincide with the direction of the applied field. This is so because the direction of chemical bonds in the molecule also affects the direction of polarization. Thus, instead of Eq. 5.6, we have the relationships:

$$\begin{aligned} P_x &= \alpha_{xx} E_x + \alpha_{xy} E_y + \alpha_{xz} E_z \\ P_y &= \alpha_{yx} E_x + \alpha_{yy} E_y + \alpha_{yz} E_z \\ P_z &= \alpha_{zx} E_x + \alpha_{zy} E_y + \alpha_{zz} E_z \end{aligned} \tag{5.7}$$

* A more complete form of this equation is

$$P = \alpha E + \tfrac{1}{2}\beta E^2 + \cdots$$

Here $\beta \ll \alpha$ and β is called the hyperpolarizability. The second term becomes significant only when E is large ($\sim 10^9$ V m^{-1}). In this case, hyper Raman spectra, $2\nu \pm \nu_i$, are observed. For a discussion of nonlinear Raman spectroscopy, see Ref. 23.

In matrix form, Eq. 5.7 is written as

$$\begin{bmatrix} P_x \\ P_y \\ P_z \end{bmatrix} = \begin{bmatrix} \alpha_{xx} & \alpha_{xy} & \alpha_{xz} \\ \alpha_{yx} & \alpha_{yy} & \alpha_{yz} \\ \alpha_{zx} & \alpha_{zy} & \alpha_{zz} \end{bmatrix} \begin{bmatrix} E_x \\ E_y \\ E_z \end{bmatrix} \tag{5.8}$$

and the first matrix on the right-hand side is called the *polarizability tensor.* In normal Raman scattering, the tensor is symmetric; $\alpha_{xy} = \alpha_{yx}$, $\alpha_{yz} = \alpha_{zy}$, and $\alpha_{xz} = \alpha_{zx}$. This is not so, however, in the case of resonance Raman scattering (Sec. I-21).

According to quantum mechanics, the vibration is Raman active if one of these six components of the polarizability changes during the vibration. Similarly, it is infrared active if one of the three components of the dipole moment (μ_x, μ_y, and μ_z) changes during the vibration. Changes in dipole moment or polarizability are not obvious from inspection of the normal modes of vibration in most polyatomic molecules. As will be shown later, application of group theory gives a clear-cut solution to this problem.

In simple molecules, however, the activity of a vibration may be determined by inspection of the normal mode. For example, it is obvious that the vibration in a homopolar diatomic molecule is not infrared active but is Raman active, whereas the vibration in a heteropolar diatomic molecule is both infrared and Raman active. It is also obvious that all three vibrations of H_2O and ν_2 and ν_3 of CO_2 are infrared active. Except for ν_1 of CO_2, the Raman activity is not easy to predict even for such simple molecules.

The polarizability tensor can be visualized easily if we draw a *polarizability ellipsoid* by plotting $1/\sqrt{\alpha}$ in any direction from the origin. This gives a three-dimensional surface such as is shown in Fig. I-8. If we orient this ellipsoid with its principal axes along the *X, Y, Z* axes of the coordinate system, Eq. 5.8 is simplified to

$$\begin{bmatrix} P_X \\ P_Y \\ P_Z \end{bmatrix} = \begin{bmatrix} \alpha_{XX} & 0 & 0 \\ 0 & \alpha_{YY} & 0 \\ 0 & 0 & \alpha_{ZZ} \end{bmatrix} \begin{bmatrix} E_X \\ E_Y \\ E_Z \end{bmatrix} \tag{5.9}$$

These three axes are called the *principal axes of polarizability.* In terms of the polarizability ellipsoid, the Raman selection rule can be stated as follows: *The vibration is Raman active if the ellipsoid changes in size, shape, or orientation during the vibration.* Consider the ν_1 vibration of CO_2. As shown in Fig. I-8, the ellipsoid changes its size during this vibration (or α_{xx}, α_{yy}, and α_{zz} change during the vibration). Thus it is Raman active. Although the size of the ellipsoid changes during the ν_3 vibration, they are identical in two extreme positions, as shown in Fig. I-8. If we consider a limiting case where the nuclei undergo very small displacements, there is effectively no change in the polarizability; hence the ν_3 vibration is not Raman active. The same is true for the ν_2 vibration.

In the case of H_2O, both the ν_1 and ν_2 vibrations are Raman active because the size and the shape of the ellipsoid change during these vibrations. The ν_3

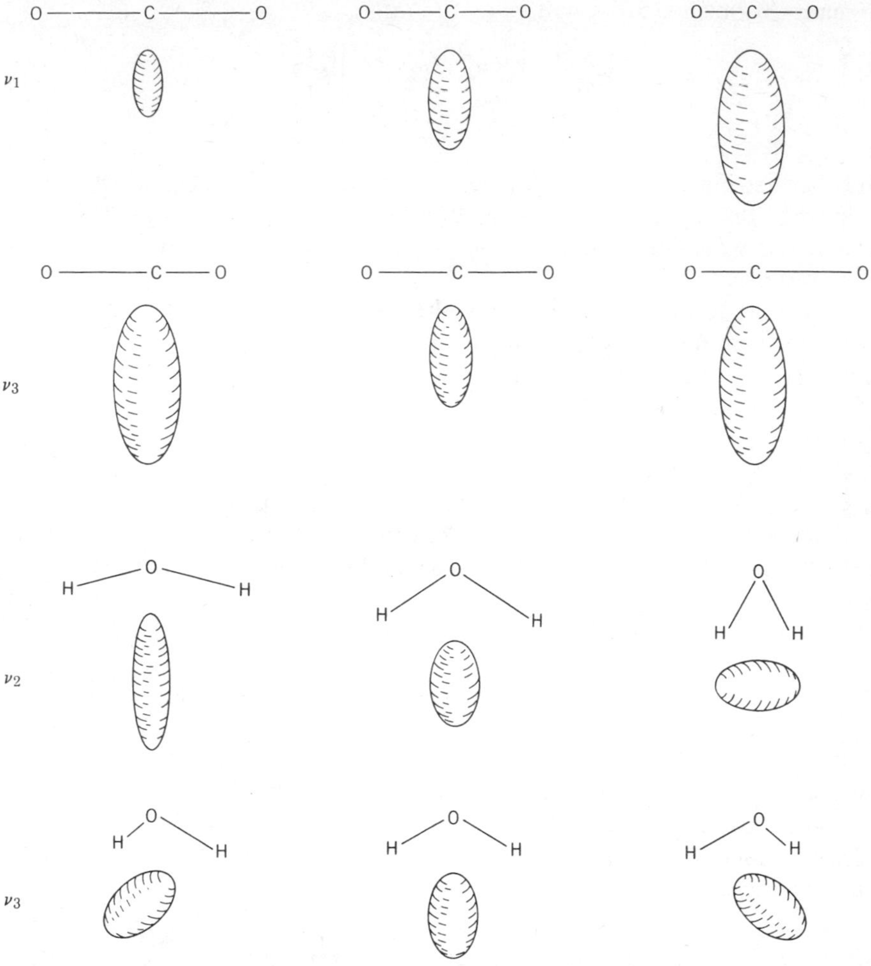

Fig. I-8. Changes of polarizability ellipsoids during the normal vibrations of CO_2 and H_2O.

vibration of H_2O is different from the other vibrations in that the orientation of the ellipsoid changes (α_{yz} changes) during the vibration. Thus all three normal vibrations of H_2O are Raman active.

It should be noted that in CO_2 the vibration symmetric with respect to the center of symmetry (ν_1) is Raman active and not infrared active, whereas the vibrations antisymmetric with respect to the center of symmetry (ν_2 and ν_3) are infrared active but not Raman active. In a polyatomic molecule having a center of symmetry, the vibrations symmetric with respect to the center of symmetry (*g* vibrations*) are Raman active and not infrared active, but the vibrations antisymmetric with respect to the center of symmetry (*u* vibrations*)

* The symbols *g* and *u* stand for *gerade* and *ungerade* (German), respectively.

are infrared active and not Raman active. This rule is called the *mutual exclusion rule.* It should be noted, however, that in polyatomic molecules having several symmetry elements in addition to the center of symmetry, the vibrations that should be active according to this rule may not necessarily be active, because of the presence of other symmetry elements. An example is seen in a square-planar XY_4-type molecule of $\mathbf{D}_{4h}$ symmetry, where the A_{2g} vibrations are not Raman active and the A_{1u}, B_{1u}, and B_{2u} vibrations are not infrared active (see Sec. II-6 and Appendix I).

I-6. INTRODUCTION TO GROUP THEORY[12–17]

In Sec. I-4, the symmetry and the point group allocation of a given molecule were discussed. To understand the symmetry and selection rules of normal vibrations in polyatomic molecules, however, a knowledge of group theory is required. The minimum amount of group theory needed for this purpose is given here.

Consider a pyramidal XY_3 molecule (Fig. I-9) for which the symmetry operations I, C_3^+, C_3^-, σ_1, σ_2, and σ_3 are applicable. Here C_3^+ and C_3^- denote rotation through 120° in the clockwise and counterclockwise directions, respectively, and σ_1, σ_2, and σ_3 indicate the symmetry planes that pass through X and Y_1, X and Y_2, and X and Y_3, respectively. For simplicity, let these symmetry operations be denoted by I, A, B, C, D, and E, respectively. Other symmetry operations are possible, but each is equivalent to some one of the operations mentioned. For instance, a clockwise rotation through 240° is identical with operation B. It may also be shown that two successive applications of any one of these operations is equivalent to some single operation of the group mentioned. Let operation C be applied to the original figure. This interchanges Y_2 and Y_3. If operation A is applied to the resulting figure, the net result is the same as application of the single operation D to the original figure. This is written as $CA = D$. If all the possible multiplicative combinations are made, Table I-2, in which the operation applied first is written across the top, is obtained. This is called the *multiplication table* of the group.

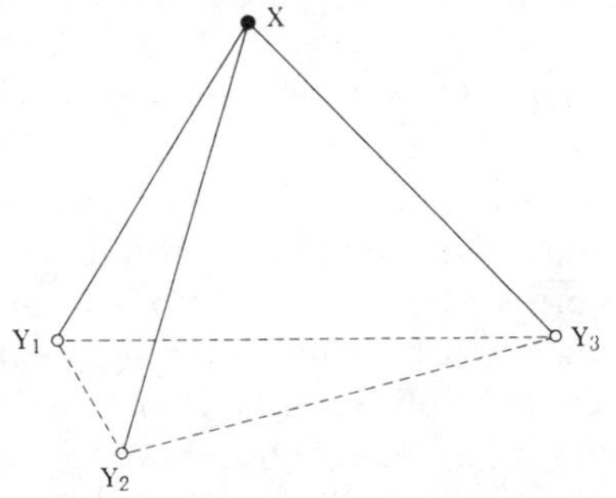

Fig. I-9. Pyramidal XY_3 molecule.

TABLE I-2

	I	A	B	C	D	E
I	I	A	B	C	D	E
A	A	B	I	D	E	C
B	B	I	A	E	C	D
C	C	E	D	I	B	A
D	D	C	E	A	I	B
E	E	D	C	B	A	I

It is seen that a group consisting of the mathematical elements (symmetry operations) I, A, B, C, D, and E satisfies the following conditions:

1. The product of any two elements in the set is another element in the set.
2. The set contains the identity operation that satisfies the relation $IP = PI = P$, where P is any element in the set.
3. The associative law holds for all the elements in the set, that is, $(CB)A = C(BA)$, for example.
4. Every element in the set has its reciprocal, X, which satisfies the relation $XP = PX = I$, where P is any element in the set. This reciprocal is usually denoted by P^{-1}.

These are necessary and sufficient conditions for a set of elements to form a *group*. It is evident that operations I, A, B, C, D, and E form a group in this sense. It should be noted that the commutative law of multiplication does not necessarily hold. For example, Table I-2 shows that $CD \neq DC$.

The six elements can be classified into three types of operations: the identity operation I, the rotations C_3^+ and C_3^-, and the reflections σ_1, σ_2, and σ_3. Each of these sets of operations is said to form a *class*. More precisely, two operations, P and Q, which satisfy the relation $X^{-1}PX = P$ or Q, where X is any operation of the group and X^{-1} is its reciprocal, are said to belong to the same class. It can easily be shown that C_3^+ and C_3^-, for example, satisfy the relation. Thus the six elements of the point group $\mathbf{C}_{3v}$ are usually abbreviated as I, $2C_3$, and $3\sigma_v$.

The relations between the elements of the group are shown in the multiplication table, Table I-2. Such a tabulation of a group is, however, awkward to handle. The essential features of the table may be abstracted by replacing the elements by some analytical function that reproduces the multiplication table. Such an analytical expression may be composed of a simple integer, an exponential function, or a matrix. Any set of such expressions that satisfies the relations given by the multiplication table is called a *representation* of the group and is designated by Γ. The representations of the point group $\mathbf{C}_{3v}$ discussed above are indicated in Table I-3. It is easily proved that each representation in the table satisfies the multiplication table.

TABLE I-3

$\mathbf{C}_{3v}$	I	A	B	C	D	E
$A_1(\Gamma_1)$	1	1	1	1	1	1
$A_2(\Gamma_2)$	1	1	1	-1	-1	-1
$E(\Gamma_3)$	$\begin{pmatrix} 1 & 0 \\ 0 & 1 \end{pmatrix}$	$\begin{pmatrix} -\frac{1}{2} & \frac{\sqrt{3}}{2} \\ -\frac{\sqrt{3}}{2} & -\frac{1}{2} \end{pmatrix}$	$\begin{pmatrix} -\frac{1}{2} & -\frac{\sqrt{3}}{2} \\ \frac{\sqrt{3}}{2} & -\frac{1}{2} \end{pmatrix}$	$\begin{pmatrix} -1 & 0 \\ 0 & 1 \end{pmatrix}$	$\begin{pmatrix} \frac{1}{2} & -\frac{\sqrt{3}}{2} \\ -\frac{\sqrt{3}}{2} & -\frac{1}{2} \end{pmatrix}$	$\begin{pmatrix} \frac{1}{2} & \frac{\sqrt{3}}{2} \\ \frac{\sqrt{3}}{2} & -\frac{1}{2} \end{pmatrix}$

In addition to the three representations in Table I-3, it is possible to write an infinite number of other representations of the group. If a set of six matrices of the type $S^{-1}R(K)S$ is chosen, where $R(K)$ is a representation of the element K given in Table I-3, S ($|S| \neq 0$) is any matrix of the same order as R, and S^{-1} is the reciprocal of S, this set also satisfies the relations given by the multiplication table. The reason is obvious from the relation

$$S^{-1}R(K)SS^{-1}R(L)S = S^{-1}R(K)R(L)S = S^{-1}R(KL)S$$

Such a transformation is called a *similarity transformation.* Thus it is possible to make an infinite number of representations by means of similarity transformations.

On the other hand, this statement suggests that a given representation may be broken into simpler ones. If each representation of the symmetry element K is transformed into the form

$$R(K) = \begin{vmatrix} Q_1(K) & 0 & 0 & 0 \\ 0 & Q_2(K) & 0 & 0 \\ 0 & 0 & Q_3(K) & 0 \\ 0 & 0 & 0 & Q_3(K) \end{vmatrix} \tag{6.1}$$

by a similarity transformation, $Q_1(K)$, $Q_2(K)$, ... are simpler representations. In such a case, $R(K)$ is called *reducible.* If a representation cannot be simplified any further, it is said to be *irreducible.* The representations Γ_1, Γ_2, and Γ_3 in Table I-3 are all irreducible representations. It can be shown generally that the number of irreducible representations is equal to the number of classes. Thus only three irreducible representations exist for the point group $\mathbf{C}_{3v}$. These representations are entirely independent of each other. Furthermore, the sum of the squares of the dimensions (l) of the irreducible representations of a group is always equal to the total number of the symmetry elements, namely, the *order of the group* (h). Thus

$$\sum l_i^2 = l_1^2 + l_2^2 + \cdots = h \tag{6.2}$$

In the point group $\mathbf{C}_{3v}$, it is seen that

$$1^2 + 1^2 + 2^2 = 6$$

TABLE I-4. THE CHARACTER TABLE OF THE POINT GROUP $\mathbf{C}_{3v}$

$\mathbf{C}_{3v}$	I	$2C_3(z)$	$3\sigma_v$
$A_1(\chi_1)$	1	1	1
$A_2(\chi_2)$	1	1	−1
$E(\chi_3)$	2	−1	0

A point group is classified into *species* according to its irreducible representations. In the point group $\mathbf{C}_{3v}$, the species having the irreducible representations Γ_1, Γ_2, and Γ_3 are called the A_1, A_2, and E species, respectively.*

The sum of the diagonal elements of a matrix is called the *character* of the matrix and is denoted by χ. It is to be noted in Table I-3 that the character of each of the elements belonging to the same class is the same. Thus, using the character, Table I-3 can be simplified to Table I-4. Such a table is called the *character table* of the point group $\mathbf{C}_{3v}$.

That the *character* of a matrix is not changed by a similarity transformation can be proved as follows. If a similarity transformation is expressed by $T=S^{-1}RS$, then

$$\chi_T=\sum_i (S^{-1}RS)_{ii}=\sum_{i,j,k}(S^{-1})_{ij}R_{jk}S_{ki}=\sum_{j,k,i}S_{ki}(S^{-1})_{ij}R_{jk}$$
$$=\sum_{j,k}\delta_{kj}R_{jk}=\sum_k R_{kk}=\chi_R$$

where δ_{kj} is Kronecker's delta (0 for $k\neq j$ and 1 for $k=j$). Thus any reducible representation can be reduced to its irreducible representations by a similarity transformation that leaves the character unchanged. Therefore the character of the reducible representation, $\chi(K)$, is written as

$$\chi(K)=\sum_m a_m\chi_m(K) \tag{6.3}$$

where $\chi_m(K)$ is the character of $Q_m(K)$, and a_m is a positive integer that indicates the number of times $Q_m(K)$ appears in the matrix of Eq. 6.1. Hereafter the character will be used rather than the corresponding representation because a 1:1 correspondence exists between these two, and the former is sufficient for vibrational problems.

It is important to note that the following relation holds in Table I-4:

$$\sum_K \chi_i(K)\chi_j(K)=h\delta_{ij} \tag{6.4}$$

If Eq. 6.3 is multiplied by $\chi_i(K)$ on both sides, and the summation is taken

* For the labeling of the irreducible representations (species), see Appendix I.

over all the symmetry operations, then

$$\sum_K \chi(K)\chi_i(K) = \sum_K \sum_m a_m\chi_m(K)\chi_i(K)$$
$$= \sum_m \sum_K a_m\chi_m(K)\chi_i(K)$$

For a fixed m, we have

$$\sum_K a_m\chi_m(K)\chi_i(K) = a_m \sum_K \chi_m(K)\chi_i(K) = a_m h\delta_{im}$$

If we consider the sum of such a term over m, only the sum in which $m = i$ remains. Thus

$$\sum_K \chi(K)\chi_m(K) = ha_m$$

or

$$a_m = \frac{1}{h}\sum_K \chi(K)\chi_m(K) \tag{6.5}$$

This formula is written more conveniently as

$$\boxed{a_m = \frac{1}{h}\sum n\chi(K)\chi_m(K)} \tag{6.6}$$

where n is the number of symmetry elements in any one class, and the summation is made over the different classes. As Sec. I-7 will show, this formula is very useful in determining the number of normal vibrations belonging to each species.

I-7. THE NUMBER OF NORMAL VIBRATIONS FOR EACH SPECIES

As shown in Sec. I-5, the $3N-6$ (or $3N-5$) normal vibrations of an N-atom molecule can be classified into various species according to their symmetry properties. The number of normal vibrations in each species can be calculated by using the general equations given in Appendix II. These equations were derived from consideration of the vibrational degrees of freedom contributed by each set of identical nuclei for each symmetry species.[1] As an example, let us consider the NH_3 molecule belonging to the $\mathbf{C}_{3v}$ point group. The general equations are as follows:

A_1 species: $3m+2m_v+m_0-1$

A_2 species: $3m+m_v-1$

E species: $6m+3m_v+m_0-2$

N (total number of atoms) $=6m+3m_v+m_0$

From the definitions given in the footnotes of Appendix II, it is obvious that $m=0$, $m_0=1$, and $m_v=1$ in this case. To check these numbers, we calculate the total number of atoms from the equation for N given above. Since the result is 4, these assigned numbers are correct. Then the number of normal vibrations in each species can be calculated by inserting these numbers in the general equations given above: 2, 0, and 2, respectively, for the A_1, A_2, and E species. Since the E species is doubly degenerate, the total number of vibrations is counted as 6, which is expected from the $3N-6$ rule.

A more general method of finding the number of normal vibrations in each species can be developed by using group theory. The principle of the method is that all the representations are irreducible if normal coordinates are used as the basis for the representations. For example, the representations for the symmetry operations based on three normal coordinates, Q_1, Q_2, and Q_3, which correspond to the ν_1, ν_2, and ν_3 vibrations in the H_2O molecule of Fig. I-7, are as follows:

$$I\begin{bmatrix} Q_1 \\ Q_2 \\ Q_3 \end{bmatrix} = \begin{bmatrix} 1 & 0 & 0 \\ 0 & 1 & 0 \\ 0 & 0 & 1 \end{bmatrix}\begin{bmatrix} Q_1 \\ Q_2 \\ Q_3 \end{bmatrix}, \qquad C_2(z)\begin{bmatrix} Q_1 \\ Q_2 \\ Q_3 \end{bmatrix} = \begin{bmatrix} 1 & 0 & 0 \\ 0 & 1 & 0 \\ 0 & 0 & -1 \end{bmatrix}\begin{bmatrix} Q_1 \\ Q_2 \\ Q_3 \end{bmatrix}$$

$$\sigma_v(xz)\begin{bmatrix} Q_1 \\ Q_2 \\ Q_3 \end{bmatrix} = \begin{bmatrix} 1 & 0 & 0 \\ 0 & 1 & 0 \\ 0 & 0 & -1 \end{bmatrix}\begin{bmatrix} Q_1 \\ Q_2 \\ Q_3 \end{bmatrix}, \qquad \sigma_v(yz)\begin{bmatrix} Q_1 \\ Q_2 \\ Q_3 \end{bmatrix} = \begin{bmatrix} 1 & 0 & 0 \\ 0 & 1 & 0 \\ 0 & 0 & 1 \end{bmatrix}\begin{bmatrix} Q_1 \\ Q_2 \\ Q_3 \end{bmatrix}$$

Let a representation be written with the $3N$ rectangular coordinates of an N-atom molecule as its basis. If it is decomposed into its irreducible components, the basis for these irreducible representations must be the normal coordinates, and the number of appearances of the same irreducible representation must be equal to the number of normal vibrations belonging to the species represented by this irreducible representation. As stated previously, however, the $3N$ rectangular coordinates involve six (or five) coordinates, which correspond to the translational and rotational motions of the molecule as a whole. Therefore the representations that have such coordinates as their basis must be subtracted from the result obtained above. Use of the character of the representation, rather than the representation itself, yields the same result.

For example, consider a pyramidal XY_3 molecule that has six normal vibrations. At first, the representations for the various symmetry operations must be written with the 12 rectangular coordinates in Fig. I-10 as their basis. Consider pure rotation C_p^+. If the clockwise rotation of the point (x, y, z) around the z axis by the angle θ brings it to the point denoted by the coordinates (x', y', z'), the relations between these two sets of coordinates are given by

$$\begin{aligned} x' &= x\cos\theta + y\sin\theta \\ y' &= -x\sin\theta + y\cos\theta \\ z' &= z \end{aligned} \tag{7.1}$$

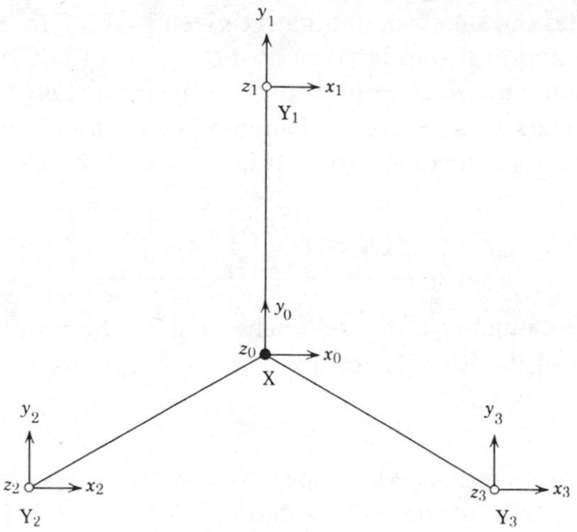

Fig. I-10. Rectangular coordinates in a pyramidal XY_3 molecule (z axis is perpendicular to the paper plane).

By using matrix notation, this can be written as

$$\begin{bmatrix} x' \\ y' \\ z' \end{bmatrix} = C_\theta^+ \begin{bmatrix} x \\ y \\ z \end{bmatrix} = \begin{bmatrix} \cos\theta & \sin\theta & 0 \\ -\sin\theta & \cos\theta & 0 \\ 0 & 0 & 1 \end{bmatrix} \begin{bmatrix} x \\ y \\ z \end{bmatrix} \tag{7.2}$$

Then the character of the matrix is given by

$$\chi(C_\theta^+) = 1 + 2\cos\theta \tag{7.3}$$

The same result is obtained for $\chi(C_\theta^-)$. If this symmetry operation is applied to all the coordinates of the XY_3 molecule, the result is

$$C_\theta \begin{bmatrix} x_0 \\ y_0 \\ z_0 \\ x_1 \\ y_1 \\ z_1 \\ x_2 \\ y_2 \\ z_2 \\ x_3 \\ y_3 \\ z_3 \end{bmatrix} = \left[\begin{array}{c:c:c:c} \mathbf{A} & 0 & 0 & 0 \\ \hdashline 0 & 0 & 0 & \mathbf{A} \\ \hdashline 0 & \mathbf{A} & 0 & 0 \\ \hdashline 0 & 0 & \mathbf{A} & 0 \end{array}\right] \begin{bmatrix} x_0 \\ y_0 \\ z_0 \\ x_1 \\ y_1 \\ z_1 \\ x_2 \\ y_2 \\ z_2 \\ x_3 \\ y_3 \\ z_3 \end{bmatrix} \tag{7.4}$$

where **A** denotes the small square matrix given by Eq. 7.2. Thus the character of this representation is simply given by Eq. 7.3. It should be noted in Eq. 7.4 that only the small matrix **A**, related to the nuclei unchanged by the symmetry operation, appears as a diagonal element. Thus a more general form of the character of the representation for rotation around the axis by θ is

$$\boxed{\chi(R) = N_R(1+2\cos\theta)} \tag{7.5}$$

where N_R is the number of nuclei unchanged by the rotation. In the present case, $N_R = 1$ and $\theta = 120°$. Therefore

$$\chi(C_3) = 0 \tag{7.6}$$

Identity (I) can be regarded as a special case of Eq. 7.5 in which $N_R = 4$ and $\theta = 0°$. The character of the representation is

$$\chi(I) = 12 \tag{7.7}$$

Pure rotation and identity are called *proper rotation.*

It is evident from Fig. I-10 that a symmetry plane such as σ_1 changes the coordinates from (x_i, y_i, z_i) to $(-x_i, y_i, z_i)$. The corresponding representation is therefore written as

$$\sigma_1 \begin{bmatrix} x \\ y \\ z \end{bmatrix} = \begin{bmatrix} -1 & 0 & 0 \\ 0 & 1 & 0 \\ 0 & 0 & 1 \end{bmatrix} \begin{bmatrix} x \\ y \\ z \end{bmatrix} \tag{7.8}$$

The result of such an operation on all the coordinates is

$$\sigma_1 \begin{bmatrix} x_0 \\ y_0 \\ z_0 \\ x_1 \\ y_1 \\ z_1 \\ x_2 \\ y_2 \\ z_2 \\ x_3 \\ y_3 \\ z_3 \end{bmatrix} = \left[\begin{array}{c|c|c|c} \mathbf{B} & 0 & 0 & 0 \\ \hline 0 & \mathbf{B} & 0 & 0 \\ \hline 0 & 0 & 0 & \mathbf{B} \\ \hline 0 & 0 & \mathbf{B} & 0 \end{array}\right] \begin{bmatrix} x_0 \\ y_0 \\ z_0 \\ x_1 \\ y_1 \\ z_1 \\ x_2 \\ y_2 \\ z_2 \\ x_3 \\ y_3 \\ z_3 \end{bmatrix} \tag{7.9}$$

where **B** denotes the small square matrix of Eq. 7.8. Thus the character of this representation is calculated as $2 \times 1 = 2$. It is noted again that the matrix on the diagonal is nonzero only for the nuclei unchanged by the operation.

More generally, a reflection at a plane (σ) is regarded as $\sigma = i \times C_2$. Thus the general form of Eq. 7.8 may be written as

$$\begin{bmatrix} -1 & 0 & 0 \\ 0 & -1 & 0 \\ 0 & 0 & -1 \end{bmatrix} \begin{bmatrix} \cos\theta & \sin\theta & 0 \\ -\sin\theta & \cos\theta & 0 \\ 0 & 0 & 1 \end{bmatrix} = \begin{bmatrix} -\cos\theta & -\sin\theta & 0 \\ \sin\theta & -\cos\theta & 0 \\ 0 & 0 & -1 \end{bmatrix}$$

Then

$$\chi(\sigma) = -(1 + 2\cos\theta)$$

As a result, the character of the large matrix shown in Eq. 7.9 is given by

$$\boxed{\chi(R) = -N_R(1 + 2\cos\theta)} \qquad (7.10)$$

In the present case, $N_R = 2$ and $\theta = 180°$. This gives

$$\chi(\sigma_v) = 2 \qquad (7.11)$$

Symmetry operations such as i and S_p are regarded as

$$i = i \times I, \qquad \theta = 0°$$

$$S_3 = i \times C_6, \qquad \theta = 60°$$

$$S_4 = i \times C_4, \qquad \theta = 90°$$

$$S_6 = i \times C_3, \qquad \theta = 120°$$

Therefore the characters of these symmetry operations can be calculated by Eq. 7.10 with the values of θ defined above. Operations such as σ, i, and S_p are called *improper rotations.* Thus the character of the representation based on 12 rectangular coordinates is as follows:

I	$2C_3$	$3\sigma_v$
12	0	2

(7.12)

To determine the number of normal vibrations belonging to each species, the $\chi(R)$ thus obtained must be resolved into the $\chi_i(R)$ of the irreducible representations of each species in Table I-4. First, however, the characters corresponding to the translational and rotational motions of the molecule must be subtracted from the result shown in Eq. 7.12.

The characters for the translational motion of the molecule in the x, y, and z directions (denoted by T_x, T_y, and T_z) are the same as those obtained in Eqs. 7.5 and 7.10. They are as follows:

$$\boxed{\chi_t(R) = \pm(1 + 2\cos\theta)} \qquad (7.13)$$

where the + and − signs are for proper and improper rotations, respectively. The characters for the rotations around the x, y, and z axes (denoted by R_x, R_y, and R_z) are given by

$$\boxed{\chi_r(R) = +(1 + 2\cos\theta)} \tag{7.14}$$

for both proper and improper rotations. This is due to the fact that a rotation of the vectors in the plane perpendicular to the x, y, and z axes can be regarded as a rotation of the components of angular momentum, M_x, M_y, and M_z, about the given axes. If p_x, p_y, and p_z are the components of linear momentum in the x, y, and z directions, the following relations hold:

$$M_x = yp_z - zp_y$$

$$M_y = zp_x - xp_z$$

$$M_z = xp_y - yp_x$$

Since (x, y, z) and (p_x, p_y, p_z) transform as shown in Eq. 7.2, it follows that

$$C_\theta \begin{bmatrix} M_x \\ M_y \\ M_z \end{bmatrix} = \begin{bmatrix} \cos\theta & \sin\theta & 0 \\ -\sin\theta & \cos\theta & 0 \\ 0 & 0 & 1 \end{bmatrix} \begin{bmatrix} M_x \\ M_y \\ M_z \end{bmatrix}$$

Then a similar relation holds for R_x, R_y, and R_z:

$$C_\theta \begin{bmatrix} R_x \\ R_y \\ R_z \end{bmatrix} = \begin{bmatrix} \cos\theta & \sin\theta & 0 \\ -\sin\theta & \cos\theta & 0 \\ 0 & 0 & 1 \end{bmatrix} \begin{bmatrix} R_x \\ R_y \\ R_z \end{bmatrix}$$

Thus the characters for the proper rotations are given by Eq. 7.14. The same result is obtained for the improper rotation if the latter is regarded as $i \times$ (proper rotation). Therefore the character for the vibration is obtained from

$$\boxed{\chi_v(R) = \chi(R) - \chi_t(R) - \chi_r(R)} \tag{7.15}$$

It is convenient to tabulate the foregoing calculations as in Table I-5. By using the formula in Eq. 6.6 and the character of the irreducible representations in Table I-4, a_m can be calculated as follows:

$$a_m(A_1) = \tfrac{1}{6}[(1)(6)(1) + (2)(0)(1) + (3)(2)(1)] = 2$$

$$a_m(A_2) = \tfrac{1}{6}[(1)(6)(1) + (2)(0)(1) + (3)(2)(-1)] = 0$$

$$a_m(E) = \tfrac{1}{6}[(1)(6)(2) + (2)(0)(-1) + (3)(2)(0)] = 2$$

and

$$\chi_v = 2\chi_{A_1} + 2\chi_E \tag{7.16}$$

TABLE I-5

Symmetry operation	I	$2C_3$	$3\sigma_v$
Kind of rotation	proper		improper
θ	0°	120°	180°
$\cos\theta$	1	$-\frac{1}{2}$	-1
$1+2\cos\theta$	3	0	-1
N_R	4	1	2
χ, $\pm N_R(1+2\cos\theta)$	12	0	2
χ_t, $\pm(1+2\cos\theta)$	3	0	1
χ_r, $+(1+2\cos\theta)$	3	0	-1
χ_v, $\chi-\chi_t-\chi_r$	6	0	2

In other words, the six normal vibrations of a pyramidal XY_3 molecule are classified into two A_1 and two E species.

This procedure is applicable to any molecule. As another example, a similar calculation is shown in Table I-6 for an octahedral XY_6 molecule. By use of Eq. 6.6 and the character table in Appendix I, the a_m are obtained as

$$\begin{aligned} a_m(A_{1g}) = \tfrac{1}{48}[&(1)(15)(1)+(8)(0)(1)+(6)(1)(1)+(6)(1)(1)\\ &+(3)(-1)(1)+(1)(-3)(1)+(6)(-1)(1)+(8)(0)(1)\\ &+(3)(5)(1)+(6)(3)(1)]\\ \doteq 1& \end{aligned}$$

$$\begin{aligned} a_m(A_{1u}) = \tfrac{1}{48}[&(1)(15)(1)+(8)(0)(1)+(6)(1)(1)+(6)(1)(1)\\ &+(3)(-1)(1)+(1)(-3)(-1)+(6)(-1)(-1)+(8)(0)(-1)\\ &+(3)(5)(-1)+(6)(3)(-1)]\\ =0& \\ \vdots \end{aligned}$$

and therefore

$$\chi_v = \chi_{A_{1g}} + \chi_{E_g} + 2\chi_{F_{1u}} + \chi_{F_{2g}} + \chi_{F_{2u}}$$

TABLE I-6

Symmetry operation	I	$8C_3$	$6C_2$	$6C_4$	$3C_4^2 \equiv C_2''$	$S_2 \equiv i$	$6S_4$	$8S_6$	$3\sigma_h$ [a]	$6\sigma_d$ [a]
Kind of rotation	proper					improper				
θ	0°	120°	180°	90°	180°	0°	90°	120°	180°	180°
$\cos\theta$	1	$-\frac{1}{2}$	-1	0	-1	1	0	$-\frac{1}{2}$	-1	-1
$1+2\cos\theta$	3	0	-1	1	-1	3	1	0	-1	-1
N_R	7	1	1	3	3	1	1	1	5	3
χ, $\pm N_R(1+2\cos\theta)$	21	0	-1	3	-3	-3	-1	0	5	3
χ_t, $\pm(1+2\cos\theta)$	3	0	-1	1	-1	-3	-1	0	1	1
χ_r, $+(1+2\cos\theta)$	3	0	-1	1	-1	3	1	0	-1	-1
χ_v, $\chi-\chi_t-\chi_r$	15	0	1	1	-1	-3	-1	0	5	3

[a] σ_h = horizontal plane of symmetry; σ_d = diagonal plane of symmetry.

I-8. INTERNAL COORDINATES

In Sec. I-3, the potential and the kinetic energies were expressed in terms of rectangular coordinates. If, instead, these energies are expressed in terms of *internal coordinates* such as increments of the bond length and bond angle, the corresponding force constants have clearer physical meanings than those expressed in terms of rectangular coordinates, since these force constants are characteristic of the bond stretching and the angle deformation involved. The number of internal coordinates must be equal to, or greater than, $3N-6$ (or $3N-5$), the degrees of vibrational freedom of an N-atom molecule. If more than $3N-6$ (or $3N-5$) coordinates are selected as the internal coordinates, this means that these coordinates are not independent of each other. Figure I-11 illustrates the internal coordinates for various types of molecules.

In linear XYZ (a), bent XY_2 (b), and pyramidal XY_3 (c) molecules, the number of internal coordinates is the same as the number of normal vibrations. In a nonplanar X_2Y_2 molecule (d) such as H_2O_2, the number of internal coordinates is the same as the number of vibrations if the twisting angle around the central bond ($\Delta\tau$) is considered. In a tetrahedral XY_4 molecule (e), however, the number of internal coordinates exceeds the number of normal vibrations by one. This is due to the fact that the six angle coordinates around the central atom are not independent of each other, that is, they must satisfy the relation

$$\Delta\alpha_{12}+\Delta\alpha_{23}+\Delta\alpha_{31}+\Delta\alpha_{41}+\Delta\alpha_{42}+\Delta\alpha_{43}=0 \qquad (8.1)$$

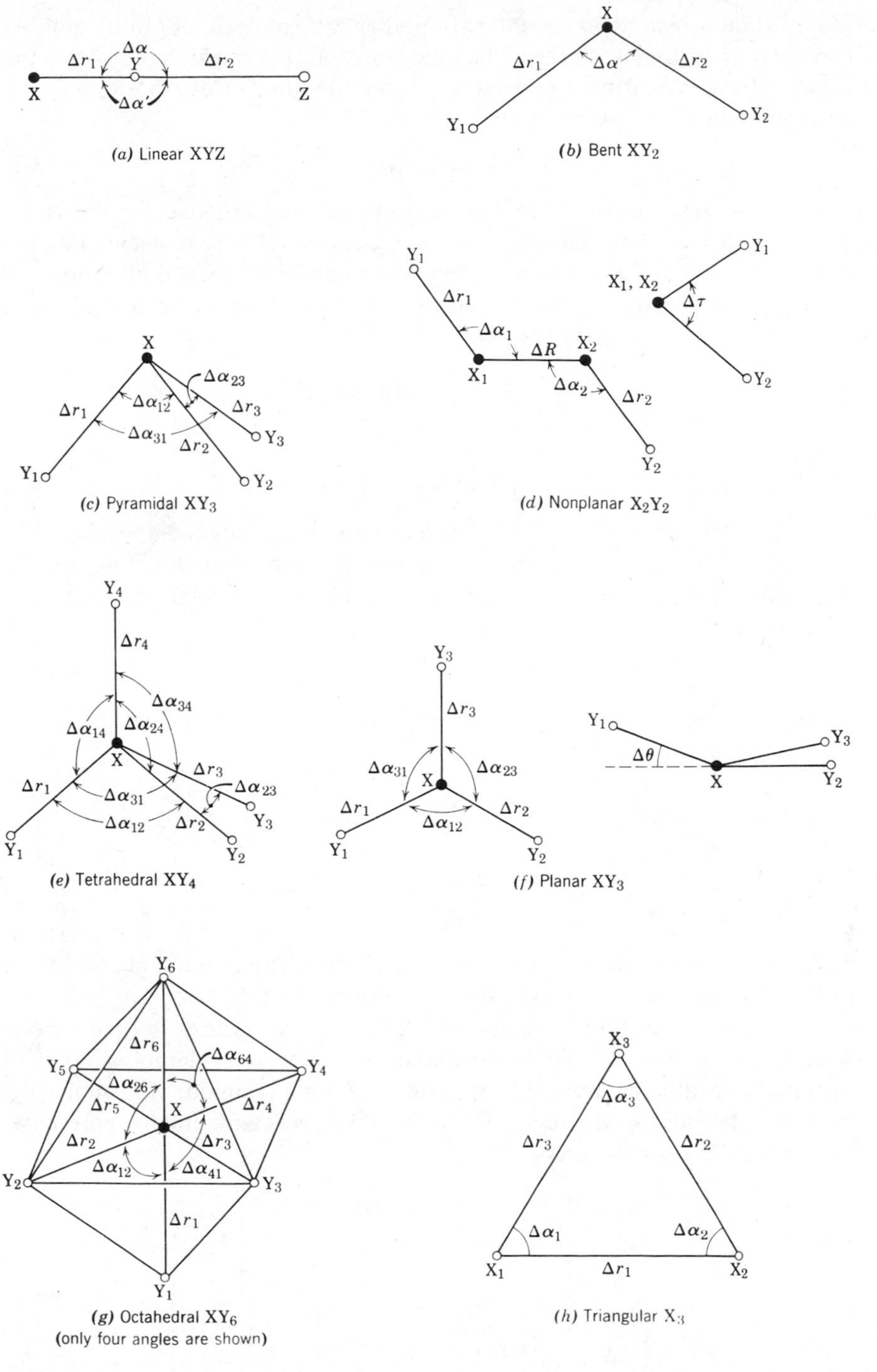

Fig. I-11. Internal coordinates for various molecules.

This is called a *redundant condition.* In planar XY_3 molecule (f), the number of internal coordinates is seven when the coordinate $\Delta\theta$, which represents the deviation from planarity, is considered. Since the number of vibrations is six, one redundant condition such as

$$\Delta\alpha_{12}+\Delta\alpha_{23}+\Delta\alpha_{31}=0 \tag{8.2}$$

must be involved. Such redundant conditions always exist for the angle coordinates around the central atom. In an octahedral XY_6 molecule (g), the number of internal coordinates exceeds the number of normal vibrations by three. This means that, of the 12 angle coordinates around the central atom, three redundant conditions are involved:

$$\begin{aligned} &\Delta\alpha_{12}+\Delta\alpha_{26}+\Delta\alpha_{64}+\Delta\alpha_{41}=0\\ &\Delta\alpha_{15}+\Delta\alpha_{56}+\Delta\alpha_{63}+\Delta\alpha_{31}=0\\ &\Delta\alpha_{23}+\Delta\alpha_{34}+\Delta\alpha_{45}+\Delta\alpha_{52}=0 \end{aligned} \tag{8.3}$$

The redundant conditions are more complex in ring compounds. For example, the number of internal coordinates in a triangular X_3 molecule (h) exceeds the number of vibrations by three. One of these redundant conditions (A_1' species) is

$$\Delta\alpha_1+\Delta\alpha_2+\Delta\alpha_3=0 \tag{8.4}$$

The other two redundant conditions (E' species) involve bond stretching and angle deformation coordinates such as

$$\begin{aligned} &(2\Delta r_1-\Delta r_2-\Delta r_3)+\frac{r}{\sqrt{3}}(\Delta\alpha_1+\Delta\alpha_2-2\Delta\alpha_3)=0\\ &(\Delta r_2-\Delta r_3)-\frac{r}{\sqrt{3}}(\Delta\alpha_1-\Delta\alpha_2)=0 \end{aligned} \tag{8.5}$$

where r is the equilibrium length of the X–X bond. The redundant conditions mentioned above can be derived by using the method described in Sec. I-11.

The procedure for finding the number of normal vibrations in each species was described in Sec. I-7. This procedure is, however, considerably simplified if internal coordinates are used. Again consider a pyramidal XY_3 molecule. Using the internal coordinates shown in Fig. I-11*c*, we can write the representation for the C_3^+ operation as

$$C_3^+\begin{bmatrix}\Delta r_1\\ \Delta r_2\\ \Delta r_3\\ \Delta\alpha_{12}\\ \Delta\alpha_{23}\\ \Delta\alpha_{31}\end{bmatrix}=\begin{bmatrix}0&0&1&0&0&0\\1&0&0&0&0&0\\0&1&0&0&0&0\\0&0&0&0&0&1\\0&0&0&1&0&0\\0&0&0&0&1&0\end{bmatrix}\begin{bmatrix}\Delta r_1\\ \Delta r_2\\ \Delta r_3\\ \Delta\alpha_{12}\\ \Delta\alpha_{23}\\ \Delta\alpha_{31}\end{bmatrix} \tag{8.6}$$

Thus $\chi(C_3^+) = 0$, as does $\chi(C_3^-)$. Similarly, $\chi(I) = 6$ and $\chi(\sigma_v) = 2$. This result is exactly the same as that obtained in Table I-5 using rectangular coordinates. *When using internal coordinates, however, the character of the representation is simply given by the number of internal coordinates unchanged by each symmetry operation.*

If this procedure is made separately for stretching (Δr) and bending ($\Delta\alpha$) coordinates, it is readily seen that

$$\begin{aligned} \chi^r(R) &= \chi_{A_1} + \chi_E \\ \chi^\alpha(R) &= \chi_{A_1} + \chi_E \end{aligned} \tag{8.7}$$

Thus it is found that both A_1 and E species have one stretching and one bending vibration, respectively. No consideration of the translational and rotational motions is necessary if the internal coordinates are taken as the basis for the representation.

Another example, for an octahedral XY_6 molecule, is given in Table I-7. Using Eq. 6.6 and the character table in Appendix I, we find that these characters are resolved into

$$\chi^r(R) = \chi_{A_{1g}} + \chi_{E_g} + \chi_{F_{1u}} \tag{8.8}$$

$$\chi^\alpha(R) = \chi_{A_{1g}} + \chi_{E_g} + \chi_{F_{1u}} + \chi_{F_{2g}} + \chi_{F_{2u}} \tag{8.9}$$

Comparison of this result with that obtained in Sec. I-7 immediately suggests that three redundant conditions are included in these bending vibrations (one in A_{1g} and one in E_g). Therefore $\chi^\alpha(R)$ for genuine vibrations becomes

$$\chi^\alpha(R) = \chi_{F_{1u}} + \chi_{F_{2g}} + \chi_{F_{2u}} \tag{8.10}$$

Thus it is concluded that six stretching and nine bending vibrations are distributed as indicated in Eqs. 8.8 and 8.10, respectively. Although the method given above is simpler than that of Sec. I-7, caution must be exercised with respect to the bending vibrations whenever redundancy is involved. In such a case, comparison of the results obtained from both methods is useful in finding the species of redundancy.

I.9. SELECTION RULES FOR INFRARED AND RAMAN SPECTRA

According to quantum mechanics,[4,5] the selection rule for the infrared spectrum is determined by the integral:

$$[\mu]_{v'v''} = \int \psi_{v'}(Q_a)\mu\psi_{v''}(Q_a)\, dQ_a \tag{9.1}$$

Here μ is the dipole moment in the electronic ground state, ψ is the vibrational eigenfunction given by Eq. 2.7, and v' and v'' are the vibrational quantum numbers before and after the transition, respectively. The activity of the normal vibration whose normal coordinate is Q_a is being determined. By resolving

TABLE I-7

	I	$8C_3$	$6C_2$	$6C_4$	$3C_4^2 \equiv C_2''$	$S_2 \equiv i$	$6S_4$	$8S_6$	$3\sigma_h$	$6\sigma_d$
$\chi^r(R)$	6	0	0	2	2	0	0	0	4	2
$\chi^\alpha(R)$	12	0	2	0	0	0	0	0	4	2

the dipole moment into the three components in the x, y, and z directions, we obtain the result

$$
\begin{aligned}
[\mu_y]_{v'v''} &= \int \psi_{v'}(Q_a)\mu_x\psi_{v''}(Q_a)\, dQ_a \\
[\mu_y]_{v'v''} &= \int \psi_{v'}(Q_a)\mu_y\psi_{v''}(Q_a)\, dQ_a \\
[\mu_z]_{v'v''} &= \int \psi_{v'}(Q_a)\mu_z\psi_{v''}(Q_a)\, dQ_a
\end{aligned}
\tag{9.2}
$$

If one of these integrals is not zero, the normal vibration associated with Q_a is infrared active. If all the integrals are zero, the vibration is infrared inactive.

Similarly, the selection rule for the Raman spectrum is determined by the integral:

$$
[\alpha]_{v'v''} = \int \psi_{v'}(Q_a)\alpha\psi_{v''}(Q_n)\, dQ_a \tag{9.3}
$$

As shown in Sec. I-5, α consists of six components, α_{xx}, α_{yy}, α_{zz}, α_{xy}, α_{yz}, and α_{xz}. Thus Eq. 9.3 may be resolved into six components:

$$
\begin{aligned}
[\alpha_{xx}]_{v'v''} &= \int \psi_{v'}(Q_a)\alpha_{xx}\psi_{v''}(Q_a)\, dQ_a \\
[\alpha_{yy}]_{v'v''} &= \int \psi_{v'}(Q_a)\alpha_{yy}\psi_{v''}(Q_a)\, dQ_a \\
&\vdots
\end{aligned}
\tag{9.4}
$$

If one of these integrals is not zero, the normal vibration associated with Q_a is Raman active. If all the integrals are zero, the vibration is Raman inactive.

It is possible to decide whether the integrals of Eqs. 9.2 and 9.4 are zero or nonzero from a consideration of symmetry. As stated in Sec. I-1, the vibrations of interest are the fundamentals in which transitions occur from $v'=0$ to $v''=1$. It is evident from the form of the vibrational eigenfunction (Eq. 2.8) that $\psi_0(Q_a)$ is invariant under any symmetry operation, whereas the symmetry of $\psi_1(Q_a)$ is the same as that of Q_a. Thus the integral does not vanish when the symmetry of μ_x, for example, is the same as that of Q_a. If the symmetry properties of μ_x and Q_a differ in even one symmetry element of the group, the integral becomes zero. In other words, for the integral to be nonzero, Q_a must belong to the same species as μ_x. More generally, the normal vibration associated with Q_a becomes infrared active when at least one of the components of the dipole moments belongs to the same species as Q_a. Similar conclusions are obtained for the Raman spectrum.

Since the species of the normal vibration can be determined by the methods described in Secs. I-7 and I-8, it is necessary only to determine the species of the components of the dipole moment and polarizability of the molecule. This can be done as follows. The components of the dipole moment, μ_x, μ_y, and μ_z, transform as do those of translational motion, T_x, T_y, and T_z, respectively. These were discussed in Sec. I-7. Thus the character of the dipole moment is given by Eq. 7.13, which is

$$\boxed{\chi_\mu(R) = \pm(1+2\cos\theta)} \tag{9.5}$$

where + and − have the same meaning as before. In a pyramidal XY_3 molecule, Eq. 9.5 gives

	I	$2C_3$	$3\sigma_v$
$\chi_\mu(R)$	3	0	1

Using Eq. 6.6, we resolve this into $A_1 + E$. It is obvious that μ_z belongs to A_1. Then μ_x and μ_y must belong to E. In fact, the pair, μ_x and μ_y, transforms as follows:

$$I\begin{bmatrix}\mu_x\\ \mu_y\end{bmatrix} = \begin{bmatrix}1 & 0\\ 0 & 1\end{bmatrix}\begin{bmatrix}\mu_x\\ \mu_y\end{bmatrix}, \qquad C_3^+\begin{bmatrix}\mu_x\\ \mu_y\end{bmatrix} = \begin{bmatrix}-\frac{1}{2} & \frac{\sqrt{3}}{2}\\ -\frac{\sqrt{3}}{2} & -\frac{1}{2}\end{bmatrix}\begin{bmatrix}\mu_x\\ \mu_y\end{bmatrix}$$

$$\chi(I) = 2, \qquad \chi(C_3^+) = -1$$

$$\sigma_1\begin{bmatrix}\mu_x\\ \mu_y\end{bmatrix} = \begin{bmatrix}-1 & 0\\ 0 & 1\end{bmatrix}\begin{bmatrix}\mu_x\\ \mu_y\end{bmatrix}$$

$$\chi(\sigma_1) = 0$$

Thus it is found that μ_z belongs to A_1 and (μ_x, μ_y) belongs to E.

The character of the representation of the polarizability is given by

$$\boxed{\chi_\alpha(R) = 2\cos\theta(1+2\cos\theta)} \tag{9.6}$$

for both proper and improper rotations. This can be derived as follows. The polarizability in the x, y, and z directions is related to that in X, Y, and Z

coordinates by

$$
\begin{bmatrix} \alpha_{XX} & \alpha_{XY} & \alpha_{XZ} \\ \alpha_{YX} & \alpha_{YY} & \alpha_{YZ} \\ \alpha_{ZX} & \alpha_{ZY} & \alpha_{ZZ} \end{bmatrix}
= \begin{bmatrix} C_{Xx} & C_{Xy} & C_{Xz} \\ C_{Yx} & C_{Yy} & C_{Yz} \\ C_{Zx} & C_{Zy} & C_{Zz} \end{bmatrix}
\begin{bmatrix} \alpha_{xx} & \alpha_{xy} & \alpha_{xz} \\ \alpha_{yx} & \alpha_{yy} & \alpha_{yz} \\ \alpha_{zx} & \alpha_{zy} & \alpha_{zz} \end{bmatrix}
\begin{bmatrix} C_{Xx} & C_{Yx} & C_{Zx} \\ C_{Xy} & C_{Yy} & C_{Zy} \\ C_{Xz} & C_{Yz} & C_{Zz} \end{bmatrix}
$$

where C_{Xx}, C_{Xy}, and so forth, denote the direction cosines between the two axes subscripted. If a rotation through θ around the Z axis superimposes the X, Y, and Z axes on the x, y, and z axes, the preceding relation becomes

$$
C_\theta \begin{bmatrix} \alpha_{xx} & \alpha_{xy} & \alpha_{xz} \\ \alpha_{yx} & \alpha_{yy} & \alpha_{yz} \\ \alpha_{zx} & \alpha_{zy} & \alpha_{zz} \end{bmatrix}
= \begin{bmatrix} \cos\theta & \sin\theta & 0 \\ -\sin\theta & \cos\theta & 0 \\ 0 & 0 & 1 \end{bmatrix}
\begin{bmatrix} \alpha_{xx} & \alpha_{xy} & \alpha_{xz} \\ \alpha_{yx} & \alpha_{yy} & \alpha_{yz} \\ \alpha_{zx} & \alpha_{zy} & \alpha_{zz} \end{bmatrix}
\begin{bmatrix} \cos\theta & -\sin\theta & 0 \\ \sin\theta & \cos\theta & 0 \\ 0 & 0 & 1 \end{bmatrix}
$$

This can be written as

$$
C_\theta \begin{bmatrix} \alpha_{xx} \\ \alpha_{yy} \\ \alpha_{zz} \\ \alpha_{xy} \\ \alpha_{xz} \\ \alpha_{yz} \end{bmatrix}
= \begin{bmatrix}
\cos^2\theta & \sin^2\theta & 0 & 2\sin\theta\cos\theta & 0 & 0 \\
\sin^2\theta & \cos^2\theta & 0 & -2\sin\theta\cos\theta & 0 & 0 \\
0 & 0 & 1 & 0 & 0 & 0 \\
-\sin\theta\cos\theta & \sin\theta\cos\theta & 0 & 2\cos^2\theta - 1 & 0 & 0 \\
0 & 0 & 0 & 0 & \cos\theta & \sin\theta \\
0 & 0 & 0 & 0 & -\sin\theta & \cos\theta
\end{bmatrix}
\begin{bmatrix} \alpha_{xx} \\ \alpha_{yy} \\ \alpha_{zz} \\ \alpha_{xy} \\ \alpha_{xz} \\ \alpha_{yz} \end{bmatrix}
$$

Thus the character of this representation is given by Eq. 9.6. The same results are obtained for improper rotations if they are regarded as the product $i \times$(proper rotation). For a pyramidal XY_3 molecule, Eq. 9.6 gives

	I	$2C_3$	$3\sigma_v$
$\chi_\alpha(R)$	6	0	2

Using Eq. 6.6, this is resolved into $2A_1 + 2E$. Again, it is immediately seen that the component α_{zz} belongs to A_1, and the pair α_{zx} and α_{zy} belongs to E since

$$\begin{bmatrix} zx \\ zy \end{bmatrix} = z \begin{bmatrix} x \\ y \end{bmatrix} \approx A_1 \times E = E$$

It is more convenient to consider the components $\alpha_{xx} + \alpha_{yy}$ and $\alpha_{xx} - \alpha_{yy}$ than a_{xx} and a_{yy}. If a vector of unit length is considered, the relation

$$x^2 + y^2 + z^2 = 1$$

holds. Since α_{zz} belongs to A_1, $\alpha_{xx} + \alpha_{yy}$ must belong to A_1. Then the pair $\alpha_{xx} - \alpha_{yy}$ and α_{xy} must belong to E. As a result, the character table of the point group $\mathbf{C}_{3v}$ is completed as in Table I-8. Thus it is concluded that, in the point group $\mathbf{C}_{3v}$, both the A_1 and the E vibrations are infrared as well as Raman active, while the A_2 vibrations are inactive.

Complete character tables like Table I-8 have already been worked out for all the point groups. Therefore no elaborate treatment such as that described in this section is necessary in practice. Appendix I gives complete character tables for the point groups that appear frequently in this book. From these tables, the selection rules for the infrared and Raman spectra are obtained immediately: *The vibration is infrared or Raman active if it belongs to the same species as one of the components of the dipole moment or polarizability, respectively.* For example, the character table of the point group $\mathbf{O}_h$ signifies immediately that only the F_{1u} vibrations are infrared active and only the A_{1g}, E_g, and F_{2g} vibrations are Raman active, for the components of the dipole moment or the polarizability belong to these species in this point group. It is to be noted in these character tables that (1) a totally symmetric vibration is Raman active in any point group, and (2) the infrared and Raman-active vibrations always belong to u and g types, respectively, in point groups having a center of symmetry.

As stated in Sec. I-2, some overtones and combination bands are observed weakly because the actual vibrations are not harmonic and some of them are allowed by symmetry selection rules. For the symmetry selection rules of these nonfundamental vibrations, see Refs. 3, 7, and 8.

TABLE I-8. CHARACTER TABLE OF THE POINT GROUP $\mathbf{C}_{3v}$

$\mathbf{C}_{3v}$	I	$2C_3$	$3\sigma_v$		
A_1	+1	+1	+1	μ_2	$\alpha_{xx} + \alpha_{yy}$, α_{zz}
A_2	+1	+1	−1		
E	+2	−1	0	(μ_x, μ_y)[a]	$(\alpha_{xz}, \alpha_{yz})$,[a] $(\alpha_{xx} - \alpha_{yy}, \alpha_{xy})$[a]

[a] A doubly degenerate pair is represented by two terms in parentheses.

I-10. STRUCTURE DETERMINATION

Suppose that a molecule has several probable structures each of which belongs to a different point group. Then the number of infrared and Raman-active fundamentals should be different for each structure. Therefore the most probable model can be selected by comparing the observed number of infrared- and Raman-active fundamentals with that predicted theoretically for each model.

Consider the XeF_4 molecule as an example. It may be tetrahedral or square-planar. By use of the methods described in the preceding sections, the number of infrared or Raman-active fundamentals can be found easily for each structure. Tables I-9*a* and I-9*b* summarize the results. It is seen that the tetrahedral structure predicts two infrared-active fundamentals (one stretching and one bending), whereas the square-planar structure predicts three infrared-active fundamentals (one stretching and two bendings). The infrared spectrum of XeF_4 in the vapor phase exhibits one XeF stretching at 586 cm^{-1} and two FXeF bendings at 291 and 161 cm^{-1} (Ref. II-635). Thus the square-planar structure of $\mathbf{D}_{4h}$ symmetry is preferable to the tetrahedral structure of $\mathbf{T}_d$ symmetry.* As is seen in Tables I-9*a* and I-9*b*, group theory predicts two Raman-active XeF stretchings for both structures, but two Raman-active bendings for the tetrahedral and one Raman-active bending for the square-planar structure. The observed Raman spectrum (554, 524, and 218 cm^{-1}) again confirms the square-planar structure.

This method is widely used for elucidation of the molecular structure of inorganic, organic, and coordination compounds. In Part II and Appendix III, the number of infrared- and Raman-active fundamentals is compared for XY_3 (planar, $\mathbf{D}_{3h}$, and pyramidal, $\mathbf{C}_{3v}$), XY_4 (square-planar, $\mathbf{D}_{4h}$, and tetrahedral, $\mathbf{T}_d$), XY_5 (trigonal-bipyramidal, $\mathbf{D}_{3h}$, and tetragonal-pyramidal, $\mathbf{C}_{4v}$) and other molecules. Recently, the structures of various metal carbonyl compounds (Sec. III-16) have been determined by this simple technique.

It should be noted, however, that this method does not give a clear-cut answer if the predicted numbers of infrared- and Raman-active fundamentals are similar for various probable structures. Furthermore, a practical difficulty arises in determining the number of fundamentals from the observed spectrum, since the intensities of overtone and combination bands are sometimes comparable to those of fundamentals when they appear as satellite bands of the fundamental. This is particularly true when overtone and combination bands are enhanced anomalously by *Fermi resonance* (accidental degeneracy). For example, the frequency of the first overtone of the ν_2 vibration of CO_2 (667 cm^{-1}) is very close to that of the ν_1 vibration (1337 cm^{-1}). Since these two vibrations belong to the same symmetry species (Σ_g^+), they interact with

---

* This conclusion may be drawn directly from observation of the mutual exclusion rule, which holds for $\mathbf{D}_{4h}$ but not for $\mathbf{T}_d$.

TABLE I-9*a*. NUMBER OF FUNDAMENTALS FOR TETRAHEDRAL XeF_4

$\mathbf{T}_d$	Activity	Number of Fundamentals	XeF Stretching	FXeF Bending
A_1	R	1	1	0
A_2	ia	0	0	0
E	R	1	0	1
F_1	ia	0	0	0
F_2	IR, R	2	1	1
Total	IR	2	1	1
	R	4	2	2

TABLE I-9*b*. NUMBER OF FUNDAMENTALS FOR SQUARE-PLANAR XeF_4

$\mathbf{D}_{4h}$	Activity	Number of Fundamentals	XeF Stretching	FXeF Bending
A_{1g}	R	1	1	0
A_{1u}	ia	0	0	0
A_{2g}	ia	0	0	0
A_{2u}	IR	1	0	1
B_{1g}	R	1	1	0
B_{1u}	ia	0	0	0
B_{2g}	R	1	0	1
B_{2u}	ia	1	0	1
E_g	R	0	0	0
E_u	IR	2	1	1
Total	IR	3	1	2
	R	3	2	1

each other and give rise to two strong Raman lines at 1388 and 1286 cm^{-1}. Fermi resonances similar to the resonance observed for CO_2 may occur for a number of other molecules. It is to be noted also that the number of observed bands depends on the resolving power of the instrument used. Finally, the molecular symmetry in the isolated state is not necessarily the same as that in the crystalline state (Sec. I-23). Therefore this method must be applied with caution to spectra obtained for compounds in the crystalline state.

I-11. PRINCIPLE OF THE GF MATRIX METHOD*

As described in Sec. I-3, the frequency of the normal vibration is determined by the kinetic and potential energies of the system. The kinetic energy is determined by the masses of the individual atoms and their geometrical arrangement in the molecule. On the other hand, the potential energy arises from interaction between the individual atoms and is described in terms of the force constants. Since the potential energy provides valuable information about the nature of interatomic forces, it is highly desirable to obtain the force constants from the observed frequencies. This is usually done by calculating the frequencies, assuming a suitable set of force constants. If the agreement between the calculated and observed frequencies is satisfactory, this particular set of the force constants is adopted as a representation of the potential energy of the system.

To calculate the vibrational frequencies, it is necessary first to express both the potential and the kinetic energies in terms of some common coordinates (Sec. I-3). Internal coordinates (Sec. I-8) are more suitable for this purpose than rectangular coordinates, since (1) force constants expressed in terms of internal coordinates have clearer physical meanings than those expressed in terms of rectangular coordinates, and (2) a set of internal coordinates does not involve translational and rotational motion of the molecule as a whole.

Using the internal coordinates, R_i, we write the potential energy as

$$2V = \tilde{\mathbf{R}}\mathbf{F}\mathbf{R} \tag{11.1}$$

For a bent Y_1XY_2 molecule such as that in Fig. I-11*b*, **R** is a column matrix of the form

$$\mathbf{R} = \begin{bmatrix} \Delta r_1 \\ \Delta r_2 \\ \Delta\alpha \end{bmatrix}$$

$\tilde{\mathbf{R}}$ is its transpose:

$$\tilde{\mathbf{R}} = [\Delta r_1 \quad \Delta r_2 \quad \Delta\alpha]$$

and **F** is a matrix whose components are the force constants:

$$\mathbf{F} = \begin{bmatrix} f_{11} & f_{12} & r_1 f_{13} \\ f_{21} & f_{22} & r_2 f_{23} \\ r_1 f_{31} & r_2 f_{32} & r_1 r_2 f_{33} \end{bmatrix} \equiv \begin{bmatrix} F_{11} & F_{12} & F_{13} \\ F_{21} & F_{22} & F_{23} \\ F_{31} & F_{32} & F_{33} \end{bmatrix} \qquad (11.2)\dagger$$

Here r_1 and r_2 are the equilibrium lengths of the $X–Y_1$ and $X–Y_2$ bonds, respectively.

---

* For details, see Refs. 3–8. The term "normal coordinate analysis" is almost synonymous with the **GF** matrix method, since most of the normal coordinate calculations are carried out by using this method.

†Here f_{11} and f_{22} are the stretching force constants of the $X–Y_1$ and $X–Y_2$ bonds, respectively, and f_{33} is the bending force constant of the Y_1XY_2 angle. The other symbols represent interaction force constants between stretching and stretching or between stretching and bending vibrations. To make the dimensions of all the force constants the same, f_{13} (or f_{31}), f_{23} (or f_{32}), and f_{33} are multiplied by r_1, r_2, and r_1r_2, respectively.

The kinetic energy is not easily expressed in terms of the same internal coordinates. Wilson[62] has shown, however, that the kinetic energy can be written as

$$2T = \tilde{\dot{\mathbf{R}}}\mathbf{G}^{-1}\dot{\mathbf{R}} \tag{11.3}*$$

where $\mathbf{G}^{-1}$ is the reciprocal of the **G** matrix, which will be defined later.

If Eqs. 11.1 and 11.3 are combined with Newton's equation,

$$\frac{d}{dt}\left(\frac{\partial T}{\partial \dot{R}_k}\right) + \frac{\partial V}{\partial R_k} = 0 \tag{3.6}$$

the following secular equation, which is similar to Eq. 3.19, is obtained:[3,7]

$$\begin{vmatrix} F_{11}-(G^{-1})_{11}\lambda & F_{12}-(G^{-1})_{12}\lambda & \cdots \\ F_{21}-(G^{-1})_{21}\lambda & F_{22}-(G^{-1})_{22}\lambda & \cdots \\ \vdots & \vdots & \end{vmatrix} \equiv |\mathbf{F}-\mathbf{G}^{-1}\lambda| = 0 \tag{11.4}$$

By multiplying by the determinant of **G**

$$\begin{vmatrix} G_{11} & G_{12} & \cdots \\ G_{21} & G_{22} & \cdots \\ \vdots & \vdots & \end{vmatrix} \equiv |\mathbf{G}| \tag{11.5}$$

from the left-hand side of Eq. 11.4, the following equation is obtained:

$$\begin{vmatrix} \sum G_{1t}F_{t1}-\lambda & \sum G_{1t}F_{t2} & \cdots \\ \sum G_{2t}F_{t1} & \sum G_{2t}F_{t2}-\lambda & \cdots \\ \vdots & \vdots & \end{vmatrix} \equiv |\mathbf{GF}-\mathbf{E}\lambda| = 0 \tag{11.6}$$

Here **E** is the unit matrix, and λ is related to the wave number $\tilde{\nu}$ by the relation $\lambda = 4\pi^2c^2\tilde{\nu}^2$.† The order of the equation is equal to the number of internal coordinates used.

The **F** matrix can be written by assuming a suitable set of force constants. If the **G** matrix is constructed by the following method, the vibrational frequencies are obtained by solving Eq. 11.6. The **G** matrix is defined as

$$\mathbf{G} = \mathbf{B}\mathbf{M}^{-1}\tilde{\mathbf{B}} \tag{11.7}$$

Here $\mathbf{M}^{-1}$ is a diagonal matrix whose components are μ_i, where μ_i is the reciprocal of the mass of the ith atom. For a bent XY_2 molecule,

$$\mathbf{M}^{-1} = \begin{bmatrix} \mu_1 & & & & 0 \\ & \mu_1 & & & \\ & & \mu_1 & & \\ & & & \ddots & \\ 0 & & & & \mu_3 \end{bmatrix}$$

*Appendix IV gives the derivation of Eq. 11.3.

†Here λ should not be confused with λ_w (wavelength).

where μ_3 and μ_1 are the reciprocals of the masses of the X and Y atoms, respectively. The **B** matrix is defined as

$$\mathbf{R} = \mathbf{B}\mathbf{X} \tag{11.8}$$

where **R** and **X** are column matrices whose components are the internal and rectangular coordinates, respectively. For a bent XY_2 molecule, Eq. 11.8 is written as

$$\begin{bmatrix} \Delta r_1 \\ \Delta r_2 \\ \Delta\alpha \end{bmatrix} = \left[\begin{array}{ccc|ccc|ccc} -s & -c & 0 & 0 & 0 & 0 & s & c & 0 \\ 0 & 0 & 0 & s & -c & 0 & -s & c & 0 \\ -c/r & s/r & 0 & c/r & s/r & 0 & 0 & -2s/r & 0 \end{array}\right] \begin{bmatrix} \Delta x_1 \\ \Delta y_1 \\ \Delta z_1 \\ \hline \Delta x_2 \\ \Delta y_2 \\ \Delta z_2 \\ \hline \Delta x_3 \\ \Delta y_3 \\ \Delta z_3 \end{bmatrix} \tag{11.9}$$

where $s = \sin(\alpha/2)$, $c = \cos(\alpha/2)$, and r is the equilibrium distance between X and Y (see Fig. I-12).

If unit vectors such as those in Fig. I-12 are considered, Eq. 11.9 can be written in a more compact form using vector notation:

$$\begin{bmatrix} \Delta r_1 \\ \Delta r_2 \\ \Delta\alpha \end{bmatrix} = \begin{bmatrix} \mathbf{e}_{31} & 0 & -\mathbf{e}_{31} \\ 0 & \mathbf{e}_{32} & -\mathbf{e}_{32} \\ \mathbf{p}_{31}/r & \mathbf{p}_{32}/r & -(\mathbf{p}_{31}+\mathbf{p}_{32})/r \end{bmatrix} \begin{bmatrix} \boldsymbol{\rho}_1 \\ \boldsymbol{\rho}_2 \\ \boldsymbol{\rho}_3 \end{bmatrix} \tag{11.10}$$

Here $\boldsymbol{\rho}_1$, $\boldsymbol{\rho}_2$, and $\boldsymbol{\rho}_3$ are the displacement vectors of atoms 1, 2, and 3, respectively. Thus Eq. 11.10 can be written simply as

$$\mathbf{R} = \mathbf{S} \cdot \boldsymbol{\rho} \tag{11.11}$$

where the dot represents the scalar product of the two vectors. Here **S** is called the **S** matrix, and its components (**S** vector) can be written according to the following formulas: (1) bond stretching,

$$\Delta r_1 = \Delta r_{31} = \mathbf{e}_{31} \cdot \boldsymbol{\rho}_1 - \mathbf{e}_{31} \cdot \boldsymbol{\rho}_3 \tag{11.12}$$

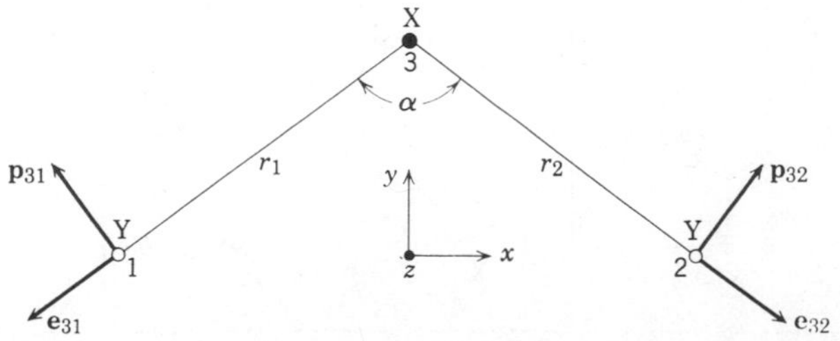

Fig. I-12. Unit vectors in a bent XY_2 molecule.

and (2) angle bending,

$$\Delta\alpha = \Delta\alpha_{132} = [\mathbf{p}_{31}\cdot\boldsymbol{\rho}_1 + \mathbf{p}_{32}\cdot\boldsymbol{\rho}_2 - (\mathbf{p}_{31}+\mathbf{p}_{32})\cdot\boldsymbol{\rho}_3]/r \tag{11.13}$$

It is seen that the direction of the **S** vector is the direction in which a given displacement of the *i*th atom will produce the greatest increase in Δr or $\Delta\alpha$. Formulas for obtaining the **S** vectors of other internal coordinates such as those of out-of-plane ($\Delta\theta$) and torsional ($\Delta\tau$) vibrations are also available.[3]

By using the **S** matrix, Eq. 11.7 is written as

$$\mathbf{G} = \mathbf{S}\mathbf{m}^{-1}\tilde{\mathbf{S}} \tag{11.14}$$

For a bent XY_2 molecule, this becomes

$$\mathbf{G} = \begin{bmatrix} \mathbf{e}_{31} & 0 & -\mathbf{e}_{31} \\ 0 & \mathbf{e}_{32} & -\mathbf{e}_{32} \\ \mathbf{p}_{31}/r & \mathbf{p}_{32}/r & -(\mathbf{p}_{31}+\mathbf{p}_{32})/r \end{bmatrix} \begin{bmatrix} \mu_1 & 0 & 0 \\ 0 & \mu_1 & 0 \\ 0 & 0 & \mu_3 \end{bmatrix}$$

$$\times \begin{bmatrix} \mathbf{e}_{31} & 0 & \mathbf{p}_{31}/r \\ 0 & \mathbf{e}_{32} & \mathbf{p}_{32}/r \\ -\mathbf{e}_{31} & -\mathbf{e}_{32} & -(\mathbf{p}_{31}+\mathbf{p}_{32})/r \end{bmatrix}$$

$$= \begin{bmatrix} (\mu_3+\mu_1)\mathbf{e}_{31}^2 & \mu_3\mathbf{e}_{31}\cdot\mathbf{e}_{32} & \frac{\mu_1}{r}\mathbf{e}_{31}\cdot\mathbf{p}_{31} + \frac{\mu_3}{r}\mathbf{e}_{31}\cdot(\mathbf{p}_{31}+\mathbf{p}_{32}) \\ & (\mu_3+\mu_1)\mathbf{e}_{32}^2 & \frac{\mu_1}{r}\mathbf{e}_{32}\cdot\mathbf{p}_{32} + \frac{\mu_3}{r}\mathbf{e}_{32}\cdot(\mathbf{p}_{31}+\mathbf{p}_{32}) \\ & & \frac{\mu_1}{r^2}\mathbf{p}_{31}^2 + \frac{\mu_1}{r^2}\mathbf{p}_{32}^2 + \frac{\mu_3}{r^2}(\mathbf{p}_{31}+\mathbf{p}_{32})^2 \end{bmatrix}$$

Considering

$$\mathbf{e}_{31}\cdot\mathbf{e}_{31} = \mathbf{e}_{32}\cdot\mathbf{e}_{32} = \mathbf{p}_{31}\cdot\mathbf{p}_{31} = \mathbf{p}_{32}\cdot\mathbf{p}_{32} = 1, \qquad \mathbf{e}_{31}\cdot\mathbf{p}_{31} = \mathbf{e}_{32}\cdot\mathbf{p}_{32} = 0$$

$$\mathbf{e}_{31}\cdot\mathbf{e}_{32} = \cos\alpha, \qquad \mathbf{e}_{31}\cdot\mathbf{p}_{32} = \mathbf{e}_{32}\cdot\mathbf{p}_{31} = -\sin\alpha$$

$$(\mathbf{p}_{31}+\mathbf{p}_{32})^2 = 2(1-\cos\alpha)$$

we find that the **G** matrix is calculated as

$$\mathbf{G} = \begin{bmatrix} \mu_3+\mu_1 & \mu_3\cos\alpha & -\frac{\mu_3}{r}\sin\alpha \\ & \mu_3+\mu_1 & -\frac{\mu_3}{r}\sin\alpha \\ & & \frac{2\mu_1}{r^2} + \frac{2\mu_3}{r^2}(1-\cos\alpha) \end{bmatrix} \tag{11.15}$$

If the **G** matrix elements obtained are written for each combination of internal coordinates, there results

$$G(\Delta r_1, \Delta r_1) = \mu_3 + \mu_1$$

$$G(\Delta r_2, \Delta r_2) = \mu_3 + \mu_1$$

$$G(\Delta r_1, \Delta r_2) = \mu_3 \cos \alpha$$

$$G(\Delta \alpha, \Delta \alpha) = \frac{2\mu_1}{r^2} + \frac{2\mu_3}{r^2}(1 - \cos \alpha) \tag{11.16}$$

$$G(\Delta r_1, \Delta \alpha) = -\frac{\mu_3}{r} \sin \alpha$$

$$G(\Delta r_2, \Delta \alpha) = -\frac{\mu_3}{r} \sin \alpha$$

If such calculations are made for several types of molecules, it is immediately seen that the **G** matrix elements themselves have many regularities. Decius[63] developed general formulas for writing **G** matrix elements.* Some of them are as follows:

$$G_{rr}^2 = \mu_1 + \mu_2$$

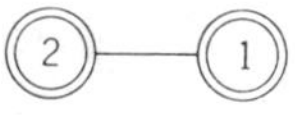

$$G_{rr}^1 = \mu_1 \cos \phi$$

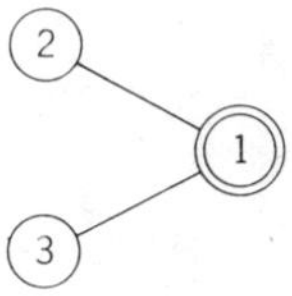

$$G_{r\phi}^2 = -\rho_{23}\mu_2 \sin \phi$$

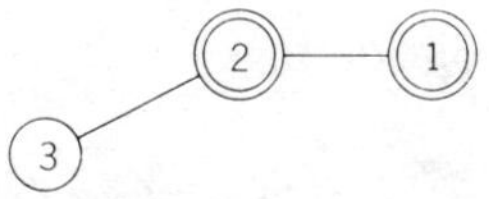

$$G_{r\phi}^1 \begin{pmatrix} 1 \\ 1 \end{pmatrix} = -(\rho_{13} \sin \phi_{213} \cos \psi_{234} + \rho_{14} \sin \phi_{214} \cos \psi_{243})\mu_1$$

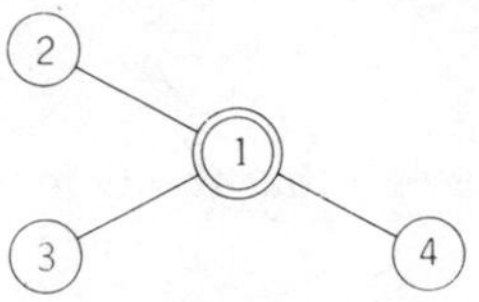

$$G_{\phi\phi}^3 = \rho_{12}^2\mu_1 + \rho_{23}^2\mu_3 + (\rho_{12}^2 + \rho_{23}^2 - 2\rho_{12}\rho_{23} \cos \phi)\mu_2$$

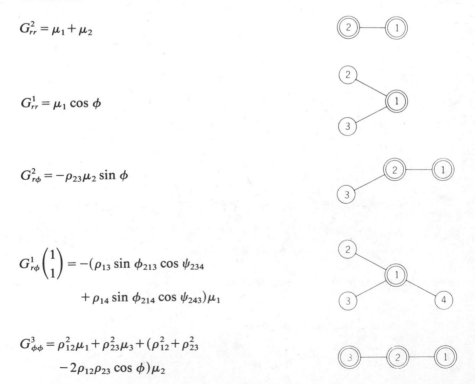

* See also Refs. 3 and 64.

$$G^2_{\phi\phi}\begin{pmatrix}1\\1\end{pmatrix} = (\rho_{12}^2 \cos \psi_{314})\mu_1 + [(\rho_{12} - \rho_{23} \cos \phi_{123} - \rho_{24} \cos \phi_{124})\rho_{12} \cos \psi_{314} + (\sin \phi_{123} \sin \phi_{124} \sin^2 \psi_{314} + \cos \phi_{324} \cos \psi_{314})\rho_{23}\rho_{24}]\mu_2$$

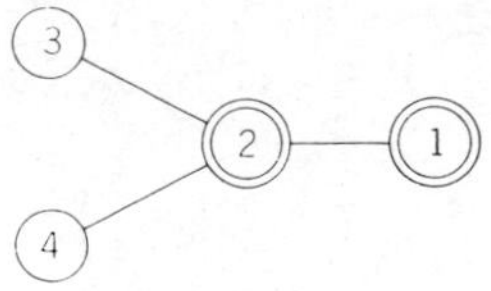

Here the atoms surrounded by a double circle are those common to both coordinates. The symbols μ and ρ denote the reciprocals of mass and bond distance, respectively. The solid angle $\psi_{\alpha\beta\gamma}$ in Fig. I-13 is defined as

$$\cos \psi_{\alpha\beta\gamma} = \frac{\cos \phi_{\alpha\delta\gamma} - \cos \phi_{\alpha\delta\beta} \cos \phi_{\beta\delta\gamma}}{\sin \phi_{\alpha\delta\beta} \sin \phi_{\beta\delta\gamma}} \tag{11.17}$$

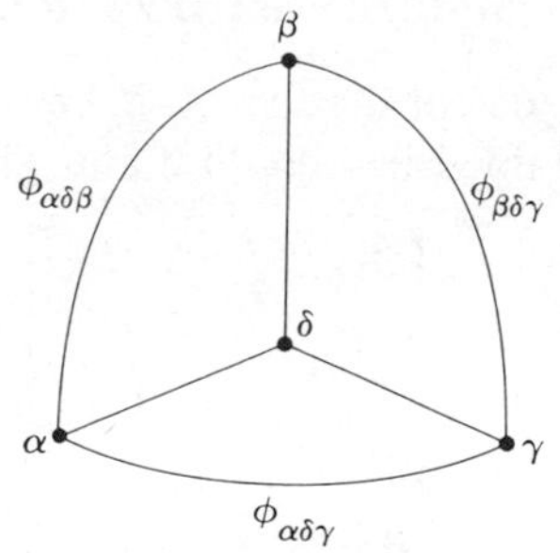

Fig. I-13. Solid angles involving atomic positions α, β, γ, and δ.

The correspondence between the Decius formulas and the results obtained in Eq. 11.16 is evident.

With the Decius formulas, the **G** matrix elements of a pyramidal XY_3 molecule have been calculated and are shown in Table I-10.

TABLE I-10

	Δr_1	Δr_2	Δr_3	$\Delta\alpha_{23}$	$\Delta\alpha_{31}$	$\Delta\alpha_{12}$
Δr_1	A	B	B	C	D	D
Δr_2	—	A	B	D	C	D
Δr_3	—	—	A	D	D	C
$\Delta\alpha_{23}$	—	—	—	E	F	F
$\Delta\alpha_{31}$	—	—	—	—	E	F
$\Delta\alpha_{12}$	—	—	—	—	—	E

$$A = G_{rr}^2 = \mu_X + \mu_Y$$

$$B = G_{rr}^1 = \mu_X \cos\alpha$$

$$C = G_{r\phi}^1\begin{pmatrix}1\\1\end{pmatrix} = -\frac{2}{r}\frac{\cos\alpha(1-\cos\alpha)\mu_X}{\sin\alpha}$$

$$D = G_{r\phi}^2 = -\frac{\mu_X}{r}\sin\alpha$$

$$E = G_{\phi\phi}^3 = \frac{2}{r^2}[\mu_Y + \mu_X(1-\cos\alpha)]$$

$$F = G_{\phi\phi}^2\begin{pmatrix}1\\1\end{pmatrix} = \frac{\mu_Y}{r^2}\frac{\cos\alpha}{1+\cos\alpha} + \frac{\mu_X}{r^2}\frac{(1+3\cos\alpha)(1-\cos\alpha)}{1+\cos\alpha}$$

I-12. UTILIZATION OF SYMMETRY PROPERTIES

In view of the equivalence of the two X–Y bonds of a bent XY_2 molecule, the **F** and **G** matrices obtained in Eqs. 11.2 and 11.15 are written as

$$\mathbf{F} = \begin{bmatrix} f_{11} & f_{12} & rf_{13} \\ f_{12} & f_{11} & rf_{13} \\ rf_{13} & rf_{13} & r^2 f_{33} \end{bmatrix} \tag{12.1}$$

$$\mathbf{G} = \begin{bmatrix} \mu_3+\mu_1 & \mu_3\cos\alpha & -\frac{\mu_3}{r}\sin\alpha \\ \mu_3\cos\alpha & \mu_3+\mu_1 & -\frac{\mu_3}{r}\sin\alpha \\ -\frac{\mu_3}{r}\sin\alpha & -\frac{\mu_3}{r}\sin\alpha & \frac{2\mu_1}{r^2}+\frac{2\mu_3}{r^2}(1-\cos\alpha) \end{bmatrix} \tag{12.2}$$

Both of these matrices are of the form

$$\begin{bmatrix} A & C & D \\ C & A & D \\ D & D & B \end{bmatrix} \tag{12.3}$$

The appearance of the same elements is evidently due to the equivalence of the two internal coordinates, Δr_1 and Δr_2. Such symmetrically equivalent sets of internal coordinates are seen in many other molecules, such as those in Fig. I-11. In these cases, it is possible to reduce the order of the **F** and **G** matrices (and hence the order of the secular equation resulting from them) by a coordinate transformation.

Let the internal coordinates R be transformed by

$$\mathbf{R}^s = \mathbf{UR} \tag{12.4}$$

Then

$$2T = \tilde{\dot{\mathbf{R}}}\mathbf{G}^{-1}\dot{\mathbf{R}} = \tilde{\dot{\mathbf{R}}}^s\tilde{\mathbf{U}}^{-1}\mathbf{G}^{-1}\mathbf{U}^{-1}\dot{\mathbf{R}}^s$$
$$= \tilde{\dot{\mathbf{R}}}^s\mathbf{G}_s^{-1}\dot{\mathbf{R}}^s$$
$$2V = \tilde{\mathbf{R}}\mathbf{F}\mathbf{R} = \tilde{\mathbf{R}}^s\tilde{\mathbf{U}}^{-1}\mathbf{F}\mathbf{U}^{-1}\mathbf{R}^s$$
$$= \tilde{\mathbf{R}}^s\mathbf{F}_s\mathbf{R}^s$$

where

$$\mathbf{G}_s^{-1} = \tilde{\mathbf{U}}^{-1}\mathbf{G}^{-1}\mathbf{U}^{-1} \quad \text{or} \quad \mathbf{G}_s = \mathbf{U}\mathbf{G}\tilde{\mathbf{U}}$$
$$\mathbf{F}_s = \tilde{\mathbf{U}}^{-1}\mathbf{F}\mathbf{U}^{-1} \qquad (12.5)$$

If $\mathbf{U}$ is an orthogonal matrix ($\mathbf{U}^{-1} = \tilde{\mathbf{U}}$), Eq. 12.5 is written as

$$\mathbf{F}_s = \mathbf{U}\mathbf{F}\tilde{\mathbf{U}} \quad \text{and} \quad \mathbf{G}_s = \mathbf{U}\mathbf{G}\tilde{\mathbf{U}} \qquad (12.6)$$

Both $\mathbf{GF}$ and $\mathbf{G}_s\mathbf{F}_s$ give the same roots, since

$$|\mathbf{G}_s\mathbf{F}_s - \mathbf{E}\lambda| = |\mathbf{U}\mathbf{G}\tilde{\mathbf{U}}\tilde{\mathbf{U}}^{-1}\mathbf{F}\mathbf{U}^{-1} - \mathbf{E}\lambda|$$
$$= |\mathbf{U}\mathbf{G}\mathbf{F}\mathbf{U}^{-1} - \mathbf{E}\lambda|$$
$$= |\mathbf{U}||\mathbf{G}\mathbf{F} - \mathbf{E}\lambda||\mathbf{U}^{-1}| \qquad (12.7)$$
$$= |\mathbf{G}\mathbf{F} - \mathbf{E}\lambda|$$

If we choose a proper $\mathbf{U}$ matrix from symmetry consideration, it is possible to factor the original $\mathbf{G}$ and $\mathbf{F}$ matrices into smaller ones. This, in turn, reduces the order of the secular equation to be solved, thus facilitating their solution. Their new coordinates $\mathbf{R}^s$ are called *symmetry coordinates.*

The $\mathbf{U}$ matrix is constructed by using the equation

$$\mathbf{R}^s = N \sum_K \chi_i(K)K(\Delta r_1) \qquad (12.8)$$

Here K is a symmetry operation, and the summation is made over all symmetry operations. Also, $\chi_i(K)$ is the character of the representation to which R^s belongs. Called a generator, Δr_1 is, by symmetry operation K, transformed into $K(\Delta r_1)$, which is another coordinate of the same symmetrically equivalent set. Finally, N is a normalizing factor.

As an example, consider a bent XY_2 molecule in which Δr_1 and Δr_2 are equivalent. Using Δr_1 as a generator, we obtain

	I	$C_2(z)$	$\sigma(xz)$	$\sigma(yz)$
$K(\Delta r_1)$	Δr_1	Δr_2	Δr_2	Δr_1
$\chi_{A_1}(K)$	1	1	1	1
$\chi_{B_2}(K)$	1	−1	−1	1

Thus

$$R^s_{A_1} = N \sum \chi_{A_1}(K) K(\Delta r_1) = 2N(\Delta r_1 + \Delta r_2)$$

$$N = \frac{1}{2\sqrt{2}} \quad \text{since } (2N)^2 + (2N)^2 = 1$$

Then

$$R^s_{A_1} = \frac{1}{\sqrt{2}}(\Delta r_1 + \Delta r_2) \tag{12.9}$$

Similarly,

$$R^s_{B_2} = \frac{1}{\sqrt{2}}(\Delta r_1 - \Delta r_2) \tag{12.10}$$

The remaining internal coordinate, $\Delta\alpha$, belongs to the A_1 species. Thus the complete **U** matrix is written as

$$\begin{bmatrix} R^s_1(A_1) \\ R^s_2(A_1) \\ R^s_3(B_2) \end{bmatrix} = \begin{bmatrix} \frac{1}{\sqrt{2}} & \frac{1}{\sqrt{2}} & 0 \\ 0 & 0 & 1 \\ \frac{1}{\sqrt{2}} & \frac{-1}{\sqrt{2}} & 0 \end{bmatrix} \begin{bmatrix} \Delta r_1 \\ \Delta r_2 \\ \Delta\alpha \end{bmatrix} \tag{12.11}$$

If the **G** and **F** matrices of type 12.3 are transformed by relations 12.6, where **U** is given by the matrix of Eq. 12.11, they become

$$\mathbf{F_s}, \mathbf{G_s} = \left[\begin{array}{cc|c} A+C & \sqrt{2}D & 0 \\ \sqrt{2}D & B & 0 \\ \hline 0 & 0 & A-C \end{array}\right] \tag{12.12}$$

or, more explicitly,

$$\mathbf{F_s} = \left[\begin{array}{cc|c} f_{11}+f_{12} & r\sqrt{2}f_{13} & 0 \\ r\sqrt{2}f_{13} & r^2 f_{33} & 0 \\ \hline 0 & 0 & f_{11}-f_{12} \end{array}\right] \tag{12.13}$$

$$\mathbf{G_s} = \left[\begin{array}{cc|c} \mu_3(1+\cos\alpha)+\mu_1 & -\frac{\sqrt{2}}{r}\mu_3 \sin\alpha & 0 \\ -\frac{\sqrt{2}}{r}\mu_3 \sin\alpha & \frac{2\mu_1}{r^2}+\frac{2\mu_3}{r^2}(1-\cos\alpha) & 0 \\ \hline 0 & 0 & \mu_3(1-\cos\alpha)+\mu_1 \end{array}\right] \tag{12.14}$$

In a pyramidal XY_3 molecule (Fig. I-11*c*), Δr_1, Δr_2, and Δr_3 are the equivalent set; so are $\Delta\alpha_{23}$, $\Delta\alpha_{31}$, and $\Delta\alpha_{12}$. It is already known from Eq. 8.7 that one A_1 and one E vibration are involved both in the stretching and in the bending vibrations. Using Δr_1 as a generator, we obtain from Eq. 12.8

	I	C_3^+	C_3^-	σ_1	σ_2	σ_3
$K(\Delta r_1)$	Δr_1	Δr_2	Δr_3	Δr_1	Δr_3	Δr_2
$\chi_{A_1}(K)$	1	1	1	1	1	1
$\chi_E(K)$	2	-1	-1	0	0	0

Then

$$R_{A_1}^s = \frac{1}{\sqrt{3}}(\Delta r_1 + \Delta r_2 + \Delta r_3) \tag{12.15}$$

$$R_{E_1}^s = \frac{1}{\sqrt{6}}(2\Delta r_1 - \Delta r_2 - \Delta r_3) \tag{12.16}$$

To find a coordinate that forms a degenerate pair with Eq. 12.16, we repeat the same procedure, using Δr_2 and Δr_3 as the generators. The results are

$$R_{E_2}^s = N(2\Delta r_2 - \Delta r_3 - \Delta r_1)$$

$$R_{E_3}^s = N(2\Delta r_3 - \Delta r_1 - \Delta r_2)$$

If we take a linear combination, $R_{E_2}^s + R_{E_3}^s$, we obtain Eq. 12.16. If we take $R_{E_2}^s - R_{E_3}^s$, we obtain

$$R_{E_4}^s = \frac{1}{\sqrt{2}}(\Delta r_2 - \Delta r_3) \tag{12.17}$$

Since Eqs. 12.16 and 12.17 are mutually orthogonal, these two coordinates are taken as a degenerate pair. Similar results are obtained for three angle-bending coordinates. Thus the complete **U** matrix is written as

$$\begin{bmatrix} R_1^s(A_1) \\ R_2^s(A_1) \\ R_{3a}^s(E) \\ R_{4a}^s(E) \\ R_{3b}^s(E) \\ R_{4b}^s(E) \end{bmatrix} = \begin{bmatrix} 1/\sqrt{3} & 1/\sqrt{3} & 1/\sqrt{3} & 0 & 0 & 0 \\ 0 & 0 & 0 & 1/\sqrt{3} & 1/\sqrt{3} & 1/\sqrt{3} \\ 2/\sqrt{6} & -1/\sqrt{6} & -1/\sqrt{6} & 0 & 0 & 0 \\ 0 & 0 & 0 & 2/\sqrt{6} & -1/\sqrt{6} & -1/\sqrt{6} \\ 0 & 1/\sqrt{2} & -1/\sqrt{2} & 0 & 0 & 0 \\ 0 & 0 & 0 & 0 & 1/\sqrt{2} & -1/\sqrt{2} \end{bmatrix} \begin{bmatrix} \alpha r_1 \\ \Delta r_2 \\ \Delta r_3 \\ \Delta\alpha_{23} \\ \Delta\alpha_{31} \\ \Delta\alpha_{12} \end{bmatrix} \tag{12.18}$$

The **G** matrix of a pyramidal XY_3 molecule has already been calculated (see Table I-10). By using Eq. 12.6 the new $\mathbf{G_s}$ matrix becomes

$$\mathbf{G_s} = \left[\begin{array}{cc:cc:cc} A+2B & C+2D & & & & \\ C+2D & E+2F & & 0 & & 0 \\ \hdashline & & A-B & C-D & & \\ & 0 & C-D & E-F & & 0 \\ \hdashline & & & & A-B & C-D \\ & 0 & & 0 & C-D & E-F \end{array}\right] \tag{12.19}$$

Here A, B, and so forth, denote the elements in Table I-10. The **F** matrix transforms similarly. Therefore it is necessary only to solve two quadratic equations for the A_1 and E species.

For the tetrahedral XY_4 molecule shown in Fig. I-11*e*, group theory (Secs. I-7 and I-8) predicts one A_1 and one F_1 stretching, and one E and one F_2 bending, vibration. The **U** matrix for the four stretching coordinates is

$$\begin{bmatrix} R_1^s(A_1) \\ R_{2a}^s(F_2) \\ R_{2b}^s(F_2) \\ R_{2c}^s(F_2) \end{bmatrix} = \begin{bmatrix} 1/2 & 1/2 & 1/2 & 1/2 \\ 1/\sqrt{6} & 1/\sqrt{6} & -2/\sqrt{6} & 0 \\ 1/\sqrt{12} & 1/\sqrt{12} & 1/\sqrt{12} & -3/\sqrt{12} \\ -1/\sqrt{2} & 1/\sqrt{2} & 0 & 0 \end{bmatrix} \begin{bmatrix} \Delta r_1 \\ \Delta r_2 \\ \Delta r_3 \\ \Delta r_4 \end{bmatrix} \quad (12.20)$$

whereas the **U** matrix for the six bending coordinates becomes

$$\begin{bmatrix} R_1^s(A_1) \\ R_{2a}^s(E) \\ R_{2b}^s(E) \\ R_{3a}^s(F_2) \\ R_{3b}^s(F_2) \\ R_{3c}^s(F_2) \end{bmatrix} = \begin{bmatrix} 1/\sqrt{6} & 1/\sqrt{6} & 1/\sqrt{6} & 1/\sqrt{6} & 1/\sqrt{6} & 1/\sqrt{6} \\ 2/\sqrt{12} & -1/\sqrt{12} & -1/\sqrt{12} & -1/\sqrt{12} & -1/\sqrt{12} & 2/\sqrt{12} \\ 0 & 1/2 & -1/2 & 1/2 & -1/2 & 0 \\ 2/\sqrt{12} & -1/\sqrt{12} & -1/\sqrt{12} & 1/\sqrt{12} & 1/\sqrt{12} & -2/\sqrt{12} \\ 1/\sqrt{6} & 1/\sqrt{6} & 1/\sqrt{6} & -1/\sqrt{6} & -1/\sqrt{6} & -1/\sqrt{6} \\ 0 & 1/2 & -1/2 & -1/2 & 1/2 & 0 \end{bmatrix} \begin{bmatrix} \Delta\alpha_{12} \\ \Delta\alpha_{23} \\ \Delta\alpha_{31} \\ \Delta\alpha_{14} \\ \Delta\alpha_{24} \\ \Delta\alpha_{34} \end{bmatrix} \quad (12.21)$$

The symmetry coordinate $R_1^s(A_1)$ in Eq. 12.21 represents a *redundant coordinate* (see Eq. 8.1). In such a case, a coordinate transformation reduces the order of the matrix by one, since all the **G** matrix elements related to this coordinate become zero. Conversely, this result provides a general method of finding redundant coordinates. Suppose that the elements of the **G** matrix are calculated in terms of internal coordinates such as those in Table I-10. If a suitable combination of internal coordinates is made so that $\sum_j G_{ij} = 0$ (where j refers to all the equivalent internal coordinates), such a combination is a redundant coordinate. By using the **U** matrices in Eqs. 12.20 and 12.21, the problem of solving a tenth-order secular equation for a tetrahedral XY_4 molecule is reduced to that of solving two first-order (A_1 and E) and one quadratic (F_2) equation.

I-13. POTENTIAL FIELDS AND FORCE CONSTANTS

Using Eqs. 11.1 and 12.1, we write the potential energy of a bent XY_2 molecule as

$$2V = f_{11}(\Delta r_1)^2 + f_{11}(\Delta r_2)^2 + f_{33}r^2(\Delta\alpha)^2 + 2f_{12}(\Delta r_1)(\Delta r_2)$$
$$+ 2f_{13}r(\Delta r_1)(\Delta\alpha) + 2f_{13}r(\Delta r_2)(\Delta\alpha) \qquad (13.1)$$

This type of potential field is called a *generalized valence force* (GVF) field.* It consists of stretching and bending force constants, as well as the interaction force constants between them. When using such a potential field, four force constants are needed to describe the potential energy of a bent XY_2 molecule. Since only three vibrations are observed in practice, it is impossible to determine all four force constants simultaneously. One method used to circumvent this difficulty is to calculate the vibrational frequencies of isotopic molecules (e.g., D_2O and HDO for H_2O), assuming the same set of force constants.† This method is satisfactory, however, only for simple molecules. As molecules become more complex, the number of interaction force constants in the GVF field becomes too large to allow any reliable evaluation.

In another approach, Shimanouchi[65] introduced the *Urey-Bradley force* (UBF) field, which consists of stretching and bending force constants, as well as repulsive force constants between nonbonded atoms. The general form of the potential field is given by

$$V = \sum_i \left(\frac{1}{2}K_i(\Delta r_i)^2 + K'_i r_i(\Delta r_i)\right) + \sum_i \left(\frac{1}{2}H_i r_{i\alpha}^2(\Delta\alpha_i)^2 + H'_i r_{i\alpha}^2(\Delta\alpha_i)\right)$$
$$+ \sum_i \left(\frac{1}{2}F_i(\Delta q_i)^2 + F'_i q_i(\Delta q_i)\right) \qquad (13.2)$$

Here Δr_i, $\Delta\alpha_i$, and Δq_i are the changes in the bond lengths, bond angles, and distances between nonbonded atoms, respectively. The symbols K_i, K'_i, H_i, H'_i, and F_i, F'_i represent the stretching, bending, and repulsive force constants, respectively. Furthermore, r_i, $r_{i\alpha}$, and q_i are the values of the distances at the equilibrium positions and are inserted to make the force constants dimensionally similar.

Using the relation

$$q_{ij}^2 = r_i^2 + r_j^2 - 2r_ir_j \cos \alpha_{ij} \qquad (13.3)$$

---

* A potential field consisting of stretching and bending force constants only is called *simple valence force* field.

† In addition to isotope frequency shifts, mean amplitudes of vibration, Coriolis coupling constants, centrifugal distortion constants, and so forth may be used to refine the force constants of small molecules (see Ref. 66).

and considering that the first derivatives can be equated to zero in the equilibrium case, we can write the final form of the potential field as

$$V=\frac{1}{2}\sum_i\left[K_i+\sum_{j(\neq i)}(t_{ij}^2F'_{ij}+s_{ij}^2F_{ij})\right](\Delta r_i)^2$$
$$+\frac{1}{2}\sum_{i<j}(H_{ij}-s_{ij}s_{ji}F'_{ij}+t_{ij}t_{ji}F_{ij})(\sqrt{r_ir_j}\,\Delta\alpha_{ij})^2$$
$$+\sum_{i<j}(-t_{ij}t_{ji}F'_{ij}+s_{ij}s_{ji}F_{ij})(\Delta r_i)(\Delta r_j)$$
$$+\sum_{i\neq j}(t_{ij}s_{ji}F'_{ij}+t_{ji}s_{ij}F_{ij})\left(\frac{r_j}{r_i}\right)^{1/2}(\Delta r_i)(\sqrt{r_ir_j}\,\Delta\alpha_{ij}) \qquad (13.4)^*$$

Here

$$s_{ij}=\frac{r_i-r_j\cos\alpha_{ij}}{q_{ij}}$$
$$s_{ji}=\frac{r_j-r_i\cos\alpha_{ij}}{q_{ij}}$$
$$t_{ij}=\frac{r_j\sin\alpha_{ij}}{q_{ij}} \qquad (13.5)$$
$$t_{ji}=\frac{r_i\sin\alpha_{ij}}{q_{ij}}$$

In a bent XY_2 molecule, Eq. 13.4 becomes

$$V=\tfrac{1}{2}(K+t^2F'+s^2F)[(\Delta r_1)^2+(\Delta r_2)^2]+\tfrac{1}{2}(H-s^2F'+t^2F)(r\,\Delta\alpha)^2$$
$$+(-t^2F'+s^2F)(\Delta r_1)(\Delta r_2)+ts(F'+F)(\Delta r_1)(r\,\Delta\alpha)$$
$$+ts(F'+F)(\Delta r_2)(r\,\Delta\alpha) \qquad (13.6)$$

where

$$s=\frac{r(1-\cos\alpha)}{q}$$
$$t=\frac{r\sin\alpha}{q}$$

Comparing Eqs. 13.6 and 13.1, we obtain the following relations between the force constants of the generalized valence force field and those of the

---

* In the case of tetrahedral molecules, a term

$$\sum_{i\neq j\neq k}\left(\frac{\kappa}{\sqrt{2}}\right)r_{ij}r_{ik}(r_{ij}\,\Delta\alpha_{ij})(r_{ik}\,\Delta\alpha_{ik})$$

must be added, where κ is called the internal tension.

Urey-Bradley force field:

$$
\begin{aligned}
f_{11} &= K + t^2F' + s^2F \\
r^2f_{33} &= (H - s^2F' + t^2F)r^2 \\
f_{12} &= -t^2F' + s^2F \\
rf_{13} &= ts(F' + F)r
\end{aligned}
\qquad (13.7)
$$

Although the Urey-Bradley field has four force constants, F' is usually taken as $-\frac{1}{10}F$, on the assumption that the repulsive energy between nonbonded atoms is proportional to $1/r$.[9]* Thus only three force constants, K, H, and F, are needed to construct the **F** matrix. The *orbital valence force* (OVF) field developed by Heath and Linnett[67] is similar to the UBF field. The OVF field uses the angle ($\Delta\beta$) which represents the distortion of the bond from the axis of the bonding orbital instead of the angle between two bonds ($\Delta\alpha$).

The number of force constants in the Urey-Bradley field is, in general, much smaller than that in the generalized valence force field. In addition, the UBF field has the advantage that (1) the force constants have clearer physical meanings than those of the GVF field, and (2) they are often transferable from molecule to molecule. For example, the force constants obtained for $SiCl_4$ and $SiBr_4$ can be used for $SiCl_3Br$, $SiCl_2Br_2$, and $SiClBr_3$. Mizushima, Shimanouchi, and their co-workers[66] and Overend and Scherer[68] have given many examples that demonstrate the transferability of the force constants in the UBF field. This property of the Urey-Bradley force constants is highly useful in calculations for complex molecules. It should be mentioned, however, that ignorance of the interactions between nonneighboring stretching vibrations and between bending vibrations in the Urey-Bradley field sometimes causes difficulties in adjusting the force constants to fit the observed frequencies. In such a case, it is possible to improve the results by introducing more force constants.[68,69]

Evidently, the values of force constants depend on the force field initially assumed. Thus a comparison of force constants between molecules should not be made unless they are obtained by using the same force field. The normal coordinate analysis developed in Secs. I-11 to I-13 has already been applied to a number of molecules of various structures. In Parts II, III, and IV, references are cited for each type. Appendix V lists the **G** and **F** matrix elements for typical molecules.

I-14. SOLUTION OF THE SECULAR EQUATION

Once the **G** and **F** matrices are obtained, the next step is to solve the matrix secular equation:

$$
|\mathbf{GF} - \mathbf{E}\lambda| = 0 \qquad (11.6)
$$

* This assumption does not cause serious error in final results, since F' is small in most cases.

In diatomic molecules, $\mathbf{G} = G_{11} = 1/\mu$ and $\mathbf{F} = F_{11} = K$. Then $\lambda = G_{11}F_{11}$ and $\tilde{\nu} = \sqrt{\lambda}/2\pi c = \sqrt{K/\mu}/2\pi c$ (Eq. 2.6). If the units of mass and force constant are atomic weight and mdyn/Å (or 10^5 dyn/cm), respectively,* λ is related to $\tilde{\nu}(\text{cm}^{-1})$ by

$$\tilde{\nu} = 1302.83\sqrt{\lambda}$$

or

$$\lambda = 0.58915\left(\frac{\tilde{\nu}}{1000}\right)^2 \qquad (14.1)$$

As an example, for the HF molecule $\mu = 0.9573$ and $K = 9.65$ in these units. Then, from Eqs. 2.6 and 14.1, $\tilde{\nu}$ is 4139 cm^{-1}.

The **F** and **G** matrix elements of a bent XY_2 molecule are given in Eqs. 12.15 and 12.16, respectively. The secular equation for the A_1 species is quadratic:

$$|\mathbf{GF} - \mathbf{E}\lambda| = \begin{vmatrix} G_{11}F_{11} + G_{12}F_{21} - \lambda & G_{11}F_{12} + G_{12}F_{22} \\ G_{21}F_{11} + G_{22}F_{21} & G_{21}F_{12} + G_{22}F_{22} - \lambda \end{vmatrix} = 0 \qquad (14.2)$$

If this is expanded into an algebraic equation, the following result is obtained:

$$\lambda^2 - (G_{11}F_{11} + G_{22}F_{22} + 2G_{12}F_{12})\lambda + (G_{11}G_{22} - G_{12}^2)(F_{11}F_{22} - F_{12}^2) = 0 \qquad (14.3)$$

For the H_2O molecule,

$$\mu_1 = \mu_H = \frac{1}{1.008} = 0.99206$$

$$\mu_3 = \mu_O = \frac{1}{15.995} = 0.06252$$

$$r = 0.96\ (\text{Å}), \qquad \alpha = 105°$$

$$\sin\alpha = \sin 105° = 0.96593$$

$$\cos\alpha = \cos 105° = -0.25882$$

Then the **G** matrix elements of Eq. 12.16 are

$$G_{11} = \mu_1 + \mu_3(1 + \cos\alpha) = 1.03840$$

$$G_{12} = -\frac{\sqrt{2}}{r}\mu_3 \sin\alpha = -0.08896$$

$$G_{22} = \frac{1}{r^2}[2\mu_1 + 2\mu_3(1 - \cos\alpha)] = 2.32370$$

---

* Although the bond distance is involved in both the **G** and **F** matrices, it is canceled during multiplication of the **G** and **F** matrix elements. Therefore any unit can be used for the bond distance.

If the force constants in terms of the generalized valence force field are selected as

$$f_{11}=8.428, \qquad f_{12}=-0.105$$

$$f_{13}=0.252, \qquad f_{33}=0.768$$

the **F** matrix elements of Eq. 12.15 are

$$F_{11}=f_{11}+f_{12}=8.32300$$

$$F_{12}=\sqrt{2}rf_{13}=0.35638$$

$$F_{22}=r^2f_{33}=0.70779$$

Using these values, we find that Eq. 14.3 becomes

$$\lambda^2-10.22389\lambda+13.86234=0$$

The solution of this equation gives

$$\lambda_1=8.61475, \qquad \lambda_2=1.60914$$

If these values are converted to $\tilde{\nu}$ through Eq. 14.1, we obtain

$$\tilde{\nu}_1=3824\ \text{cm}^{-1}, \qquad \tilde{\nu}_2=1653\ \text{cm}^{-1}$$

With the same set of force constants, the frequency of the B_2 vibration is calculated as

$$\lambda_3=G_{33}F_{33}=[\mu_1+\mu_3(1-\cos\alpha)](f_{11}-f_{12})$$

$$=9.13681$$

$$\tilde{\nu}_3=3938\ \text{cm}^{-1}$$

The observed frequencies corrected for anharmonicity are as follows: $\omega_1=3825\ \text{cm}^{-1}$, $\omega_2=1654\ \text{cm}^{-1}$, and $\omega_3=3936\ \text{cm}^{-1}$.

If the secular equation is third order, it gives rise to a cubic equation:

$$\begin{aligned}
&\lambda^3-(G_{11}F_{11}+G_{22}F_{22}+G_{33}F_{33}+2G_{12}F_{12}+2G_{13}F_{13}+2G_{23}F_{23})\lambda^2\\
&+\left\{\begin{vmatrix} G_{11} & G_{12} \\ G_{21} & G_{22}\end{vmatrix}\begin{vmatrix} F_{11} & F_{12} \\ F_{21} & F_{22}\end{vmatrix}+\begin{vmatrix} G_{12} & G_{13} \\ G_{22} & G_{23}\end{vmatrix}\begin{vmatrix} F_{12} & F_{13} \\ F_{22} & F_{23}\end{vmatrix}\right.\\
&+\begin{vmatrix} G_{11} & G_{13} \\ G_{21} & G_{23}\end{vmatrix}\begin{vmatrix} F_{11} & F_{13} \\ F_{21} & F_{23}\end{vmatrix}+\begin{vmatrix} G_{11} & G_{12} \\ G_{31} & G_{32}\end{vmatrix}\begin{vmatrix} F_{11} & F_{12} \\ F_{31} & F_{32}\end{vmatrix}\\
&+\begin{vmatrix} G_{12} & G_{13} \\ G_{32} & G_{33}\end{vmatrix}\begin{vmatrix} F_{12} & F_{13} \\ F_{32} & F_{33}\end{vmatrix}+\begin{vmatrix} G_{11} & G_{13} \\ G_{31} & G_{33}\end{vmatrix}\begin{vmatrix} F_{11} & F_{13} \\ F_{31} & F_{33}\end{vmatrix}\\
&+\begin{vmatrix} G_{21} & G_{22} \\ G_{31} & G_{32}\end{vmatrix}\begin{vmatrix} F_{21} & F_{22} \\ F_{31} & F_{32}\end{vmatrix}+\begin{vmatrix} G_{22} & G_{23} \\ G_{32} & G_{33}\end{vmatrix}\begin{vmatrix} F_{22} & F_{23} \\ F_{32} & F_{33}\end{vmatrix}\\
&\left.+\begin{vmatrix} G_{21} & G_{23} \\ G_{31} & G_{33}\end{vmatrix}\begin{vmatrix} F_{21} & F_{23} \\ F_{31} & F_{33}\end{vmatrix}\right\}\lambda-\begin{vmatrix} G_{11} & G_{12} & G_{13} \\ G_{21} & G_{22} & G_{23} \\ G_{31} & G_{32} & G_{33}\end{vmatrix}\begin{vmatrix} F_{11} & F_{12} & F_{13} \\ F_{21} & F_{22} & F_{23} \\ F_{31} & F_{32} & F_{33}\end{vmatrix}=0
\end{aligned} \tag{14.4}$$

Thus it is possible to solve the secular equation by expanding it into an algebraic equation. If the order of the secular equation is higher than three, however, direct expansion such as that just shown becomes too cumbersome. There are several methods of calculating the coefficients of an algebraic equation using indirect expansion.[3] The use of an electronic computer greatly reduces the burden of calculation. Excellent programs written by Schachtschneider[70] and other workers are available for the vibrational analysis of polyatomic molecules.

I-15. VIBRATIONAL FREQUENCIES OF ISOTOPIC MOLECULES

As stated in Sec. I-13, the vibrational frequencies of isotopic molecules are very useful in refining a set of force constants in vibrational analysis. For large molecules, isotopic substitution is indispensable in making band assignments, since only vibrations involving the motion of the isotopic atom will be shifted by isotopic substitution.

Two important rules hold for the vibrational frequencies of isotopic molecules. The first, called the *product rule*, can be derived as follows.

Let $\lambda_1, \lambda_2, \ldots, \lambda_n$ be the roots of the secular equation $|\mathbf{GF}-\mathbf{E}\lambda|=0$. Then

$$\lambda_1\lambda_2\cdots\lambda_n=|\mathbf{G}|\,|\mathbf{F}| \qquad (15.1)$$

holds for a given molecule. Since the isotopic molecule has exactly the same $|\mathbf{F}|$ as that in Eq. 15.1, a similar relation

$$\lambda_1'\lambda_2'\cdots\lambda_n'=|\mathbf{G}'|\,|\mathbf{F}|$$

holds for this molecule. It follows that

$$\frac{\lambda_1\lambda_2\cdots\lambda_n}{\lambda_1'\lambda_2'\cdots\lambda_n'}=\frac{|\mathbf{G}|}{|\mathbf{G}'|} \qquad (15.2)$$

Since

$$\tilde{\nu}=\frac{1}{2\pi c}\sqrt{\lambda}$$

Eq. 15.2 can be written as

$$\frac{\tilde{\nu}_1\tilde{\nu}_2\cdots\tilde{\nu}_n}{\tilde{\nu}_1'\tilde{\nu}_2'\cdots\tilde{\nu}_n'}=\sqrt{\frac{|\mathbf{G}|}{|\mathbf{G}'|}} \qquad (15.3)$$

This rule has been confirmed by using pairs of molecules such as H_2O and D_2O, CH_4 and CD_4. The rule is also applicable to the product of vibrational frequencies belonging to a single symmetry species.

A more general form of Eq. 15.3 is given by the *Redlich–Teller product rule*:[1]

$$\frac{\tilde{\nu}_1\tilde{\nu}_2\cdots\tilde{\nu}_n}{\tilde{\nu}_1'\tilde{\nu}_2'\cdots\tilde{\nu}_n'}=\sqrt{\left(\frac{m_1'}{m_1}\right)^{\alpha}\left(\frac{m_2'}{m_2}\right)^{\beta}\cdots\left(\frac{M}{M'}\right)^{t}\left(\frac{I_x}{I_x'}\right)^{\delta_x}\left(\frac{I_y}{I_y'}\right)^{\delta_y}\left(\frac{I_z}{I_z'}\right)^{\delta_z}} \qquad (15.4)$$

Here $m_1, m_2, \ldots$ are the masses of the representative atoms of the various sets of equivalent nuclei (atoms represented by m, m_0, $m_{xy}, \ldots$ in the tables given in Appendix II); $\alpha, \beta, \ldots$ are the coefficients of m, m_0, $m_{xy}, \ldots$; M is the total mass of the molecule; t is the number of T_x, T_y, T_z in the symmetry type considered; I_x, I_y, I_z are the moments of inertia about the x, y, z axes, respectively, which go through the center of the mass; and δ_x, δ_y, δ_z are 1 or 0, depending on whether or not R_x, R_y, R_z belong to the symmetry type considered. A degenerate vibration is counted only once on both sides of the equation.

Another useful rule in regard to the vibrational frequencies of isotopic molecules, called the *sum rule*, can be derived as follows. It is obvious from Eqs. 14.3 and 14.4 that

$$\lambda_1+\lambda_2+\cdots+\lambda_n=\sum_n \lambda=\sum_{i,j} G_{ij}F_{ij} \tag{15.5}$$

Let σ_k denote $\sum_{ij} G_{ij}F_{ij}$ for k different isotopic molecules, all of which have the same **F** matrix. If a suitable combination of molecules is taken, so that

$$\begin{aligned}\sigma_1+\sigma_2+\cdots+\sigma_k &= (\sum G_{ij}F_{ij})_1+(\sum G_{ij}F_{ij})_2+\cdots+(\sum G_{ij}F_{ij})_k\\ &= [(\sum G_{ij})_1+(\sum G_{ij})_2+\cdots+(\sum G_{ij})_k](\sum F_{ij})\\ &= 0\end{aligned}$$

then it follows that

$$(\sum\lambda)_1+(\sum\lambda)_2+\cdots+(\sum\lambda)_k=0 \tag{15.6}$$

This rule has been verified for such combinations as H_2O, D_2O, and HDO, where

$$2\sigma(\text{HDO})-\sigma(\text{H}_2\text{O})-\sigma(\text{D}_2\text{O})=0$$

Such relations between the frequencies of isotopic molecules are highly useful in making band assignments.

I-16. METAL-ISOTOPE SPECTROSCOPY[71,72]

As a first approximation, vibrational spectra of coordination compounds can be classified into ligand vibrations which occur in the high-frequency region (4000–600 cm^{-1}) and metal–ligand vibrations which appear in the low-frequency region (below 600 cm^{-1}). The former provide information about the effect of coordination on the electronic structure of the ligand while the latter provide direct information about the structure of the coordination sphere and the nature of the metal–ligand bond. Since the main interest of coordination chemistry is the coordinate bond, it is the metal–ligand vibrations that have held the interest of inorganic vibrational spectroscopists. It is difficult, however, to make unequivocal assignments of metal–ligand vibrations since the interpretation of the low-frequency spectrum is complicated by the appearance of ligand vibrations as well as lattice vibrations in the case of solid-state spectra.

Conventional methods which have been used to assign metal–ligand vibrations are:

1. Comparison of spectra between a free ligand and its metal complex; the metal–ligand vibration should be absent in the spectrum of the free ligand. This method often fails to give a clear-cut assignment since some ligand vibrations activated by complex formation may appear in the same region as the metal–ligand vibrations.
2. The metal–ligand vibration should be metal sensitive and be shifted by changing the metal or its oxidation state. This method is applicable only when a series of metal complexes have exactly the same structure, with only the central metal being different. Also, it does not provide definitive assignments since some ligand vibrations (such as chelate ring deformations) are also metal sensitive.
3. The metal–ligand stretching vibration should appear in the same frequency region if the metal is the same and the ligands are similar. For example, the ν(Zn–N) (ν: stretching) of Zn(II) pyridine complexes are expected to be similar to those of Zn(II) α-picoline complexes. This method is applicable only when the metal–ligand vibration is known for one parent compound.
4. The metal–ligand vibration exhibits an isotope shift if the ligand is isotopically substituted. For example, the ν(Ni–N) of $[Ni(NH_3)_6]Cl_2$ at 334 cm^{-1} is shifted to 318 cm^{-1} upon deuteration of the ammonia ligands. The observed shift (16 cm^{-1}) is in good agreement with that predicted theoretically for this mode. This method was used to assign the metal–ligand vibrations of chelate compounds such as oxamido($^{14}N/^{15}N$) and acetylacetonato($^{16}O/^{18}O$) complexes. However, isotopic substitution of the α-atom (atom directly bonded to the metal) causes shifts of not only metal–ligand vibrations but also of ligand vibrations involving the motion of the α-atom. Thus, this method alone cannot provide an unequivocal assignment of the metal–ligand vibration.
5. The frequency of a metal–ligand vibration may be predicted if the metal–ligand stretching and other force constants are known *a priori*. At present, this method is not practical since only a very limited amount of information is available on the force constants of coordination compounds.

It is obvious that none of the above methods is perfect in assigning metal–ligand vibrations. Furthermore, these methods encounter more difficulties as the structure of the complex (and hence the spectrum) becomes more complicated. Fortunately, the "metal isotope technique" which was developed in 1969 may be used to obtain reliable metal–ligand assignments.[73] Isotope pairs such as (H/D) and ($^{16}O/^{18}O$) had been used routinely by many spectroscopists. However, isotopic pairs of heavy metals such as ($^{58}Ni/^{62}Ni$) and ($^{104}Pd/^{110}Pd$) were not employed until 1969 when the first report on the assignments of the Ni–P vibrations of *trans*-$Ni(PEt_3)_2X_2$ (X = Cl and Br) was made. The delay in their use was probably due to two reasons:

1. It was thought that the magnitude of isotope shifts arising from metal isotope substitution might be too small to be of practical value.
2. Pure metal isotopes were too expensive to use routinely in the laboratory.

Nakamoto and co-workers[71,72] have shown, however, that the magnitudes of metal isotope shifts are generally of the order of 2–10 cm^{-1} for stretching modes and 0–2 cm^{-1} for bending modes, and that the experimental error in measuring the frequency could be as small as ± 0.2 cm^{-1} if proper precautions are taken. They have also shown that this technique is financially feasible (see Table I-11) if the compounds are prepared on a milligram scale. Normally, the vibrational spectrum of a compound can be obtained with a sample less than 10 mg.

TABLE I-11. SOME STABLE METAL ISOTOPES

Element	Atomic Weight	Inventory Form	Isotope	Natural Abundance (%)	Purity (%)	Price ($/mg)[a]
Chromium	51.996	Cr_2O_3	^{50}Cr	4.31	96.4–96.8	5.73
			^{52}Cr	83.76	99.74–99.90	0.82
			^{53}Cr	9.55	96.98	2.64
Iron	55.847	Fe_2O_3	^{54}Fe	5.82	97.08	1.89
			^{56}Fe	91.66	99.93–99.97	0.09
			^{57}Fe	2.19	86.06–90.24	11.02
Nickel	58.71	Ni	^{58}Ni	67.88	99.76	0.35
			^{60}Ni	26.23	99.07–99.62	0.23
			^{62}Ni	3.66	96.64–98.7	9.28
Copper	63.54	CuO	^{63}Cu	69.09	99.89	0.14
			^{65}Cu	30.91	99.69	0.29
Zinc	65.37	ZnO	^{64}Zn	48.89	99.69–99.85	0.55
			^{66}Zn	27.81	98.22–98.69	0.70
			^{68}Zn	18.57	99.26–99.34	2.50
Molybdenum	95.94	Mo	^{92}Mo	15.84	97.37–98.27	0.30
			^{95}Mo	15.72	96.47	0.28
			^{97}Mo	9.46	92.44–94.25	0.47
			^{100}Mo	9.63	97.27–97.42	0.49
Zirconium	91.22	ZrO_2	^{90}Zr	51.46	99.36	0.94
			^{94}Zr	17.40	98.58	5.12
Palladium	106.4	Pd	^{104}Pd	10.97	95.25	2.20
			^{108}Pd	26.71	—	—[b]
Tin	118.69	SnO_2	^{116}Sn	14.30	95.60	0.37
			^{120}Sn	32.85	98.05–98.39	0.39
			^{124}Sn	5.94	96.71–96.96	0.54

[a] Oak Ridge National Laboratory, 1984.
[b] Not available.

It should be noted that the central atom of a highly symmetrical molecule ($\mathbf{T}_d$, $\mathbf{O}_h$, etc.) does not move during the totally symmetric vibration. Thus, no metal-isotope shifts are expected in these cases.[74] When the central atom is coordinated by several different donor atoms, multiple isotope labeling is necessary to distinguish different coordinate bond-stretching vibrations. For example, complete assignments of bis(glycino)nickel(II) require $^{14}N/^{15}N$ and/or $^{16}O/^{18}O$ isotope shift data as well as $^{58}Ni/^{62}Ni$ isotope shift data.[75]

Parts II–V of this book include typical results obtained by using metal-isotope techniques. These results show that metal-isotope data are indispensable not only in assigning the metal–ligand vibrations but also in refining metal–ligand stretching force constants in normal coordinate analysis. The presence of vibrational coupling between metal–ligand and other vibrations can also be detected by combining metal-isotope data with normal coordinate calculations (Sec. I-17) since both experimental and theoretical isotope shift values would be smaller when such couplings occur.

The metal-isotope techniques become more important as the molecules become larger and complex. In biological molecules such as heme proteins, structural and bonding information about the active site (iron porphyrin) can be obtained through definitive assignments of coordinate bond-stretching vibrations around the iron atom. Using resonance Raman techniques (Sec. I-21), it is possible to observe iron porphyrin and iron-axial ligand vibrations without interference from peptide chain vibrations. Thus, these vibrations can be assigned by comparing resonance Raman spectra of a natural heme protein with that of a ^{54}Fe-reconstituted heme protein. Part V includes several examples of applications of metal-isotope techniques to bioinorganic compounds [hemoglobin (^{54}Fe, ^{57}Fe), oxy-hemocyanin (^{63}Cu, ^{65}Cu), etc.]. An example of a multiple labeling is seen in the case of cytochrome P-450$_{cam}$ having the axial Fe–S linkage in addition to four Fe–N (porphyrin) bonds; Champion et al.[76] were able to assign its Fe–S stretching vibration (351 cm^{-1}) by combining ^{32}S–^{34}S and ^{54}Fe–^{56}Fe isotope shift data.

I-17. GROUP FREQUENCIES AND BAND ASSIGNMENTS

From observation of the infrared spectra of a number of compounds having a common group of atoms, it is found that, regardless of the rest of the molecule, this common group absorbs over a narrow range of frequencies, called the *group frequency.* For example, the group frequencies of the methyl group are 3000–2860, 1470–1400, 1380–1200, and 1200–800 cm^{-1}. Group frequencies have been found for a number of organic and inorganic groups, and they have been summarized as *group frequency charts,*[33,34] which are highly useful in identifying the atomic groups from infrared spectra. Group frequency charts for inorganic and coordination compounds are given in Appendix VI as well as in Figs. II-21 and II-22.

The concept of group frequency rests on the assumption that the vibrations of a particular group are relatively independent of those of the rest of the molecule. As stated in Sec. I-3, however, all the nuclei of the molecule perform their harmonic oscillations in a normal vibration. Thus an *isolated vibration*, which the group frequency would have to be, cannot be expected in polyatomic molecules. If, however, a group includes relatively light atoms such as hydrogen (OH, NH, NH_2, CH, CH_2, CH_3, etc.) or relatively heavy atoms such as the halogens (CCl, CBr, CI, etc.), as compared to other atoms in the molecule, the idea of an isolated vibration may be justified, since the amplitudes (or velocities) of the harmonic oscillation of these atoms are relatively larger or smaller than those of the other atoms in the same molecule. Vibrations of groups having multiple bonds (C≡C, C≡N, C=C, C=N, C=O, etc.) may also be relatively independent of the rest of the molecule if the groups do not belong to a conjugated system.

If atoms of similar mass are connected by bonds of similar strength (force constant), the amplitude of oscillation is similar for each atom of the whole system. Therefore it is not possible to isolate the group frequencies in a system like the following:

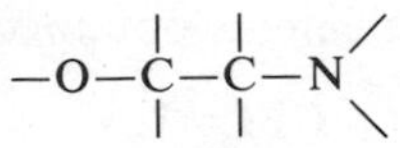

A similar situation may occur in a system in which resonance effects average out the single and multiple bonds by conjugation. Examples of this effect are seen in the metal chelate compounds of β-diketones, α-diimines, and oxalic acid (discussed in Part III). When the group frequency approximation is permissible, the mode of vibration corresponding to this frequency can be inferred empirically from the band assignments obtained theoretically for simple molecules. If *coupling* between various group vibrations is serious, it is necessary to make a theoretical analysis for each individual compound, using a method like the following one.

As stated in Sec. I-3, the generalized coordinates are related to the normal coordinates by

$$q_k = \sum_i B_{ki} Q_i \tag{3.10}$$

In matrix form, this is written as

$$\mathbf{q} = \mathbf{B}_q \mathbf{Q} \tag{17.1}$$

It can be shown[3] that the internal coordinates are also related to the normal coordinates by

$$\mathbf{R} = \mathbf{LQ} \tag{17.2}$$

This is written more explicitly as

$$\begin{aligned} R_1 &= l_{11}Q_1 + l_{12}Q_2 + \cdots + l_{1N}Q_N \\ R_2 &= l_{21}Q_1 + l_{22}Q_2 + \cdots + l_{2N}Q_N \\ &\vdots \\ R_i &= l_{i1}Q_1 + l_{i2}Q_2 + \cdots + l_{iN}Q_N \end{aligned} \tag{17.3}$$

In a normal vibration in which the normal coordinate Q_N changes with frequency ν_N, all the internal coordinates, $R_1, R_2, \ldots, R_i$, change with the same frequency. The amplitude of oscillation is, however, different for each internal coordinate. The relative ratio of the amplitudes of the internal coordinates in a normal vibration associated with Q_N is given by

$$l_{1N}:l_{2N}:\cdots:l_{iN} \tag{17.4}$$

If one of these elements is relatively large compared to the others, the normal vibration is said to be predominantly due to the vibration caused by the change of this coordinate.

The ratio of l's given by Eq. 17.4 can be obtained as a column matrix (or eigenvector) $\boldsymbol{l}_N$, which satisfies the relation[3,4]

$$\mathbf{GF}\boldsymbol{l}_N = \boldsymbol{l}_N\lambda_N \tag{17.5}$$

It consists of i elements, $l_{1N}, l_{2N}, \ldots, l_{iN}$, i being the number of internal coordinates, and can be calculated if the **G** and **F** matrices are known. An assembly by columns of the l elements obtained for each λ gives the relation

$$\mathbf{GFL} = \mathbf{L\Lambda} \tag{17.6}$$

where $\mathbf{\Lambda}$ is a diagonal matrix whose elements consist of λ values.

As an example, calculate the **L** matrix of the H_2O molecule, using the results obtained in Sec. I-14. The **G** and **F** matrices for the A_1 species are as follows:

$$\mathbf{G} = \begin{bmatrix} 1.03840 & -0.08896 \\ -0.08896 & 2.32370 \end{bmatrix}, \qquad \mathbf{F} = \begin{bmatrix} 8.32300 & 0.35638 \\ 0.35638 & 0.70779 \end{bmatrix}$$

with $\lambda_1 = 8.61475$ and $\lambda_2 = 1.60914$. The **GF** product becomes

$$\mathbf{GF} = \begin{bmatrix} 8.61090 & 0.30710 \\ 0.08771 & 1.61299 \end{bmatrix}$$

The **L** matrix can be calculated from Eq. 17.6:

$$\begin{bmatrix} 8.61090 & 0.30710 \\ 0.08771 & 1.61299 \end{bmatrix} \begin{bmatrix} l_{11} & l_{12} \\ l_{21} & l_{22} \end{bmatrix} = \begin{bmatrix} l_{11} & l_{12} \\ l_{21} & l_{22} \end{bmatrix} \begin{bmatrix} 8.61475 & 0 \\ 0 & 1.60914 \end{bmatrix}$$

However, this equation gives only the ratios $l_{11}:l_{21}$ and $l_{12}:l_{22}$. To determine their values, it is necessary to use the following normalization condition:

$$\mathbf{L\tilde{L}} = \mathbf{G} \tag{17.7*}$$

Then the final result is

$$\begin{bmatrix} l_{11} & l_{12} \\ l_{21} & l_{22} \end{bmatrix} = \begin{bmatrix} 1.01683 & -0.06686 \\ 0.01274 & 1.52432 \end{bmatrix}$$

* This equation can be derived as follows. According to Eq. 11.3, $2T = \tilde{\dot{\mathbf{R}}}\mathbf{G}^{-1}\dot{\mathbf{R}}$. On the other hand, Eq. 17.2 gives $\dot{\mathbf{R}} = \mathbf{L}\dot{\mathbf{Q}}$ and $\tilde{\dot{\mathbf{R}}} = \tilde{\dot{\mathbf{Q}}}\tilde{\mathbf{L}}$. Thus $2T = \tilde{\dot{\mathbf{Q}}}\tilde{\mathbf{L}}\mathbf{G}^{-1}\mathbf{L}\dot{\mathbf{Q}}$. Comparing this with $2T = \tilde{\dot{\mathbf{Q}}}\mathbf{E}\dot{\mathbf{Q}}$ (matrix form of Eq. 3.11), we obtain $\tilde{\mathbf{L}}\mathbf{G}^{-1}\mathbf{L} = \mathbf{E}$ or $\mathbf{L\tilde{L}} = \mathbf{G}$.

This result indicates that, in the normal vibration Q_1, the relative ratio of amplitudes of two internal coordinates, R_1 (symmetric OH stretching) and R_2 (HOH bending), is 1.0168:0.0127. Therefore this vibration (3824 cm^{-1}) is assigned to an almost pure OH stretching mode. The relative ratio of amplitudes for the Q_2 vibration is $-0.0669:1.5243$. Thus this vibration is assigned to an almost pure HOH bending mode.

In other cases, the l values do not provide the band assignments that are expected empirically. This occurs because the dimension of l for a stretching coordinate is different from that for a bending coordinate. Morino and Kuchitsu[77] proposed that the potential energy distribution of a normal vibration Q_N, defined by

$$V(Q_N) = \frac{1}{2} Q_N^2 \sum_{ij} F_{ij} l_{iN} l_{jN} \tag{17.8}^*$$

gives a better measure for making band assignments. In general, the value of $F_{ij} l_{iN} l_{jN}$ is large when $i = j$. Therefore the $F_{ii} l_{iN}^2$ terms are most important in determining the distribution of the potential energy. Thus the ratios of the $F_{ii} l_{iN}^2$ terms provide a measure of the relative contribution of each internal coordinate R_i to the normal coordinate Q_N. If any $F_{ii} l_{iN}^2$ term is exceedingly large compared with the others, the vibration is assigned to the mode associated with R_i. If $F_{ii} l_{iN}^2$ and $F_{jj} l_{jN}^2$ are relatively large compared with the others, the vibration is assigned to a mode associated with both R_i and R_j (coupled vibration).

As an example, let us calculate the potential energy distribution for the H_2O molecule. Using the **F** and **L** matrices obtained previously, we find that the $\tilde{\mathbf{L}}\mathbf{FL}$ matrix is calculated to be

$$\begin{bmatrix} \begin{pmatrix} l_{11}^2 F_{11} & + & l_{21}^2 F_{22} & + & 2 l_{21} l_{11} F_{12} \\ 8.60551 & & 0.00011 & & 0.00923 \end{pmatrix} & 0 \\ 0 & \begin{pmatrix} l_{12}^2 F_{11} & + & l_{22}^2 F_{22} & + & 2 l_{12} l_{22} F_{12} \\ 0.03721 & & 1.64459 & & -0.07264 \end{pmatrix} \end{bmatrix}$$

Then the potential energy distribution in each normal vibration ($F_{ii} l_{iN}^2$) is given by

$$\begin{array}{c} \\ R_1 \\ R_2 \end{array} \begin{array}{cc} \lambda_1 & \lambda_2 \\ \left[\begin{array}{c} 8.60551 \\ 0.00011 \end{array}\right. & \left.\begin{array}{c} 0.03721 \\ 1.64459 \end{array}\right] \end{array}$$

---

* According to Eq. 11.1, the potential energy is written as $2V = \tilde{\mathbf{R}}\mathbf{FR}$. Using Eq. 17.2, we can write this as $2V = \tilde{\mathbf{Q}}\tilde{\mathbf{L}}\mathbf{FLQ}$. On the other hand, Eq. 3.12 can be written as $2V = \tilde{\mathbf{Q}}\mathbf{\Lambda Q}$. A comparison of these two expressions gives $\mathbf{\Lambda} = \tilde{\mathbf{L}}\mathbf{FL}$. If this is written for one normal vibration whose frequency is λ_N, we have

$$\lambda_N = \sum_{ij} \tilde{l}_{Ni} F_{ij} l_{jN} = \sum_{ij} F_{ij} l_{iN} l_{jN}$$

Then the potential energy due to this vibration is expressed by Eq. 17.8.

More conveniently, the result is expressed by calculating $(F_{ii}l_{iN}^2/\sum F_{ii}l_{iN}^2) \times 100$ for each coordinate:

$$\begin{array}{c} \\ R_1 \\ R_2 \end{array} \begin{array}{c} \begin{array}{cc} \lambda_1 & \lambda_2 \end{array} \\ \begin{bmatrix} 99.99 & 2.21 \\ 0.01 & 97.79 \end{bmatrix} \end{array}$$

In this case, the final results are the same whether the band assignments are based on the **L** matrix or on the potential energy distribution: Q_1 is the symmetric OH stretching and Q_2 is the HOH bending. In other cases, different results may be obtained, depending on which criterion is used for band assignments.

A more rigorous method of determining the vibrational mode is to draw the displacements of individual atoms in terms of rectangular coordinates. As in Eq. 17.2, the relationship between the rectangular and normal coordinates is given by

$$\mathbf{X} = \mathbf{L}_x\mathbf{Q} \tag{17.9}$$

The $\mathbf{L}_x$ matrix can be obtained from the relationship[78]

$$\mathbf{L}_x = \mathbf{M}^{-1}\tilde{\mathbf{B}}\mathbf{G}^{-1}\mathbf{L} \qquad (17.10)^*$$

The matrices on the right-hand sides have already been defined.

Three-dimensional drawings of normal modes such as those shown in Part II can be made from the cartesian displacement calculations obtained above. However, hand plotting of these data is laborious and complicated. Use of computer plotting programs greatly facilitates this process.[79]

I-18. INTENSITY OF INFRARED ABSORPTION[52]

The absorption of strictly monochromatic light (ν) is expressed by the Lambert–Beer law:

$$I_\nu = I_{0,\nu}\, e^{-\alpha_\nu pl} \tag{18.1}$$

where I_ν is the intensity of the light transmitted by a cell of length l containing a gas at pressure p, $I_{0,\nu}$ is the intensity of the incident light, and α_ν is the absorption coefficient for unit pressure. The true integrated absorption coefficient A is defined by

$$A = \int_{\text{band}} \alpha_\nu\, d\nu = \frac{1}{pl}\int_{\text{band}} \ln\left(\frac{I_{0,\nu}}{I_\nu}\right) d\nu \tag{18.2}$$

where the integration is carried over the entire frequency region of a band.

---

* By combining Eqs. 11.8 and 17.9, we have $\mathbf{R} = \mathbf{BX} = \mathbf{BL}_x\mathbf{Q}$. Since $\mathbf{R} = \mathbf{LQ}$ (Eq. 17.2), it follows that $\mathbf{LQ} = \mathbf{BL}_x\mathbf{Q}$ or $\mathbf{L} = \mathbf{BL}_x$. The kinetic energy is written as $2T = \tilde{\dot{\mathbf{X}}}\mathbf{M}\dot{\mathbf{X}}$. In terms of internal coordinates, it is written as $2T = \tilde{\dot{\mathbf{R}}}\mathbf{G}^{-1}\dot{\mathbf{R}} = \tilde{\dot{\mathbf{X}}}\tilde{\mathbf{B}}\mathbf{G}^{-1}\mathbf{B}\dot{\mathbf{X}}$. By comparing these two expressions, we have $\mathbf{M} = \tilde{\mathbf{B}}\mathbf{G}^{-1}\mathbf{B}$. Then we can write $\mathbf{L}_x = \mathbf{M}^{-1}\mathbf{ML}_x = \mathbf{M}^{-1}\tilde{\mathbf{B}}\mathbf{G}^{-1}\mathbf{BL}_x = \mathbf{M}^{-1}\tilde{\mathbf{B}}\mathbf{G}^{-1}\mathbf{L}$.

In practice, I_ν and $I_{0,\nu}$ cannot be measured accurately, since no spectrophotometers have infinite resolving power. Therefore we measure instead the apparent intensity T_ν:

$$T_\nu = \int_{\text{slit}} I(\nu) g(\nu, \nu')\, d\nu \tag{18.3}$$

where $g(\nu, \nu')$ is a function indicating the amount of light of frequency ν when the spectrophotometer reading is set at ν'. Then the apparent integrated absorption coefficient B is defined by

$$B = \frac{1}{pl} \int_{\text{band}} \ln \frac{\int_{\text{slit}} I_0(\nu) g(\nu, \nu')\, d\nu}{\int_{\text{slit}} I(\nu) g(\nu, \nu')\, d\nu}\, d\nu' \tag{18.4}$$

It can be shown that

$$\lim_{pl \to 0} (A - B) = 0 \tag{18.5}$$

if I_0 and α_ν are constant within the slit width used. (This condition is approximated by using a narrow slit.) In practice, we plot B/pl against pl, and extrapolate the curve to $pl \to 0$. To apply this method to gaseous molecules, it is necessary to broaden the vibrational-rotational bands by adding a high-pressure inert gas (pressure broadening).

For liquids and solutions, p and α in preceding equations are replaced by M (molar concentration) and ε (molar absorption coefficient), respectively. However, the extrapolation method just described is not applicable, since experimental errors in determining B values become too large at low concentration or at small cell length. The true integrated absorption coefficient of a liquid can be calculated if we assume that the shape of an absorption band is represented by the Lorentz equation and that the slit function is triangular.[80]

Theoretically, the true integrated absorption coefficient A_N of the Nth normal vibration is given by[3]

$$A_N = \frac{n\pi}{3c}\left[\left(\frac{\partial \mu_x}{\partial Q_N}\right)_0^2 + \left(\frac{\partial \mu_y}{\partial Q_N}\right)_0^2 + \left(\frac{\partial \mu_z}{\partial Q_N}\right)_0^2\right] \tag{18.6}$$

where n is the number of molecules per cubic centimeter, and c is the velocity of light. As shown by Eq. 17.2, an internal coordinate R_i is related to a set of normal coordinates by

$$R_i = \sum_N L_{iN} Q_N \tag{18.7}$$

If the additivity of the bond dipole moment is assumed, it is possible to write

$$\begin{aligned}\frac{\partial \mu}{\partial Q_N} &= \sum_i \left(\frac{\partial \mu}{\partial R_i}\right)\left(\frac{\partial R_i}{\partial Q_N}\right) \\ &= \sum_i \left(\frac{\partial \mu}{\partial R_i}\right) L_{iN}\end{aligned} \tag{18.8}$$

Then Eq. 18.6 is written as

$$A_N = \frac{n\pi}{3c}\left[\left(\sum_i \frac{\partial \mu_x}{\partial R_i} L_{iN}\right)_0^2 + \left(\sum_i \frac{\partial \mu_y}{\partial R_i} L_{iN}\right)_0^2 + \left(\sum \frac{\partial \mu_z}{\partial R_i} L_{iN}\right)_0^2\right]$$

$$= \frac{n\pi}{3c}\sum_i \left[\left(\frac{\partial \mu_x}{\partial R_i}\right)_0^2 + \left(\frac{\partial \mu_y}{\partial R_i}\right)_0^2 + \left(\frac{\partial \mu_z}{\partial R_i}\right)_0^2\right](L_{iN})^2 \qquad (18.9)$$

This equation shows that the intensity of an infrared band depends on the values of the $\partial\mu/\partial R$ terms as well as of the L matrix elements.

Equation 18.9 has been applied to relatively small molecules to calculate the $\partial\mu/\partial R$ terms from the observed intensity and known L_{iN} values.[7] However, the additivity of the bond dipole moment does not strictly hold, and the results obtained are often inconsistent and conflicting. Thus far, very few studies have been made on infrared intensities of large molecules because of these difficulties.

I-19. DEPOLARIZATION OF RAMAN LINES

As stated in Secs. I-7 and I-8, it is possible, by using group theory, to classify the normal vibration into various symmetry species. Experimentally, measurements of the infrared dichroism and polarization properties of Raman lines of an orientated crystal provide valuable information about the symmetry of normal vibrations (Sec. I-23). Here we consider the polarization properties of Raman lines in liquids and solutions in which molecules or ions take completely random orientations.*

Suppose that we irradiate a molecule fixed at the origin of a space-fixed coordinate system with natural light from the positive-y direction, and observe the Raman scattering in the x direction as shown in Fig. I-14. The incident light vector E may be resolved into two components, E_x and E_z, of equal magnitude ($E_y = 0$). Both components give induced dipole moments, P_x, P_y, and P_z. However, only P_y and P_z contribute to the scattering along the x axis, since an oscillating dipole cannot radiate in its own direction. Then, from Eq. 5.7. we have

$$P_y = \alpha_{yx}E_x + \alpha_{yz}E_z \qquad (19.1)\dagger$$

$$P_z = \alpha_{zx}E_x + \alpha_{zz}E_z \qquad (19.2)$$

The intensity of the scattered light is proportional to the sum of squares of the individual $\alpha_{ij}E_j$ terms. Thus the ratio of the intensities in the y and z

---

* It is possible to obtain approximate depolarization ratios of fine powders where the molecules or ions take pseudorandom orientations (see Ref. 81).

† In the case of Raman scattering, it is the $\partial\alpha/\partial Q$ term that should be used in Eq. 19.1 and the following equations.

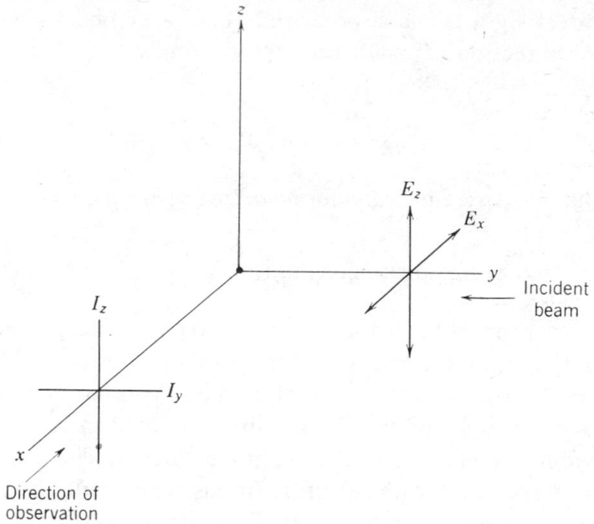

Fig. I-14. Schematic representation of experimental condition for measuring depolarization ratios.

directions is

$$\rho_n = \frac{I_y}{I_z} = \frac{\alpha_{yx}^2 E_x^2 + \alpha_{yz}^2 E_z^2}{\alpha_{zx}^2 E_x^2 + \alpha_{zz}^2 E_z^2} \tag{19.3}$$

where ρ_n is called the *depolarization ratio for natural light* (n).

In a homogeneous liquid or gas, the molecules are randomly orientated, and we must consider the polarizability components averaged over all molecular orientations. The results are expressed in terms of two quantities: $\bar{\alpha}$ (*mean value*) and γ (*anisotropy*):

$$\bar{\alpha} = \tfrac{1}{3}(\alpha_{xx} + \alpha_{yy} + \alpha_{zz}) \tag{19.4}$$

$$\gamma^2 = \tfrac{1}{2}[(\alpha_{xx} - \alpha_{yy})^2 + (\alpha_{yy} - \alpha_{zz})^2 + (\alpha_{zz} - \alpha_{xx})^2 + 6(\alpha_{xy}^2 + \alpha_{yz}^2 + \alpha_{zx}^2)] \tag{19.5}$$

These two quantities are invariant to any coordinate transformation. It can be shown[3] that the average values of the squares of α_{ij} are

$$\overline{(\alpha_{xx})^2} = \overline{(\alpha_{yy})^2} = \overline{(\alpha_{zz})^2} = \tfrac{1}{45}[45(\bar{\alpha})^2 + 4\gamma^2] \tag{19.6}$$

$$\overline{(\alpha_{xy})^2} = \overline{(\alpha_{yz})^2} = \overline{(\alpha_{zx})^2} = \tfrac{1}{15}\gamma^2 \tag{19.7}$$

Since $E_x = E_z = E$, Eq. 19.3 can be written as

$$\rho_n = \frac{I_y}{I_z} = \frac{6\gamma^2}{45(\bar{\alpha})^2 + 7\gamma^2} \tag{19.8}$$

The total intensity I_n is given by

$$I_n = I_y + I_z = \text{const}\,\{\tfrac{1}{45}[45(\bar{\alpha})^2 + 13\gamma^2]\}E^2 \tag{19.9}$$

If the incident light is plane polarized (e.g. laser beam), with its electric vector in the z direction ($E_x = 0$), Eq. 19.8 becomes

$$\rho_p = \frac{I_y}{I_z} = \frac{3\gamma^2}{45(\bar{\alpha})^2 + 4\gamma^2} \qquad (19.10)$$

where ρ_p is the *depolarization ratio for polarized light* (p). In this case, the total intensity is given by

$$I_p = I_y + I_z = \text{const}\,\{\tfrac{1}{45}[45(\bar{\alpha})^2 + 7\gamma^2]\}E^2 \qquad (19.11)$$

The symmetry property of a normal vibration can be determined by measuring the depolarization ratio. From an inspection of character tables (Appendix I), it is obvious that $\bar{\alpha}$ is nonzero only for totally symmetric vibrations. Then Eq. 19.8 gives $0 \leq \rho_n < \frac{6}{7}$, and the Raman lines are said to be *polarized*. For all nontotally symmetric vibrations, $\bar{\alpha}$ is zero, and $\rho_n = \frac{6}{7}$. Then the Raman lines are said to be *depolarized*. If the exciting line is plane polarized, these criteria must be changed according to Eq. 19.10. Thus $0 \leq \rho_p < \frac{3}{4}$ for totally symmetric vibrations, and $\rho_p = \frac{3}{4}$ for nontotally symmetric vibrations. Figure I-15 shows the Raman spectra of CCl_4 (500–150 cm^{-1}) in two directions of polarization obtained with the 488 nm excitation. The three bands at 459, 314, and 218 cm^{-1} give ρ_p values of approximately 0.02, 0.75, and 0.75, respectively. Thus it is concluded that the 459 cm^{-1} band is polarized (A_1), whereas the two bands at 314 (F_2) and 218 (E) cm^{-1} are depolarized.

As stated in Sec. I-5, the polarizability tensors are symmetric in normal Raman scattering. If the exciting frequency approaches that of an electronic absorption, some scattering tensors become antisymmetric,* and resonance Raman scattering can occur (Sec. I-21). In this case, Eq. 19.10 must be written in a more general form:[82]

$$\rho_p = \frac{3g^s + 5g^a}{10g^o + 4g^s} \qquad (19.12)$$

where

$$\begin{aligned} g^o &= 3(\bar{\alpha})^2 \\ g^s &= \tfrac{1}{3}[(\alpha_{xx} - \alpha_{yy})^2 + (\alpha_{xx} - \alpha_{zz})^2 + (\alpha_{yy} - \alpha_{zz})^2] \\ &\quad + \tfrac{1}{2}[(\alpha_{xy} + \alpha_{yx})^2 + (\alpha_{xz} + \alpha_{zx})^2 + (\alpha_{yz} + \alpha_{zy})^2] \\ g^a &= \tfrac{1}{2}[(\alpha_{xy} - \alpha_{yx})^2 + (\alpha_{xz} - \alpha_{zx})^2 + (\alpha_{yz} - \alpha_{zy})^2] \end{aligned} \qquad (19.13)$$

If we define

$$\gamma_s^2 = \tfrac{3}{2}g^s \quad \text{and} \quad \gamma_{as}^2 = \tfrac{3}{2}g^a \qquad (19.14)$$

---

* A tensor is called antisymmetric if $\alpha_{xx} = \alpha_{yy} = \alpha_{zz} = 0$ and $\alpha_{xy} = -\alpha_{yx}$, $\alpha_{yz} = -\alpha_{zy}$, and $\alpha_{zx} = -\alpha_{xz}$.

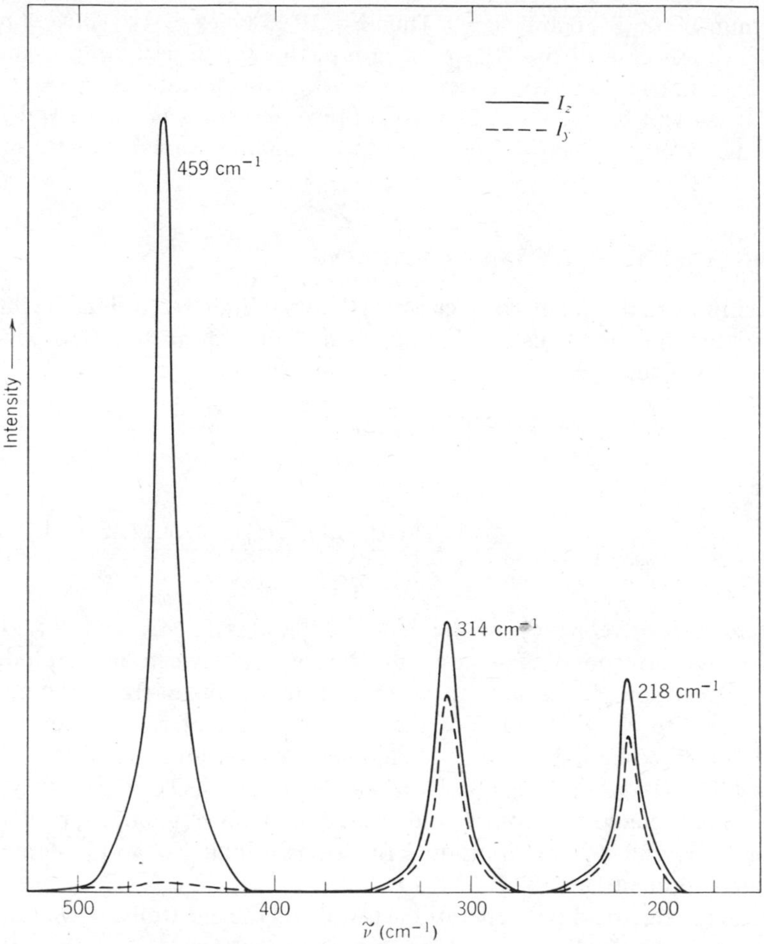

Fig. I-15. Raman spectra of CCl_4 (500–150 cm^{-1}) in two directions of polarization (488 nm excitation).

Eq. 19.12 can be written as

$$\rho_p = \frac{3\gamma_s^2 + 5\gamma_{as}^2}{45(\bar{\alpha})^2 + 4\gamma_s^2} \tag{19.15}$$

In normal Raman scattering, $\gamma_s^2 = \gamma^2$ and $\gamma_{as}^2 = 0$. Then Eq. 19.15 is reduced to Eq. 19.10.

The symmetry properties of resonance Raman lines can be predicted on the basis of Eq. 19.15. For totally symmetric vibrations, $\bar{\alpha} \neq 0$ and $\gamma_{as} = 0$. Then Eq. 19.15 gives $0 \leq \rho_p < \frac{3}{4}$. Nontotally symmetric vibrations ($\bar{\alpha} = 0$) are classified into two types: those which have symmetric scattering tensors, and those which have antisymmetric scattering tensors. If the tensor is symmetric, $\gamma_{as} = 0$ and $\gamma_s \neq 0$. Then Eq. 19.15 gives $\rho_p = \frac{3}{4}$ (depolarized). If the tensor is antisymmetric, $\gamma_{as} \neq 0$ and $\gamma_s = 0$. Then Eq. 19.15 gives $\rho_p = \infty$ (*inverse polarization*). In the case of the $\mathbf{D}_{4h}$ point group, the B_{1g} and B_{2g} representations belong to the former type, whereas the A_{2g} representations belongs to the latter.[83] As will be shown in Sec. I-21, Spiro and Strekas[82] observed for the first time inversely polarized bands in the resonance Raman spectra of heme proteins.

I-20. INTENSITY OF RAMAN SCATTERING

According to the quantum mechanical theory of light scattering, the intensity per unit solid angle of scattered light arising from a transition between states m and n is given by

$$I_{n \leftarrow m} = \text{const}\,(\nu_0 + \nu_{mn})^4 \sum_{\rho\sigma} |(P_{\rho\sigma})_{mn}|^2 \tag{20.1}$$

where

$$(P_{\rho\sigma})_{mn} = (\alpha_{\rho\sigma})_{mn} E = \frac{1}{h} \sum_r \left[\frac{(M_\rho)_{rn}(M_\sigma)_{mr}}{\nu_{rm} - \nu_0} + \frac{(M_\rho)_{mr}(M_\sigma)_{rn}}{\nu_{rn} + \nu_0} \right] E \tag{20.2}$$

Here ν_0 is the frequency of the incident light: ν_{rm}, ν_{rn}, and ν_{mn} are the frequencies corresponding to the energy differences between subscripted states; terms of the type $(M_\sigma)_{mr}$ are the cartesian components of transition moments such as $\int \Psi_r^* \mu_\sigma \Psi_m \, d\tau$; and E is the electric vector of the incident light. It should be noted here that the states denoted by m, n, and r represent vibronic states $\psi_g(\xi, Q)\phi_i^g(Q)$, $\psi_g(\xi, Q)\phi_j^g(Q)$, and $\psi_e(\xi, Q)\phi_v^e(Q)$, respectively, where ψ_g and ψ_e are electronic ground- and excited-state wave functions, respectively, and ϕ_i^g, ϕ_j^g, and ϕ_v^e are vibrational functions. Finally, σ and ρ denote x, y, and z components.

Since the electric dipole operator acts only on the electronic wave functions, the $(\alpha_{\rho\sigma})_{mn}$ term in Eq. 20.2 can be written in the form[22]

$$(\alpha_{\rho\sigma})_{mn} = \frac{1}{h} \int \phi_j^g (\alpha_{\rho\sigma})_{gg} \phi_i^g \, dQ \tag{20.3}$$

where

$$(\alpha_{\rho\sigma})_{gg} = \sum_e \left(\frac{\int \psi_g^* \mu_\sigma \psi_e \, d\tau \cdot \int \psi_e^* \mu_\rho \psi_g \, d\tau}{\bar{\nu}_{eg} - \nu_0} + \frac{\int \psi_g^* \mu_\rho \psi_e \, d\tau \cdot \int \psi_e^* \mu_\sigma \psi_g \, d\tau}{\bar{\nu}_{eg} + \nu_0} \right)$$

Here $\bar{\nu}_{eg}$ corresponds to the energy of a pure electronic transition between the ground and excited states.

To discuss the Raman scattering, we expand the $(\alpha_{\rho\sigma})_{gg}$ term as a Taylor series with respect to the normal coordinate Q:

$$(\alpha_{\rho\sigma})_{gg} = (\alpha_{\rho\sigma})_{gg}^0 + \left[\frac{\partial (\alpha_{\rho\sigma})_{gg}}{\partial Q} \right]_0 Q + \cdots \tag{20.4}$$

Then, we write Eq. 20.3 as

$$(\alpha_{\rho\sigma})_{mn} = \frac{1}{h}(\alpha_{\rho\sigma})^0_{gg} \int \phi^g_j \phi^g_i \, dQ + \frac{1}{h}\left[\frac{\partial(\alpha_{\rho\sigma})_{gg}}{\partial Q}\right]_0 \int \phi^g_j Q \phi^g_i \, dQ \quad (20.5)$$

The first term on the right-hand side is zero unless $i = j$. This term is responsible for Rayleigh scattering. The second term determines the activity of fundamental vibrations in Raman scattering; it vanishes for a harmonic oscillator unless $j = i \pm 1$.

If we consider a Stokes transition, $v \to v+1$, Eq. 20.5 is written as[3]

$$(\alpha_{\rho\sigma})_{v,v+1} = \frac{1}{h}\left[\frac{\partial(\alpha_{\rho\sigma})_{gg}}{\partial Q}\right]_0 \sqrt{\frac{(v+1)h}{8\pi^2 \mu \nu}} \quad (20.6)$$

where μ and ν are the reduced mass and the Stokes frequency. Then Eq. 20.1 is written as

$$I = \text{const}\,(\nu_0 - \nu)^4 \frac{E^2}{h^2}\left[\frac{\partial(\alpha_{\rho\sigma})_{gg}}{\partial Q}\right]_0^2 \frac{(v+1)h}{8\pi^2 \mu \nu} \quad (20.7)$$

In Sec. I-19, we derived a classical equation for Raman intensity:

$$I_n = \text{const}\left(\frac{\partial \alpha}{\partial Q}\right)^2 E^2$$

$$= \text{const}\left\{\frac{1}{45}[45(\bar{\alpha})^2 + 13\gamma^2]\right\} E^2 \quad (19.9)$$

By replacing the $\partial\alpha/\partial Q$ term of Eq. 20.7 with the term enclosed in braces in Eq. 19.9, we obtain

$$I_n = \text{const}\,(\nu_0 - \nu)^4 \frac{(v+1)}{8\pi^2 \mu \nu} \frac{E^2}{h^2}\left\{\frac{1}{45}[45(\bar{\alpha})^2 + 13\gamma^2]\right\} \quad (20.8)$$

At room temperature, most of the scattering molecules are in the $v = 0$ state, but some are in higher vibrational states. Using the Maxwell-Boltzmann distribution law, we find that the fraction of molecules f_v with vibrational quantum number v is given by

$$f_v = \frac{e^{-[v+(1/2))h_\nu/kT}}{\sum_v e^{-[v+(1/2)]h\nu/kT}} \quad (20.9)$$

Then the total intensity is proportional to $\sum_v f_v(v+1)$, which is equal to $(1 - e^{-h\nu/kT})^{-1}$ (see Ref. 18). Hence we can rewrite Eq. 20.8 in the form

$$I_n = K I_0 \frac{(\nu_0 - \nu)^4}{\mu\nu(1 - e^{-h\nu/kT})}[45(\bar{\alpha})^2 + 13\gamma^2] \quad (20.10)$$

Here I_0 is the incident light intensity which is proportional to E^2, and K summarizes all other constant terms.

If the incident light is polarized, the form of Eq. 20.10 is slightly modified:

$$I_p = KI_0 \frac{(\nu_0 - \nu)^4}{\mu\nu(1 - e^{-h\nu/kT})}[45(\bar{\alpha})^2 + 7\gamma^2] \qquad (20.11)$$

As shown in Sec. I-19, the degree of depolarization ρ_p is

$$\rho_p = \frac{3\gamma^2}{45(\bar{\alpha})^2 + 4\gamma^2} \quad \text{or} \quad \gamma^2 = \frac{45(\bar{\alpha})^2 \rho_p}{3 - 4\rho_p} \qquad (20.12)$$

Since $\rho_p = \frac{3}{4}$ for nontotally symmetric vibrations, Eq. 20.12 holds only for totally symmetric vibrations. Then Eq. 20.11 is written as

$$I_p = K'I_0 \frac{(\nu_0 - \nu)^4}{\mu\nu(1 - e^{-h\nu/kT})}\left(\frac{1 + \rho_p}{3 - 4\rho_p}\right)(\bar{\alpha})^2 \qquad (20.13)$$

In the case of a solution, the intensity is proportional to the molar concentration C. Then Eq. 20.13 is written as

$$I_p = K''I_0 \frac{C(\nu_0 - \nu)^4}{\mu\nu(1 - e^{-h\nu/kT})}\left(\frac{1 + \rho_p}{3 - 4\rho_p}\right)(\bar{\alpha})^2 \qquad (20.14)$$

If we compare the intensities of totally symmetric vibrations (A_1 mode) of two tetrahedral XY_4-type molecules, the intensity ratio is given by

$$\frac{I_1}{I_2} = \frac{C_1}{C_2}\left(\frac{\tilde{\nu}_0 - \tilde{\nu}_1}{\tilde{\nu}_0 - \tilde{\nu}_2}\right)^4 \frac{\tilde{\nu}_2\mu_2}{\tilde{\nu}_1\mu_1} \frac{(1 - e^{-hc\tilde{\nu}_2/kT})(\bar{\alpha}_1)^2}{(1 - e^{-hc\tilde{\nu}_1/kT})(\bar{\alpha}_2)^2} \qquad (20.15)$$

In this case, the ρ_p term drops out, since $\gamma^2 = 0$ for isotropic molecules such as tetrahedral XY_4 and octahedral XY_6 types. By using CCl_4 as the standard, it is possible to determine the relative value of the $\partial\alpha/\partial Q$ term, which provides information about the degree of covalency and the bond order.[18]

I-21. RESONANCE RAMAN SPECTRA

In normal Raman spectroscopy, the exciting frequency lies in the region where the compound has no electronic absorption band (Sec. I-1). In resonance Raman spectroscopy, the exciting frequency falls within the electronic band. In the gaseous phase, this tends to cause resonance fluorescence since the rotational-vibrational levels are discrete. In the liquid and solid states, however, these levels are no longer discrete because of molecular collisions and/or intermolecular interactions. If such a broad vibronic band is excited, it tends to give resonance Raman rather than resonance fluorescence spectra.[84,85]

As an example, Fig. I-16 shows the resonance Raman spectra of TiI_4 in cyclohexane obtained by Clark and Mitchell.[86] Figure I-17 shows the positions of the exciting frequencies relative to the electronic absorption spectrum. As the exciting wavelength is changed from 647.1 to 514.5 nm, the intensity of the $\nu_1(A_1)$ band is enhanced relative to the 806 cm^{-1} band of cyclohexane (the internal standard). It is seen that the intensity of the 161 cm^{-1} band (ν_1) is

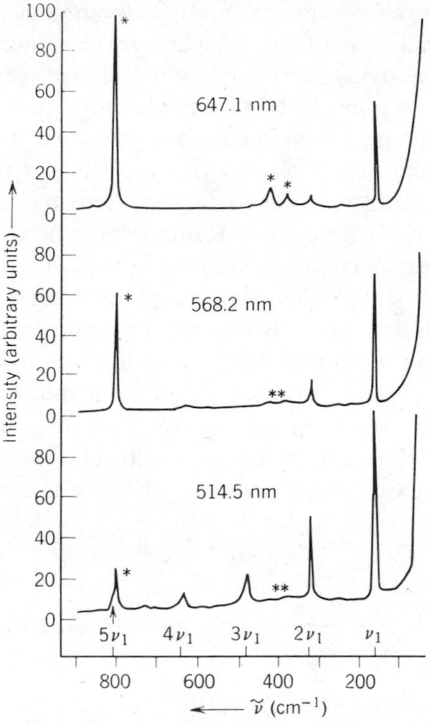

Fig. I-16. Raman spectra of titanium tetraiodide in cyclohexane obtained with 647.1, 568.2, and 514.5 nm excitation. Solvent peaks are marked with an asterisk, and overtones as $n\nu_1$.

maximized when the exciting frequency is near the absorption maximum (515 nm). In one experiment, a series of overtones up to $13\nu_1$ was observed, although Fig. I-16 shows the series up to $5\nu_1$. These overtone frequencies have been used to calculate the anharmonicity constant with great accuracy.[86]

Resonance Raman spectroscopy is particularly suited to the study of biological macromolecules such as heme proteins because only a dilute solution (biological condition) is needed to observe the spectrum and only vibrations

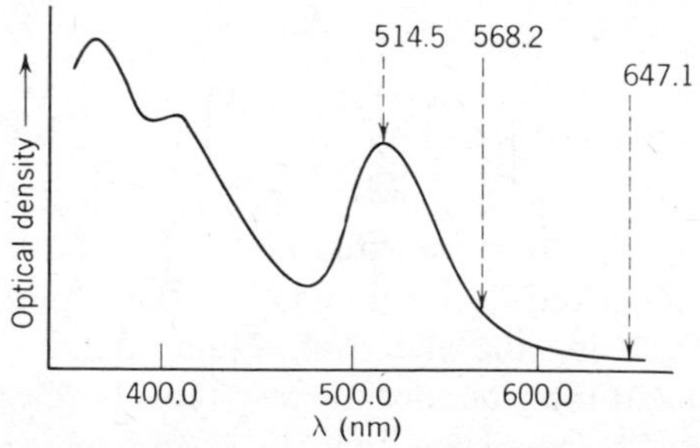

Fig. I-17. Electronic spectrum of titanium tetraiodide.

localized within the chromophoric group are enhanced when the exciting frequency approaches that of the relevant chromophore. This *selectivity* is highly important in studying the theoretical relationship between the electronic transition and the vibrations to be resonance enhanced.

The origin of resonance Raman enhancement is explained in terms of Eq. 20.2. In normal Raman spectroscopy, ν_0 is chosen in the region that is far from the electronic absorption. Then $\nu_{rm} \gg \nu_0$, and $\alpha_{\rho\sigma}$ is independent of the exciting frequency ν_0. In resonance Raman spectroscopy, the denominator, $\nu_{rm} - \nu_0$, becomes very small as ν_0 approaches ν_{rm}. Thus the first term in the square brackets of Eq. 20.2 dominates all other terms and results in striking enhancement of Raman lines. However, Eq. 20.2 cannot account for the selectivity of resonance Raman enhancement since it is not specific about the states of the molecule. Albrecht[87] derived a more specific equation for the initial and final states of resonance Raman scattering by introducing the Herzberg–Teller expansion of electronic wave functions into the Kramers–Heisenberg dispersion formula. The results are as follows:

$$(\alpha_{\rho\sigma})_{gi,gj} = A + B + C \tag{21.1}$$

$$A = \sum_{e\neq g}{}' \sum_{v} \left[\frac{(g^0|R_\sigma|e^0)(e^0|R_\rho|g^0)}{E_{ev} - E_{gi} - E_0} + (\text{nonresonance term}) \right] \langle i|v\rangle\langle v|j\rangle \tag{21.2}$$

$$\begin{aligned} B = \sum_{e\neq g}{}' \sum_{v} \sum_{s\neq e}{}' \sum_{a} \Bigg\{ \Bigg[&\frac{(g^0|R_\sigma|e^0)(e^0|h_a|s^0)(s^0|R_\rho|g^0)}{E_{ev} - E_{gi} - E_0} + (\text{nonresonance term}) \Bigg] \\ &\times \frac{\langle i|v\rangle\langle v|Q_a|j\rangle}{E_e^0 - E_s^0} + \Bigg[\frac{(g^0|R_\sigma|s^0)(s^0|h_a|e^0)(e^0|R_\rho|g^0)}{E_{ev} - E_{gi} - E_0} \\ &+ (\text{nonresonance term}) \Bigg] \times \frac{\langle i|Q_a|v\rangle\langle v|j\rangle}{E_e^0 - E_s^0} \Bigg\} \end{aligned} \tag{21.3}$$

$$\begin{aligned} C = \sum_{e\neq g}{}' \sum_{t\neq g}{}' \sum_{v} \sum_{a} \Bigg\{ \Bigg[&\frac{(e^0|R_\rho|g^0)(g^0|h_a|t^0)(t^0|R_\sigma|e^0)}{E_{ev} - E_{gi} - E_0} + (\text{nonresonance term}) \Bigg] \\ &\times \frac{\langle i|v\rangle\langle v|Q_a|j\rangle}{E_g^0 - E_t^0} + \Bigg[\frac{(e^0|R_\rho|t^0)(t^0|h_a|g^0)(g^0|R_\sigma|e^0)}{E_{ev} - E_{gi} - E_0} \\ &+ (\text{nonresonance term}) \Bigg] \times \frac{\langle i|Q_a|v\rangle\langle v|j\rangle}{E_g^0 - E_t^0} \Bigg\} \end{aligned} \tag{21.4}$$

The notations *g, i, j, e,* and *v* were explained in Sec. I-20. Other notations are as follows: *s*, another excited electronic state; h_a, the vibronic coupling operator $\partial\mathcal{H}/\partial Q_a$, $\mathcal{H}$ and Q_a being the electronic Hamiltonian and the *a*th normal coordinate of the electronic ground state, respectively; E_{gi} and E_{ev}, the energies of states *gi* and *ev*, respectively; $|g^0)$, $|e^0)$, and $|s^0)$, the electronic wave functions for the equilibrium nuclear positions of the ground and excited states; E_e^0 and E_s^0, the corresponding energies of the electronic states, e^0 and s^0, respectively; and E_0, the energy of the exciting light. The nonresonance terms are similar to the preceding terms except that the denominator is $(E_{ev} - E_{gj} + E_0)$ instead of $(E_{ev} - E_{gi} - E_0)$ and that R_σ and R_ρ in the numerator are interchanged. These terms can be neglected under the strict resonance condition since the resonance terms become very large. The C term is usually neglected because its components are denominated by $E_g^0 - E_t^0$, where t refers to an excited state which is much higher in energy than the first excited state.

In the case of totally symmetric modes, the product of the integrals $\langle i|v\rangle$ and $\langle v|j\rangle$ in Eq. 21.2 is finite due to nonorthogonality of the vibrational wave functions at the electronic ground and excited states. On the other hand, these wave functions are nearly orthogonal for nontotally symmetric vibrations. Thus, only totally symmetric modes can derive Raman intensities via the A term. Nontotally symmetric and totally symmetric modes can gain Raman intensity via the B term as long as the $\langle i|v\rangle\langle v|Q_a|j\rangle/(E_e^0 - E_s^0)$ in Eq. 21.3 is nonzero. In general, the B term contribution is small relative to the A term due in part to the additional denominator $E_e^0 - E_s^0$.

Spiro and co-workers[88] carried out an extensive study on resonance Raman spectra of various heme proteins. As is shown in Fig. I-18, ferrocytochrome c exhibits two electronic transitions referred to as the Q_0 (or α) and B (or Soret) bands along with a vibronic side band (Q_v or β) in the 400–600 cm^{-1} region. According to MO (molecular orbital) calculations on the porphine core of $\mathbf{D}_{4h}$ symmetry, the Q_0 and B transitions result from strong interaction between the $a_{1u}(\pi) \rightarrow e_g(\pi^*)$ and $a_{2u}(\pi) \rightarrow e_g(\pi^*)$ transitions which have similar energies and the same excited-state symmetry (E_u). This is an ideal situation for B-term resonance. According to Eq. 21.3, any normal modes which give nonzero values for the integral $(e^0|h_a|s^0)$ are enhanced via the B term. These vibrations must belong to $E_u \times E_u = A_{1g} + B_{1g} + B_{2g} + A_{2g}$.

Figure I-19 shows the resonance Raman spectra of cytochrome c obtained by the Q_v excitation (514.5 nm).[82] As discussed in Sec. I-19, the A_{1g}, B_{1g} and

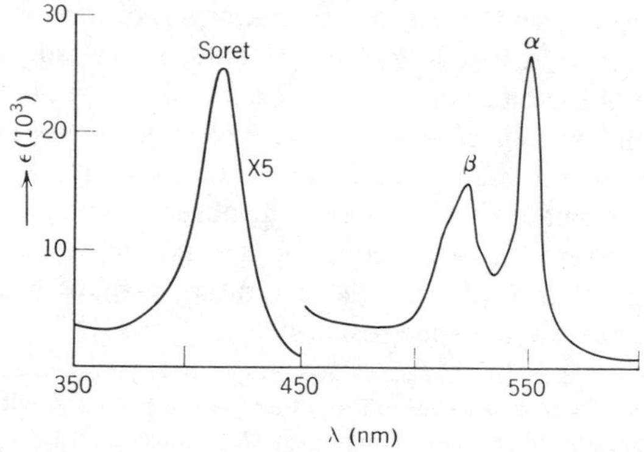

Fig. I-18. Electronic spectrum of ferrocytochrome c.[88]

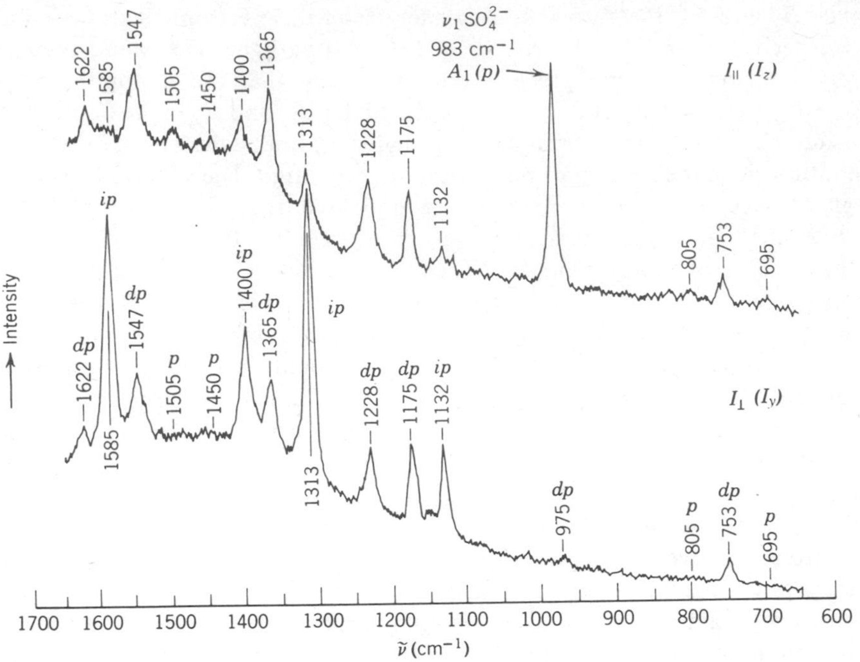

Fig. I-19. Resonance Raman spectra of ferrocytochrome *c* (514.5 nm excitation).[82] The scattering geometry is shown in Fig. I-14.

B_{2g}, and A_{2g} vibrations are expected to be polarized (*p*), depolarized (*dp*), and inversely polarized (*ip*),* respectively. These polarization properties, together with their vibrational frequencies, were used by Spiro and his co-workers to make complete assignments of vibrational spectra of the Fe-porphin skeletons of a series of heme proteins. They showed that the resonance Raman spectrum may be used to predict the oxidation and spin states of the Fe atom in heme proteins. For example, the Fe atom in oxyhemoglobin has been shown to be low-spin Fe(III). It should be noted that the A_{2g} mode, which is normally Raman inactive, is observed under the resonance condition. Although the A_{1g} modes are rather weak in Fig. I-19, these vibrations are enhanced markedly and exclusively by the excitation near the *B* band since the *A*-term resonance is predominant under such condition. The majority of compounds studied thus far exhibit the *A*-term rather than the *B*-term resonance. A more complete study of resonance Raman spectra involves the observation of *excitation profiles* (Raman intensity plotted as a function of the exciting frequency for each mode), and the simulation of observed excitation profiles based on various theories of resonance Raman spectroscopy.[89]

* For the A_{2g} modes, $\frac{3}{4} < \rho_p < \infty$ was observed rather than $\rho_p = \infty$, as predicted by the theory. This discrepancy was attributed to the lowering of the $\mathbf{D}_{4h}$ symmetry of the complex due to the presence of the peripheral groups in cytochrome *c* (Ref. 82).

I-22. VIBRATIONAL SPECTRA IN GASEOUS PHASE[90] AND INERT GAS MATRICES[40,41]

As distinct from molecules in condensed phases, those in the gaseous phase are free from intermolecular interactions. If the molecules are relatively small, vibrational spectra of gases exhibit rotational fine structure (see Fig. I-4) from which moments of inertia and hence internuclear distances and bond angles can be calculated.[1,2] Furthermore, detailed analysis of rotational fine structure provides information about the magnitude of rotation–vibration interaction (Coriolis coupling), centrifugal distortion, anharmonicity, and even nuclear spin statistics in some cases. In the past, infrared spectroscopy was the main tool in measuring gas-phase vibrational spectra. Recently, Raman spectroscopy has been playing a significant role because of the development of powerful laser sources and high-resolution spectrophotometers. For example, Clark and Rippon[91] measured gas-phase Raman spectra of Group IV tetrahalides, and calculated the Coriolis coupling constants from the observed band contours. For gas-phase Raman spectra of other inorganic compounds, see Ref. 90. Unfortunately, the majority of inorganic and coordination compounds exist as solids at room temperature. Although some of these compounds can be vaporized at high temperatures without decomposition, it is rather difficult to measure their spectra by the conventional method. Furthermore, high-temperature spectra are difficult to interpret because of the increased importance of rotational and vibrational hot bands.

In 1954, Pimentel and his co-workers[92] developed the *matrix isolation technique* to study the infrared spectra of unstable and stable species. In this method, solute molecules and inert gas molecules such as Ar and N_2 are mixed at a ratio of 1:500 or greater and deposited on an IR window such as a CsI crystal cooled to 10–15 K. Since the solute molecules trapped in an inert gas matrix are completely isolated from each other, the matrix isolation spectrum is similar to the gas-phase spectrum; no crystal field splittings and no lattice modes are observed. However, the former spectrum is simpler than the latter because, except for a few small hydride molecules, no rotational transitions are observed because of the rigidity of the matrix environment at low temperatures. The lack of rotational structure and intermolecular interactions results in a sharpening of the solute band so that even very closely located metal isotope peaks can be resolved in a matrix environment. This technique is also applicable to a compound which is not volatile at room temperature. For example, matrix isolation spectra of metal halides can be measured by vaporizing these compounds at high temperatures in a Knudsen cell and co-condensing their vapors with inert gas molecules on a cold window.[93] The recent development of closed-cycle helium refrigerators greatly facilitated the experimental technique. The matrix isolation technique has now been applied to a number of inorganic and coordination compounds to obtain structural and bonding information. Some important applications of this technique are described below.

(1) Vibrational Spectra of Radicals

Highly reactive radicals can be produced *in situ* in inert gas matrices by photolysis and other techniques. Since these radicals are stabilized in matrix environments, their spectra can be measured by routine spectroscopic techniques. For example, the spectrum of the HOO radical[94] was obtained by measuring the spectrum of the photolysis product of a mixture of HI and O_2 in an Ar matrix at ~4 K. Part II lists the vibrational frequencies of many other radicals, such as NH, OF, and FCE, obtained by similar methods.

(2) Vibrational Spectra of High-Temperature Species

Alkali halide vapors produced at high temperatures consist mainly of monomers and dimers. The vibrational spectra of these salts at high temperatures are difficult to measure and difficult to interpret because of the presence of hot bands. The matrix isolation technique utilizing a Knudsen cell has solved this problem. The vibrational frequencies of some of these high-temperature species are listed in Part II.

(3) Isotope Splittings

As stated before, it is possible to observe individual peaks due to heavy metal isotopes in inert gas matrices since the bands are extremely sharp (half-band width, 1.5–1.0 cm^{-1}) under these conditions. Figure I-20*a* shows the infrared spectrum of the ν_7 band (coupled vibration between CrC stretching and CrCO bending modes) of $Cr(CO)_6$ in a N_2 matrix.[95] The bottom curve shows a computer simulation using the measured isotope shift of 2.5 cm^{-1} per atomic mass unit, a 1.2 cm^{-1} half-band width, and the percentages of natural abundance of Cr isotopes: ^{50}Cr (4.31%), ^{52}Cr (83.76%), ^{53}Cr (9.55%), and ^{54}Cr (2.38%). These isotope frequencies are highly important in calculating force constants and anharmonicity corrections. In Parts II and III, isotope frequencies will be given for some representative compounds.

(4) Chemical Synthesis

Recently, the matrix co-condensation technique has been used to synthesize a number of unstable and transient coordination compounds. For example, a series of nickel carbonyls of the type $Ni(CO)_x$, where $x = 1$, 2, 3, and 4, have been synthesized by allowing metal vapor to react with CO diluted in Ar on a cold window and warming the matrix carefully. Figure I-21 shows the result obtained by DeKock.[96] The structures of $Ni(CO)_2$ and $Ni(CO)_3$ were concluded to be linear and trigonal-planar, respectively, since these compounds exhibit only one CO stretching band in the infrared. Similar methods have been applied to the synthesis of a number of coordination compounds ML_x, where M is Pt, Pd, Ni, and so on, and L is CO, N_2, O_2, PF_3, and so on.[41] More detailed discussions of individual compounds will be given in Part III.

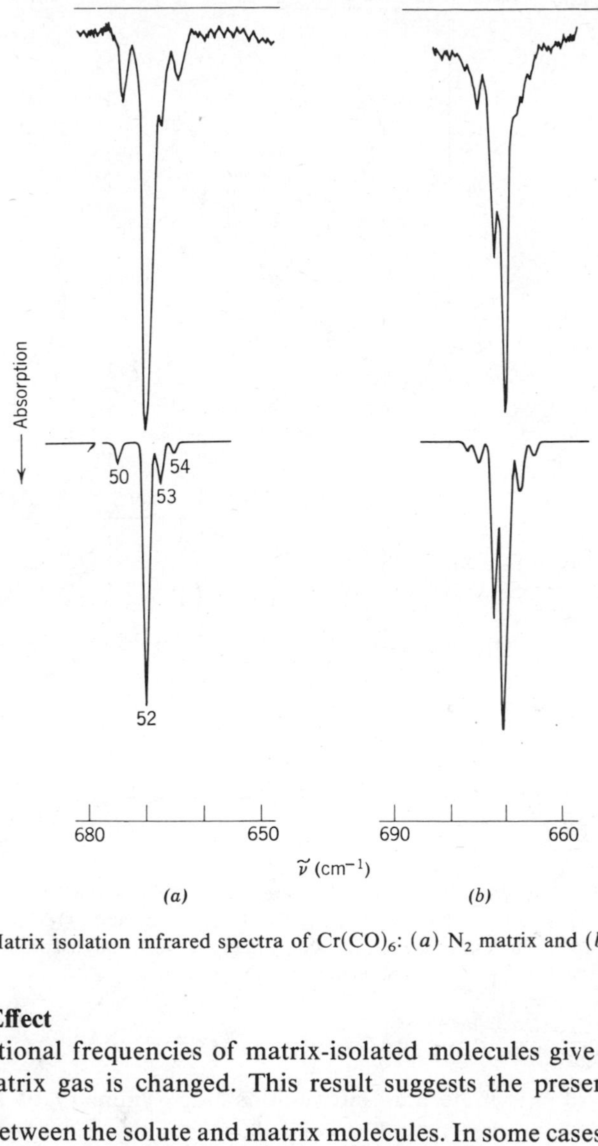

Fig. I-20. Matrix isolation infrared spectra of $Cr(CO)_6$: (*a*) N_2 matrix and (*b*) Ar matrix.

(5) Matrix Effect

The vibrational frequencies of matrix-isolated molecules give slight shifts when the matrix gas is changed. This result suggests the presence of weak interaction between the solute and matrix molecules. In some cases, the spectra are complicated by the presence of more than one trapping site. For example, the infrared spectrum of $Cr(CO)_6$ in an Ar matrix (Fig. I-20*b*) is markedly different from that in a N_2 matrix (Fig. I-20*a*).[95] The former spectrum can be interpreted by assuming two different sites in an Ar matrix. The computer-simulation spectrum (bottom curve) was obtained by assuming that these two sites are populated in a 2:1 ratio, the frequency separation of the corresponding peaks being 2 cm^{-1}. Thus it is always desirable to obtain the matrix isolation spectra in several different environments.

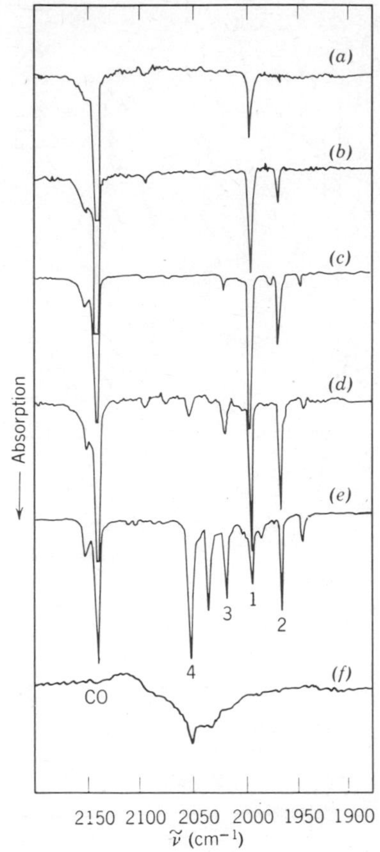

Fig. I-21. Infrared spectra of Ni atoms deposited in a 500:1 Ar/CO matrix and subsequent annealing: (*a*) original, (*b*) 17 K, (*c*) 18 K, (*d*) 19 K, (*e*) 26 K, and (*f*) 35 K (temperatures are relative). The arabic numerals refer to the relative rate of growth and disappearance of the bands and hence to *n* in $Ni(CO)_n$.

I-23. VIBRATIONAL SPECTRA OF CRYSTALS[42–46]

Because of intermolecular interactions, the symmetry of a molecule is generally lower in the crystalline state than in the gaseous (isolated) state.* This change in symmetry may split the degenerate vibrations and activate infrared- (or Raman-) inactive vibrations. In addition, the spectra obtained in the crystalline state exhibit *lattice modes*—vibrations due to translatory and rotatory motions of a molecule in the crystalline lattice. Although their frequencies are usually lower than 300 cm^{-1}, they may appear in the high-frequency region as the combination bands with internal modes (see Fig. II-10, for

* The symmetry of a molecule may be the lowest in solution (or liquid) because it interacts with randomly oriented molecules.

example). Thus the vibrational spectra of crystals must be interpreted with caution, especially in the low-frequency region.

To analyze the spectra of crystals, it is necessary to carry out the site group or factor group analysis described in the following subsection.

(1) Site Group Analysis

According to Halford,[97] the vibrations of a molecule in the crystalline state are governed by a new selection rule derived from *site symmetry*—a local symmetry around the center of gravity of a molecule in a unit cell. The site symmetry can be found by using the following two conditions: (1) the site group must be a subgroup of both the space group of the crystal and the molecular point group of the isolated molecule, and (2) the number of equivalent sites must be equal to the number of molecules in the unit cell. Halford derived a complete table that lists possible site symmetries and the number of equivalent sites for 230 space groups.* Suppose that the space group of the crystal, the number of molecules in the unit cell (Z), and the point group of the isolated molecule are known. Then the site symmetry can be found from Halford's table. In general, the site symmetry is lower than the molecular symmetry in an isolated state.

The vibrational spectra of calcite and aragonite crystals are markedly different, although both have the same composition (Sec. II-4). This result can be explained if we consider the difference in site symmetry of the CO_3^{2-} ion between these crystals. According to X-ray analysis, the space group of calcite is $\mathbf{D}_{3d}^6$ and Z is two. Halford's table gives

$$\mathbf{D}_3(2), \quad \mathbf{C}_{3i}(2), \quad \infty\mathbf{C}_3(4), \quad \mathbf{C}_i(6), \quad \infty\mathbf{C}_2(6)$$

as possible site symmetries for space group $\mathbf{D}_{3d}^6$ (the number in front of point group notation indicates the number of distinct sets of sites, and that in parenthesis denotes the number of equivalent sites for each distinct set). Rule 2 eliminates all but $\mathbf{D}_3(2)$ and $\mathbf{C}_{3i}(2)$. Rule 1 eliminates the latter since $\mathbf{C}_{3i}$ is not a subgroup of $\mathbf{D}_{3h}$. Thus the site symmetry of the CO_3^{2-} ion in calcite must be $\mathbf{D}_3$. On the other hand, the space group of aragonite is $\mathbf{D}_{2h}^{16}$ and Z is four. Halford's table gives

$$2\mathbf{C}_i(4), \quad \infty\mathbf{C}_s(4)$$

Since $\mathbf{C}_i$ is not a subgroup of $\mathbf{D}_{3h}$, the site symmetry of the CO_3^{2-} ion in aragonite must be $\mathbf{C}_s$. Thus the $\mathbf{D}_{3h}$ symmetry of the CO_3^{2-} ion in an isolated state is lowered to $\mathbf{D}_3$ in calcite and to $\mathbf{C}_s$ in aragonite. Then the selection rules are changed as shown in Table I-12.

There is no change in the selection rule in going from the free CO_3^{2-} ion to calcite. In aragonite, however, ν_1 becomes infrared active, and ν_3 and ν_4 each

*For a more complete table, see Ref. 98.

TABLE I-12. CORRELATION TABLE FOR $\mathbf{D}_{3h}$, $\mathbf{D}_3$, $\mathbf{C}_{2v}$, AND $\mathbf{C}_s$

Point Group	ν_1	ν_2	ν_3	ν_4
$\mathbf{D}_{3h}$	A_1'(R)	A_2''(I)	E'(I, R)	E'(I, R)
$\mathbf{D}_3$	A_1(R)	A_2(I)	E(I, R)	E(I, R)
$\mathbf{C}_{2v}$	A_1(I, R)	B_1(I, R)	A_1(I, R) + B_2(I, R)	A_1(I, R) + B_2(I, R)
$\mathbf{C}_s$	A'(I, R)	A''(I, R)	A'(I, R) + A'(I, R)	A'(I, R) + A'(I, R)

split into two bands. The observed spectra of calcite and aragonite are in good agreement with these predictions (see Table II-4*b*).

(2) Factor Group Analysis

A more complete analysis including lattice modes can be made by the method of factor group analysis developed by Bhagavantam and Venkatarayudu.[99] In this method, we consider all the normal vibrations for an entire Bravais primitive cell.* Figure I-22 illustrates the Bravais cell of calcite, which consists of the following symmetry elements: I, $2S_6$, $2S_6^2 \equiv 2C_3$, $S_6^3 \equiv i$, $3C_2$, and $3\sigma_v$ (glide plane). These elements are exactly the same as those of the point group $\mathbf{D}_{3d}$, although the last element is a glide plane rather than a plane of symmetry in a single molecule.

It is possible to derive the 230 space groups by combining operations possessed by the 32 crystallographic point groups† with operations such as pure translation, screw rotation (translation + rotation), and glide plane reflection (translation + reflection). If we regard the translations that carry a point in a unit cell into the equivalent point in another cell as identity, we define the 230 factor groups that are the subgroups of the corresponding space groups. In the case of calcite, the factor group consists of the symmetry elements described above, and is denoted by the same notation as that used for the space group ($\mathbf{D}_{3d}^6$). The site group discussed previously is a subgroup of a factor group.

Since the Bravais cell contains 10 atoms, it has $3 \times 10 - 3 = 27$ normal vibrations, excluding three translational motions of the cell as a whole.‡ These 27 vibrations can be classified into various symmetry species of the factor group $\mathbf{D}_{3d}^6$, using a procedure similar to that described in Sec. I-7 for internal vibrations. First, we calculate the characters of representations corresponding to the entire freedom possessed by the Bravais primitive cell [$\chi_R(N)$], translational motions of the whole cell [$\chi_R(T)$], translatory lattice modes [$\chi_R(T')$],

* Every molecule (or ion) in a Bravais primitive cell can be superimposed on that of the neighboring cell by simple translation. In the case of calcite, the Bravais primitive cell is the same as the crystallographic unit cell. However, this is not always the case.

† In crystals, the number of point groups is limited to 32 since only C_1, C_2, C_3, C_4, and C_6 are possible due to the space-filling requirements of the crystal lattice.

‡ These three motions give "acoustic modes" that propagate sound waves through the crystal.

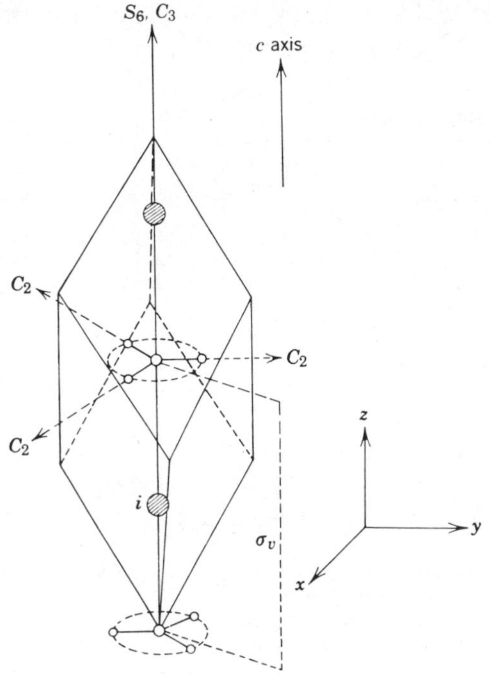

Fig. I-22. The Bravais primitive cell of calcite.

rotatory lattice modes [$\chi_R(R')$], and internal modes [$\chi_R(n)$], using the equations given in Table I-13. Then each of these characters is resolved into the symmetry species of the point group, $\mathbf{D}_{3d}$. The final results show that three internal modes (A_{2u} and two E_u), three translatory modes (A_{2u} and two E_u), and two rotatory modes (A_{2u} and E_u) are infrared active, and three internal modes (A_{1g} and two E_g), one translatory mode (E_g), and one rotatory mode (E_g) are Raman active. As will be shown in the following subsections, these predictions are in perfect agreement with the observed spectra.

The correlation method developed by Fateley et al.[15] is simpler than factor group analysis and gives the same results. A complete normal coordinate analysis on the whole Bravais cell of calcite-type crystals was carried out by Nakagawa and Walter.[100] Similar treatments have been extended to nitro and aquo complexes.[101] For a review of lattice vibrations of inorganic and coordination compounds, see Ref. 101.

(3) Infrared Dichroism

Suppose that we irradiate a single crystal of calcite with polarized infrared radiation whose electric vector vibrates along the c axis (z direction) in Fig. I-22. Then the infrared spectrum shown by the solid curve of Fig. I-23 is obtained.[102] According to Table I-13, only the A_{2u} vibrations are activated

TABLE I-13. FACTOR GROUP ANALYSIS OF CALCITE CRYSTAL[a]

D_{3d}^6	I	$2S_6$	$2C_3 \equiv 2S_6^2$	$i \equiv S_6^3$	$3C_2$	$3\sigma_v$	Number of Vibrations N	T	T'	R'	n		
A_{1g}	1	1	1	1	1	1	1	0	0	0	1		$\alpha_{xx}+\alpha_{yy}, \alpha_{zz}$
A_{1u}	1	−1	1	−1	1	−1	2	0	1	0	1		
A_{2g}	1	1	1	1	−1	−1	3	0	1	1	1		
A_{2u}	1	−1	1	−1	−1	1	4	1	1	1	1	T_z	
E_g	2	−1	−1	2	0	0	4	0	1	1	2		$(\alpha_{xx}-\alpha_{yy}, \alpha_{xy}), (\alpha_{xz}, \alpha_{yz})$
E_u	2	1	−1	−2	0	0	6	1	2	1	2	(T_x, T_y)	
$N_R(p)$	10	2	4	2	4	0	↑	↑	↑	↑	↑		
$N_R(s)$	4	2	4	2	2	0							
$N_R(s-v)$	2	0	2	0	2	0							
$\chi_R(N)$	30	0	0	−6	−4	0							
$\chi_R(T)$	3	0	0	−3	−1	1							
$\chi_R(T')$	9	0	0	−3	−1	−1							
$\chi_R(R')$	6	0	0	0	−2	0							
$\chi_R(n)$	12	0	0	0	0	0							

[a] p, total number of atoms in the Bravais primitive cell.
s, total number of molecules (ions) in the primitive cell.
v, total number of monoatomic molecules (ions) in the primitive cell.
$N_R(p)$, number of atoms unchanged by symmetry operation R.
$N_R(s)$, number of molecules (ions) whose center of gravity is unchanged by symmetry operation R.
$N_R(s-v)$, $N_R(s)$ minus number of monoatomic molecules (ions) unchanged by symmetry operation R.
$\chi_R(N)=N_R(p)[\pm(1+2\cos\theta)]$, character of representation for entire freedom possessed by the primitive cell.
$\chi_R(T)=\pm(1+2\cos\theta)$, character of representation for translational motions of the whole primitive cell.
$\chi_R(T')=\{N_R(s)-1\}\{\pm(1+2\cos\theta)\}$, character of representation for translatory lattice modes.
$\chi_R(R')=N_R(s-v)\{\pm(1+2\cos\theta)\}$, character of representation for rotary lattice modes.
$\chi_R(n)=\chi_R(N)-\chi_R(T)-\chi_R(T')-\chi_R(R')$, character of representation for internal modes.
Note that + and − signs are for proper and improper rotations, respectively. The symbol θ should be taken as defined in Sec. I-7.

under such conditions. Thus the three bands observed at 885(v), 357(t) and 106(r) cm^{-1} are assigned to the A_{2u} species. The spectrum shown by the dotted curve is obtained if the direction of polarization is perpendicular to the c axis (x, y plane). In this case, only the E_u vibrations should be infrared active. Therefore the five bands observed at 1484(v), 706(v), 330(t), 182(t), and 106(r) cm^{-1} are assigned to the E_u species. As shown above, polarized infrared studies of single crystals provide valuable information about the symmetry properties of normal vibrations if the crystal structure is known from other sources.

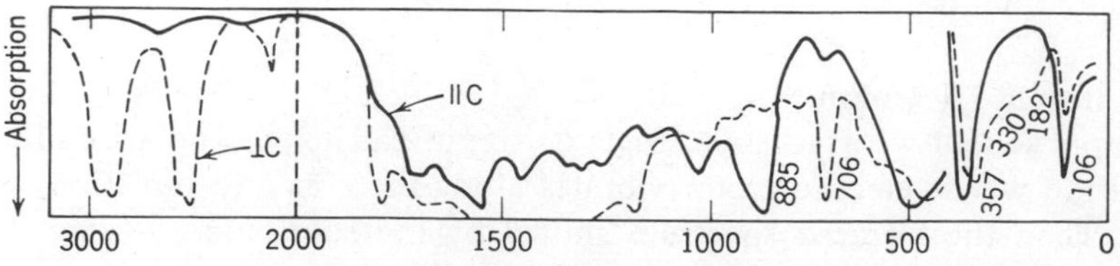

Fig. I-23. Infrared dichroism of calcite.[102]

The ratio of the absorption intensities in the directions parallel and perpendicular to a crystal axis, called the *dichroic ratio*, is defined by

$$R = \frac{\int_{\text{band}} \varepsilon_{\parallel}(\tilde{\nu})\, d\tilde{\nu}}{\int_{\text{band}} \varepsilon_{\perp}(\tilde{\nu})\, d\tilde{\nu}} \tag{23.1}$$

Here $\varepsilon_{\parallel}$ and $\varepsilon_{\perp}$ denote the absorption coefficients for radiation whose electric vector vibrates in the directions parallel and perpendicular, respectively, to the crystal axis. In the case of calcite, the maximum dichroic ratio is expected, since the carbonate ions are oriented parallel to each other. The dichroic ratio will be smaller if the molecules are not parallel to each other in a crystal lattice.

(4) Polarized Raman Spectra

Polarized Raman spectra provide more information about the symmetry properties of normal vibrations than do polarized infrared spectra.[103] Again consider a single crystal of calcite. According to Table I-13, the A_{1g} vibrations become Raman active if any one of the polarizability components, α_{xx}, α_{yy}, and α_{zz}, is changed. Suppose that we irradiate a calcite crystal from the y direction, using polarized radiation whose electric vector vibrates parallel to the z axis (see Fig. I-24), and observe the Raman scattering in the x direction with its polarization in the z direction. This condition is abbreviated as $y(zz)x$. In this case, Eq. 5.7 is simplified to $P_z = \alpha_{zz}E_z$ because $E_x = E_y = 0$ and $P_x = P_y = 0$. Since α_{zz} belongs to the A_{1g} species, only the A_{1g} vibrations are observed under this condition. Figure I-25c illustrates the Raman spectrum obtained with this condition. Thus the strong Raman line at 1088 cm^{-1}(v) is assigned to the A_{1g} species. Both the A_{1g} and E_g vibrations are observed if the $z(xx)y$ condition is used. The Raman spectrum (Fig. I-25a) shows that five Raman

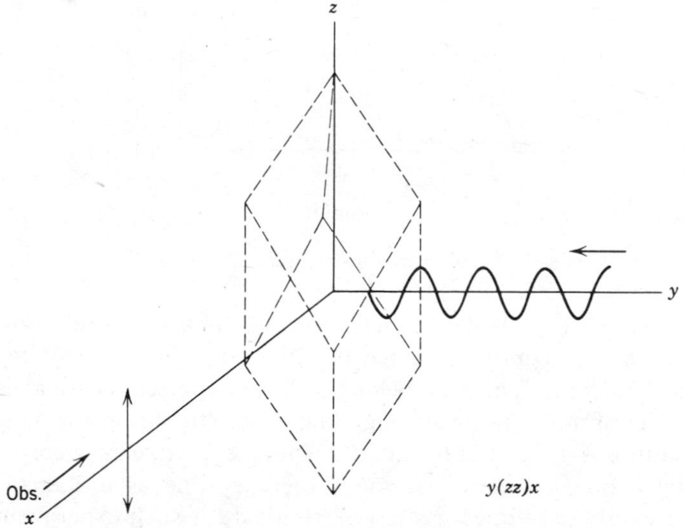

Fig. I-24. Schematic representation of experimental condition used for the measurement of depolarization ratios of calcite crystal.

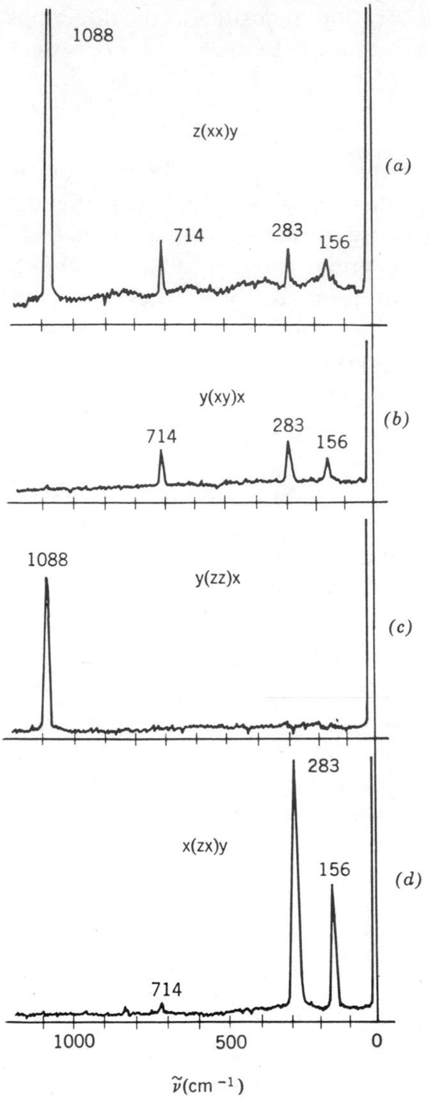

Fig. I-25. Polarized Raman spectra of calcite.[103]

lines [1088(v), 714(v), 283(r), 156(t), and 1434(v) cm^{-1} (not shown)] are observed under this condition. Since the 1088 cm^{-1} line belongs to the A_{1g} species, the remaining four must belong to the E_g species. These assignments can also be confirmed by measuring Raman spectra using the $y(xy)x$ and $x(zx)y$ conditions (Fig. I-25b and d). These experiments were originally performed by Bhagavantam,[104] using a mercury line as a Raman source. However, recently developed gas lasers are ideal for such experiments, since they provide strong and completely polarized radiation. Figure I-26 shows the vibrational modes of calcite crystals obtained by Nakagawa and Walter.[100]

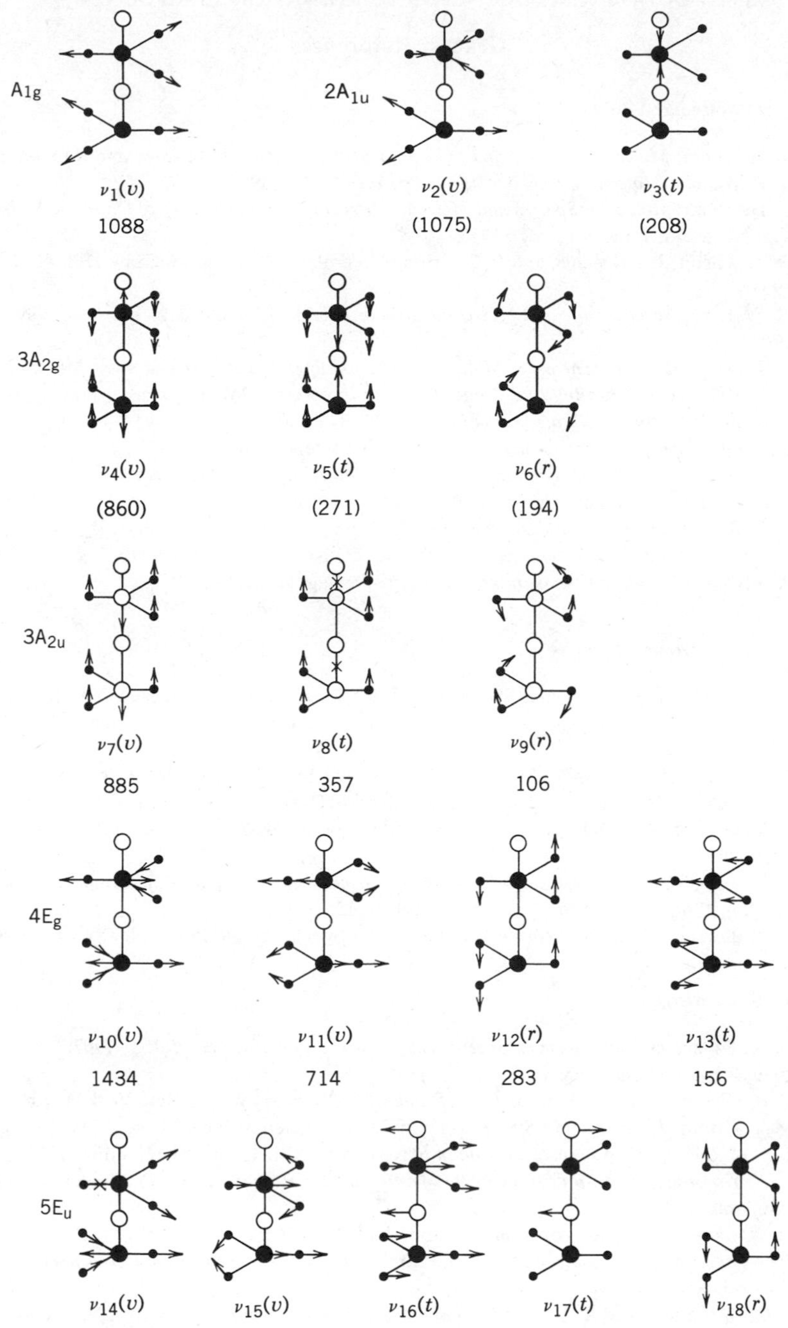

Fig. I-26. Vibrational modes of calcite.[100] The observed and calculated (in parentheses) are listed under each mode.

General References

Theory of Molecular Vibrations

1. G. Herzberg, *Molecular Spectra and Molecular Structure.* Vol. II: *Infrared and Raman Spectra of Polyatomic Molecules,* Van Nostrand, Princeton, N.J., 1945.
2. G. Herzberg, *Molecular Spectra and Molecular Structure.* Vol. I: *Spectra of Diatomic Molecules,* Van Nostrand, Princeton, N.J., 1950.
3. E. B. Wilson, J. C. Decius, and P. C. Cross, *Molecular Vibrations,* McGraw-Hill, New York, 1955.
4. G. W. King, *Spectroscopy and Molecular Structure,* Holt, Rinehart, and Winston, New York, 1964.
5. C. J. H. Schutte, *The Theory of Molecular Spectroscopy,* Vol. I: *The Quantum Mechanics and Group Theory of Vibrating and Rotating Molecules,* North Holland, Amsterdam, 1976.
6. P. Gans, *Vibrating Molecules,* Chapman and Hall, London, 1971.
7. D. Steele, *Theory of Vibrational Spectroscopy,* Saunders, London, 1971.
8. L. A. Woodward, *Introduction to the Theory of Molecular Vibrations and Vibrational Spectroscopy,* Oxford University Press, London, 1972.
9. C. N. Banwell, *Fundamentals of Molecular Spectroscopy,* 2nd ed., McGraw-Hill, London, 1972.
10. S. J. Cyvin, *Molecular Structures and Vibrations,* Elsevier, Amsterdam, 1972.
11. A. Fadini, *Molekülkraftkonstanten,* Steinkopff-Verlag, Darmstadt, 1976.

Symmetry and Group Theory

12. L. H. Hall, *Group Theory and Symmetry in Chemistry,* McGraw-Hill, New York, 1969.
13. F. A. Cotton, *Chemical Application of Group Theory,* 2nd ed., Wiley-Interscience, New York, 1971.
14. M. Orchin and H. H. Jaffé, *Symmetry, Orbitals and Spectra,* Wiley, New York, 1971.
15. W. G. Fateley, F. R. Dollish, N. T. McDevitt, and F. F. Bentley, *Infrared and Raman Selection Rules for Molecular and Lattice Vibrations: The Correlation Method,* Wiley-Interscience, New York, 1972.
16. J. R. Ferraro and J. S. Ziomek, *Introductory Group Theory and its Application to Molecular Structure,* 2nd ed., Plenum Press, New York, 1975.
17. P. R. Bunker, *Molecular Symmetry and Spectroscopy,* Academic Press, New York, 1979.

Raman Spectroscopy

18. H. A. Szymanski, ed., *Raman Spectroscopy: Theory and Practice,* Vol. 1, 1967, and Vol. 2, 1970, Plenum Press, New York.
19. T. R. Gilson and P. J. Hendra, *Laser Raman Spectroscopy,* Wiley, New York, 1970.
20. M. C. Tobin, *Laser Raman Spectroscopy,* Wiley-Interscience, New York, 1971.
21. J. P. Mathieu, ed., *Advances in Raman Spectroscopy,* Vol. 1, Heyden, London, 1973.
22. J. A. Koningstein, *Introduction to the Theory of the Raman Effect,* D. Reidel, Dordrecht (Holland), 1973.
23. D. A. Long, *Raman Spectroscopy,* McGraw-Hill, New York, 1977.
24. J. G. Grasselli, M. K. Snavely, and B. J. Bulkin, *Chemical Applications of Raman Spectroscopy,* Wiley, New York, 1981.

24a. D. P. Strommen and K. Nakamoto, *Laboratory Raman Spectroscopy,* Wiley, New York, 1984.

Vibrational Spectra of Inorganic, Coordination, and Organometallic Compounds

25. H. Siebert, *Anwendungen der Schwingungsspektroskopie in der Anorganischen Chemie,* Springer-Verlag, Berlin, 1966.

26. D. M. Adams, *Metal-Ligand and Related Vibrations*, Edward Arnold, London, 1967.
27. J. R. Ferraro, *Low-frequency Vibrations of Inorganic and Coordination Compounds*, Plenum Press, New York, 1971.
28. L. H. Jones, *Inorganic Spectroscopy*, Vol. 1, Marcel Dekker, New York, 1971.
29. R. A. Nyquist and R. O. Kagel, *Infrared Spectra of Inorganic Compounds*, Academic Press, New York, 1971.
30. S. D. Ross, *Inorganic Infrared and Raman Spectra*, McGraw-Hill, New York, 1972.
31. N. N. Greenwood, E. J. F. Ross, and B. P. Straughan, *Index of Vibrational Spectra of Inorganic and Organometallic Compounds*, Vol. 1 (1972) and Vol. 2 (1975), Butterworths, London.
32. E. Maslowsky, Jr., *Vibrational Spectra of Organometallic Compounds*, Wiley, New York, 1976.

Vibrational Spectra of Organic Compounds

33. L. J. Bellamy, *The Infrared Spectra of Complex Molecules*, 3rd ed., Vol. 1 (1975) and Vol. 2 (1980), Chapman and Hall, London.
34. N. B. Colthup, L. H. Daly, and S. E. Wiberley, *Introduction to Infrared and Raman Spectroscopy*, 2nd ed., Academic Press, New York, 1975.
35. F. R. Dollish, W. G. Fateley, and F. F. Bentley, *Characteristic Raman Frequencies of Organic Compounds*, Wiley, New York, 1974.
36. F. S. Parker, *Application of Infrared, Raman and Resonance Raman Spectroscopy in Biochemistry*, Plenum Press, New York, 1983.
37. A. T. Tu, *Raman Spectroscopy in Biology*, Wiley, New York, 1982.
38. P. R. Carey, *Biochemical Applications of Raman and Resonance Raman Spectroscopies*, Academic Press, New York, 1982.

Low-Temperature and Matrix Isolation Spectroscopy

39. B. Meyer, *Low-Temperature Spectroscopy*, Elsevier, Amsterdam, 1971.
40. H. E. Hallam, ed., *Vibrational Spectroscopy of Trapped Species*, Wiley, New York, 1973.
41. M. Moskovits and G. A. Ozin, eds., *Cryochemistry*, Wiley, New York, 1976.

Time-Resolved Spectroscopy

41a. G. H. Atkinson, *Time-Resolved Vibrational Spectroscopy*, Academic Press, New York, 1983.

High-Pressure Spectroscopy

41b. J. R. Ferraro, *Vibrational Spectroscopy at High External Pressures—The Diamond Anvil Cell*, Academic Press, New York, 1984.

Vibrational Spectra of Crystals and Minerals

42. G. Turrell, *Infrared and Raman Spectra of Crystals*, Academic Press, New York, 1972.
43. J. C. Decius and R. M. Hexter, *Molecular Vibrations in Crystals*, McGraw-Hill, New York, 1977.
44. V. C. Farmer, *The Infrared Spectra of Minerals*, Mineralogical Society, London, 1974.
45. J. A. Gadsden, *Infrared Spectra of Minerals and Related Inorganic Compounds*, Butterworths, London, 1975.
46. C. Karr, ed., *Infrared and Raman Spectroscopy of Lunar and Terrestrial Minerals*, Academic Press, New York, 1975.

Infrared Spectra of Adsorbed Species

47. L. H. Little, *Infrared Spectra of Adsorbed Species*, Academic Press, New York, 1967.
48. M. L. Hair, *Infrared Spectroscopy in Surface Chemistry*, Marcel Dekker, New York, 1967.
49. A. T. Bell and M. L. Hair, eds., "Vibrational Spectroscopies for Adsorbed Species," American Chemical Society, Washington, D.C., 1980.

Fourier-Transform Spectroscopy

50. R. J. Bell, *Introductory Fourier Transform Spectroscopy*, Academic Press, New York, 1972.
51. J. R. Ferraro and L. J. Basile, eds., *Fourier Transform Infrared Spectroscopy*, Vol. 1 (1978) to present, Academic Press, New York.

Vibrational Intensities

52. W. B. Person and G. Zerbi, eds., *Vibrational Intensities in Infrared and Raman Spectroscopy*, Elsevier, Amsterdam, 1982.

Advances Series

53. *Spectroscopic Properties of Inorganic and Organometallic Compounds*, Vol. 1 to present, The Chemical Society, London.
54. *Molecular Spectroscopy—Specialist Periodical Reports*, Vol. 1 to present, The Chemical Society, London.
55. R. J. H. Clark and R. E. Hester, eds., *Advances in Infrared and Raman Spectroscopy*, Vol. 1 to present, Heyden, London.
56. J. Durig, ed., *Vibrational Spectra and Structure*, Vol. 1 to present, Elsevier, Amsterdam.
57. C. B. Moore, ed., *Chemical and Biochemical Applications of Lasers*, Vol. 1 to present, Academic Press, New York.
58. *Structure and Bonding*, Vol. 1 to present, Springer-Verlag, New York.

References

59. W. Holzer, W. F. Murphy, and H. J. Bernstein, *J. Chem. Phys.*, **52,** 399 (1970).
60. C. F. Shaw, III, *J. Chem. Educ.*, **58,** 343 (1981).
61. P. Pulay, *Mol. Phys.*, **17,** 197 (1969); **18,** 473 (1970); P. Pulay and W. Meyer, *J. Mol. Spectrosc.*, **40,** 59 (1971).
61a. Y. Nishimura, M. Tsuboi, S. Kato, and K. Morokuma, *J. Am. Chem. Soc.*, **103,** 1354 (1981).
62. E. B. Wilson, *J. Chem. Phys.*, **7,** 1047 (1939); **9,** 76 (1941).
63. J. C. Decius, *J. Chem. Phys.*, **16,** 1025 (1948).
64. T. Shimanouchi, *J. Chem. Phys.*, **25,** 660 (1956).
65. T. Shimanouchi, *J. Chem. Phys.*, **17,** 245, 734, 848 (1949).
66. T. Shimanouchi, "The Molecular Force Field," in D. Henderson, ed., *Physical Chemistry: An Advanced Treatise*, Vol. 4, Academic Press, New York, 1970.
67. D. F. Heath and J. W. Linnett, *Trans. Faraday Soc.*, **44,** 556, 873, 878, 884 (1948); **45,** 264 (1949).
68. J. Overend and J. R. Scherer, *J. Chem. Phys.*, **32,** 1289, 1296, 1720 (1960); **33,** 446 (1960); **34,** 547 (1961); **36,** 3308 (1962).
69. T. Shimanouchi, *Pure Appl. Chem.*, **7,** 131 (1963).
70. J. H. Schachtschneider, "Vibrational Analysis of Polyatomic Molecules," Pts. V and VI, Tech. Rept. 231-64 and 53-65, Shell Development Co., Emeryville, Calif., 1964 and 1965.
71. K. Nakamoto, *Angew. Chem.*, **11,** 666 (1972).

72. N. Mohan, K. Nakamoto, and A. Müller, "The Metal Isotope Effect on Molecular Vibrations," in R. J. H. Clark and R. E. Hester, eds., *Advances in Infrared and Raman Spectroscopy*, Vol. 1, Heyden, London, 1976.
73. K. Nakamoto, K. Shobatake, and B. Hutchinson, *Chem. Commun.*, 1451 (1969).
74. J. Takemoto and K. Nakamoto, *Chem. Commun.*, 1017 (1970).
75. J. R. Kincaid and K. Nakamoto, *Spectrochim. Acta*, **32A,** 277 (1976).
76. P. M. Champion, B. R. Stallard, G. C. Wagner, and I. C. Gunsalus, *J. Am. Chem. Soc.*, **104,** 5469 (1982).
77. Y. Morino and K. Kuchitsu, *J. Chem. Phys.*, **20,** 1809 (1952).
78. B. L. Crawford and W. H. Fletcher, *J. Chem. Phys.*, **19,** 141 (1951).
79. P. LaBonville and J. M. Williams, *Appl. Spectrosc.*, **25,** 672 (1971).
80. D. A. Ramsay, *J. Am. Chem. Soc.*, **74,** 72 (1952).
81. D. P. Strommen and K. Nakamoto, *Appl. Spectrosc.*, **37,** 436 (1983).
82. T. G. Spiro and T. C. Strekas, *Proc. Natl. Acad. Sci.*, **69,** 2622 (1972).
83. W. M. McClain, *J. Chem. Phys.*, **55,** 2789 (1971).
84. W. Kiefer, *Appl. Spectrosc.*, **28,** 115 (1974).
85. M. Mingardi and W. Siebrand, *J. Chem. Phys.*, **62,** 1074 (1975).
86. R. J. H. Clark and P. D. Mitchell, *J. Am. Chem. Soc.*, **95,** 8300 (1973).
87. A. C. Albrecht, *J. Chem. Phys.*, **34,** 1476 (1961).
88. T. G. Spiro, *Acc. Chem. Res.*, **7,** 339 (1974).
89. F. Inagaki, M. Tasumi, and T. Miyazawa, *J. Mol. Spectrosc.*, **50,** 286 (1974).
90. G. A. Ozin, "Single Crystal and Gas Phase Raman Spectroscopy in Inorganic Chemistry," in S. J. Lippard, ed., *Progress in Inorganic Chemistry*, Vol. 14, Wiley-Interscience, New York, 1971.
91. R. J. H. Clark and D. M. Rippon, *J. Mol. Spectrosc.*, **44,** 479 (1972).
92. E. Whittle, D. A. Dows, and G. C. Pimentel, *J. Chem. Phys.*, **22,** 1943 (1954).
93. M. J. Linevsky, *J. Chem. Phys.*, **34,** 587 (1961).
94. D. E. Milligan and M. E. Jacox, *J. Chem. Phys.*, **38,** 2627 (1963).
95. D. Tevault and K. Nakamoto, *Inorg. Chem.*, **14,** 2371 (1975).
96. R. L. DeKock, *Inorg. Chem.*, **10,** 1205 (1971).
97. R. S. Halford, *J. Chem. Phys.*, **14,** 8 (1946).
98. W. G. Fateley, *Appl. Spectrosc.*, **27,** 395 (1973); W. G. Fateley, N. J. McDevitt, and F. F. Bentley, *Appl. Spectrosc.*, **25,** 155 (1971).
99. S. Bhagavantam and T. Venkatarayudu, *Proc. Indian Acad. Sci.*, **9A,** 224 (1939); "Theory of Groups and Its Application to Physical Problems," Andhra University, Waltair, 1951.
100. I. Nakagawa and J. L. Walter, *J. Chem. Phys.*, **51,** 1389 (1969).
101. I. Nakagawa, *Coord. Chem. Rev.*, **4,** 423 (1969).
102. M. Tsuboi, *Infrared Absorption Spectra*, Vol. 6, Nankodo, Tokyo, 1958, p. 41.
103. S. P. Porto, J. A. Giordmaine, and T. C. Damen, *Phys. Rev.*, **147,** 608 (1966).
104. S. Bhagavantam, *Proc. Indian Acad. Sci.*, **11A,** 62 (1940).

Inorganic Compounds

Part II

II-1. DIATOMIC MOLECULES

As shown in Sec. I-2, diatomic molecules have only one vibration along the chemical bond; its frequency is given by

$$\tilde{\nu} = \frac{1}{2\pi c}\sqrt{\frac{K}{\mu}}$$

where K is the force constant, μ the reduced mass, and c the velocity of light. In homopolar XX molecules ($\mathbf{D}_{\infty h}$), the vibration is not infrared active but is Raman active, whereas it is both infrared- and Raman-active in heteropolar XY molecules ($\mathbf{C}_{\infty v}$). Table II-1*a* lists the frequencies corrected for anharmonicity (ω_e) and anharmonicity constants. The force constants can be calculated directly from these ω_e values.

Tables II-1*b* and II-1*c* lists the observed frequencies of a number of diatomic molecules, ions, and radicals as reported recently. The matrix isolation technique was employed extensively to observe the spectra of unstable and reactive species (Sec. I-22). Resonance Raman spectra of colored diatomic species exhibit a series of overtone frequencies which can be used to calculate anharmonicity constants (Sec. I-21).

Tables II-1*a*, II-1*b*, and II-1*c* show several interesting trends in frequencies and force constants. In the same family of the periodic table, we find the following:

	F_2		Cl_2		Br_2		I_2
$\tilde{\nu}$ (cm^{-1})	892	>	546	>	319	>	215
K (mdyn/Å)	4.45	>	3.19	>	2.46	>	1.76

A similar trend is seen for $Li_2 > Na_2 > K_2 > Rb_2$. For a series of hydrogen halides, we find:

	HF		HCl		HBr		HI
$\tilde{\nu}$ (cm^{-1})	4138.5	>	2991	>	2649.7	>	2309.5
K (mdyn/Å)	9.66	>	5.16	>	4.12	>	3.12

The same trend is seen for HBe > HMg > HCa > HSr. Across the periodic table, a series such as the following:

	HC		HN		HO		HF
$\tilde{\nu}$ (cm^{-1})	2861.6	<	(3300)	<	3735.2	<	4138.5
K (mdyn/Å)	4.49	<	6.03	<	7.79	<	9.66

TABLE II-1*a*. HARMONIC FREQUENCIES AND ANHARMONICITY CONSTANTS OF SOME DIATOMIC MOLECULES AND IONS (cm^{-1})[a,b]

Molecule	ω_e	$x_e\omega_e$	Molecule	ω_e	$x_e\omega_e$
HH	4395.2	117.91	^{7}LiF (4)	906.2	7.90
$[HH]^+$	2297	62	$^{7}LiCl$ (4)	641.1	4.2
HD	3817.1	94.96	$^{7}LiBr$ (4)	563.2	3.53
DD	3118.5	64.10	^{7}LiI (4)	498.2	3.39
$H^{7}Li$	1405.7	23.20	$^{23}Na^{23}Na$	159.2	0.73
$H^{23}Na$	1172.2	19.72	^{23}NaK	123.3	0.40
$H^{39}K$	985.0	14.65	$^{23}NaRb$	106.6	0.46
HRb	936.8	14.15	$^{23}Na^{19}F$ (5)	536.1	3.83
$H^{138}Cs$	890.7	12.6	$^{23}NaCl$ (6)	366	2.05
$H^{9}Be$	2058.6	35.5	$^{23}NaBr$ (6)	302	1.50
$H^{24}Mg$	1495.7	31.5	^{23}NaI (6)	258	1.08
$H^{40}Ca$	1299	19.5	$^{39}K^{39}K$	92.6	0.35
HSr	1206.2	17.0	KF (7)	426	2.4
HBa	1172	16	KCl (6)	281	1.30
$H^{11}B$	(2366)	(49)	KBr (6)	213	0.80
$H^{27}Al$	1682.6	29.15	KI	212	0.70
$H^{115}In$	1474.7	24.7	$^{85}Rb^{85}Rb$	57.3	0.96
HTl	1390.7	22.7	$Rb^{133}Cs$	49.4	—
$H^{12}C$	2861.6	64.3	RbF (7)	376	1.9
$D^{12}C$	2101.0	34.7	RbCl (6)	228	0.92
$H^{28}Si$(2)	2042.5	35.67	$^{133}Cs^{133}Cs$	42.0	0.08
HSn	(1580)	—	CsF (7)	353	1.7
HPb	1564.1	29.75	CsCl (6)	209	0.75
$H^{14}N$	(3300)	—	$^{133}CsBr$	(194)	(2.0)
$H^{31}P$	(2380)	—	$^{133}Cs^{127}I$	142	(1.2)
HBi	1698.9	31.6	$^{9}Be^{19}F$	1265.6	9.12
$H^{16}O$	3735.2	82.81	$^{9}Be^{35}Cl$	846.6	5.11
$D^{16}O$	2720.9	44.2	$^{9}Be^{16}O$	1487.3	11.83
$H^{19}F$	4138.5	90.07	$^{24}Mg^{19}F$	717.6	3.84
$D^{19}F$	2998.3	45.71	$^{24}Mg^{35}Cl$	465.4	2.05
$H^{35}Cl$ (3)	2991.0	52.85	$^{24}Mg^{79}Br$	373.8	1.34
$D^{35}Cl$ (3)	2145.2	27.18	$Mg^{127}I$	[312]	—
HBr	2649.7	45.21	$^{24}Mg^{16}O$	785.1	5.18
HI	2309.5	39.73	MgS	525.2	2.93
$H^{63}Cu$	1940.4	37.0	$^{40}Ca^{19}F$	587.1	2.74
HAg	1760.0	34.05	$Ca^{35}Cl$	369.8	1.31
$H^{197}Au$	2305.0	43.12	$Ca^{79}Br$	285.3	0.86
HZn	1607.6	55.14	$Ca^{127}I$	242.0	0.64
HCd	1430.7	46.3	$Ca^{16}O$	650	6.6
HHg	1387.1	83.01	$Sr^{19}F$	500.1	2.21
$H^{55}Mn$	[1490.6]	—	$Sr^{35}Cl$	302.3	0.95
HCo	(1890)	—	$Sr^{79}Br$	216.5	0.51
HNi	[1926.6]	—	$Sr^{127}I$	173.9	0.42
$^{7}Li^{7}Li$	351.4	2.59	$Sr^{16}O$	653.5	4.0

TABLE II-1*a* (*Continued*)

Molecule	ω_e	$x_e\omega_e$	Molecule	ω_e	$x_e\omega_e$
$Ba^{19}F$	468.9	1.79	$Ge^{19}F$	665	2.79
$^{138}Ba^{35}Cl$	279.3	0.89	$^{74}Ge^{35}Cl$	407.6	1.36
$Ba^{79}Br$	193.8	0.42	GeBr	296.6	0.9
$Ba^{16}O$	669.8	2.05	$^{74}Ge^{16}O$	985.7	4.30
$^{11}B^{11}B$	1051.3	9.4	$^{74}Ge^{32}S$	575.8	1.80
$^{11}B^{19}F$	1399.8	11.3	$^{74}Ge^{80}Se$	406.8	1.2
$^{11}B^{35}Cl$	839.1	5.11	$^{74}Ge^{130}Te$	323.4	1.0
$^{11}B^{79}Br$	684.3	3.52	$Sn^{19}F$	582.9	2.69
$^{11}B^{14}N$	1514.6	12.3	$Sn^{35}Cl$	352.5	1.06
$^{11}B^{16}O$	1885.4	11.77	SnBr	247.7	0.62
$^{27}Al^{19}F$	814.5	8.1	$Sn^{16}O$	822.4	3.73
$^{27}Al^{35}Cl$	481.3	1.95	SnS	487.7	1.34
$^{27}Al^{79}Br$	378.0	1.28	SnSe	331.2	0.74
$^{27}Al^{127}I$	316.1	1.0	SnTe	259.5	0.50
$^{27}Al^{16}O$	978.2	7.12	PbPb	256.5	2.96
$^{69}Ga^{35}Cl$	365.0	1.1	$Pb^{19}F$	507.2	2.30
$^{69}Ga^{81}Br$	263.0	0.81	$Pb^{35}Cl$	303.8	0.88
$^{69}Ga^{127}I$	216.4	0.5	$Pb^{79}Br$	207.5	0.50
$Ga^{16}O$	767.7	6.34	$Pb^{127}I$	160.5	0.25
$^{115}In^{35}Cl$	317.4	1.01	$Pb^{16}O$	721.8	3.70
$^{115}In^{81}Br$	221.0	0.65	$^{208}Pb^{32}S$	428.1	1.20
$^{115}In^{127}I$	177.1	0.4	PbSe	277.6	0.51
$In^{16}O$	703.1	3.71	PbTe	211.8	0.12
$Tl^{19}F$	475.0	1.89	$^{14}N^{14}N$	2359.6	14.46
$Tl^{35}Cl$	287.5	1.24	$[^{14}N^{14}N]^+$	2207.2	16.14
$Tl^{81}Br$	192.1	0.39	$^{14}N^{16}O$	1903.9	13.97
$Tl^{127}I$	150	—	^{14}NS	1220.0	7.75
$^{12}C^{12}C$	1641.4	11.67	^{14}NBr	693	5.0
$^{12}C^{35}Cl$	846	1.0	$^{14}N^{31}P$	1337.2	6.98
$^{12}C^{14}N$	2068.7	13.14	$^{14}N^{75}As$	1068.0	5.36
$^{12}C^{31}P$	1239.7	6.86	^{14}NSb	942.0	5.6
$^{12}C^{16}O$	2170.2	13.46	$^{31}P^{31}P$	780.4	2.80
$[^{12}C^{16}O]^+$	2214.24	15.16	$^{31}P^{16}O$	1230.6	6.52
$^{12}C^{32}S$	1285.1	6.5	$^{75}As^{75}As$	429.4	1.12
^{12}CSe	1036.0	4.8	$[^{75}As^{75}As]^+$	(314.8)	(1.25)
SiSi	(750)	—	$^{75}As^{16}O$	967.4	5.3
$^{28}Si^{19}F$	856.7	4.7	SbSb	269.9	0.59
$^{28}Si^{35}Cl$	535.4	2.20	$Sb^{209}Bi$	220.0	0.50
SiBr	425.4	1.5	$Sb^{19}F$	614.2	2.77
$^{28}Si^{14}N$	1151.7	6.56	$Sb^{35}Cl$	369.0	0.92
$^{28}Si^{16}O$	1242.0	6.05	$Sb^{14}N$	942.0	5.6
$^{28}Si^{32}S$	749.5	2.56	$Sb^{16}O$	817.2	5.30
$^{28}SiSe$	580.0	1.78	$^{209}Bi^{209}Bi$	172.7	0.32
$^{28}SiTe$	481.2	1.30	$^{209}Bi^{19}F$	510.8	2.05

TABLE II-1*a* (*Continued*)

Molecule	ω_e	$x_e\omega_e$	Molecule	ω_e	$x_e\omega_e$
$^{209}Bi^{35}Cl$	308.0	0.96	$^{197}Au^{35}Cl$	382.8	1.30
$^{209}Bi^{79}Br$	209.3	0.47	$Zn^{19}F$	(630)	(3.5)
$^{209}Bi^{127}I$	163.9	0.31	$Zn^{35}Cl$	390.5	1.55
$^{209}Bi^{16}O$	702.1	5.20	ZnBr	(220)	—
$^{16}O^{16}O$	1580.4	12.07	$^{64}Zn^{127}I$	223.4	0.75
$[OO]^+$	1876.4	16.53	$Cd^{19}F$	(535)	—
^{16}OCl	(780)	—	$Cd^{35}Cl$	330.5	1.2
^{16}OBr	713	7	CdBr	230.0	0.50
^{16}OI	(687)	(5)	$Cd^{127}I$	178.5	0.63
$^{16}O^{32}S$	1123.7	6.12	$Hg^{19}F$	490.8	4.05
$^{32}S^{32}S$	725.7	2.85	$Hg^{35}Cl$	292.6	1.60
$Se^{16}O$	907.1	4.61	$^{202}Hg^{81}Br$	186.3	0.98
$^{80}Se^{80}Se$	391.8	1.06	$Hg^{127}I$	125.6	1.09
$Te^{16}O$	796.0	3.50	HgTl	26.9	0.69
TeTe	251	0.55	$^{45}Sc^{16}O$	971.6	3.95
$^{19}F^{19}F$	[892.1]	—	$^{89}Y^{16}O$	852.5	2.45
$^{19}F^{35}Cl$	793.2	9.9	$^{139}La^{16}O$	811.6	2.23
$^{19}F^{79}Br$	671	3	$Ce^{16}O$	865.0	2.99
$^{35}Cl^{35}Cl$	564.9	4.0	$^{141}Pr^{16}O$	818.9	1.20
$[^{35}Cl^{35}Cl]^+$	645.3	2.90	$Ge^{16}O$	841.0	3.70
ClBr (8)	442.5	1.5	$Lu^{16}O$	841.7	4.07
$^{79}Br^{79}Br$ (9)	325.4	1.098	YbCl	293.6	1.23
$^{79}Br^{81}Br$	323.2	1.07	$^{48}Ti^{35}Cl$	456.4	6.3
$^{127}I^{35}Cl$	384.2	1.47	$^{48}Ti^{16}O$	1008.4	4.61
$^{127}I^{79}Br$	268.4	0.78	$^{90}Zr^{16}O$	936.6	3.45
$^{127}I^{127}I$	214.6	0.61	$V^{16}O$	1012.7	4.9
$^{63}Cu^{19}F$	622.7	3.95	$Cr^{16}O$	898.8	6.5
$^{63}Cu^{35}Cl$	416.9	1.57	^{55}MnF	618.8	3.01
$^{63}Cu^{79}Br$	314.1	0.87	$^{55}Mn^{35}Cl$	384.9	1.4
$^{63}Cu^{127}I$	264.8	0.71	$^{55}MnBr$	289.7	0.9
$Cu^{16}O$	628	3	$^{55}Mn^{16}O$	840.7	4.89
$^{107}Ag^{35}Cl$	343.6	1.16	$Fe^{35}Cl$	406.6	1.2
$^{109}Ag^{81}Br$	247.7	0.68	$Fe^{16}O$	880	5
$^{107}Ag^{127}I$	206.2	0.43	CoCl	421.2	0.74
$Ag^{16}O$	493.2	4.10	NiCl	419.2	1.04

[a] The compounds are listed in the order of the periodic table: IA, IIA, . . . , VIIIA, IB, IIB, . . . , VIIIB.

[b] Most of these values (gas-phase spectra) were taken from Herzberg.[1] Data obtained from other sources are indicated by reference numbers in parentheses after chemical formulas. Values in parentheses are not certain, and those in square brackets are observed frequencies without anharmonicity correction.

TABLE II-1*b*. OBSERVED FREQUENCIES OF HOMOPOLAR DIATOMIC MOLECULES, IONS, AND RADICALS (cm^{-1})

Species	State[a]	$\tilde{\nu}$	Refs.	Species	State[a]	$\tilde{\nu}$	Refs.
TT[b]	Liquid	2458	10	ClCl[c]	Mat	546	17
SnSn	Mat	188	11	$[ClCl]^-$ [c]	Mat	247	18
PbPb	Mat	112	11	$[BrBr]^+$	Sol'n	360	19
PP	Gas	775	12	BrBr	Gas	319	20
AsAs	Gas	421	12	$[II]^+$	Sol'n	238	21
$[OO]^-$	Mat	1097	13	II	Gas	213	20
$[OO]^{2-}$	Solid	794 738	13a	$[II]^-$	Mat	~115	22
SS	Mat	718	14	AgAg	Mat	194	23
$[SS]^-$	Solid[d]	594	15	ZnZn	Mat	80	24
$[SeSe]^-$	Solid[d]	325	15	CdCd	Mat	58	24
$[FF]^-$	Mat	475	16				

[a] Mat = inert gas matrix.
[b] T = ^{3}H(tritium).
[c] Cl = ^{35}Cl.
[d] Doped in KI.

is found. Other interesting series are these:

	N_2		CO		NO		O_2
$\tilde{\nu}$ (cm^{-1})	2359.6	>	2138	>	1880	>	1580
K (mdyn/Å)	22.98	>	18.47	>	15.55	>	11.77

and these:

	O_2^+		O_2		O_2^-		O_2^{2-}
$\tilde{\nu}$ (cm^{-1})	1865	>	1580	>	1097	>	~766
K (mdyn/Å)	16.39	>	11.76	>	5.67	>	~2.76

As stated in Sec. I-2, the force constant is not directly related to the dissociation energy. For a series of similar molecules, however, there is an approximate linear relationship between the two. The last series above is of particular interest since it shows the effect of charge on the frequency and bond energy. Here $\nu(O_2^+)$ is higher than $\nu(O_2)$ because O_2^+ is formed by losing one electron from an antibonding orbital of O_2, while $\nu(O_2^-)$ is lower than $\nu(O_2)$ because one extra electron of O_2^- enters an antibonding orbital of O_2. The O_2^+ ion was found in compounds such as $O_2^+[MF_6]^-$ (M = As, Sb, etc.) and $O_2^+[M_2F_{11}]^-$ (M = Nb, Ta, etc.),[79] whereas the O_2^- ion was observed in a triangular $M^+O_2^-$ complex formed by the reaction of alkali metals with O_2 in an argon matrix.[13]

TABLE II-1c. OBSERVED FREQUENCIES OF HETEROPOLAR DIATOMIC MOLECULES, IONS, AND RADICALS (cm^{-1})

Species[a]	State[b]	$\tilde{\nu}$	Refs.	Species[a]	State[b]	$\tilde{\nu}$	Refs.
AlH	Mat	1593	25	CdO	Mat	719	59
SiH	Mat	1967	26	TiO	Mat	1005	60
NH	Mat	3131.6	27	ZrO	Mat	975	60
OH	Mat	3452.3	28	NbO	Mat	~968	61
		3428.2		TaO	Mat	1020	62
$[OH]^-$	Solid	3637.4	29	WO	Mat	1050.9	63
SH	Mat	2540.8	30	^{56}FeO	Mat	873.1	64
CrH	Mat	1548	31	CoO	Mat	846.4	64
NiH	Mat	1906	25	^{58}NiO	Mat	825.7	64
CuH	Mat	1882	25	UO	Mat	820	65
$[CN]^-$	Sol'n	2080	32	PuO	Mat	~820	66
CN	Mat	2046	33	CS	Mat	1274	67
PN	Mat	1323	34	^{28}SiS	Mat	736.0	34
FN	Mat	~1117	35	^{74}GeS	Mat	566.6	68
ClN	Mat	818.5	35	$^{74}Ge^{80}Se$	Mat	397.9	68
		825		$^{74}GeTe$	Mat	317.6	68
BrN	Mat	691	35	^{120}SnS	Mat	480.5	68
IN	Mat	590	36	$^{120}Sn^{80}Se$	Mat	325.2	68
TiN	Mat	1037	37	PbS	Mat	423.1	68
NbN	Mat	1002.5	38	$Pb^{80}Se$	Mat	275.1	68
ThN	Mat	934.6	39	$[SSe]^-$	Solid[c]	464	69
UN	Mat	~995	40	LiF	Mat	~885	70
PuN	Mat	855.7	41	LiCl	Mat	~575	71
^{7}LiO	Mat	745	42, 43	LiBr	Mat	~510	71
^{26}MgO	Mat	815.4	44	LiI	Mat	433	71
^{40}CaO	Mat	707.0	45	NaF	Mat	515	70
CO	Mat	2138.4	46	MgF	Mat	738	72
^{28}SiO	Mat	1225.9	47	AlF	Mat	793	73
^{74}GeO	Mat	973.4	48	TlF	Mat	441	74
SnO	Mat	816.1	49	TlCl	Mat	261	74
PbO	Mat	718.4	50	TlBr	Mat	179	74
$[NO]^+$	Solid	2273	51	TlI	Mat	143	74
NO	Mat	1880	52	CF	Mat	1279	75
$[NO]^-$	Mat	1358–1374	53	SCl	Mat	617	76
$[NO]^{2-}$	Mat	886	54	SBr	Mat	518	76
SO	Mat	1136.7	55	SI	Mat	443	76
FO	Mat	1028.7	56	$[ClF]^+$	Sol'n	819	77
ClO	Mat	850	57	BrCl	Gas	440	78
BrO	Mat	729.9	58	ICl	Gas	381	78
ZnO	Mat	802	59	IBr	Gas	265	78

[a] C, N, O, S, and Cl denote ^{12}C, ^{14}N, ^{16}O, ^{32}S, and ^{35}Cl, respectively.
[b] Mat = inert gas matrix
[c] Doped in KI.

The effect of changing the charge on the vibrational frequency is also seen in other series: $N_2 > N_2^+$, $NO^+ > NO > NO^- > NO^{2-}$, $CO^+ > CO$, $F_2 > F_2^-$, and so on. As will be shown in Part III, these frequency trends provide valuable information about the nature of the metal–ligand bond involving these diatomic species.

Alkali halide vapor consists mainly of monomers and dimers. As stated in Sec. I-22, Linevsky[80] developed a technique to isolate these monomers and dimers, produced at high temperature, in inert gas matrices. This technique has been used extensively to study the infrared spectra and structures of a number of inorganic salts. Some references on metal halide dimers are: $(LiF)_2$,[81] $(LiCl)_2$,[71,82] $(LiBr)_2$,[71,82] and $(NaX)_2$ (X = F, Cl, Br, and I).[83] These dimers are known to be cyclic-planar ($\mathbf{D}_{2h}$). On the other hand, $(TlX)_2$(X = F and Cl) are linear and symmetrical ($\mathbf{D}_{\infty h}$, X–Tl–Tl–X).[74] Such structures have been well known for $(HgX)_2$ (X = Cl, Br, and I).[84] The dimer $(LiO)_2$ is also cyclic-planar.[42] However, $(NH)_2$ is *trans*-planar ($\mathbf{C}_{2h}$) in a N_2 matrix[85] but takes the *cis* structure upon complex formation with the $Cr(CO)_5$ group.[86]

Hydrogen halides polymerize in the condensed phases; hydrogen fluoride polymerizes even in the gaseous phase.[87] The HX stretching bands are shifted markedly to lower frequencies by polymerization. For example, the monomer frequencies of HF (3962 cm^{-1}), HCl (2886 cm^{-1}), HBr (2558 cm^{-1}), and HI (2230 cm^{-1}) in the gaseous phase are lowered to 3420–3060,[88] 2746–2704,[89] 2438–2404,[89] and 2120 cm^{-1},[89] respectively, in the solid phase. The infrared spectra of monomeric and polymeric hydrogen halides in inert gas matrices have been reported.[90]

Halogens form molecular compounds with organic solvents. For example, the band at 213 cm^{-1} of gaseous I_2 is shifted to 201 cm^{-1} in benzene solution,[91] and the band at 381.5 cm^{-1} of gaseous ICl is shifted to 275 cm^{-1} in pyridine solution.[92] The Raman spectra of I_2, Br_2, and ICl have been studied in many solvents.[93] These frequencies are much lower than the corresponding gas-phase frequencies because of charge-transfer interaction with solvent molecules.

The hydroxyl ion $[OH]^-$ is characterized by a sharp band at 3700–3500 cm^{-1}. For example, $LiOH \cdot H_2O$ exhibits a sharp OH stretching band at 3574 cm^{-1} and a broad OH_2 stretching band in the 3200–2800 cm^{-1} region.[94] The cyanide ion $[CN]^-$ exhibits a relatively sharp band in the 2250–2050 cm^{-1} region. The CN stretching bands are at 2080 cm^{-1} for ionic cyanides such as Na[CN] and K[CN][95] and at 2170–2250 cm^{-1} for covalent cyanides such as Cu[CN] and Au[CN], in which two metals are bridged by the CN groups.[96] The vibrational frequencies of cyanogen, N≡C—C≡N, have been determined and its harmonic frequencies and anharmonicity constants calculated using ^{13}C and ^{15}N isotope data.[97] The NO stretching frequencies of nitric oxide (NO) in an Ar matrix are 1883 (monomer), 1862 and 1768 (*cis*-dimer), and 1740 (*trans*-dimer) cm^{-1} (Ref. 98).

Finally, many compounds containing the M=O groups, such as V=O, Nb=O, Ta=O, Mo=O, W=O, Re=O, Ru=O, and Os=O, exhibit the M=O stretching bands in the 1050–800 cm^{-1} region.

II-2. TRIATOMIC MOLECULES

The three normal modes of linear X_3- ($\mathbf{D}_{\infty h}$) and YXY- ($\mathbf{D}_{\infty h}$) type molecules were shown in Fig. I-7; ν_1 is Raman active but not infrared active, whereas ν_2 and ν_3 are infrared active but not Raman active (mutual exclusion rule). However, all three vibrations become infrared as well as Raman active in linear XYZ-type molecules ($\mathbf{C}_{\infty v}$), shown in Fig. II-1. The three normal modes of bent X_3- ($\mathbf{C}_{2v}$) and YXY- ($\mathbf{C}_{2v}$) type molecules were also shown in Fig. I-7. In this case, all three vibrations are both infrared and Raman active. The same holds for bent XXY- and XYZ- ($\mathbf{C}_s$) type molecules. In the following, the vibrational frequencies of a number of triatomic molecules are listed for each class of compounds.

Table II-2*a* lists the vibrational frequencies of XY_2-type metal halides. Most of these data were obtained in inert gas matrices. Although the structures of these halides are classified as either linear or bent, it should be noted that the bond angles in the latter type range from 95° to 170°.[99] Thus, some bent molecules are almost linear. These two types can be distinguished by the infrared activity of the ν_1 mode; it is active for bent but not active for linear molecules. However, this simple criterion has led to conflicting results in some cases.[100] The bond angle of the YXY-type molecule can be determined by the metal (X atom) isotope frequencies of the ν_3 modes observed in inert gas matrices. If a pair of $\tilde{\nu}_3$ frequencies is determined, the bond angle (2α) can be calculated from the equation[101]

$$\left(\frac{\tilde{\nu}'_3}{\tilde{\nu}_3}\right)^2 = \frac{M_X}{M_{X'}}\left(\frac{M_{X'}+2M_Y\sin^2\alpha}{M_X+2M_Y\sin^2\alpha}\right)$$

where M denotes the mass of the atom subscripted, and the prime indicates an isotope. (For the derivation of this equation, see Part I.) Figure II-2 shows the ν_3 spectra of NiF_2 in Ne and Ar matrices.[101] Using these isotope frequencies,

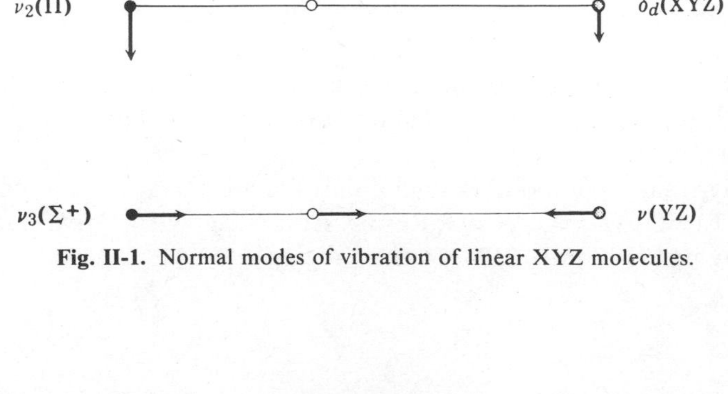

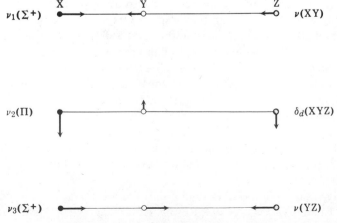

Fig. II-1. Normal modes of vibration of linear XYZ molecules.

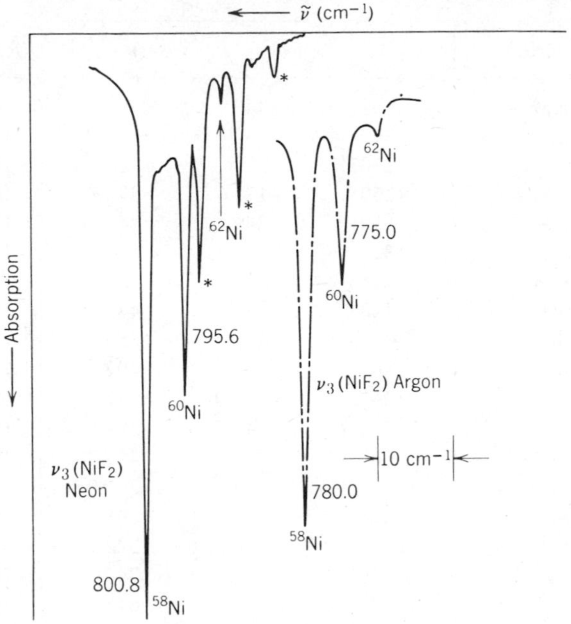

Fig. II-2. Infrared spectrum (ν_3) of NiF_2 in Ne and Ar matrices. Matrix splitting, indicated by asterisks, is present in the Ne matrix spectrum.

the FNiF angle was calculated to be 154°–167°. It should be noted in this table and others that the majority of compounds follow the trend $\nu_3 > \nu_1$ and that the exceptions to this rule occur in some bent molecules.

The structure of the dimeric species $(MX_2)_2$ is known to be cyclic-planar:

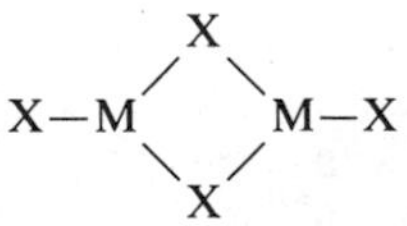

($\mathbf{D}_{2h}$)[134,135]

although an exception is reported for $(GeF_2)_2$:[136]

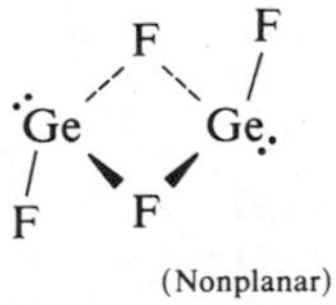

($\mathbf{C}_{2h}$)

Table II-2*b* lists the vibrational frequencies of triatomic oxides, sulfides, and selenides. Most of these data were obtained in inert gas matrices. The dioxo groups, such as VO_2^+, MoO_2^{2+}, WO_2^{2+}, ReO_2^+, and UO_2^{2+}, exhibit strong bands in the 1000–850 cm^{-1} region. *Trans*-dioxo groups exhibit only one M=O

TABLE II-2a. VIBRATIONAL FREQUENCIES OF XY_2-TYPE METAL HALIDES (cm^{-1})

Compound[a]	Structure	ν_1	ν_2	ν_3	Refs.
BeF_2	Linear	(680)	345	1555	102
$BeCl_2$	Linear	(390)	250	1135	102
$BeBr_2$	Linear	(230)	220	1010	102
BeI_2	Linear	(160)	(175)	873	102
MgF_2	Linear	550	249	842	103
$MgCl_2$	Linear	327	93	601	103
$MgBr_2$	Linear	198	82	497	103
MgI_2	Linear	148	56	445	103
$^{40}CaF_2$	Bent	484.8	163.4	553.7	104
$^{40}Ca^{35}Cl_2$	Linear	—	63.6	402.3	105
$^{86}SrF_2$	Bent	441.5	82.0	443.4	104
$^{88}Sr^{35}Cl_2$	Bent	269.3	43.7	299.5	105
BaF_2	Bent	389.6	(64)	413.2	104
$BaCl_2$	Bent	255.2	—	260.0	105
$InCl_2^-$	Bent	328	177	291	106
CF_2	Bent	1102	668	1222	107
$C^{35}Cl_2$	Bent	719.5	—	745.7	108
CBr_2	Bent	595.0	—	640.5	109
SiF_2	Bent	851.5	(345)	864.6	110
$^{28}Si^{35}Cl_2$	Bent	513	—	502	111
$^{28}SiBr_2$	Bent	402.6	—	399.5	112
GeF_2	Bent	692	263	663	113
$GeCl_2$	Bent	398	—	373	114
SnF_2	Bent	592.7	197	570.9	115
$SnCl_2$	Bent	354	(120)	334	114
$SnBr_2$	Bent	237	84	223	116
PbF_2	Bent	531.2	165	507.2	115
$PbCl_2$	Bent	297	—	321	114
$PbBr_2$ (g)	Bent	200	64	—	117
NF_2	Bent	1069.6	573.4	930.7	118
PF_2	Bent	834.0	—	843.5	119
PCl_2	Bent	452.0	—	524.8	120
PBr_2	Bent	369.0	—	410.0	120
$O^{35}Cl_2$	Bent	630.7	296.4	670.8	121
SF_2	Bent	825.4	357.5	799.2	122a
SCl_2	Bent	518	208	526	122
SBr_2	Bent	405	—	418	122
SI_2	Bent	368	—	376	122
$SeCl_2$ (g)	Bent	415	153	377	123
$TeCl_2$ (g)	Bent	377	125	—	123
KrF_2 (g)	Linear	449	233	596, 580	124
XeF_2	Linear	497	213.2	555	125
$Xe^{35}Cl_2$	Linear	—	—	314.1	126
$^{63}CuF_2$	Bent	—	183.0	743.9	101

TABLE II-2*a* (*Continued*)

Compound[a]	Structure	ν_1	ν_2	ν_3	Refs.
$[CuCl_2]^-$(s)	Linear	300	109	405	127
$[CuBr_2]^-$(s)	Linear	193	81	322	127
$[CuI_2]^-$(s)	Linear	148	65	279	127
$[AgCl_2]^-$(s)	Linear	268	88	333	127
$[AgBr_2]^-$(s)	Linear	170	61	253	127
$[AgI_2]^-$(s)	Linear	133	49	215	127
$[AuCl_2]^-$(s)	Linear	329	120, 112	350	128
$[AuBr_2]^-$(s)	Linear	209	79, 75	254	128
$[AuI_2]^-$(s)	Linear	158	67, 59	210	128
ZnF_2	Linear	596	150	754	129
$ZnCl_2$	Linear	352	103, 100	503	130
$ZnBr_2$	Linear	223	71	404	130
ZnI_2	Linear	163	61	346	130
CdF_2	Linear	555	121	660	129
$CdCl_2$	Linear	(327)	88	419	131
$CdBr_2$	Linear	(205)	62	319	131
CdI_2	Linear	(149)	(50)	269	131
HgF_2	Linear	568	170	642	129
$HgCl_2$	Linear	(348)	107	405	131
$HgBr_2$	Linear	(219)	73	294	131
HgI_2	Linear	(158)	63	237	131
TiF_2	Bent	665	~180	766	132
VF_2	Bent	—	—	733.2	99
CrF_2	Linear	—	155.4	654.5	99
$CrCl_2$	Linear	—	—	493.5	133
MnF_2	Linear	—	124.8	700.1	99
$MnCl_2$	Linear	—	83	476.8	133
FeF_2	Linear	—	141.0	731.3	99
$FeCl_2$	Linear	—	88	493.2	133
CoF_2	Bent	—	151.0	723.5	99
$CoCl_2$	Linear	—	94.5	493.4	133
NiF_2	Bent	—	139.7	779.6	99
$NiCl_2$	Linear	(350)	85	520.6	133
$NiBr_2$	Linear	—	69	414.2	133

[a] All data were obtained in inert gas matrices except those for which the physical state is indicated as g (gas), l (liquid), or s (solid).

stretching band, whereas *cis*-dioxo groups show two such bands.[175] For example, *cis*-VO_2^+ groups show one band at 907–876 (antisymmetric) and the other at 922–910 cm^{-1} (symmetric).[176] Similar results are reported for *cis*-MoO_2^{2+} (Ref. 177) and *cis*-WO_2^{2+} (Ref. 178) groups. The UO_2^{2+}, NpO_2^{2+}, PuO_2^{2+}, and AmO_2^{2+} groups are linear and exhibit only one band in the 940–850 cm^{-1}

TABLE II-2*b*. VIBRATIONAL FREQUENCIES OF YXY- AND XXY-TYPE OXIDES AND RELATED COMPOUNDS (cm^{-1})

Compound[a]	Structure	ν_1[b]	ν_2	ν_3[b]	Refs.
$^6Li^6LiO$	Linear	—	118	1028.5	42
O^6LiO	Bent	—	243.4	730	43
ONaO	Bent	1080.0	390.7	332.8	137
OBO	Linear	—	—	1276	138
AlOAl	Bent	715	(120)	994	139
GaOGa	Bent	472	—	809.4	140
$O^{12}CO$ (g)	Linear	(1337)[c]	667	2349	141, 142
$O^{13}CO$	Linear	—	(649)	2284	142
SCS (g)	Linear	658	397	1533	143
SeCSe	Linear	364	313	1303	144
$O^{120}SnO$	Bent	—	—	863.1	145
$[ONO]^-$	Bent	—	—	1244	146
$O^{14}NO$	Bent	(1318)	749	1610	147
$[ONO]^+$ (s)	Linear	1396	570	2360	148
NNO	Linear	2223.8	588.7	1284.9	149
$[O_3]^-$	Bent	1016	600.9	802.3	150
O_3	Bent	1134.9	716.0	1089.2	151
S_3 (g)	Bent	585	490/310	651	152
$[S_3]^-$ (l)	Bent	535	235.5	571	153
OSO	Bent	1147	517	1351	154
SSO	Bent	672	382	1156	155
$[OSO]^-$	Bent	984.8	495.6	1042.0	156
$O^{80}SeO$	Bent	992.0	372.5	965.6	157
OTeO	Bent	831.7	294	848.3	158
FOO	Bent	~1500	586.4	376.0	159
FOF	Bent	925, 915	461	821	160
ClOCl (s)	Bent	630.7	296.4	670.8	161
OOCl	Bent	1441	407	373	162
$[OClO]^-$ (s)	Bent	790	400	840	163
OClO (g)	Bent	943.2			
$[OClO]^+$ (s)	Bent	~1040	~520	~1290	165
BrOBr (s)	Bent	504	197	587	166
$[OBrO]^+$	Bent	865	375	932	167
$[OBrO]^-$	Bent	775	400	800	163
OCeO	Bent	757.0	—	736.8	168
OTbO	Bent	758.7	—	718.8	168
OThO	Bent	787.4	—	735.3	168
OPrO	Bent	(7))	—	730.4	168
$O^{48}TiO$	Bent	—	—	934.8	169
OTaO	Bent	971	—	912	170
OWO	Bent	992	—	928	171
OUO	Linear	(765.4)	—	776.1	172
$[OUO]^{2+}$ (s)	Linear	(856)	—	931	173
OPuO	Linear	—	—	794.3	174

[a] All data were obtained in inert gas matrices except those for which the physical state is indicated as g (gas), l (liquid), or s (solid).
[b] For XXY molecules, ν_1 and ν_2 are XX and XY stretching modes, respectively.
[c] Fermi resonance with $2\nu_2$ (see Sec. I-10).

region.[179] Jones[180] derived an equation which relates the U=O stretching force constant to the U=O distance. McGlynn et al.[181] noted that, in a series of $K_xUO_2L_y(NO_3)_2$-type compounds, the U=O stretching frequencies decrease as L is changed in the order of the spectrochemical series

$$[CN]^- > en > NH_3 > [NCS]^- > [ONO]^- > py > H_2O > F^- > [NO_3]^-$$

The relationship between the UO_2 frequency and the U=O distance has been studied by several other investigators.[182–184]

Metal dioxides produced by the sputtering technique[185] in inert gas matrices take linear or bent O–M–O structures. Metal dinitrides produced by the same technique also take linear N–M–N (M = U [186] and Pu [187]) structures although $Th(N_2)$ is triangular ($\mathbf{C}_{2v}$).[188] These structures are markedly different from those of molecular oxygen and nitrogen complexes of various metals produced by the conventional matrix co-condensation techniques (see Sec. III-13).

Table II-2*c* lists the vibrational frequencies of triatomic interhalogeno compounds. The resonance Raman spectrum of the I_3^- ion gives a series of overtones of the ν_1 vibration.[211,212] The resonance Raman spectra of the I_2Br^- and IBr_2^- ions and their complexes with amylose have been studied.[213] The same table also lists the vibrational frequencies of XHY-type (X, Y: halogens) compounds. All these species are linear except the $ClHCl^-$ ion, which was found to be bent in an inelastic neutron scattering (INS) and Raman spectral study.[205] Salt-molecule reactions such as $CsF + F_2 = Cs[F_3]$ have been utilized to produce a number of novel triatomic and other anions in inert gas matrices.[208,210]

Table II-2*d* lists the vibrational frequencies of bent XH_2-type molecules. The XH stretching frequencies are lower and the XH_2 bending frequencies are higher in condensed phases than in the vapor phase because of hydrogen bonding in the former. This trend is also seen for H_2O and H_2S dissolved in organic solvents such as pyridine and dioxane.[228,229] According to Walrafen,[219] liquid H_2O exhibits ν_1 and ν_3 at 3450 and 3615 cm^{-1}, respectively. On the other hand, Senior and Thompson[230] assign both modes at 3450 cm^{-1} and interpret the 3615-cm^{-1} band as a combination $\nu_1 + \nu_0$ band where ν_0 is the low-frequency stretching mode of the O—H······O system. The spectrum of ice is complicated because of its polymorphism. Hornig et al.[220] assigned the spectrum of ice I as shown in Table II-2*d*. They also assigned some librational and translational modes. Bertie and Whalley[231] studied the vibrational spectra of ice in other phases and found the spectra to be consistent with reported crystal structures in each phase. For crystal water and aquo complexes, see Sec. III-6. The vibrational frequencies of H_2O, D_2O, HDO and their dimers in argon matrices have been reported.[232]

Table II-2*e* lists the vibrational frequencies of linear XYZ-type molecules. Some of these vibrations split into two because of Fermi resonance or the crystal-field effect. Vibrational spectra of coordination compounds containing pseudohalide ions such as NCS^-, NCO^-, and N_3^- are discussed in Sec. III-15.

TABLE II-2c. VIBRATIONAL FREQUENCIES OF TRIATOMIC HALOGENO COMPOUNDS (cm^{-1})

Compound[a]	Structure[b]	ν_1[c]	ν_2	ν_3[c]	Refs.	Compound[a]	Structure[b]	ν_1[c]	ν_2	ν_3[c]	Refs.
$FClF^+$ (s)	b	807	387	830	189	$BrICl^-$ (s)	l	203	—	180	201
FClF (m)	b	500	242	578, 570	190	$BrII^-$ (s)	l	117	84	168	201
$FClF^-$ (s)	l	510, 478	—	636	191	$BrII^+$ (s)	b	258	—	198	199
$FClCl^+$ (s)	b	744	299, 293	535, 528	192	I_3^- (sl)	l	114	52	145	197
$FBrF^+$ (s)	b	706	362	702	193	HFH^+ (s)	b	2970	1680	3080	202
$FBrF^-$ (s)	l	442	198	596, 562	194	FHF^- (s)	l	596, 603	1233	1450	203
$ClFCl^+$ (s)	b	~529	293, 258	586, 593	195	$ClHF^-$ (s)	l	275	863, 823	2710	204
Cl_3^+ (s)	b	493, 485	225	508	192	$ClHCl^-$ (s)	b	199	660, 602	1670	205
Cl_3^- (m)	b	253	—	340	196	$ClHBr^-$ (s)	l	—	508	1705	206
$ClBrCl^-$ (sl)	l	278	~135	225	197	$ClHI^-$ (s)	l	—	485	2200	206
$ClICl^+$ (s)	b	371	147	364	198	$BrHF^-$ (s)	l	220	740	2900	204
$ClICl^-$ (sl)	l	269	127	226	197	$BrHBr^-$ (s)	l	126	1038	1420	207
$ClII^+$ (s)	b	360	126 (?)	197	199	$BrHBr^-$ (m)	l	168	—	670	208
Br_3^- (sl)	l	164	53	191	197	IHF^- (s)	l	180	635	3145	204
Br_3 (m)	l	190	—	—	200	IHI (m)	l	(120.7)	—	682.1	209
$BrIBr^+$ (s)	l	256	124	256	199	F_3^- (m)	l	461	—	550	210

[a] m = inert gas matrix; sl = solution; s = solid.
[b] b = bent; l = linear.
[c] For XYZ type, ν_1 and ν_3 correspond to ν(XY) and ν(YZ), respectively.

TABLE II-2*d*. VIBRATIONAL FREQUENCIES OF BENT XH_2-TYPE MOLECULES (cm^{-1})

Molecule	State	ν_1	ν_2	ν_3	Refs.
GeH_2	Matrix	1887	920	1864	214
NH_2	Matrix	—	1499	3220	215
	Surface	3290	1610	3380	216
$[NH_2]^-$	Solid	3270	1556	3323	217
$^{16}OH_2$	Gas	3657	1595	3756	218
	Liquid	3450	1640	3615	219
	Solid	3400	1620	3220	220
^{16}OHD[a]	Gas	2727	1402	3707	218
	Solid	2416	1490	3275	220
$^{16}OD_2$	Gas	2671	1178	2788	218
	Solid	2495, 2336	1210	2432	220
$^{18}OD_2$	Gas	2657	1169	2764	221
OT_2[b]	Gas	—	996	2370	222
SH_2	Gas	2615	1183	2627	223
	Matrix	2619.5	—	2632.6	224
	Solid	2532, 2523	1186, 1171	2544	225
SD_2	Gas	1892	934	2000	226
	Solid	1843, 1835	857, 847	1854	225
SeH_2	Gas	2345	1034	2358	227
SeD_2	Gas	1687	741	1697	227

[a] Here ν_1 and ν_3 denote ν(OD) and ν(OH), respectively.
[b] $T = {}^3H$ (tritium).

Table II-2*f* lists the vibrational frequencies of bent XYZ-type molecules. The spectra of most of these compounds were measured in inert gas matrices.

II-3. PYRAMIDAL FOUR-ATOM MOLECULES

(1) XY_3 Molecules (C_{3v})

The four normal modes of vibrations of a pyramidal XY_3 molecule are shown in Fig. II-3. These four vibrations are both infrared and Raman active. Table II-3*a* lists the fundamental frequencies of XH_3-type molecules. Several bands marked by an asterisk are split into two by *inversion doubling.* As is shown in Fig. II-4, two configurations of the XH_3 molecule are equally probable. If the potential barrier between them is small, the molecule may resonate between the two structures. As a result, each vibrational level splits into two levels (positive and negative).[281] Transitions between levels of different sign are allowed in the infrared spectrum, whereas those between levels of the same sign are allowed in the Raman spectrum. The transition between the two levels at $v = 0$ is also observed in the microwave region ($\tilde{\nu} = 0.79$ cm^{-1}). If the

TABLE II-2*e*. VIBRATIONAL FREQUENCIES OF LINEAR XYZ AND X_3 MOLECULES (CM^{-1})

XYZ	State	$\nu_1(XY)$	$\nu_2(\delta)$	$\nu_3(YZ)$	Refs.
HCN	Gas	3311	712	2097	233
	Matrix	3306	721	—	234
DCN	Gas	2630	569	1925	233
TCN	Gas	2460	513	1724	235
FCN	Gas	1077	449	2290	236
ClCN	Gas	714	380	2219	237
	Matrix	718	384, 387[a]	—	238
BrCN	Gas	574	342.5	2200	237
	Matrix	575	349, 351[a]	—	238
ICN	Gas	485	304	2188	239
NaCN	Matrix	382	170	2044	240
HNC	Matrix	3620	477	2029	241
^6LiNC	Matrix	722.9	121.7	2080.5	242
$H^{11}BO$	Matrix	(2849)	754	1817	243
CCO	Matrix	1074	381	1978	244
SCO	Gas	859	520	2062	245
SCSe	Gas	1435	(355)	506	246
SCTe	Sol'n	1347	(377)	423	246
$[CNO]^-$	Solid	2096	471	1106	247
$[NNF]^+$	Solid	2371	391	1057	248
NCN	Matrix	—	423	1475	249
NNO	Gas	2277	596.5	1300.3	250
NNC	Matrix	1235	394	2824	251
NCO	Matrix	1275	487	1922	252
$[NCO]^-$	Solid	2160, 2135	629	1282, 1202[b]	253
$[NNN]^-$	Solid	1344	645	2041	254
$[NC^{32}S]^-$	Solid	2053	486, 471	748	255
$[NCSe]^-$	Solid	2070	424, 416	558	256
$[NCTe]^-$	Solid	2073	366	450	253a

[a] Splitting due to matrix environment.
[b] Fermi resonance between ν_3 and $2\nu_2$.

potential barrier is sufficiently high and if the three Y groups are not identical, optical isomers may be anticipated.

As is seen in Table II-3*a*, ν_1 and ν_3 overlap or are close in most compounds. The presence of the hydronium (H_3O^+) ion in hydrated acids has been confirmed by observing its characteristic frequencies. For example, it was shown from infrared spectra that $H_2PtCl_6 \cdot 2H_2O$ should be formulated as $(H_3O)_2[PtCl_6]$.[292] For normal coordinate analysis of pyramidal XH_3 molecules, see Refs. 293–295.

Table II-3*b* lists the vibrational frequencies of pyramidal XY_3 halogeno compounds. Clark and Rippon[306] have measured the Raman spectra of a

TABLE II-2*f*. VIBRATIONAL FREQUENCIES OF BENT XYZ-TYPE MOLECULES (CM^{-1})

XYZ	State	ν_1(XY)	$\nu_2(\delta)$	ν_3(YZ)	Refs.
HCO	Matrix	2483	1087	1863	257
$H^{12}CF$	Matrix	—	1405	1181.5	75
HNO	Matrix	3450	1110	1563	258
$H^{14}NF$	Matrix	—	1432	1000	35
HOO	Matrix	3414	1389	1101	259
FOO	Matrix	376.0	586.4	1499.7	260
ClOO	Matrix	407	373	1441	261
BrOO	Matrix	—	—	1487	262
HOF	Matrix	3537.1	1359.0	886.0	263
HOCl	Matrix	3578	1239	728	264
HOBr	Matrix	3589	1164	626	264
HOI	Matrix	3597	1103	575	265
$H^{72}GeBr$	Matrix	1858	701	283	266
NOF	Matrix	1886.2	734.9	492.2	267
NSCl	Gas	1325	273	414	268
ONF	Gas	1843.5	765.8	519.9	269
$ON^{35}Cl$	Gas	1799.7	595.8	331.9	270
ONBr	Gas	1799.0	542.0	266.4	271
ONI	Matrix	1809	470	216	272
OPCl	Gas	1218	—	780	273
OCF	Matrix	1855	626	1018	274
OCCl	Matrix	1880	281	570	275
$O^{35}ClF$	Matrix	1038.0	593.5	315.2	276
SNF	Gas	1372	640	366	277
$^{35}ClCF$	Matrix	742	—	1146	278
$^{35}Cl^{32}SN$	Matrix	1327.3	403.8	267.4	279
ClSnBr	Matrix	328	—	228	280
ClPbBr	Matrix	295	—	200	280
$Br^{32}SN$	Matrix	1312.9	346.1	226.2	279

number of these compounds in the gaseous phase. The compounds show a $\nu_1 > \nu_3$ and $\nu_2 > \nu_4$ trend, whereas the opposite trend holds for the neutral XH_3 molecules listed in Table II-3*a*. In some cases, two stretching frequencies (ν_1 and ν_3) are too close to be distinguished empirically. This is also true for two

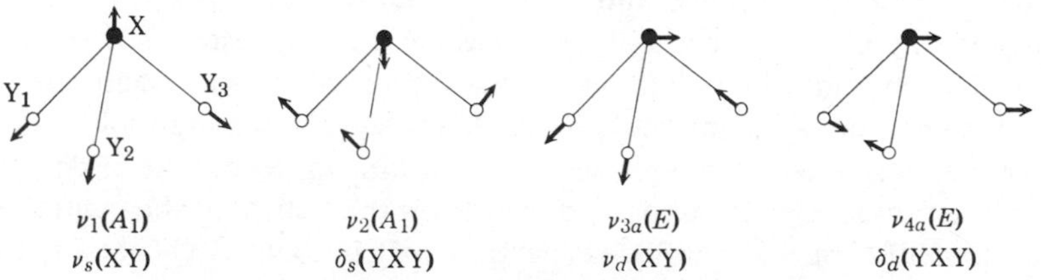

Fig. II-3. Normal modes of vibration of pyramidal XY_3 molecules.

TABLE II-3*a*. VIBRATIONAL FREQUENCIES OF PYRAMIDAL XH_3 MOLECULES (CM^{-1})

Molecule	State	$\nu_1(A_1)$	$\nu_2(A_1)$	$\nu_3(E)$	$\nu_4(E)$	Refs.
NH_3	Gas	3335.9, 3337.5^a	931.6, 968.1^a	3414	1627.5	281
	Solid	3223	1060	3378	1646	282
ND_3	Gas	2419	748.6, 749.0^a	2555	1191.0	281
	Solid	2318	815	2500	1196	282
$^{15}NH_3$	Gas	3335	926, 961^a	3335	1625	283
NT_3	Gas	2016	647	2163	1000	284
PH_3	Gas	2327	990, 992^a	2421	1121	285
PD_3	Gas	1694	730	(1698)	806	285
AsH_3	Gas	2122	906	2185	1005	285
AsD_3	Gas	1534	660	—	714	285
SbH_3	Gas	1891	782	1894	831	286
SbD_3	Gas	1359	561	1362	593	286
$[OH_3]^+SbCl_6^-$	Sol'n	3560	1095	3510	1600	287
$[OH_3]^+ClO_4^-$	Solid	3285	1175	3100	1577	288
$[OH_3]^+NO_3^-$	Solid	2780	1135	2780	1680	289
$[OH_3]^+$	Liquid SO_2	3385	—	3470, 3400	1700, 1635	290
$[OD_3]^+$	Liquid SO_2	2490	—	2660, 2580	1255	290
$[GeH_3]^-$	Liquid NH_3	1740	809	—	886	291

^{a} Splitting due to Fermi resonance.

bending bands (ν_2 and ν_4). The symmetry of the SnX_3^- ion in $[As(Ph)_4][SnX_3]$ (X = Br and I) is lowered to $\mathbf{C}_s$ in the solid state, so that ν_3 and ν_4 each split into two bands.[302] Normal coordinate analyses on $[SnX_3]^-$ (X = F, Cl, Br, and I)[302] and Group VB trihalides[312] have been carried out.

Table II-3*c* lists the vibrational frequencies of pyramidal XO_3-type compounds. Rocchiciolli[325] has measured the infrared spectra of a number of sulfites, selenites, chlorates, and bromates. Dasent and Waddington[326] also measured the infrared spectra of metal iodates and suggested that extra bands at 480–420 cm^{-1} may be due to the metal-oxygen vibrations. Again the ν_3 and ν_4 vibrations may split into two bands because of lowering of symmetry in the crystalline state. Although $\nu_2 > \nu_4$ holds in all cases, the order of two stretching frequencies (ν_1 and ν_3) depends on the nature of the central metal. Figure II-5 illustrates the infrared spectra of $KClO_3$ and KIO_3 obtained in the crystalline state.

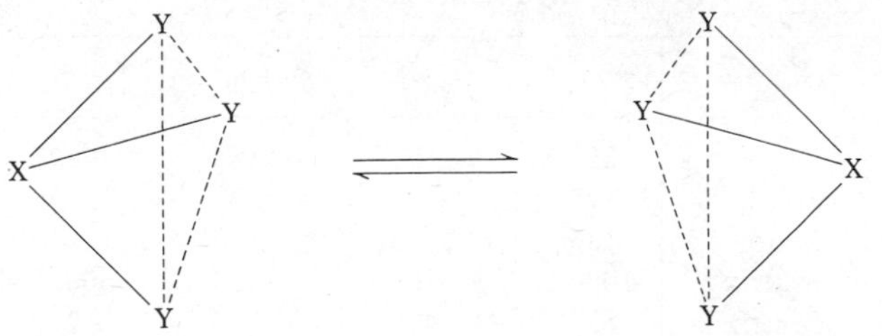

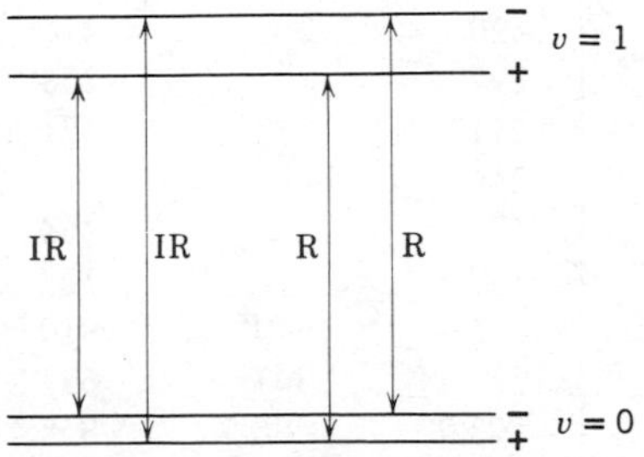

Fig. II-4. Inversion doubling of XY_3-type molecules.

(2) $ZXY_2(C_s)$ and ZXYW (C_1) Molecules

Substitution of one of the Y atoms of a pyramidal XY_3 molecule by a Z atom lowers the symmetry from $\mathbf{C}_{3v}$ to $\mathbf{C}_s$. Then the degenerate vibrations split into two bands, and all six vibrations become infrared and Raman active. The relationship between $\mathbf{C}_{3v}$ and $\mathbf{C}_s$ is shown in Table II-3*d*. Table II-3*e* lists the vibrational frequencies of pyramidal ZXY_2 molecules. Simon and Paetzold[346] made an extensive study of the vibrational spectra of selenium compounds. The ZXYW-type molecule belongs to the $\mathbf{C}_1$ point group, and all six vibrations are infrared and Raman active. The vibrational spectra of OSClBr[347] and $[XSnYZ]^-$ (X, Y, Z: a halogen)[348] have been reported.

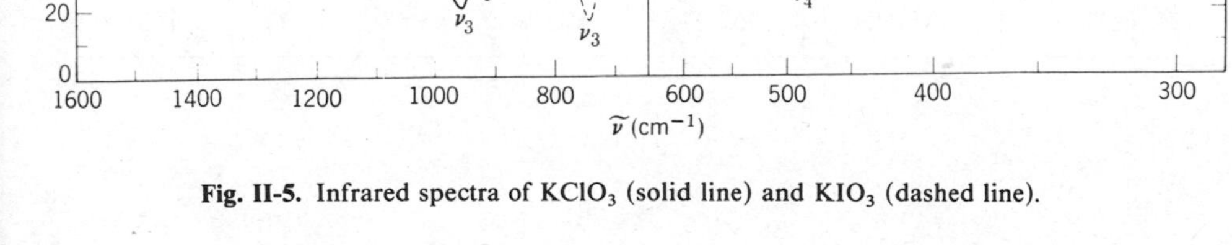

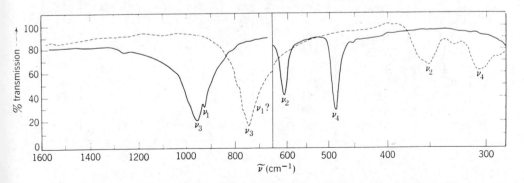

Fig. II-5. Infrared spectra of $KClO_3$ (solid line) and KIO_3 (dashed line).

TABLE II-3b. VIBRATIONAL FREQUENCIES OF PYRAMIDAL XY_3 HALOGENO COMPOUNDS (cm^{-1})

Molecule	State	ν_1	ν_2	ν_3	ν_4	Refs.
$[InCl_3]^-$	Solid	252	102	185	97	296
$[InBr_3]^-$	Solid	177	74	149	46	296
$[InI_3]^-$	Solid	136	78	110	40	296
CF_3	Matrix	1084	703	1250	600, 500	297
$^{28}SiF_3$	Matrix	832	406	954	290	298
$^{28}SiCl_3$	Matrix	470.2	—	582.0	—	299
$GeCl_3$	Matrix	388	—	362	—	300
$[GeCl_3]^-$	Solid	303	—	285	—	301
$[SnF_3]^-$	Solid	520	280	477	224	302
$[SnCl_3]^-$	Sol'n	297	128	256	103	302
$[SnBr_3]^-$	Sol'n	211	83	181	65	302
$[PbCl_3]^-$	Sol'n	249	—	—	—	303
$[PbBr_3]^-$	Sol'n	176	—	164	58	303
$[PbI_3]^-$	Sol'n	137	—	127	30 ~ 45	303
NF_3	Gas	1035	649	910	500	304
NCl_3	Sol'n	535	347	637	254	305
PF_3	Gas	893.2	486.5	858.4	345.6	306
PCl_3	Gas	515.0	258.3	504.0	186.0	306
PBr_3	Gas	390.0	159.9	384.4	112.8	306
PI_3	Solid	303	111	325	79	307
AsF_3	Gas	738.5	336.8	698.8	262.0	306
$AsCl_3$	Gas	416.5	192.5	391.0	150.2	306
$AsBr_3$	Melt	272	128	287	99	308
AsI_3	Gas	212.0	89.6	(201)	63.9	306
SbF_3	Matrix	654	259	624	—	309
$SbCl_3$	Gas	380.7	150.8	358.9	121.8	306
$SbBr_3$	Sol'n	254	101	245	81	308
SbI_3	Gas	186.5	74.0	(147)	54.3	306
$BiCl_3$	Gas	342	123	322	107	310
$BiBr_3$	Gas	220	77	214	63	311
BiI_3	Solid	145.5	90.2	115.2	71.0	312
$[SF_3]^+$	Melt	943	690	922	356	314
$[SCl_3]^+$	Solid	498	276	533, 521	215, 208	313
$[SBr_3]^+$	Solid	375	175	429, 414	128	315
$[SeF_3]^+$	Melt	781	381	743	275	314
$[SeCl_3]^+$	Sol'n	430	206	415	172	316
$[SeBr_3]^+$	Sol'n	291	138	298	108	308
$[TeCl_3]^+$	Sol'n	362	186	347	(150)	317
$[TeBr_3]^+$	Sol'n	265	112	266	92	308
$[MnBr_3]^-$	Solid	280	110	150	80	318
$FeCl_3$	Matrix	363.0	68.7	460.2	113.8	319

TABLE II-3*c*. VIBRATIONAL FREQUENCIES OF PYRAMIDAL XO_3 MOLECULES (CM^{-1})

Molecule	State	$\nu_1(A_1)$	$\nu_2(A_1)$	$\nu_3(E)$	$\nu_4(E)$	Refs.
$[SO_3]^{2-}$	Sol'n	967	620	933	469	320
$[SeO_3]^{2-}$	Sol'n	807	432	737	374	321
$[TeO_3]^{2-}$	Sol'n	758	364	703	326	321
$[ClO_3]^-$	Sol'n	933	608	977	477	322
	Solid	939	614	971	489	323
$[BrO_3]^-$	Sol'n	805	418	805	358	322
	Solid	810	428	790	361	323
$[IO_3]^-$	Sol'n	805	358	775	320	322
	Solid	796	348	745	306	323
XeO_3	Sol'n	780	344	833	317	324

TABLE II-3*d*. RELATIONSHIP BETWEEN C_{3v} AND C_s

C_{3v} XY_3	$\nu_1(A_1)$ $\nu_s(XY)$	$\nu_2(A_1)$ $\delta_s(YXY)$	$\nu_3(E)$ $\nu_d(XY)$		$\nu_4(E)$ $\delta_d(YXY)$	
	↓	↓	↙	↘	↙	↘
C_s ZXY_2	$\nu_1(A')$ $\nu_s(XZ)$	$\nu_3(A')$ $\delta_s(YXZ)$	$\nu_2(A')$ $\nu_s(XY)$	$\nu_5(A'')$ $\nu_a(XY)$	$\nu_4(A')$ $\delta_s(YXY)$	$\nu_6(A'')$ $\delta_a(YXZ)$

II-4. PLANAR FOUR-ATOM MOLECULES

(1) XY_3 Molecules (D_{3h})

The four normal modes of vibration of planar XY_3 molecules are shown in Fig. II-6; ν_2, ν_3, and ν_4 are infrared active, and ν_1, ν_3, and ν_4 are Raman active. This case should be contrasted with pyramidal XY_3 molecules, for which all four vibrations are both infrared and Raman active.

Table II-4*a* lists the vibrational frequencies of planar XY_3 molecules. As stated above, pyramidal and planar structures can be distinguished easily on the basis of the difference in selection rules. In some cases, however, this approach leads to conflicting results. For example, rare-earth trifluorides in inert gas matrices exhibit infrared spectra that are consistent with planar structures except for PrF_3, which was thought to be pyramidal since it showed two stretching bands at 542 and 458 cm^{-1}.[365] Later, a Raman study of matrix-isolated PrF_3 showed that the infrared band at 542 cm^{-1} was not a PrF stretching fundamental. Thus the structure of PrF_3 was concluded to be planar in agreement with the result of an electron diffraction study in the gaseous phase.[365] The CH_3 radical produced by the reaction of Li atoms with CH_3Br or CH_3I is planar.[362] For dimeric species such as Al_2F_6 and Fe_2Cl_6, see Sec. II-10. Normal coordinate analyses of planar XY_3 molecules have been carried out by many investigators.[366–371,358]

TABLE II-3*e*. VIBRATIONAL FREQUENCIES OF PYRAMIDAL ZXY_2-TYPE METAL HALIDES (CM^{-1})

Z—X(—Y)(—Y)	$\nu_1(A')$ $\nu(XZ)$	$\nu_2(A')$ $\nu_s(XY)$	$\nu_3(A')$ $\delta_s(YXZ)$	$\nu_4(A')$ $\delta(YXY)$	$\nu_5(A'')$ $\nu_a(XY)$	$\nu_6(A'')$ $\delta_a(YXZ)$	Refs.
HNF_2	3193	972	500	1307	888	1424	327
$HNCl_2$	3279	687	—	1002	695	1295	328
$ClNH_2$	686	—	1553	1032	3380	—	328
$ClNF_2$	692	918	552	366	842	382	329
$ClPF_2$	545	864	411	(302)	852	260	330
$BrPF_2$	459	858	244	391	849	215	330
$FPCl_2$	838	525	328	203	525	267	330
$FPBr_2$	824	398	258	123	423	221	330
IPF_2	375	851	198	413	846	204	331
$[BrOF_2]^+$	1062	655	365	290	630	315	332
OSF_2	1333	808	530	(410)	748	390	333
$OSCl_2$	1251	492	194	344	455	284	334
$OSBr_2$	1121	405	120	267	379	223	335
$[FSO_2]^-$	598	1100	378	496	1170	280	336
$[ClSO_2]^-$	214	1120	172	526	1312	(103)	336
$[BrSO_2]^-$	203	1117	115	530	1308	—	336
$[ISO_2]^-$	184	1112	(55)	530	1300	—	336
$[FSeO_2]^-$	~430	903	283	324	888	238	337
$OSeF_2$	1049	667	362	282	637	253	338
$OSeCl_2$	995	388	161	279	347	255	339
$OSe(OH)_2$[a]	831	702	430	336	690	364	340, 341
$FClO_2$	602	1097	398	533	1253	351	342
$FBrO_2$	506	908	305	394	953	271	343
$[OClF_2]^+$	1333	731	513	384	695	404	344, 345

[a] The OH group was assumed to be a single atom.

Table II-4*b* lists the vibrational frequencies of planar XO_3-type compounds. The infrared spectra of rare-earth orthoborates have been studied by Laperches and Tarte.[381] For metaborates, see Sec. II-12. As discussed in Sec. I-23, the spectra of calcite and aragonite are markedly different because of the difference in crystal structure. Recent normal coordinate calculations[382] on the CO_3 radical prefer the *trans*-C_s structure (A) rather than the three-membered ring C_{2v} structure (B) suggested originally.[383]

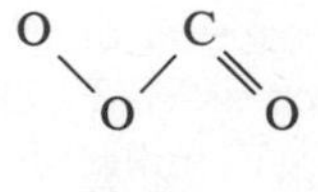

(A)

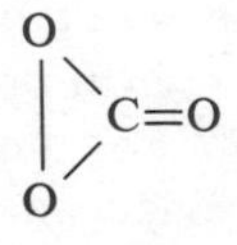

(B)

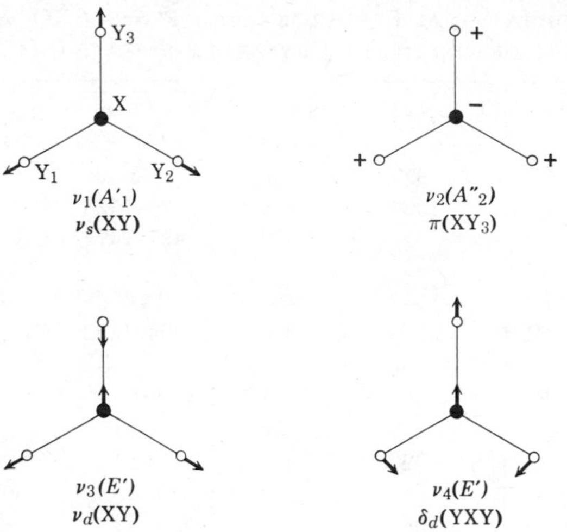

Fig. II-6. Normal modes of vibration of planar XY_3 molecules.

TABLE II-4*a*. VIBRATIONAL FREQUENCIES OF PLANAR XY_3 MOLECULES (CM^{-1})

Molecule	State[a]	$\nu_1(A'_1)$	$\nu_2(A''_2)$	$\nu_3(E')$	$\nu_4(E')$	Refs.
$^{10}BH_3$	Mat	(2623)	1132	2820	1610	349
$^{10}BF_3$	Gas	888	718	1505	482	350, 351
$^{11}BCl_3$	Liquid	472.7	—	950.7	253.7	352, 353
$^{11}BBr_3$	Liquid	278	374	802	150	352, 354
$^{11}BI_3$	Liquid	192	—	691.8	101.0	352, 354
AlF_3	Mat	—	286.2	909.4	276.9	355
$AlCl_3$	Mat	393.5	—	618.8	150	356
$AlBr_3$	Gas	228	107	450–500	93	357, 358
AlI_3	Gas	156	77	370–410	64	357, 358
$^{69}GaCl_3$	Mat	386.2	—	469.3	132	356
$GaBr_3$	Gas	219, 237[b]	95	—	84	358, 359
GaI_3	Gas	147	63	275	50	358, 359
$InCl_3$	Mat	359.0	—	400.5	119	356, 360
$InBr_3$	Gas	212	74	280	62	357, 358
InI_3	Gas	151	56	200–230	44	357, 358
$TlBr_3$	Sol'n	190	125	220	51	361
CH_3	Mat	—	730.3	—	1383	362
$[CdCl_3]^-$	Sol'n	265	—	287	90	363
$[CdBr_3]^-$	Sol'n	168	—	184	58	363
$[CdI_3]^-$	Sol'n	124	—	161	51	363
$[HgCl_3]^-$	Solid	273	113	263	100	364
PrF_3	Mat	526, 542	86	458	99	365

[a] Mat = inert gas matrix.
[b] Fermi resonance.

TABLE II-4*b*. VIBRATIONAL FREQUENCIES OF PLANAR XO_3 AND RELATED COMPOUNDS IN THE CRYSTALLINE STATE (CM^{-1})

Compound		$\nu_1(A_1')$	$\nu_2(A_2'')$	$\nu_3(E')$	$\nu_4(E')$	Refs.
$La[^{10}BO_3]$	IR	939	740.5	1330.0	606.2	372
$H_3[BO_3]$	IR	1060	668, 648	1490–1428	545	373
$Ca[CO_3]$ (calcite)	IR	—	879	1429–1492	706	374
	R	1087	—	1432	714	374
$Ca[CO_3]$ (aragonite)	IR	1080	866	1504, 1492	711, 706	374
	R	1084	852	1460	704	374
$Ba[CS_3]$	R	510	516[b]	920	314	375, 376
$Ba[CSe_3]$	IR	290	420	802	185	377, 376
$Na[NO_3]$	IR	—	831	1405	692	378
	R	1068	—	1385	724	378
$K[NO_3]$	IR	—	828	1370	695	378
	R	1049	—	1390	716	378
SO_3[a]	IR	—	484	1391	536	379
	R	1065	497.5	1390	530.2	380

[a] Gaseous state.
[b] Infrared.

Crystals of $LiNO_3$, $NaNO_3$, and KNO_3 take the calcite structure (Sec. I-23). Nakagawa and Walter[378] carried out normal coordinate analyses on the whole Bravais lattices of these crystals. The spectra of anhydrous metal nitrates such as $Zn(NO_3)_2$[384] and $UO_2(NO_3)_2$[385] can be interpreted in terms of $\mathbf{C}_{2v}$ symmetry since the NO_3 group is covalently bonded to the metal (see Sec. III-11). Raman spectra of metal nitrates in the molten state[386,387] indicate that the degeneracy of the ν_3 vibration is lost and the Raman-inactive ν_2 vibration appears. Apparently, the $\mathbf{D}_{3h}$ selection rule is violated because of cation–anion interaction. The infrared spectrum of monomeric lithium nitrate ($LiNO_3$) in an inert gas matrix shows a large splitting of the ν_3 vibration (240 cm^{-1}) due to distortion of the NO_3 group.[388] Like CO_3^{2-} and NO_3^- ions, CS_3^{2-} and CSe_3^{2-} ions act as chelating ligands (see Sec. III-22). Normal coordinate analyses of planar XY_3 molecules have been made by many investigators. For some relatively recent work, see Refs. 389 and 390.

Some XY_3-type halides take the unusual T-shaped structure of $\mathbf{C}_{2v}$ symmetry shown below. This geometry is derived from a trigonal-bipyramidal structure in which two equatorial positions are occupied by two lone-pair electrons. Typical examples are ClF_3 and BrF_3. With the equatorial Y atom represented as Y′, the following assignments have been made for these molecules:[391] $\nu(XY')$, A_1, 754 and 672; $\nu(XY)$, B_1, 683.2 and 597; $\nu(XY)$, A_1, 523 and 547; δ, A_1, 328 and 235; δ, B_1, 431 and 347; π, B_2, 332 and 251.5 cm^{-1} (for each

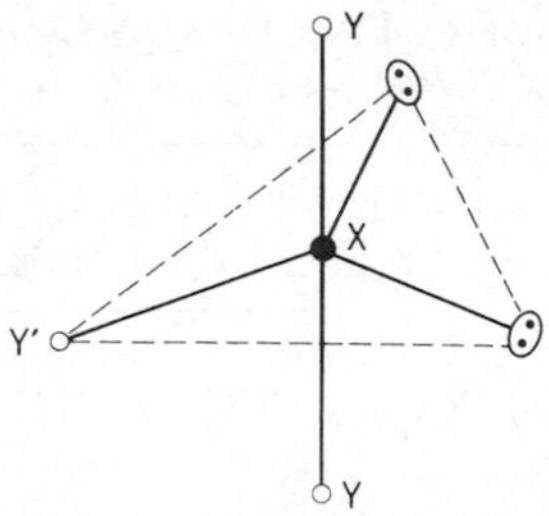

mode, the former value is for ClF_3 and the latter is for BrF_3). The Raman spectrum of XeF_4 in SbF_5 exhibits two strong polarized bands at 643 and 584 cm^{-1}, which were assigned to the T-shaped XeF_3^+ ion.[392] $XeOF_2$ takes a similar T structure.[393] In an inert gas matrix UO_3 gives an infrared spectrum consistent with the T-shaped structure.[394]

(2) ZXY_2 (C_{2v}) and ZXYW(C_s) Molecules

If one of the Y atoms of a planar XY_3 molecule is replaced by a Z atom, the symmetry is lowered to $\mathbf{C}_{2v}$. If two of the Y atoms are replaced by two different atoms, W and Z, the symmetry is lowered to $\mathbf{C}_s$. As a result, the selection rules are changed, as already shown in Table I-12. In both cases, all six vibrations become active in infrared and Raman spectra. Table II-4*c* lists the vibrational frequencies of planar ZXY_2 and ZXYW molecules. Although not listed in this table, the infrared spectra of binary mixed halides of boron[410] and aluminum[411] have been measured. The frequencies listed for the formate and acetate ions were obtained in aqueous solution. These frequencies are important when we discuss the vibrational spectra of metal salts of these anions (Sec. III-7).

II-5. OTHER FOUR-ATOM MOLECULES

(1) X_2Y_2 Molecules

Molecules like O_2H_2 take the nonplanar $\mathbf{C}_2$ structure (twisted about the O–O bond by ca. 90°), whereas N_2F_2 and $[N_2O_2]^{2-}$ exist in two forms: *trans*-planar ($\mathbf{C}_{2h}$) and *cis*-planar ($\mathbf{C}_{2v}$). Figure II-7 shows the six normal modes of vibration for the $\mathbf{C}_{2v}$ and $\mathbf{C}_2$ structures. The selection rules for these two structures are different only in the ν_6 vibration, which is infrared inactive and Raman depolarized in the planar model but infrared active and Raman polarized in the nonplanar model.

Table II-5*a* lists the vibrational frequencies of X_2Y_2-type compounds. In a N_2 matrix $^{14}N_2O_2$ exists in three forms all containing the N–N linkage; their NO stretching frequencies (cm^{-1}) are as follows: *trans*, 1764; *cis* I, 1870, 1776; and *cis* II, 1870, 1785.[423] Bands characteristic of diazene (N_2H_2) are: *trans*, 3109, 1333; *cis*, 3116, 3025, 1347, 1304.[424] Ketelaar et al.[425] observed the infrared frequencies of S_2Cl_2 and S_2Br_2 from *simultaneous transitions* in mixtures of each of these compounds with CS_2. These vibrations are similar to the combina-

TABLE II-4*c*. VIBRATIONAL FREQUENCIES OF PLANAR ZXY_2 AND ZXYW MOLECULES (CM^{-1})

$XY_3(\mathbf{D}_{3h})$	$\nu_1(A_1')$ $\nu_s(XY)$	$\nu_2(A_2'')$ $\pi(XY_3)$	$\nu_3(E')$ $\nu_d(XY)$		$\nu_4(E')$ $\delta_d(YXY)$		
$ZXY_2(\mathbf{C}_{2v})$	$\nu_1(A_1)$ $\nu(XZ)$	$\nu_6(B_1)$ $\pi(ZXY_2)$	$\nu_2(A_1)$ $\nu_s(XY)$	$\nu_4(B_2)$ $\nu_a(XY)$	$\nu_3(A_1)$ $\delta_s(ZXY)$	$\nu_5(B_2)$ $\delta_a(ZXY)$	
$ZXYW(\mathbf{C}_s)$	$\nu_1(A')$ $\nu(XZ)$	$\nu_6(A'')$ π	$\nu_2(A')$ $\nu(XY)$	$\nu_4(A')$ $\nu(XW)$	$\nu_3(A')$ $\delta(ZXY)$	$\nu_5(A')$ $\delta(ZXW)$	Refs.
$(HO)-NO_2$	886	765	1320	1710	—	583	395
$F-NO_2$	822.4	742.0	1309.6	1791.5	559.6	567.8	396
$Cl-NO_2$	792.6	652.3	1267.1	1684.6	408.1	369.6	396
$[(HO)-CO_2]^-$	960	835	1338	1697	712	579	397
$[H-CO_2]^-$	2803	1069	1351	1585	760	1383	398
$[(CH_3)-CO_2]^-$	926	621	1413	1556	650	471	398
$O=CF_2$	1930	767.4	965.6	1243.7	582.9	619.9	399
$O=CCl_2$	1827	580	569	849	285	440	400
$O=CBr_2$	1828	512	425	757	181	350	400
$O=CClF$	1868	667	776	1095	501	415	400
$O=CBrCl$	1828	547	517	806	240	372	400
$O=CBrF$	1874	620	721	1068	398	335	400
$O=CHF$	1837	—	2981	1065	1343	663	401
$S=CH_2$	1063	993	2970	~3028	1550	1437	402
$S=CF_2$	1368	622	787	1189	526	417	403
$S=CCl_2$	1120	472	~500	812	296	~302	404
$Se=CF_2$	1280	560	710	1208	432	352	405
$[Se-CS_2]^-$	442	485	433	925	284	265	406
$[O=NF_2]^+$	1862	715	897	1163	569	647	407, 408
$[O=NCl_2]^+$	1650	—	635	747	220	420	409

tion bands in one molecule, but here the vibrations of two molecules are combined because of transient molecular collisions. The Raman spectra of Se_2Cl_2 and Se_2Br_2 in CS_2 and CCl_4 solutions indicate that the former takes the $\mathbf{C}_2$ structure whereas the latter takes the $\mathbf{C}_{2v}$ structure.[426]

Normal coordinate analyses of O_2H_2,[427] *trans*-$N_2O_2^{2-}$,[428] and S_2X_2 (X: a halogen)[422,428] have been made. Other $(XY)_2$-type compounds are known for dimeric metal halides such as $(LiF)_2$, which takes a planar-ring structure (Sec. II-1).

(2) Other Planar Molecules ($\mathbf{C}_s$)

Planar four-atom molecules of the WXYZ, XYZY, and XY_3 types have six normal modes of vibration, as shown in Fig. II-8. All these vibrations are both infrared and Raman active. In WXYZ compounds, the XYZ skeleton may be linear (HNCO, HOCN, HSCN) or nonlinear (HONO, HNSO). In the latter case, the whole molecule may take the *cis* or *trans* structure. Table II-5*b* lists

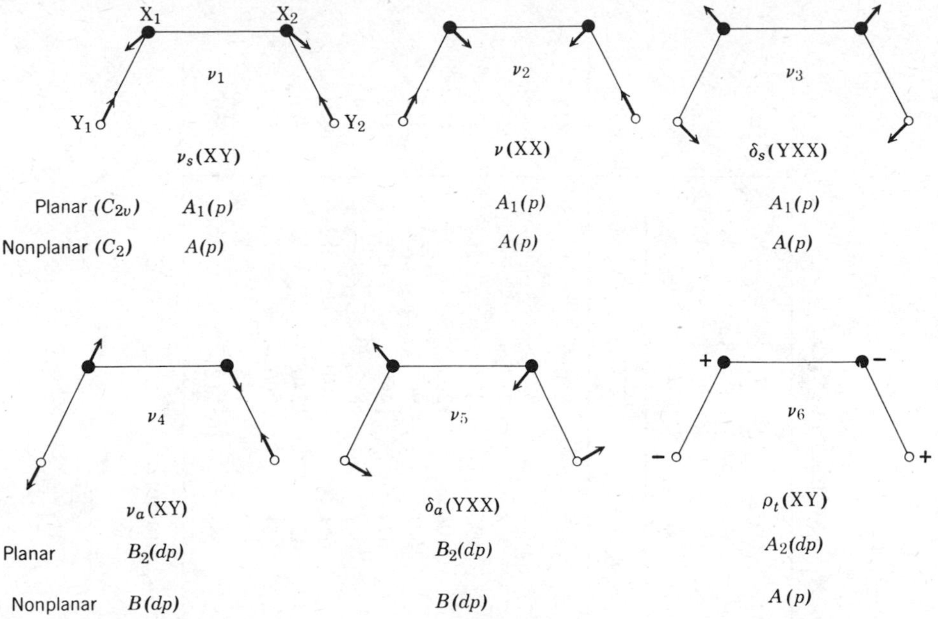

Fig. II-7. Normal modes of vibration of nonlinear X_2Y_2 molecules (p: polarized; dp: depolarized).[403]

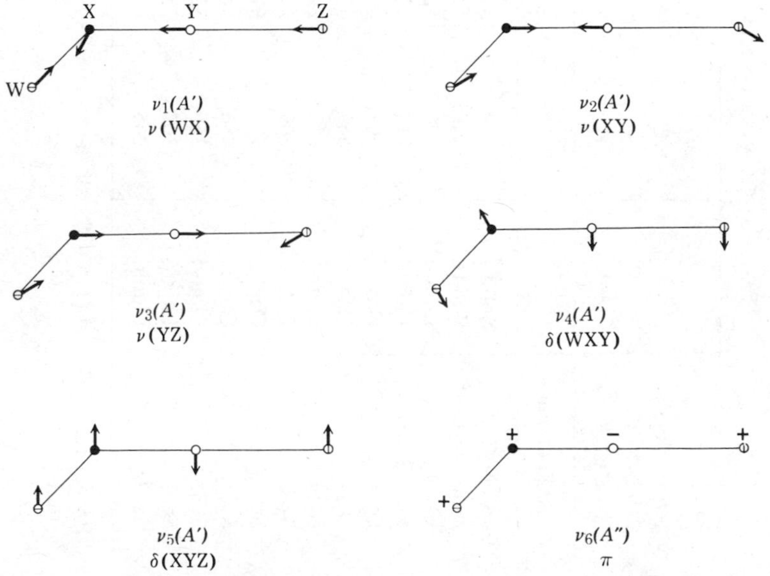

Fig. II-8. Normal modes of vibration of nonlinear WXYZ molecules.

TABLE II-5a. VIBRATIONAL FREQUENCIES OF X_2Y_2 MOLECULES[a] (CM^{-1})

trans-X_2Y_2	C_{2h}	ν_1 A_g	ν_2 A_g	ν_3 A_g	ν_4 B_u	ν_5 B_u	ν_6 A_u	
cis-X_2Y_2	C_{2v}	A_1	A_1	A_1	B_2	B_2	A_2	
Twisted X_2Y_2	C_2	A	A	A	B	B	A	
Assignment		$\nu_s(XY)$	$\nu(XX)$	$\delta_s(YXX)$	$\nu_a(XY)$	$\delta_a(YXX)$	π	Refs.
cis-$[N_2O_2]^{2-}$	Solid	830	1314	584	1047	330	—	412
trans-$[N_2O_2]^{2-}$	Solid	(1419)	(1121)	(696)	1031	371	492	413, 414
cis-N_2F_2	Liquid	896	1525	341	952	737	(550)	415
trans-N_2F_2	Liquid	1010	1522	600	990	423	364	416
O_2H_2	Gas	3607	1394	864	3608	1266	317	417
O_2D_2	Gas	2669	1029	867	2661	947	230	417
O_2F_2	Matrix	611	1290	366	624	459	(202)	418
S_2H_2	Liquid	2509	509	(868)[b]	2557[c]	882	—	419
S_2F_2	Gas	717	615	320	681	301	183	420
S_2Cl_2	Gas	466	546	202	457	240	92	421
S_2Br_2	Liquid	365	529	172	351	200	66	422
Se_2Cl_2	Liquid	367	288	130	367	146	87	422
Se_2Br_2	Liquid	265	292	107	265	118	50	422

[a] Except for N_2F_2 and $[N_2O_2]^{2-}$, all the molecules listed take the C_2 or C_{2v} structure.
[b] Solid.
[c] Gas.

TABLE II-5*b*. VIBRATIONAL FREQUENCIES OF PLANAR FOUR-ATOM MOLECULES (cm^{-1})

Molecule WXYZ	State[a]	ν_1 ν(WX)	ν_2 ν(XY)	ν_3 ν(YZ)	ν_4 δ(WXY)	ν_5 δ(XYZ)	ν_6 π	Refs.
HCNO	Gas	3336	1256	2198	331	538	—	429
DCNO	Gas	2580	1254	2066	—	—	—	429
HCNS	Gas	3539	1989	857	615	469	539	430
DCNS	Gas	2645	1944	851	549	366	481	430
HCNO	Gas	3531	2274	1527	777	660	578	431
DCNO	Gas	2635	2235	1310	460	767	603	431
HNCS	Mat	3505	1979	988	577	461	—	432
DNCS	Mat	2623	1938	—	548	366	—	432
HNNN	Mat	3324	2150	1273	1168	527	588	433
DNNN	Mat	2466	—	1198	964	493	—	433
cis-HNSO	Mat	3309	1083	1249	900	447	755	434
trans-HNSO	Mat	3308	982	1381	878	496	651	435
cis-HONO	Mat	3412	1633	1265	850	610	637	436
trans-HONO	Mat	3558	1684	1298	815	625	583	437
HOCN	Mat	3506	1098	2294	1241	460	438	438
DOCN	Mat	2590	1093	2292	957	437	—	438
FNSO	Liquid	825	995	1230	228	600	395	439
ClNSO	Liquid	526	989	1221	187	672	359	439
BrNSO	Liquid	451	1000	1214	161	624	342	439
INSO	Liquid	372	1028	1247	154	602	330	439
ClSCN	Sol'n	520	678	2162	—	353	—	440
BrSCN	Sol'n	451	676	2157	—	369	—	440
ISCN	Sol'n	372	700	2130	—	362	—	440

[a] Mat = inert gas matrix.

the vibrational frequencies of molecules belonging to these types. Normal coordinate analyses have been made of HN_3[441] and HONO.[436]

II-6. TETRAHEDRAL AND SQUARE-PLANAR FIVE-ATOM MOLECULES

(1) Tetrahedral XY_4 Molecules (T_d)

Figure II-9 illustrates the four normal modes of vibration of a tetrahedral XY_4 molecule. All four vibrations are Raman active, whereas only ν_3 and ν_4 are infrared active. Fundamental frequencies of XH_4-type molecules are listed in Table II-6*a*. The trends $\nu_3 > \nu_1$ and $\nu_2 > \nu_4$ hold for the majority of the compounds. The XH stretching frequencies may be lowered whenever the XH_4 ions form hydrogen bonds with counterions. In the same family of the periodic table, the XH stretching frequency decreases as the mass of the X atom increases. Shirk and Shriver[444] noted, however, that the ν_1 frequency and the corresponding force constant show an unusual trend in Group IIIA:

	$[BH_4]^-$	$[GaH_4]^-$	$[AlH_4]^-$
$\tilde{\nu}_1$ (cm^{-1})	2270	1807	1757
F_{11} (mdyn/Å)	3.07	1.94	1.84

Figure II-10 shows the infrared spectrum of NH_4Cl, measured by Hornig et al.[456] They noted that the combination band between ν_4 (F_2) and ν_6 (rotatory lattice mode) is observed for NH_4F, NH_4Cl, and NH_4Br because the NH_4^+ ion does not rotate freely in these crystals. In NH_4I (phase I), however, this band is not observed because the NH_4^+ ion rotates freely.

Table II-6*b* lists the vibrational frequencies of a number of tetrahalogeno compounds. Except for $[TlBr_4]^-$ and UF_4, the trends $\nu_3 > \nu_1$ and $\nu_4 > \nu_2$ hold for all the compounds. The latter trend is opposite to that found for MH_4 compounds. In the solid state, ν_3 and ν_4 may split into two or three bands because of the site effect. In some cases, the MX_4 ions are distorted to a flattened tetrahedron ($\mathbf{D}_{2d}$) or a structure of lower symmetry ($\mathbf{C}_s$).[484,498] According to X-ray analysis[499] the unit cell of $[(CH_3)_2CHNH_3]_2$ $[CuCl_4]$ contains two square-planar and four distorted tetrahedral $[CuCl_4]^{2-}$ ions.

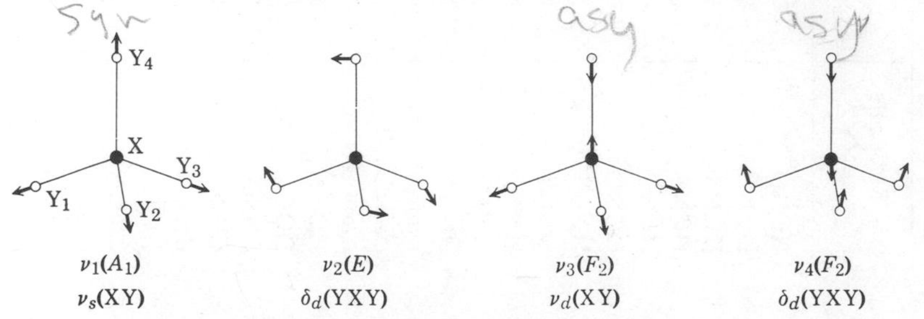

Fig. II-9. Normal modes of vibration of tetrahedral XY_4 molecules.

TABLE II-6*a*. VIBRATIONAL FREQUENCIES OF TETRAHEDRAL XH_4 MOLECULES (CM^{-1})

Molecule	ν_1	ν_2	ν_3	ν_4	Refs.
$[^{10}BH_4]^-$	2270	1208	2250	1093	442, 443
$[^{10}BD_4]^-$	1604	856	1707	827	442, 443
$[AlH_4]^-$	1757	772	1678	760 or 766	444
$[AlD_4]^-$	1256	549	1220	560 or 556	444
$[GaH_4]^-$	1807	—	—	—	444
CH_4	2917	1534	3019	1306	445
CD_4	2085	1092	2259	996	446, 447
SiH_4	2180	970	2183	910	445, 448
SiH_4	—	—	2192	913	449
SiD_4	(1545)	(689)	1597	681	448, 450
GeH_4	2106	931	2114	819	445, 451
GeD_4	1504	665	1522	596	451
SnH_4	—	758	1901	677	452
SnD_4	—	539	1368	487	452
$[^{14}NH_4]^+$	3040	1680	3145	1400	445
$[^{15}NH_4]^+$	—	(1646)	3137	1399	453
$[ND_4]^+$	2214	1215	2346	1065	445
$[NT_4]^+$	—	976	2022	913	453
$[PH_4]^+$	2295	1086, 1026	2366, 2272	974, 919	454
$[PD_4]^+$	1654	772, 725	1732	677	454
$[AsH_4]^+$	2080	949	2142	818, 813	455

Willett et al.[500] demonstrated by using infrared spectroscopy that distorted tetrahedral ions can be pressed to square-planar ions under high pressure. In nitromethane, $(Et_4N)_2[CuCl_4]$ exhibits two bands at 278 and 237 cm^{-1}, indicating the distortion in solution.[501] A solution of $(Et_3NH)[GaCl_4]$ in 1,2-dichloroethylene exhibits three bands at 390, 383, and 359 cm^{-1} due to lowering of symmetry caused by the NH···Cl hydrogen bonding.[502] Work and Good[503] studied the infrared spectra of long chain tertiary and quaternary ammonium

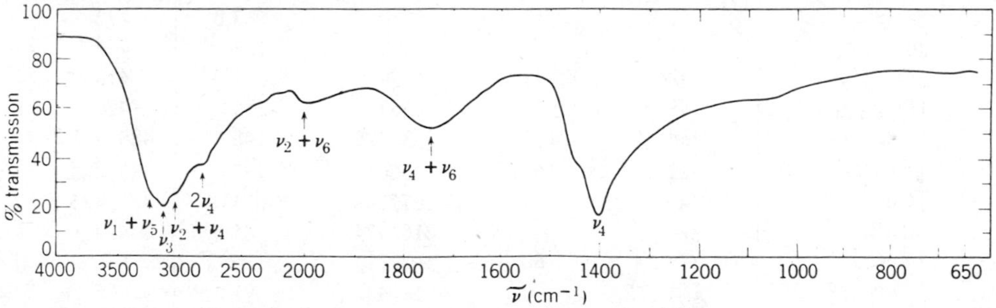

Fig. II-10. Infrared spectrum of NH_4Cl (ν_5, ν_6: lattice modes).

TABLE II-6*b*. VIBRATIONAL FREQUENCIES OF TETRAHEDRAL HALOGENO COMPOUNDS (CM^{-1})

Molecule	ν_1	ν_2	ν_3	ν_4	Refs.
$[BeF_4]^{2-}$	547	255	800	385	457
$[MgCl_4]^{2-}$	252	100	330	142	458
$[MgBr_4]^{2-}$	150	61	290	90	458
$[MgI_4]^{2-}$	107	42	259	60	458
$[BF_4]^-$	777	360	1070	533	459
$[BCl_4]^-$	405	190	670	274	460
$[BBr_4]^-$	243	117	605	166	461
$[AlF_4]^-$	622	210	760	322	462
$[AlCl_4]^-$	348	119	498	182	463
$[AlBr_4]^-$	212	98	394	114	464
$[AlI_4]^-$	146	51	336	82	465
$[GaCl_4]^-$	343	120	370	153	466
$[GaBr_4]^-$	210	71	278	102	467
$[GaI_4]^-$	145	52	222	73	468
$[InCl_4]^-$	321	89	337	112	469
$[InBr_4]^-$	197	55	239	79	470
$[InI_4]^-$	139	42	185	58	468
$[TlCl_4]^-$	312	—	293	93, 110	471
$[TlBr_4]^-$	192	—	173, 185	78	471
$[TiI_4]^-$	130	—	146	60	471
CF_4	908.4	434.5	1283.0	631.2	472
CCl_4	460.0	214.2	792, 765	313.5	472
CBr_4	267	123	672	183	473
CI_4	178	90	555	123	474
SiF_4	800.8	264.2	1029.6	388.7	472
$SiCl_4$	423.1	145.2	616.5	220.3	472
$SiBr_4$	246.7	84.8	494.0	133.6	472
SiI_4	168.1	62	405.5	90.5	475
GeF_4	738	205	800	260	476
$GeCl_4$	396.9	125.0	459.1	171.0	472
$GeBr_4$	235.7	74.7	332.0	111.1	472
GeI_4	158.7	~60	264.1	79.0	475
$SnCl_4$	369.1	95.2	408.2	126.1	472
$SnBr_4$	222.1	59.4	284.0	85.9	472
SnI_4	147.7	42.4	210	63.0	472
$PbCl_4$	331	90	352	103	477
$[NF_4]^+$	848	443	1159	611	478
$[PCl_4]^+$	458	178	662	255	479
$[PBr_4]^+$	254	116	503, 496	148	480, 481
$[AsCl_4]^+$	422	156	500	187	482, 483
$[CuCl_4]^{2-}$	—	77	267, 248	136, 118	484, 485
$[CuBr_4]^{2-}$	—	—	216, 174	85	484
$[ZnCl_4]^{2-}$	276	80	277	126	486
$[ZnBr_4]^{2-}$	171	—	204	91	486

TABLE II-6*b* (*Continued*)

Molecule	ν_1	ν_2	ν_3	ν_4	Refs.
$[ZnI_4]^{2-}$	118	—	164	—	486
$[CdCl_4]^{2-}$	261	84	249, 240	98	487
$[CdBr_4]^{2-}$	161	49	177	75, 61	487
$[CdI_4]^{2-}$	116	39	—	50	487
$[HgCl_4]^{2-}$	267	180	276	192	488
$[HgI_4]^{2-}$	126	35	140	41	489
TiF_4	712	185	793	209	490
$TiCl_4$	389	114	498	136	491
$TiBr_4$	231.5	68.5	393	88	491
TiI_4	162	51	323	67	491
ZrF_4	(725 ~ 600)	(200 ~ 150)	668	190	492
$ZrCl_4$	377	98	418	113	491
$ZrBr_4$	225.5	60	315	72	491
ZrI_4	158	43	254	55	491
$HfCl_4$	382	101.5	390	112	491
$HfBr_4$	235.5	63	273	71	491
HfI_4	158	55	224	63	491
VCl_4	383	128	475	128, 150	493
$CrCl_4$	373	116	486	126	494
$CrBr_4$	224	60	368	71	494
$[MnCl_4]^{2-}$	256	—	278, 301	120	486, 495
$[MnBr_4]^{2-}$	195	65	209, 221	89	486, 495
$[MnI_4]^{2-}$	108	46	188, 193	56	486, 495
$[FeCl_4]^-$	330	114	~378	~136	486
$[FeBr_4]^-$	200	—	290	95	486
$[FeCl_4]^{2-}$	266	82	286	119	486
$[FeBr_4]^{2-}$	162	—	219	84	486
$[FeI_4]^{2-}$	—	—	186	—	484
$[NiCl_4]^{2-}$	264	—	294, 280	119	484, 496
$[NiBr_4]^{2-}$	—	—	228	81	484, 496
$[NiI_4]^{2-}$	105	—	191	—	484
$[CoCl_4]^{2-}$	269	—	311, 291	135	496
$[CoBr_4]^{2-}$	166	—	231, 222	96	496
$[CoI_4]^{2-}$	118	—	202, 194	56	484
UF_4	614	340	420	180	497

salts of $[GaCl_4]^-$ and $[GaBr_4]^-$ ions in benzene solution, and found that the degree of distortion of these ions depends on the nature of the cation and the concentration. Distortion of $[MCl_4]^{2-}$ ions (M = Fe, Co, Ni and Zn) is also reported for their cesium and rubidium salts.[504]

Table II-6*b* includes a number of data obtained by Clark et al. from their gas-phase Raman studies.[472,491] Some halides such as TiI_4,[505] SnI_4,[475,506] and

VCl_4[507] have strong electronic absorption in the visible region, and their resonance Raman spectra have been measured in solution. As discussed in Sec. I-21, these compounds exhibit a series of ν_1 overtones. Two important trends in frequency are noted in Table II-6*b*. First, the MX stretching frequency decreases as the halogen is changed in the order F > Cl > Br > I. The average values of ν(MBr)/ν(MCl) and ν(MI)/ν(MCl) calculated from all the compounds listed in Table II-6*b* are 0.76 and 0.62, respectively, for ν_3, and 0.61 and 0.42, respectively, for ν_1. These values are very useful when we assign the MX stretching bands of halogeno complexes (see Sec. III-19). Second, the effect of changing the oxidation state on the MX stretching frequency is seen in a pair such as $[FeX_4]^-$ and $[FeX_4]^{2-}$ (X = Cl and Br);[486] the MX stretching frequency increases as the oxidation state of the metal becomes higher. The ratio $\nu_3(FeX_4^-)/\nu_3(FeX_4^{2-})$ is 1.32 in this case.

A tetrahedral MCl_4 molecule in which M is isotopically pure and Cl is in natural abundance consists of five isotopic species because of the mixing of the ^{35}Cl (75.4%) and ^{37}Cl (24.6%) isotopes. Table II-6*c* lists their symmetries, percentages of natural abundance, and symmetry species of infrared-active modes corresponding to the ν_3 vibration of the $\mathbf{T}_d$ molecule. It has been established[508] that these nine bands overlap partially to give a "five-peak chlorine isotope pattern" whose relative intensity is indicated by the vertical lines shown in Fig. II-11*b*. If M is isotopically mixed, the spectrum is too complicated to assign by the conventional method. For example, tin is a mixture of 10 isotopes, none of which is predominant. Thus 50 bands are expected to appear in the ν_3 region of $SnCl_4$. It is almost impossible to resolve all these peaks, even in an inert gas matrix at 10 K. Königer and Müller,[509] therefore, prepared $^{116}SnCl_4$ and $^{116}Sn^{35}Cl_4$ on a milligram scale and measured their infrared spectra in Ar matrices. As expected, the former gave a "five-peak chlorine isotope pattern," whereas the latter showed a single peak at 409.8 cm^{-1}. This work was extended to $GeCl_4$, which consists of two Cl and five Ge isotopes. In this case, 25 peaks are expected to appear in the ν_3 region. However, the observed spectrum (Fig. II-11*a*) shows about 10 bands. Königer et al.,[510] therefore, prepared $^{74}GeCl_4$ and $Ge^{35}Cl_4$ and measured their spectra in Ar

TABLE II-6*c*. INFRARED-ACTIVE VIBRATIONS OF $M^{35}Cl_n{}^{37}Cl_{4-n}$-TYPE MOLECULES

Species	Symmetry	Abundance (%)	IR-Active Modes
$M^{35}Cl_4$	$\mathbf{T}_d$	32.5	F_2
$M^{35}Cl_3^{37}Cl$	$\mathbf{C}_{3v}$	42.2	A_1, E
$M^{35}Cl_2^{37}Cl_2$	$\mathbf{C}_{2v}$	20.5	A_1, B_1, B_2
$M^{35}Cl^{37}Cl_3$	$\mathbf{C}_{3v}$	4.4	A_1, E
$M^{37}Cl_4$	$\mathbf{T}_d$	0.4	F_2

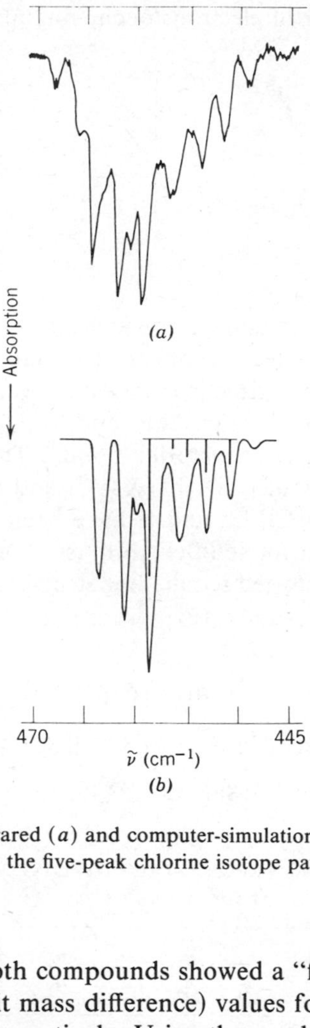

Fig. II-11. Matrix-isolation infrared (*a*) and computer-simulation spectra (*b*) of $GeCl_4$. Vertical lines in (*b*) show the five-peak chlorine isotope pattern of $^{74}GeCl_4$.

matrices. As expected, both compounds showed a "five-peak" spectrum. The *ism* (isotope shift per unit mass difference) values for Cl and Ge were found to be 3.8 and 1.2 cm^{-1}, respectively. Using these values, it is now possible to calculate the frequencies of all other isotopic molecules. Furthermore, the relative intensity of individual peaks is known from the relative concentration of each isotopic molecule. On the basis of this information, Tevault et al.[511] obtained a computer-simulation infrared spectrum of $GeCl_4$ in natural abundance (Fig. II-11*b*).

Normal coordinate analyses of tetrahedral XY_4 molecules have been carried out by a number of investigators.[512] Thus far, Basile et al.[513] have made the most complete study; they calculated the force constants of 146 compounds by using GVF, UBF, and OVF fields (Sec. I-13), and discussed several factors that influence the values of the XY stretching force constants.

It has long been known that molecules such as SF_4, SeF_4, and TeF_4 assume a distorted tetrahedral structure ($\mathbf{C}_{2v}$) derived from a trigonal-bipyramidal geometry with a lone pair of electrons occupying an equatorial position:

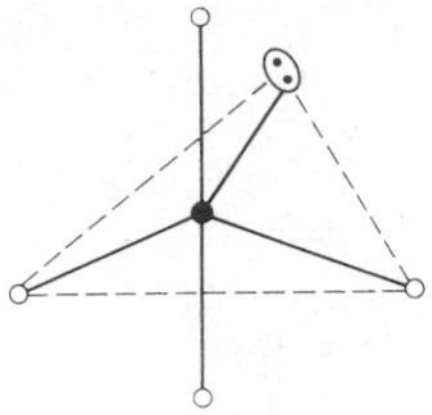

Table II-6*d* lists the vibrational frequencies of nine normal modes of such molecules. It should be noted that these compounds exhibit four stretching modes, two of which are polarized in the Raman. Adams and Downs[519] carried out normal coordinate analyses on SeF_4 and TeF_4 and found that the axial bonds are weaker than the equatorial bonds. The vibrational spectra of tetraalkylammonium salts of $[AsX_4]^-$ (X = Cl and Br), $[BiX_4]^-$ (X = Cl, Br, and I), and $[SbX_4]^-$ (X = Cl, Br, and I) have been studied in the solid state and in solution.[516] Except for solid $(Et_4N)[AsCl_4]$ and $[(n\text{-}Bu)_4N][SbI_4]$, all these ions assume the distorted tetrahedral structure shown above.

The $[ClF_4]^+$, $[BrF_4]^+$, and $[IF_4]^+$ ions were found in the following adducts:[520]

$$ClF_5 \cdot (AsF_5) = [ClF_4][AsF_6]$$

$$BrF_5 \cdot (SbF_5)_2 = [BrF_4][Sb_2F_{11}]$$

$$IF_5 \cdot (SbF_5) = [IF_4][SbF_6]$$

TABLE II-6*d*. VIBRATIONAL FREQUENCIES OF DISTORTED TETRAHEDRAL XY_4 MOLECULES[a] (CM^{-1})

C_{2v}	ν_1 A_1	ν_2 A_1	ν_3 A_1	ν_4 A_1	ν_5 A_2	ν_6 B_1	ν_7 B_1	ν_8 B_2	ν_9 B_2	Refs.
$[SbF_4]^-$	596	449	285	163	220	431	257	566	180	514
$[SbCl_4]^-$	339	296	147	—	—	321	199	246	—	514
	337	300	170	—	—	—	—	252	146	515
$[SbBr_4]^-$	228	190	—	—	—	201	140	169	—	514, 516
$[SbI_4]^-$	169	—	114	—	—	162	85	148	—	514, 516
SF_4	892	558	532	228	(437)	730	475	867	353	517
SeF_4	739	551	362	200	—	585	254	717	403	518
TeF_4	695	572	333	(152)	—	587	273	682	(185)	519
$[ClF_4]^+$	800	571	385	250	475	795	515	829	385	517
$[BrF_4]^+$	723	606	385	219	—	704	419	736	369	520
$[IF_4]^+$	728	614	345	263	—	—	388	719	311	520

[a] Whereas ν_1 and ν_8 are stretching modes of equatorial bonds, ν_2 and ν_6 are stretching modes of axial bonds. For the normal modes of bending vibrations, see Ref. 521.

It should be noted that $SeCl_4$, $SeBr_4$, $TeCl_4$, and $TeBr_4$ consist of the pyramidal XY_3^+ cation and the Y^- anion in the solid state.[316,317]

Table II-6*e* lists the vibrational frequencies of tetrahedral MO_4-, MS_4-, and MSe_4-type compounds. The rules $\nu_3 > \nu_1$ and $\nu_4 > \nu_2$ hold for the majority of the compounds. It should be noted that ν_2 and ν_4 are often too close to be observed as separate bands in Raman spectra. Weinstock et al.[535] showed that, in Raman spectra, ν_2 should be stronger than ν_4, and that ν_4 is hidden by ν_2 in $[MoO_4]^{2-}$ and $[ReO_4]^-$ ions.

Baran et al.[547,548] have found several relationships between the ν_1/ν_3 ratio and the negative charge of the anion or the mass of the central atom in a series of oxoanions listed in Table II-6*e*. These are:

1. For a given central atom, the ν_1/ν_3 ratio increases as the negative charge of the anion increases (e.g., $[MnO_4]^- < [MnO_4]^{2-} < [MnO_4]^{3-}$).
2. For anions of the same negative charge with the central atom belonging to the same group of the periodic table, the ν_1/ν_3 ratio increases with the mass of the central atom (e.g., $[PO_4]^{3-} < [AsO_4]^{3-}$).
3. For isoelectronic ions in which the mass of the central atom remains approximately constant, the ν_1/ν_3 ratio increases with the increasing negative charge of the anion (e.g., $[ReO_4]^- < [WO_4]^{2-}$).

These trends are very useful in making correct assignments of the ν_1 and ν_3 vibrations of tetraoxoanions.[548]

The infrared spectrum and normal coordinate analysis of the SO_4 radical produced by the reaction of SO_3 with atomic oxygen in inert gas matrices is suggestive of the $\mathbf{C}_s$ structure[382] rather than the $\mathbf{C}_{2v}$ structure.[549a]

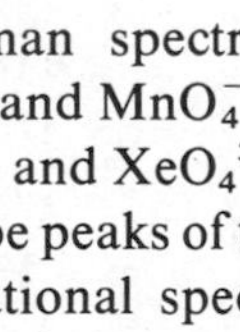

$\mathbf{C}_s$

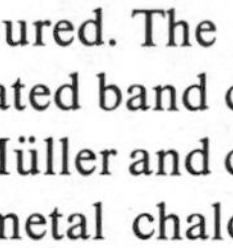

$\mathbf{C}_{2v}$

Resonance Raman spectra of highly colored ions such as CrO_4^{2-},[549] MoS_4^{2-},[550] VS_4^{3-},[551] and MnO_4^-[549] have been measured. The ν_3 (infrared) bands of gaseous RuO_4[542] and XeO_4[552] exhibit complicated band contours consisting of individual isotope peaks of the central metal. Müller and co-workers[547,553,554] reviewed the vibrational spectra of transition-metal chalcogen compounds. Basile et al.[513] carried out normal coordinate analyses of more than 60 compounds of these types.

(2) Tetrahedral ZXY_3, Z_2XY_2, and $ZWXY_2$ Molecules

If one of the Y atoms of an XY_4 molecule is replaced by a Z atom, the symmetry of the molecule is lowered to $\mathbf{C}_{3v}$. If two Y atoms are replaced, the

TABLE II-6e. VIBRATIONAL FREQUENCIES OF TETRAHEDRAL MO_4-, MS_4-, AND MSe_4-TYPE COMPOUNDS (CM^{-1})

Compound	ν_1	ν_2	ν_3	ν_4	Refs.
$[CS_4]^-$	495	353	1000, 805	—	522
$[SiO_4]^{4-}$	819	340	956	527	523, 524
$[PO_4]^{3-}$	938	420	1017	567	525, 526
$[PS_4]^{3-}$	391	282	535, 512	317, 296	527
$[NO_4]^{3-}$	843	540, 500	1012, 988	669, 651	528
$[AsO_4]^{3-}$	837	349	878	463	529
$[AsS_4]^{3-}$	386	171	419	216	529
$[SbS_4]^{3-}$	366	156	380	178	529
$[SO_4]^{2-}$	983	450	1105	611	523
$[SeO_4]^{2-}$	833	335	875	432	523
$[ClO_4]^-$	928	459	1119	625	529
$[BrO_4]^-$	801	331	878	410	530, 531
$[IO_4]^-$	791	256	853	325	532
XeO_4	775.7	267	879.2	305.9	533
$[TiO_4]^{4-}$	761	306	770	371	534
$[ZrO_4]^{4-}$	792	332	846	387	534
$[HfO_4]^{4-}$	796	325	800	379	534
$[VO_4]^{3-}$	826	336	804	(336)	535
$[VO_4]^{4-}$	818	319	780	368	534
$[VS_4]^{3-}$	404.5	193.5	470	(193.5)	536
$[VSe_4]^{3-}$	(232)	121	365	(121)	536
$[NbS_4]^{3-}$	408	163	421	(163)	536
$[NbSe_4]^{3-}$	239	100	316	(100)	536
$[TaS_4]^{3-}$	424	170	399	(170)	536
$[TaSe_4]^{3-}$	249	103	277	(103)	536
$[CrO_4]^{2-}$	846	349	890	378	535
$[CrO_4]^{3-}$	830	330	765	330	537
$[MoO_4]^{2-}$	897	317	837	(317)	535
$[MoS_4]^{2-}$	458	184	472	(184)	538
$[MoSe_4]^{2-}$	255	120	340	120	539
$[WO_4]^{2-}$	931	325	838	(325)	535
$[WS_4]^{2-}$	479	182	455	(182)	538
$[WSe_4]^{2-}$	281	107	309	(107)	536
$[MnO_4]^-$	834	346	902	386	535a
$[MnO_4]^{2-}$	812	325	820	332	534
$[MnO_4]^{3-}$	789	308	778	332	540
$[TcO_4]^-$	912	325	912	336	535
$[ReO_4]^-$	971	331	920	(331)	535
$[ReS_4]^-$	501	200	486	(200)	541
$[FeO_4]^{2-}$	832	340	790	322	534
RuO_4	885.3	~319	921	336	542

TABLE II-6e (*Continued*)

Compound	ν_1	ν_2	ν_3	ν_4	Refs.
$[RuO_4]^-$	830	339	845	312	534
$[RuO_4]^{2-}$	840	331	804	336	534
OsO_4	965.2	333.1	960.1	322.7	543
$[CoO_4]^{4-}$	670	320	633	320	544
$[B(OH)_4]^{-}$ [a]	754	379	945	533	545
$[Al(OH)_4]^{-}$ [a]	615	310	(720)	(310)	546
$[Zn(OH)_4]^{2-}$ [a]	470	300	(570)	(300)	546

[a] Only MO_4 skeletal vibrations are listed for this ion.

symmetry becomes $\mathbf{C}_{2v}$. In $ZWXY_2$ and ZWXYU types (X: central atom), the symmetry is further lowered to $\mathbf{C}_1$. As a result, the selection rules are changed as shown in Table II-6*f*. The number of infrared-active vibrations is six for ZXY_3 and eight for Z_2XY_2. Table II-6*g* lists the vibrational frequencies of ZXY_3-type molecules. The SO stretching frequency of the $[OSF_3]^+$ ion in $[OSF_3]SbF_6$ (1536 cm^{-1}) is the highest that has been observed, and corresponds to a force constant of 14.7 mdyn/Å.[570] It is also interesting to note that the structure of the $[OXeF_3]^+$ ion is $\mathbf{C}_s$[392] whereas that of the $[OXeF_3]^-$ ion is $\mathbf{C}_{2v}$[393] shown below:

$\mathbf{C}_s$ $\mathbf{C}_{2v}$

Vibrational spectra have been reported for a number of mixed halogeno complexes. Some references are as follows: $[AlCl_nBr_{4-n}]^-$ (594), SiF_nCl_{4-n} (595), $SiCl_nBr_{4-n}$ (596), and $[FeCl_nBr_{4-n}]^-$ (597). It is interesting to note that the SiFClBrI molecule exhibits the SiF, SiCl, SiBr, and SiI stretching bands at 910, 587, 486, and 333 cm^{-1}, respectively.[598] The vibrational spectrum of $OClF_3$ suggests a trigonal-bipyramidal structure which is similar to that of the $[OXeF_3]^+$ ion shown above.[599]

TABLE II-6*f*. CORRELATION TABLE FOR $\mathbf{T}_d$, $\mathbf{C}_{3v}$, $\mathbf{C}_{2v}$, AND $\mathbf{C}_1$

Point Group	ν_1	ν_2	ν_3	ν_4
$\mathbf{T}_d$	A_1(R)	E(R)	F_2(IR, R)	F_2(IR, R)
$\mathbf{C}_{3v}$	A_1(IR, R)	E(IR, R)	A_1(IR, R)+E(IR, R)	A_1(IR, R)+E(IR, R)
$\mathbf{C}_{2v}$	A_1(IR, R)	A_1(IR, R)+A_2(R)	A_1(IR, R)+B_1(IR, R) +B_2(IR, R)	A_1(IR, R)+B_1(IR, R) +B_2(IR, R)
$\mathbf{C}_1$	A(IR, R)	$2A$(IR, R)	$3A$(IR, R)	$3A$(IR, R)

TABLE II-6g. VIBRATIONAL SPECTRA OF ZXY_3 MOLECULES (cm^{-1})

C_{3v} ZXY_3	$\nu_1(A_1)$ $\nu(XY_3)$	$\nu_2(A_1)$ $\nu(XZ)$	$\nu_3(A_1)$ $\delta(XY_3)$	$\nu_4(E)$ $\nu(XY_3)$	$\nu_5(E)$ $\delta(XY_3)$	$\nu_6(E)$ $\rho_r(XY_3)$	Refs.
$[OCF_3]^-$	813	1560	595	960	422	576	555
$FCCl_3$	537.6	1080.4	351.3	847.8	242.8	395.3	556
$BrCCl_3$	419.9	730.7	246.5	785.2	290.2	188.0	556
$ICCl_3$	390	684	224	755	284	188	557
$H^{28}SiF_3$	855.8	2315.6	425.3	997.8	843.6	306.2	558
$FSiCl_3$	465	948	239	640	282	167	559
$FSiBr_3$	318	912	163	520	226	110	559
$FSiI_3$	242	894	115	424	194	71	559
$BrSiF_3$	858	505	288	940	338	200	559
$HGeCl_3$	418.4	2155.7	181.8	708.6	454	145.0	560
ONF_3	743	1691	528	883	558	400	561
$[ClPBr_3]^+$	285	587	149	500	172	120	562
$[BrPCl_3]^+$	399	582	217	657	235	159	562
OPF_3	873	1415	473	990	485	345	563
$OPCl_3$	486	1290	267	581	337	193	564
$OPBr_3$	340	1261	173	488	267	118	565, 564
$[FPO_3]^-$	1001	794	534	1125	—	382	566
SPF_3	975.3	693.4	439.2	947.0	405.4	273.7	567
$SPCl_3$	435	753	250	542	250	167	568
$SPBr_3$	299	718	165	438	179	115	569
$[OSF_3]^+$	909	1540, 1532	535	1063	508	387	570
NSF_3	772.6	1522.9	524.6	814.6	432.3	346.2	571
$ISCl_3$	482	293	242	493	140	255	572
$[HSO_3]^-$	1038	2588	629	1200	509	1123	573
$[FSO_3]^-$	1142	862	571	1302	619	424	574
$[ClSO_3]^-$	1042	381	601	1300	553	312	575
$[SSO_3]^-$	995	446	669	1123	541	335	576
$[FSeO_3]^-$	896	580, 603	392	968, 974	409	301	577
$FClO_3$	1062.6	716.3	549.7	1317.9	589.8	403.9	578
$FBrO_3$	875.2	605.0	(354)	974	(376)	(296)	579
$[NClO_3]^{2-}$	815	1256	594	870	623	457	580
OVF_3	722	1058	258	806	308	204	581
$OVCl_3$	408	1035	165	504	249	129	582
$OVBr_3$	271	1025	120	400	83	212	583
$ONbCl_3$	395	997	106	448	225	110	582
$[FCrO_3]^-$	911	635	338	955	370	261	584
$[ClCrO_3]^-$	907	438	295	954	365	209	585
$[BrCrO_3]^-$	906	395	242	948	364	200	584a
$[OMoS_3]^{2-}$	461	862	183	470	183	263	585
$[OMoSe_3]^{2-}$	293	858	120	355	120	188	585a
$[SMoO_3]^{2-}$	900	472	318	846	318	239	586
$[SMoSe_3]^{2-}$	—	471	121	342	121	—	553
$[SeMoS_3]^{2-}$	349	458	—	473	150	183	587
$[OWS_3]^{2-}$	474	878	182	451	182	264	585

TABLE II-6g (*Continued*)

C_{3v} ZXY_3	$\nu_1(A_1)$ $\nu(XY_3)$	$\nu_2(A_1)$ $\nu(XZ)$	$\nu_3(A_1)$ $\delta(XY_3)$	$\nu_4(E)$ $\nu(XY_3)$	$\nu_5(E)$ $\delta(XY_3)$	$\nu_6(E)$ $\rho_r(XY_3)$	Refs.
$[OWSe_3]^{2-}$	292	878	(120)	312	(120)	194	585a
$[SWSe_3]^{2-}$	468	281	108	311	150	108	587
$FMnO_3$	903.6	715.5	339.0	950.6	380.0	264.0	588
$ClMnO_3$	889.9	458.9	305	951.9	365	~200	586
$FTcO_3$	962	696	317	951	347	231	589
$ClTcO_3$	948	451	300	932	342	197	590
$FReO_3$	1013.2	701	305.5	978.3	346.9	234.2	591
$ClReO_3$	994	436, 427	291	963	337	192	591
$BrReO_3$	997	350	195	963	332	168	584
$[NReO_3]^{2-}$	878	1022	315	830	273	380	592
$[SReO_3]^-$	948	528	322	906	322	(240)	584
$[NOsO_3]^-$	892.5	1026.2	310.0	872.0	299.7	371.5	593

Table II-6*h* lists the vibrational frequencies of tetrahedral Z_2XY_2 molecules. The vibrational spectrum of O_2XeF_2 can be interpreted on the basis of a trigonal-bipyramidal structure in which two F atoms are axial and two O atoms and a pair of electrons are equatorial.[612] The structures of $[O_2ClF_2]^-$ [610] and $[O_2BrF_2]^-$ [611] are similar to that of O_2XeF_2, but that of $[O_2ClF_2]^+$ [620] is pseudotetrahedral. The gas-phase Raman spectrum of Cl_2TeBr_2 at 310°C

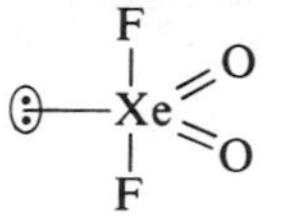

C_{2v}

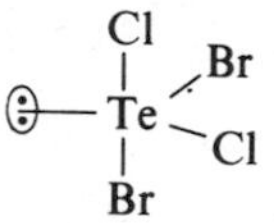

C_1

is indicative of the C_1 symmetry shown in the above diagram.[621] Table II-6*i* lists the vibrational frequencies of tetrahedral $ZWXY_2$ molecules. Other references are as follows: $SFPCl_2$ (628), $FOPCl_2$ (629), $FOPBr_2$ (630), $ClOPBr_2$ and $BrOPCl_2$ (564), $FBrSO_2$ (631), and $FClCrO_2$ (632).

(3) Square-Planar XY_4 Molecules (D_{4h})

Figure II-12 shows the seven normal modes of vibration of square-planar XY_4 molecules. Vibrations ν_3, ν_6, and ν_7 are infrared active, whereas ν_1, ν_2, and ν_4 are Raman active. Table II-6*j* lists the vibrational frequencies of some ions belonging to this group; XeF_4 (Sec. I-10) is an unusual example of a neutral molecule which takes a square-planar structure. Bosworth and Clark[638] measured the relative intensities of Raman-active fundamentals of some of these ions and calculated their bond polarizability derivatives and bond anisotropies. They[639] also measured the resonance Raman spectra of several $[AuBr_4]^-$ salts in the solid state, and observed progressions such as $n\nu_1$

TABLE II-6*h*. VIBRATIONAL FREQUENCIES OF Z_2XY_2 MOLECULES (cm^{-1})

Z_2XY_2	$\nu_1(A_1)$ $\nu(XY)$	$\nu_2(A_1)$ $\nu(XZ)$	$\nu_3(A_1)$ $\delta(XY_2)$	$\nu_4(A_1)$ $\delta(XZ_2)$	$\nu_5(A_2)$ $\rho_t(XY_2)$	$\nu_6(B_1)$ $\nu(XY)$	$\nu_7(B_1)$ $\rho_w(XY_2)$	$\nu_8(B_2)$ $\nu(XZ)$	$\nu_9(B_2)$ $\rho_r(XY_2)$	Refs.
$[F_2NH_2]^+$	2637	~1060	1543	528	—	2790	1176	1036	1487	600
$[O_2PH_2]^-$	2365	1046	1160	470	930	2308	820	1180	1093	601
F_2CI_2	605	1064	272	112	251	740	200	1110	278	602
H_2SiCl_2	942	2221	514	188	710	868	566	2221	602	603
F_2SiBr_2	414	891	270	115	187	540	241	974	257	604
H_2GeF_2	860	2155	720	(270)	(664)	814	720	2174	596	605
H_2GeCl_2	840	2132	404	163	648	772	420	2155	533	603
H_2GeBr_2	848	2122	298	104	105	757	324	2138	492	605
H_2GeI_2	821	2090	220	96	628	706	294	2110	451	605
$[Cl_2PBr_2]^+$	326	584	191	132	150	518	173	616	201	584
$[O_2PF_2]^-$	910	1179	269	567	—	962	492	1269	528	606
O_2SF_2	849	1270	385	553	—	886	544	1503	540	607
O_2SCl_2	405	1182	218	560	388	362	380	1414	282	608
$[O_2SeF_2]^-$	396	859	241	445	—	—	—	833	304	609
$[O_2ClF_2]^-$	363	1070	198	559	480	510	~337	1221	~337	610
$[O_2BrF_2]^-$	374	885	201	429	405	448	339	912	314	611
O_2XeF_2	490	845	198	333	—	578	~313	902	~313	612
$[O_2VF_2]^-$	664	970	—	330	—	631	295	962	295	613
$[O_2VCl_2]^-$	438	970	—	332	—	431	230	959	199	614
$[O_2NbS_2]^{3-}$	464	897	246	356	—	514	(297)	872	(271)	615
O_2CrF_2	727	1006	208	364	(259)	789	304	1016	274	616
O_2CrCl_2	475	995	140	356	(224)	500	257	1002	215	617
O_2MoBr_2	262	995	147	373	—	338	161	970	184	618
$[O_2{}^{92}MoS_2]^{2-}$	473	819	200	307	267	506	267	801	246	619
$[O_2MoSe_2]^{2-}$	283	864	114	339	251	353	251	834	—	615
$[O_2WS_2]^{2-}$	454	886	196	310	280	442	280	848	235	615
$[O_2WSe_2]^{2-}$	282	888	116	319	235	329	235	845	156	615

TABLE II-6*i*. VIBRATIONAL FREQUENCIES OF $ZWXY_2$ MOLECULES (CM^{-1})

$ZWXY_2$[a]	$\nu_1(A')$ $\nu_s(XY_2)$	$\nu_2(A')$ $\nu(XW)$	$\nu_3(A')$ $\nu(XZ)$	$\nu_4(A')$ $\delta_s(XY_2)$	$\nu_5(A')$ $\delta_s(WXY)$	$\nu_6(A')$ $\delta_s(ZXY_2)$	$\nu_7(A'')$ $\nu_a(XY_2)$	$\nu_8(A'')$ $\delta(ZXW)$	$\nu_9(A'')$ $\delta_a(ZXY_2)$	Refs.
$OClPF_2$	900	623	1384	(419)	(274)	412	960	274	419	622, 330
$SClPF_2$	949	549	735	394	363	209	925	252	316	622, 330
$OBrPF_2$	884	561	1380	(413)	(240)	316	947	240	413	622, 330
$SBrPF_2$	938	477	719	389	288	175	911	231	297	622, 330
$OFPCl_2$	546	907	1358	205	330	382	626	253	374	622, 330
$SFPCl_2$	479	915	750	193	328	268	574	193	317	622, 330
$OFPBr_2$	472	888	1337	133	273	304	536	220	290	622, 330
$SFPBr_2$	377	887	713	129	274	218	470	162	254	622, 330
$OBrPCl_2$	545	432	1285	242	172	285	580	161	327	623
$SBrPCl_2$	493	372	743	(230)	150	206	536	150	230	623
$OClPBr_2$	391	552	1275	130	209	291	492	157	271	623
$SClPBr_2$	333	500	729	121	196	190	436	136	205	623
$[OSeMoS_2]^{2-}$	478[b]	355	869	190	—	273	467[b]	—	273	624, 625
$[OSeWS_2]^{2-}$	473	320	879	190	—	265[b]	458	—	255[b]	626, 625
$[OSMoSe_2]^{2-}$	360[b]	461	865	—	—	—	320[b]	—	—	627
$[OSWSe_2]^{2-}$	317[b]	459	882	—	—	—	312[b]	—	—	627

[a] X denotes the central atom.
[b] These assignments may be interchanged.

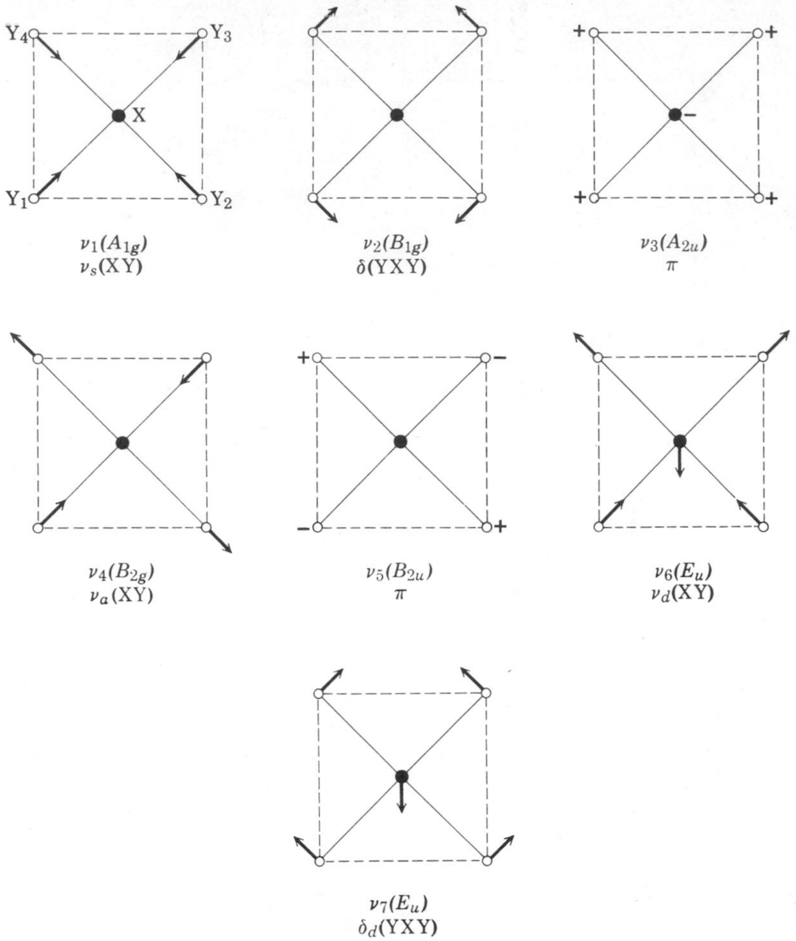

Fig. II-12. Normal modes of vibration of square-planar XY_4 molecules.

($n=1$-9) and $\nu_2+n\nu_1$ ($n=1$-5). For normal coordinate analyses of square-planar XY_4 molecules, see Refs. 634, 640, and 641.

II-7. TRIGONAL-BIPYRAMIDAL AND TETRAGONAL-PYRAMIDAL XY_5 AND RELATED MOLECULES

An XY_5 molecule may be a trigonal bipyramid ($\mathbf{D}_{3h}$) or a tetragonal pyramid ($\mathbf{C}_{4v}$). If it is trigonal-bipyramidal, only two stretching vibrations (A_2'' and E') are infrared active. If it is tetragonal-pyramidal, three stretching vibrations (two A_1 and E) are infrared active. As discussed in Sec. I-10, however, it is not always possible to make clear-cut distinctions of these structures based on selection rules since practical difficulties arise in counting the number of fundamental vibrations in infrared and Raman spectra.

TABLE II-6*j*. VIBRATIONAL FREQUENCIES OF SQUARE-PLANAR XY_4 MOLECULES (CM^{-1})[a]

XY_4	$\nu_1(A_{1g})$ $\nu_s(XY)$	$\nu_2(B_{1g})$ $\delta(XY_2)$	$\nu_3(A_{2u})$ π	$\nu_4(B_{2g})$ $\nu_a(XY)$	$\nu_6(E_u)$ $\nu_d(XY)$	$\nu_7(E_u)$ $\delta_d(XY_2)$	Refs.
$[ClF_4]^-$	505	288	425	417	680–500	—	633
$[BrF_4]^-$	523	246	317	449	580–410	(194)	633
$[ICl_4]^-$	288	128	—	261	266	—	634
XeF_4	554.3	218	291	524	586	(161)	635
$[AuCl_4]^-$	347	171	—	324	350	179	636
$[AuBr_4]^-$	212	102	—	196	252[b]	~110[b]	636
$[AuI_4]^-$	148	75	—	110	192	113	636
$[PdCl_4]^{2-}$	303	164	150	275	321	161	637
$[PdBr_4]^{2-}$	188	102	114	172	243	104	636, 637
$[PtCl_4]^{2-}$	330	171	147	312	313	165	637
$[PtBr_4]^{2-}$	208	106	105	194	227	112	637
$[PtI_4]^{2-}$	155	85	105	142	180	127	636, 637

[a] For these molecules ν_5 is inactive. The designations B_{1g} and B_{2g} may be interchanged, depending on the definition of symmetry axes involved.
[b] From Ref. 639.

(1) Trigonal-Bipyramidal XY_5 Molecules (D_{3h})

Figure II-13 shows the eight normal vibrations of a trigonal-bipyramidal XY_5 molecule. Six of these eight (A_1', E', and E'') are Raman active and five (A_2'' and E') are infrared active. Three stretching vibrations (ν_1, ν_2, and ν_5) are allowed in the Raman, whereas two (ν_3 and ν_5) are allowed in the infrared. Table II-7*a* lists the observed frequencies and band assignments of trigonal-bipyramidal XY_5 molecules.

The majority of the compounds sho
All the data for $[XY_5]^{n-}$ were obtaine
where the ions are monomeric. Most
polymerized in the condensed phases.
$TaCl_5$,[659] and WCl_5[660] are dimeric in t
NbF_5 and TaF_5 are known to be tetra

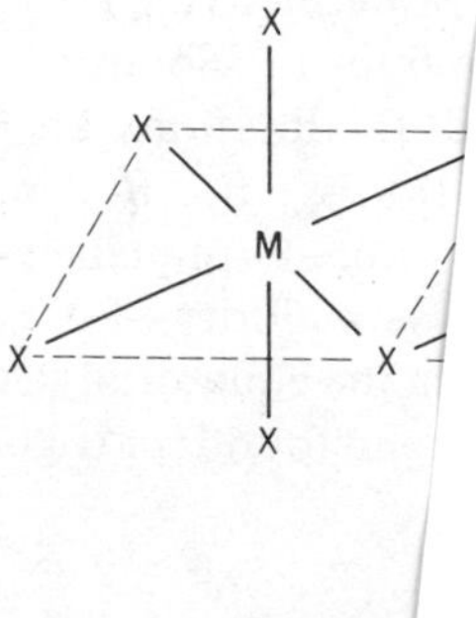

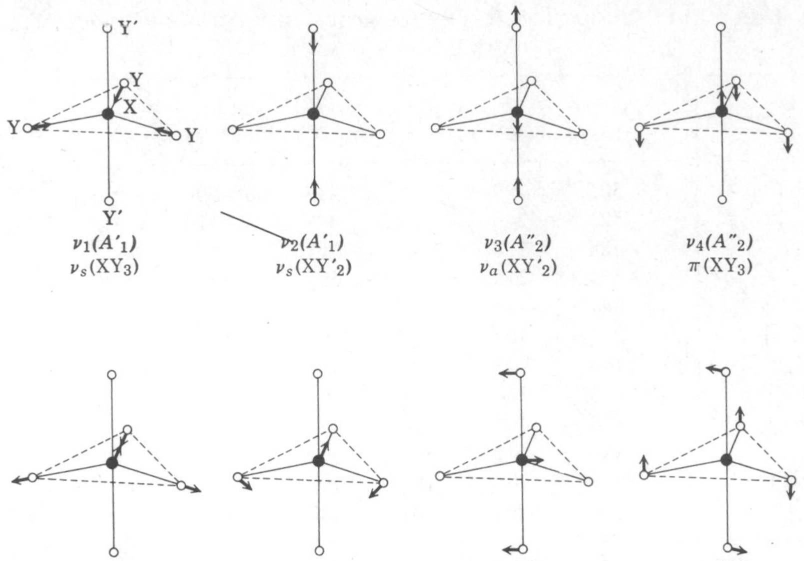

Fig. II-13. Normal modes of vibration of trigonal-bipyramidal XY_5 molecules.

these molecules are polymerized even in the gaseous phase. For example, SbF_5 is monomeric ($\mathbf{D}_{3h}$) at 350°C but polymeric at 140°C in the gaseous phase,[662] and NbF_5 and TaF_5 are polymeric in the gaseous phase if the temperature is below 350°C.[663] Although PCl_5 exists as a $\mathbf{D}_{3h}$ molecule in the gaseous and liquid states, it has an ionic structure consisting of $[PCl_4]^+[PCl_6]^-$ units in the crystalline state, as proved by Raman spectroscopy.[664] The importance of the ν_7 vibration in the intramolecular conversion of pentacoordinate molecules has been discussed by Holmes.[665]

Normal coordinate analyses on trigonal-bipyramidal XY_5 molecules[666-668] have been carried out by several investigators. These calculations show that equatorial bonds are stronger than axial bonds. Vibrational spectra of mixed halogeno compounds such as PF_nCl_{5-n}[669] and PF_3X_2 (X = Cl and Br)[670] have been assigned. The structure of OSF_4 is trigonal-bipyramidal with two fluorines and one oxygen occupying the three equatorial positions ($\mathbf{C}_{2v}$).[671]

etragonal-Pyramidal XY_5 and ZXY_4 Molecules ($\mathbf{C}_{4v}$)

e II-14 shows the nine normal modes of vibration of a tetragonal-
ZXY_4 molecule. Only A_1 and E vibrations are infrared active,
ne vibrations belonging to the A_1, B_1, B_2, and E species are
able II-7*b* lists the vibrational frequencies of tetragonal-
ZXY_4 molecules. In the majority of XY_5 molecules, the
cy (ν_1) is higher than the equatorial stretching frequen-
pposite to the trend found for trigonal-bipyramidal

TABLE II-7a. VIBRATIONAL FREQUENCIES OF TRIGONAL-BIPYRAMIDAL XY_5 MOLECULES (cm^{-1})

Molecule	Phase	ν_1	ν_2	ν_3	ν_4	ν_5	ν_6	ν_7	ν_8	Refs.
$[SiF_5]^-$	Sol'n[a]	708	519	785	481	874	449	—	—	642
$[SiCl_5]^-$	Sol'n[a]	372	—	395	271	550	250	—	—	643
$[GeCl_5]^-$	Solid	348	236	310	200	395	200	—	—	644
$[SnCl_5]^-$	Solid	340	—	314	160	350	150	66	169	645
$[SnBr_5]^-$	Solid	—	—	208	106	256	111	—	—	646
PF_5	Gas	817	640	944	575	1026	532	300	514	647
PCl_5	Sol'n[a]	394	385	444	299	580	278	98	261	648
AsF_5	Gas	733	642	784	400	809	366	123	388	649
$AsCl_5$	Sol'n[a]	369	295	385	184	437	220	83	213	650
$SbCl_5$	Gas	355	309	—	—	400	173	58	120	651
$[CuCl_5]^{3-}$	Solid	260	—	268	—	170[b]	95[c]	—	—	652
$[CdCl_5]^{3-}$	Solid	251	—	236	—	157[b]	98[c]	—	—	652
$[TiCl_5]^-$	Sol'n[a]	348	302	355	178	411	190	66	166	653
$[TiCl_5]^-$	Solid	—	—	346	170	385	212	(83)	—	646
VF_5	Gas	719	608	784	331	810	282	(200)	350	654
$NbCl_5$	Matrix	(349)	(293)	396	126	444	159	99	(139)	655
$NbBr_5$	Gas	234	178	—	—	—	119	67	101	651
$TaCl_5$	Gas	406	324	—	—	—	181	54	127	651
$TaBr_5$	Gas	240	182	—	—	—	110	70	93	651
MoF_5	Matrix	—	—	683	—	713	261	112	—	656
$MoCl_5$	Gas	390	313	—	—	418	200	100	175	651

[a] Nonaqueous solution.
[b] May be assigned to ν_7.
[c] May be assigned to ν_8.

XY_5 molecules, discussed in the preceding section. For normal coordinate analyses on these compounds, see Refs. 675, 676, 688, and 689.

II-8. OCTAHEDRAL MOLECULES

(1) Octahedral XY_6 Molecules (O_h)

Figure II-15 illustrates the six normal modes of vibration of an octahedral XY_6 molecule. Vibrations ν_1, ν_2, and ν_5 are Raman active, whereas only ν_3 and ν_4 are infrared active. Since ν_6 is inactive in both, its frequency is estimated from an analysis of combination and overtone bands.

Table II-8a lists the vibrational frequencies of a number of hexahalogeno compounds. In general, the order of the stretching frequencies is $\nu_1 > \nu_3 \gg \nu_2$ or $\nu_1 < \nu_3 \gg \nu_2$, depending on the compound. The order of the bending frequencies is $\nu_4 > \nu_5 > \nu_6$ in most cases. In the same family of the periodic table, the stretching frequencies decrease as the mass of the central atom increases, for

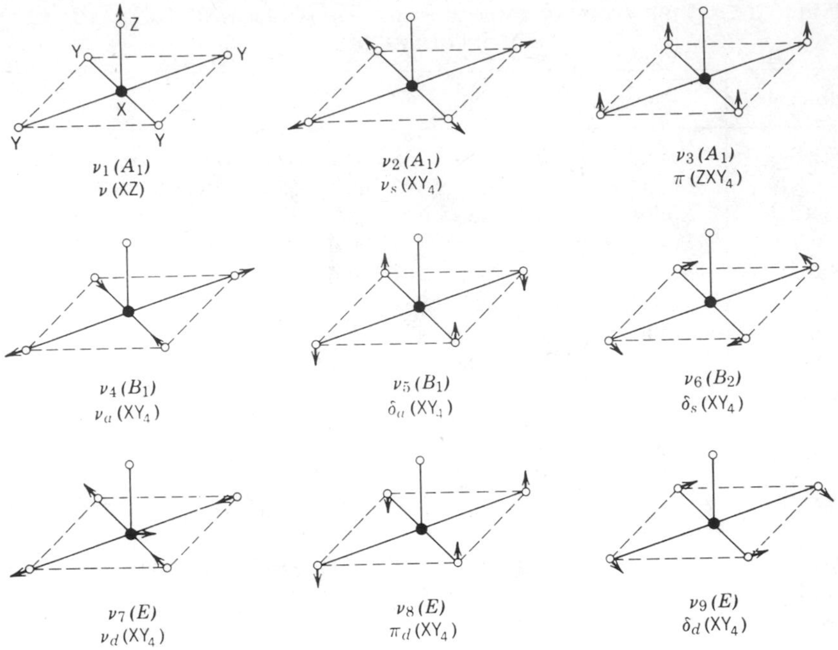

Fig. II-14. Normal modes of vibration of tetragonal-pyramidal ZXY_4 molecules.

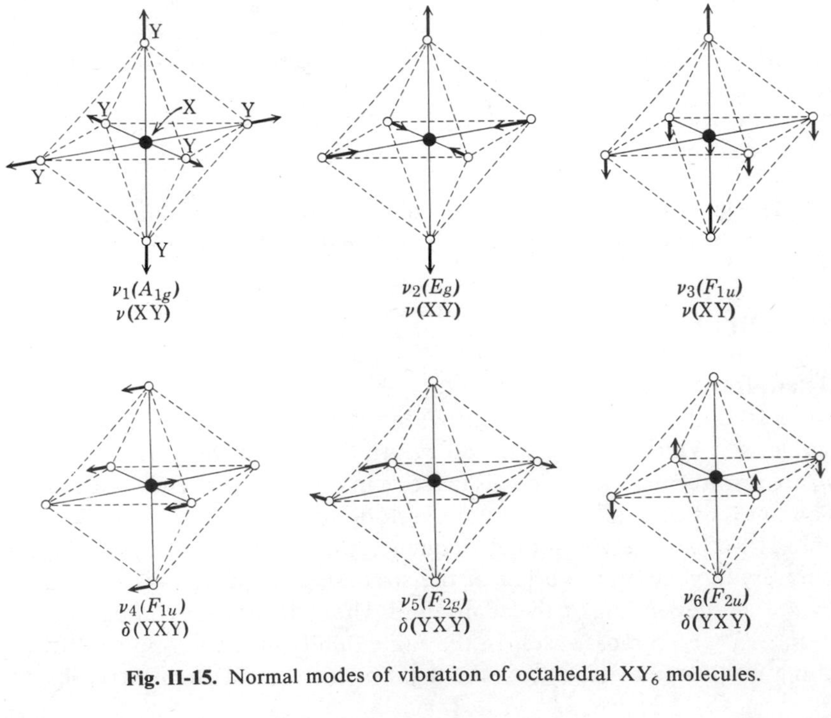

Fig. II-15. Normal modes of vibration of octahedral XY_6 molecules.

TABLE II-7*b*. VIBRATIONAL FREQUENCIES OF TETRAGONAL-PYRAMIDAL XY_5 AND ZXY_4 MOLECULES (CM^{-1})

XY_5 or ZXY_4	ν_1	ν_2	ν_3	ν_4	ν_5	ν_6	ν_7	ν_8	ν_9	Refs.
$[InCl_5]^{2-}$	294	283	140	287	193	165	274	108	143	672
$[SbF_5]^{2-}$	557	427	278	388	—	220	375, 347	307	142	673
$[SbCl_5]^-$	445	285	180	420	—	117	300	255	90	674
$[SF_5]^-$	796	522	469	(435)	269	342	590	(435)	241	675
$[SeF_5]^-$	666	515	332	460	236	282	480	399	202	675
$[TeF_5]^-$	624	517	291	579	—	243	488	345	—	673
ClF_5	709	538	480	480	(346)	375	732	—	296	676, 677
BrF_5	682	570	365	535	(281)	312	644	414	237	676, 678
IF_5	698	593	315	575	(257)	273	640	374	189	673, 676
NbF_5	740	686	513	—	—	—	729	261	103	679
$[OTeF_4]^{2-}$	837	461	—	390	—	190	335	—	129	680
$[OClF_4]^-$	1216	462	339	350	—	283	600, 550	415, 394	213	681
$[OBrF_4]^-$	932	525	312	459	236	248	506 } 483 }	421 } 407 } 398 }	194 164	682
$[OIF_4]^-$	888	533	273	475	—	214	485	365	124	683
$OXeF_4$	920	567	285	527	(230)	233	608	365	161	676
$OMoF_4$	1048	714	264	—	—	—	720	294	236	684, 685
$OMoCl_4$	1015	450	143	400	148	220	396	256	172	686, 687
$[OMoCl_4]^-$	1008	354	184	327	158	167	364	240	114	688
OWF_4	1055	733	248	631	328	291	698	298	236	684, 685
$[NRuCl_4]^-$	1092	346	197	304	154	172	378	267	163	688
$[NRuBr_4]^-$	1088	224	156	187	103	128	304	211	98	688
$[NOsCl_4]^-$	1123	358	184	352	149	174	365	271	132	688
$[NOsBr_4]^-$	1119	162	122	156	110	120	220	273	98	688

TABLE II-8a. VIBRATIONAL FREQUENCIES OF OCTAHEDRAL XY_6 MOLECULES (cm^{-1})

Molecule	ν_1	ν_2	ν_3	ν_4	ν_5	ν_6 [a]	Refs.
$[AlF_6]^{3-}$	541	(400)	568	387	322	(228)	690, 691
$[GaF_6]^{3-}$	535	(398)	481	298	281	(198)	690, 691
$[InF_6]^{3-}$	497	(395)	447	226	229	(162)	690, 691
$[InCl_6]^{3-}$	277	193	250	157	(149)	—	692
$[TlF_6]^{3-}$	478	387	412	202	209	(148)	690, 691
$[TlCl_6]^{3-}$	280	262	294	222, 246	155	(136)	693
$[TlBr_6]^{3-}$	161	153	190, 195	134, 156	95	(80)	693
$[SiF_6]^{2-}$	663	477	741	483	408	—	694
$[GeF_6]^{2-}$	624	471	603	339, 359	335	—	694
$[GeCl_6]^{2-}$	318	213	310	213	191	—	695
$[SnF_6]^{2-}$	592	477	559	300	252	—	694
$[SnCl_6]^{2-}$	311	229	303	166	158	—	696
$[SnBr_6]^{2-}$	190	144	224	118	109	—	697
$[SnI_6]^{2-}$	122	93	161	84	78	—	698
$[PbCl_6]^{2-}$	281	209	262	142	139	—	696
$[PF_6]^{-}$	756	585, 570	865, 835	559, 530	480 ~ 468	—	699
$[PCl_6]^{-}$	360	283	444	285	238	—	695
$[AsF_6]^{-}$	689	573	700	385	375	(252)	694
$[AsCl_6]^{-}$	337	289	333	220	202	—	695
$[SbF_6]^{-}$	668	558	669	350	294	—	694
$[SbCl_6]^{-}$	330	282	353	180	175	—	700
$[SbCl_6]^{3-}$	327	274	—	—	137	—	701
$[SbBr_6]^{-}$	192	169	239, 224	119	103, 78	—	702
$[SbBr_6]^{3-}$	180	153	180	107	73	—	703, 704
$[SbI_6]^{3-}$	107	96	108	82	54	—	703, 704
$[BiF_6]^{-}$	579	526	—	—	231/241	—	705
$[BiCl_6]^{3-}$	259	215	172	130	115	—	703, 704
$[BiBr_6]^{3-}$	156	130	128	75	62	—	703, 704
$[BiI_6]^{3-}$	114	103	96	(59)	54	—	703, 704
SF_6	775	643	939	614	524	(347)	706-708
SeF_6	708	658	780	437	403	(264)	706, 707
$[SeCl_6]^{2-}$	299	255	280	160-140	165	—	692
$[SeBr_6]^{2-}$	179	157	225	122	105	—	709
TeF_6	698	672	752	325	312	(197)	706, 707
$[TeCl_6]^{2-}$	298	250	250	136	140	—	710
$[TeCl_6]^{3-}$	264	192	230	146	(135)	—	692
$[TeBr_6]^{2-}$	174	153	198	—	75	—	710
$[ClF_6]^{+}$	679	630	890	582	513	—	711
$[BrF_6]^{+}$	658	660	—	—	405	—	712
$[BrF_6]^{-}$	568	454	400	204, 184	250	(138)	713
$[AuF_6]^{-}$	595	530	—	—	225	—	714
$[ScF_6]^{3-}$	498	390	176	90	230	—	715
$[YF_6]^{3-}$	476	382	160	74	194	—	715
$[LaF_6]^{3-}$	443	334	130	63	171	—	715

TABLE II-8*a* (*Continued*)

Molecule	ν_1	ν_2	ν_3	ν_4	ν_5	ν_6 [a]	Refs.
$[GaF_6]^{3-}$	473	380	140	72	185	—	715
$[YbF_6]^{3-}$	491	370	156	70	196	—	715
$[CeCl_6]^{2-}$	295	265	268	117	120	(86)	716
$[TiF_6]^{2-}$	618	(440)	615, 600	315, 281	308, 300	—	717
$[TiCl_6]^{2-}$	320	271	316	183	173	—	718
$[TiCl_6]^{3-}$	322	278	304, 290	—	175	—	719
$[TiBr_6]^{2-}$	192	—	244	119	115	—	718
$[ZrF_6]^{2-}$	589	(416)	537, 522	241, 192	258, 244	—	720
$[ZrCl_6]^{2-}$	327	237	290	150	153	—	721
$[ZrBr_6]^{2-}$	194	144	223	106	99	—	722
$[HfF_6]^{2-}$	572	(389)	448, 490	217, 184	259, 247	—	717
$[HfCl_6]^{2-}$	326	257	275	145	156	(80)	696
$[HfBr_6]^{2-}$	197	142	189	102	101	—	722
$[VF_6]^{-}$	676	538	646	300	330	—	723
$[VF_6]^{2-}$	584	—	578	273	—	—	723
$[VF_6]^{3-}$	533	—	511	292	—	—	723
$[VCl_6]^{2-}$	—	—	355, 305	—	—	—	724
$[NbF_6]^{-}$	683	562	602	244	280	—	725
$[NbCl_6]^{-}$	368	288	333	162	183	—	696
$[NbCl_6]^{2-}$	—	—	314	165	—	—	726
$[NbBr_6]^{-}$	—	—	240 ~ 216	—	—	—	727
$[NbBr_6]^{2-}$	—	—	236	112	—	—	726
$[NbI_6]^{-}$	—	—	180	70, 66	—	—	728
$[TaF_6]^{-}$	692	581	560	240	272	(192)	729
$[TaCl_6]^{-}$	378	298	330	158	180	—	696
$[TaCl_6]^{2-}$	—	—	297	160	—	—	696
$[TaBr_6]^{-}$	—	—	234 ~ 223	—	—	—	727
$[TaBr_6]^{2-}$	—	—	217	109	—	—	726
$[TaI_6]^{-}$	—	—	160	80	—	—	728
CrF_6	(720)	(650)	790	(266)	(309)	(110)	730
$[CrCl_6]^{3-}$	286	237	315	199	162	182	731
$^{92}MoF_6$	741.8	652.0	749.5	265.7	317	117	732
$[MoF_6]^{2-}$	685	598	653	250	274	—	733
$[MoCl_6]^{-}$	356	—	327	162	—	—	734
$[MoCl_6]^{2-}$	329	—	308	168	154	—	734
$[MoCl_6]^{3-}$	305	—	268, 286, 302	167, 187	150	—	734
WF_6	770	676	711	258	321	(127)	706, 707
WCl_6	437	331	373	160	182	—	696
$[WCl_6]^{-}$	382	—	332, 312	157	168	—	734
$[WCl_6]^{2-}$	341	—	293	166, 150	—	—	734

TABLE II-8a (*Continued*)

Molecule	ν_1	ν_2	ν_3	ν_4	ν_5	ν_6 [a]	Refs.
$[MnF_6]^{2-}$	592	508	620	335	308	—	735
TcF_6	713	(639)	748	275	(297)	(145)	707
$[ReF_6]^+$	797	734	783	353	359	—	736
ReF_6	754	(671)	715	257	(295)	(147)	707
$[ReF_6]^{2-}$	611	530	535	249	221	(181)	737
$[ReCl_6]^-$	—	—	318	161	—	—	738
$[ReCl_6]^{2-}$	346	(275)	313	172	159	—	696, 739
$[ReBr_6]^{2-}$	213	(174)	217	118	104	—	740
$[NiF_6]^{2-}$	555	512	648	332	307, 298	—	741
$[PdCl_6]^{2-}$	318	289	346	200	178	—	742
$[PdBr_6]^{2-}$	198	176	253	130	100	—	743
PtF_6	656	(601)	705	273	(242)	(211)	707
$[PtF_6]^{2-}$	611	576	571	281	210	(143)	744
$[PtCl_6]^{2-}$	348	318	342	183	171	(88)	697
$[PtBr_6]^{2-}$	213	190	243	146	137	—	742
$[PtI_6]^{2-}$	—	—	186	46	—	—	745
$[FeF_6]^{3-}$	538	374	—	—	253	—	746
RuF_6	(675)	(624)	735	275	(283)	(186)	707
$[RuCl_6]^{2-}$	—	—	346	188	—	—	747
OsF_6	731	(668)	720	268	(276)	(205)	707
$[OsCl_6]^{2-}$	352	—	304	174	177	—	739, 748
$[OsBr_6]^{2-}$	218	162	211	—	—	—	739, 748
RhF_6	(634)	(595)	724	283	(269)	(192)	707
$[RhCl_6]^{2-}$	—	—	329	187	—	—	749
IrF_6	702	645	719	276	267	(206)	707
$[IrCl_6]^{2-}$	352	(225)	333	184	190	—	696
$[IrCl_6]^{3-}$	—	—	296	200	—	—	745
$[IrBr_6]^{2-}$	—	—	235	82	—	—	749
$[ThCl_6]^{2-}$	294	255	259	—	114	—	750
UF_6	666	530	619	184	200	—	751
$[UF_6]^-$	—	—	525	173	—	—	752
$[UCl_6]^-$	343	273	310	122	136	—	752, 753
$[UCl_6]^{2-}$	299	237	262	114	121	(80)	750
$[UBr_6]^-$	—	—	214	87	—	—	752
NpF_6	654	535	624	199	208	(164)	707
$[NpCl_6]^{2-}$	310	—	265	117	128	—	754
PuF_6	(628)	(523)	616	206	(211)	(173)	707

[a] The value of ν_6 can also be estimated by the relation $\nu_6 = \nu_5/\sqrt{2}$ (Refs. 755 and 756).

example:

	$[AlF_6]^{3-}$		$[GaF_6]^{3-}$		$[InF_6]^{3-}$		$[TlF_6]^{3-}$
$\tilde{\nu}_1$ (cm^{-1})	541	>	535	>	497	>	478
$\tilde{\nu}_3$ (cm^{-1})	568	>	481	>	447	>	412

The trend in ν_1 directly reflects the trend in the stretching force constant (and bond strength) since the central atom is not moving in this mode. In ν_3, however, both X and Y atoms are moving, and the mass effect of the X atom cannot be ignored completely. Across the periodic table, the stretching frequencies increase as the oxidation state of the central atom becomes higher. Thus we have:

	$[AlF_6]^{3-}$		$[SiF_6]^{2-}$		$[PF_6]^{-}$		SF_6
$\tilde{\nu}_1$ (cm^{-1})	541	<	663	<	745	<	774
$\tilde{\nu}_3$ (cm^{-1})	568	<	741	<	840	<	939

The effect of lowering the oxidation state is clearly seen in a series such as $[VF_6]^{n-}$ ($n = 1$, 2, and 3) and $[WCl_6]^{n-}$ ($n = 0$, 1, and 2):

	$[VF_6]^{-}$		$[VF_6]^{2-}$		$[VF_6]^{3-}$
$\tilde{\nu}_1$ (cm^{-1})	676	>	584	>	533
$\tilde{\nu}_3$ (cm^{-1})	646	>	578	>	511

As in many other cases, the higher the oxidation state, the higher the frequency. The bending frequencies do not exhibit clear-cut trends. The effect of changing the halogen is seen in a number of series, for example:

	$[SnF_6]^{2-}$		$[SnCl_6]^{2-}$		$[SnBr_6]^{2-}$		$[SnI_6]^{2-}$
$\tilde{\nu}_1$ (cm^{-1})	592	>	311	>	190	>	127
$\tilde{\nu}_3$ (cm^{-1})	559	>	303	>	224	>	161

The stretching force constants also follow the same order. The ratios ν(MBr)/ν(MCl) and ν(MI)/ν(MCl) are about 0.61 and 0.42, respectively, for ν_1, and about 0.76 and 0.62, respectively, for ν_3.

In the $[MCl_6]^{3-}$ series, ν_3 and ν_4 change as follows:

	$Cr^{3+}(d^3)$	$Mn^{3+}(d^4)$	$Fe^{3+}(d^5)$	In^{3+}
$\tilde{\nu}_3$ (cm^{-1})	315	342	248	248
$\tilde{\nu}_4$ (cm^{-1})	200	183	184	161

All these metals are in the high-spin states. For $Fe^{3+}(t_{2g}^3e_g^2)$, occupation of the antibonding orbitals lowers ν_3 drastically in relation to the Cr^{3+} complex; its ν_3 is comparable to that of the In^{3+} complex, whose ν_3 is lowered because of the increased mass of the metal. On the other hand, the ν_3 of $[MnCl_6]^{3-}$ is higher than that of $[CrCl_6]^{3-}$ because the static Jahn–Teller effect of the Mn^{3+} ion causes a tetragonal distortion.[716] Table II-8*a* includes several XY_6-type ions which exhibit splitting of degenerate vibrations due to lowering of symmetry in the crystalline state.

Coriolis coupling constants have been calculated from the vapor-phase Raman spectra of some XY_6-type molecules.[706] The Raman intensity of an XY_6 molecule normally follows the order $I(\nu_1) > I(\nu_2) > I(\nu_5)$. Adams and Downs[757] noted that $I(\nu_2)/I(\nu_1)$ is 0.5–1 for $[TeCl_6]^{2-}$ and $[TeBr_6]^{2-}$, although it normally ranges from 0.05 to 0.1. Furthermore, they perceived ν_3, which is not allowed in Raman spectra. From these and other items of evidence, they proposed that the $\mathbf{O}_h$ selection rule breaks down in $[TeX_6]^{2-}$ because less symmetrical electronic excited states perturb the $\mathbf{O}_h$ ground state. They also noted the distortion of $[SbX_6]^{3-}$ ions to $\mathbf{C}_{3v}$ symmetry from their Raman spectra in solution. Woodward and Creighton[758] noted that $I(\nu_2) > I(\nu_1)$ holds in the aqueous Raman spectra of Na_2PtX_6 (X = Cl and Br) and Na_2PdCl_6, and attributed this unusual trend to the presence of six nonbonding *d*-electrons in the valence shell. Bosworth and Clark[701] carried out a Raman study on 17 $[XY_6]^{n-}$-type metal halides, and discussed the results in terms of the preresonance Raman effect. Hamaguchi et al.[759] observed anomalous polarization (Sec. I-21) for all the Raman bands of the $IrCl_6{}^{2-}$ ion measured under resonance conditions. It was suggested that Jahn–Teller effect in electronic excited states is responsible for this anomaly.

Weinstock et al.[760] noted that the combination bands $(\nu_1+\nu_3)$ and $(\nu_2+\nu_3)$ appear with similar frequencies, intensities, and shapes in the infrared spectra of $MoF_6(d^0)$ and $RhF_6(d^3)$. As is shown in Fig. II-16, however, $(\nu_2+\nu_3)$ was very broad and weak in $TcF_6(d^1)$, $ReF_6(d^1)$, $RuF_6(d^2)$, and $OsF_6(d^2)$. This anomaly was attributed to a dynamic Jahn–Teller effect. The static Jahn–Teller effect does not seem to operate in these compounds since no splittings of the triply degenerate fundamentals were observed. Perhaps the most fascinating

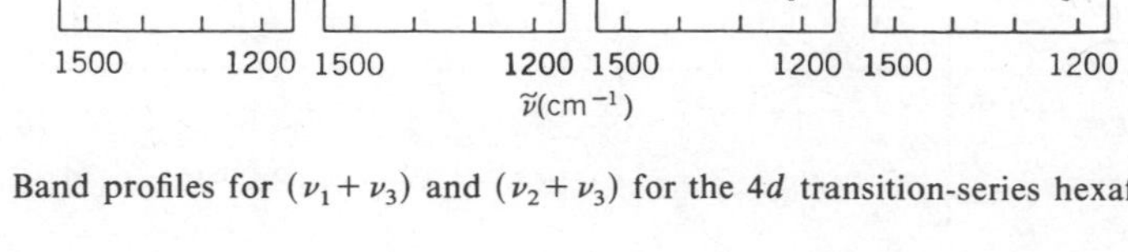

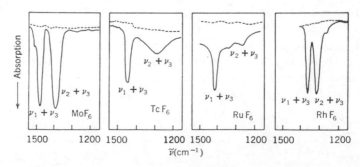

Fig. II-16. Band profiles for $(\nu_1+\nu_3)$ and $(\nu_2+\nu_3)$ for the 4*d* transition-series hexafluorides.[760]

XY_6-type molecule is XeF_6. In their earlier work, Claassen et al.[761] suggested the distortion of XeF_6 from $\mathbf{O}_h$ symmetry since they observed two stretching bands in infrared and three stretching bands in Raman spectra. It was not possible, however, to determine the precise structure of XeF_6 until they[762] carried out a detailed infrared, Raman and electronic spectral study of XeF_6 vapor as a function of temperature. They were then able to show that XeF_6 consists of the three electronic isomers shown in Fig. II-17, and to explain subtle differences in spectra at different temperatures as a shift of equilibrium among these three isomers. As expected, the ^{235}U-^{238}U isotope shift of the ν_3 band of gaseous UF_6 is only 0.65 ± 0.1 cm^{-1}.[763]

The ^{35}Cl NQR spectra provide information about the σ and π contributions to the covalent M–Cl bonding in $[MCl_6]^{n-}$-type ions, and these can be correlated with the force constants obtained from infrared and Raman studies.[696] Both infrared and NQR spectra suggest low-site symmetry of the $[MCl_6]^{3-}$ ion in $K_3[MCl_6]\cdot H_2O$ (M = Ir and Rh) crystals.[764]

Normal coordinate analyses on octahedral hexahalogeno compounds have been made by a number of investigators. Kim et al.[765] calculated the force constants of 15 metal fluorides using the UBF and OVF fields (see Sec. I-13), and found that the latter is better than the former. LaBonville et al.[766] calculated the force constants of 62 metal halides by using the UBF, OVF, GVF, modified UBF, and modified OVF fields, and found that the modified OVF field gives the best overall agreement with the observed frequencies. They also discussed the dependence of force constants on the mass of the halogen, the oxidation state of the metal, the number of nonbonding electrons in the valence shell, and the crystal-field stabilization energy.

Table II-8*b* lists the vibrational frequencies of MO_6-type ions. Hauck and Fadini calculated the force constants of these ions.[767,768]

(2) Octahedral XY_nZ_{6-n} Molecules

The XY_5Z molecule belongs to the $\mathbf{C}_{4v}$ point group, and its 11 normal vibrations are classified into $4A_1$, $2B_1$, B_2, and $4E$ modes, of which only A_1

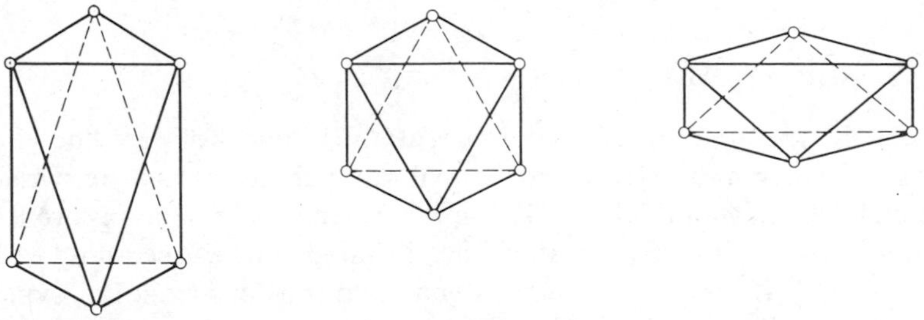

Prolate $\mathbf{D}_{3d}$	Regular $\mathbf{O}_h$	Oblate $\mathbf{D}_{3d}$
599 (A_{2u})	619 (F_{1u})	630 (A_{2u})
520 (E_u)		507 (E_u)

Fig. II-17. Structures of three isomers of XeF_6 and their IR-active XeF stretching frequencies (cm^{-1}). The Xe atom at the center is not shown.

TABLE II-8*b*. VIBRATIONAL FREQUENCIES OF OCTAHEDRAL MO_6 MOLECULES (cm^{-1})

Compound	ν_1	ν_2	ν_3	ν_4	ν_5	Refs.
$Li_6[TeO_6]$	700	540	640	470	355	767
$Li_6[WO_6]$	740	450	620	425	360	767
α-$Li_6[ReO_6]$	680	505	620	425	360	767
$Ca_4[PtO_6]$	—	530	575	425	345	768
α-$Li_6[TeO_6]$	700	540	640	470	355	768
$Ca_5[IO_6]_2$	771	490, 538	765, 695	451, 435	—	768

and E are infrared active; all are Raman active. Table II-8*c* lists the observed frequencies of XY_5Z-type molecules.

The XY_4Z_2 molecule may be *cis* ($\mathbf{C}_{2v}$) or *trans* ($\mathbf{D}_{4h}$). The *cis*-isomer is expected to give four XY stretching ($2A+B_1+B_2$) and two XZ stretching (A_1+B_1) modes, all of which are infrared as well as Raman active. The *trans*-isomer is expected to give three XY stretching ($A_{1g}+B_{1g}+E_u$) and two XZ stretching ($A_{1g}+A_{2u}$) modes, of which E_u and A_{2u} are infrared active and A_{1g} and B_{1g} are Raman active. The selection rules for other XY_nZ_{6-n} molecules are tabulated in Appendix III. These selection rules can be used to distinguish the structures of stereo isomers on the basis of their vibrational spectra. For example, Clark et al.[718] measured the infrared and Raman spectra of mixed halogeno ions of the $[MX_4Y_2]^{2-}$-type (M = Ti and Sn; X = Cl, Br, and I) and concluded that most and probably all of these ions take the *cis* structure in the solid state.

Preetz and co-workers isolated a series of mixed-halogeno complexes of the $[MX_nY_{6-n}]^{2-}$ ($n=1\sim5$) type and assigned their infrared and Raman spectra. Their compounds include: $[OsF_nCl_{6-n}]^{2-}$ (Ref. 786), $[IrF_nCl_{6-n}]^{2-}$ (Ref. 787), $[PtCl_nBr_{6-n}]^{2-}$ (Ref. 788), and $[MF_nCl_{6-n}]^{2-}$ (M = Pt and Ir, Ref. 789).

II-9. XY_7 AND XY_8 MOLECULES

The XY_7-type molecules are very rare. Both IF_7 and ReF_7 are known to be pentagonal-bipyramidal ($\mathbf{D}_{5h}$), and their vibrational spectra have been assigned completely, as shown in Table II-9. The A_1', E_1'', and E_2' vibrations are Raman active, and the A_2'' and E_1' vibrations are infrared active. According to Eysel and Seppelt,[791] IF_7 undergoes minor dynamic distortions from $\mathbf{D}_{5h}$ symmetry which cause violation of the $\mathbf{D}_{5h}$ selection rules for combination bands but not for the fundamentals. Normal coordinate analysis[791] shows that the axial bonds are definitely stronger and shorter than the equatorial ones. Table II-9 also lists the frequencies and assignments of the $[UO_2F_5]^{2-}$ ion of $\mathbf{D}_{5h}$ symmetry.[792]

TABLE II-8c. VIBRATIONAL FREQUENCIES OF OCTAHEDRAL XY_5Z AND XY_4WZ MOLECULES (CM^{-1})

XY_5Z or XY_4WZ	$\nu_1(A_1)$ $\nu(XZ)$	$\nu_2(A_1)$ $\nu(XW)^a$	$\nu_3(A_1)$ $\nu(XY_4)$	$\nu_4(A_1)$ $\pi(XY_4)$	$\nu_5(B_1)$ $\nu(XY_4)$	$\nu_6(B_1)$ $\pi(XY_4)$	$\nu_7(B_2)$ $\delta(XY_4)$	$\nu_8(E)$ $\nu(XY_4)$	$\nu_9(E)$ $\rho_w(XW)^a$	$\nu_{10}(E)$ $\rho_w(XZ)$	$\nu_{11}(E)$ $\delta(XY_4)$	Refs.
$[SbCl_5Br]^-$	219	308	334	151	287	—	—	344	—	—	—	769, 770
$[SbBr_5Cl]^-$	305	206	192	—	186	—	—	239	—	—	—	769
SF_5Cl	402	855	707	602	625	271	505	909	597	399	441	771
SF_5Br	272	848	691	586	620	—	500	898	575	222	419	772
$[SF_5O]^-$	1153	722	697	506	541	472	452	780	607	530	325	773
SeF_5Cl	729	654	440	384	636	—	380	745	421	334	213	774
$[SeF_5O]^-$	919	559	649	—	556	—	—	639	—	—	—	775
TeF_5Cl	708	662	312	410.5 404.0	651	—	302	726	324.6	259	167	776
IF_5O	927	680	640	363	647	307	330	710	372	343	205	777
$[TiF_5O]^{3-}$	920	379	520	290	—	—	—	520	138	335	235	778
$[VF_5O]^{3-}$	943	383	525	317	—	—	—	525	139	342	237	778
$[NbCl_5Br]^-$	210	310	365	181	285	120	134	352	161	153	75	779
$[TaCl_5Br]^-$	204	318	368	183	300	(120)	168	325	151	143	73	779
$[TaBr_5Cl]^-$	323	231	187	110	180	(73)	96	214	123	144	76	779
$[MoCl_5O]^{2-}$	998	318	331	168	336	159	164	321	233	137	147	780
$[MoF_5O]^-$	973	662	492	300	580	—	—	580	324	252	—	781
$[WF_5O]^-$	987	686	507	286	594	—	—	608	329	242	—	781
WF_5Cl	407	744	703	257	644	182	377	661	290	227	307	782
ReF_5O	990	739	643	309	652	234	334	713	260	365	125	777
$[RuCl_5N]^-$	1048	284	318	192	307	168	184	337	233	154	174	780
$[RuBr_5N]^-$	1046	201	207	156	181	136	147	204	257	110	144	780
OsF_5O	963	716	644	281	644	210	332	701	263	367	164	777
$[OsCl_5N]^{2-}$	1084	324	348	189	334	169	181	336	264	146	172	780
$[OsBr_5N]^{2-}$	1085	192	198	156	172	136	149	234	217	115	144	780
$[UF_5O]^-$	820	593	602	182	445	(161)	(283)	480	248	209	201	783
$[XeF_5O]^-$	883	524	420	361	544	177	390	473 468 435	410 396	384 365	293 274	784
$[MoCl_4OBr]^{2-}$	235	964	301	149	288	—	—	320	229	92	162	785
$[WCl_4OBr]^{2-}$	233	960	326	149	—	—	187	298	230	92	162	785

[a] For XY_5Z, W is regarded as Y *trans* to Z.

TABLE II-9. VIBRATIONAL FREQUENCIES[a] OF XY_7 AND XY_5Z_2 MOLECULES (CM^{-1})

D_{5h}	ReF_7 [790]	IF_7 [790]	IF_7 [791]	$UF_5O_2^{3-}$ [792]	Assignment[b] [791]
$\nu_1(A_1')$	736	676	675	(668)	$\nu_s(MF_a)$
$\nu_2(A_1')$	645	635	629	816	$\nu_s(MF_e)$
$\nu_3(A_2'')$	703	670	672	873	$\nu_a(MF_a)$
$\nu_4(A_2'')$	299	365	257	380	$\delta(F_eMF_a)$
$\nu_5(E_1')$	703	746	746	740	$\nu_d(MF_e)$
$\nu_6(E_1')$	353	425	425	425	$\delta_d(F_eMF_e)$
$\nu_7(E_1')$	217	257	363	240	$\delta_d(F_aMF_a)$
$\nu_8(E_1'')$	597	510	308	—	$\delta_d(F_eMF_a)$
$\nu_9(E_2')$	489	352	509	—	$\nu_d(MF_e)$
$\nu_{10}(E_2')$	352	310	342	—	$\delta_d(F_eMF_e)$

[a] In these molecules $\nu_{11}(E_2'')$ is inactive.
[b] F_a and F_e denote the axial and equatorial F atoms, respectively.

The XY_8-type molecule may take the form of (I) a cube ($\mathbf{O}_h$), (II) an archimedean antiprism ($\mathbf{D}_{4d}$), (III) a dodecahedron ($\mathbf{D}_{2d}$), or (IV) a face-centered trigonal prism ($\mathbf{C}_{2v}$). Although XY_8 molecules are rare, X-ray analysis indicates that $[TaF_8]^{3-}$ and $[CrO_8]^{3-}$ ions take structures II and III, respectively.[793,794] The infrared and Raman spectra of crystalline $Na_3[TaF_8]$ are in accord with structure II, proposed by X-ray analysis.[795] For normal coordinate analyses of a cubic and an archimedean antiprism XY_8 molecule, see Refs. 796 and 797, respectively.

II-10. X_2Y_4 AND X_2Y_6 MOLECULES

(1) X_2Y_4 Molecules

Depending upon the twisting angle (τ) between the two XY_2 planes, the symmetry of the $Y_2X{-}XY_2$ molecule may be $\mathbf{D}_{2h}$ ($\tau=0°$, planar), $\mathbf{D}_{2d}$ ($\tau=90°$, staggered), or $\mathbf{D}_2$ ($0°<\tau<90°$, intermediate).

The $\mathbf{D}_{2h}$ structure may be confirmed if the infrared and Raman mutual exclusion rule holds. The $\mathbf{D}_{2d}$ and $\mathbf{D}_2$ structures can be distinguished by comparing the number of fundamentals with that predicted for each structure: 8 for $\mathbf{D}_2$ and 5 for $\mathbf{D}_{2d}$ in the infrared, and 12 for $\mathbf{D}_2$ and 9 for $\mathbf{D}_{2d}$ in the Raman.

B_2F_4[798] and B_2Cl_4[799] are staggered in the gaseous and liquid phases and planar in the solid state, whereas B_2Br_4[800] is staggered in all phases. Apparently, steric hindrance plays a main role in determining the conformation. Both B_2F_4 and B_2Cl_4 are also staggered in Ar matrices.[801] The vibrational spectra of crystalline N_2O_4 have been assigned on the basis of $\mathbf{D}_{2h}$ symmetry.[802] In a N_2 matrix, N_2O_4 is a mixture of the $\mathbf{D}_{2h}$, $\mathbf{D}_{2d}$, and $ONO{-}NO_2$ isomers.[98] The vibrational spectra of the oxalato ion ($C_2O_4^{2-}$) have been assigned based on $\mathbf{D}_{2d}$,[803] $\mathbf{D}_{2h}$,[804] and $\mathbf{D}_2$[805] symmetry.

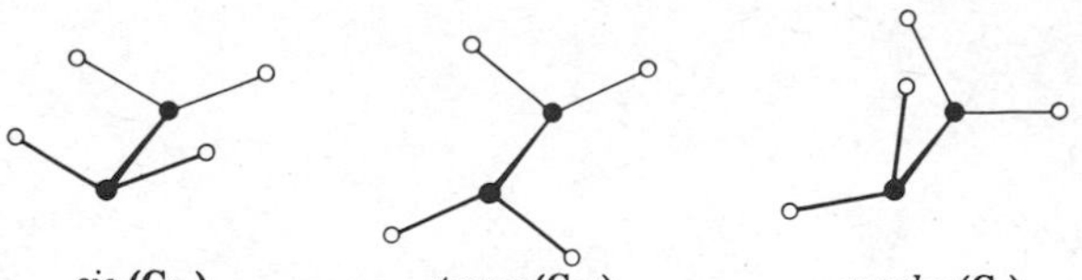

Fig. II-18. Various conformations of hydrazine.

Molecules like N_2H_4 and N_2F_4 take the *trans* (C_{2h}), *gauche* (C_2), or *cis* (C_{2v}) structure (Fig. II-18), depending on the angle of internal rotation. Evidently, the presence of lone-pair electrons on the X atom is responsible for the deviation of the X–XY_2 plane from the planarity. Most of these compounds exist as the *trans*- or *gauche*-isomer or as a mixture of both. The *trans*-isomer shows 6, whereas the *gauche*-isomer shows 12, fundamentals in the infrared.

In the gaseous, liquid, and solid states, N_2F_4 is a mixture of the *trans*- and *gauche*-isomers, and complete vibrational assignments have been made on each isomer.[806] N_2H_4 is pure *gauche* in all physical states.[807] The infrared spectrum of P_2H_4 in the gaseous state has been assigned on the basis of the *gauche* structure.[808] However, it is *trans* in the solid state.[809] The *trans* structure has been deduced from the vibrational spectra of P_2F_4 (all states),[810] P_2Cl_4 (all states),[811] and P_2I_4 (solid and solution).[812]

The infrared spectra of dimeric metal halides $(MX_2)_2$ isolated in inert gas matrices have been assigned on the basis of a planar-cyclic ring structure (Sec. II-2). Durig and co-workers reviewed the vibrational spectra of X_2Y_4-type molecules.[813]

(2) Bridged X_2Y_6 Molecules (D_{2h})

Figure II-19 illustrates the 18 normal modes of vibration[814] and band assignments for nonplanar bridging X_2Y_6-type molecules. The A_g, B_{1g}, B_{2g}, and B_{3g} vibrations are Raman active, whereas the B_{1u}, B_{2u}, and B_{3u} vibrations are infrared active. Table II-10*a* lists the vibrational frequencies of molecules belonging to this type. In most compounds, the ν_1, ν_8, ν_{11}, and ν_{16} vibrations are largely due to the terminal XY_2 stretching motions, and their frequencies are higher than those of ν_2, ν_6, ν_{13}, and ν_{17}, which are mainly due to the vibrations of the bridging $X_2Y'_2$ group. Normal coordinate analyses of these molecules have been made by several investigators.[818–821] Adams and Churchill[818] showed that, except for Ga_2I_6 and Al_2I_6, the values of the bridging stretching force constants are only 40–45% of those of the terminal force constants. It should be noted that In_2Cl_6 and In_2Br_6 are polymeric and In_2I_6 is dimeric in the crystalline state, as shown by their spectra.[822]

Table II-10*b* lists eight stretching frequencies of planar X_2Y_6 ions and molecules. In the $[M_2X_6]^{2-}$ (M = Pd and Pt; X = Cl, Br, and I) series, Goggin[823] showed that the distinction between terminal and bridging vibrations is meaningless except for X = Cl, since these vibrations couple so strongly with each

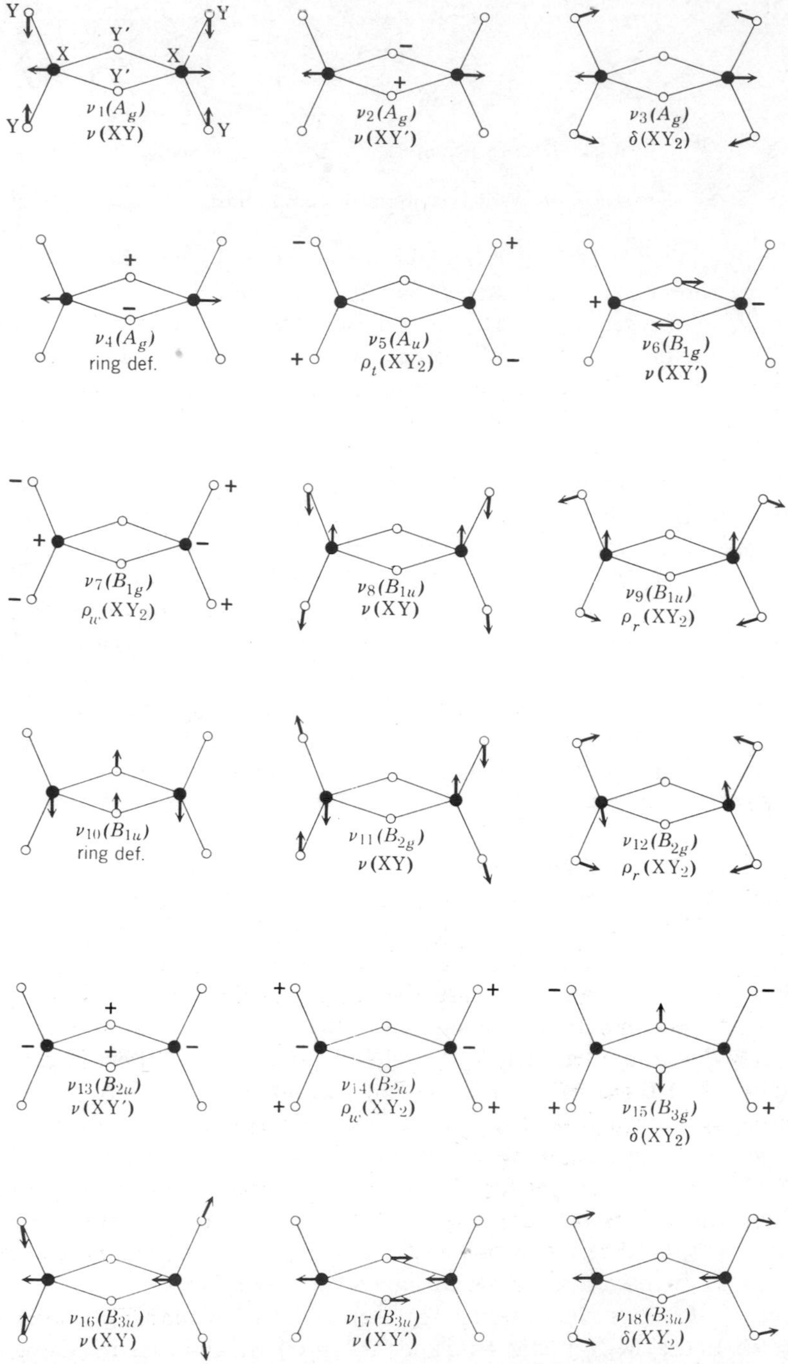

Fig. II-19. Normal modes of vibration of bridged X_2Y_6 molecules.[814]

Table II-10*a*. Vibrational Frequencies of Nonplanar Bridging X_2Y_6 Molecules (cm^{-1})

		$^{11}B_2H_6$	Al_2F_6	Al_2Cl_6	Al_2Br_6	Al_2I_6	Ga_2Cl_6	Ga_2Br_6	Ga_2I_6	In_2I_6	Fe_2Cl_6
A_g	ν_1	2537	—	523	410	348	413	291	229	187	426
	ν_2	2088	—	342	212	148	318	204	143	134	315
	ν_3	1187	—	219	142	95	167	119	85	69	115
	ν_4	818	—	107	70	56	100	64	50	40	108
A_u	ν_5	(833)	—	—	—	—	—	—	—	—	(50)
B_{1g}	ν_6	1760	—	284	354	—	243	241	195	114	315
	ν_7	874	—	166	82	82	125	85	64	55	(130)
B_{1u}	ν_8	2623	995	622	500	415	464	347	273	228	468
	ν_9	954	340	174	—	—	—	102	—	—	119
	ν_{10}	369	—	—	—	—	—	—	—	—	24
B_{2g}	ν_{11}	2616	—	612	491	405	462	339	265	232	459
	ν_{12}	929	—	121	116	63	117	74	55	49	(66)
B_{2u}	ν_{13}	(1939)	600	422	341	291	318	232	189	158	328
	ν_{14}	972	—	135	90	64	114	82	61	49	99
B_{3g}	ν_{15}	1020	—	—	—	54	215	158	68	44	(96)
B_{3u}	ν_{16}	2525	805	485	373	320	390	269	213	178	406
	ν_{17}	1607	575	320	198	140	282	188	134	125	280
	ν_{18}	1179	300	144	110	81	156	90	77	59	116
Refs.		815	816	817	818	818	818, 819	818	818	818	820, 319

TABLE II-10*b*. VIBRATIONAL FREQUENCIES[a] OF PLANAR BRIDGING X_2Y_6 MOLECULES (cm^{-1})

		$[Pd_2Cl_6]^{2-}$	$[Pd_2Br_6]^{2-}$	$[Pd_2I_6]^{2-}$	$[Pt_2Cl_6]^{2-}$	$[Pt_2Br_6]^{2-}$	$[Pt_2I_6]^{2-}$	I_2Cl_6	Au_2Cl_6
A_g	ν_1, $\nu(XY_t)$	346	262	219	349	241	196	344	379
	ν_2, $\nu(XY_b)$	302	194	143	316	211	160	198	328
B_{1g}	ν_6, $\nu(XY_t)$	328	253	—	333	238	196	314	366
	ν_7, $\nu(XY_b)$	265	173	130	294	193	145	142	289
B_{2u}	ν_{12}, $\nu(XY_t)$	335	257	218	330	236	196	327	364
	ν_{13}, $\nu(XY_b)$	262	178	—	300	192	147	170	309
B_{3u}	ν_{16}, $\nu(XY_t)$	343	264	218	341	239	196	340	374
	ν_{17}, $\nu(XY_b)$	297	192	140	312	210	157	205	309
Refs.		823	823	823	823	823	823	824	824

[a] XY_t and XY_b denote terminal and bridging XY stretching modes, respectively. When these two modes couple strongly, distinction between them is not clear (see Ref. 823).

other. According to Forneris et al.,[824] the terminal and bridging stretching force constants of Au_2Cl_6 are 2.22 and 1.15 mdyn/Å, respectively, and those of I_2Cl_6 are 1.70 and 0.40 mdyn/Å, respectively (modified UBF). On the other hand, Adams and Churchill[818] report values of 2.419 and 1.482 mdyn/Å, respectively, for the terminal and bridging force constants of Au_2Cl_6 (GVF).

(3) Ethane-Type X_2Y_6 Molecules ($\mathbf{D}_{3d}$)

The ethane-type X_2Y_6 molecule may be staggered ($\mathbf{D}_{3d}$), eclipsed ($\mathbf{D}_{3h}$), or *gauche* ($\mathbf{D}_3$). Figure II-20 shows the 12 normal modes of vibration of the staggered X_2Y_6 molecule. The A_{1g} and E_g vibrations are Raman active, and the A_{2u} and E_u vibrations are infrared active. Table II-10*c* lists the observed frequencies and band assignments based on $\mathbf{D}_{3d}$ symmetry. It should be noted that neutral Ga_2X_6 (X: a halogen) molecules take the bridging $\mathbf{D}_{2h}$ structure (Table II-10*a*), whereas $[Ga_2X_6]^{2-}$ ions take the ethane-like $\mathbf{D}_{3d}$ structure.

The structure of Si_2Cl_6 has been controversial; Griffiths[834] prefers the $\mathbf{D}_{3h}$ or $\mathbf{D}'_{3h}$ structure* for liquid Si_2Cl_6, whereas Ozin[835] favors the $\mathbf{D}_{3d}$ model for

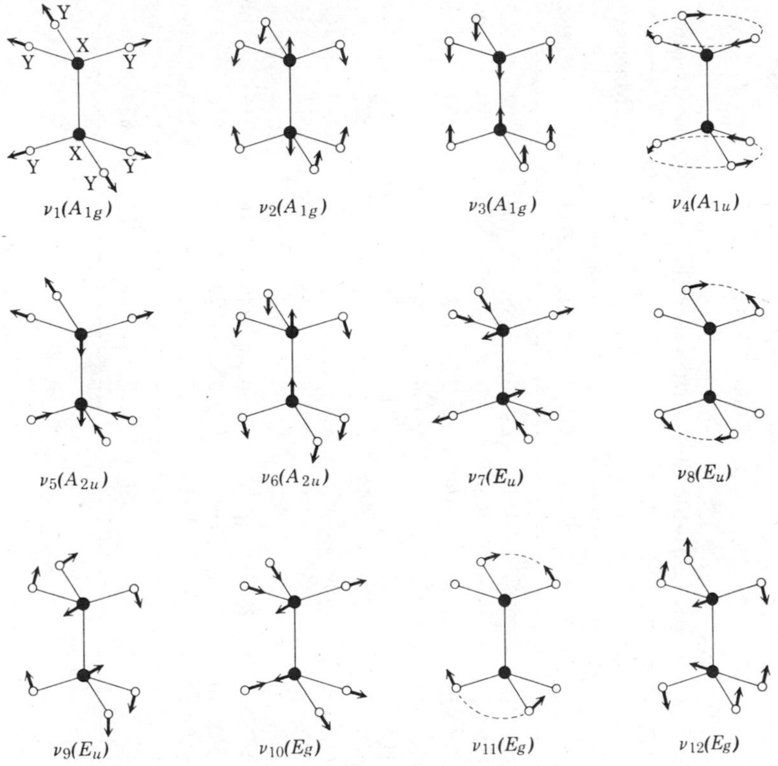

Fig. II-20. Normal modes of vibration of ethane-type X_2Y_6 molecules.

* If free rotation about the SiSi bond occurs, the symmetry plane (σ_h) is lost from $\mathbf{D}_{3h}$, and the molecule belongs to $\mathbf{D}'_{3h}$. Its selection rules are essentially the same as those of $\mathbf{D}_{3h}$.

TABLE II-10c. VIBRATIONAL FREQUENCIES OF ETHANE-TYPE X_2Y_6 MOLECULES (CM^{-1})

		Molecule ($\mathbf{D}_{3d}$)												
		C_2H_6	Si_2H_6	Ge_2H_6	$N_2H_6^{2+}$	$P_2O_6^{4-}$	$S_2O_6^{2-}$	$Ga_2Cl_6^{2-}$	$Ga_2Br_6^{2-}$	$Ga_2I_6^{2-}$	Si_2F_6	Si_2Cl_6	Si_2Br_6	Si_2I_6
A_{1g}	ν_1, $\nu(XY_3)$	2899	2152	(2070)	2650	1062	1102	375	316	285	910	351	223	154
	ν_2, $\delta(XY_3)$	(1375)	909	765	1524	670	710	106	70	42, 48	220	127	80	(51)
	ν_3, $\nu(XX)$	993	434	229	1027	275	293	233	164	118	541	624	562	510
A_{1u}	ν_4, $\rho_t(XY_3)$	275	—	144	455	—	—	—	—	—	—	—	—	—
A_{2u}	ν_5, $\nu(XY_3)$	2954	2154	2078	2600	942	1000	302	201	155	819	460	329	255
	ν_6, $\delta(XY_3)$	1379	844	755	1485	562	577	151	110	88	403	241	168	116
E_u	ν_7, $\nu(XY_3)$	2994	2179	2114	2739	1085	1240	327	237, 228	200	971	603	479	388
	ν_8, $\delta(XY_3)$	1486	940	898	1613	494	516	141	92	74	340	178	114	81
	ν_9, $\rho_r(XY_3)$	821	379	407	1096	200	204	89-66	64	(50)	203	74	50	(31)
E_g	ν_{10}, $\nu(XY_3)$	2963	2155	2150	2745	1168	1216	314	228	184	985	590	473	398
	ν_{11}, $\delta(XY_3)$	1460	929	875	1599	508	556	116	84	75	306	211	139	94
	ν_{12}, $\rho_r(XY_3)$	(1155)	625	417	1105	323	320	146	102	84	135	132	89	(53)
Refs.		825	826	827	828	829	830	831	831	831	832	833	833	833

all phases. The $\mathbf{D}_{3h}$ and $\mathbf{D}_{3d}$ selection rules are similar except that the E_u modes of $\mathbf{D}_{3d}$ which are infrared active become both infrared and Raman active (E') in $\mathbf{D}_{3h}$. The SiSi stretching mode was assigned at 354 cm^{-1} by Griffiths and at 627 cm^{-1} by Ozin. According to Höfler et al.,[833] the MM stretching force constant increases in the order $Si_2H_6 < Si_2I_6 < Si_2Br_6 < Si_2Cl_6$. Normal coordinate analyses have also been made on Si_2Cl_6[835] and $[Ga_2Cl_6]^{2-}$.[831] The stretching frequencies of the $[In_2X_6]^{2-}$ ions (X = Cl, Br, and I) have been assigned.[836]

II-11. X_2Y_7, X_2Y_8, X_2Y_9, AND X_2Y_{10} MOLECULES

(1) X_2Y_7 Molecules

The X_2Y_7- (XY_3—Y—XY_3) type molecule belongs to the $\mathbf{C}_s$, $\mathbf{C}_{2v}$, or $\mathbf{C}_1$ point group, depending on the relative orientation of the two XY_3 groups:

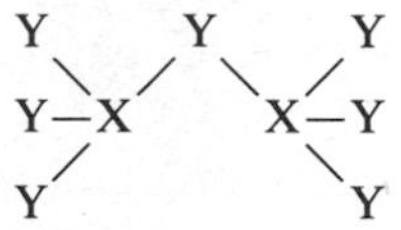

Seventeen vibrations are infrared active in $\mathbf{C}_{2v}$, while all 21 vibrations are infrared active in $\mathbf{C}_s$ and $\mathbf{C}_1$ symmetry. The 21 normal vibrations of the X_2Y_7 molecule may be classified into in-phase and out-of-phase coupling motions of terminal XY_3 group vibrations and the skeletal vibrations of the XYX bridge. Table II-11 lists the observed frequencies of these bridging vibrations.

On the basis of normal coordinate analyses, Brown and Ross[838] have made complete assignments of the vibrational spectra of the $S_2O_7^{2-}$, $Se_2O_7^{2-}$, $V_2O_7^{4-}$, and $Cr_2O_7^{2-}$ ions ($\mathbf{C}_{2v}$ symmetry). The molecule Re_2O_7 is monomeric in the

TABLE II-11. YXY BRIDGING FREQUENCIES OF X_2Y_7 MOLECULES (CM^{-1})

Compound	$\nu_a(YX_2)$	$\nu_s(YX_2)$	$\delta(YX_2)$	Refs.
$Na_4[P_2O_7]$	915	730	—	837
$Na_4[As_2O_7]$	735	550	245	837
$Na_2[S_2O_7]$	825	725	182 or 116	838
$Na_2[Se_2O_7]$	707	556	—	838
$Na_4[V_2O_7]$	710	533	200	838
$Na_2[Cr_2O_7]$	770	565, 554	220	838
Re_2O_7	804	456	50	839
Cl_2O_7	785	700	(165)	840
$[Ga_2Cl_7]^-$	286	276	—	841
$[Ga_2Br_7]^-$	222	195	—	841

gaseous and liquid states and polymeric in the solid state. Beattie and Ozin[839] assigned the spectra of gaseous Re_2O_7 on the basis of $\mathbf{C}_{2v}$ symmetry. Vibrational analysis of Cl_2O_7 has been made by assuming $\mathbf{C}_2$[842] or $\mathbf{C}_{2v}$[840] symmetry. On the assumption of a linear OPO bridge, the vibrational assignments of divalent metal pyrophosphates ($M_2P_2O_7$) have been made in terms of $\mathbf{D}_{3h}$ or $\mathbf{D}_{3d}$ symmetry.[843,844] According to Wing and Callahan,[845] the MOM bridge angle is always larger than 115°, and ν_a(MO) is at least 215 cm^{-1} higher than ν_s(MO). This separation increases as the MOM angle increases. In contrast to the $[Ga_2Cl_7]^-$ ion, the Ga-I-Ga bond of the $[Ga_2I_7]^-$ ion in the molten and crystalline states is linear.[846]

(2) X_2Y_8 Molecules

The symmetry of the X_2Y_8 ($Y_3X—Y—Y—XY_3$) molecule may be low enough to activate all 24 normal vibrations in both infrared and Raman spectra. Thus far, the XYYX bridging frequencies have been assigned at 988 [ν_a(XYYX)], 784 [ν_s(XYYX)], 890 [ν(YY)], and 397 and 328 [δ(XYYX)] for the $[P_2O_8]^{4-}$ ion, and at 1062, 834, 854, and 328 and 236 cm^{-1}, respectively, for the $[S_2O_8]^{4-}$ ion.[847]

In another type of X_2Y_8 ions, short, multiple X-X (metal-metal) bonds link two XY_4 units so that the overall symmetry becomes $\mathbf{D}_{4h}$ (eclipsed form). Such ions have been found in $[Mo_2Y_8]^{4-}$ and $[Re_2Y_8]^{2-}$ (Y = Cl and Br), and so on. Thus far, vibrational studies have been concentrated on their resonance Raman spectra which exhibit long overtone series involving ν(X-X) vibrations [Sec. III-17(3)]. Some references are: $[Mo_2Cl_8]^{4-}$ (848), $[Mo_2Br_8]^{4-}$ (849), $[Re_2X_8]^{2-}$ (X = Cl and Br) (850), and $[Te_2Cl_8]^{2-}$ (851).

(3) X_2Y_9 Molecules

The X_2Y_9 molecule shown below belongs to the point group $\mathbf{D}_{3h}$, and its

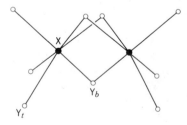

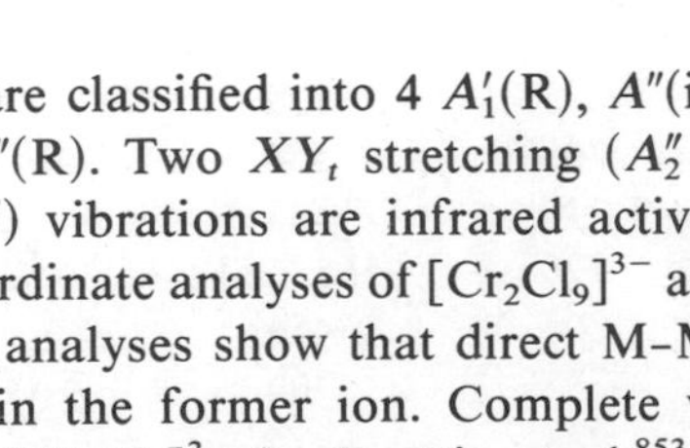

27 normal vibrations are classified into 4 A_1'(R), A''(i.a.), A_2'(i.a.), 3 A_2''(IR), 5 E'(IR, R), and 4 E''(R). Two XY_t stretching (A_2'' and E') and two XY_b stretching (A_2'' and E') vibrations are infrared active. Ziegler and Risen[852] carried out normal coordinate analyses of $[Cr_2Cl_9]^{3-}$ and $[W_2Cl_9]^{3-}$ ions. Both X-ray and vibrational analyses show that direct M-M interaction is present in the latter but not in the former ion. Complete vibrational assignments are also reported for $[Tl_2Cl_9]^{3-}$ by Beattie et al.[853] and for $[Cr_2Cl_9]^{3-}$ by Black et al.[854]

(4) X_2Y_{10} Molecules

The X_2Y_{10} molecule may take either one of the following structures:

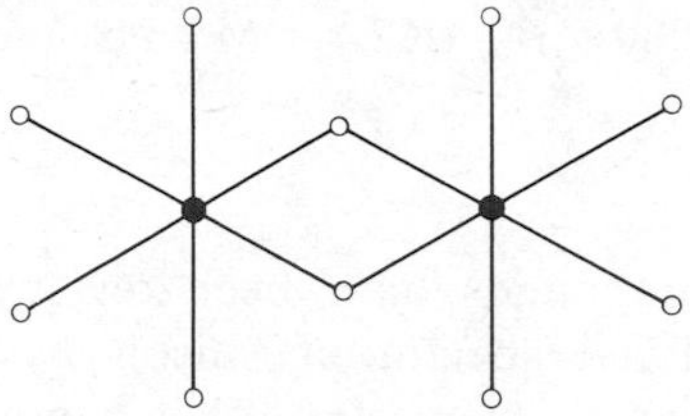

$\mathbf{D}_{2h}$

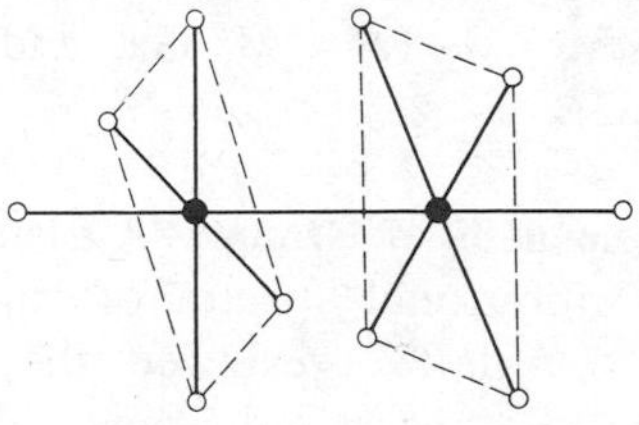

$\mathbf{D}_{4d}$

The 30 normal vibrations of the $\mathbf{D}_{2h}$ molecule are classified into 6 A_g, 2 A_u, 4 B_{1g}, 4 B_{1u}, 3B_{2g}, 4 B_{2u}, 2 B_{3g}, and 5 B_{3u}, of which 13 vibrations (B_{1u}, B_{2u}, and B_{3u}) are infrared active. These include four terminal and two bridging stretching modes. Beattie et al.[853] carried out normal coordinate analysis to assign the vibrational spectra of Nb_2X_{10} and Ta_2X_{10} (X = Cl and Br) based on $\mathbf{D}_{2h}$ symmetry. The infrared frequencies of Re_2Cl_{10},[855] Nb_2Cl_{10},[855] and Os_2Cl_{10}[856] have been reported.

The symmetry of the single-bridge XY_5—XY_5 type molecule may be $\mathbf{D}_{4d}$ (staggered) or $\mathbf{D}_{4h}$ (eclipsed), and these two structures cannot be distinguished by the selection rules. Under $\mathbf{D}_{4d}$ symmetry, the 30 normal vibrations are classified into 4 A_1, B_1, 3 B_2, 4 E_1, 3 E_2, and 4 E_3 species, of which B_2 and E_1 are infrared active and A_1, E_2, and E_3 are Raman active. Jones and Ekberg[857] made complete assignments of infrared and Raman spectra of S_2F_{10} vapor based on $\mathbf{D}_{4d}$ symmetry.

II-12. COMPOUNDS OF OTHER TYPES

Many compounds do not belong to any of the types discussed in the preceding sections. For these, the reader should consult *Spectroscopic Properties of Inorganic and Organometallic Compounds,* (1968–present), published by the Chemical Society, London, and other reference books cited at the end of Part I. Here, the references are given for some representative compounds which are relatively simple and for which complete band assignments are available.

(1) Compounds of Group IIIA Elements

Several review articles are available on the vibrational spectra of boron compounds. Lehmann and Shapiro[858] reviewed ν(BH), $\delta(BH_2)$, ν(BC), ν(BB), etc. of alkylboranes, and Bellamy et al.[859] discussed ν(BN), ν(BH), ν(BX), ν(BC), etc. Meller[860] lists ν(BN), ν(BH), ν(BX), and ν(BC) of a number of organoboron-nitrogen compounds. For vibrational spectra of boron compounds containing the B–P bond, see a review by Verkade.[861] The group frequency charts shown in Appendix VI include ν(BH), ν(BO), and ν(BX).

References of individual molecules include: B_2O_2, B_2O_3 (862), $(BO_2^-)_3$ (863), $(B_2O_5)^{4-}$ (864), $B_3(OH)_3$ (865), $B_3O_3(OH)_3$ (866), $H_2B_2O_3$ (867), B_5H_9

(868), $B_{10}H_{14}$ (869), $B_3N_3H_6$(Borazine) (870), $B_3N_3H_3X_3$ (871), B_4Cl_4 (872), $Zr(BH_4)_4$ (873), $U(BH_4)_4$ (874), BF_3-$(CH_3OH)_{1,2}$ (875), BX_3-phosphine (876), BX_3-NMe_3 (877), $BX_n(NCS)_{3-n}$ (878), GaX_3-L (L = PH_3, PMe_3, etc.) (879), $M[N(SiMe_3)_2]_3$ (M = Al, Ga, and In) (880), $M(AlCl_4)_2$ (M = Pd and Cu) (881, 882).

(2) Compounds of Group IVA Elements

The vibrational spectra of silicon compounds have been reviewed by Smith.[883] Aylett[884] reviewed the vibrational spectra of silicon hydrides. Campbell-Ferguson and Ebsworth[885] reviewed the spectra of halogenosilane-amine adducts, and Schumann[886] reviewed ν(GeP) and ν(SnP) of a large number of compounds. Appendix VI gives group frequency charts for ν(SiH), ν(GeH), ν(SiO), ν(SiX), and ν(GeX). For individual compounds, only references are cited: C_3O_2 (887), $[C_3O_3]^{2-}$ (888), C_3S_2 (889), Si_5H_{10} (890), Si_6H_{12} (891), quartz (892), silicates (893, 894), $(SiH_3)_2O$ (895), $(SiH_3)_4N_2$ (896), $(SiH_3)_2S$ (897), $(SiH_3)PH_2$ (898), $(SiCl_3)_2O$ (899), $Si_2(NCO)_6$ (900), $(SiH_3)_3N$ (901), $[Ge_4S_{10}]^{4-}$ (902), germanates (894), $(GeH_3)_3M$ (M = As, Sb) (903).

(3) Compounds of Group VA Elements

The vibrational spectra of nitrogen oxides[904] and phosphorus compounds[905–910] have been reviewed. Figure II-21 shows a group frequency chart based on these reviews. Appendix VI also gives group frequency chart for ν(PH), ν(PO), and ν(PX) (X: a halogen).

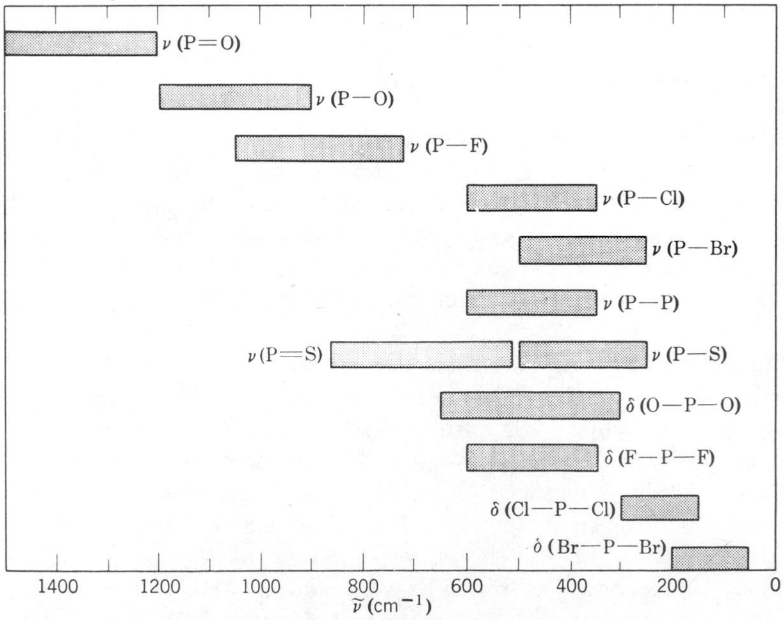

Fig. II-21. Characteristic frequencies of phosphorus compounds.

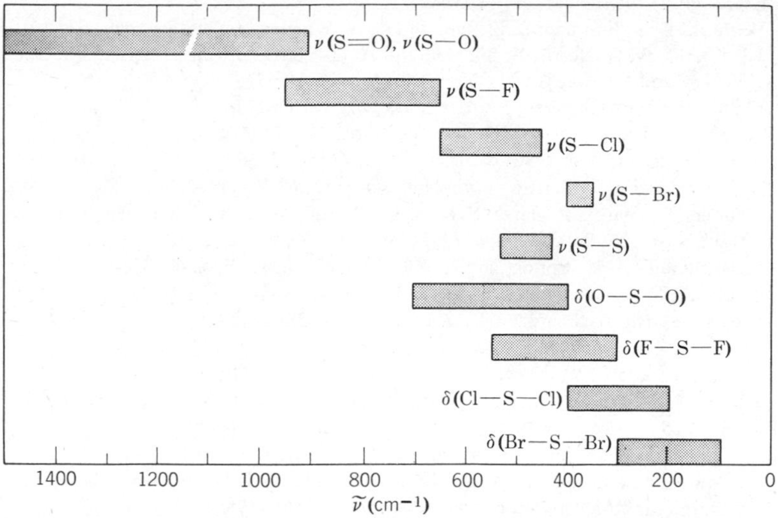

Fig. II-22. Characteristic frequencies of sulfur compounds.

For individual compounds, only references are given: N_2O_3 (911), N_2O_5 (912), HNO_3 (913), XNO_2 and $XONO_2$ (X = F and Cl) (914, 915), NH_2OH (916), $[NH_3OH]Cl$ (917), $[N_2F_3]^+$ (918), $N(CF_3)_3$ (919), P_4 (920), $[P_3O_9]^{3-}$ (921), $[P_4O_{12}]^{4-}$ (922), P_4S_n (n = 3, 5, 7, 10) (923), P_4S_4 (924), $As_{4-n}P_n$ (n = 1 ~ 3) (925), $[HPO_4]^{2-}$ and $[H_2PO_4]^-$ (926), $[HPO_3]^{2-}$ and $[H_2PO_2]^-$ (927), $[H_2PF_4]^-$ (928), HPF_4 and H_2PF_3 (929), PF_3X_2 (X = Cl, Br) (930), $(PNCl_2)_3$ (931), $(PNBr_2)_3$ (932), $[PO_2NH]_4{}^{4-}$ (933), $P(CN)_3$ (934), As_4 and As_4O_6 (935), $[As_3O_9]^{3-}$ (936), $(As_2O_3)_n$ (937), $As(CN)_3$ (934), $[BiX_4]^-$, $[BiX_5]^{2-}$, and $[Bi_2X_9]^{3-}$ (X = Cl, Br) (938).

(4) Compounds of Group VIA Elements

As stated in Sec. II-6(A), Müller and co-workers[547,553,554] reviewed the vibrational spectra of thio and seleno compounds of transition-metal ions. The group frequency chart of sulfur compounds shown in Fig. II-22 is based on the literature appearing in this book. For individual compounds, only references are cited: $[S_n]^{2-}$ (n = 3 ~ 6) (939), S_6 (940), S_7 (941), S_8 (942), $H_2S_{3,4}$ (943), $[S_2O_5]^{2-}$ (944), H_2SO_4 (945), S_4N_4 (946), $[SNS_3]^-$ (947), $ClOSO_2F$ (948), $XNSF_2$ (X = F, Cl, Br) (949), $[S_3N_3]^-$ (950), $[Se_4]^{2+}$ (951).

References

1. G. Herzberg, *Molecular Spectra and Molecular Structure,* Vol. I: *Spectra of Diatomic Molecules,* Van Nostrand, Princeton, N.J., 1950; K. P. Huber and G. Herzberg, *Molecular Spectra and Molecular Structure,* Vol. IV: *Constants of Diatomic Molecules,* Van Nostrand, Princeton, N.J., 1979.

2. A. E. Douglas, *Can. J. Phys.*, **35**, 71 (1957).
3. D. H. Rank, D. P. Eastman, B. S. Rao, and T. A. Wiggins, *J. Opt. Soc. Am.*, **52**, 1 (1962).
4. W. Klemperer, W. G. Norris, A. Büchler, and A. G. Emslie, *J. Chem. Phys.*, **33**, 1534 (1960).
5. S. E. Veazey and W. Gordy, *Phys. Rev.* **138**, *A*, 1303 (1965).
6. S. A. Rice and W. Klemperer, *J. Chem. Phys.*, **27**, 573 (1957).
7. V. I. Baikov and K. P. Vasilevskii, *Opt. Spectrosc.*, **22**, 198 (1967).
8. W. V. F. Brooks and B. Crawford, Jr., *J. Chem. Phys.*, **23**, 363 (1955).
9. J. A. Horsley and R. F. Barrow, *Trans. Faraday Soc.*, **63**, 32 (1967).
10. P. C. Souers, D. Fearon, R. Garza, E. M. Kelly, P. E. Roberts, R. H. Sanborn, R. T. Tsugawa, J. L. Hunt, and J. D. Poll, *J. Chem. Phys.*, **70**, 1581 (1979).
11. R. A. Teichman III, M. Epting, and E. R. Nixon, *J. Chem. Phys.*, **68**, 336 (1978).
12. G. A. Ozin, *Chem. Commun.* 1325 (1969).
13. L. Andrews and R. R. Smardzewski, *J. Chem. Phys.*, **58**, 2258 (1973).
13a. J. C. Evans, *Chem. Commun.*, **682** (1969).
14. A. G. Hopkins and C. W. Brown, *J. Chem. Phys.*, **62**, 1598 (1975).
15. W. Holzer, W. F. Murphy, and H. J. Bernstein, *J. Mol. Spectrosc.*, **32**, 13 (1969).
16. W. F. Howard Jr. and L. Andrews, *J. Am. Chem. Soc.*, **95**, 3045 (1973).
17. M. R. Clarke and G. Mamantov, *Inorg. Nucl. Chem. Lett.*, **7**, 993 (1971).
18. W. F. Howard Jr. and L. Andrews, *J. Am. Chem. Soc.*, **95**, 2056 (1973).
19. R. J. Gillespie and M. J. Morton, *Chem. Commun.*, 1565 (1968).
20. W. Holzer, W. F. Murphy, and H. J. Bernstein, *J. Chem. Phys.*, **52**, 399 (1970).
21. R. J. Gillespie and M. J. Morton, *J. Mol. Spectrosc.*, **30**, 178 (1969).
22. W. F. Howard Jr. and L. Andrews, *J. Am. Chem. Soc.*, **97**, 2956 (1975).
23. W. Schulze, H. U. Becker, R. Minkwitz, and K. Manzel, *Chem. Phys. Lett.*, **55**, 59 (1978).
24. A. Givan and A. Loewenschuss, *Chem. Phys. Lett.*, **62**, 592 (1979).
25. R. B. Wright, J. K. Bates, and D. M. Gruen, *Inorg. Chem.*, **17**, 2275 (1978).
26. D. E. Milligan and M. E. Jacox, *J. Chem. Phys.*, **52**, 2594 (1970).
27. K. Rosengren and G. C. Pimentel, *J. Chem. Phys.*, **43**, 507 (1965).
28. N. Acquista, L. J. Schoen, and D. R. Lide, Jr., *J. Chem. Phys.*, **48**, 1534 (1968).
29. W. R. Busing, *J. Chem. Phys.*, **23**, 933 (1955).
30. N. Acquista and L. J. Schoen, *J. Chem. Phys.*, **53**, 1290 (1970).
31. R. J. Van Zee, T. C. DeVore, and W. Weltner, Jr., *J. Chem. Phys.*, **71**, 2051 (1979).
32. R. A. Penneman and L. H. Jones, *J. Chem. Phys.*, **24**, 293 (1956).
33. D. E. Milligan and M. E. Jacox, *J. Chem. Phys.*, **47**, 278 (1967).
34. R. M. Atkins and P. L. Timms, *Spectrochim. Acta*, **33A**, 853 (1977).
35. M. E. Jacox and D. E. Milligan, *J. Chem. Phys.*, **46**, 184 (1967); **40**, 2461 (1964).
36. R. Minkwitz and F. W. Froben, *Chem. Phys. Lett.*, **39**, 473 (1976).
37. F. W. Froben and F. Rogge, *Chem. Phys. Lett.*, **78**, 264 (1981).
38. D. W. Green, W. Korfmacher, and D. M. Gruen, *J. Chem. Phys.*, **58**, 404 (1973).
39. D. W. Green and G. T. Reedy, *J. Mol. Spectrosc.*, **74**, 423 (1979).
40. D. W. Green and G. T. Reedy, *J. Chem. Phys.*, **65**, 2921 (1976).
41. D. W. Green and G. T. Reedy, *J. Chem. Phys.*, **69**, 552 (1978).
42. D. White, K. S. Seshadri, D. F. Dever, D. E. Mann, and M. J. Linevsky, *J. Chem. Phys.*, **39**, 2463 (1963).
43. K. S. Seshadri, D. White, and D. E. Mann, *J. Chem. Phys.*, **45**, 4697 (1966).
44. L. Andrews, E. S. Prochaska, and B. S. Ault, *J. Chem. Phys.*, **69**, 556 (1978).
45. L. Andrews and B. S. Ault, *J. Mol. Spectrosc.*, **68**, 114 (1977).
46. H. Dubost and A. Abouaf-Marguin, *Chem. Phys. Lett.*, **17**, 269 (1972).
47. J. S. Anderson and J. S. Ogden, *J. Chem. Phys.*, **51**, 4189 (1969).
48. J. S. Ogden and M. J. Ricks, *J. Chem. Phys.*, **52**, 352 (1970).
49. J. S. Ogden and M. J. Ricks, *J. Chem. Phys.*, **53**, 896 (1970).
50. J. S. Ogden and M. J. Ricks, *J. Chem. Phys.*, **56**, 1658 (1972).
51. V. P. Babaeva and V. Y. Rosolovskii, *Russ. J. Inorg. Chem.*, **16**, 471 (1971).
52. W. A. Guillory and C. E. Hunter, *J. Chem. Phys.*, **50**, 3516 (1969).

53. D. E. Tevault and L. Andrews, *J. Phys. Chem.*, **77**, 1646 (1973).
54. D. E. Tevault and L. Andrews, *J. Phys. Chem.*, **77**, 1640 (1973).
55. A. G. Hopkins and C. W. Brown, *J. Chem. Phys.*, **62**, 2511 (1975).
56. L. Andrews and J. I. Raymond, *J. Chem. Phys.*, **55**, 3078 (1971).
57. F. K. Chi and L. Andrews, *J. Phys. Chem.*, **77**, 3062 (1973).
58. D. E. Tevault, N. Walker, R. R. Smardzewski, and W. B. Fox, *J. Phys. Chem.*, **82**, 2733 (1978).
59. E. S. Prochaska and L. Andrews, *J. Chem. Phys.*, **72**, 6782 (1980).
60. W. Weltner, Jr. and D. McLeod, Jr., *J. Phys. Chem.*, **69**, 3488 (1965).
61. D. W. Green, W. Korfmacher, and D. M. Gruen, *J. Chem. Phys.*, **58**, 404 (1973).
62. W. Weltner, Jr. and D. McLeod, Jr., *J. Chem. Phys.*, **42**, 882 (1965).
63. D. W. Green and K. M. Ervin, *J. Mol. Spectrosc.*, **89**, 145 (1981).
64. D. W. Green, G. T. Reedy, and J. G. Kay, *J. Mol. Spectrosc.*, **78**, 257 (1979).
65. S. D. Gabelnick, G. T. Reedy, and M. G. Chasanov, *J. Chem. Phys.*, **58**, 4468 (1973).
66. D. W. Green and G. T. Reedy, *J. Chem. Phys.*, **69**, 544 (1978).
67. R. R. Steudel, *Z. Anorg. Allg. Chem.*, **361**, 180 (1968).
68. C. P. Marino, J. D. Guerin, and E. R. Nixon, *J. Mol. Spectrosc.*, **51**, 160 (1974).
69. W. Holzer, W. F. Murphy, and H. J. Bernstein, *J. Mol. Spectrosc.*, **32**, 13 (1969).
70. A. Snelson and K. S. Pitzer, *J. Phys. Chem.*, **67**, 882 (1963).
71. S. Schlick and O. Schnepp, *J. Chem. Phys.*, **41**, 463 (1964).
72. D. E. Mann, G. V. Calder, K. S. Seshadri, D. White, and M. J. Linevsky, *J. Chem. Phys.*, **46**, 1138 (1967).
73. A. Snelson, *J. Phys. Chem.*, **71**, 3202 (1967).
74. J. M. Brom, Jr. and H. F. Franzen, *J. Chem. Phys.*, **54**, 2874 (1971).
75. M. E. Jacox and D. E. Milligan, *J. Chem. Phys.*, **50**, 3252 (1969).
76. M. Feuerhahn, R. Minkwitz, and G. Vahl, *Spectrochim. Acta*, **36A**, 183 (1980).
77. G. A. Olah and M. B. Comisarow, *J. Am. Chem. Soc.*, **91**, 2172 (1969).
78. W. Holzer, W. F. Murphy, and H. J. Bernstein, *J. Chem. Phys.*, **52**, 399 (1970).
79. A. J. Edwards, W. E. Falconer, J. E. Griffiths, W. A. Sunder, and M. J. Vasile, *J. Chem. Soc., Dalton Trans.*, 1129 (1974).
80. M. J. Linevsky, *J. Chem. Phys.*, **34**, 587 (1961).
81. A. Snelson, *J. Chem. Phys.*, **46**, 3652 (1967).
82. M. Freiberg, A. Ron., and O. Schnepp, *J. Phys. Chem.*, **72**, 3526 (1968).
83. T. P. Martin and H. Schaber, *J. Chem. Phys.*, **68**, 4299 (1978).
84. J. R. Durig, K. K. Lau, G. Nagarajan, M. Walker, and J. Bragin, *J. Chem. Phys.*, **50**, 2130 (1969).
85. V. E. Bondybey and J. W. Nibler, *J. Chem. Phys.*, **58**, 2125 (1973).
86. D. Sellmann, A. Brandl, and R. Endell, *Angew. Chem.*, Int. Ed., **12**, 1019 (1973).
87. J. L. Hollenberg, *J. Chem. Phys.*, **46**, 3271 (1967); D. F. Smith, *J. Chem. Phys.*, **48**, 1429 (1968).
88. P. A. Giguère and N. Zengin, *Can. J. Chem.*, **36**, 1013 (1958).
89. D. F. Hornig and W. E. Osberg, *J. Chem. Phys.*, **23**, 662 (1955); G. L. Hiebert and D. F. Hornig, *J. Chem. Phys.*, **28**, 316 (1958).
90. A. J. Barnes, H. E. Hallam, and G. F. Scrimshaw, *Trans. Faraday Soc.*, **65**, 3150, 3159, 3172 (1969).
91. V. Lorenzelli, *Compt. Rend.*, **258**, 5386 (1964).
92. W. B. Person, R. E. Humphrey, W. A. Deskin, and A. I. Popov, *J. Am. Chem. Soc.*, **80**, 2049 (1958); W. B. Person, R. E. Erickson, and R. E. Buckles, *J. Am. Chem. Soc.*, **82**, 29 (1960); A. I. Popov, R. E. Humphrey, and W. B. Person, *J. Am. Chem. Soc.*, **82**, 1850 (1960).
93. P. Klaboe, *J. Am. Chem. Soc.*, **89**, 3667 (1967).
94. L. H. Jones, *J. Chem. Phys.*, **22**, 217 (1954).
95. W. D. Stalleup and D. Williams, *J. Chem. Phys.*, **10**, 199 (1942).
96. R. A. Penneman and L. H. Jones, *J. Chem. Phys.*, **28**, 169 (1958).
97. L. H. Jones, *J. Mol. Spectrosc.*, **45**, 55 (1973); **49**, 82 (1974).
98. W. G. Fateley, H. A. Bent, and B. Crawford, Jr., *J. Chem. Phys.*, **31**, 204 (1959).
99. J. W. Hastie, R. H. Hauge, and J. L. Margrave, *Chem. Commun.*, 1452 (1969).
100. I. Eliezer and A. Reger, *Coord. Chem. Rev.*, **9**, 189 (1972–73).

101. J. W. Hastie, R. H. Hauge, and J. L. Margrave, *High Temp. Sci.*, **1,** 76 (1969).
102. A. Snelson, *J. Phys. Chem.*, **72,** 250 (1968).
103. M. L. Lesiecki and J. W. Nibler, *J. Chem. Phys.*, **64,** 871 (1976).
104. G. V. Calder, D. E. Mann, K. S. Seshadri, M. Allavena, and D. White, *J. Chem. Phys.*, **51,** 2093 (1969).
105. D. White, G. V. Calder, S. Hemple, and D. E. Mann, *J. Chem. Phys.*, **59,** 6645 (1973).
106. J. J. Habeeb and D. G. Tuck, *J. Chem. Soc., Dalton Trans.*, 866 (1976).
107. D. E. Milligan and M. E. Jacox, *J. Chem. Phys.*, **48,** 2265 (1968).
108. L. Andrews, *J. Chem. Phys.*, **48,** 979 (1968).
109. L. Andrews and T. G. Carver, *J. Chem. Phys.*, **49,** 896 (1968).
110. J. W. Hastie, R. H. Hauge, and J. L. Margrave, *J. Am. Chem. Soc.*, **91,** 2536 (1969).
111. D. E. Milligan and M. E. Jacox, *J. Chem. Phys.*, **49,** 1938 (1968).
112. G. Maass, R. H. Hauge, and J. L. Margrave, *Z. Anorg. Allg. Chem.*, **392,** 295 (1972).
113. J. W. Hastie, R. H. Hauge, and J. L. Margrave, *J. Phys. Chem.*, **72,** 4492 (1968).
114. L. Andrews and D. L. Frederick, *J. Am. Chem. Soc.*, **92,** 775 (1970).
115. R. H. hauge, J. W. Hastie, and J. L. Margrave, *J. Mol. Spectrosc.*, **45,** 420 (1973).
116. G. A. Ozin and A. Vander Voet, *J. Chem. Phys.*, **56,** 4768 (1972).
117. I. R. Beattie and R. O. Perry, *J. Chem. Soc., A*, 2429 (1970).
118. M. D. Harmony and R. J. Myers, *J. Chem. Phys.*, **37,** 636 (1962).
119. J. K. Burdett, L. Hodges, V. Dunning, and J. H. Current, *J. Phys. Chem.*, **74,** 4053 (1970).
120. L. Andrews and D. L. Frederick, *J. Phys. Chem.*, **73,** 2774 (1969).
121. M. M. Rochkind and G. C. Pimentel, *J. Chem. Phys.*, **42,** 1361 (1965).
122. M. Feuerhahn and G. Vahl, *Inorg. Nucl. Chem. Lett.*, **16,** 5 (1980).
122a. A. Haas and H. Willner, *Spectrochim. Acta*, **34A,** 541 (1978).
123. G. A. Ozin and A. Vander Voet, *Chem. Commun.*, 896 (1970).
124. H. H. Claassen, G. L. Goodman, J. G. Malm, and F. Schreiner, *J. Chem. Phys.*, **42,** 1229 (1965).
125. P. A. Agron, G. M. Begun, H. A. Levy, A. A. Mason, C. G. Jones, and D. F. Smith, *Science*, **139,** 842 (1963).
126. L. Y. Nelson and G. C. Pimentel, *Inorg. Chem.*, **6,** 1758 (1967).
127. D. N. Waters and B. Basak, *J. Chem. Soc. A*, 2733 (1971).
128. P. Braunstein and R. J. H. Clark, *J. Chem. Soc., Dalton Trans.*, 1845 (1973).
129. A. Givan and A. Loewenschuss, *J. Chem. Phys.*, **72,** 3809 (1980).
130. A. Givan and A. Loewenschuss, *J. Chem. Phys.*, **68,** 2228 (1978).
131. A. Loewenschuss, A. Ron, and O. Schnepp, *J. Chem. Phys.*, **50,** 2502 (1969).
132. J. W. Hastie, R. Hauge, and J. L. Margrave, *J. Chem. Phys.*, **51,** 2648 (1969).
133. M. E. Jacox and D. E. Milligan, *J. Chem. Phys.*, **51,** 4143 (1969).
134. K. R. Thompson and K. D. Carlson, *J. Chem. Phys.*, **49,** 4379 (1968).
135. D. L. Cocke, C. A. Chang, and K. A. Gingerich, *Appl. Spectrosc.*, **27,** 260 (1973).
136. H. Huber, E. P. Kündig, G. A. Ozin, and A. Vander Voet, *Can. J. Chem.*, **52,** 95 (1974).
137. L. Andrews, *J. Phys. Chem.*, **73,** 3922 (1969).
138. A. Sommer, D. White, M. J. Linevsky, and D. E. Mann, *J. Chem. Phys.*, **38,** 87 (1963).
139. A. Snelson, *J. Phys. Chem.*, **74,** 2574 (1970).
140. A. J. Hinchcliffe and J. S. Ogden, *J. Phys. Chem.*, **77,** 2537 (1973); *Chem. Commun.*, 1053 (1969).
141. J. H. Taylor, W. S. Benedict, and J. Strong, *J. Chem. Phys.*, **20,** 1884 (1952).
142. A. H. Nielsen and R. J. Lagemann, *J. Chem. Phys.*, **22,** 36 (1954).
143. T. Wentink, Jr., *J. Chem. Phys.*, **29,** 188 (1958).
144. G. W. King and K. Srikameswaran, *J. Mol. Spectrosc.*, **29,** 491 (1969).
145. J. S. Anderson, A. Bos, and J. S. Ogden, *Chem. Commun.*, 1381 (1971).
146. D. E. Milligan and M. E. Jacox, *J. Chem. Phys.*, **55,** 3404 (1971).
147. R. V. St. Louis and B. L. Crawford, Jr., *J. Chem. Phys.*, **42,** 857 (1965).
148. J. W. Nebgen, A. D. McElroy, and H. F. Klodowski, *Inorg. Chem.*, **4,** 1796 (1965).
149. D. F. Smith, Jr., J. Overend, R. C. Spiker, and L. Andrews, *Spectrochim. Acta*, **28A,** 87 (1972).
150. L. Andrews and R. C. Spiker, *J. Chem. Phys.*, **59,** 1863 (1973); D. M. Thomas and L. Andrews, *J. Mol. Spectrosc.*, **50,** 220 (1974).

151. A. Barbe, C. Secroun, and P. Jouve, *J. Mol. Spectrosc.*, **49,** 171 (1974).
152. A. G. Hopkins, S. Tang, and C. W. Brown, *J. Am. Chem. Soc.*, **95,** 3486 (1973).
153. R. J. H. Clark and D. G. Cobbold, *Inorg. Chem.*, **17,** 3169 (1978).
154. D. Maillard, M. Allavena, and J. P. Perchard, *Spectrochim. Acta*, **31A,** 1523 (1975).
155. A. G. Hopkins, F. P. Daly, and C. W. Brown, *J. Phys. Chem.*, **79,** 1849 (1975).
156. D. E. Milligan and M. E. Jacox, *J. Chem. Phys.*, **55,** 1003 (1971).
157. S. N. Cesaro, M. Spoliti, A. J. Hinchcliffe, and J. S. Ogden, *J. Chem. Phys.*, **55,** 5834 (1971).
158. M. Spoliti, S. N. Cesaro, and E. Coffari, *J. Chem. Thermodyn.*, **4,** 507 (1972).
159. R. D. Spratley, J. J. Turner, and G. C. Pimentel, *J. Chem. Phys.*, **44,** 2063 (1966).
160. J. S. Ogden and J. J. Turner, *J. Chem. Soc. A*, 1483 (1967).
161. M. M. Rochkind and G. C. Pimentel, *J. Chem. Phys.*, **42,** 1361 (1965).
162. A. Arkell and I. Schwager, *J. Am. Chem. Soc.*, **89,** 5999 (1967).
163. B. Tanguy, B. Frit, G. Turrell, and P. Hagenmuller, *Compt. Rend.*, **264C,** 301 (1967).
164. A. H. Nielsen and P. J. H. Woltz, *J. Chem. Phys.*, **20,** 1878 (1952).
165. K. O. Christe, C. J. Schack, D. Pilipovich, and W. Sawodny, *Inorg. Chem.*, **8,** 2489 (1969).
166. C. Campbell, J. P. M. Jones, and J. J. Turner, *Chem. Commun.*, 888 (1968).
167. E. Jacob, *Angew. Chem. Int. Ed.*, **15,** 158 (1976).
168. S. D. Gabelnick, G. T. Reedy, and M. G. Chasanov, *J. Chem. Phys.*, **60,** 1167 (1974).
169. N. S. McIntyre, K. R. Thompson, and W. Weltner, *J. Phys. Chem.*, **75,** 3243 (1971).
170. W. Weltner and D. McLeod, *J. Chem. Phys.*, **42,** 882 (1965).
171. W. Weltner and D. McLeod, *J. Mol. Spectrosc.*, **17,** 276 (1965).
172. S. D. Gabelnick, G. T. Reedy, and M. G. Chasanov, *J. Chem. Phys.*, **58,** 4468 (1973).
173. L. H. Jones, *J. Chem. Phys.*, **23,** 2105 (1955).
174. D. W. Green and G. T. Reedy, *J. Chem. Phys.*, **69,** 544 (1978).
175. W. P. Griffith and J. D. Wickins, *J. Chem. Soc. A*, 400 (1968).
176. G. Pausewang and K. Dehnicke, *Z. Anorg. Allg. Chem.*, **369,** 265 (1969).
177. F. W. Moore and R. E. Rice, *Inorg. Chem.*, **7,** 2510 (1968).
178. B. Šoptrajanov, A. Nikolovski, and I. Petrov, *Spectrochim. Acta*, **24A**, 1617 (1968).
179. L. H. Jones and R. A. Penneman, *J. Chem. Phys.*, **21,** 542 (1953).
180. L. H. Jones, *Spectrochim. Acta*, **11,** 409 (1959).
181. S. P. McGlynn, J. K. Smith, and W. C. Neely, *J. Chem. Phys.*, **35,** 105 (1961).
182. H. R. Hoekstra, *Inorg. Chem.*, **2,** 492 (1963).
183. J. I. Bullock, *J. Chem. Soc. A*, **781** (1969).
184. K. Ohwada, *Spectrochim. Acta*, **24A,** 595 (1968).
185. D. W. Green and G. T. Reedy, in J. R. Ferraro and L. J. Basile, eds., *Fourier Transform Infrared Spectroscopy: Applications to Chemical Systems*, Vol. 1, Academic Press, New York, 1978.
186. D. W. Green and G. T. Reedy, *J. Chem. Phys.*, **65,** 2921 (1976).
187. D. W. Green and G. T. Reedy, *J. Chem. Phys.*, **69,** 552 (1978).
188. D. W. Green and G. T. Reedy, *J. Mol. Spectrosc.*, **74,** 423 (1979).
189. R. J. Gillespie and M. J. Morton, *Inorg. Chem.*, **9,** 616 (1970).
190. E. S. Prochaska and L. Andrews, *Inorg. Chem.*, **16,** 339 (1977).
191. K. O. Christe, W. Sawodny, and J. P. Guertin, *Inorg. Chem.*, **6,** 1159 (1967).
192. R. J. Gillespie and M. J. Morton, *Inorg. Chem.*, **9,** 811 (1970).
193. K. O. Christe and C. J. Schack, *Inorg. Chem.*, **9,** 2296 (1970).
194. T. Surles, L. A. Quaterman, and H. H. Hyman, *J. Inorg. Nucl. Chem.*, **35,** 668 (1973).
195. K. O. Christe and W. Sawodny, *Inorg. Chem.*, **8,** 212 (1969).
196. B. S. Ault and L. Andrews, *J. Chem. Phys.*, **64,** 4853 (1976).
197 W. Gabes and H. Gerding, *J. Mol. Struct.*, **14,** 267 (1972).
198. R. Forneris and Y. Tavares-Forneris, *J. Mol. Struct.*, **23,** 241 (1974).
199. W. W. Wilson and F. Aubke, *Inorg. Chem.*, **13,** 326 (1974).
200. D. H. Boal and G. A. Ozin, *J. Chem. Phys.*, **55,** 3598 (1971).
201. A. G. Maki and R. Forneris, *Spectrochim. Acta*, **23A,** 867 (1967).
202. M. Couzi, J. C. Cornut, and P. V. Huong, *J. Chem. Phys.*, **56,** 426 (1972).
203. J. J. Rush, L. W. Schroeder, and A. J. Melveger, *J. Chem. Phys.*, **56,** 2793 (1972).

204. J. C. Evans and G. Y-S. Lo, *J. Phys. Chem.*, **70,** 543 (1966).
205. G. C. Stirling, C. J. Ludman, and T. C. Waddington, *J. Chem. Phys.*, **52,** 2730 (1970).
206. J. W. Nibler and G. C. Pimentel, *J. Chem. Phys.*, **47,** 710 (1967).
207. J. C. Evans and G. Y-S. Lo, *J. Phys. Chem.*, **71,** 3942 (1967).
208. B. S. Ault and L. Andrews, *J. Chem. Phys.*, **64,** 1986 (1976).
209. P. N. Noble, *J. Chem. Phys.*, **56,** 2088 (1972).
210. B. S. Ault and L. Andrews, *Inorg. Chem.*, **16,** 2024 (1977).
211. K. Kaya, N. Mikami, Y. Udagawa, and M. Ito, *Chem. Phys. Lett.*, **16,** 151 (1972).
212. W. Kiefer and H. J. Bernstein, *Chem. Phys. Lett.*, **16,** 5 (1972).
213. M. E. Heyde, L. Rimai, R. G. Kilponen, and D. Gill, *J. Am. Chem. Soc.*, **94,** 5222 (1972).
214. G. R. Smith and W. A. Guillory, *J. Chem. Phys.*, **56,** 1423 (1972).
215. D. E. Milligan and M. E. Jacox, *J. Chem. Phys.*, **43,** 4487 (1965).
216. T. Nakata and S. Matsushita, *J. Phys. Chem.*, **72,** 458 (1968).
217. A. Müller, R. Kebabcioglu, B. Krebs, P. Bouclier, J. Portier, and P. Hagenmuller, *Z. Anorg. Allg. Chem.*, **368,** 31 (1969).
218. W. S. Benedict, N. Gailar, and E. K. Plyler, *J. Chem. Phys.*, **24,** 1139 (1956).
219. G. E. Walrafen, *J. Chem. Phys.*, **40,** 3249 (1964).
220. C. Haas and D. F. Hornig, *J. Chem. Phys.*, **32,** 1763 (1960); D. F. Hornig, H. F. White, and F. P. Reding, *Spectrochim. Acta*, **12,** 338 (1958).
221. S. Pinchas and M. Halmann, *J. Chem. Phys.*, **31,** 1692 (1959).
222. P. A. Staats, H. W. Morgan, and J. H. Goldstein, *J. Chem. Phys.*, **24,** 916 (1956).
223. H. C. Allen and E. K. Plyler, *J. Chem. Phys.*, **25,** 1132 (1956).
224. A. J. Tursi and E. R. Nixon, *J. Chem. Phys.*, **53,** 518 (1970).
225. F. P. Reding and D. F. Hornig, *J. Chem. Phys.*, **27,** 1024 (1957).
226. A. H. Nielsen and H. H. Nielsen, *J. Chem. Phys.*, **5,** 277 (1937).
227. D. M. Cameron, W. C. Sears, and H. H. Nielsen, *J. Chem. Phys.*, **7,** 994 (1939).
228. E. Greinacher, W. Lüttke, and R. Mecke, *Z. Elektrochem.*, **59,** 23 (1955).
229. M. L. Josien and P. Saumagne, *Bull. Soc. Chim. Fr.*, 937 (1956).
230. W. A. Senior and W. K. Thompson, *Nature*, **205,** 170 (1965).
231. J. E. Bertie, H. J. Labbé, and E. Whalley, *J. Chem. Phys.*, **49,** 775, 2141 (1968); J. E. Bertie and E. Whalley, *J. Chem. Phys.*, **40,** 1637 (1964).
232. G. P. Ayers and A. D. E. Pullin, *Spectrochim. Acta*, **32A,** 1629 (1976).
233. H. C. Allen, E. D. Tidwell, and E. K. Plyler, *J. Chem. Phys.*, **25,** 302 (1956).
234. J. Pacansky and G. V. Calder, *J. Phys. Chem.*, **76,** 454 (1972).
235. P. A. Staats, H. W. Morgan, and J. H. Goldstein, *J. Chem. Phys.*, **25,** 582 (1956).
236. R. E. Dodd and R. Little, *Spectrochim. Acta*, **16,** 1083 (1960).
237. W. O. Freitag and E. R. Nixon, *J. Chem. Phys.*, **24,** 109 (1956).
238. T. B. Freedman and E. R. Nixon, *J. Chem. Phys.*, **56,** 698 (1972).
239. S. Hemple and E. R. Nixon, *J. Chem. Phys.*, **47,** 4273 (1967).
240. Z. K. Ismail, R. H. Hauge, and J. L. Margrave, *J. Mol. Spectrosc.*, **45,** 304 (1973).
241. D. E. Milligan and M. E. Jacox, *J. Chem. Phys.*, **47,** 278 (1967).
242. Z. K. Ismail, R. H. Hauge, and J. L. Margrave, *J. Chem. Phys.*, **57,** 5137 (1972).
243. E. R. Lory and R. F. Porter, *J. Am. Chem. Soc.*, **93,** 6301 (1971).
244. M. E. Jacox, D. E. Milligan, N. G. Mall, and W. E. Thompson, *J. Chem. Phys.*, **43,** 3734 (1965).
245. A. Maki, E. K. Plyler, and E. D. Tidwell, *J. Res. Natl. Bur. Stds.*, **66A,** 163 (1962).
246. T. Wentink, Jr., *J. Chem. Phys.*, **29,** 188 (1958).
247. W. Beck, *Chem. Ber.*, **95,** 341 (1962); **98,** 298 (1965).
248. K. O. Christe, R. D. Wilson, and W. Sawodny, *J. Mol. Struct.*, **8,** 245 (1971).
249. D. E. Milligan, M. E. Jacox, and A. M. Bass, *J. Chem. Phys.*, **43,** 3149 (1965).
250. G. M. Begun and W. H. Fletcher, *J. Chem. Phys.*, **28,** 414 (1958).
251. V. E. Bondybey and J. H. English, *J. Chem. Phys.*, **67,** 664 (1979).
252. D. E. Milligan and M. E. Jacox, *J. Chem. Phys.*, **47,** 5157 (1967).
253. O. H. Ellestad, P. Klaeboe, E. E. Tucker, and J. Songstad, *Acta Chem. Scand.*, **26,** 1721 (1972).
253a. O. H. Ellestad, P. Klaeboe and J. Songstad, *Acta Chem. Scand.*, **26,** 1724 (1972).

254. P. Gray and T. C. Waddington, *Trans. Faraday Soc.*, **53**, 901 (1957).
255. P. O. Kinell and B. Strandberg, *Acta Chem. Scand.*, **13**, 1607 (1959).
256. H. W. Morgan, *J. Inorg. Nucl. Chem.*, **16**, 368 (1960).
257. D. E. Milligan and M. E. Jacox, *J. Chem. Phys.*, **51**, 277 (1969).
258. J. F. Ogilvie, *Spectrochim. Acta*, **23A**, 737 (1967).
259. D. E. Milligan and M. E. Jacox, *J. Chem. Phys.*, **38**, 2627 (1963); *J. Mol. Spectrosc.*, **42**, 495 (1972).
260. R. D. Spratley, J. J. Turner, and G. C. Pimentel, *J. Chem. Phys.*, **44**, 2063 (1966).
261. A. Arkell and I. Schwager, *J. Am. Chem. Soc.*, **89**, 5999 (1967).
262. D. E. Tevault and R. R. Smardzewski, *J. Am. Chem. Soc.*, **100**, 3955 (1978).
263. J. A. Goleb, H. H. Claassen, M. H. Studier, and E. H. Appelman, *Spectrochim. Acta*, **28A**, 65 (1972).
264. I. Schwager and A. Arkell, *J. Am. Chem. Soc.*, **89**, 6006 (1967).
265. N. Walker, D. E. Tevault, and R. R. Smardzewski, *J. Chem. Phys.*, **69**, 564 (1978).
266. R. J. Isabel and W. A. Guillory, *J. Chem. Phys.*, **57**, 1116 (1972).
267. R. R. Smardzewski and W. B. Fox, *J. Am. Chem. Soc.*, **96**, 304 (1974); *J. Chem. Phys.*, **60**, 2104 (1974).
268. A. Müller, G. Nagarajan, O. Glemser, and J. Wegener, *Spectrochim. Acta*, **23A**, 2683 (1967); A. Müller, N. Mohan, S. J. Cyvin, N. Weinstock, and O. Glemser, *J. Mol. Spectrosc.*, **59**, 161 (1976).
269. R. R. Ryan and L. H. Jones, *J. Chem. Phys.*, **50**, 1492 (1969); L. H. Jones, L. B. Asprey, and R. R. Ryan, *J. Chem. Phys.*, **47**, 3371 (1967).
270. L. H. Jones, R. R. Ryan, and L. B. Asprey, *J. Chem. Phys.*, **49**, 581 (1968).
271. J. Laane, L. H. Jones, R. R. Ryan, and L. B. Asprey, *J. Mol. Spectrosc.*, **30**, 485 (1969).
272. M. Feuerhahn, W. Hilbig, R. Minkwitz, and U. Engelhardt, *Spectrochim. Acta*, **34A**, 1065 (1978).
273. R. D. Verma and S. Nagaraj, *J. Mol. Spectrosc.*, **58**, 301 (1975).
274. D. E. Milligan, M. E. Jacox, A. M. Bass, J. J. Comeford, and D. E. Mann, *J. Chem. Phys.*, **42**, 3187 (1965).
275. M. E. Jacox and D. E. Milligan, *J. Chem. Phys.*, **43**, 866 (1965).
276. L. Andrews, F. K. Chi, and A. Arkell, *J. Am. Chem. Soc.*, **96**, 1997 (1974).
277. H. Richert and O. Glemser, *Z. Anorg. Allg. Chem.*, **307**, 328 (1961).
278. C. E. Smith, D. E. Milligan, and M. E. Jacox, *J. Chem. Phys.*, **54**, 2780 (1971).
279. S. C. Peake and A. J. Downs, *J. Chem. Soc., Dalton Trans.*, 859 (1974).
280. G. A. Ozin and A. Vander Voet, *J. Chem. Phys.*, **56**, 4768 (1972).
281. G. Herzberg, *Infrared and Raman Spectra of Polyatomic Molecules*, Van Nostrand, Princeton, N.J., 1945, p. 295.
282. F. P. Reding and D. F. Hornig, *J. Chem. Phys.*, **19**, 594 (1951); **22**, 1926 (1954).
283. H. W. Morgan, P. A. Staats, and J. H. Goldstein, *J. Chem. Phys.*, **27**, 1212 (1957).
284. S. Sundaram and F. F. Cleveland, *J. Mol. Spectrosc.*, **5**, 61 (1960).
285. E. Lee and C. K. Wu, *Trans. Faraday Soc.*, **35**, 1366 (1939).
286. W. H. Haynie and H. H. Nielsen, *J. Chem. Phys.*, **21**, 1839 (1953).
287. P. V. Huong and B. Desbat, *J. Raman Spectrosc.*, **2**, 373 (1974).
288. R. C. Taylor and G. L. Vidale, *J. Am. Chem. Soc.*, **78**, 5999 (1956).
289. R. Savoie and P. A. Giguère, *J. Chem. Phys.*, **41**, 2698 (1964).
290. P. A. Giguère and C. Madec, *Chem. Phys. Lett.*, **37**, 569 (1976).
291. T. Birchall and I. Drummond, *J. Chem. Soc. A*, 3162 (1971).
292. R. D. Gillard and G. Wilkinson, *J. Chem. Soc.*, 1640 (1964).
293. S. Sundaram, F. Suszek, and F. F. Cleveland, *J. Chem. Phys.*, **32**, 251 (1960).
294. G. DeAlti, G. Costa, and V. Galasso, *Spectrochim. Acta*, **20**, 965 (1964).
295 M. Pariseau, E. Wu, and J. Overend, *J. Chem. Phys.*, **39**, 217 (1963).
296. J. G. Contreras, J. S. Poland, and D. G. Tuck, *J. Chem. Soc., Dalton Trans.*, 922 (1973).
297. D. E. Milligan, M. E. Jacox, and J. J. Comeford, *J. Chem. Phys.*, **44**, 4058 (1966).
298. D. E. Milligan, M. E. Jacox, and W. A. Guillory, *J. Chem. Phys.*, **49**, 5330 (1968).

299. M. E. Jacox and D. E. Milligan, *J. Chem. Phys.*, **49**, 3130 (1968).
300. W. A. Guillory and C. E. Smith, *J. Chem. Phys.*, **53**, 1661 (1970).
301. P. S. Poskozim and A. L. Stone, *J. Inorg. Nucl. Chem.*, **32**, 1391 (1970).
302. I. Wharf and D. F. Shriver, *Inorg. Chem.*, **8**, 914 (1969).
303. B. Basak, *Inorg. Chim. Acta*, **45**, L47 (1980).
304. J. Shamir and H. H. Hyman, *Spectrochim. Acta*, **23A**, 1899 (1967).
305. P. J. Hendra and J. R. Mackenzie, *Chem. Commun.*, 760 (1968).
306. R. J. H. Clark and D. M. Rippon, *J. Mol. Spectrosc.*, **52**, 58 (1974).
307. H. Stammreich, R. Forneris, and Y. Tavares, *J. Chem. Phys.*, **25**, 580 (1956).
308. W. V. F. Brooks, J. Passmore, and E. K. Richardson, *Can. J. Chem.*, **57**, 3230 (1979).
309. C. J. Adams and A. J. Downs, *J. Chem. Soc. A*, 1534 (1971).
310. E. Denchik, S. C. Nyburg, G. A. Ozin, and J. T. Szymanski, *J. Chem. Soc. A*, 3157 (1971).
311. V. A. Maroni and P. T. Cunningham, *Appl. Spectrosc.*, **27**, 428 (1973).
312. T. R. Manley and D. A. Williams, *Spectrochim. Acta*, **21**, 1773 (1965).
313. H. E. Doorenbos, J. C. Evans, and R. O. Kagel, *J. Phys. Chem.*, **74**, 3385 (1970).
314. J. A. Evans and D. A. Long, *J. Chem. Soc. A*, 1688 (1968).
315. J. Passmore, E. K. Richardson, and P. Taylor, *Inorg. Chem.*, **17**, 1681 (1978).
316. E. A. Robinson and J. A. Ciruna, *Can. J. Chem.*, **46**, 3197 (1968).
317. D. M. Adams and P. J. Lock, *J. Chem. Soc. A*, 145 (1967).
318. J. Kouinis and A. G. Galinos, *Monatsh. Chem.*, **108**, 835 (1977).
319. A. Givan and A. Loewenschuss, *J. Raman Spectrosc.*, **6**, 84 (1977).
320. J. C. Evans and H. J. Bernstein, *Can. J. Chem.*, **33**, 1270 (1955).
321. H. Siebert, *Z. Anorg. Allg. Chem.*, **275**, 225 (1955).
322. D. J. Gardiner, R. B. Girling, and R. E. Hester, *J. Mol. Struct.*, **13**, 105 (1972).
323. W. Sterzel and W. D. Schnee, *Z. Anorg. Allg. Chem.*, **383**, 231 (1971).
324. H. H. Claassen and G. Knapp, *J. Am. Chem. Soc.*, **86**, 2341 (1964).
325. C. Rocchiciolli, *Compt. Rend.*, **242**, 2922 (1956); **244**, 2704 (1957); **247**, 1108 (1958); **249**, 236 (1959).
326. W. E. Dasent and T. C. Waddington, *J. Chem. Soc.*, 2429, 3350 (1960).
327. J. J. Comeford, D. E. Mann, L. J. Schoen, and D. R. Lide, *J. Chem. Phys.*, **38**, 461 (1963).
328. G. E. Moore and R. M. Badger, *J. Am. Chem. Soc.*, **74**, 6076 (1952).
329. J. J. Comeford, *J. Chem. Phys.*, **45**, 3463 (1966); J. J. Comeford, D. E. Mann, L. J. Schoen, and D. R. Lide, *J. Chem. Phys.*, **38**, 461 (1963).
330. A. Müller, E. Niecke, B. Krebs, and O. Glemser, *Z. Naturforsch.*, **23b**, 588 (1968); A. Müller, K. Königer, S. J. Cyvin, and A. Fadini, *Spectrochim. Acta*, **29A**, 219 (1973).
331. C. R. S. Dean, A. Finch, and P. N. Crates, *J. Chem. soc., Dalton Trans.*, 1384 (1972).
332. M. Adelhelm and E. Jacob, *Angew. Chem. Int. Ed.*, **16**, 461 (1977).
333. J. K. O'Loane and M. K. Wilson, *J. Chem. Phys.*, **23**, 1313 (1955).
334. D. E. Martz and R. T. Lagemann, *J. Chem. Phys.*, **22**, 1193 (1954).
335. H. Stammreich, R. Forneris, and Y. Tavares, *J. Chem. Phys.*, **25**, 1277 (1956).
336. D. F. Burow, *Inorg. Chem.*, **11**, 573 (1972).
337. R. J. Gillespie, P. H. Spekkens, J. B. Milne, and D. M. Moffett, *J. Fluorine Chem.*, **7**, 43 (1976).
338. L. E. Alexander and I. R. Beattie, *J. Chem. Soc., Dalton Trans.*, 1745 (1972).
339. J. A. Rolfe and L. A. Woodward, *Trans. Faraday Soc.*, **51**, 779 (1955).
340. A. Simon and R. Paetzold, *Z. Anorg. Allg. Chem.*, **301**, 246 (1959); *Naturwissenschaften*, **44**, 108 (1957).
341. M. Falk and P. A. Giguère, *Can. J. Chem.*, **34**, 1680 (1958).
342. D. F. Smith, G. M. Begun, and W. H. Fletcher, *Spectrochim. Acta*, **20**, 1763 (1964).
343. R. J. Gillespie and P. H. Spekkens, *J. Chem. Soc., Dalton Trans.*, 1539 (1977).
344. K. O. Christe, E. C. Curtis, and C. J. Schack, *Inorg. Chem.*, **11**, 2212 (1972).
345. R. Bougon, J. Isabey, and P. Plurien, *Compt. Rend.*, **273C**, 415 (1971).
346. A. Simon and R. Paetzold, *Z. Anorg. Allg. Chem.*, **303**, 39, 46, 53, 72, 79 (1960); *Z. Elektrochem.*, **64**, 209 (1960).
347. R. Steudel and D. Lautenbach, *Z. Naturforsch.*, **24b**, 350 (1969).

348. M. Goldstein and G. C. Tok, *J. Chem. Soc. A*, 2303 (1971).
349. A. Kaldor and R. F. Porter, *J. Am. Chem. Soc.*, **93,** 2140 (1971).
350. J. Vanderryn, *J. Chem. Phys.*, **30,** 331 (1959).
351. D. A. Dows, *J. Chem. Phys.*, **31,** 1637 (1959).
352. R. J. H. Clark and P. D. Mitchell, *J. Chem. Phys.*, **56,** 2225 (1972).
353. D. A. Dows and G. Bottger, *J. Chem. Phys.*, **34,** 689 (1961).
354. T. Wentink, Jr. and V. H. Tiensuu, *J. Chem. Phys.*, **28,** 826 (1958).
355. Y. S. Yang and J. S. Shirk, *J. Mol. Spectrosc.*, **54,** 39 (1975).
356. I. R. Beattie, H. E. Blayden, S. M. Hall, S. N. Jenny, and J. S. Ogden, *J. Chem. Soc., Dalton Trans.*, 666 (1976).
357. I. R. Beattie and J. R. Horder, *J. Chem. Soc. A*, 2655 (1969).
358. D. F. Wolfe and G. L. Humphrey, *J. Mol. Struct.*, **3,** 293 (1969).
359. G. K. Selivanov and A. A. Mal'tsev, *Zh. Strukt. Khim.*, **14,** 943 (1973).
360. R. G. Pong, A. E. Shirk, and J. S. Shirk, *J. Mol. Spectrosc.*, **66,** 35 (1977).
361. J. E. D. Davies and D. A. Long, *J. Chem. Soc. A*, 2050 (1968).
362. L. Andrews and G. C. Pimentel, *J. Chem. Phys.*, **47,** 3637 (1967).
363. J. E. D. Davies and D. A. Long, *J. Chem. Soc. A*, 2054 (1968).
364. P. Biscarini, L. Fusina, G. Nivellini, and G. Pelizzi, *J. Chem. Soc., Dalton Trans.*, 664 (1977).
365. R. D. Wesley and C. W. DeKock, *J. Chem. Phys.*, **55,** 3866 (1971); M. Lesiecki, J. W. Nibler, and C. W. DeKock, *J. Chem. Phys.*, **57,** 1352 (1972).
366. K. Shimizu and H. Shingu, *Spectrochim. Acta*, **22,** 1999 (1966).
367. S. Konaka, Y. Murata, K. Kuchitsu, and Y. Morino, *Bull. Chem. Soc. Jpn.*, **39,** 1134 (1966).
368. L. Beckmann, L. Gutjahr, and R. Mecke, *Spectrochim. Acta*, **21,** 141 (1965).
369. I. W. Levin and S. Abramowitz, *J. Chem. Phys.*, **43,** 4213 (1965).
370. J. L. Duncan, *J. Mol. Spectrosc.*, **13,** 338 (1964).
371. C. D. Bass, L. Lynds, T. Wolfram, and R. E. DeWames, *J. Chem. Phys.*, **40,** 3611 (1964).
372. W. C. Steele and J. C. Decius, *J. Chem. Phys.*, **25,** 1184 (1956).
373. P. E. Bethell and N. Sheppard, *Trans. Faraday Soc.*, **51,** 9 (1959).
374. S. Bhagavantum and T. Venkatarayudu, *Proc. Indian Acad. Sci.*, **9A,** 224 (1939).
375. A. Müller and M. Stockburger, *Z. Naturforsch.*, **20A,** 1242 (1965).
376. A. Müller, N. Mohan, P. Cristophliemk, I. Tossidis, and M. Dräger, *Spectrochim. Acta*, **29A,** 1345 (1973).
377. A. Müller, G. Gattow, and H. Seidel, *Z. Anorg. Allg. Chem.*, **347,** 24 (1966).
378. I. Nakagawa and J. L. Walter, *J. Chem. Phys.*, **51,** 1389 (1969).
379. K. Stopperka, *Z. Anorg, Allg. Chem.*, **345,** 277 (1966).
380. A. Kalder, A. G. Maki, A. J. Dorney, and I. M. Mills, *J. Mol. Spectrosc.*, **45,** 247 (1973).
381. J. P. Laperches and P. Tarte, *Spectrochim. Acta*, **22,** 1201 (1966).
382. P. LaBonville, R. Kugel, and J. R. Ferraro, *J. Chem. Phys.*, **67,** 1477 (1977).
383. M. E. Jacox and D. E. Milligan, *J. Chem. Phys.*, **54,** 919 (1971).
384. C. C. Addison and B. M. Gatehouse, *J. Chem. Soc.*, 613 (1960).
385. J. R. Ferraro and A. Walker, *J. Chem. Phys.*, **45,** 550 (1966).
386. G. E. Walrafen and D. E. Irish, *J. Chem. Phys.*, **40,** 911 (1964).
387. G. J. Janz and T. R. Kozlowski, *J. Chem. Phys.*, **40,** 1699 (1964).
388. D. Smith, D. W. James, and J. P. Devlin, *J. Chem. Phys.*, **54,** 4437 (1971).
389. C. J. Peacock, A. Müller, and R. Kebabcioglu, *J. Mol. Struct.*, **2,** 163 (1968).
390. P. Thirugnanasambandam and G. J. Srinivasan, *J. Chem. Phys.*, **50,** 2467 (1969).
391. R. A. Frey, R. L. Redington, and A. L. K. Aljibury, *J. Chem. Phys.*, **54,** 344 (1971).
392. R. J. Gillespie, B. Landa, and G. J. Schrobilgen, *Inorg. Chem.*, **15,** 1256 (1976).
393. R. J. Gillespie and G. J. Schrobilgen, *Chem. Commun.*, 595 (1977).
394. D. W. Green, G. T. Reedy, and S. D. Gabelnick, *J. Chem. Phys.*, **73,** 4207 (1980).
395. H. Cohn, C. K. Ingold, and H. G. Poole, *J. Chem. Soc.*, 4272 (1952).
396. D. L. Bernitt, R. H. Miller, and I. C. Hisatsune, *Spectrochim. Acta*, **23A,** 237 (1967).
397. D. L. Bernitt, K. O. Hartman, and I. C. Hisatsune, *J. Chem. Phys.*, **42,** 3553 (1965).
398. K. Itoh and H. J. Bernstein, *Can. J. Chem.*, **34,** 170 (1956).

399. P. D. Mallinson, D. C. McKean, J. H. Holloway, and I. A. Oxton, *Spectrochim. Acta*, **31A**, 143 (1975).
400. J. Overend and J. C. Evans, *Trans. Faraday Soc.*, **55**, 1817 (1959).
401. R. F. Stratton and A. H. Nielsen, *J. Mol. Spectrosc.*, **4**, 373 (1960).
402. M. E. Jacox and D. E. Milligan, *J. Mol. Spectrosc.*, **58**, 142 (1975).
403. A. J. Downs, *Spectrochim. Acta*, **19**, 1165 (1963).
404. I. S. Butler and A. M. English, *Spectrochim. Acta*, **33A**, 545 (1977).
405. A. Haas, B. Koch, N. Welcman, and H. Willner, *Spectrochim. Acta*, **32A**, 497 (1976).
406. A. Müller, N. Mohan, P. Cristophliemk, I. Tossidis, and M. Dräger, *Spectrochim. Acta*, **29A**, 1345 (1973).
407. C. A. Wamser, W. B. Fox, B. Sukornick, J. R. Holmes, B. B. Stewart, R. Juurik, N. Vanderkooi, and D. Gould, *Inorg. Chem.*, **8**, 1249 (1969).
408. A. Allan, J. L. Duncan, J. H. Holloway, and D. C. McKean, *J. Mol. Spectrosc.*, **31**, 368 (1969).
409. K. Dehnicke, H. Aeissen, M. Kölmel, and J. Strähle, *Angew. Chem. Inst. Ed.*, **16**, 547 (1977).
410. D. F. Wolfe and G. L. Humphrey, *J. Mol. Struct.*, **3**, 293 (1969).
411. R. G. S. Pong, A. E. Shirk, and J. S. Shirk, *Ber. Bunsenges. Phys. Chem.*, **82**, 79 (1978).
412. J. Goubeau and K. Laitenberger, *Z. Anorg. Allg. Chem.*, **320**, 78 (1963).
413. J. E. Rauch and J. C. Decius, *Spectrochim. Acta*, **22**, 1963 (1966).
414. G. E. McGraw, D. L. Bernitt, and I. C. Hisatsune, *Spectrochim. Acta*, **23A**, 25 (1967).
415. S. T. King and J. Overend, *Spectrochim. Acta*, **23A**, 61 (1967).
416. S. T. King and J. Overend, *Spectrochim. Acta*, **22**, 689 (1966).
417. P. A. Giguère and T. K. K. Srinivasan, *J. Raman Spectrosc.*, **2**, 125 (1974).
418. D. J. Gardiner, N. J. Lawrence, and J. J. Turner, *J. Chem. Soc. A*, 400 (1971).
419. N. Zengin and P. A. Giguère, *Can. J. Chem.*, **37**, 632 (1959).
420. R. D. Brown and G. P. Pez, *Spectrochim. Acta*, **26A**, 1375 (1970).
421. S. G. Frankiss and D. S. Harrison, *Spectrochim. Acta*, **31A**, 161 (1975).
422. R. Forneris and C. E. Hennies, *J. Mol. Struct.*, **5**, 449 (1970).
423. W. A. Guillory and C. E. Hunter, *J. Chem. Phys.*, **50**, 3516 (1969).
424. N. Wiberg, G. Fischer, and H. Bachhuber, *Angew. Chem. Int. Eds.*, **16**, 780 (1977).
425. J. A. A. Ketelaar, F. N. Hooge, and G. Blasse, *Rec. Trav. Chim.*, **75**, 220 (1956): F. N. Hooge and J. A. A. Ketelaar, *Rec. Trav. Chim.*, **77**, 902 (1958).
426. W. Kiefer, *Spectrochim. Acta*, **27A**, 1285 (1971).
427. P. A. Giguère and O. Bain, *J. Phys. Chem.*, **56**, 340 (1952).
428. C. A. Frenzel and K. E. Blick, *J. Chem. Phys.*, **55**, 2715 (1971).
429. W. Beck and K. Feldl, *Agnew. Chem.*, **78**, 746 (1966); W. Beck, P. Swoboda, K. Feldl, and R. S. Tobias, *Chem. Ber.*, **104**, 533 (1971).
430. G. R. Draper and R. L. Werner, *J. Mol. Spectrosc.*, **50**, 369 (1974).
431. D. A. Dows and G. C. Pimental, *J. Chem. Phys.*, **23**, 1258 (1955).
432. J. R. Durig and D. W. Wertz, *J. Chem. Phys.*, **46**, 3069 (1967).
433. C. B. Moore and K. Rosengren, *J. Chem. Phys.*, **44**, 4108 (1966).
434. P. O. Tchir and R. D. Spratley, *Can. J. Chem.*, **53**, 2311 (1975).
435. P. O. Tchir and R. D. Spratley, *Can. J. Chem.*, **53**, 2331 (1975).
436. W. A. Guillory and C. E. Hunter, *J. Chem. Phys.*, **54**, 598 (1971).
437. R. T. Hall and G. C. Pimentel, *J. Chem. Phys.*, **38**, 1889 (1963).
438. M. E. Jacox and D. E. Milligan, *J. Chem. Phys.*, **40**, 2457 (1964).
439. H. H. Eysel, *J. Mol. Struct.*, **5**, 275 (1970).
440. M. J. Nielsen and A. D. E. Pullin, *J. Chem. Soc.*, 604 (1960).
441. W. T. Thompson and W. H. Fletcher, *Spectrochim. Acta*, **22**, 1907 (1966).
442. A. R. Emery and R. C. Taylor, *J. Chem. Phys.*, **28**, 1029 (1958).
443. C. J. H. Schutte, *Spectrochim. Acta*, **16**, 1054 (1960); J. A. A. Ketelaar and C. J. H. Schutte, *Spectrochim. Acta*, **17**, 1240 (1961).
444. A. E. Shirk and D. F. Shriver, *J. Am. Chem. Soc.*, **95**, 5904 (1973).
445. Landolt-Börnstein, *Physikalisch-chemische Tabellen*, Vol. 2, 1951.
446. G. E. MacWood and H. C. Urey, *J. Chem. Phys.*, **4**, 402 (1936).

447. H. M. Kaylor and A. H. Nielsen, *J. Chem. Phys.*, **23,** 2139 (1955).
448. I. F. Kovalev, *Opt. Spektrosk.*, **2,** 310 (1957).
449. R. E. Wilde, T. K. K. Srinivasan, R. W. Harral, and S. G. Sankar, *J. Chem. Phys.*, **55,** 5681 (1971).
450. J. H. Meal and M. K. Wilson, *J. Chem. Phys.*, **24,** 385 (1956).
451. L. P. Lindemann and M. K. Wilson, *J. Chem. Phys.*, **22,** 1723 (1954).
452. I. W. Levin and H. Ziffer, *J. Chem. Phys.*, **43,** 4023 (1965).
453. H. W. Morgan, P. A. Staats, and J. H. Goldstein, *J. Chem: Phys.*, **27,** 1212 (1957).
454. J. R. Durig, D. J. Antion, and F. G. Baglin, *J. Chem. Phys.*, **49,** 666 (1968).
455. J. R. Durig, C. B. Pate and Y. S. Li, *J. Chem. Phys.*, **54,** 1033 (1971).
456. E. L. Wagner and D. F. Hornig, *J. Chem. Phys.*, **18,** 296, 305 (1950); R. C. Plumb and D. F. Hornig, *J. Chem. Phys.*, **21,** 366 (1953); **23,** 947 (1955); W. Vedder and D. F. Hornig, *J. Chem. Phys.*, **35,** 1560 (1961).
457. A. S. Quist, J. B. Bates, and G. E. Boyd, *J. Phys. Chem.*, **76,** 78 (1972).
458. V. A. Maroni, *J. Chem. Phys.*, **55,** 4789 (1971).
459. A. S. Quist, J. B. Bates, and G. E. Boyd, *J. Chem. Phys.*, **54,** 4896 (1971).
460. M. C. Dhamelincourt and M. Migeon, *Compt. Rend.*, **281,** C79 (1975).
461. J. A. Creighton, *J. Chem. Soc.*, 6589 (1965).
462. B. Gilbert, G. Mamantov, and G. M. Begun, *Inorg. Nucl. Chem. Lett.*, **10,** 1123 (1974).
463. E. Rytter and H. A. Øye, *J. Inorg. Nucl. Chem.*, **35,** 4311 (1973).
464. D. H. Brown and D. T. Stewart, *Spectrochim. Acta*, **26A,** 1344 (1970).
465. G. M. Begun, C. R. Boston, G. Torsi, and G. Mamantov, *Inorg. Chem.*, **10,** 886 (1971).
466. H. A. Øye and W. Bues, *Inorg. Nucl. Chem. Lett.*, **8,** 31 (1972).
469. L. A. Woodward and A. A. Nord, *J. Chem. Soc.*, 2655 (1955).
468. L. A. Woodward and G. H. Singer, *J. Chem. Soc.*, 716 (1958).
469. L. A. Woodward and M. J. Taylor, *J. Chem. Soc.*, 4473 (1960).
470. L. A. Woodward and P. T. Bill, *J. Chem. Soc.*, 1699 (1955).
471. D. M. Adams and D. M. Morris, *J. Chem. Soc. A*, 694 (1968).
472. R. J. H. Clark and D. M. Rippon, *Chem. Commun.*, 1295 (1971); R. J. H. Clark and P. D. Mitchell, *J. Chem. Soc., Faraday Trans. 2*, **71,** 515 (1975).
473. R. R. Haun and W. D. Harkins, *J. Am. Chem. Soc.*, **54,** 3917 (1932).
474. H. Stammreich, Y. Tavares, and D. Bassi, *Spectrochim. Acta*, **17,** 661 (1961).
475. R. J. H. Clark and T. J. Dines, *Inorg. Chem.*, **19,** 1681 (1980).
476. A. D. Caunt, L. N. short, and L. A. Woodward, *Trans. Faraday Soc.*, **48,** 873 (1952); *Nature*, **168,** 557 (1951).
477. R. J. H. Clark and B. K. Hunter, *J. Mol. Struct.*, **9,** 354 (1971).
478. K. O. Christe, *Spectrochim. Acta*, **36A,** 921 (1980).
479. P. Van Huong and B. Desbat, *Bull. Soc. Chim. Fr.*, 2631 (1972).
480. M. Delahaye, P. Dhamelincourt, and J. C. Merlin, *Compt. Rend.*, **272B,** 370 (1971).
481. W. Gabes and H. Gerding, *Rec. Trav. Chim.*, **90,** 157 (1971); W. Gabes, K. Olie, and H. Gerding, *Rec. Trav. Chim.*, **91,** 1367 (1972).
482. A. Müller and A. Fadini, *Z. anorg. Allg. Chem.*, **349,** 164 (1967).
483. J. Weidlein and K. Dehnicke, *Z. Anorg. Allg. Chem.*, **337,** 113 (1965).
484. A. Sabatini and L. Sacconi, *J. Am. Chem. Chem. Soc.*, **86,** 17 (1964).
485. I. R. Beattie, T. R. Gilson, and G. A. Ozin, *J. Chem. Soc. A*, 534 (1969).
486. J. S. Avery, C. D. Burbridge, and D. M. L. Goodgame, *Spectrochim. Acta*, **24A,** 1721 (1968).
487. P. L. Goggin, R. J. Goodfellow, and K. Kessler, *J. Chem. Soc., Dalton Trans.*, 1914 (1977).
488. G. J. Janz and D. W. James, *J. Chem. Phys.*, **38,** 905 (1963).
489. D. A. Long and J. Y. H. Chau, *Trans. Faraday Soc.*, **58,** 2325 (1962).
490. L. E. Alexander and I. R. Beattie, *J. Chem. Soc., Dalton Trans.*, 1745 (1972).
491. R. J. H. Clark, B. K. Hunter, and D. M. Rippon, *Inorg. Chem.*, **11,** 56 (1972); R. J. H. Clark and D. M. Rippon, *J. Mol. Spectrosc.*, **44,** 479 (1972).
492. A. Büchler, J. B. Berkowitz-Mattuck, and D. H. Dugre, *J. Chem. Phys.*, **34,** 2202 (1961).
493. M. F. A. Dove, J. A. Creighton, and L. A. Woodward, *Spectrochim Acta*, **18,** 267 (1962).

494. B. Cuoni, F. P. Emmenegger, C. Rohrbasser, C. W. Schläpfer, and P. Studer, *Spectrochim. Acta*, **34A**, 247 (1978).
495. H. G. M. Edwards, M. J. Ware, and L. A. Woodward, *Chem. Commun.*, 540 (1968).
496. H. G. M. Edwards, L. A. Woodward, M. J. Gall, and M. J. Ware, *Spectrochim. Acta*, **26A**, 287 (1970).
497. W. Krasser and H. W. Nürnberg, *Spectrochim. Acta*, **26A**, 1059 (1970).
498. D. M. Adams and P. J. Lock, *J. Chem. Soc. A*, 620 (1967).
499. D. N. Anderson and R. D. Willett, *Inorg. Chim. Acta*, **8**, 167 (1974).
500. R. D. Willett, J. R. Ferraro, and M. Choca, *Inorg. Chem.*, **13**, 2919 (1974).
501. D. Forster, *Chem. Commun.*, 113 (1967).
502. P. L. Goggin and T. G. Buick, *Chem. Commun.*, 290 (1967).
503. R. A. Work, III, and M. L. Good, *Spectrochim. Acta*, **28A**, 1537 (1972).
504. J. T. R. Dunsmuir and A. P. Lane, *J. Chem. Soc. A*, 404, 2781 (1971).
505. R. J. H. Clark and P. D. Mitchell, *J. Am. Chem. Soc.*, **95**, 8300 (1973); *J. Raman Spectrosc.*, **2**, 399 (1974).
506. R. J. H. Clark and P. D. Mitchell, *Chem. Commun.*, 762 (1973).
507. T. Kamisuki and S. Maeda, *Chem. Phys. Lett.*, **21**, 330 (1973).
508. S. T. King, *J. Chem. Phys.*, **49**, 1321 (1968).
509. F. Königer and A. Müller, *J. Mol. Spectrosc.*, **56**, 200 (1975).
510. F. Königer, A. Müller, and K. Nakamoto, *Z. Naturforsch.*, **30b**, 456 (1975).
511. D. Tevault, J. D. Brown, and K. Nakamoto, *Appl. Spectrosc.*, **30**, 461 (1976).
512. A. Müller and B. Krebs, *J. Mol. Spectrosc.*, **24**, 180 (1967); A. Müller and A. Fadini, *Z. Anorg. Allg. Chem.*, **349**, 164 (1967).
513. L. J. Basile, J. R. Ferraro, P. LaBonville, and M. C. Wall, *Coord, Chem. Rev.*, **11**, 21 (1973).
514. C. J. Adams and A. J. Downs, *J. Chem. Soc. A*, 1534 (1971).
515. J. Milne, *Can. J. Chem.*, **53**, 888 (1975).
516. G. Y. Ahlijah and M. Goldstein, *J. Chem. Soc. A*, 326 (1970); *Chem. Commun.*, 1356 (1968).
517. K. O. Christe, H. Willner, and W. Sawodny, *Spectrochim. Acta*, **35A**, 1347 (1979).
518. K. Seppelt, *Z. Anorg. Allg. Chem.*, **416**, 12 (1975).
519. C. J. Adams and A. J. Downs, *Spectrochim. Acta*, **28A**, 1841 (1972).
520. K. O. Christe and W. Sawodny, *Inorg. Chem.*, **12**, 2879 (1973).
521. L. E. Alexander and I. R. Beattie, *J. Chem. Soc., Dalton Trans.*, 1745 (1972).
522. M. Robineau and D. Zins, *Compt. Rend.*, **280**, C759 (1975).
523. Landolt-Börnstein, *Physikalisch-Chemische Tabellen*, Vol. 2, 1951.
524. D. Fortnum and J. O. Edwards, *J. Inorg. Nucl. Chem.*, **2**, 264 (1956).
525. E. Steger and K. Herzog, *Z. Anorg. Allg. Chem.*, **331**, 169 (1964).
526. E. Steger and W. Schmidt, *Ber. Bunsenges. Phys. Chem.*, **68**, 102 (1964).
527. O. Sala and M. L. A. Temperini, *Chem. Phys. Lett.*, **36**, 652 (1975).
528. M. Jansen, *Angew. Chem. Int. Ed.*, **16**, 534 (1977).
529. H. Siebert, *Z. Anorg. Allg. Chem.*, **275**, 225 (1954).
530. L. C. Brown, G. M. Begun, and G. E. Boyd, *J. Am. Chem. Soc.*, **91**, 2250 (1969).
531. E. H. Appelman, *Inorg. Chem.*, **8**, 223 (1969).
532. H. Siebert, *Z. Anorg. Allg. Chem.*, **273**, 21 (1953).
533. R. S. McDowell and L. B. Asprey, *J. Chem. Phys.*, **57**, 3062 (1972).
534. F. Gonzalez-Vilchez and W. P. Griffith, *J. Chem. Soc., Dalton Trans.*, 1416 (1972).
535. N. Weinstock, H. Schulze, and A. Müller, *J. Chem. Phys.*, **59**, 5063 (1973).
535a. H. Homborg and W. Preetz, *Spectrochim. Acta*, **32A**, 709 (1976).
536. A. Müller, K. H. Schmidt, K. H. Tytko, J. Bouwma, and F. Jellinek, *Spectrochim. Acta*, **28A**, 381 (1972).
537. A. Müller, R. Kebabcioglu, M. J. F. Leroy, and G. Kaufmann, *Z. Naturforsch.*, **23b**, 740 (1968).
538. A. Müller, N. Weinstock, and H. Schulze, *Spectrochim. Acta*, **28A**, 1075 (1972); K. H. Schmidt and A. Müller, *Spectrochim. Acta*, **28A**, 1829 (1972).
539. A. Müller, B. Krebs, R. Kebabcioglu, M. Stockburger, and O. Glemser, *Spectrochim. Acta*, **24A**, 1831 (1968).

540. E. J. Baran and S. G. Manca, *Spectrosc. Lett.*, **15,** 455 (1982).
541. A. Müller, E. Diemann, and U. V. K. Rao, *Chem. Ber.*, **103,** 2961 (1970).
542. R. S. McDowell, L. B. Asprey, and L. C. Hoskins, *J. Chem. Phys.*, **56,** 5712 (1972).
543. R. S. McDowell, *Inorg. Chem.*, **6,** 1759 (1967); R. S. McDowell and M. Goldblatt, *Inorg. Chem.*, **10,** 625 (1971).
544. E. J. Baran, *Z. Anorg. Allgem. Chem.*, **399,** 57 (1973).
545. J. O. Edwards, G. C. Morrison, V. F. Ross, and J. W. Schultz, *J. Am. Chem. Soc.*, **77,** 266 (1955).
546. E. R. Lippincott, J. A. Psellos, and M. C. Tobin, *J. Chem. Phys.*, **20,** 536 (1952).
547. A. Müller, E. J. Baran, and R. O. Carter, *Struct. Bonding* (*Berlin*), **26,** 81 (1976).
548. E. J. Baran, *Inorg. Chem.*, **20,** 4453 (1981).
549. W. Kiefer and H. J. Bernstein, *Mol. Phys.*, **23,** 835 (1972).
549a. R. Kugel and H. Taube, *J. Phys. Chem.*, **79,** 2130 (1975).
550. A. Ranade and M. Stockburger, *Chem. Phys. Lett.*, **22,** 257 (1973).
551. A. Ranade, W. Krasser, A. Müller, and E. Ahlborn, *Spectrochim. Acta*, **30A,** 1341 (1974).
552. R. S. McDowell and L. B. Asprey, *J. Chem. Phys.*, **57,** 3062 (1972).
553. K. H. Schmidt and A. Müller, *Coord. Chem. Rev.*, **14,** 115 (1974).
554. A. Müller, E. Diemann, R. Jostes, and H. Bögge, *Angew. Chem. Int. Ed.*, **20,** 934 (1981).
555. K. O. Christe, E. C. Curtis, and C. J. Schack, *Spectrochim. Acta*, **31A**, 1035 (1975).
556. R. J. H. Clark and O. H. Ellestad, *J. Mol. Spectrosc.*, **56,** 386 (1975).
557. R. H. Mann and P. M. Harris, *J. Mol. Spectrosc.*, **45,** 65 (1973).
558. H. Bürger, S. Biedermann, and A. Ruoff, *Spectrochim. Acta*, **30A,** 1655 (1974).
559. J. Goubeau, F. Haenschke, and A. Ruoff, *Z. Anorg. Allg. Chem.*, **366,** 113 (1969).
560. A. Ruoff, H. Bürger, S. Biedermann, and J. Cichon, *Spectrochim. Acta*, **30A,** 1647 (1974).
561. E. C. Curtis, D. Philipovich, and W. H. Maberly, *J. Chem. Phys.*, **46,** 2904 (1967).
562. A. Finch, P. N. Gates, F. J. Ryan, and F. F. Bentley, *J. Chem. Soc.*, *Dalton Trans.*, 1863 (1973).
563. H. S. Gutowsky and A. D. Liehr, *J. Chem. Phys.*, **26,** 329 (1957).
564. M. L. Delwaulle and F. François, *Compt. Rend.*, **220,** 817 (1945).
565. H. Gerding and M. van Driel, *Rec. Trav. Chim.*, **61,** 419 (1942).
566. J. Durand, L. Beys, P. Hillaire, S. Aleonard, and L. Cot, *Spectrochim. Acta*, **34A,** 123 (1978).
567. F. Königer and A. Müller, *Spectrochim. Acta*, **33A,** 971 (1977).
568. H. Gerding and R. Westrik, *Rec. Trav. Chim.*, **61,** 842 (1942).
569. M. L. Delwaulle and F. François, *Compt. Rend.*, **226,** 896 (1948).
570. M. Brownstein, P. A. W. Dean, and R. J. Gillespie, *Chem. Commun.*, 9 (1970)
571. F. Königer, A. Müller, and O. Glemser, *J. Mol. Struct.*, **46,** 29 (1978).
572. Y. Tavares-Forneris and R. Forneris, *J. Mol. Struct.*, **24,** 205 (1965).
573. I. C. Hisatsune and J. Heicklen, *Can. J. Chem.*, **53,** 2646 (1975).
574. C. S. Alleyne, K. O. Mailer, and R. C. Thompson, *Can. J. Chem.*, **52,** 336 (1974).
575. D. J. Stufkens and H. Gerding, *Rec. Trav. Chim.*, **89,** 417 (1970).
576. E. Steger, I. C. Ciurea, and A. Fadini, *Z. Anorg. Allg. Chem.*, **350,** 225 (1967).
577. M. Černík and K. Dostál, *Z. Anorg. Allg. Chem.*, **425,** 37 (1976).
578. W. F. Murphy and H. Katz, *J. Raman Spectrosc.*, **7,** 76 (1978).
579. H. H. Claassen and E. H. Appelman, *Inorg. Chem.*, **9,** 622 (1970).
580. J. Goubeau, E. Kilcioglu, and E. Jacob, *Z. Anorg. Allg. Chem.*, **357,** 190 (1968).
581. H. Selig and H. H. Claassen, *J. Chem. Phys.*, **44,** 1404 (1966).
582. G. A. Ozin and D. J. Reynolds, *Chem. Commun.*, 884 (1969).
583. F. A. Miller and W. K. Baer, *Spectrochim. Acta*, **17,** 114 (1961).
584. H. Stammreich, O. Sala, and D. Bassi, *Spectrochim. Acta*, **19,** 593 (1963).
584a. A. Müller, K. H. Schmidt, E. Ahlborn, and C. J. L. Lock, *Spectrochim. Acta*, **29A,** 1773 (1973).
585. H. Stammreich, O. Sala, and K. Kawai, *Spectrochim. Acta*, **17,** 226 (1961).
585a. K. H. Schmidt and A. Müller, *Spectrochim. Acta*, **28A,** 1829 (1972).
586. A. Müller, N. Mohan, H. Dornfeld, and C. Tellez, *Spectrochim. Acta*, **34A,** 561 (1978).
587. A. Müller, K. H. Schmidt, and U. Zint, *Spectrochim. Acta*, **32A,** 901 (1976).
588. E. L. Varetti, R. R. Filgueira, and A. Müller, *Spectrochim. Acta*, **37A,** 369 (1981).
589. J. Binenboym, U. El-Gad, and H. Selig, *Inorg. Chem.*, **13,** 319 (1974).

590. A. Guest, H. E. Howard-Lock, and C. J. L. Lock, *J. Mol. Spectrosc.*, **43,** 273 (1972).
591. I. R. Beattie, R. A. Crocombe, and J. S. Ogden, *J. Chem. Soc., Dalton Trans.*, 1481 (1977).
592. A. Müller, B. Krebs, and W. Höltje, *Spectrochim. Acta,* **23A,** 2753 (1967).
593. K. H. Schmidt, V. Flemming, and A. Müller, *Spectrochim. Acta,* **31A,** 1913 (1975).
594. R. H. Bradley, P. N. Brier, and D. E. H. Jones, *J. Chem. Soc. A,* 1397 (1971).
595. K. Hamada, G. A. Ozin, and E. A. Robinson, *Bull. Chem. Soc. Jpn.*, **44,** 2555 (1971).
596. F. Höfler, *Z. Naturforsch.*, **26a,** 547 (1971).
597. C. A. Clausen and M. L. Good, *Inorg. Chem.*, **9,** 220 (1970).
598. F. Höfler and W. Veigl, *Agnew. Chem. Int. Ed.*, **10,** 919 (1971).
599. K. O. Christe and E. C. Curtis, *Inorg. Chem.*, **11,** 2196 (1972).
600. K. O. Christe, *Inorg. Chem.*, **14,** 2821 (1975).
601. M. Abenoza and V. Tabacik, *J. Mol. Struct.*, **26,** 95 (1975).
602. I. McAlpine and H. Sutcliffe, *Spectrochim. Acta,* **25A,** 1723 (1969).
603. J. E. Drake, C. Riddle, and D. E. Rogers, *J. Chem. Soc. A,* 910 (1969).
604. M.-L. Dubois, M.-B. Delhaye, and F. Wallart, *Compt. Rend.*, **269B,** 260 (1969).
605. J. E. Drake and C. Riddle, *J. Chem. Soc. A,* 2114 (1969).
606. J. Weidlein, *Z. Anorg. Allg. Chem.*, **358,** 13 (1968).
607. S. Sportouch, R. J. H. Clark, and R. Gaufres, *J. Raman Spectrosc.*, **2,** 153 (1974).
608. D. E. Martz and R. T. Lagemann, *J. Chem. Phys.*, **22,** 1193 (1954).
609. R. J. Gillespie, J. B. Milne, D. Moffett, and P. Spekkens, *J. Fluorine Chem.*, **7,** 43 (1976).
610. K. O. Christe and E. C. Curtis, *Inorg. Chem.*, **11,** 35 (1972).
611. R. Bougon, P. Joubert, and G. Tantot, *J. Chem. Phys.*, **66,** 1562 (1977).
612. H. H. Claassen, E. L. Gasner, H. Kim, and J. L. Huston, *J. Chem. Phys.*, **49,** 253 (1968).
613. E. Ahlborn, E. Diemann, and A. Müller, *Chem. Commun.*, 378 (1972).
614. V. D. Fenske, A.-F. Shihada, H. Schwab, and K. Dehnicke, *Z. Anorg. Allg. Chem.*, **471,** 140 (1980).
615. M. Muller, M. J. F. Leroy, and R. Rohmer, *Compt. Rend.*, **270C,** 1458 (1970).
616. S. D. Brown, G. L. Gard, and T. M. Loehr, *J. Chem. Phys.*, **64,** 1219 (1976).
617. M. Spoliti, J. H. Thirtle, and T. M. Dunn, *J. Mol. Spectrosc.*, **52,** 146 (1974).
618. V. V. Kovba and A. A. Mal'tsev, *Russ. J. Inorg. Chem.*, **20,** 11 (1975).
619. A. Müller, N. Weinstock, K. H. Schmidt, K. Nakamoto, and C. W. Schläpfer, *Spectrochim. Acta,* **28A,** 2289 (1972).
620. K. O. Christe, R. D. Wilson, and E. C. Curtis, *Inorg. Chem.*, **12,** 1358 (1973).
621. G. A. Ozin and A. Vander Voet, *Chem. Commun.*, 1489 (1970).
622. A. Müller, B. Krebs, E. Niecke, and A. Ruoff, *Ber. Bunsenges, Phys. Chem.*, **71,** 571 (1967).
623. M. L. Delwaulle and F. François, *J. Chim. Phys.*, **46,** 87 (1949); *Compt. Rend.*, **226,** 894 (1948).
624. A. Müller and E. Diemann, *Z. Naturforsch.*, **B24,** 353 (1969).
625. A. Müller and E. Diemann, *Chem. Ber.*, **102,** 2603 (1969).
626. E. Diemann and A. Müller, *Inorg. Nucl. Chem. Lett.*, **5,** 339 (1969).
627. A. Müller and E. Diemann, *Z. Anorg. Allg. Chem.*, **373,** 57 (1970).
628. J. R. Durig and J. W. Clark, *J. Chem. Phys.*, **46,** 3057 (1967).
629. M. L. Delwaulle and F. François, *Compt. Rend.*, **222,** 1193 (1946).
630. A. Müller, E. Niecke, and O. Glemser, *Z. Anorg. Allg. Chem.*, **350,** 246 (1967).
631. T. T. Crow and R. T. Lagemann, *Spectrochim. Acta,* **12,** 143 (1958).
632. G. D. Flesch and H. J. Svec, *J. Am. Chem. Soc.*, **80,** 3189 (1958).
633. K. O. Christe and C. J. Schack, *Inorg. Chem.*, **9,** 1852 (1970).
634. H. Stammreich, and R. Forneris, *Spectrochim. Acta,* **16,** 363 (1960).
635. P. Tsao, C. C. Cobb, and H. A. Claassen, *J. Chem. Phys.*, **54,** 5247 (1971).
636. P. J. Hendra, *J. Chem. Soc. A,* 1298 (1967); *Spectrochim. Acta,* **23A,** 2871 (1967).
637. P. L. Goggin and J. Mink, *J. Chem. Soc., Dalton Trans.*, 1479 (1974).
638. Y. M. Bosworth and R. J. H. Clark, *Inorg. Chem.*, **14,** 170 (1975).
639. Y. M. Bosworth and R. J. H. Clark, *J. Chem. Soc., Dalton Trans.*, 381 (1975).
640. J. Hiraishi and T. Shimanouchi, *Spectrochim. Acta,* **22,** 1483 (1966).
641. A. N. Pandey and U. P. Verma, *J. Mol. Struct.*, **42,** 171 (1977).

642. H. C. Clark, K. R. Dixon, and J. G. Nicolson, *Inorg. Chem.* **8,** 450 (1969).
643. I. R. Beattie and K. M. Livingston, *J. Chem. Soc. A,* 859 (1969).
644. I. R. Beattie, T. Gilson, K. Livingston, V. Fawcett, and G. A. Ozin, *J. Chem. Soc. A,* 712 (1967).
645. J. I. Bullock, N. J. Taylor, and F. W. Parrett, *J. Chem. Soc., Dalton Trans.,* 1843 (1972).
646. J. A. Creighton and J. H. S. Green, *J. Chem. Soc. A,* 808 (1968).
647. I. R. Beattie, K. M. S. Livingston, and D. J. Reynolds, *J. Chem. Phys.,* **51,** 4269 (1969).
648. P. van Huong and B. Desbat, *Bull. Soc. Chim. Fr.,* 2631 (1972).
649. L. C. Hoskins and R. C. Lord, *J. Chem. Phys.,* **46,** 2402 (1967).
650. K. Seppelt, *Z. Anorg. Allg. Chem.,* **434,** 5 (1977).
651. I. R. Beattie and G. A. Ozin, *J. Chem. Soc. A,* 1691 (1969).
652. T. V. Long, A. W. Herlinger, E. F. Epstein, and I. Bernal, *Inorg. Chem.,* **9,** 459 (1970).
653. C. S. Creaser and J. A. Creighton, *J. Chem. Soc., Dalton Trans.,* 1402 (1975).
654. H. H. Claassen and H. Selig, *J. Chem. Phys.,* **44,** 4039 (1965).
655. R. D. Werder, R. A. Frey, and H. Günthard, *J. Chem. Phys.,* **47,** 4159 (1967).
656. N. Acquista and S. Abramowitz, *J. Chem. Phys.,* **58,** 5484 (1973).
657. D. E. Sands and A. Zalkin, *Acta Crystallogr.,* **12,** 723 (1959).
658. A. Zalkin and D. E. Sands, *Acta Crystallogr.,* **11,** 615 (1958).
659. R. A. Walton and B. J. Brisdon, *Spectrochim. Acta,* **23A,** 2489 (1967).
660. P. M. Boorman, N. N. Greenwood, M. A. Hildon, and H. J. Whitfield, *J. Chem. Soc. A,* 2017 (1967).
661. A. J. Edwards, *J. Chem. Soc.,* 3714 (1964).
662. L. E. Alexander and I. R. Beattie, *J. Chem. Phys.,* **56,** 5829 (1972).
663. L. E. Alexander, I. R. Beattie, and P. J. Jones, *J. Chem. Soc., Dalton Trans.,* 210 (1972).
664. H. Gerding and H. Houtgraaf, *Rec. Trav. Chim.,* **74,** 5 (1955).
665. R. R. Holmes, *Acc. Chem. Res.,* **5,** 296 (1972).
666. R. R. Holmes, R. M. Deiters, and J. A. Golen, *Inorg. Chem.,* **8,** 2612 (1969).
667. I. W. Levin, *J. Mol. Spectrosc.,* **33,** 61 (1970).
668. H. Selig, J. H. Holloway, J. Tyson, and H. H. Claassen, *J. Chem. Phys.,* **53,** 2559 (1970).
669. R. R. Holmes, *J. Chem. Phys.,* **46,** 3718, 3724, 3730 (1967).
670. J. A. Salthouse and T. C. Waddington, *Spectrochim. Acta,* **23A,** 1069 (1967).
671. K. O. Christe, C. J. Schack, and E. C. Curtis, *Spectrochim. Acta,* **33A,** 323 (1977).
672. D. M. Adams and R. R. Smardzewski, *J. Chem. Soc. A,* 714 (1971).
673. L. E. Alexander and I. R. Beattie, *J. Chem. Soc. A,* 3091 (1971).
674. H. A. Szymanski, R. Yelin, and L. Marabella, *J. Chem. Phys.,* **47,** 1877 (1967).
675. K. O. Christe, E. C. Curtis, C. J. Schack, and D. Pilipovich, *Inorg. Chem.,* **11,** 1679 (1972).
676. G. M. Begun, W. H. Fletcher, and D. F. Smith, *J. Chem. Phys.,* **42,** 2236 (1965).
677. K. O. Christe, *Spectrochim. Acta,* **27A,** 631 (1971).
678. R. A. Frey, R. L. Redington, and A. L. Khidir Aljibury, *J. Chem. Phys.,* **54,** 344 (1971).
679. N. Acquista and S. Abramowitz, *J. Chem. Phys.,* **56,** 5221 (1972).
680. J. B. Milne and D. Moffett, *Inorg. Chem.,* **12,** 2240 (1973).
681. K. O. Christe and E. C. Curtis, *Inorg. Chem.,* **11,** 2209 (1972).
682. R. Bougon, T. B. Huy, P. Charpin, and G. Tantot, *Compt. Rend.,* **283,** C71 (1976).
683. J. B. Milne and D. M. Moffett, *Inorg. Chem.,* **15,** 2165 (1976).
684. L. E. Alexander, I. R. Beattie, A. Bukovszky, P. J. Jones, C. J. Marsden, and G. J. Van Schalkwyk, *J. Chem. Soc., Dalton Trans.,* 81 (1974).
685. R. T. Paine and R. S. McDowell, *Inorg. Chem.,* **13,** 2366 (1974).
686. I. R. Beattie, K. M. S. Livingston, D. J. Reynolds, and G. A. Ozin, *J. Chem. Soc. A,* 1210 (1970).
687. K. Iijima and S. Shibata, *Bull. Chem. Soc., Jpn.,* **48,** 666 (1975).
688. R. J. Collin, W. P. Griffith, and D. Pawson, *J. Mol. Struct.,* **19,** 531 (1973).
689. M. G. Krishna Pillai and P. Parameswaran Pillai, *Can. J. Chem.,* **46,** 2393 (1968).
690. M. J. Reisfeld, *Spectrochim. Acta,* **29A,** 1923 (1973).
691. E. J. Baran and A. E. Lavat, *Z. Naturforsch.,* **36a,** 677 (1981).
692. T. Barrowcliffe, I. R. Beattie, P. Day, and K. Livingston, *J. Chem. Soc. A,* 1810 (1967).
693. T. G. Spiro, *Inorg. Chem.,* **6,** 569 (1967).

694. C. Naulin and R. Bougon, *J. Chem. Phys.*, **64,** 4155 (1976).
695. I. R. Beattie, T. Gilson, K. Livingston, V. Fawcett, and G. A. Ozin, *J. Chem. Soc. A*, 712 (1967).
696. T. L. Brown, W. G. McDugle, Jr., and L. G. Kent, *J. Am. Chem. Soc.*, **92,** 3645 (1970).
697. M. Debeau and M. Krauzman, *Compt. Rend.*, **264B,** 1724 (1967).
698. I. Wharf and D. F. Shriver, *Inorg. Chem.*, **8,** 914 (1969).
699. A. M. Heyns, *Spectrochim. Acta*, **33A,** 315 (1977).
700. M. Burgard and J. MacCordick, *Inorg. Nucl. Chem. Lett.*, **6,** 599 (1970).
701. Y. M. Bosworth and R. J. H. Clark, *J. Chem. Soc., Dalton Trans.*, 1749 (1974).
702. R. J. H. Clark and M. L. Duarte, *J. Chem. Soc., Dalton Trans.*, 790 (1977).
703. M. A. Hooper and D. W. James, *Aust. J. Chem.*, **26,** 1401 (1973).
704. M. A. Hooper and D. W. James, *J. Inorg. Nucl. Chem.*, **35,** 2335 (1973).
705. T. Surles, L. A. Quaterman, and H. H. Hyman, *J. Inorg. Nucl. Chem.*, **35,** 670 (1973).
706. Y. M. Bosworth, R. J. H. Clark, and D. M. Rippon, *J. Mol. Spectrosc.*, **46,** 240 (1973).
707. H. H. Claassen, G. L. Goodman, J. H. Holloway, and H. Selig, *J. Chem. Phys.*, **53,** 341 (1970).
708. A. Aboumajd, H. Berger, and R. Saint-Loup, *J. Mol. Spectrosc.*, **78,** 486 (1979).
709. P. J. Hendra and Z. Jović, *J. Chem. Soc. A*, 600 (1968).
710. R. W. Berg, F. W. Poulsen, and N. J. Bjerrum, *J. Chem. Phys.*, **67,** 1829 (1977).
711. K. O. Christe, *Inorg. Chem.*, **12,** 1580 (1973).
712. R. J. Gillespie and G. J. Schrobilgen, *Inorg. Chem.*, **13,** 1230 (1974).
713. R. Bougon, P. Charpin, and J. Soriano, *Compt. Rend.*, **272C,** 565 (1971).
714. N. Bartlett and K. Leary, *Rev. Chim. Minérale*, **13,** 82 (1976).
715. R. Becker, A. Lentz, and W. Sawodny, *Z. Anorg. Allg. Chem.*, **420,** 210 (1976).
716. D. M. Adams and D. M. Morris, *J. Chem. Soc. A*, 694 (1968).
717. I. W. Forrest and A. P. Lane, *Inorg. Chem.*, **15,** 265 (1976).
718. R. J. H. Clark, L. Maresca, and R. J. Puddephatt, *Inorg. Chem.*, **7,** 1603 (1968).
719. P. C. Crouch, G. W. A. Fowles, and R. A. Walton, *J. Chem. Soc. A*, 972 (1969).
720. P. A. W. Dean and D. F. Evans, *J. Chem. Soc. A*, 698 (1967).
721. D. M. Adams and D. C. Newton, *J. Chem. Soc. A*, 2262 (1968).
722. W. von Bronswyk, R. J. H. Clark, and L. Maresca, *Inorg. Chem.*, **8,** 1395 (1969).
723. R. Becker and W. Sawodny, *Z. Naturforsch.*, **28b,** 360 (1973).
724. R. A. Walton and B. J. Brisdon, *Spectrochim. Acta*, **23A,** 2222 (1967).
725. O. L. Keller, *Inorg. Chem.*, **2,** 783 (1963).
726. S. M. Horner, R. J. H. Clark, B. Crociani, D. B. Copley, W. W. Horner, F. N. Collier, and S. Y. Tyree, *Inorg. Chem.*, **7,** 1859 (1968).
727. D. Brown and P. J. Jones, *J. Chem. Soc. A*, 247 (1967).
728. G. A. Ozin, G. W. A. Fowles, D. J. Tidmarsh, and R. A. Walton, *J. Chem. Soc. A*, 642 (1969).
729. O. L. Keller and A. Chetham-Strode, *Inorg. Chem.*, **5,** 367 (1966).
730. B. Weinstock and G. L. Goodman, *Adv. Chem. Phys.*, **11,** 169 (1965).
731. H. H. Eysel, *Z. Anorg. Allg. Chem.*, **390,** 210 (1972).
732. R. S. McDowell, R. J. Sherman, L. B. Asprey, and R. C. Kennedy, *J. Chem. Phys.*, **62,** 3974 (1975).
733. R. R. Smardzewski, R. E. Noftle, and W. B. Fox, *J. Mol. Spectrosc.*, **62,** 449 (1976).
734. J. A. Creighton and T. J. Sinclair, *Spectrochim. Acta*, **35A,** 507 (1979).
735. C. D. Flint, *J. Mol. Spectrosc.*, **37,** 414 (1971).
736. E. Jacob and M. Fähnle, *Angew. Chem. Int. Ed.*, **15,** 159 (1976).
737. J. A. LoMenzo, S. Strobridge, H. H. Patterson, and E. Engstrom, *J. Mol. Spectrosc.*, **66,** 150 (1977).
738. P. W. Frais, C. J. L. Lock, and A. Guest, *Chem. Commun.*, 1612 (1970).
739. G. L. Bottger and C. V. Damsgard, *Spectrochim. Acta*, **28A,** 1631 (1972).
740. L. A. Woodward and M. J. Ware, *Spectrochim. Acta*, **20,** 711 (1964).
741. K. O. Christe, *Inorg. Chem.*, **16,** 2238 (1977).
742. M. Debeau and H. Poulet, *Spectrochim. Acta*, **25A,** 1553 (1969).
743. P. J. Hendra and P. J. D. Park, *Spectrochim. Acta*, **23A,** 1635 (1967).
744. L. A. Woodward and M. J. Ware, *Spectrochim. Acta*, **19,** 775 (1963).

745. D. M. Adams and H. A. Gebbie, *Spectrochim. Acta*, **19,** 925 (1963).
746. K. Wieghardt and H. H. Eysel, *Z. Naturforsch.*, **25b**, 105 (1970).
747. J. M. Fletcher, W. E. Gardner, A. C. Fox, and G. Topping, *J. Chem. Soc. A*, 1038 (1967).
748. D. A. Kelly and M. L. Good, *Spectrochim. Acta*, **28A,** 1529 (1972).
749. M. Debeau, *Spectrochim. Acta*, **25A,** 1311 (1969).
750. L. A. Woodward and M. J. Ware, *Spectrochim. Acta*, **24A,** 921 (1968).
751. R. T. Paine, R. S. McDowell, L. B. Asprey, and L. H. Jones, *J. Chem. Phys.*, **64,** 3081 (1976).
752. J. L. Ryan, *J. Inorg. Nucl. Chem.*, **33,** 153 (1971).
753. E. Stumpp and G. Piltz, *Z. Anorg. Allg. Chem.*, **409,** 53 (1974).
754. B. W. Berringer, J. B. Gruber, T. M. Loehr, and G. P. O'Leary, *J. Chem. Phys.*, **55,** 4608 (1971).
755. D. M. Yost, C. S. Steffens, and S. T. Gross, *J. Chem. Phys.*, **2,** 311 (1934).
756. A. Fadini and S. Kemmler-Sack, *Spectrochim. Acta*, **34A,** 853 (1978).
757. C. J. Adams and A. J. Downs, *Chem. Commun.*, 1699 (1970).
758. L. A. Woodward and J. A. Creighton, *Spectrochim. Acta*, **17,** 594 (1961).
759. H. Hamaguchi, I. Harada, and T. Shimanouchi, *Chem. Phys. Lett.*, **32,** 103 (1975).
760. B. Weinstock, H. H. Claassen, and C. L. Chernick, *J. Chem. Phys.*, **38,** 1470 (1963).
761. H. Kim, H. H. Claassen, and E. Pearson, *Inorg. Chem.*, **7,** 616 (1968).
762. H. H. Claassen, G. L. Goodman, and H. Kim, *J. Chem. Phys.*, **56,** 5042 (1972).
763. D. M. Cox and J. Elliott, *Spectrosc. Lett.*, **12,** 275 (1979).
764. P. J. Cresswell, J. E. Fergusson, B. R. Penfold, and D. E. Scaife, *J. Chem. Soc., Dalton Trans.*, 254 (1972).
765. H. Kim, P. A. Souder, and H. H. Claassen, *J. Mol. Spectrosc.*, **26,** 46 (1968).
766. P. LaBonville, J. R. Ferraro, M. C. Wall, and L. J. Basile, *Coord. Chem. Rev.*, **7,** 257 (1972).
767. J. Hauck and A. Fadini, *Z. Naturforsch.*, **25b,** 422 (1970).
768. J. Hauck, *Z. Naturforsch.*, **25b**, 224, 468, 647 (1970).
769. C. J. Adams and A. J. Downs, *J. Inorg. Nucl. Chem.*, **34,** 1829 (1972).
770. G. Goetz, M. Deneux, and M. J. F. Leroy, *Bull. Soc. Chim. Fr.*, 29 (1971).
771. J. E. Griffith, *Spectrochim. Acta*, **23A,** 2145 (1967).
772. K. O. Christe, E. C. Curtis, and C. J. Schack, *Spectrochim. Acta*, **33A,** 69 (1977).
773. K. O. Christe, C. J. Schack, D. Pilipovich, E. C. Curtis, and W. Sawodny, *Inorg. Chem.*, **12,** 620 (1973).
774. K. O. Christe, C. J. Schack, and E. C. Curtis, *Inorg. Chem.*, **11,** 583 (1972).
775. K. Seppelt, *Z. Arnog. Allg. Chem.*, **399,** 87 (1973).
776. W. V. F. Brooks, M. E. Eshaque, C. Lau, and J. Passmore, *Can. J. Chem.*, **54,** 817 (1976).
777. J. H. Holloway, H. Selig, and H. H. Claassen, *J. Chem. Phys.*, **54,** 4305 (1971).
778. K. Dehnicke, G. Pausewang, and W. Rüdorff, *Z. Anorg. Allg. Chem.*, **366,** 64 (1969).
779. G. A. Ozin, G. W. A. Fowles, D. J. Tidmarsh, and R. A. Walton, *J. Chem. Soc. A*, 642 (1969).
780. R. J. Collin, W. P. Griffith, and D. Pawson, *J. Mol. Struct.*, **19,** 531 (1973).
781. A. Beuter and W. Sawodny, *Z. Anorg. Allg. Chem.*, **427,** 37 (1976).
782. D. M. Adams, G. W. Fraser, D. M. Morris, and R. D. Peacock, *J. Chem. Soc. A*, 1131 (1968).
783. P. Joubert, R. Bougon, and B. Gaudreau, *Can. J. Chem.*, **56,** 1874 (1978).
784. G. J. Schrobilgen, *Chem. Commun.*, 894 (1980).
785. J. P. Brunnette and M. J. F. Leroy, *J. Inorg. Nucl. Chem.*, **36,** 289 (1974).
786. W. Preetz, D. Ruf, and D. Tensfeldt, *Z. Naturforsch.*, **39b,** 1100 (1984).
787. D. Tensfeldt and W. Preetz, *Z. Naturforsch.*, **39b,** 1185 (1984).
788. W. Preetz and G. Rimkus, *Z. Naturforsch.*, **37b,** 579 (1982).
789. W. Preetz and H. Kühl, *Z. Anorg. Allg. Chem.*, **425,** 97 (1976).
790. H. H. Claassen, E. L. Gasner, and H. Selig, *J. Chem. Phys.*, **49,** 1803 (1968).
791. H. H. Eysel and K. Seppelt, *J. Chem. Phys.*, **56,** 5081 (1972).
792. Nguyen-Quy-Dao, *Bull. Soc. Chim. Fr.*, 3976 (1968).
793. J. L. Hoard, W. G. Martin, M. E. Smith, and J. E. Whitney, *J. Am. Chem. Soc.*, **76,** 3820 (1954).
794. R. Stomberg and C. Brosset, *Acta Chem. Scand.*, **14,** 441 (1960).
795. K. O. Hartman and F. A. Miller, *Spectrochim. Acta*, **24A,** 669 (1968).
796. C. W. F. T. Pistorius, *Bull. Soc. Chim. Belg.*, **68,** 630 (1959).

797. H. L. Schlaefer and H. F. Wasgestian, *Theor. Chim. Acta*, **1,** 369 (1963).
798. J. R. Durig, J. W. Thompson, J. D. Witt, and J. D. Odom, *J. Chem. Phys.*, **58,** 5339 (1973).
799. D. E. Mann and L. Fano, *J. Chem. Phys.*, **26,** 1665 (1957).
800. J. D. Odom, J. E. Saunders, and J. R. Durig, *J. Chem. Phys.*, **56,** 1643 (1972).
801. L. A. Minon, K. S. Seshadri, R. C. Taylor, and D. White, *J. Chem. Phys.*, **53,** 2416 (1970).
802. B. Andrews and A. Anderson, *J. Chem. Phys.*, **74,** 1534 (1981).
803. G. M. Begun and W. H. Fletcher, *Spectrochim. Acta*, **19,** 1343 (1963).
804. H. Murata and K. Kawai, *J. Chem. Phys.*, **25,** 589, 796 (1956).
805. R. E. Hester and R. A. Plane, *Inorg. Chem.*, **3,** 513 (1964).
806. D. N. Shchepkin, L. A. Zhygula, and L. P. Belozerskaya, *J. Mol. Structure*, **49,** 265 (1978).
807. J. R. Durig, S. F. Bush, and E. E. Mercer, *J. Chem. Phys.*, **44,** 4238 (1966).
808. E. R. Nixon, *J. Phys. Chem.*, **60,** 1054 (1956).
809. S. G. Frankiss, *Inorg. Chem.*, **7,** 1931 (1968).
810. K. H. Rhee, A. M. Snider, Jr., and F. A. Miller, *Spectrochim. Acta*, **29A,** 1029 (1973).
811. S. G. Frankiss and F. A. Miller, *Spectrochim. Acta*, **21,** 1235 (1965).
812. S. G. Frankiss, F. A. Miller, H. Stammreich, and Th. T. Sans, *Spectrochim. Acta*, **23A,** 543 (1967).
813. J. R. Durig, B. M. Gimarc, and J. D. Odom, in J. R. Durig, ed., *Vibrational Spectra and Structure*, Vol. 2, Marcel Dekker, New York, 1975, p. 1.
814. R. P. Bell and H. C. Longuet-Higgins, *Proc. Roy. Soc. London, Ser. A*, **183,** 357 (1945).
815. J. L. Duncan, D. C. McKean, and I. Torto, *J. Mol. Spectrosc.*, **85,** 16 (1981).
816. A. Snelson, *J. Phys. Chem.*, **71,** 3202 (1967).
817. M. Tranquille and M. Fouassier, *J. Chem. Soc., Faraday Trans. 2,* **76,** 26 (1980).
818. D. M. Adams and R. G. Churchill, *J. Chem. Soc. A*, 697 (1970).
819. I. R. Beattie, T. Gilson, and P. Cocking, *J. Chem. Soc. A*, 702 (1967).
820. R. A. Frey, R. D. Werder, and H. H. Günthard, *J. Mol. Spectrosc.*, **35,** 260 (1970).
821. T. Onishi and T. Shimanouchi, *Spectrochim. Acta*, **20,** 721 (1964).
822. N. N. Greenwood, D. J. Prince, and B. P. Straughan, *J. Chem. Soc. A*, 1694 (1968).
823. P. L. Goggin, *J. Chem. Soc., Dalton Trans.*, 1483 (1974).
824. R. Forneris, J. Hiraishi, F. A. Miller, and M. Uehara, *Spectrochim. Acta*, **26A,** 581 (1970).
825. B. L. Crawford, W. H. Avery, and J. W. Linnett, *J. Chem. Phys.*, **6,** 682 (1938).
826. G. W. Bethke and M. K. Wilson, *J. Chem. Phys.*, **26,** 1107 (1957).
827. D. A. Dows and R. M. Hexter, *J. Chem. Phys.*, **24,** 1029 (1956).
828 R. G. Snyder and J. C. Decius, *Spectrochim. Acta*, **13,** 280 (1959).
829. W. G. Palmer, *J. Chem. Soc.*, 1552 (1961).
830. K. Buijs, *J. Chem. Phys.*, **36,** 861 (1962).
831. C. A. Evans, K. H. Tan, S. P. Tapper, and M. J. Taylor, *J. Chem. Soc., Dalton Trans.*, 988 (1973).
832. F. Höfler, S. Waldhör, and E. Hengge, *Spectrochim. Acta*, **28A,** 29 (1972).
833. F. Höfler, W. Sawodny, and E. Hengge, *Spectrochim. Acta*, **26A,** 819 (1970).
834. J. E. Griffiths, *Spectrochim. Acta*, **25A,** 965 (1969).
835. G. A. Ozin, *J. Chem. Soc. A*, 2952 (1969).
836. B. H. Freeland, J. H. Hencher, D. G. Tuck, and J. G. Contreras, *Inorg. Chem.*, **15,** 2144 (1976).
837. W. Bues, K. Buchler, and P. Kuhnle, *Z. Anorg. Allg. Chem.*, **325,** 8 (1963).
838. R. G. Brown and S. D. Ross, *Spectrochim. Acta*, **28,** 1263 (1972).
839. I. R. Beattie and G. A. Ozin, *J. Chem. Soc. A*, 2615 (1969).
840. J. Roziere, J.-L. Pascal, and A. Potier, *Spectrochim. Acta*, **29A,** 169 (1973).
841. A. Grodzicki and A. Potier, *J. Inorg. Nucl. Chem.*, **35,** 61 (1973).
842. J. D. Witt and R. M. Hammaker, *J. Chem. Phys.*, **58,** 303 (1973).
843. A. Hezel and S. D. Ross, *Spectrochim. Acta*, **23A,** 1583 (1967).
844. R. W. Mooney and R. L. Goldsmith, *J. Inorg. Nucl. Chem.*, **31,** 933 (1969).
845. R. M. Wing and K. P. Callahan, *Inorg. Chem.*, **8,** 871 (1969).
846. D. Mascherpa-Corral and A. Potier, *J. Inorg. Nucl. Chem.*, **38,** 211 (1976).
847. A. Simon and H. Richter, *Z. Anorg. Allg. Chem.*, **304,** 1 (1960); **315,** 196 (1962).
848. R. J. H. Clark and M. L. Franks, *J. Am. Chem. Soc.*, **97,** 2691 (1975).

849. R. J. H. Clark and N. R. D'Urso, *J. Am. Chem. Soc.*, **100,** 3088 (1978).
850. R. J. H. Clark and M. L. Franks, *J. Am. Chem. Soc.*, **98,** 2763 (1976).
851. K. Schwochau, K. Hedwig, H. J. Schenk, and O. Greis, *Inorg. Nucl. Chem. Lett.*, **13,** 77 (1977).
852. R. J. Ziegler and W. M. Risen, Jr., *Inorg. Chem.*, **11,** 2796 (1972).
853. I. R. Beattie, T. R. Gilson, and G. A. Ozin, *J. Chem. Soc.*, 2765 (1968).
854. J. D. Black, J. T. R. Dunsmuir, I. W. Forrest, and A. P. Lane, *Inorg. Chem.*, **14,** 1257 (1975).
855. D. A. Edwards and R. T. Ward, *J. Chem. Soc. A*, 1617 (1970).
856. R. C. Burns and T. A. O'Donnell, *Inorg. Chem.*, **18,** 3081 (1979).
857. L. H. Jones and S. A. Ekberg, *Spectrochim. Acta*, **36A,** 761 (1980).
858. W. J. Lehmann and I. Shapiro, *Spectrochim. Acta*, **17,** 396 (1961).
859. L. J. Bellamy, W. Gerrard, M. F. Lappert, and R. L. Williams, *J. Chem. Soc.*, 2412 (1958).
860. A. Meller, *Organometal. Chem. Rev.*, **2,** 1 (1967).
861. J. G. Verkade, *Coord. Chem. Rev.*, **9,** 1 (1972).
862. W. Weltner, Jr. and J. R. W. Warn, *J. Chem. Phys.*, **37,** 292 (1962).
863. I. Hisatsune and N. A. Suarez, *Inorg. Chem.*, **3,** 168 (1964).
864. W. Bues, G. Foerster, and R. Schmitt, *Z. Anorg. Allg. Chem.*, **344,** 148 (1966).
865. A. Kaldor and R. F. Porter, *Inorg. Chem.*, **10,** 775 (1971).
866. J. L. Parsons, *J. Chem. Phys.*, **33,** 1860 (1960).
867. F. A. Grimm and R. F. Porter, *Inorg. Chem.*, **8,** 731 (1969).
868. V. F. Kalasinsky, *J. Phys. Chem.*, **83,** 3239 (1979).
869. W. E. Keller and H. L. Johnston, *J. Chem. Phys.*, **20,** 1749 (1952).
870. B. Roussel, A. Chapput, and G. Fleury, *J. Mol. Struct.*, **31,** 371 (1976).
871. K. E. Blick, K. Niedenzu, W. Sawodny, T. Takasuka, T. Totani, and H. Watanabe, *Inorg. Chem.*, **10,** 1133 (1971).
872. F. R. Brown, F. A. Miller, and C. Sourisseau, *Spectrochim. Acta*, **32A,** 125 (1976).
873. B. E. Smith, H. F. Shurvell, and B. D. James, *J. Chem. Soc., Dalton Trans.*, 711 (1978).
874. R. T. Paine, R. W. Light, and M. Nelson, *Spectrochim. Acta*, **35A,** 213 (1979).
875. J. Derouault, T. Dziembowska, and M.-T. Forel, *Spectrochim. Acta*, **35A,** 773 (1979).
876. J. E. Drake, J. L. Hencher, and L. N. Khasrou, *Can. J. Chem.*, **59,** 2898 (1981).
877. P. H. Laswick and R. C. Taylor, *J. Mol. Struct.*, **34,** 197 (1976).
878. J. Dazord, H. Mongeot, H. Atchekzai, and J. P. Tuchagues, *Can. J. Chem.*, **54,** 2135 (1976).
879. A. Balls, N. N. Greenwood, and B. P. Straughan, *J. Chem. Soc. A*, 753 (1968).
880. H. Burger, J. Cichon, V. Goetze, U. Wannagat, and H. J. Wismar, *J. Organometal. Chem.*, **33,** 1 (1971).
881. G. N. Papatheodorou and M. A. Capote, *J. Chem. Phys.*, **69,** 2067 (1978).
882. C. W. Schläpfer and C. Rohrbasser, *Inorg. Chem.*, **17,** 1623 (1978).
883. A. L. Smith, *Spectrochim. Acta*, **16,** 87 (1960).
884. B. J. Aylett, *Adv. Inorg. Chem. Radiochem.*, **11,** 262 (1968).
885. H. J. Campbell-Ferguson and E. A. V. Ebsworth, *J. Chem. Soc. A*, 705 (1967).
886. H. Schumann, *Angew. Chem. Int. Ed.*, **8,** 937 (1969).
887. W. H. Weber, P. D. Maker, and C. W. Peters, *J. Chem. Phys.*, **64,** 2149 (1976).
888. R. West, D. Eggerding, J. Perkins, D. Handy, and E. C. Tuazon, *J. Am. Chem. Soc.*, **101,** 1710 (1979).
889. J. B. Bates and W. H. Smith, *Chem. Phys. Lett.*, **14,** 362 (1972).
890. F. Höfler, G. Bauer, and E. Hengge, *Spectrochim. Acta*, **32A,** 1435 (1976).
891. K. Hassler, E. Hengge, and D. Kovar, *Spectrochim. Acta*, **34A,** 1193 (1978).
892. J. B. Bates and A. S. Quist., *J. Chem. Phys.*, **56,** 1528 (1972).
893. J. Etchepare, *Spectrochim. Acta*, **26A,** 2147 (1970).
894. P. Tarte, M. J. Pottier, and A. M. Procès, *Spectrochim. Acta*, **29A,** 1017 (1973).
895. J. R. Durig, M. J. Flanagan, and V. F. Kalasinsky, *J. Chem. Phys.*, **66,** 2775 (1977).
896. J. R. Durig, K. S. Kalasinsky, and V. F. Kalasinsky, *J. Mol. Struct.*, **35,** 201 (1976).
897. J. R. Durig, M. J. Flanagan, and V. F. Kalasinsky, *Spectrochim. Acta*, **34A,** 63 (1978).
898. J. R. Durig, Y. S. Li, M. M. Chen, and J. D. Odom, *J. Mol. Spectrosc.*, **59,** 74 (1976).
899. K. Hamada, *J. Mol. Struct.*, **48,** 191 (1978).

900. F. Höfler and W. Peter, *Z. Naturforsch.*, **30b**, 282 (1975).
901. F. A. Miller, J. Perkins, G. A. Gibbon, and B. A. Swisshelm, *J. Raman Spectrosc.*, **2**, 93 (1974).
902. S. Pohl and B. Krebs, *Z. Anorg. Allg. Chem.*, **424**, 265 (1976).
903. E. A. V. Ebsworth, D. W. H. Rankin, and G. M. Sheldrick, *J. Chem. Soc. A*, 2828 (1968).
904. J. Laane and J. R. Ohlsen, *Progr. Inorg. Chem.*, **27**, 465 (1980).
905. D. E. C. Corbridge, in M. Grayson and E. J. Griffith, eds., *Topics in Phosphorus Chemistry*, Vol. 6, Interscience, New York, 1969, p. 235.
906. D. E. C. Corbridge, *The Structural Chemistry of Phosphorus*, Elsevier, Amsterdam, 1974.
907. L. C. Thomas, *Interpretation of the Infrared Spectra of Organophosphorus Compounds*, Heyden, London, 1974.
908. E. Steger, *Z. Chem.*, **12**, 52 (1972).
909. L. C. Thomas and R. A. Chittenden, *Spectrochim. Acta*, **26A**, 781 (1970).
910. J. Goubeau, *Pure Appl. Chem.*, **44**, 393 (1975).
911. C. H. Bibart and G. E. Ewing, *J. Chem. Phys.*, **61**, 1293 (1974).
912. I. C. Hisatsune, J. P. Devlin, and Y. Wada, *Spectrochim. Acta*, **18**, 1641 (1962).
913. W. A. Guillory and M. L. Bernstein, *J. Chem. Phys.*, **62**, 1058 (1975).
914. K. O. Christe, C. J. Schack, and R. D. Wilson, *Inorg. Chem.*, **13**, 2811 (1974).
915. D. Tevault and R. R. Smardzewski, *J. Phys. Chem.*, **82**, 375 (1978).
916. R. E. Nightingale and E. L. Wagner, *J. Chem. Phys.*, **22**, 203 (1954).
917. D. L. Frasco and E. L. Wagner, *J. Chem. Phys.*, **30**, 1124 (1959).
918. K. O. Christe and C. J. Schack, *Inorg. Chem.*, **17**, 2749 (1978).
919. H. Bürger, H. Niepel, G. Pawelke, and H. Oberhammer, *J. Mol. Struct.*, **54**, 159 (1979).
920. Y. M. Bosworth, R. J. H. Clark, and D. M. Rippon, *J. Mol. Spectrosc.*, **46**, 240 (1973).
921. W. P. Griffith and K. J. Rutt, *J. Chem. Soc. A*, 2331 (1968).
922. E. Steger and A. Simon, *Z. Anorg. Allg. Chem.*, **291**, 76 (1957); **294**, 1 (1958); **294**, 147 (1958).
923. M. Gardner, *J. Chem. Soc., Dalton Trans.*, 691 (1973).
924. W. Bues, M. Somer, and W. Brockner, *Z. Naturforsch.*, **36a**, 842 (1981).
925. G. A. Ozin, *J. Chem. Soc. A*, 2307 (1970).
926. A. C. Chapman and L. E. Thirlwell, *Spectrochim. Acta*, **20**, 937 (1964).
927. M. Tsuboi, *J. Am. Chem. Soc.*, **79**, 1351 (1957).
928. K. O. Christe, C. J. Schack, and E. C. Curtis, *Inorg. Chem.*, **15**, 843 (1976).
929. R. R. Holmes and C. J. Hora, Jr., *Inorg. Chem.*, **11**, 2506 (1972).
930. J. A. Salthouse and T. C. Waddington, *Spectrochim. Acta*, **23A**, 1069 (1967).
931. I. C. Hisatsune, *Spectrochim. Acta*, **21**, 18 (1965).
932. T. R. Manley and D. A. Williams, *Spectrochim. Acta*, **23A**, 149 (1967).
933. E. Steger and K. Lunkwitz, *J. Mol. Struct.*, **3**, 67 (1969).
934. H. G. M. Edwards, J. S. Ingman, and D. A. Long, *Spectrochim. Acta*, **32A**, 731 (1976).
935. S. B. Brumbach and G. M. Rosenblatt, *J. Chem. Phys.*, **56**, 3110 (1972).
936. W. P. Griffith, *J. Chem. Soc. A*, 905 (1967).
937. R. Mercier and C. Sourisseau, *Spectrochim. Acta*, **34A**, 337 (1978).
938. R. A. Work and M. L. Good, *Spectrochim. Acta*, **29A**, 1547 (1973).
939. G. J. Janz, J. W. Coutts, J. R. Downey, and E. Roduner, *Inorg. Chem.*, **15**, 1755 (1976).
940. L. A. Nimon, V. D. Neff, R. E. Cantley, and R. O. Buttlar, *J. Mol. Spectrosc.*, **22**, 105 (1967).
941. R. Steudel and F. Schuster, *J. Mol. Struct.*, **44**, 143 (1978).
942. G. A. Ozin, *J. Chem. Soc. A*, 116 (1969).
943. H. Wieser, P. J. Krueger, E. Muller, and J. B. Hyne, *Can. J. Chem.*, **47**, 1633 (1969).
944. A. W. Herlinger and T. V. Long, *Inorg. Chem.*, **8**, 2661 (1969).
945. K. Stopperka and F. Kilz, *Z. Anorg. Allg. Chem.*, **370**, 49 (1969).
946. A. Turowski, R. Appel, W. Sawodny, and K. Molt, *J. Mol. Struct.*, **48**, 313 (1978).
947. T. Chivers and I. Drummond, *Inorg. Chem.*, **13**, 1222 (1974).
948. K. O. Christe, C. J. Schack, and E. C. Curtis, *Spectrochim. Acta*, **26A**, 2367 (1970).
949. R. Kebabcioglu, R. Mews, and O. Glemser, *Spectrochim. Acta*, **28A**, 1593 (1972).
950. J. Bojes, T. Chivers, W. G. Laidlaw, and M. Trsic, *J. Am. Chem. Soc.*, **101**, 4517 (1979).
951. R. J. Gillespie and G. P. Pez, *Inorg. Chem.*, **8**, 1229 (1969).

Coordination Compounds

Part III

III-1. AMMINE, AMIDO, AND RELATED COMPLEXES

(1) Ammine (NH_3) Complexes

Vibrational spectra of metal ammine complexes have been studied extensively, and these are reviewed by Schmidt and Müller.[1] Figure III-1 shows the infrared spectra of typical hexammine complexes in the high-frequency region. To assign these NH_3 group vibrations, it is convenient to use the six normal modes of vibration of a simple 1:1 (metal/ligand) complex model such as that shown in Fig. III-2. Table III-1 lists the infrared frequencies and band assignments of hexammine complexes. It is seen that the antisymmetric and symmetric NH_3 stretching, NH_3 degenerate deformation, NH_3 symmetric deformation, and NH_3 rocking vibrations appear in the regions of 3400–3000, 1650–1550, 1370–1000, and 950–590 cm^{-1}, respectively. These assignments have been confirmed by NH_3/ND_3 and $NH_3/^{15}NH_3$ isotope shifts.

The NH_3 stretching frequencies of the complexes are lower than those of the free NH_3 molecule for two reasons.[18] One is the effect of coordination. Upon coordination, the N–H bond is weakened and the NH_3 stretching frequencies are lowered. The stronger the M–N bond, the weaker is the N–H bond and the lower are the NH_3 stretching frequencies if other conditions are equal. Thus the NH_3 stretching frequencies may be used as a rough measure of the M–N bond strength. The other reason is the effect of the counterion. The NH_3 stretching frequencies of the chloride are much lower than those of the perchlorate, for example. This is attributed to the weakening of the N–H bond due to the formation of the N–H . . . Cl-type hydrogen bond in the former.

The effects of coordination and hydrogen bonding mentioned above shift the NH_3 deformation and rocking modes to higher frequencies. Among them, the NH_3 rocking mode is most sensitive, and the degenerate deformation is least sensitive, to these effects. Thus the NH_3 rocking frequency is often used to compare the strength of the M–N bond in a series of complexes of the same type and anion.[18] It is interesting to note that a genuine 1:1 complex such as shown in Fig. III-2 can be prepared by reacting alkali halide with NH_3 in Ar

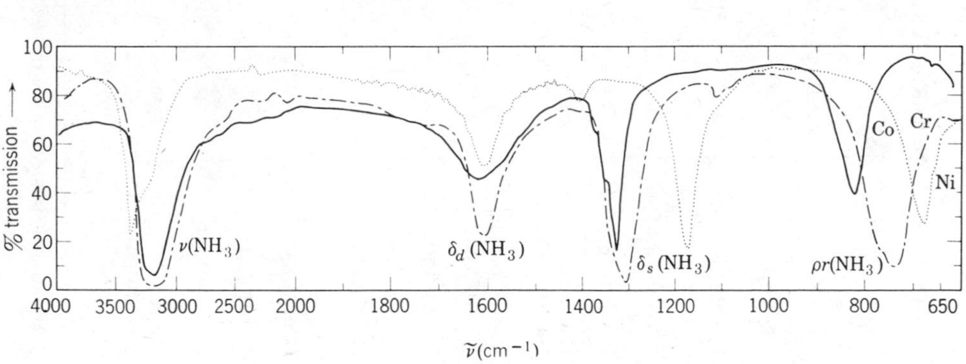

Fig. III-1. Infrared spectra of hexammine complexes: $[Co(NH_3)_6]Cl_3$ (solid line), $[Cr(NH_3)_6]Cl_3$ (dot-dashed line), and $[Ni(NH_3)_6]Cl_2$ (dotted line).

TABLE III-1. INFRARED FREQUENCIES OF OCTAHEDRAL HEXAMMINE COMPLEXES (CM^{-1})[a]

Complex	$\nu_a(NH_3)$	$\nu_s(NH_3)$	$\delta_a(HNH)$	$\delta_s(HNH)$	$\rho_r(NH_3)$	$\nu(MN)$ IR	$\nu(MN)$ Raman	$\delta(NMN)$	Refs.
$[Mg(NH_3)_6]Cl_2$	3353	3210	1603	1170	660	363	335 (A_{1g})	198	2
							243 (E_g)		
$[Cr(NH_3)_6]Cl_3$	3257	3185	1630	1307	748	495	465 (A_{1g})	—	3, 4
		3130				473	412 (E_g)		
						456			
$[^{50}Cr(NH_3)_6](NO_3)_3$	3310	3250	1627	1290	770	471	—	270	5
		3190							
$[Mn(NH_3)_6]Cl_2$	3340	3160	1608	1146	592	302	330 (A_{1g})	165	1, 6
$[Fe(NH_3)_6]Cl_2$	3335	3175	1596	1156	633	315	—	170	1, 6
$[Ru(NH_3)_6]Cl_2$	3315	3210	1612	1220	763	409	—	—	7
$[Ru(NH_3)_6]Cl_3$	3077		1618	1368	788	463	500 (A_{1g})	283	8
				1342			475 (E_g)	263	
$[Os(NH_3)_6]OsBr_6$	3125		1595	1339	818	452	—	256	8
$[Co(NH_3)_6]Cl_2$	3330	3250	1602	1163	654	325	357 (A_{1g})	92	6, 9, 10
							255 (E_g)		
$[Co(NH_3)_6]Cl_3$	3240	3160	1619	1329	831	498	500 (A_{1g})	331	11–13
						477	445 (E_g)		
						449			
$[Co(ND_3)_6]Cl_3$	2440	2300	1165	1020	667	462	—	294	5
						442			
						415			
$[Rh(NH_3)_6]Cl_3$	3200		1618	1352	845	472	515 (A_{1g})	302	8, 14
							480 (E_g)		
$[Ir(NH_3)_6]Cl_3$	3155		1587	1350	857	475	527 (A_{1g})	279	8, 14
				1323			500 (E_g)	264	
$[^{58}Ni(NH_3)_6]Cl_2$	3345	3190	1607	1176	685	335	370 (A_{1g})	217	11, 15
							265 (E_g)		
$[Zn(NH_3)_6]Cl_2$	3350	3220	1596	1145	645	300	—	—	1, 10
$[Cd(NH_3)_6]Cl_2$	—	—	1585	1091	613	298	342 (A_{1g})	—	6, 10
$[Pt(NH_3)_6]Cl_4$	3150	3050	1565	1370	950	530	569 (A_{1g})	318	16, 17
						516	545 (E_g)		

[a] All infrared frequencies are those of the F_{1u} species.

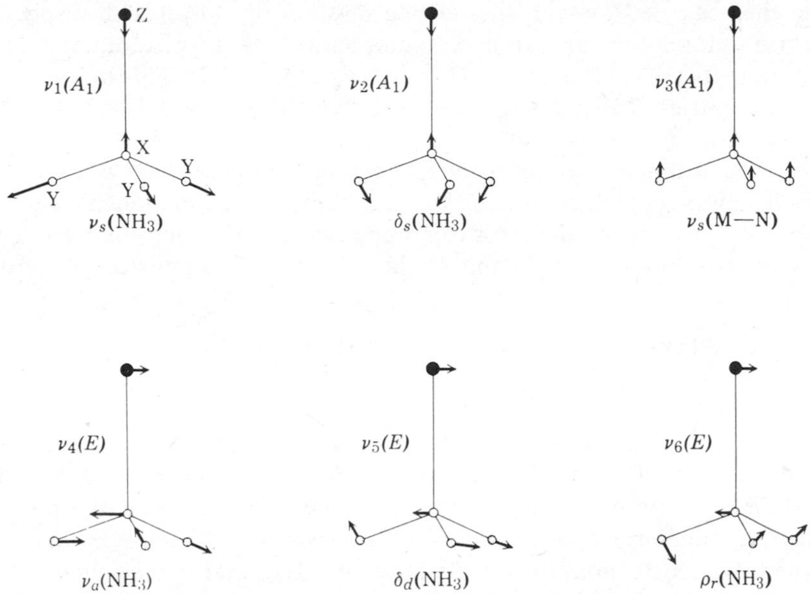

Fig. III-2. Normal modes of vibration of tetrahedral ZXY_3 molecules. (The band assignment is given for an $M\text{-}NH_3$ group.)

matrices.[19] For example, $[H_3N\text{-}K]^+Cl^-$ exhibits the NH_3 rocking band near 460 cm^{-1}, reflecting a rather weak $K\text{-}NH_3$ linkage.

To assign the skeletal modes such as the MN stretching and NMN bending modes, it is necessary to consider the normal modes of the octahedral MN_6 skeleton ($\mathbf{O}_h$ symmetry). The MN stretching mode in the low-frequency region is of particular interest since it provides direct information about the structure of the MN skeleton and the strength of the M-N bond. The octahedral MN_6 skeleton exhibits two ν(M-N)(A_{1g} and E_g) in Raman and one ν(M-N)(F_{1u}) in infrared spectra (Sec. II-8). Most of these vibrations have been assigned based on observed isotope shifts (including metal isotopes, NH_3/ND_3 and $NH_3/^{15}NH_3$) and normal coordinate calculations. Although the assignment of the ν(Co-N) in the infrared spectrum of $[Co(NH_3)_6]Cl_3$ had been controversial, Schmidt and Müller[5] confirmed the original assignments made by Nakamoto et al.; the three weak bands at 498, 477, and 449 cm^{-1} are the split components of the triply degenerate F_{1u} mode (Fig. III-3).[20] According to Nakagawa and Shimanouchi,[21] the intensity of the MN stretching mode in the infrared increases as the M-N bond becomes more ionic and as the MN stretching frequency becomes lower. Relative to the Co(III)-N bond of the $[Co(NH_3)_6]^{3+}$ ion, the Co(II)-N bond of the $[Co(NH_3)_6]^{2+}$ ion is more ionic, and its stretching frequency is much lower (325 cm^{-1}). This may be responsible for the strong appearance of the Co(II)-N stretching band in the infrared.

As listed in Table III-1, two Raman-active MN stretching modes (A_{1g} and E_g) are observed for the octahedral hexammine salts. In general, $\nu(A_{1g})$ is higher than $\nu(E_g)$. However, the relative position of $\nu(F_{1u})$ with respect to these two vibrations changes from one compound to another. Another obvious trend in ν(MN) is $\nu(M^{4+}—N) > \nu(M^{3+}—N) > \nu(M^{2+}—N)$. This holds for all symmetry species. Table III-1 shows that the NH_3 rocking frequency also follows the same trend as above.

Normal coordinate analyses on metal ammine complexes have been carried out by many investigators. Among them, Nakagawa, Shimanouchi, and co-workers[9,17,21] have made the most comprehensive study, using the UBF field. The MN stretching force constants of the hexammine complexes follow this order:

$$\begin{array}{ccccccccc} \text{Pt(IV)} & \gg & \text{Co(III)} & > & \text{Cr(III)} & > & \text{Ni(II)} & \approx & \text{Co(II)} \\ 2.13 & & 1.05 & & 0.94 & & 0.34 & & 0.33 \text{ mdyn/Å} \end{array}$$

Terrasse et al.[14] report a value of 1.6 mdyn/Å for the Rh–N stretching force constant of the $[Rh(NH_3)_6]^{3+}$ ion in the UBF field. On the other hand, Schmidt and Müller[5,6] and other workers[11] calculated the GVF constants of a number of ammine complexes by using the point mass model (i.e., the NH_3 ligand is regarded as a single atom having the mass of NH_3), and refined their values with isotope shift data (H–D, ^{14}N–^{15}N, and metal isotopes). For the hexammine series, they obtained the following order:

$$\begin{array}{ccccccccccc} Pt^{4+} & > & Ir^{3+} & > & Os^{3+} & > & Rh^{3+} & > & Ru^{3+} & > & Co^{3+} > \\ 2.75 & & 2.28 & & 2.13 & & 2.10 & & 2.01 & & 1.86 \end{array}$$

$$\begin{array}{ccccccccccccc} Cr^{3+} & > & Ni^{2+} & > & Co^{2+} & > & Fe^{2+} & \sim & Cd^{2+} & > & Zn^{2+} & \gtrsim & Mn^{2+} \\ 1.66 & & 0.85 & & 0.80 & & & 0.73 & & & 0.69 & & 0.67 \text{ mdyn/Å} \end{array}$$

For a series of divalent metals, the above order is parallel to the Irving–Williams series ($Mn^{2+} < Fe^{2+} < Co^{2+} < Ni^{2+} < Cu^{2+} > Zn^{2+}$). Schmidt and Müller[1] discussed the relationship between the MN stretching force constant and the stability constant or the bond energy.

Table III-2 lists the observed infrared frequencies and band assignments of tetrahedral, square-planar, and linear metal ammine complexes. The Raman-active MN stretching frequencies are also included in Table III-2. Normal coordinate analyses have been made by Nakagawa et al.[9,17,21] by using the UBF field; the following values were obtained for the MN stretching force constants:

$$\begin{array}{ccccccc} Hg^{2+} & > & Pt^{2+} & > & Pd^{2+} & > & Cu^{2+} \\ 2.05 & & 1.92 & & 1.71 & & 0.84 \text{ mdyn/Å} \end{array}$$

On the other hand, Schmidt and Müller[6] obtained the following values by using the GVF field and the point mass approximation:

$$\begin{array}{ccccccccccc} Pt^{2+} & > & Pd^{2+} & \gg & Co^{2+} & \sim & Zn^{2+} & \sim & Cu^{2+} & > & Cd^{2+} \\ 2.54 & & 2.15 & & 1.44 & & 1.43 & & 1.42 & & 1.24 \text{ mdyn/Å} \end{array}$$

Table III-2. Infrared Frequencies of Other Ammine Complexes (cm^{-1})

Complex	$\nu_a(NH_3)$	$\nu_s(NH_3)$	$\delta_a(HNH)$	$\delta_s(HNH)$	$\rho_r(NH_3)$	$\nu(MN)$ IR	$\nu(MN)$ Raman	$\delta(NMN)$	Refs.
Tetrahedral									
$[Co(NH_3)_4](ReO_4)_2$	3340	3260	1610	1240	693	430	405 (A_1)	195	22
$[^{64}Zn(NH_3)_4]I_2$	3275	3150	1596	1253	685	—	432 (A_1)	156	23
	3233			1239			412 (F_2)		
$[Cd(NH_3)_4](ReO_4)_2$	3354	3267	1617	1176	670	370	—	166	1
								160	
Square-planar									
$[^{104}Pd(NH_3)_4]Cl_2 \cdot H_2O$	3270	3170	1630	1279	849	495	510 (A_{1g})	291	5, 24
					802		468 (B_{1g})	238	
$[Pt(NH_3)_4]Cl_2$	3236	3156	1563	1325	842	510	524 (A_{1g})	297	17, 25
							508 (B_{1g})	235	
$[Cu(NH_3)_4]SO_4 \cdot H_2O$	3327	3169	1669	1300	735	426	420 (A_{1g})	256	5, 26
	3253		1639	1283			375 (B_{1g})	227	
$[Au(NH_3)_4](NO_3)_3$	3490	3105	1571	1331	936	555	566	327	27
	3220				914		544	307	
Linear									
$[Ag(NH_3)_2]_2SO_4$	3320	3150	1642	1236	740	476	372 (A_1)	[illegible]	[illegible]
	3230		1626	1222	703	400			
$[Hg(NH_3)_2]Cl_2$	3265	3197	1605	1268	719	513	412	[illegible]	[illegible]

Vibrational spectra of metal ammine complexes in the crystalline state exhibit lattice vibrations below 200 cm^{-1}. Assignments of lattice modes have been made for the hexammine salts of Mg(II),[2] Co(II),[10,32] Ni(II),[10,21,31,32] $[Co(NH_3)_6][Co(CN)_6]$,[33] and $[Pt(NH_3)_4]Cl_2$.[34] The Magnus green salt, $[Pt(NH_3)_4][PtCl_4]$, is of particular interest. Originally, Hiraishi et al.[17] assigned the infrared band at 200 cm^{-1} to a lattice mode which corresponds to the stretching mode of the Pt—Pt—Pt chain. This high frequency was justified on the basis of the strong Pt–Pt interaction in this salt. Adams and Hall,[35] on the other hand, assigned this mode at 81 cm^{-1}, and the 201 cm^{-1} band to a NH_3 torsion. In fact, the latter is shifted to 158 cm^{-1} by the deuteration of NH_3 ligands.[36] Lattice modes and low-frequency internal modes of hexammine complexes have also been studied by Janik et al.,[37,38] using the inelastic neutron scattering technique. Attempts have been made to determine the vibrational frequencies in the electronic excited states by the analysis of vibronic spectra; the absorption,[4,39] emission,[40] and luminescence[41] spectra of $[Cr(NH_3)_6]^{3+}$ salts ($^2E_g \leftarrow {}^4A_{2g}$ transition) have yielded the NH_3 rocking and NCrN bending frequencies of the F_{2u} vibrations, which are forbidden in both infrared and Raman spectra of octahedral hexammine complexes.

(2) Halogenoammine Complexes

If the NH_3 groups of a hexammine complex are partly replaced by other groups, the degenerate vibrations are split because of lowering of symmetry, and new bands belonging to other groups appear. Here we discuss only halogenoammine complexes. The infrared spectra of $[Co(NH_3)_5X]^{2+}$- and *trans*-$[Co(NH_3)_4X_2]^+$-type complexes have been studied by Nakagawa and Shimanouchi.[9,21] Table III-3 lists the observed frequencies and band assignments obtained by these workers. The infrared spectra of some of these complexes in the CoN stretching region are shown in Fig. III-3. Normal coordinate analyses on these complexes[9] have yielded the following UBF stretching force constants (mdyn/Å): K(Co—N), 1.05; K(Co—F), 0.99; K(Co—Cl), 0.91; K(Co—Br), 1.03; and K(Co—I), 0.62. Raman spectra of some chloroammine Co(III) complexes have been assigned.[42] For halogenoammine complexes of Cr(III), see Refs. 43 and 3. Detailed vibrational assignments have been made for halogenoammine complexes of Os(III)[44] and of Ru(III), Rh(III), Os(III), and Ir(III).[45]

In regard to $M(NH_3)_4X_2$- and $M(NH_3)_3X_3$-type complexes, the main interest has been the distinction of stereoisomers by vibrational spectroscopy. As shown in Appendix III, *trans*-MN_4X_2 ($\mathbf{D}_{4h}$) exhibits one MN stretching (E_u) and one MX stretching (A_{2u}), while *cis*-MN_4X_2 ($\mathbf{C}_{2v}$) shows four MN stretching (two A_1, B_1, and B_2) and two MX stretching (A_1 and B_1) vibrations in the infrared. For *mer*-MN_3X_3 ($\mathbf{C}_{2v}$), three MN stretching and three MX stretching vibrations are infrared active, whereas only two MN stretching and two MX stretching vibrations are infrared active for *fac*-MN_3X_3 ($\mathbf{C}_{3v}$). Nolan and James[16] have measured and assigned the Raman spectra of a series of $[Pt(NH_3)_nCl_{6-n}]^{(n-2)+}$-type complexes.

TABLE III-3. SKELETAL VIBRATIONS OF PENTAMMINE AND *trans*-TETRAMMINE Co(III) COMPLEXES (cm^{-1})[9,21]

Complex		ν(CoN)	ν(CoX)	Skeletal Bending
Pentammine ($\mathbf{C}_{4v}$ symmetry)				
$[Co(NH_3)_5F]^{2+}$	A_1	480, 438	343	308
	E	498	—	345, 290, 219
$[Co(NH_3)_5Cl]^{2+}$	A_1	476, 416	272	310
	E	498	—	292, 287, 188
$[Co(NH_3)_5Br]^{2+}$	A_1	475, 410	215	287
	E	497	—	290, 263, 146
$[Co(NH_3)_5I]^{2+}$	A_1	473, 406	168	271
	E	498	—	290, 259, 132
trans-Tetrammine ($\mathbf{D}_{4h}$ symmetry)				
$[Co(NH_3)_4Cl_2]^+$	A_{2u}	—	353	186
	E_u	501	—	290, 167
$[Co(NH_3)_4Br_2]^+$	A_{2u}	—	317	227
	E_u	497	—	280, 120

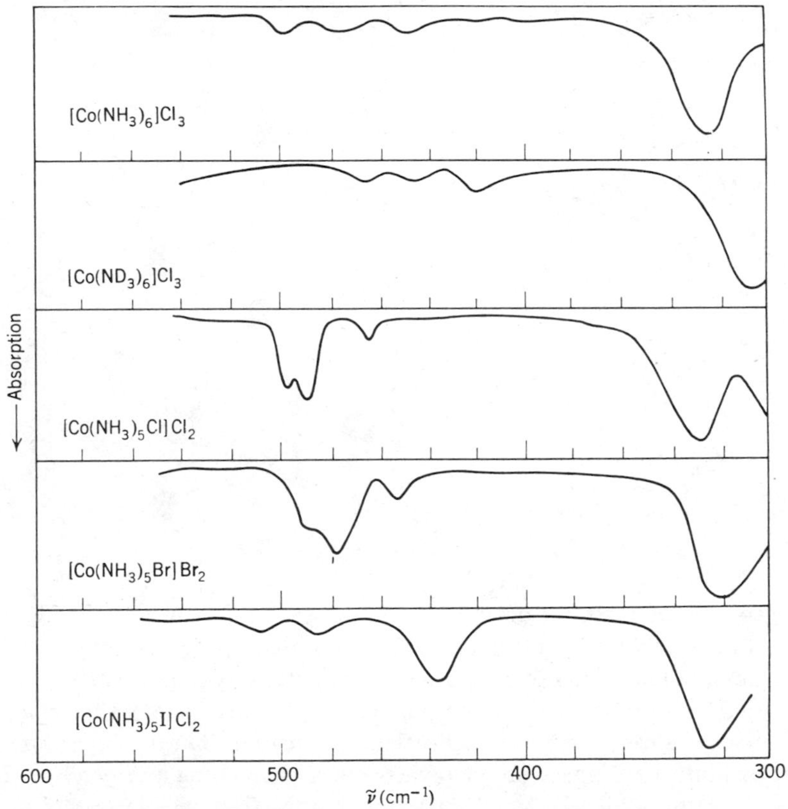

Fig. III-3. Infrared spectra of Co(III) ammine complexes.[20]

Vibrational spectra of the planar $M(NH_3)_2X_2$-type complexes [M = Pt(II) and Pd(II)] have been studied by many investigators. Table III-4 summarizes the observed frequencies and band assignments of their skeletal vibrations, including those of "*cis*-platin"—the well-known anticancer drug. Figure III-4 shows the infrared spectra of *cis*- and *trans*-$[Pd(NH_3)_2Cl_2]$ obtained by Layton et al.[49] As expected, both the PdN and PdCl stretching bands split into two in the *cis*-isomer. Durig et al.[50] found that the PdN stretching frequencies range from 528 to 436 cm^{-1}, depending on the nature of other ligands in the complex. In general, the PtN stretching band shifts to a lower frequency as a ligand of stronger *trans*-influence is introduced in the position *trans* to the Pt–N bond.[51] Using infrared spectroscopy, Durig and Mitchell[52] studied the isomerization of *cis*-$[Pd(NH_3)_2X_2]$ to its *trans*-isomer.

Mixed-valence compounds such as $Pd^{II}Pt^{IV}(NH_3)_4Cl_6$ and $Pd^{II}Pd^{IV}(NH_3)_4Cl_6$ take the form of a chain structure as shown below:

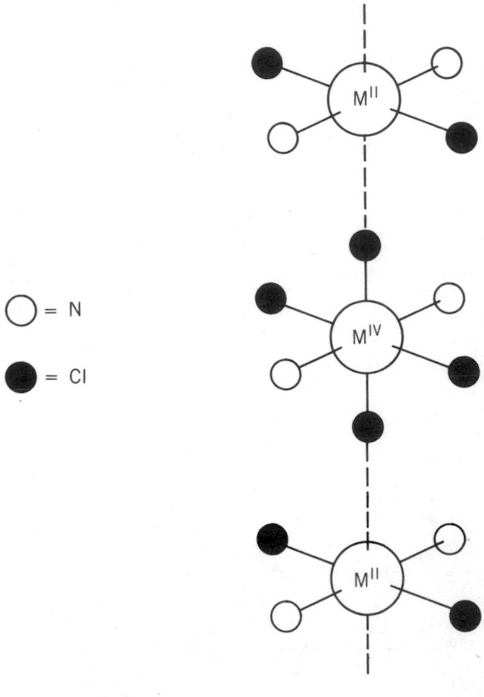

Both compounds exhibit an intense, broad absorption in the visible region. When the laser wavelength is chosen in this region, the former shows a RR spectrum involving the progressions of three totally symmetric metal–chlorine stretching vibrations of both components. Thus, the visible absorption was attributed to a metal–metal mixed-valence transition. On the other hand, the latter exhibits a RR spectrum involving several stretching and bending fundamentals of the $[Pd(NH_3)_2Cl_4]$ component only. The visible spectrum must therefore be attributed to the metal–ligand charge-transfer transitions within this component.[53]

TABLE III-4. SKELETAL FREQUENCIES OF SQUARE-PLANAR $M(NH_3)_2X_2$-TYPE COMPLEXES (cm^{-1})[a]

Complex	$\nu(MN)$	$\nu(MX)$	Bending	Refs.
trans-$[Pd(NH_3)_2Cl_2]$				
IR	496	333	245, 222, 162, 137	46
R	494	295		
cis-$[Pd(NH_3)_2Cl_2]$				
IR	495, 476	327, 306	245, 218, 160, 135	46
trans-$[Pd(NH_3)_2Br_2]$				
IR	490	—	220, 220, 122, 101	46
cis-$[Pd(NH_3)_2Br_2]$				
IR	480, 460	258	225, 225, 120, 100	46
trans-$[Pd(NH_3)_2I_2]$				
IR	480	191	263, 218, 109	46
trans-$[Pt(NH_3)_2Cl_2]$				
IR	572	365	220, 195	47, 47a
R	529	318	—	47
cis-$[Pt(NH_3)_2Cl_2]$				
IR	510	330, 323	250, 198, 155, 123	34, 47a
trans-$[Pt(NH_3)_2Br_2]$				
IR	504	260	230	47, 47a
R	535	206	—	47
trans-$[Pt(NH_3)_2I_2]$				
R	532	153	—	47

[a] For band assignments, see also Refs. 17 and 48.

(3) Amido (NH_2) Complexes

The vibrational spectra of amido complexes may be interpreted in terms of the normal vibrations of a pyramidal ZXY_2-type molecule. Mizushima et al.[54] and Niwa et al.[55] carried out normal coordinate analysis on the $[Hg(NH_2)_2]^+_\infty$ ion (infinite-chain polymer); the results of the latter workers are given in Table III-5. Brodersen and Becher[56] studied the infrared spectra of a number of compounds containing Hg–N bonds and assigned the HgN stretching bands at 700–400 cm^{-1}. The infrared spectrum of the NH_2^- ion in alkali-metal salts has been measured.[57] Alkylamido complexes of the type $M(NR_2)_{4,5}$ (M = Ti, Zr, Hf, V, Nb, and Ta) exhibit their MN stretching bands in the 700–530 cm^{-1} region.[58]

(4) Alkylamine Complexes

Infrared spectra of methylamine complexes, $[Pt(CH_3NH_2)_2X_2]$ (X: a halogen), have been studied by Watt et al.[59] and Kharitonov et al.[60] Far-infrared spectra of $[M(R_2NH)_2X_2]$- [M = Zn(II) or Cd(II); R = ethyl or *n*-propyl; X = Cl or Br] type complexes have also been reported.[61] Chatt and co-workers[62] studied the effect of hydrogen bonding on the NH stretching frequencies of

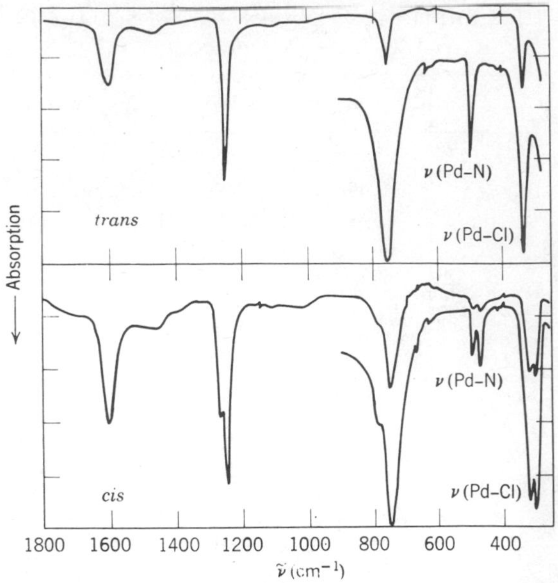

Fig. III-4. Infrared spectra of *trans*- and *cis*-$[Pd(NH_3)_2Cl_2]$.[49]

trans-$[Pt(RNH_2)Cl_2L]$-type complexes (R = Me, Et, etc.; L = C_2H_4, PEt_3, etc.) in organic solvents such as chloroform and dioxane. Their study revealed that the complexes of primary amines have a strong tendency to associate through intermolecular hydrogen bonds of the NH···Cl type, whereas those of secondary amines have little tendency to associate. Later, this difference was explained on the basis of steric repulsion and intramolecular interaction between the NH hydrogen and the nonbonding *d*-electrons of the metal.[63]

(5) Complexes of Hydroxylamine and Hydrazine

The vibrational spectra of hydroxylamine (NH_2OH) complexes have been studied by Kharitonov et al.[64] Sacconi and Sabatini[65] reported the infrared spectra of hydrazine (NH_2NH_2) complexes of the type $[M(N_2H_4)_2Cl_2]$, where M is a divalent metal, and assigned the MN stretching bands between 440

TABLE III-5. INFRARED FREQUENCIES AND BAND ASSIGNMENTS OF AMIDO COMPLEXES (CM^{-1})[55]

Compound	$\nu(NH_2)$	$\delta(NH_2)$	$\rho_w(NH_2)$	$\rho_r(NH_2)$	$\nu(HgN)$
$[Hg(NH_2)]^+_\infty(Cl)^-_\infty$	3200, 3175	1540	1025	673	573
$[Hg(NH_2)]^+_\infty(Br)^-_\infty$	3220, 3180	1525	1008	652	560

and 330 cm^{-1}. Infrared spectra of hydrazine complexes of Hg [66] and Ni [67,68] have been reported.

III-2. COMPLEXES OF ETHYLENEDIAMINE AND RELATED COMPOUNDS

As is shown in Fig. III-5, 1,2-disubstituted ethane may exist in the *cis*, *trans*, or *gauche* form, depending on the angle of internal rotation. The *cis* form may not be stable in the free ligand because of steric repulsion between two X groups. The *trans* form belongs to point group $\mathbf{C}_{2h}$, in which only the *u* vibrations (antisymmetric with respect to the center of symmetry) are infrared active. On the other hand, both *gauche* forms belong to point group $\mathbf{C}_2$, in which all the vibrations are infrared active. Thus the gauche form exhibits more bands than the *trans* form. Mizushima and co-workers[69] have shown that 1,2-dithiocyanatoethane ($NCS—CH_2—CH_2—SCN$) in the crystalline state definitely exists in the *trans* form, because no infrared frequencies coincide with Raman frequencies (mutual exclusion rule). By comparing the spectrum of the crystal with that of a $CHCl_3$ solution, they concluded that several extra bands observed in solution can be attributed to the *gauche* form. Table III-6 summarizes the infrared frequencies and band assignments obtained by Mizushima et al. It is seen that the CH_2 rocking vibration provides the most clear-cut diagnosis of conformation: one band (A_u) at 749 cm^{-1} for the *trans* form, and two bands (A and B) at 918 and 845 cm^{-1} for the *gauche* form.

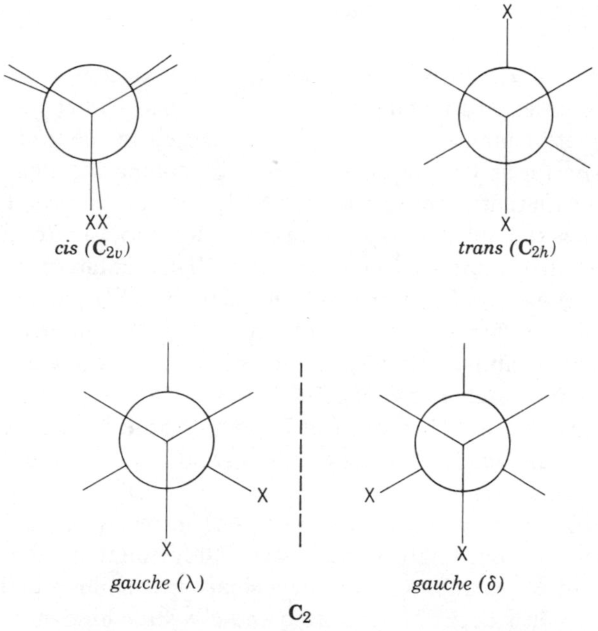

Fig. III-5. Rotational isomers of 1,2-disubstituted ethane.

TABLE III-6. INFRARED SPECTRA OF 1,2-DITHIOCYANATOETHANE AND ITS PT(II) COMPLEX (CM^{-1})[69]

Ligand			
Crystal (*trans*)	$CHCl_3$ Solution (*gauche* + *trans*)	Pt Complex (*gauche*)	Assignment
—	2170 (*g*)	2165 (*g*)	$\nu(C{\equiv}N)$
2155 (*t*)	2170 (*t*)	—	
1423 (*t*)	1423 (*t*)	—	$\delta(CH_2)$
—	1419 (g)	1410 (g)	
1291[a] (t)	—	—	$\rho_w(CH_2)$
—	1285 (g)	1280 (g)	
1220 (*t*)	1215 (*t*)	—	
1145 (*t*)	1140 (*t*)	—	$\rho_t(CH_2)$
—	1100 (*g*)	1110 (*g*)	
—	— (*g*)[b]	1052 (*g*)	$\nu(CC)$
1037[a]	—	—	
—	918 (*g*)	929 (*g*)	$\rho_r(CH_2)$
—	845 (*g*)	847 (*g*)	
749 (*t*)	—[b]	—	
680 (*t*)	677 (*t*)	—	$\nu(CS)$
660 (*t*)	660 (*t*)	—	

[a] Raman frequencies in the crystalline state.
[b] Hidden by $CHCl_3$ absorption.

The compound 1,2-dithiocyanatoethane may take the *cis* or *gauche* form when it coordinates to a metal through the S atoms. The chelate ring formed will be completely planar in the *cis*, and puckered in the *gauche*, form. The *cis* and *gauche* forms can be distinguished by comparing the spectrum of a metal chelate with that of the ligand in $CHCl_3$ solution (*gauche* + *trans*). Table III-6 compares the infrared spectrum of 1,2-dithiocyanatoethanedichloroplatinum(II) with that of the free ligand in a $CHCl_3$ solution. Only the bands characteristic of the *gauche* form are observed in the Pt(II) complex. This result definitely indicates that the chelate ring in the Pt(II) complex is *gauche*. The method described above has also been applied to the metal complexes of 1,2-dimethylmercaptoethane ($CH_3S—CH_2—CH_2—SCH_3$).[70] In this case, the free ligand exhibits one CH_2 rocking at 735 cm^{-1} in the crystalline state (*trans*), whereas the metal complex always exhibits two CH_2 rockings at 920-890 and 855-825 cm^{-1} (*gauche*).

The conformation of ethylenediamine (en) in metal complexes is of particular interest in coordination chemistry. Unfortunately, the CH_2 rocking mode discussed above does not provide a clear-cut diagnosis in this case, since it couples strongly with the NH_2 rocking and CN stretching modes that appear in the same frequency region. However, X-ray analyses on *trans*-

$[Co(en)_2Cl_2]Cl \cdot HCl \cdot 2H_2O$[71] and other complexes indicate that the chelating ethylenediamine takes the *gauche* conformation without exception.

The $[M(en)_3]^{n+}$ ion can take eight different conformations if we consider all possible combinations of conformations of the three chelate rings (λ or δ) around the chiral metal center. They are designated as $\Lambda(\delta\delta\delta)$, $\Lambda(\delta\delta\lambda)$, $\Lambda(\delta\lambda\lambda)$, $\Lambda(\lambda\lambda\lambda)$, $\Delta(\lambda\lambda\lambda)$, $\Delta(\lambda\lambda\delta)$, $\Delta(\lambda\delta\delta)$, and $\Delta(\delta\delta\delta)$. According to Cramer and Huneke,[72] it is possible to distinguish some of these conformers based on the number of infrared-active C–C stretching vibrations. For example, $[Cr(en)_3]Cl_3 \cdot 3.5H_2O$ [$\Lambda(\delta\delta\delta)$, $\mathbf{D}_3$ symmetry] exhibits only one band at 1003 cm^{-1} whereas $[Cr(en)_3][Ni(CN)_5] \cdot 1.5H_2O$ [$\Lambda(\delta\delta\lambda, \delta\lambda\lambda)$, $\mathbf{C}_2$ symmetry] exhibits three bands at 1008, 1002 (shoulder), and 995 cm^{-1}. Gouteron has shown[73] that racemic(*dl*) and optically active(*d*) forms of $[Co(en)_3]Cl_3$ can be distinguished in the crystalline state by comparing vibrational spectra below 200 cm^{-1} if they do not belong to the same crystal system. Borch et al.[74] carried out the most complete vibrational study on the $[Rh(en)_3]^{3+}$ ion [$\Lambda(\delta\delta\delta)$, $\mathbf{D}_3$ symmetry]. Their work includes ^{15}N, ND_2, CD_2 isotope shift data as well as a complete normal coordinate analysis on the whole ion. Flint and Matthews[75] found from their electronic spectral studies that the vibrational modes of the $[M(en)_3]^{3+}$ ion (M = Cr and Co) in the 320–280 cm^{-1} region have at least as much M–N stretching character as those in the 600–450 cm^{-1} region although previous Raman studies assign the former to the NMN bending and the latter to the M–N stretching mode.

The M–N stretching vibrations have been assigned in the 410–380 cm^{-1} region for the $[M(en)_3]^{2+}$ ion (M = Zn, Cd, Fe, Ni, Co, and Mn)[76] and in the 550–400 cm^{-1} region for the $[M(en)_2]^{2+}$ ion (M = Cu, Pd, and Pt).[77]

Lever and Mantovani[78] assigned the MN stretching bands of $M(N{-}N)_2X_2$-[M = Cu(II), Co(II), and Ni(II); N—N = en, dimethyl-en, etc.; X = Cl, Br, etc.] type complexes by using the metal isotope technique. For these compounds, the CoN and NiN stretching bands have been assigned to 400–230 cm^{-1} [78] and the CuN stretching vibrations have been located in the 420–360 cm^{-1} range.[79] A straight-line relationship between the square of the CuN stretching frequency and the energy of the main electronic *d-d* band was found,[80] with some exceptions.[79]

The infrared spectra of *cis*- and *trans*-$[M(en)_2X_2]^+$ [M = Co(III), Cr(III), Ir(III), and Rh(III); X = Cl, Br, etc.] have been studied extensively.[81–84] These isomers can be distinguished by comparing the spectra in the regions of 1700–1500 (NH_2 bending), 950–850 (CH_2 rocking), and 610–500 cm^{-1} (MN stretching). The infrared spectra of Magnus-type $[Pt(en)_2][PtCl_4]$ and its analogs have been reported.[85,86] So are the resonance Raman spectra[87] of $[Pt(en)_2][Pt(en)_2Cl_2]$ $(ClO_4)_4$ which takes a linear Pt(II)—Cl—Pt(IV) chain structure shown previously [Sec. III-1(2)].

Ethylenediamine takes the *trans* form when it functions as a bridging group between two metal atoms. Powell and Sheppard[88] were the first to suggest that ethylenediamine in $(C_2H_4)Cl_2Pt(en)PtCl_2(C_2H_4)$ is likely to be *trans*, since the infrared spectrum of this compound is simpler than that of other complexes in which ethylenediamine is *gauche*. Similar results have been obtained for $(AgCl)_2en$,[89] $(AgSCN)_2en$,[90] $(AgCN)_2en$,[90] $Hg(en)Cl_2$,[91] and $M(en)Cl_2$ (M = Zn or Cd).[89] The structure of these complexes may be depicted as follows:

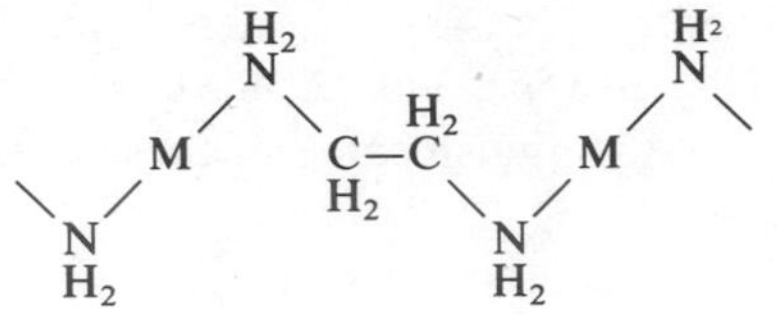

A more complete study, including the infrared and Raman spectra, of $M(en)X_2$-type complexes [M = Zn(II), Cd(II), and Hg(II); X = Cl, Br, and SCN] has been made by Iwamoto and Shriver.[92] Mutual exclusion of infrared and Raman spectra, along with other evidence, supports the $\mathbf{C}_{2h}$ bridging structure of the en ligand in the Cd and Hg complexes (see Fig. III-6).

The infrared spectra of diethylenetriamine (dien) complexes have been reported for [Pd(dien)X]X (X = Cl, Br, and I)[93] and $[Co(dien)(en)Cl]^{2+}$.[94] The latter exists in the four isomeric forms shown in Fig. III-7. Their infrared spectra revealed that the ω- and κ-isomers contain dien in the *mer*-configuration; the π- and ε-isomers contain dien in the *fac*-configuration. The *mer*- and *fac*-isomers of $[M(dien)X_3]$ [M = Cr(III), Co(III), and Rh(III); X: a halogen] can also be distinguished by infrared spectra.[95]

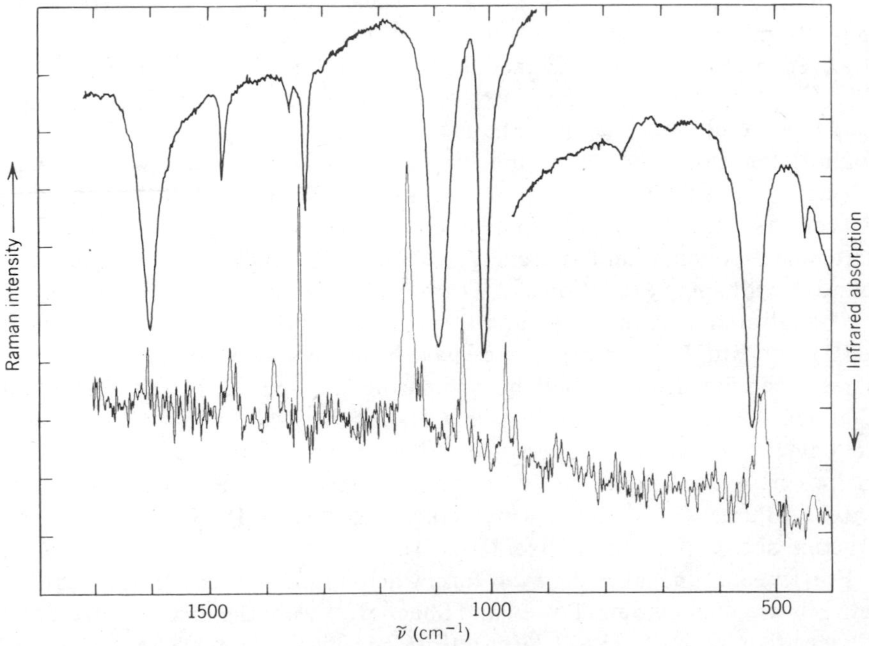

Fig. III-6. Infrared (top) and Raman (bottom) spectra of $[Cd(en)Br_2]$.[92]

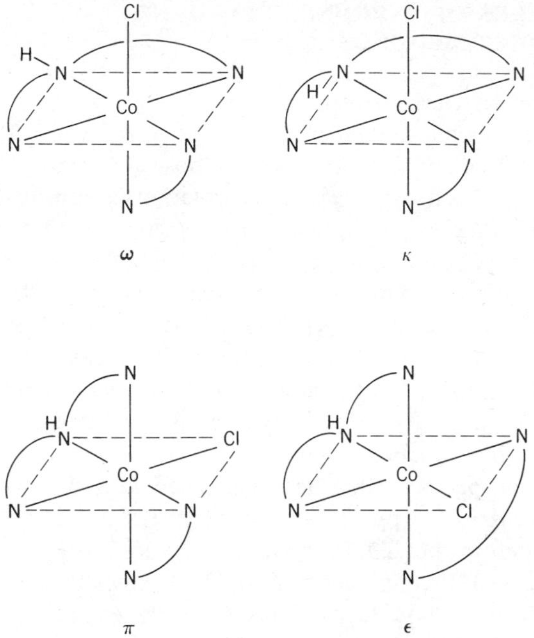

Fig. III-7. Structures of $[Co(dien)(en)Cl]^{2+}$.

The infrared spectra of β, β′, β″-triaminotriethylamine (tren) complexes with Co(III)[96] and lanthanides[97] have been reported. Buckingham and Jones[98] measured the infrared spectra of $[M(trien)X_2]^+$, where trien is triethylenetetramine, M is Co(III), Cr(III), or Rh(III), and X is a halogen or an acido anion. These compounds give three isomers (Fig. III-8) which can be distinguished, for example, by the CH_2 rocking vibrations in the 920–869 cm^{-1} region. For $[Co(trien)Cl_2]$ ClO_4, *cis-α*-isomer exhibits two strong bands at 905 and 871 cm^{-1} and *cis-β*-isomer shows four bands at 918, 898, 868, and 862 cm^{-1}; *trans*-isomer gives only one band at 874 cm^{-1} with a weak band at 912 cm^{-1}. Far-infrared spectra of some of these trien complexes have been reported.[99]

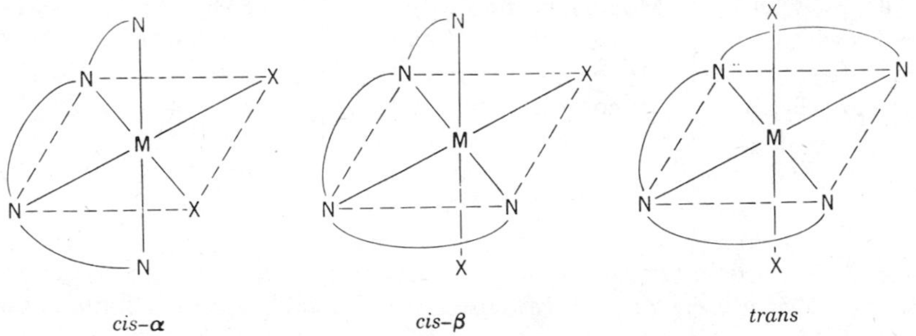

Fig. III-8. Structures of $[M(trien)X_2]^+$-type ions.

III-3. COMPLEXES OF PYRIDINE, BIPYRIDINE, AND RELATED COMPOUNDS

(1) Pyridine and Its Derivatives

Upon complex formation, the pyridine (py) vibrations in the high-frequency region are not shifted appreciably, whereas those at 604 (in-plane ring deformation) and 405 cm^{-1} (out-of-plane ring deformation) are shifts to higher frequencies. Clark and Williams[100] carried out an extensive far-infrared study on metal pyridine complexes. Table III-7 lists the observed frequencies of these metal-sensitive py vibrations and metal—py stretching vibrations. Clark and Williams showed that ν(M—py) and ν(MX) (X: a halogen) are very useful in elucidating the stereochemistry of these py complexes. For example, *fac*-[$Rh(py)_3Cl_3$] exhibits two ν(Rh—py) ($\mathbf{C}_{3v}$ symmetry), whereas *mer*-$Rh(py)_3Cl_3$ shows three ν(Rh—py) ($\mathbf{C}_{2v}$ symmetry) near 250 cm^{-1}. The infrared spectra of these two compounds are given in Fig. III-54.

The metal isotope technique has been used to assign the ν(M—py) and ν(MX) vibrations of $Zn(py)_2X_2$[101] and $Ni(py)_4X_2$.[102] The former vibrations have been located in the 225–160 and 250–225 cm^{-1} regions, respectively, for the Zn(II) and Ni(II) complexes. Figure III-9 shows the infrared and Raman spectra of [$^{64}Zn(py)_2Cl_2$] and its ^{68}Zn analog. As expected from its $\mathbf{C}_{2v}$ symmetry, two ν(Zn—py) and two ν(ZnCl) are metal-isotope sensitive. Far-infrared spectra of metal pyridine nitrate complexes, $M(py)_x(NO_3)_y$, have been reported.[103,104]

TABLE III-7. VIBRATIONAL FREQUENCIES OF PYRIDINE COMPLEXES (CM^{-1})[100]

Complex	Structure	py	py	ν(M—py)
$Co(py)_2Cl_2$[a]	Monomeric, tetrahedral	642	422	253[b]
$Ni(py)_2I_2$	Monomeric, tetrahedral	643	428	240
$Cr(py)_2Cl_2$	Polymeric, octahedral	640	440	219
$Cu(py)_2Cl_2$	Polymeric, octahedral	644	441	268
$Co(py)_2Cl_2$[a]	Polymeric, octahedral	631	429	243, 235[b]
mer-[$Rh(py)_3Cl_3$]	Monomeric, octahedral	650	468	265, 245, 230
fac-[$Rh(py)_3Cl_3$]	Monomeric, octahedral	643	464	266, 245
trans-[$Ni(py)_4Cl_2$]	Monomeric, octahedral	626	426	236
trans-[$Ir(py)_4Cl_2$]Cl	Monomeric, octahedral	650	469	260, (255)
cis-[$Ir(py)_4Cl_2$]Cl	Monomeric, octahedral	656	468	287, 273
trans-[$Pt(py)_2Br_2$]	Monomeric, square-planar	656	476	297
cis-[$Pt(py)_2Br_2$]	Monomeric, square-planar	659 644	448	260, 234

[a] $Co(py)_2Cl_2$ exists in two forms: blue complex (monomeric, tetrahedral) and violet complex (polymeric, octahedral). The far-infrared spectra of these two forms are shown in Fig. III-55 (Sec. III-19).

[b] Assignments made by Postmus et al.[109].

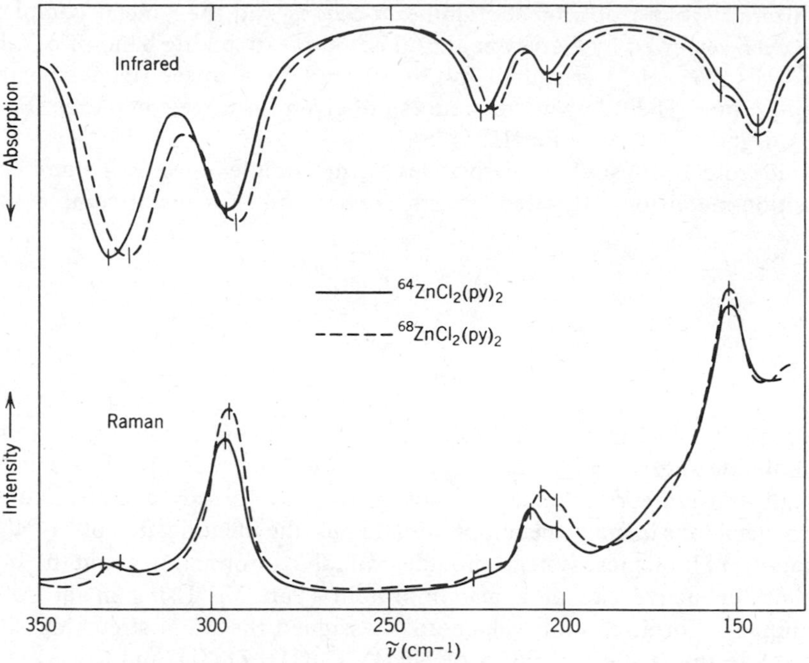

Fig. III-9. Infrared and Raman spectra of $^{64}Zn(py)_2Cl_2$ and its ^{68}Zn analog.

Fleischmann et al.[105] observed that pyridine adsorbed on a silver electrode surface exhibits three Raman bands (totally symmetric ring breathing vibrations) at 1036, 1008, and 1025 cm^{-1} whose intensities vary markedly depending upon the applied electrode potential. They assigned the former two to the pyridine which is hydrogen-bonded to water molecules adsorbed on the electrode and the latter to the pyridine directly bonded to the Lewis-acid site of the electrode. Jeanmaire and Van Duyne[106] first recognized that the molar intensities of these adsorbed pyridine molecules are enhanced by a factor of 10^5–10^6 times relative to normal pyridine. Since then, "surface-enhanced Raman scattering (SERS)" has been observed for other metal electrodes and adsorbates.[107] Thus far, no single definitive theories are available about the origin of this interesting phenomenon.[108]

The infrared spectra of metal complexes with alkyl pyridines have been studied extensively.[110-114] Using the metal isotope technique, Lever and Ramaswamy[115] assigned the M—pic stretching bands of $M(pic)_2X_2$ [M = Ni(II) and Cu(II); pic = picoline; X = Cl, Br, and I] in the 300–230 cm^{-1} region. The infrared spectra of metal complexes with halogenopyridines have been reported.[111,116,117] Infrared spectra have been used to determine whether coordination occurs through the nitrile or the pyridine nitrogen in cyanopyridine complexes. It was found that 3- and 4-cyanopyridines coordinate to the metal via the pyridine nitrogen,[118-120] whereas 2-cyanopyridine coordinates to the metal via the nitrile nitrogen.[118,121]

Infrared spectra of aromatic amine N-oxides and their metal complexes have been reviewed by Garvey et al.[122] The N=O stretching band of pyridine N-oxide (1265 cm^{-1}) is shifted by 70–30 cm^{-1} to a lower frequency upon complexation. The following references are given for three complexes: Fe(II) (123), Hg(II) (124), and Fe(III) (125).

Imidazole(ImH) and its derivatives form complexes with a number of transition-metal ions. Infrared spectra are reported for metal complexes of

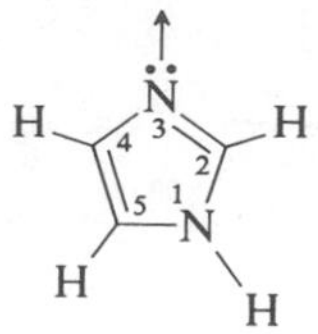

imidazole,[126–128] 2-methylimidozole,[129–130] N-methylimidazole,[131] 4- and 5-bromoimidazole,[132] and benzimidazole.[133,134] Among them, imidazole is biologically most important since imidazole nitrogens of histidyl residues coordinate to metal ions in many metalloproteins. Thus, the identification of M–N(Im) vibrations in biological systems provides valuable information about the structure of the active site of a metalloprotein (Part V). Using metal isotope techniques, Cornilsen and Nakamoto[128] assigned the M–N stretching vibrations of 16 imidazole complexes of Ni(II), Cu(II), Zn(II), and Co(II) in the 325–210 cm^{-1} region. Hodgson et al.[135] also assigned these vibrations in the same region. Salama and Spiro[136] were first to assign the Co–N stretching vibrations in resonance Raman spectra of $Co(ImH)_2Cl_2$ (274 and 232 cm^{-1}), $[Co(ImH)_4]^{2+}$ (301 cm^{-1}), and $[Co(Im^-)_4]^{2-}$ (306 cm^{-1}).

(2) Complexes of 2,2′-Bipyridine and Related Ligands

Infrared spectra of metal complexes of 2,2′-bipyridine (bipy) and 1,10-phenanthroline (phen) have been studied extensively. In general, the bands in the high-frequency region are not metal sensitive since they originate in the heterocyclic or aromatic ring of the ligand. Thus the main interest has been focused on the low-frequency region, where $\nu(MN)$ and other metal-sensitive vibrations appear. It has been difficult, however, to assign $\nu(MN)$ empirically since several ligand vibrations also appear in the same frequency region. This difficulty was overcome by using the metal isotope technique. Hutchinson et al.[137] first applied this method to the tris-bipy and phen complexes of Fe(II), Ni(II), and Zn(II). Later, this work was extended to other metals in various oxidation states.[138] Table III-8 lists $\nu(MN)$, magnetic moments, and the electronic configuration of these tris-bipy complexes. The results revealed several interesting relationships between $\nu(MN)$ and the electronic structure:

1. In terms of simple MO theory, Cr(III), Cr(II), Cr(I), Cr(0), V(II), V(0), Ti(0), Ti(-I), Fe(III), Fe(II), and Co(III) have filled or partly filled t_{2g} (bonding) and empty e_g (antibonding) orbitals. The $\nu(MN)$ of these metals (Group A) are in the 300–390 cm^{-1} region.

TABLE III-8. MN STRETCHING FREQUENCIES AND ELECTRONIC STRUCTURES IN $[M(bipy)_3]^{n+}$-TYPE COMPOUNDS (cm^{-1})[a]

	−I	0	I	II	III
d^3				374 V 335 (3.67) $(t_{2g})^3$	385 Cr 349 (3.78) $(t_{2g})^3$
d^4		374 Ti 339 (0) $(t_{2g})^4$-ls		351 Cr 343 (2.9) $(t_{2g})^4$-ls	
d^5	365 Ti 322 (1.74) $(t_{2g})^5$-ls	371 V 343 (1.68) $(t_{2g})^5$-ls	371 Cr 343 (2.0) $(t_{2g})^5$-ls	224 Mn 191 (5.95) $(t_{2g})^3(e_g)^2$-hs	384 Fe 367 (?)
d^6		382 Cr 308 (0) $(t_{2g})^6$		386 Fe 376 (0) $(t_{2g})^6$	378[b] Co 370 (0) $(t_{2g})^6$
d^7		258 Mn 227 (4.10) $(t_{2g})^5(e_g)^2$		266 Co 228 (4.85) $(t_{2g})^5(e_g)^2$	
d^8	235 Mn 184 (3.71) $(t_{2g})^6(e_g)^2$		244 Co 194 (3.3) $(t_{2g})^6(e_g)^2$	282 Ni 258 (3.10) $(t_{2g})^6(e_g)^2$	
d^9		280 Co 257 (2.23) $(t_{2g})^6(e_g)^3$		291 Cu 268 (?) $(t_{2g})^6(e_g)^3$	
d^{10}				230 Zn 184 (0) $(t_{2g})^6(e_g)^4$	

[a] The numbers at the upper right of each group indicate the MN stretching frequencies (cm^{-1}). The number in parentheses gives the observed magnetic moment in Bohr magnetons. ls = low spin; hs = high spin.
[b] Values for $[Co(phen)_3](ClO_4)_3$.

2. On the other hand, Co(II), Co(I), Co(0), Mn(II), Mn(0), Mn(-I), Ni(II), Cu(II), and Zn(II) have filled or partly filled e_g orbitals. The ν(MN) of these metals (Group B) are in the 180–290 cm^{-1} region.

3. Thus no marked changes in frequencies are seen in the Cr(III)–Cr(0) and Co(II)–Co(0) series, although a dramatic decrease in frequencies is observed in going from Co(III) to Co(II).

4. The fact that the ν(MN) do not change appreciably in the former two series indicates that the M–N bond strength remains approximately the same.

These results also suggest that, as the oxidation state is lowered, increasing numbers of electrons of the metal reside in essentially ligand orbitals which do not affect the M–N bond strength.

Other work on bipy and phen complexes includes a far-infrared study of tris-bipy complexes with low-oxidation-state metals [Cr(0), V(-I), Ti(0), etc.],[139] the assignments of infrared spectra of M(bipy)Cl_2 and M(phen)Cl_2 (M = Cu, Ni, etc.),[140] normal coordinate analysis on Pd(bipy)Cl_2 and its bipy-d_8 analog[141] and Raman spectra of phen complexes with Zn(II) and Hg(II).[142]

The $[Fe(bipy)_3]^{2+}$ ion and its analogs exhibit strong absorption near 520 nm which is due to Fe(3d)–ligand(π) CT transition. When the laser wavelength is tuned in this region, a number of bipy vibrations (all totally symmetric) are strongly resonance-enhanced as shown in Fig. III-10.[143] Excitation profile studies show that the intensities of all these bands are maximized at the main absorption maximum at 19,100 cm^{-1} (524 nm) and that no maxima are present at the side band near 20,500 cm^{-1} (488 nm). Thus, Clark et al.[143] concluded that the latter band is due to a vibronic transition.

Recently, the $[Ru(bipy)_3]^{2+}$ ion and related complexes have attracted much attention as potential compounds of solar-energy-conversion devices because of their excited-state redox properties. When solutions of this ion are irradiated with 7-ns, high-intensity pulses from the third harmonic (354.5 nm) of a Nd–YAG laser, the irradiated volume can be saturated with the long-lived (~600 ns) triplet M–L CT state (A_3) via efficient ($\phi \cong 1$) and rapid ($\tau < 10$ ps) intersystem crossing ($A_2 \rightarrow A_3$) as shown in Fig. III-11. Since the A_3–A_4 (π–π^*) transition is close to the 354.5-nm exciting line, conditions are favorable for efficient resonance Raman scattering from the A_3 state; namely, it is possible to obtain the resonance Raman spectrum of the ion in the electronic excited state: Detailed analysis of the resonance Raman spectra of this ion,[144–144b] 3,3′-d_2 bipy[144c] analog, and a related Re complex of bipy[144d] have shown unambiguously that the triplet M–L CT state (A_3) is properly formulated as $[Ru(III)(bipy)_2(bipy^{\dot{-}})]^{2+}$, that is, the electron is localized on one bipy rather than delocalized over all three ligands (at least on the vibrational time scale).

The metal isotope technique has been used to study the effect of magnetic crossover on the low-frequency spectrum of $Fe(phen)_2(NCS)_2$. This compound exists as a high-spin complex at 298 K and as a low-spin complex at 100 K. Figure III-12 shows the infrared spectra of $^{54}Fe(phen)_2(NCS)_2$ obtained by Takemoto and Hutchinson.[145] On the basis of observed isotopic shifts, along

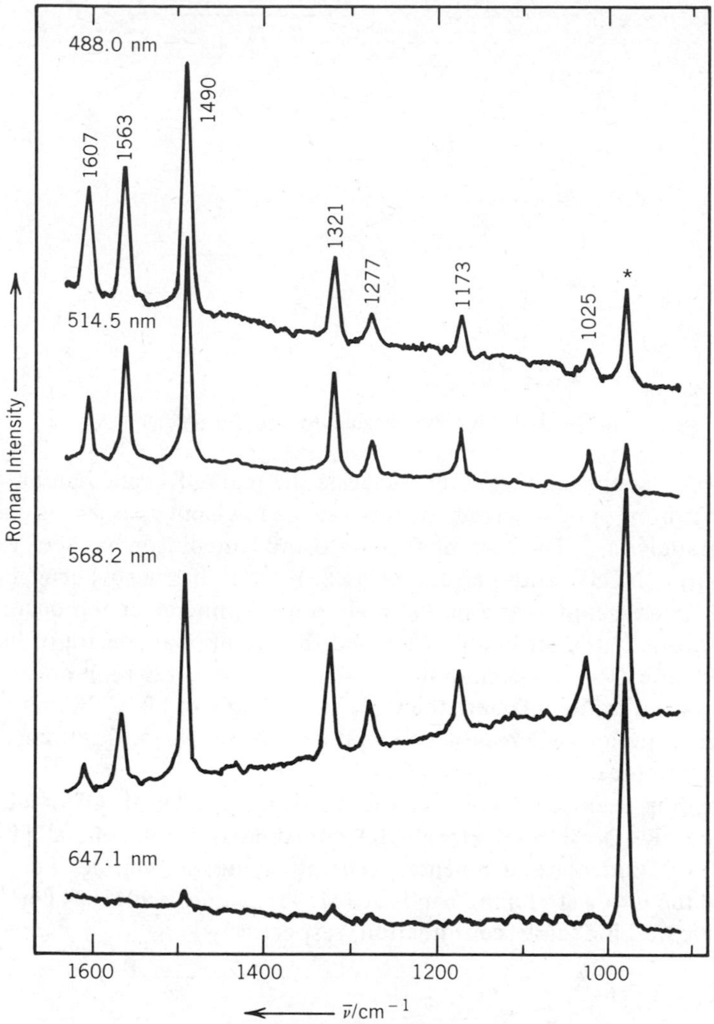

Fig. III-10. Resonance Raman spectra of the $[Fe(bipy)_3]^{2+}$ ion. The asterisk indicates the 981-cm^{-1} band of the SO_4^{2-} ion (internal standard).

with other evidence, they made the following assignments (cm^{-1}):

	ν(Fe—NCS)	ν[Fe—N(phen)]
High spin	252 (4.0)	222 (4.5)
Low spin	532.6 (1.6)	379 (5.0)
	528.5 (1.7)	371 (6.0)

The numbers in parentheses indicate the isotope shift, $\nu(^{54}Fe) - \nu(^{57}Fe)$. Both vibrations show large shifts to higher frequencies in going from the high- to

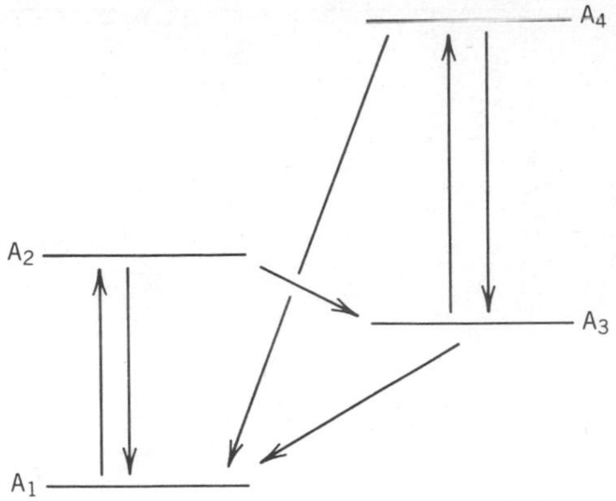

Fig. III-11. Energy-level diagram for the $[Ru(bipy)_3]^{2+}$ ion.

the low-spin complex. This result suggests the marked strengthening of these coordinate bonds in going from the high- to the low-spin complex as confirmed by X-ray analysis.[146] The work of Takemoto and Hutchinson has been extended to $Fe(bipy)_2(NCS)_2$ and $Fe(phen)_2(NCSe)_2$.[147] It has also been shown by infrared spectroscopy that a partial high-→low-spin conversion occurs under high pressure.[148] Barnard et al.[149] studied the vibrational spectra of bis[tri-(2-pyridyl)amine]Co(II) perchlorate in the high-spin (293 K) and low-spin (100 K) states. In the infrared, the CoN stretching band is at 263 cm^{-1} for the high-spin complex, whereas it splits and shifts to 312 and 301 cm^{-1} in the low-spin complex.

Depending upon the nature of the alkyl group (R), an alkyl-substituted α-diimine (R—N=CH—CH=N—R) coordinates to the metal [Pt(II) or Pd(II)] as a unidentate or bidentate (chelating) ligand. van der Poel et al.[150] observed the C=N stretching bands at 1615–1624 and 1590–1604 cm^{-1} for the unidentate and bidentate coordination, respectively.

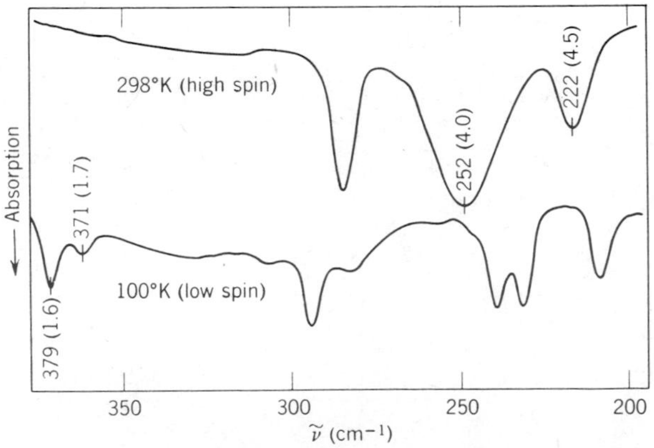

Fig. III-12. Infrared spectra of $^{54}Fe(phen)_2(NCS)_2$. The number in parentheses indicates the isotope shift, $\tilde{\nu}(^{54}Fe)-\tilde{\nu}(^{57}Fe)$.

The metal isotope technique has been used to assign the MN vibrations of metal complexes with many other ligands. For example, Takemoto[151] assigned the NiN_2 and NiN_1 stretching vibrations of $[Ni(DAPD)_2]^{2-}$ at 416–341 and 276 cm^{-1}, respectively.

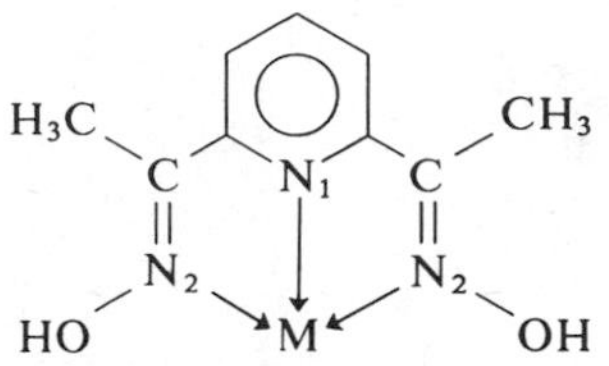

DAPD: 2,6-diacetylpyridine dioxime

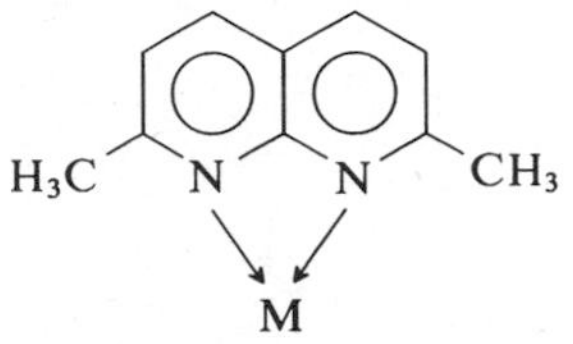

DMNAPY: 2,7-dimethyl-1,8-naphthyridine

In $Ni(DAPD)_2$, where the Ni atom is in the +IV state, the NiN_2 and NiN_1 stretching bands are located at 509.8–472.0 and 394.8 cm^{-1}, respectively. These high-frequency shifts in going from Ni(II) (d^8) to Ni(IV) (d^6, diamagnetic) have also been observed for diarsine complexes (Sec. III-21). Hutchinson and Sunderland[152] have noted that the MN stretching frequencies of the Ni(II) and Zn(II) complexes of 2,7-dimethyl-1,8-naphthyridine(DMNAPY), shown above, are lower than those of the corresponding tris-bipy complexes by 16 to 24%. This was attributed to the weakening of the M–N bond due to the strain in the four-membered chelate rings of the DMNAPY complexes. Normal coordinate analysis has been carried out on the $M(DMG)_2$ series (DMG: dimethylglyoximate ion)[153] and the MN stretching force constants (mdyn/Å) have been found to be as follows:

$$\begin{array}{ccccccc} \text{Pt(II)} & & \text{Pd(II)} & & \text{Cu(II)} & & \text{Ni(II)} \\ 3.77 & > & 2.84 & > & 1.92 & > & 1.88 \ \text{(GVF)} \end{array}$$

This work was extended to bis(glyoximato) complexes of Pt(II), Pd(II), and Ni(II).[154] The Co–N(DMG) and Co–N(py) stretching bands of $Co(DMG)_2(py)X$ (X = a halogen) were assigned at 512 and 453 cm^{-1}, respectively, based on ^{15}N and py-d_5 isotope shifts.[155] The metal isotope technique has been used to assign the MN stretching vibrations of metal complexes with 8-hydroxyquinoline[156] and 1,8-naphthyridine.[157]

III-4. METALLOPORPHYRINS AND RELATED COMPOUNDS

Vibrational spectra of metalloporphyrins and related compounds have been studied extensively because of their biological importance as prosthetic groups of heme proteins (Sec. V-1 and V-2). Figure III-13 shows the planar $\mathbf{D}_{4h}$ structure of a metalloporphyrin.

The simplest porphyrin, porphin ($R_1 \sim R_8$ = H and R′ = H), forms complexes with a variety of metal ions. It has $105(3 \times 37 - 6)$ normal vibrations which are classified as shown in Table III-9. Ogoshi et al.[158] prepared porphin complexes with ^{64}Zn, ^{68}Zn, Cu, and Ni, and carried out normal coordinate calculations

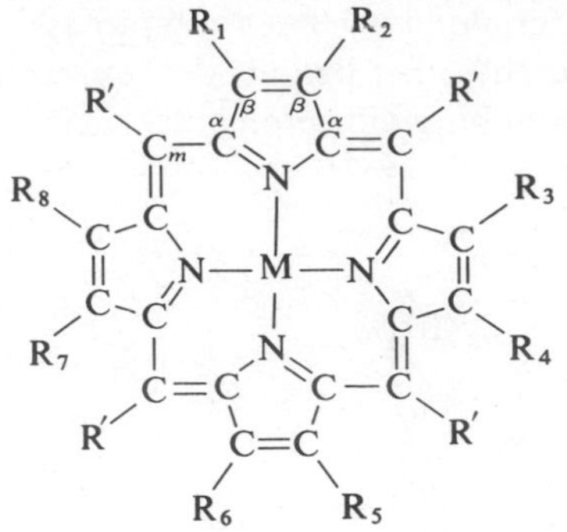

Fig. III-13. The structure of a metalloporphyrin.

to assign the 18 infrared-active in-plane vibrations. Figure III-14 illustrates the infrared (IR) spectra of these complexes in the high-frequency region. As expected, their calculations show that the bands between 1700 and 950 cm^{-1} are due to ν(CC), ν(CN), δ(CH), and δ(CCN) which are strongly coupled to each other.

As stated in Sec. I-21, metalloporphyrins are ideal for resonance Raman (RR) studies since they exhibit strong absorption bands in the visible and near-UV regions. It is well established that excitation at the α- and β-bands (see Fig. I-18) causes resonance enhancements of the A_{1g}, B_{1g}, B_{2g}, and A_{2g} vibrations of the porphyrin core whereas that at the Soret band enhances only the A_{1g} vibrations. Resonance Raman spectra of porphin complexes have been measured by several investigators.[159–161] Among them, Gladkov et al.[161] carried out the most complete study; they measured the IR and RR spectra of porphin-d_0, -d_4, -d_8, -d_{12} and their Cu(II) and Pd(II) complexes, and made band assignments based on the potential energy distribution obtained from normal coordinate analysis.

TABLE III-9. CLASSIFICATION OF NORMAL VIBRATIONS OF A METAL PORPHIN COMPLEX OF $\mathbf{D}_{4h}$ SYMMETRY[a] [158]

In-Plane Vibrations		Out-of-Plane Vibrations	
A_{1g}(R)	9	A_{1u}	3
A_{2g}[b]	8	A_{2u}(IR)	6
B_{1g}(R)	9	B_{1u}	5
B_{2g}(R)	9	B_{2u}	4
E_u(IR)	18	E_g(R)[c]	8

[a] R = Raman active; IR = IR active.
[b] The A_{2g} vibrations become Raman active under resonance conditions (see Sec. I-21).
[c] The E_g vibrations are weak even under resonance conditions.

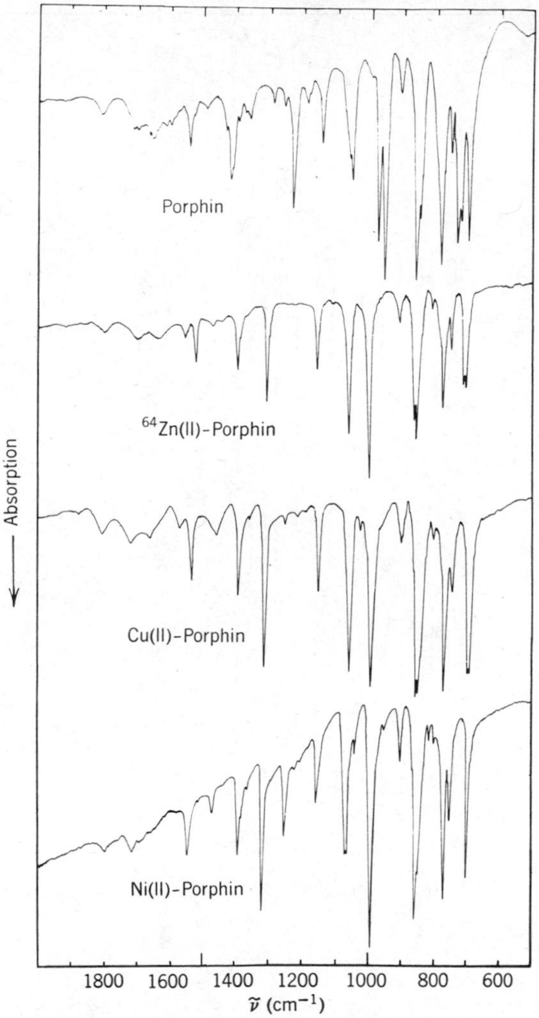

Fig. III-14. Infrared spectra of porphin and its ^{64}Zn, Cu, and Ni complexes.

The IR and RR spectra of metal complexes of octaethylporphyrin (OEP: $R_1 \sim R_8 = C_2H_5$, $R' = H$) have been studied extensively. Ogoshi et al.[162] obtained the IR spectra of seven OEP complexes, and assigned the ν(MN) vibrations in the 290–200 cm^{-1} region. Kincaid et al.[163] obtained matrix-isolation IR spectra of five OEP complexes shown in Fig. III-15. Kitagawa and co-workers[164] made the most complete study on Ni(OEP). They measured the vibrational spectra of Ni(OEP)(IR, RR), Ni(OEP-$^{15}N_4$)(RR) and Ni(OEP-d_4) (IR, RR), and made band assignments of all the in-plane porphyrin-core vibrations based on normal coordinate calculations. Figure III-16 shows the normal modes of six vibrations which are useful for structural diagnosis of iron porphyrins (see also Sec. V-I and V-2).

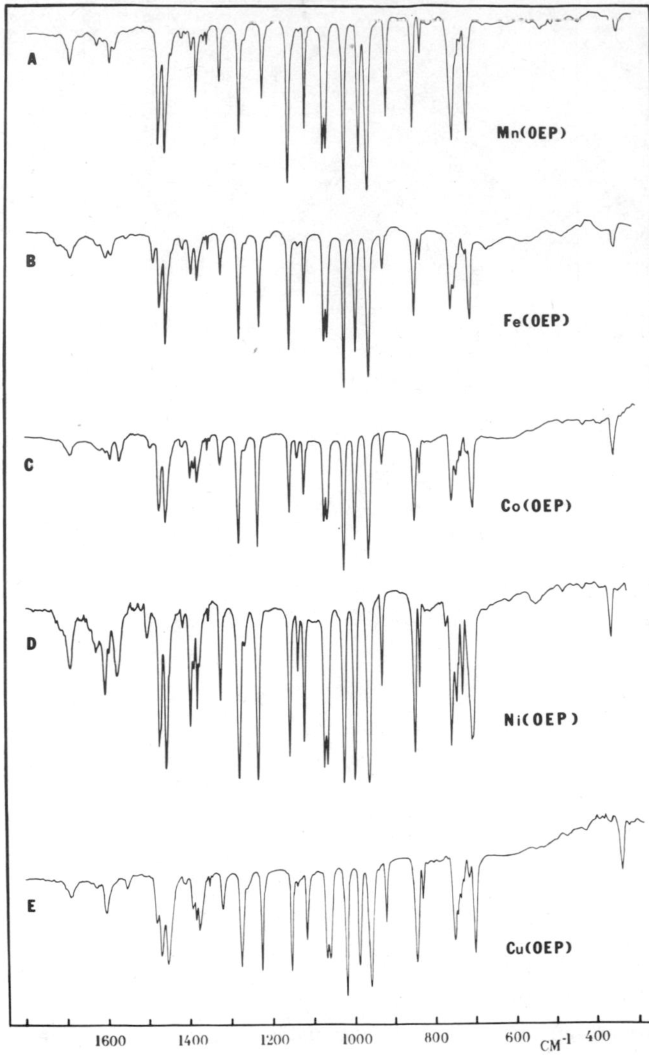

Fig. III-15. Infrared spectra of OEP complexes in inert gas matrices.

Table III-10 lists main contributors to their potential energy distribution for the above six vibrations which account for their sensitivities to the oxidation and spin states in iron porphyrins. Changes in the oxidation state cause alterations in π-back-bonding to the porphyrin ring. Thus, ν_3, ν_4, and ν_{10} which contain $\nu(C_\alpha C_\beta)$, $\nu(C_\alpha C_m)$ and/or $\nu(C_\alpha N)$ are sensitive to the oxidation state. The sensitivity of ν_4 to the change in the central metal[165] can also be accounted for on the same basis.

In going from the low- to high-spin state, the porphyrin core tends to expand or be domed, and this results in weakening of the C_α-C_m bond. Thus, ν_3, ν_{10},

TABLE III-10. POTENTIAL ENERGY DISTRIBUTIONS OF SIX STRUCTURALLY SENSITIVE VIBRATIONS OF Ni(OEP)[164]

		$\tilde{\nu}$ (cm^{-1})	Potential Energy Distribution (%)	Sensitivity
A_{1g}	$\nu_3(p)$	1519	$\nu(C_\alpha C_m)$ 41, $\nu(C_\alpha C_\beta)$ 35	Spin, oxidation
	$\nu_4(p)$	1383	$\nu(C_\alpha N)$ 53, $\delta(C_\alpha C_m)$ 21	Oxidation, metal
	$\nu_7(p)$	674[b]	$\delta(C_\beta C_\alpha N)$ 20, $\nu(C_\alpha C_\beta)$ 19	Porphyrin ring doming
B_{1g}	$\nu_{10}(dp)$	1655	$\nu(C_\alpha C_m)$ 49, $\nu(C_\alpha C_\beta)$ 17	Spin, oxidation
	$\nu_{11}(dp)$	1576	$\nu(C_\beta C_\beta)$ 57, $\nu(C_\beta Et)$ 16	Substituent
A_{2g}	$\nu_{19}(ap)$	1603	$\nu'(C_\alpha C_m)$ 67, $\nu'(C_\alpha C_\beta)^a$ 18	Spin

[a] ν' = antisymmetric stretch.
[b] Value for Ni(OEP-d_4).

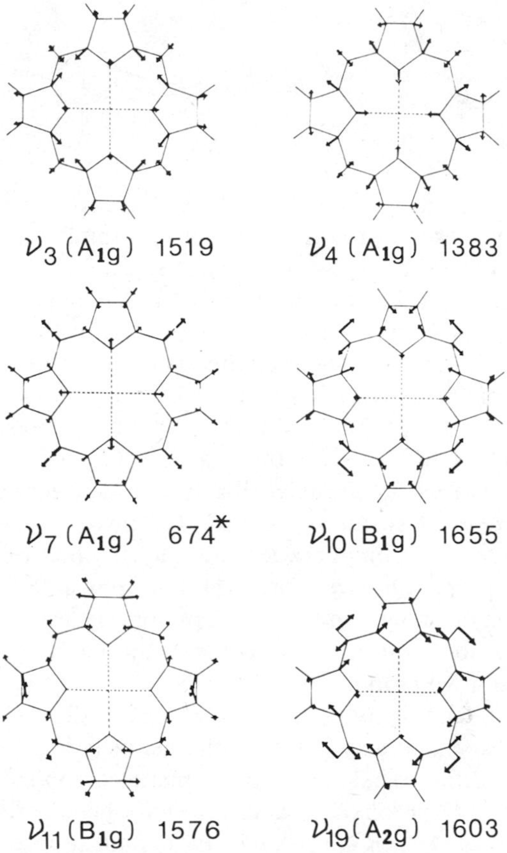

Fig. III-16. Normal modes of vibration of structure-sensitive bands. [The asterisk refers to the value obtained for Ni(OEP-d_4).]

and ν_{19} which contain $\nu(C_\alpha C_m)$ as the major contributor are sensitive to the spin state. In particular, ν_{10} is sensitive to the number and the nature of axial ligands.[165a] ν_{11} is sensitive to the nature of the peripheral substituents since it is mainly due to $\nu(C_\beta C_\beta)$. Finally, ν_7 is a 16-membered porphyrin ring breathing motion. Thus it is strong for planar complexes and weak for non-planar (or domed) complexes.[166]

Ogoshi et al.[167] assigned the ν(MN) of [Fe(OEP)X] (X = Cl, Br, etc.) and $[Fe(OEP)L_2]^+$ (L = γ-picoline, imidazole, etc.) based on ^{54}Fe-^{56}Fe isotope shifts. The ν(FeN) of the former (280–250 cm^{-1}) are lower than those of the latter (320–294 cm^{-1}). According to X-ray analysis, the iron atom in the former (high spin) is out-of-plane whereas it is on the porphyrin plane in the latter (low spin).

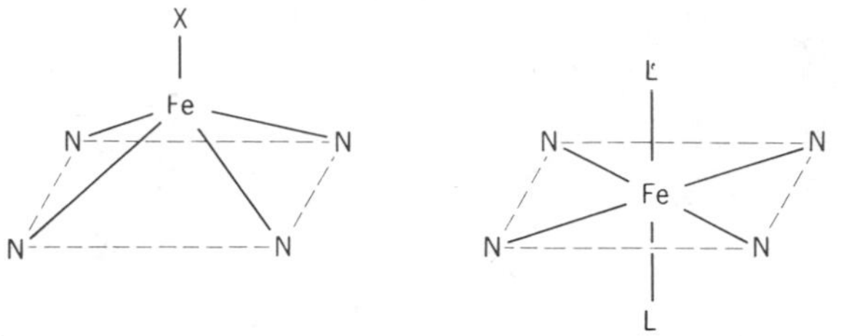

In going from high- to low-spin complexes, electrons are shifted from the antibonding ($d_{x^2-y^2}$ and d_{z^2}) to the bonding (d_π) orbitals. Removal of electrons from the antibonding orbitals would strengthen the Fe–N bonds and raise the ν(FeN). Also the splitting of the ν(FeN) band may disappear since the symmetry of the porphyrin core changes from $\mathbf{C}_{4v}$ to $\mathbf{D}_{4h}$.

Kincaid and Nakamoto[168] observed the ν(Fe–F) of ^{54}Fe(OEP)F at 595 cm^{-1} with the 514.5 nm excitation. Kitagawa et al.[166] also observed the ν(Fe–X) of Fe(OEP)X at 364 and 279 cm^{-1} for X = Cl and Br, respectively and the ν_s(L–Fe–L) of $[Fe(OEP)L_2]^+$ (L = ImH) at 290 cm^{-1} using the 488 nm excitation. These results show that the axial vibrations can be enhanced via resonance with in-plane $\pi - \pi^*$ transitions (α and β bands). According to the latter workers, vibrational coupling between these axial vibrations and totally symmetric in-plane porphyrin-core vibrations is responsible for their resonance enhancement. On the other hand, Spiro[169] prefers electronic coupling, namely, the $\pi - \pi^*$ transition induces the changes in the Fe–X (or L) distance, thus activating the axial vibration.

Complexes of tetraphenylporphyrin (TPP, $R_1 \sim R_8 = H$, $R' = C_6H_5$) have been studied most extensively due to their relatively high-yield synthesis and convenient purification. Thus, many biomimetic compounds such as "picket-fence" and "capped" porphyrins (Sec. V-1) have been derived from the basic TPP structure. However, IR spectra of TPP complexes are extremely difficult to assign because vibrations of phenyl groups are mixed with or overlapped by those of the porphyrin-core. Recently, Oshio et al.[170] prepared 20 Fe(TPP)LL′- (L, L′: axial ligands) type complexes, and located two spin-state sensitive bands (I and III) and one oxidation-state-sensitive band (II) in the regions shown in Fig. III-17. Based on shifts by TPP-d_8 and TPP-d_{20} substitutions, they proposed the following assignments: Band I [$\nu(C_\alpha C_m)$ mixed with $\nu(C_m$-phenyl)], Band II [pyrrole ring deformation coupled with $\delta(C_\beta$-H)], and Band III (porphyrin-core deformation). Kincaid and Nakamoto[171] measured the far-IR spectra of TPP complexes of five divalent metals including $^{58}Ni/^{62}Ni$, $^{63}Cu/^{65}Cu$, and $^{64}Zn/^{68}Zn$ pairs, and were able to locate their ν(MN) (coupled with other modes) in the 325–200 cm^{-1} region.

Band assignments obtained for RR spectra of OEP complexes[164] are not applicable to TPP complexes because the nature of vibrational coupling is markedly different between these two systems. Burke et al.[172] first made empirical assignments of the RR spectra of $(Fe(TPP))_2O$ and its d_8 and d_{20} analogs. They noted that, in a nearly linear Fe–O–Fe bridge, the ν_s(FeOFe)-(363 cm^{-1}) is markedly lower than the ν_{as}(FeOFe)(885 cm^{-1}). They also noted three structure-sensitive bands at ~1560(p), ~1360(p), and ~390(p) cm^{-1} in a series of Fe(TPP)LL′-type complexes; the first is mainly spin-state sensitive while the last two are oxidation-state sensitive.[173] Stong et al.[174] found linear relationships between the porphyrin-core size and the vibrational frequencies of three bands at ~1570(p), ~1540(ap), and ~1460(p) cm^{-1} in a series of M(TPP)-type complexes. Chottard et al.[175] found four structure-sensitive bands in the Fe(TPP)LL′-type complexes: Band A(p), 1370–1345 cm^{-1}; Band B(dp),

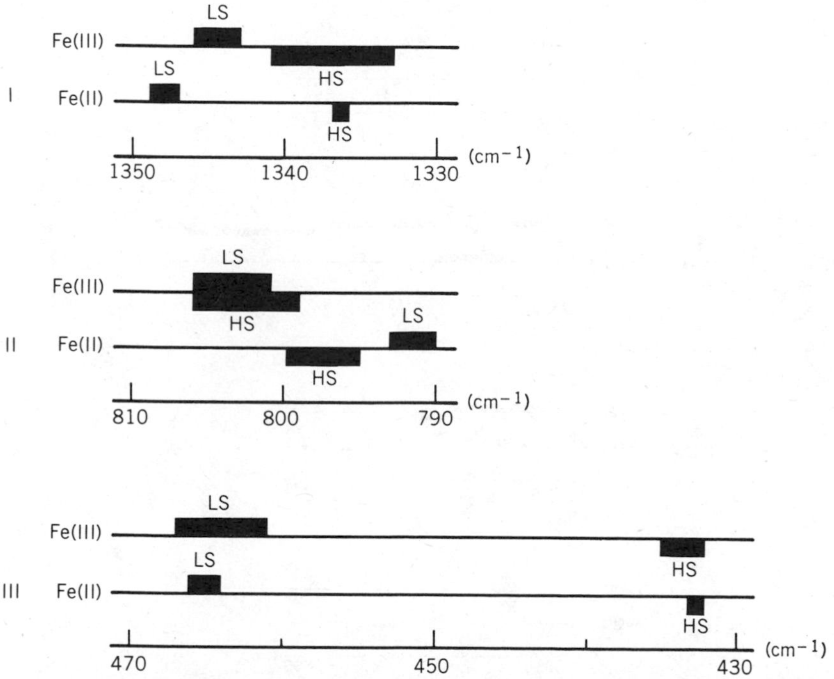

Fig. III-17. Structure-sensitive TPP vibrations in infrared spectra.

1379–1365 cm^{-1}; Band C(*ap*), 1545–1500 cm^{-1}; Band D(*p*), 1572–1542 cm^{-1}. They concluded that all these bands are essentially spin-state sensitive except Band B which is characteristic of pentacoordination. Figure III-18 shows the sensitivities to spin and oxidation states of four bands selected by Chottard et al. (A, C, and D) and Burke et al. (E) as applied to 20 Fe(TPP)LL′-type complexes studied by Oshio et al.[170]

Indirect excitation of axial vibrations has been discussed previously for the Fe(OEP)LL′-type compounds.[166,169] Direct excitation is possible if the metal–axial ligand CT transition is in the visible region. Thus, Asher and Sauer[176] observed the ν(Mn–X) of Mn(EP)X (X = F, Cl, Br, and I) with exciting lines in the 460–490 nm region where the Mn–X CT bands appear. Here EP denotes etioporphyrin ($R_1 = R_3 = R_5 = R_7 = CH_3$, $R_2 = R_4 = R_6 = R_8 = C_2H_5$ in Fig. III-13). Similarly, Wright et al.[177] were able to observe totally symmetric pyridine(py) vibrations as well as ν_s(Fe–N(py)) of $Fe(MP)(py)_2$ with exciting lines near 497 nm which are in resonance with the Fe($d\pi$)–py(π^*) CT transition. Here, MP denotes mesoporphyrin IX dimethylester ($R_1 = R_3 = R_5 = R_8 = CH_3$, $R_2 = R_4 = C_2H_5$, and $R_6 = R_7 = -CH_2-CH_2COOCH_3$ in Fig. III-13). The RR spectra of M(MP) [M = Ni(II), Co(II), and Cu(II)] are reported by Verma et al.[178]

IR spectra of divalent-metal complexes of protoporphyrin IX dimethyl ester (PPIXDME, $R_1 = R_3 = R_5 = R_8 = CH_3$, $R_2 = R_4 =$ vinyl group, and $R_6 = R_7 = CH_2-CH_2-COOCH_3$) are reported by Boucher and Katz[179] who noted several

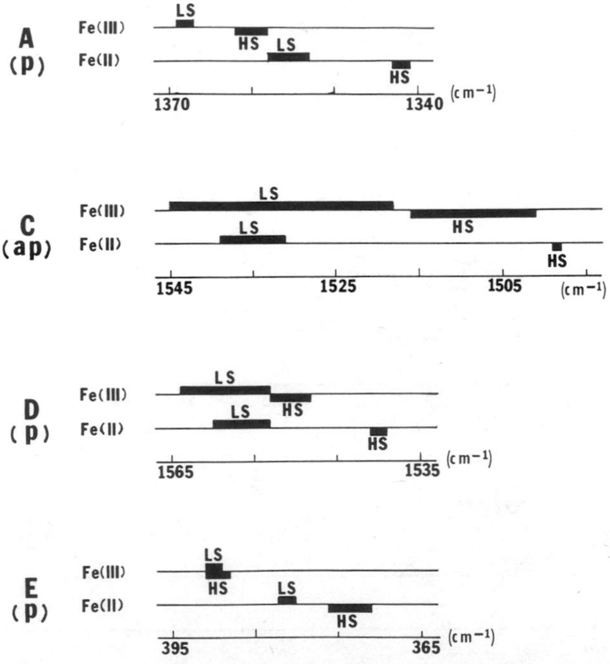

Fig. III-18. Structure-sensitive TPP vibrations in resonance Raman spectra.

metal-sensitive bands in the 970–920 and 530–500 cm^{-1} regions. Choi et al.[180] assigned the RR spectra of Ni(PPIXDME) with deuterated vinyl groups, and discussed structural correlations and vinyl influences in RR spectra of protoheme complexes and proteins.[181]

Ogoshi et al.[182] reported the IR spectra of M(OEC) (M = Zn, Cu, and Ni), $Mg(OEC)(py)_2$, and Fe(OEC)X (X = F, Cl, Br, and I), and assigned ν(M–N(OEC)) and ν(Mg–N(py)) using metal isotope techniques. Here OEC denotes octaethylchlorin. Ozaki et al.[183] report RR spectra of these and other OEC complexes. Murakami et al.[184] made empirical assignments of IR spectra of dipyrromethene and tetradehydrocorrin derivatives.[185] Infrared[186] and RR spectra[187] of metal complexes of phthalocyanine(Pc) have been reported.

Sections V-1 and V-2 describe vibrational studies of other metalloporphyrins with axial ligands which serve as model compounds of a variety of heme proteins

III-5. NITRO AND NITRITO COMPLEXES

The NO_2^- ion coordinates to a metal in a variety of ways:

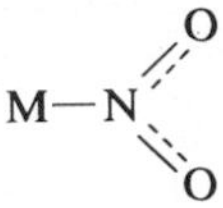

Nitro complex

I

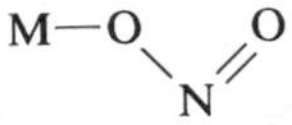

Nitrito complex

II

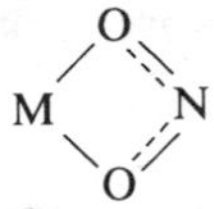

Chelating nitro complex

III

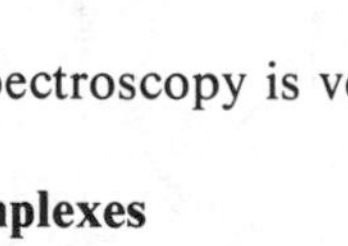

IV

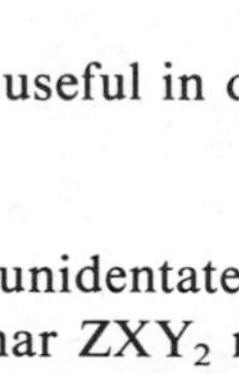

V

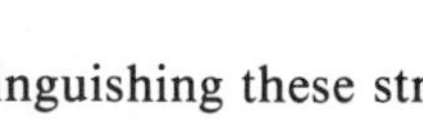

VI

Bridging nitro complexes

Vibrational spectroscopy is very useful in distinguishing these structures.

(1) Nitro Complexes

The normal vibrations of the unidentate N-bonded nitro complex may be approximated by those of a planar ZXY_2 molecule, as shown in Fig. III-19. In addition to these modes, the NO_2 twisting and skeletal modes of the whole complex may appear in the low-frequency region. Table III-11 summarizes the observed frequencies and band assignments for typical nitro complexes. It is seen that these complexes exhibit $\nu_a(NO_2)$ and $\nu_s(NO_2)$ in the 1470–1370 and 1340–1320 cm^{-1} regions, respectively. On the other hand, the free NO_2^- ion exhibits these modes at 1250 and 1335 cm^{-1}, respectively. Thus $\nu_a(NO_2)$

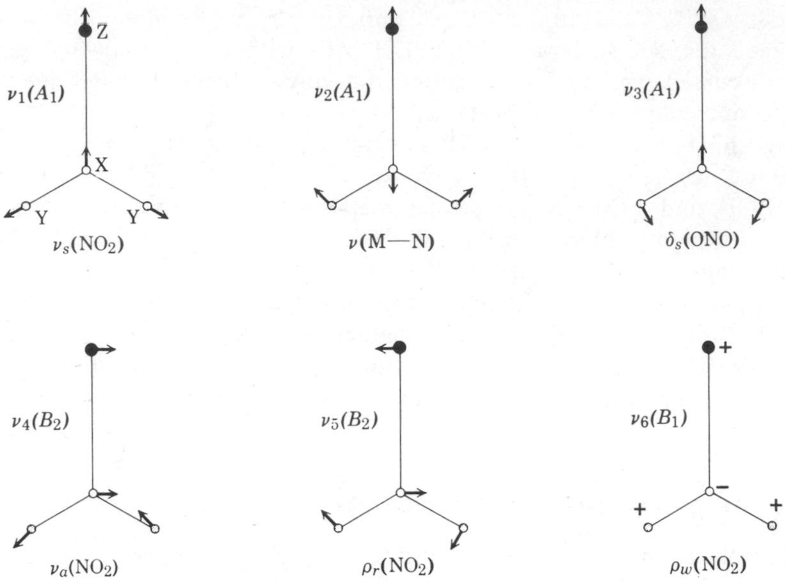

Fig. III-19. Normal modes of vibration of planar ZXY_2 molecules. (The band assignment is given for an $M-NO_2$ group.)

shifts markedly to a higher frequency, whereas $\nu_s(NO_2)$ changes very little upon coordination.

Nakagawa and Shimanouchi[31] and Nakagawa et al.[188] carried out normal coordinate analyses to assign the infrared spectra of crystalline hexanitro cobaltic salts; both internal and lattice modes were assigned completely by factor group analysis. The results indicate that the complex ion takes the $\mathbf{T}_h$ symmetry in K, Rb, and Cs salts but the $\mathbf{S}_6$ symmetry in the Na salt. Figure III-20 illustrates these unusual point groups for the $[Co(NO_2)_6]^{3-}$ ion. The infrared spectra of K and Na salts are compared in Fig. III-21.

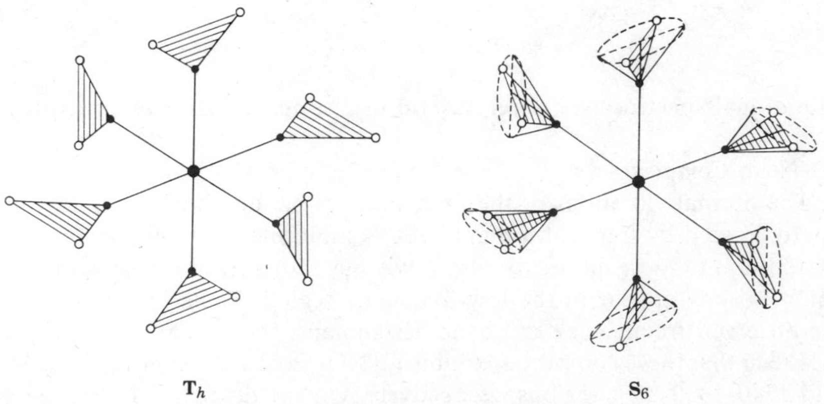

Fig. III-20. Possible structures of the $[Co(NO_2)_6]^{3-}$ ion.

TABLE III-11. OBSERVED INFRARED FREQUENCIES AND BAND ASSIGNMENTS OF NITRO COMPLEXES (CM^{-1})

Complex	$\nu_a(NO_2)$	$\nu_s(NO_2)$	$\delta(ONO)$	$\rho_w(NO_2)$	$\nu(MN)$	$\rho_r(NO_2)$[a]	Refs.
$K_3[Co(NO_2)_6]$	1386	1332	827	637	416	293	188
$Na_3[Co(NO_2)_6]$	1425	1333	854, 831	623	449, 372	276, 249	188
$K_2Ba[Ni(NO_2)_6]$	1343	1306	838	433	291	255	188
$K_3[Ir(NO_2)_6]$	1395, 1375	1330	830	657	390	300	189
$K_3[Rh(NO_2)_6]$	1395	1340	833	627	386	283	189
$K_3[Ir(NO_2)Cl_5]$	1374	1315	835	644	325	288	190
$[Pt(NO_2)_6]^{4-}$	1488, 1458	1328	834	621	368	294	191
$K_2[Pt(^{15}NO_2)_4]$	1466, 1397	1343	847, 839, 833	640, 623	421		192, 193
$[Pd(NO_2)_4]^{2-}$ [b]	1408	1364, 1320	834, 824	440	290		193
$K_2[Pt(NO_2)Cl_3]$	1401	1325	844	614	350	304	190

[a] This mode may couple with other low-frequency modes.

[b] Raman data in aqueous solution.

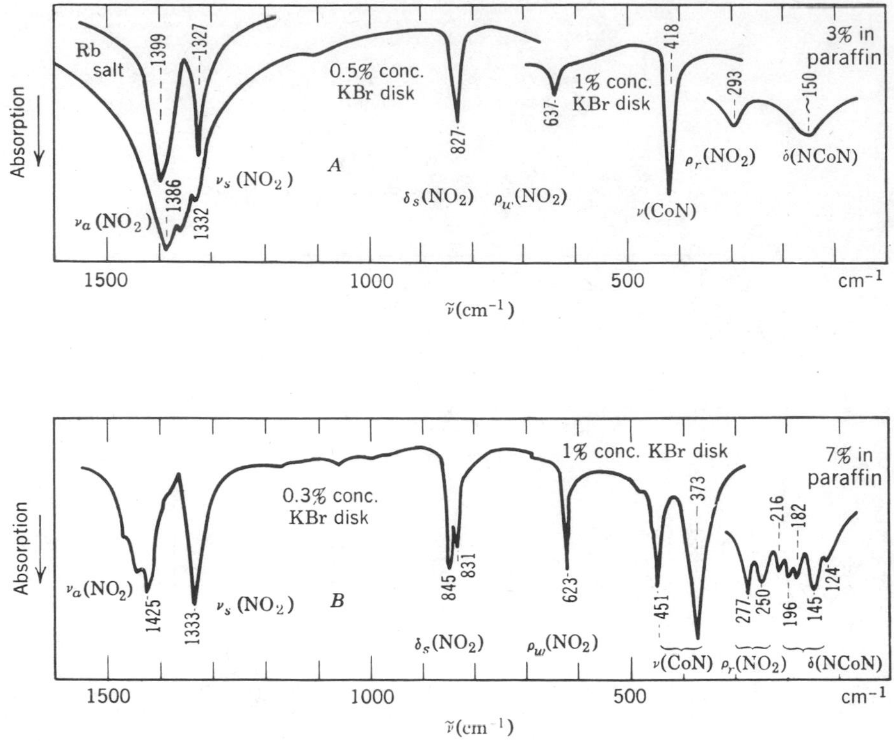

Fig. III-21. Infrared spectra of (A) $K_3[Co(NO_2)_6]$ and (B) $Na_3[Co(NO_2)_6]$.[188]

There are many nitro complexes containing other ligands such as NH_3 and Cl. In these cases, the main interest has been the distinction of stereoisomers by the symmetry selection rules and the differences in frequency between isomers. It is possible to distinguish *cis*- and *trans*-$[Co(NH_3)_4(NO_2)_2]^+$ [194] by the rule that the *cis*-isomer exhibits more bands than the *trans*-isomer, and to distinguish *fac*- and *mer*-$[Co(NH_3)_3(NO_2)_3]$[195] by the observation that $\delta(NO_2)$ and $\rho_w(NO_2)$ are higher for the *fac*-isomer ($\mathbf{C}_3$, 832 and 625 cm^{-1}, respectively) than for the *mer*-isomer ($\mathbf{C}_{2v}$, 825 and 610 cm^{-1}, respectively). Nakagawa and Shimanouchi[196] measured the infrared spectra of the $[Co(NO_2)_n(NH_3)_{6-n}]^{(3-n)+}$ series and carried out normal coordinate analysis on the mononitro and dinitro complexes. Nolan and James[197] studied the infrared and Raman spectra of $[Pt(NO_2)_nCl_{6-n}]^{2-}$-type salts in the crystalline state.

(2) Nitrito Complexes

If the NO_2 group is bonded to a metal through one of its O atoms, it is called a nitrito complex. Table III-12 lists the NO stretching frequencies of typical nitrito complexes. The two $\nu(NO_2)$ of nitrito complexes are well separated, $\nu(N{=}O)$ and $\nu(NO)$ being at 1485–1400 and 1110–1050 cm^{-1},

TABLE III-12. VIBRATIONAL FREQUENCIES OF NITRITO COMPLEXES (CM^{-1})

Complex	ν(N=O)	ν(NO)	δ(ONO)	Ref.
$[Co(NH_3)_5(ONO)]Cl_2$	1468	1065	825	198
$[Cr(NH_3)_5(ONO)]Cl_2$	1460	1048	839	198
$[Rh(NH_3)_5(ONO)]Cl_2$	1461, 1445	1063	830	190
$[Ni(py)_4(ONO)_2]$	1393	1114	825	199
trans-$[Cr(en)_2(ONO)_2]ClO_4$	1485, 1430	—	835, 825	200
$[Co(py)_4(ONO)_2](py)_2$	1405	1109	824	201

respectively. Distinction between the nitro and nitrito coordination can be made on this basis. It is to be noted that nitrito complexes lack the wagging modes near 620 cm^{-1} which appear in all nitro complexes. The ν(MO) of nitrito complexes were assigned in the 360–340 cm^{-1} region for metals such as Cr(III), Rh(III), and Ir(III).[190]

In many nitro complexes several types of nitro coordination are mixed. Goodgame, Hitchman, and their co-workers carried out an extensive study on vibrational spectra of nitro complexes containing various types of coordination. For example, all six nitro groups in $K_4[Ni(NO_2)_6]\cdot H_2O$ are coordinated through the N atom. However, its anhydrous salt exhibits the bands characteristic of nitro as well as nitrito coordination. From UV spectral evidence, Goodgame and Hitchman[202] suggested the structure $K_4[Ni(NO_2)_4(ONO)_2]$ for the anhydrous salt. Table III-13 lists the observed frequencies of two Ni(II) complexes containing both nitro and nitrito groups.

The red nitritopentammine complex, $[Co(NH_3)_5(ONO)]Cl_2$, is unstable and is gradually converted to the stable yellow nitro complex. The kinetics of this conversion can be studied by observing the disappearance of the nitrito bands.[204,205] Burmeister[206] reviewed the vibrational spectra of these and other linkage isomers.

TABLE III-13. VIBRATIONAL FREQUENCIES OF Ni(II) COMPLEXES CONTAINING NITRO AND NITRITO GROUPS (CM^{-1})

	Nitro Group			Nitrito Group		
Complex	$\nu_a(NO_2)$	$\nu_s(NO_2)$	ρ_w(ONO)	ν(N=O)	ν(NO)	Ref.
$K_4[Ni(NO_2)_6]H_2O$	1346	1319	427	—	—	202
$K_4[Ni(NO_2)_4(ONO)_2]$	1347	1325	423, 414	1387	1206	202
Ni[2-(aminomethyl)-py]$_2(NO_2)$(ONO)	1338	1318	—	1368	1251	203

(3) Chelating Nitro Groups

If the nitro group is chelating, both the antisymmetric and the symmetric NO_2 stretching frequencies are lower, and the ONO bending frequency is higher, than those of the unidentate N-bonded nitro complexes. Table III-14 lists the vibrational frequencies of chelating nitro groups. It should be noted that $\nu_a(NO_2)$ depends on the degree of asymmetry of the coordinated nitro group; it is lowest when two N–O bonds are equivalent, and becomes higher as the degree of asymmetry increases. The high $\nu_a(NO_2)$ frequencies observed for the last two compounds in Table III-14 may be accounted for on this basis.

(4) Bridging Nitro Group

The nitro group is known to form a bridge between two metal atoms. Nakamoto et al.[198] suggested that among the three possible structures, IV, V, and VI, shown before, IV is most probable for

$$[(NH_3)_3Co\!\!\begin{array}{c} \nearrow OH \searrow \\ \diagdown NO_2 \diagup \end{array}\!\!Co(NH_3)_3]^{3+}$$

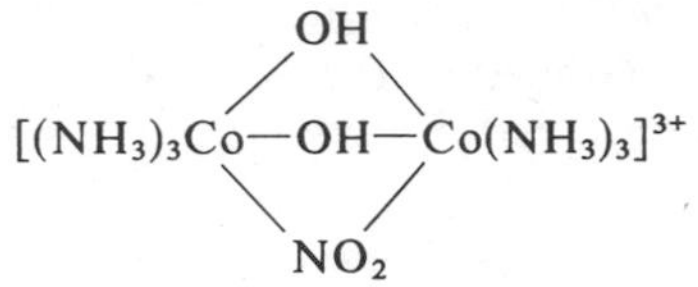

since its NO_2 stretching frequencies (1516 and 1200 cm^{-1}) are markedly different from those of other types discussed thus far. Later, this structure was found by X-ray analysis of[210]

$$[(NH_3)_4Co\!\!\begin{array}{c} \nearrow NH_2 \searrow \\ \diagdown NO_2 \diagup \end{array}\!\!Co(NH_3)_4]Cl_4\cdot 4H_2O$$

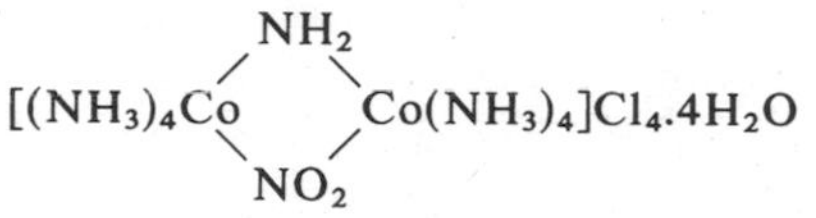

TABLE III-14. VIBRATIONAL FREQUENCIES OF CHELATING
NITRO GROUPS (CM^{-1})

Complex	$\nu_a(NO_2)$	$\nu_s(NO_2)$	$\delta(ONO)$	Ref.
Co(Ph$_3$PO)$_2$(NO$_2$)$_2$	1266	1199 1176	856	207
Ni(α-pic)$_2$(NO$_2$)$_2$	1272	1199	866 862	207
Cs$_2$[Mn(NO$_2$)$_4$]	1302	1225	841	208
Co(Me$_4$en)(NO$_2$)$_2$	1290	1207	850	201
Zn(py)$_2$(NO$_2$)$_2$	1351	1171	850	209
(o-cat)[Co(NO$_2$)$_4$]^a	1390	1191	—	208

^{a} o-cat = [o-xylylenebis(triphenylphosphonium)]$^{2+}$ ion.

This compound exhibits the NO_2 stretching bands at 1492 and 1180 cm^{-1}. Upon $^{16}O \rightarrow {}^{18}O$ substitution of the bridging oxygen, the latter is shifted by $-10\ cm^{-1}$ while the former is almost unchanged. Thus, these bands are assigned to the $\nu(N{=}O)$ (outside the bridge) and $\nu(N\text{-}O)$ (bridge), respectively.[211] The $[Co_2\{NO_2(OH)_2\}(NO_2)_6]^{3-}$ ion exhibits the NO_2 bands at 1516, 1190, and 860 cm^{-1}, indicating the presence of a bridging nitro group:[212]

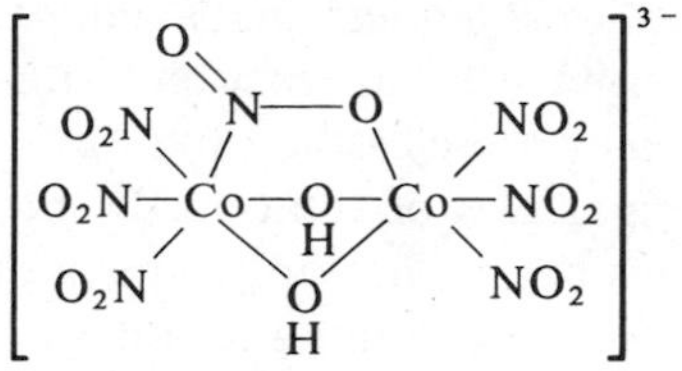

$[Ni(\beta\text{-pic})_2(NO_2)_2]_3 \cdot C_6H_6$ exhibits a number of bands due to coordinated nitro groups. Goodgame et al.[213] suggested the presence of two different types of bridging nitro groups, IV, V, and III, on the basis of the crystal structure and infrared data for this compound: Type IV absorbs at 1412 and 1236, Type V at 1460 and 1019, and Type III at 1299 and 1236 cm^{-1}. Goodgame et al.[214] also studied the infrared spectra of other bridging nitro complexes of Ni(II). For example, they found that $Ni(en)(NO_2)_2$ contains a Type-IV bridge (1429 and 1241 cm^{-1}), while $Ni(py)_2(NO_2)_2(\frac{1}{3}C_6H_6)$ is similar to that of the analogous β-picoline complex.

III-6. LATTICE WATER AND AQUO AND HYDROXO COMPLEXES

Water in inorganic salts may be classified as lattice or coordinated water. There is, however, no definite borderline between the two. The former term denotes water molecules trapped in the crystalline lattice, either by weak hydrogen bonds to the anion or by weak ionic bonds to the metal, or by both:

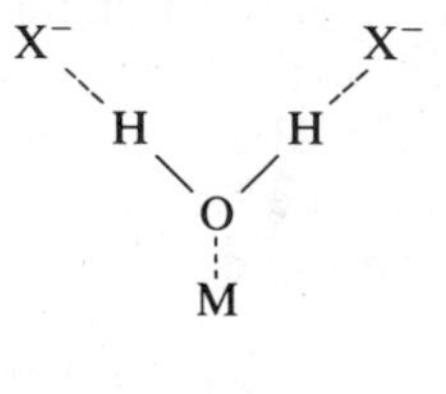

whereas the latter denotes water molecules bonded to the metal through partially covalent bonds. Although bond distances and angles obtained from X-ray and neutron-diffraction data provide direct information about the geometry of the water molecule in the crystal lattice, studies of vibrational spectra are also useful for this purpose. It should be noted, however, that the spectra of water molecules are highly sensitive to their surroundings.

(1) Lattice Water

In general, lattice water absorbs at 3550–3200 cm^{-1} (antisymmetric and symmetric OH stretchings) and at 1630–1600 cm^{-1} (HOH bending). If the spectrum is examined under high resolution, the fine structure of these bands is observed. For example, $CaSO_4 \cdot 2H_2O$ exhibits eight peaks in the 3500–3400 cm^{-1} region,[215] and its complete vibrational analysis can be made by factor group analysis (Sec. I-23). In the low-frequency region (600 ~ 200 cm^{-1}) lattice water exhibits "librational modes" that are due to rotational oscillations of the water molecule, restricted by interactions with neighboring atoms. As are shown in Fig. III-22, they are classified into three types depending upon the direction of the principal axis of rotation. It should be noted, however, that these librational modes couple not only among themselves but also with internal modes of water (HOH bending) and other ions (SO_4^{2-}, NO_3^-, etc.) in the crystal. Tayal et al.[216] reviewed librational modes of water in hydrated solids.

The presence of the hydronium (H_3O^+) ion in crystalline acid hydrates is well established, and their spectra were discussed in Sec. II-3. The existence of the $H_5O_2^+$ ion was first detected by X-ray analysis.[217] Pavia and Giguère[218] further confirmed its presence in $HClO_4 \cdot 2H_2O$ (namely, $[H_5O_2]ClO_4$) by the absence of some characteristic bands of the H_3O^+ and H_2O species. Its structure is suggested to be centrosymmetric $H_2O—H—OH_2$ of approximately $\mathbf{C}_{2h}$ symmetry. Both X-ray[71] and neutron-diffraction[219] studies suggest the presence of the $H_5O_2^+$ ion in *trans*-$[Co(en)_2Cl_2]Cl \cdot HCl \cdot 2H_2O$. Thus it should be formulated as *trans*-$[Co(en)_2Cl_2]Cl \cdot [H_5O_2]Cl$. The existence of the $H_7O_3^+$ ion in crystalline $HNO_3 \cdot 3H_2O$ and $HClO_4 \cdot 3H_2O$ was confirmed by infrared studies.[220] The spectra are consistent with a structure in which two of the hydrogens of the H_3O^+ ion are bonded to two H_2O molecules through short, asymmetrical hydrogen bonds.

(2) Aquo (H_2O) Complexes

In addition to the three fundamental modes of the free water molecule, coordinated water exhibits other modes, such as those shown in Fig. III-19. Nakagawa and Shimanouchi[221] carried out normal coordinate analyses on $[M(H_2O)_6]$- ($\mathbf{T}_h$ symmetry) and $[M(H_2O)_4]$- ($\mathbf{D}_{4h}$ symmetry) type ions to assign these low-frequency modes. Table III-15 lists the frequencies and band assignments, and Fig. III-23 illustrates the far-infrared spectra of aquo complexes obtained by these authors. Adams and Lock[222] also studied the infrared and

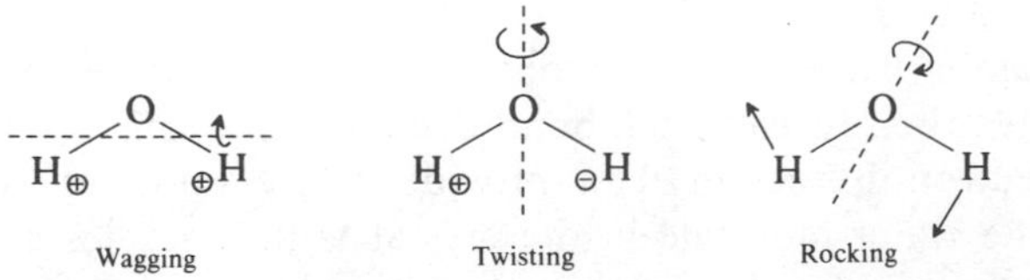

Fig. III-22. The three librational modes of water in the solid state.

TABLE III-15. OBSERVED FREQUENCIES, BAND ASSIGNMENTS, AND MO STRETCHING FORCE CONSTANTS OF AQUO COMPLEXES[221]

Compound	$\rho_r(H_2O)$	$\rho_w(H_2O)$	$\nu(MO)$	$K(M{-}O)^a$
$[Cr(H_2O)_6]Cl_3$	800	541	490	1.31
$[Ni(H_2O)_6]SiF_6$	$(755)^b$	645	405	0.84
$[Ni(D_2O)_6]SiF_6$	—	450	389	0.84
$[Mn(H_2O)_6]SiF_6$	$(655)^c$	560	395	0.80
$[Fe(H_2O)_6]SiF_6$	—	575	389	0.76
$[Cu(H_2O)_4]SO_4\cdot H_2O$	887, 855	535	440	0.67
$[Zn(H_2O)_6]SO_4\cdot H_2O$	—	541	364	0.64
$[Zn(D_2O)_6]SO_4\cdot D_2O$	467	392	358	0.64
$[Mg(H_2O)_6]SO_4\cdot H_2O$	—	460	310	0.32
$[Mg(D_2O)_6]SO_4\cdot D_2O$	474	391	—	0.32

^{a} UBF field (mdyn/Å).
^{b} $Ni(H_2O)_4Cl_2$.
^{c} $Mn(H_2O)_4Cl_2$.

Raman spectra of aquo-halogeno complexes of various types. Their assignments show that the wagging frequencies of these complexes are higher than the rocking frequencies. For example, the wagging, rocking, and MO stretching bands of solid $K_2[FeCl_5(H_2O)]$ have been assigned at 600, 460, and 390 cm^{-1}, respectively.

Vibrational spectroscopy is very useful in elucidating the structures of aquo complexes. For example, $TiCl_3\cdot 6H_2O$ should be formulated as *trans*-$[Ti(H_2O)_4Cl_2]Cl\cdot 2H_2O$ since it exhibits one TiO stretching (500 cm^{-1}, E_u) and one TiCl stretching (336 cm^{-1}, A_{2u}) mode.[223] Chang and Irish[224] showed from infrared and Raman studies that the structures of the tetrahydrates and dihydrates resulting from the dehydration of $Mg(NO_3)_2\cdot 6H_2O$ are as follows:

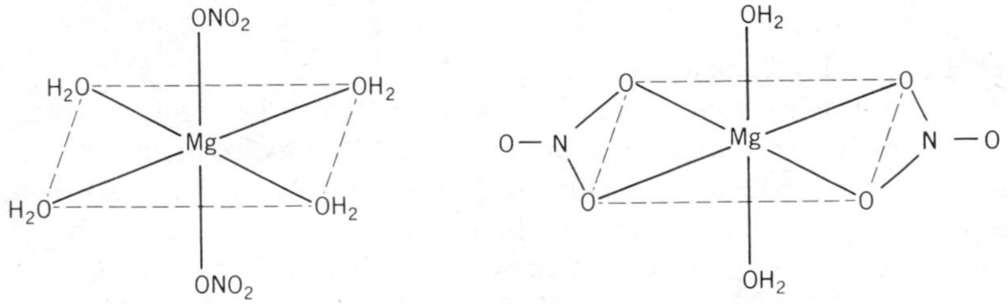

Raman spectra of aqueous solutions of inorganic salts have been studied extensively. For example, Hester and Plane[225] observed polarized Raman bands in the 400–360 cm^{-1} region for the nitrates, sulfates, and perchlorates of Zn(II), Hg(II), and Mg(II), and assigned them to the MO stretching modes of the hexacoordinated aquo complex ions.

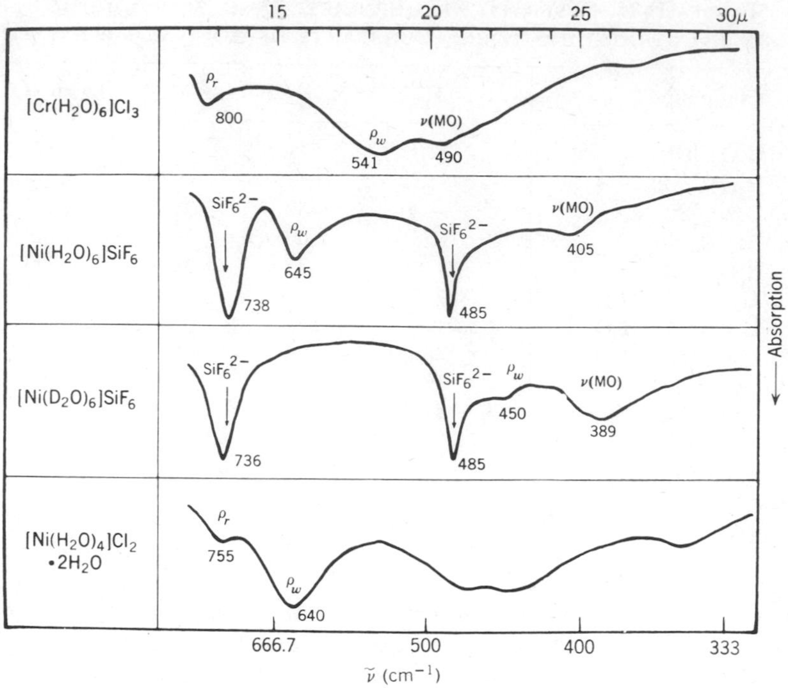

Fig. III-23. Infrared spectra of aquo complexes in the low-frequency region.[221]

A number of hydrated inorganic salts have also been studied by the inelastic neutron scattering (INS) technique.[226,227] Since the proton scattering cross section is quite large, the INS spectrum reflects mainly the motion of the protons in the crystal. Furthermore, INS spectroscopy has no selection rules involving dipole moments or polarizabilities. Thus it serves as a complementary tool to vibrational spectroscopy in studying the hydrogen vibrations of hydrated salts.

(3) Hydroxo (OH) Complexes

The spectra of hydroxo complexes are expected to be similar to those of the metal hydroxides discussed in Sec. II-1. The hydroxo group can be distinguished from the aquo group since the former lacks the HOH bending mode near 1600 cm^{-1}. Furthermore, the hydroxo complex exhibits the MOH bending mode below 1200 cm^{-1}. For example, this mode is at 1150 cm^{-1} for the $[Sn(OH)_6]^{2-}$ ion[228] and at ca. 1065 cm^{-1} for the $[Pt(OH)_6]^{2-}$ ion.[229] The OH group also forms a bridge between two metals. For example,

$$[(bipy)Cu\underset{O}{\overset{O}{\diagup\diagdown}}Cu(bipy)]\ SO_4.5H_2O$$

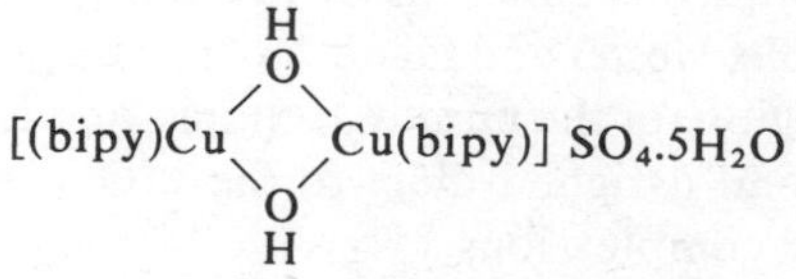

exhibits the bridging OH bending mode at 955 cm^{-1}; this is shifted to 710 cm^{-1} upon deuteration.[230] For other bridging hydroxo complexes, the following references are given: Cu(II) complexes (230, 231), Cr(III) and Fe(III) complexes (232), Co(III) complexes (233), and Pb(II) complexes (234).

III-7. COMPLEXES OF ALCOHOLS, ETHERS, KETONES, ALDEHYDES, ESTERS, AND CARBOXYLIC ACIDS

Metal alkoxides, $M(OR)_n$ (R: alkyl), exhibit $\nu(CO)$ at ca. 1000 cm^{-1} and $\nu(MO)$ at 600–300 cm^{-1}.[235] Infrared spectra have been reported for various alkoxides of Er(III)[236] and isopropoxides of rare-earth metals.[237]

The infrared spectra of alcohol complexes, $[M(EtOH)_6]Y_2$, where M is a divalent metal and Y is ClO_4^-, BF_4^-, and NO_3^-, have been measured by Van Leeuwen.[238] As expected, the anions have considerable influence on $\nu(OH)$ and $\delta(MOH)$. In ethylene glycol complexes with MX_2 (X = Cl, Br, and I), $\nu(OH)$ are shifted to lower frequencies and $\delta(CCO)$ to higher frequencies relative to those of free ligand. It was shown that ethylene glycol serves as a bidentate chelating as well as a unidentate ligand, and that the *gauche* form prevails in the complexes.[239]

The vibrational spectra of diethyl ether complexes with $MgBr_2$ and MgI_2 have been assigned completely;[240] $\nu(MgO)$ are 390–300 cm^{-1}. The solid-state Raman spectra of 1 : 1 and 1 : 2 adducts of 1,4-dioxane with metal halides show that the ligand is bridging between metals in the chair conformation.[241] Infrared spectra of alkali-metal ions complexed by cyclic polyethers have been reported.[242]

There are many coordination compounds with weakly coordinating ligands containing oxygen donors. These include ketones, aldehydes, esters, and some nitro compounds. Driessen and Groeneveld[243-245] and Driessen et al.[246] prepared metal complexes of these ligands through the reaction

$$MCl_2 + 6L + 2FeCl_3 \xrightarrow[CH_3NO_2]{} [ML_6](FeCl_4)_2$$

in a moisture-free atmosphere; CH_3NO_2 was chosen as the solvent because it is the weakest ligand available. In acetone complexes, $\nu(C{=}O)$ are lower, and $\delta(CO)$, $\pi(CO)$, and $\delta(CCC)$ are higher than those of free ligand.[243] Similar results have been obtained for complexes of acetophenone, chloracetone, and butanone.[244] In metal complexes of acetaldehyde, $\nu(C{=}O)$ are lower and $\delta(CCO)$ are higher than those of free ligand.[245] In ester complexes,[246] $\nu(C{=}O)$ shifts to lower and $\nu(C{-}O)$ to higher frequency by complex formation. When these shifts are dependent on the metal ions, the magnitudes of the shifts follow the well-known Irving–Williams order: Mn(II) < Fe(II) < Co(II) < Ni(II) < Cu(II) > Zn(II).

Extensive infrared studies have been made on metal complexes of carboxylic acids. Table III-16 gives the infrared frequencies and band assignments for the formate and acetate ions obtained by Itoh and Bernstein.[247] The carboxylate ion may coordinate to a metal in one of the following modes:

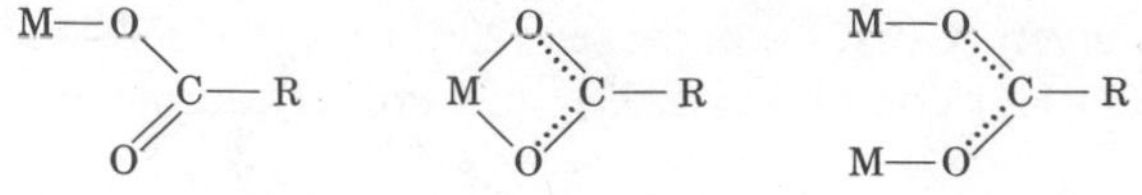

I II III

Deacon and Phillips[248a] made careful examinations of IR spectra of many acetates and trifluoroacetates having known X-ray crystal structures, and arrived at the following conclusions:

1. Unidentate complexes (structure I) exhibit the Δ values [$\nu_a(CO_2^-)-\nu_s(CO_2^-)$] which are much greater than the ionic complexes.
2. Chelating (bidentate) complexes (structure II) exhibit Δ values which are significantly less than the ionic values.
3. The Δ values for bridging complexes (structure III) are greater than those of chelating (bidentate) complexes, and close to the ionic values. These criteria are substantiated by the results shown in Table III-17 where

TABLE III-16. INFRARED FREQUENCIES AND BAND ASSIGNMENTS FOR FORMATE AND ACETATE IONS (CM^{-1})[247]

$[HCOO]^-$		$[CH_3COO]^-$			
Na Salt	Aqueous Solution	Na Salt	Aqueous Solution	C_{2v}	Band Assignment
2841	2803	2936	2935	A_1	$\nu(CH)$
—	—	—	1344		$\delta(CH_3)$
1366	1351	1414	1413		$\nu(COO)$
—	—	924	926		$\nu(CC)$
772	760	646	650		$\delta(OCO)$
—	—	—	—	A_2	$\rho_t(CH_3)$
—	—	2989	3010 or 2981	B_1	$\nu(CH)$
1567	1585	1578	1556		$\nu(COO)$
—	—	1430	1429		$\delta(CH_3)$
—	—	1009	1020		$\rho_r(CH_3)$
1377	1383	460	471		$\delta(CH)$ or $\rho_r(COO)$
—	—	2989	2981 or 3010	B_2	$\nu(CH)$
—	—	1443	1456		$\delta(CH_3)$
—	—	1042	1052		$\rho_r(CH_3)$
1073	1069	615	621		$\pi(CH)$ or $\pi(COO)$

TABLE III-17. CARBOXYL STRETCHING FREQUENCIES AND STRUCTURES OF ACETATO COMPLEXES

Compound	$\nu(C{=}O)^a$	$\nu(C{-}O)^a$	Δ^b	Structure	Ref.
CH_3COO^-, (ac)	1578	1414	164	Ionic	247
$Rh(ac)(CO)(PPh_3)_2$	1604	1376	228	Unidentate	248
$Ru(ac)_2(CO)_2(PPh_3)_2$	1613	1315	298	Unidentate	248
$RuCl(ac)(CO)(PPh_3)_2$	1507	1465	42	Bidentate	248
$RuH(ac)(PPh_3)_2$	1526	1449	77	Bidentate	248
$[Pd(ac)_2(PPh_3)]_2$	1629	1314	315	Unidentate	249
	1580	1411	169	Bridging	249
$Rh_2(ac)_2(CO)_3(PPh_2)$	1580	1440	140	Bridging	250

[a] These correspond to $\nu_a(COO^-)$ and $\nu_s(COO^-)$ of the symmetrical COO^- group.
[b] Difference between two frequencies.

$[Pd(CH_3COO)_2(PPh_3)]_2$ contains one unidentate and one bridging acetate group in one molecule. According to Stoilova et al.,[251] unidentate acetates exhibit three bands (COO deformation) at 920–720 cm^{-1} and a strong band $[\pi(CO_2)]$ at 540 cm^{-1} which are absent in bridging complexes and reduced in number in bidentate complexes. Infrared spectra of formates have been reviewed by Busca and Lorenzelli.[252]

The linkage isomerism involving the acetate group has been reported by Baba and Kawaguchi:[253]

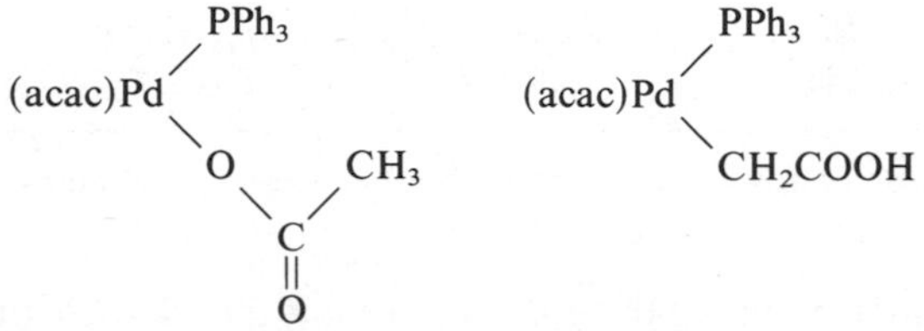

The O-isomer exhibits $\nu(C{=}O)$ at 1640 cm^{-1}, whereas the C-isomer shows $\nu(C{=}O)$ at 1670 and 1650 and $\nu(OH)$ at 2700–2500 cm^{-1}. The infrared spectra of metal glycolato complexes have been assigned by Nakamoto et al.[254]

III-8. COMPLEXES OF AMINO ACIDS, EDTA, AND RELATED COMPOUNDS

Amino acids exist as zwitterions in the crystalline state. Table III-18 gives band assignments made for the zwitterions of glycine[255] and α-alanine.[256] According to X-ray analysis, two glycino anions (gly) in $[Ni(gly)_2]\cdot 2H_2O$,[257] for example, coordinate to the metal by forming a *trans*-planar structure, and the noncoordinating C=O groups are hydrogen-bonded to the neighboring

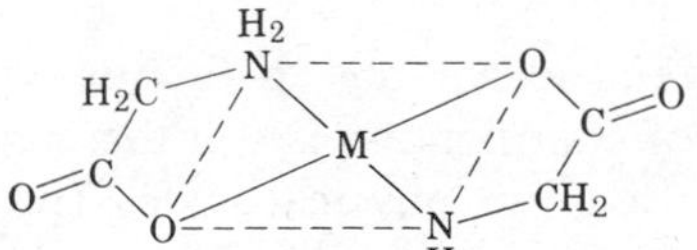

TABLE III-18. INFRARED FREQUENCIES AND BAND ASSIGNMENTS OF GLYCINE AND α-ALANINE IN THE CRYSTALLINE STATE (CM^{-1})[255,256]

Glycine	α-Alanine	Band Assignment
1610	1597	$\nu_a(COO^-)$
1585	1623	$\delta_d(NH_3^+)$
1492	1534	$\delta_s(NH_3^+)$
—	1455	$\delta_d(CH_3)$
1445	—	$\delta(CH_2)$
1413	1412	$\nu_s(COO^-)$
—	1355	$\delta_s(CH_3)$
1333	—	$\rho_w(CH_2)$
—	1308	$\delta(CH)$
1240 (R)	—	$\rho_t(CH_2)$
1131, 1110	1237, 1113	$\rho_r(NH_3^+)$[a]
1033	1148	$\nu_a(CCN)$[a]
—	1026, 1015	$\rho_r(CH_3)$[a]
910	—	$\rho_r(CH_2)$
893	918, 852	$\nu_s(CCN)$[a]
694	648	$\rho_w(COO^-)$
607	771	$\delta(COO^-)$
516	492	$\rho_t(NH_3^+)$
504	540	$\rho_r(COO^-)$

[a] These bands are coupled with other modes in α-alanine.

molecule or water of crystallization, or weakly bonded to the metal of the neighboring complex. Thus $\nu(CO_2)$ of amino acid complexes are affected by coordination as well as by intermolecular interactions.

To examine the effects of coordination and hydrogen bonding, Nakamoto et al.[258] made extensive IR measurements of the COO stretching frequencies of various metal complexes of amino acids in D_2O solution, in the hydrated crystalline state, and in the anhydrous crystalline state. The results showed that, in any one physical state, the same frequency order is found for a series of metals, regardless of the nature of the ligand. The antisymmetric frequencies increase, the symmetric frequencies decrease, and the separation between the two frequencies increases in the following order of metals:

$$\text{Ni(II)} < \text{Zn(II)} < \text{Cu(II)} < \text{Co(II)} < \text{Pd(II)} \approx \text{Pt(II)} < \text{Cr(III)}$$

Although there are several exceptions to this order, these results generally indicate that the effect of coordination is still the major factor in determining the frequency order in a given physical state. The above frequency order indicates the increasing order of the metal–oxygen interaction since the COO group becomes more asymmetrical as the metal–oxygen interaction becomes stronger.

To give theoretical band assignments on metal glycino complexes, Condrate and Nakamoto[259] carried out a normal coordinate analysis on the metal–glycino chelate ring. Figure III-24 shows the infrared spectra of bis(glycino) complexes of Pt(II), Pd(II), Cu(II), and Ni(II). Table III-19 lists the observed frequencies and theoretical band assignments. The CH_2 group frequencies are not listed, since they are not metal-sensitive. It is seen that the C=O stretching, NH_2 rocking, and MN and MO stretching bands are metal sensitive and are shifted progressively to higher frequencies as the metal is changed in the order Ni(II) < Cu(II) < Pd(II) < Pt(II). Table III-19 shows that both the MN and MO stretching force constants also increase in the same order of the metals. These results provide further support to the preceding discussion of the M–O bonds of glycino complexes.

To give definitive band assignments in the low frequency region of *bis*(glycino) complexes of Ni(II), Cu(II), and Co(II), Kincaid and Nakamoto[260] carried out H–D, ^{14}N–^{15}N, ^{58}Ni–^{62}Ni, and ^{63}Cu–^{65}Cu substitutions, and performed normal coordinate analyses on the skeletal modes of bis(glycino) complexes. Their results show that, in *trans*-$[M(gly)_2]2H_2O$, the

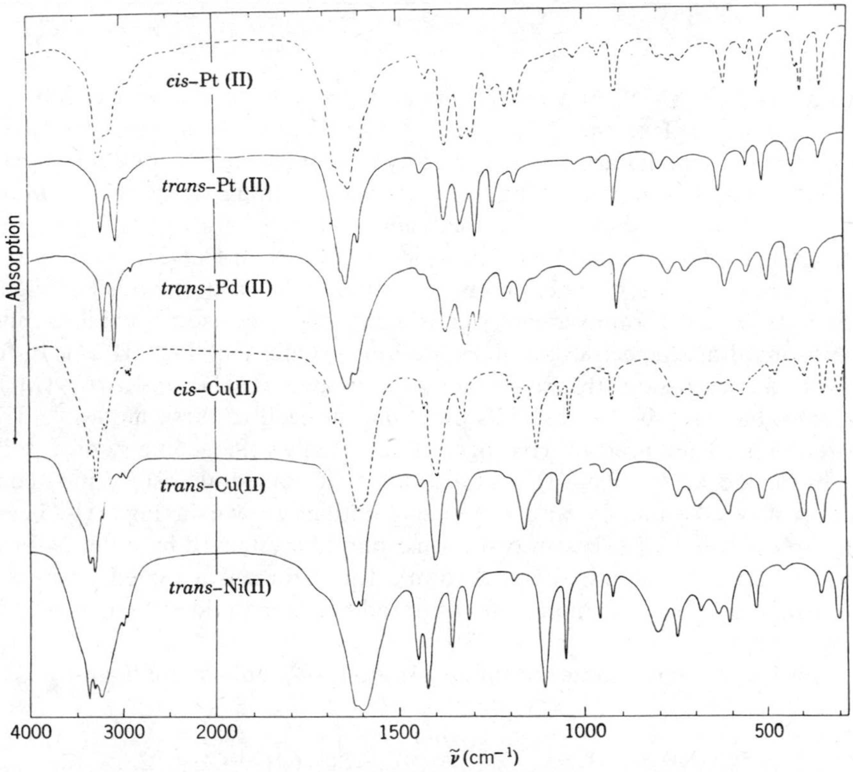

Fig. III-24. Infrared spectra of bis(glycino) complexes of divalent metals.[259]

TABLE III-19. OBSERVED FREQUENCIES AND BAND ASSIGNMENTS
OF BIS(GLYCINO) COMPLEXES $(CM^{-1})^{259}$

trans-[Pt(gly)₂]	trans-[Pd(gly)₂]	trans-[Cu(gly)₂]	trans-[Ni(gly)₂]	Band Assignment
3230 ⎫ 3090 ⎭	3230 ⎫ 3120 ⎭	3320 ⎫ 3260 ⎭	3340 ⎫ 3280 ⎭	$\nu(NH_2)$
1643	1642	1593	1589	$\nu(C{=}O)$
1610	1616	1608	1610	$\delta(NH_2)$
1374	1374	1392	1411	$\nu(C{-}O)$
1245	1218	1151	1095	$\rho_t(NH_2)$
1023	1025	1058	1038	$\rho_w(NH_2)$
792	771	644	630	$\rho_r(NH_2)$
745	727	736	737	$\delta(C{=}O)$
620	610	592	596	$\pi(C{=}O)$
549	550	439	439	$\nu(MN)$
415	420	360	290	$\nu(MO)$
2.10	2.00	0.90	0.70	$K(M{-}N)$ (mdyn/Å)^a
2.10	2.00	0.90	0.70	$K(M{-}O)$ (mdyn/Å)^a

^{a} UBF.

infrared-active $\nu(MN)$ and $\nu(MO)$ are at 483 and 337 cm^{-1}, respectively, for the Cu(II) complex, and at 442 and 289 cm^{-1}, respectively, for the Ni(II) complex. Both modes are coupled strongly with other skeletal modes however. Use of multiple isotope labeling techniques in assigning IR spectra of amino acid complexes has been extended to [Cd(gly)₂]·H₂O,[261] cis-[Ni(gly)₂(ImH)₂],[262] and [M(L–Ala)₂] [M = Ni(II) and Cu(II)].[263]

Square-planar bis(glycino) complexes can take the cis or the trans configuration. As expected from symmetry consideration, the cis-isomer exhibits more bands in infrared spectra than does the trans-isomer (see Fig. III-24). In the low-frequency region, the cis-isomer exhibits two $\nu(MN)$ and two $\nu(MO)$, whereas the trans-isomer exhibits only one for each of these modes.[259] This criterion has been used by Herlinger et al. to assign the geometry of a series of bis(amino acidato)Cu(II) complexes.[264,265] Octahedral tris(glycino) complexes may take the fac and mer configurations shown in Fig. III-25. For example, [Co(gly)₃] exists in two forms: purple crystals (dihydrate, α-form) and red crystals (monohydrate, β-form). The α-form is assigned to the mer configuration since it exhibits more infrared bands than does the β-form (fac configuration).[266]

Glycine also coordinates to the Pt(II) atom as a unidentate ligand:

$$-\overset{|}{\underset{|}{Pt}}-NH_2-CH_2-C\overset{\displaystyle O}{\underset{\displaystyle OH}{\big<}} \qquad\qquad -\overset{|}{\underset{|}{Pt}}-NH_2-CH_2-C\overset{\displaystyle O^{-1/2}}{\underset{\displaystyle O^{-1/2}}{\big<}}$$

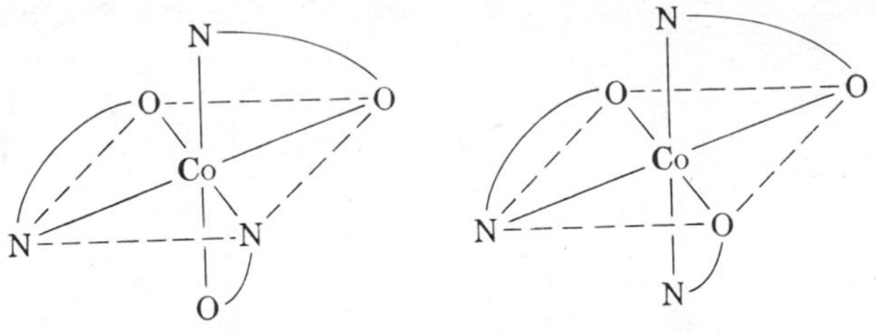

fac(C₃), β   *mer*(C₁), α

Fig. III-25. Structures of the tris(glycino) complex.

The carboxyl group is not ionized in *trans*-[Pt(glyH)$_2$X$_2$] (X: a halogen), whereas it is ionized in *trans*-[Pt(gly)$_2$(NH$_3$)$_2$]. The former exhibits the un-ionized COO stretching band near 1710 cm^{-1}, while the latter shows the ionized COO stretching band near 1610 cm^{-1}.[267]

The distinction between unidentate and bidentate glycino complexes of Pt(II) can be made readily from their infrared spectra. Figure III-26 illustrates the infrared spectra of *trans*-[Pt(glyH)$_2$Cl$_2$] and K[Pt(gly)Cl$_2$] in the COO stretching and PtO stretching regions. The bidentate (chelated) glycino group absorbs at 1643 cm^{-1}, unlike either the ionized unidentate group (1610 cm^{-1}) or the unionized unidentate group (1710 cm^{-1}). Furthermore, the bidentate glycino group exhibits the PtO stretching band at 388 cm^{-1}, whereas the unidentate glycino group has no absorption between 470 and 350 cm^{-1}. Figure III-26 also shows the spectrum of [Pt(gly)(glyH)Cl], in which both the unidentate and bidentate glycino groups are present. It is seen that the spectrum of this compound can be interpreted as a superposition of the spectra of the former two compounds.[267]

The Cu(III) complexes of tetraglycine and tetraglycineamide exhibit the N(amide)–Cu(III) CT absorption at 365 nm. Using the 363.8 nm excitation, Kincaid et al.[268] were able to resonance-enhance the ν(Cu–N) vibrations at 420 and 417 cm^{-1}, respectively.

From the infrared spectra observed in the solid state, Busch and co-workers[269] determined the coordination numbers of the metals in metal chelate compounds of EDTA and its derivatives:

$$\begin{array}{ccc} \text{HOOCH}_2\text{C} & & \text{CH}_2\text{COOH} \\ \diagdown & & \diagup \\ & \text{N--CH}_2\text{--CH}_2\text{--N} & \\ \diagup & & \diagdown \\ \text{HOOCH}_2\text{C} & & \text{CH}_2\text{COOH} \end{array}$$

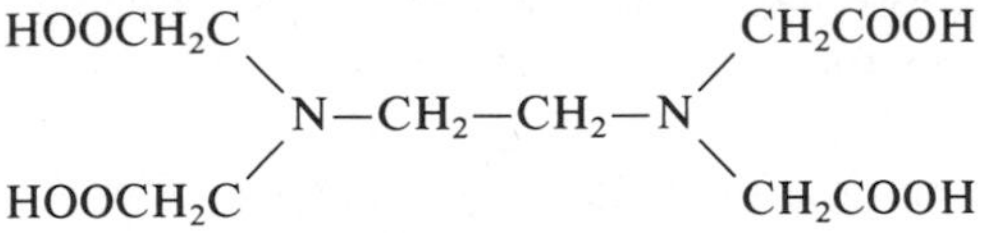

Ethylenediaminetetraacetic acid
(EDTA) or (H$_4$Y)

The method is based on the simple rule that the un-ionized and uncoordinated COO stretching band occurs at 1750–1700 cm^{-1}, whereas the ionized and coordinated COO stretching band is at 1650–1590 cm^{-1}. The latter frequency

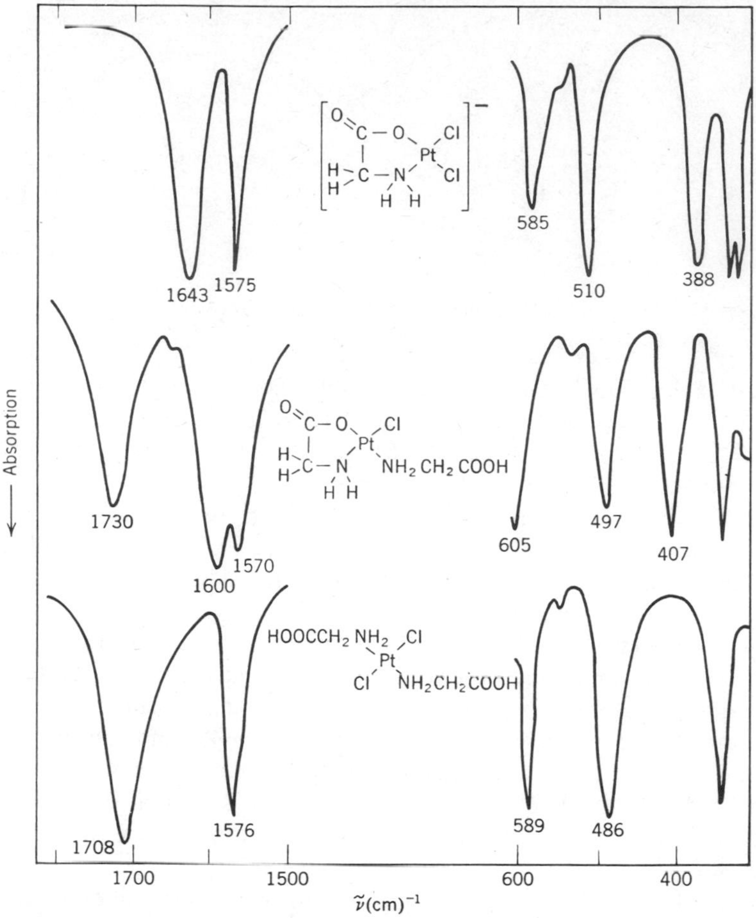

Fig. III-26. Infrared spectra of K[Pt(gly)Cl₂], [Pt(gly)(glyH)Cl], and *trans*-[Pt(glyH)₂Cl₂].[267]

depends on the nature of the metal: 1650–1620 cm⁻¹ for metals such as Cr(III) and Co(III), and 1610–1590 cm⁻¹ for metals such as Cu(II) and Zn(II). Since the free ionized COO⁻ stretching band is at 1630–1575 cm⁻¹, it is also possible to distinguish the coordinated and free COO⁻ stretching bands if a metal such as Co(III) is chosen for complex formation. Table III-20 shows the results obtained by Busch et al.

Tomita and Ueno[271] studied the infrared spectra of metal complexes of NTA, using the method described above. They concluded that NTA

$$
\begin{array}{c}
CH_2COOH \\
\diagup \\
N{-}CH_2COOH \\
\diagdown \\
CH_2COOH
\end{array}
\qquad \text{Nitrilotriacetic acid (NTA)}
$$

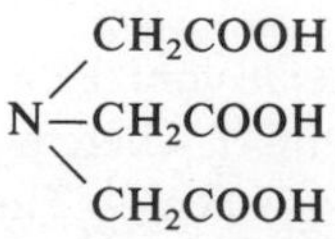

TABLE III-20. ANTISYMMETRIC COO STRETCHING FREQUENCIES AND NUMBER OF FUNCTIONAL GROUPS USED FOR COORDINATION IN EDTA COMPLEXES (CM^{-1})[269]

Compound[a]	Un-ionized COOH	Coordinated $COO^-\cdots M$	Free COO^-	Number of Coordinated Groups
$H_4[Y]$	1698[b]	—	—	
$Na_2[H_2Y]$	1668[b]	—	1637[b]	
$Na_4[Y]$	—	—	1597[b]	
$Ba[Co(Y)]_2\cdot 4H_2O$	—	1638	—	6
$Na_2[Co(Y)Cl]$	—	1648	1600	5
$Na_2[Co(Y)NO_2]$	—	1650	1604	5
$Na[Co(HY)Cl]\cdot \frac{1}{2}H_2O$	1750	1650	—	5
$Na[Co(HY)NO_2]\cdot H_2O$	1745	1650	—	5
$Ba[Co(HY)Br]\cdot 9H_2O$	1723	1628	—	5
$Na[Co(YOH)Cl]\cdot \frac{3}{2}H_2O$	—	1658	—	5
$Na[Co(YOH)Br]\cdot H_2O$	—	1654	—	5
$Na[Co(YOH)NO_2]$	—	1652	—	5
$[Pd(H_2Y)\cdot 3H_2O$	1740	1625	—	4
$[Pt(H_2Y)]\cdot 3H_2O$	1730	1635	—	4
$[Pd(H_4Y)Cl_2]\cdot 5H_2O$	1707, 1730	—	—	2
$[Pt(H_4Y)Cl_2]\cdot 5H_2O$	1715, 1730	—	—	2

[a] Y = tetranegative ion; HY = trinegative ion; H_2Y = dinegative ion; H_4Y = neutral species of EDTA; YOH = trinegative ion of HEDTA (hydroxyethylenediaminetriacetic acid).
[b] Reference 270.

acts as a quadridentate ligand in complexes of Cu(II), Ni(II), Co(II), Zn(II), Cd(II), and Pb(II), and as a tridentate in complexes of Ca(II), Mg(II), Sr(II), and Ba(II).

Krishnan and Plane[272] studied the Raman spectra of EDTA and its metal complexes in aqueous solution. They noted that ν(MN) appears strongly in the 500–400 cm^{-1} region for Cu(II), Zn(II), Cd(II), Hg(II), and so on, and that its frequency decreases with an increasing radius of the metal ion, independently of the stability of the metal complex. McConnell and Nuttall[273] assigned the ν(MN) and ν(MO) of $Na_2[M(EDTA)]2H_2O$ (M = Sn and Pb) in their Raman and infrared spectra.

III-9. INFRARED SPECTRA OF AQUEOUS SOLUTIONS

Infrared studies of aqueous solutions provide a valuable tool for elucidating the structures of complex ions in equilibria. To observe the infrared spectrum of an aqueous solution, it is necessary to use window materials such as AgCl and BaF_2, which are insoluble in water, and a thin spacer (0.02–0.01 mm) to reduce the strong absorption of water. The latter condition necessitates a solution of relatively high concentrations. Even if these conditions are met, it is still difficult to measure the spectrum of a solute in the regions at 3700–2800, 1800–1600, and below 1000 cm^{-1}, where water (H_2O) absorbs strongly. This difficulty may be overcome if water absorption is subtracted from the solution spectrum by the use of computers. A cylindrical internal reflection (CIR) cell combined with a Fourier-transform IR spectrophotometer was found to be very useful in measuring IR spectra of aqueous solutions.[273a]

The C≡N stretching band (2200–2000 cm^{-1}) can be measured in aqueous solution since it is outside of these regions. Thus the solution equilibria of cyano complexes have been studied extensively by using aqueous infrared spectroscopy (Sec. III-14). Fronaeus and Larsson[274] extended similar studies to thiocyanato complexes that exhibit the C≡N stretching bands in the same region. They[275] also studied the solution equilibria of oxalato complexes in the 1500–1200 cm^{-1} region, where the CO stretching bands of the coordinated oxalato group appear. Larsson[276] studied the infrared spectra of metal glycolato complexes in aqueous solution. In this case, the C—OH stretching band near 1060 cm^{-1} was used to elucidate the structures of the complex ions in equilibria.

If D_2O is used instead of H_2O, it is possible to observe infrared spectra in the regions 4000–2900, 2000–1300, and 1100–900 cm^{-1}. The COO stretching bands of NTA, EDTA, and their metal complexes appear between 1750 and 1550 cm^{-1} (Sec. III-8). Nakamoto and co-workers,[277] therefore, studied the solution (D_2O) equilibria of NTA, EDTA, and related ligands in this frequency region. By combining the results of potentiometric studies with the spectra obtained as a function of the pH(pD) of the solution, it was possible to establish the following COO stretching frequencies:

Type A, un-ionized carboxyl ($R_2N—CH_2COOH$), 1730–1700 cm^{-1}
Type B, α-ammonium carboxylate ($R_2N^+H—CH_2COO^-$), 1630–1620 cm^{-1}
Type C, α-aminocarboxylate ($R_2N—CH_2COO^-$), 1585–1575 cm^{-1}

As stated in Sec. III-8, the coordinated (ionized) COO group absorbs at 1650–1620 cm^{-1} for Cr(III) and Co(III), and at 1610–1590 cm^{-1} for Cu(II) and Zn(II). Thus it is possible to distinguish the coordinated COO group from those of Types B and C if a proper metal ion is selected.

Tomita et al.[278] studied the complex formation of NTA with Mg(II) by aqueous infrared spectroscopy. Figure III-27 shows the infrared spectra of equimolar mixtures of NTA and $MgCl_2$ at concentrations about 5–10% by weight. The spectra of the mixture from pD 3.2 to 4.2 exhibit a single band at 1625 cm^{-1}, which is identical to that of the free $H(NTA)^{2-}$ ion in the same pD range.[279] This result indicates that no complex formation occurs in this pD range, and that the 1625-cm^{-1} band is due to the $H(NTA)^{2-}$ ion (Type B). If the pD is raised to 4.2, a new band appears at 1610 cm^{-1}, which is not observed for the free NTA solution over the entire pD range investigated. Figure III-27 shows that this 1610-cm^{-1} band becomes stronger, and the

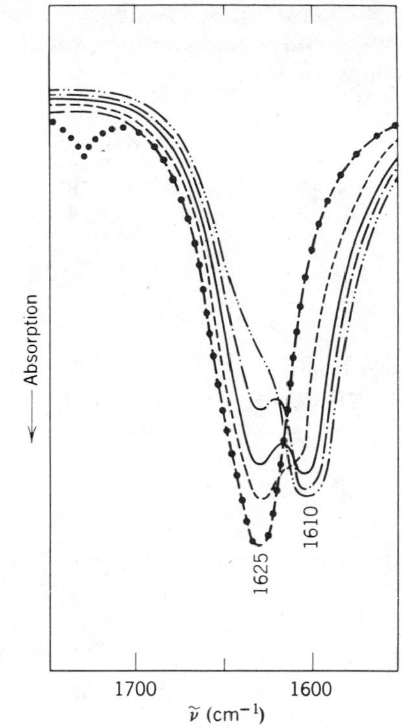

Fig. III-27. Infrared spectra of Mg–NTA complex in D_2O solutions: (······) pD 3.2; (----) pD 4.2; (----) pD 5.5; (———) pD 6.8; (–·–·–) pD 10.0; (–···–···) pD 11.6.[278]

1625-cm^{-1} band becomes weaker, as the pD increases. It was concluded that this change is due mainly to a shift of the following equilibrium in the direction of complex formation:

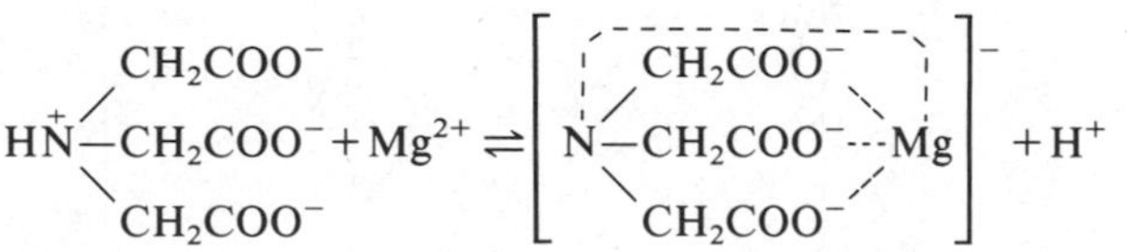

$$\begin{array}{c} CH_2COO^- \\ / \\ H\overset{+}{N}{-}CH_2COO^- + Mg^{2+} \\ \backslash \\ CH_2COO^- \end{array} \rightleftharpoons \left[\begin{array}{c} CH_2COO^- \\ / \\ N{-}CH_2COO^-\text{---}Mg \\ \backslash \\ CH_2COO^- \end{array}\right]^- + H^+$$

1625 cm^{-1} 1610 cm^{-1}
(Type B)

By plotting the intensity of these two bands as a function of pD, the stability constant of the complex ion was calculated to be 5.24. This value is in good agreement with that obtained from potentiometric titration (5.41).

Martell and Kim[280] carried out an extensive study on solution equilibria involving the formation of Cu(II) complexes with various polypeptides. As an example, the glycylglycino–Cu(II) system is discussed below.[281] Figure

III-28 illustrates the infrared spectra of free glycylglycine in D_2O solution as a function of pD. The observed spectral changes were interpreted in terms of the solution equilibria shown below:

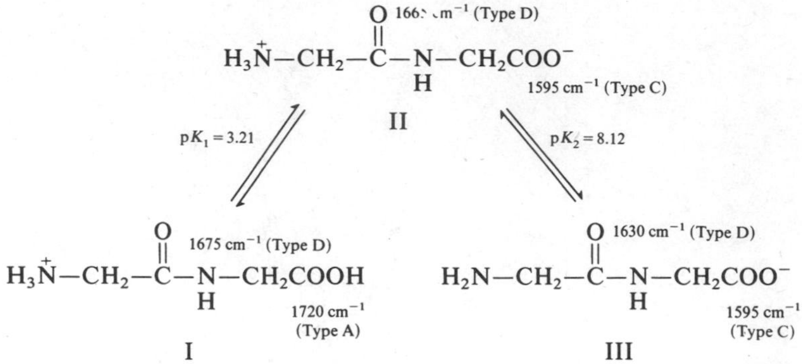

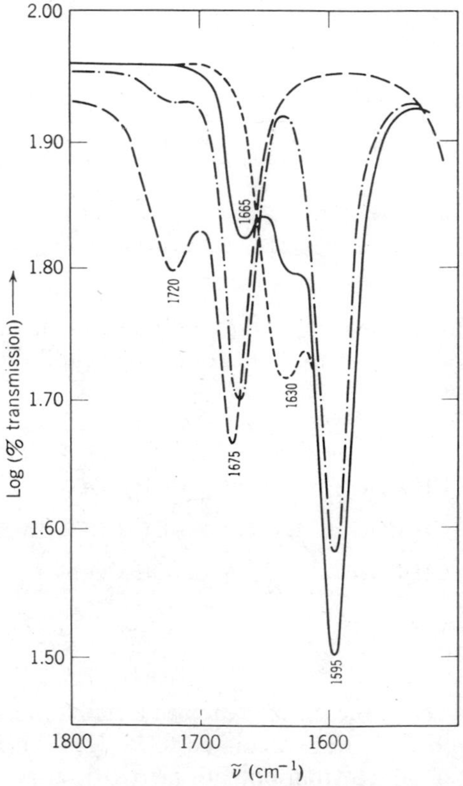

Fig. III-28. Infrared spectra of glycylglycine in D_2O solution at 0.288 *M* concentration and ionic strength 1.0 adjusted with KCl: (- - - -) pD 1.75; (–·–·–) pD 4.31; (———) pD 8.77; (----) pD 10.29.[281]

Band assignments have been made by using the criteria given previously. In addition, Type-D frequency (1680–1610 cm^{-1}) was introduced to denote the peptide carbonyl group. The exact frequency of this group depends on the nature of the neighboring groups.

Figure III-29 shows the infrared spectra of glycylglycine mixed with copper chloride at equimolar ratio in D$_2$O solution.[282] At pD = 3.58, the ligand exhibits three bands at 1720, 1675, and 1595 cm^{-1} (Fig. III-28). This result indicates that I and II are in equilibrium. At the same pD value, however, the mixture exhibits one extra band at 1625 cm^{-1}. This band was attributed to the metal complex (IV), which was formed by the following reaction:

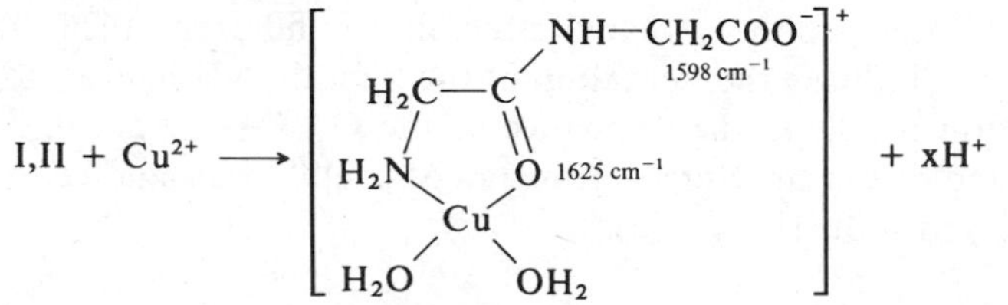

$$
\text{I,II} + \text{Cu}^{2+} \longrightarrow
\left[
\begin{array}{c}
\text{NH—CH}_2\text{COO}^- \\
{\scriptstyle 1598\ cm^{-1}} \\
\text{H}_2\text{C—C} \\
\text{H}_2\text{N}\quad\quad \text{O}\ {\scriptstyle 1625\ cm^{-1}} \\
\text{Cu} \\
\text{H}_2\text{O}\quad\quad \text{OH}_2 \\
\text{IV}
\end{array}
\right]^+
+ x\text{H}^+
$$

At pD = 5.18, the solution exhibits one broad band at about 1610 cm^{-1}. This result was interpreted as an indication that the following equilibrium was shifted almost completely to the right-hand side, and that the 1610-cm^{-1} band

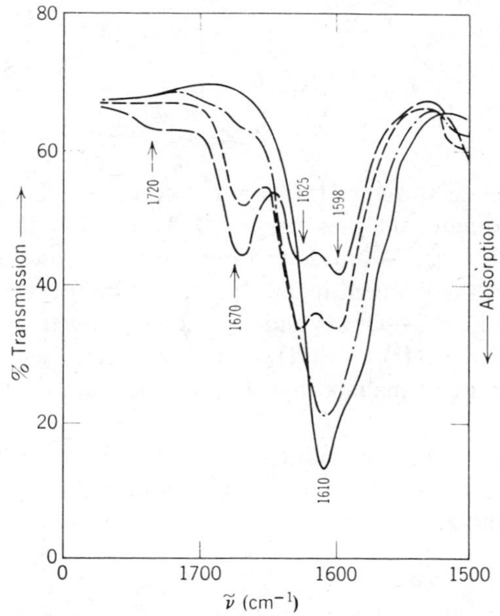

Fig. III-29. Infrared spectra of Cu(II)–glycylglycine complexes in D$_2$O solutions: (– – – –) pD 3.58; (- - - -) pD 4.24; (– · – · –) pD 5.18; (———) pD 10.65. Total concentration of the ligand and the metal is 0.2333 M, and the ionic strength is 1.0 adjusted with KCl.[282]

is an overlap of two bands at 1610 and 1598 cm^{-1}:

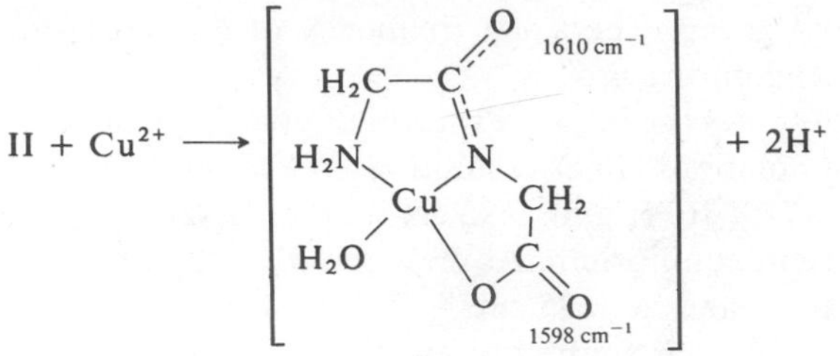

V

The shift of the peptide carbonyl stretching band from 1625 (IV) to 1610 (V) cm^{-1} may indicate the ionization of the peptide NH hydrogen, since such an ionization results in the resonance of the O—C—N system, as indicated by the dotted line in structure V. Kim and Martell[283] also studied the triglycine and tetraglycine Cu(II) systems.

III-10. COMPLEXES OF OXALIC ACID AND RELATED COMPOUNDS

The oxalato anion (ox^{2-}) coordinates to a metal as a unidentate (I) or bidentate (II) ligand:

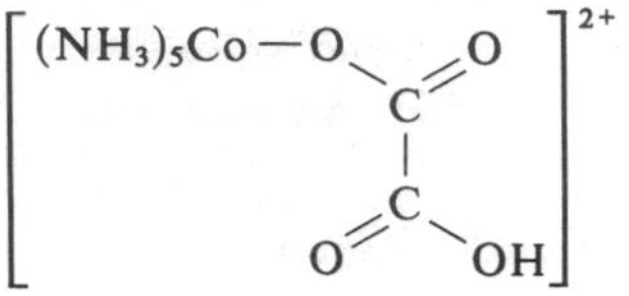

I

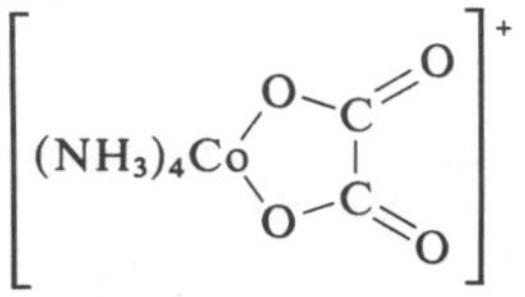

II

The bidentate chelate structure (II) is most common. Fujita et al.[284] carried out normal coordinate analyses on the 1:1 (metal-ligand) model of the $[M(ox)_2]^{2-}$ and $[M(ox)_3]^{3-}$ series, and obtained the band assignments listed in Table III-21. In the divalent metal series, $\nu(C{=}O)$ (average of ν_1 and ν_7) becomes higher, and $\nu(C{-}O)$ (ν_2 and ν_8) becomes lower, as $\nu_4(MO)$ becomes higher in the order Zn(II) < Cu(II) < Pd(II) < Pt(II) (see Fig. III-30). This relation holds in spite of the fact that ν_2, ν_4, and ν_8 are all coupled with other vibrations.

In the trivalent metal series, Hancock and Thornton[285] found that ν_{11} (MO stretching) follows the same trend as the crystal field stabilization energies (CFSE) of these metals, namely:

	Sc d^0		V d^2		Cr d^3		Mn d^4		Fe d^5		Co d^6		Ga d^{10}
$\nu(MO)(cm^{-1})$	340	<	367	<	416	>	372	>	354	<	446	>	368
CFSE (10^3 cm^{-1})	0	<	10.2	<	21.2	>	10.2	>	0	<	27.0	>	0

TABLE III-21. FREQUENCIES AND BAND ASSIGNMENTS OF VARIOUS OXALATO COMPLEXES (CM^{-1})[284]

$K_2[Zn(ox)_2]\cdot 2H_2O$	$K_2[Cu(ox)_2]\cdot 2H_2O$	$K_2[Pd(ox)_2]\cdot 2H_2O$	$K_2[Pt(ox)_2]\cdot 3H_2O$	$K_3[Fe(ox)_3]\cdot 3H_2O$	$K_3[V(ox)_3]\cdot 3H_2O$	$K_3[Cr(ox)_3]\cdot 3H_2O$	$K_3[Co(ox)_3]\cdot 3H_3O$	$K_3[Al(ox)_3]\cdot 3H_2O$	$[Cr(NH_3)_4(ox)]\cdot Cl$	Band Assignment	
1632	(1720) 1672	1698	1709	1712	1708	1708	1707	1722	1704	$\nu_a(C{=}O)$	ν_7
—	1645	1675, 1657	1674	1677, 1649	1675, 1642	1684, 1660	1670	1700, 1683	1668	$\nu_a(C{=}O)$	ν_1
1433	1411	1394	1388	1390	1390	1387	1398	1405	1393	$\nu_s(CO) + \nu(CC)$	ν_2
1302	1277	1245 (1228)	1236	1270, 1255	1261	1253	1254	1292, 1269	1258	$\nu_s(CO) + \delta(O{-}C{=}O)$	ν_8
890	886	893	900	885	893	893	900	904	914, 890	$\nu_s(CO) + \delta(O{-}C{=}O)$	ν_3
785	795	818	825	797, 785	807, 797	810, 798	822, 803	820, 803	804	$\delta(O{-}C{=}O) + \nu(MO)$	ν_9
622	593	610	—	580	581	595	—	—	—	Crystal water?	
519	541	556	575, 559	528	531	543	565	587	545	$\nu(MO) + \nu(CC)$	ν_4
519	481	469	469	498	497	485	472	436	486, 469	Ring. def. $+ \delta(O{-}C{=}O)$	ν_{10}
428, 419	420	417	405	366	368	415	446	485	366	$\nu(MO)$ + ring def.	ν_{11}
377, 364	382, 370	368	370	340	336	358	364	364	347	$\delta(O{-}C{=}O) + \nu(CC)$	ν_5
291	339	350	328	—	—	313	332	—	328	π	

Both quantities are maximized at the d^3 and d^6 configurations (d^4 and d^5 ions are in high-spin states). The IR spectra of $[Ir(ox)Cl_4]^{3-}$($\mathbf{C}_{2v}$), $[Ir(ox)_2Cl_2]^{3-}$ (*trans*, $\mathbf{D}_{2h}$; *cis*, $\mathbf{C}_2$), and $[Ir(ox)_3]^{3-}$($\mathbf{D}_3$) have been assigned by Gouteron.[286]

The oxalato anion may act as a bridging group between metal atoms. According to Scott et al.,[287] the oxalato anion can take the following four bridging structures:

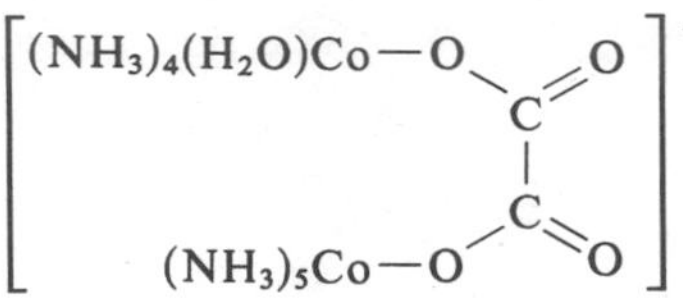

III

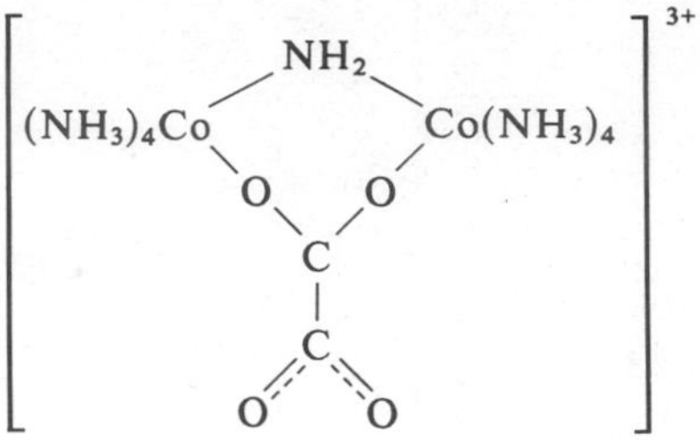

IV

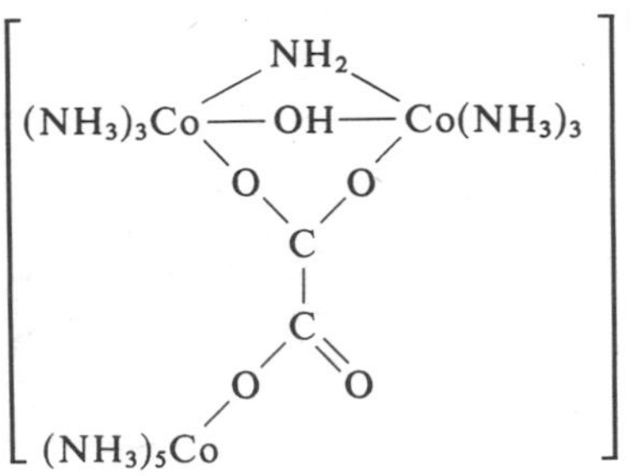

V

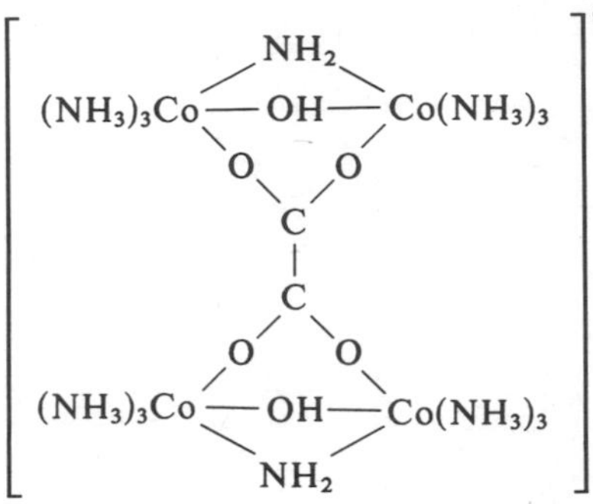

VI

Table III-22 lists the ν(CO) of each type. The spectrum of the tetradentate complex (VI) is the most simple. Because of its high symmetry [$\mathbf{D}_{2h}$ (planar) or $\mathbf{D}_{2d}$ (twisted)], it exhibits only two ν(CO). The spectra of bidentate complexes (III and IV) show four ν(CO), as expected from the $\mathbf{C}_{2v}$ symmetry. The spectrum of the tridentate complex (V) should show four ν(CO), although only three are observed.

The Raman spectra of metal oxalato complexes have also been examined to investigate the solution equilibria and the nature of the M–O bond.[288]

Vibrational assignments have been made on metal oxamido complexes of $\mathbf{V}_h$ symmetry:[289]

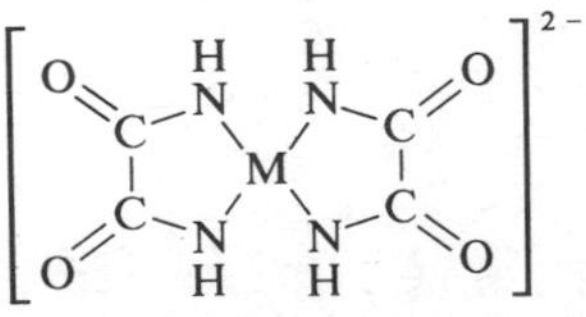

(M = Ni or Cu)

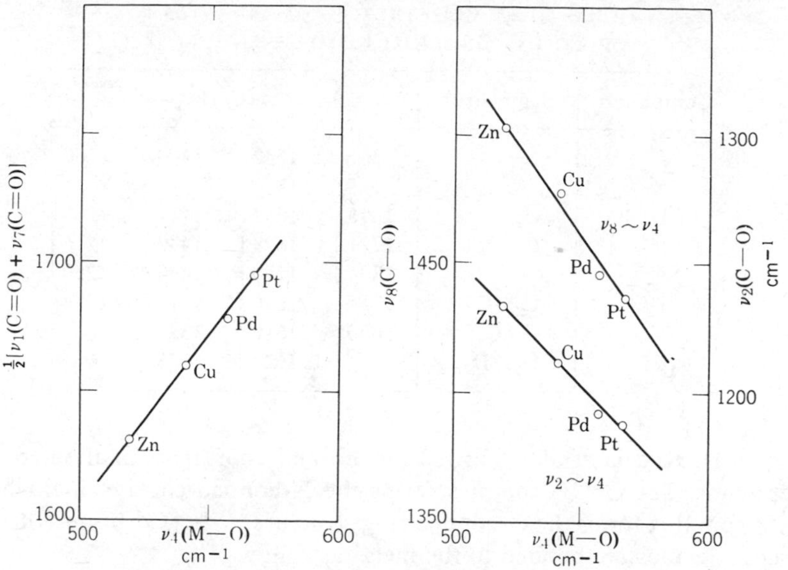

Fig. III-30. M-O stretching frequency vs C=O and C—O stretching frequencies in oxalato complexes of divalent metals.

and the *cis-* and *trans*-dimethyloxamido complexes of dimethylgallium:[290]

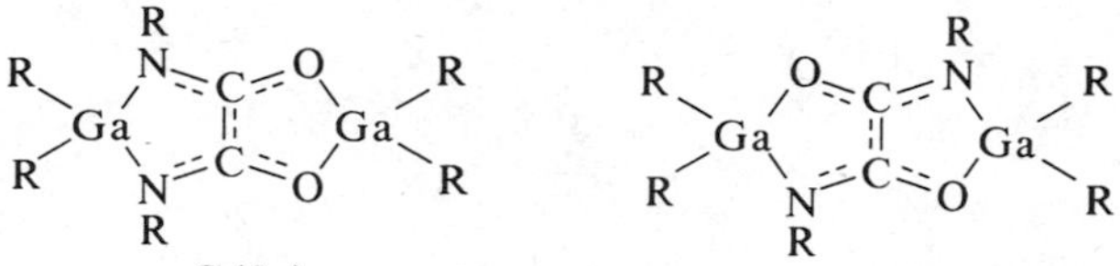

Cis(C_{2v}) *Trans*(C_{2h})

Biuret ($NH_2CONHCONH_2$) is known to form the following two types of chelate rings:

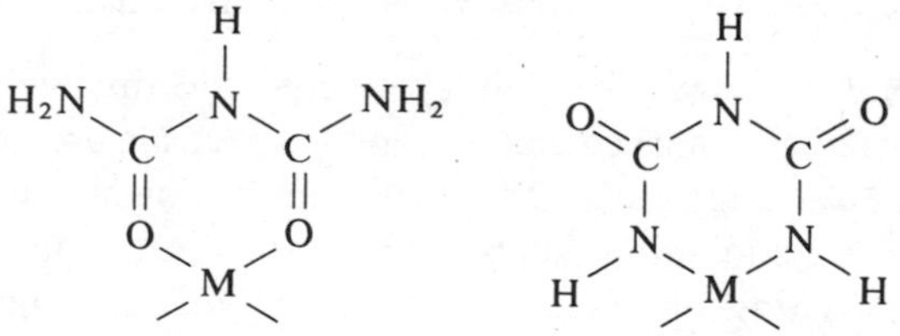

VII VIII

Violet crystals of composition $K_2[Cu(biureto)_2]\cdot 4H_2O$ are obtained when the Cu(II) ion is added to an alkaline solution of biuret, whereas pale blue-green crystals of composition $[Cu(Biuret)_2]Cl_2$ result when the Cu(II) ion is mixed with biuret in neutral (alcoholic) solution. The former contains the N-bonded chelate ring structure (VIII), while the latter consists of the O-bonded chelate

TABLE III-22. CO STRETCHING VIBRATIONS OF Co(III) OXALATO COMPLEXES (cm^{-1})

Compound	Symmetry[a]	$\nu(CO)$			
I	C_s/C_1	1761	1682 1665	1400	1260
II	C_{2v}	1696	1667	1410	1268
III	C_{2v}/C_2	1721 1701	1629 1670	1439 1430	1276 1250
IV	C_{2v}/C_2	1755	1626	1318	1284
V	C_s/C_1	1650	1610	1322	—
VI	D_{2h}/D_{2d}	—	1628	1345	—

rings (VII). Kedzia et al.[291] carried out normal coordinate analyses of both compounds. The Co(II) complex forms the N-bonded chelate ring, whereas the Zn complex forms the O-bonded ring structure.[291] In $[Cd(biuret)_2]Cl_2$, the biuret molecules are bonded to the metal as follows:[292]

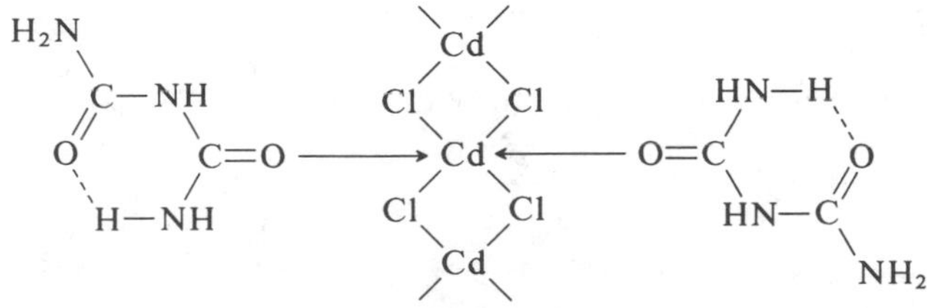

IX

Saito et al.[293] carried out normal coordinate analysis on the ligand portion of the Cd complex.

III-11. SULFATO, CARBONATO, AND OTHER ACIDO COMPLEXES

When a ligand of relatively high symmetry coordinates to a metal, its symmetry is lowered and marked changes in the spectrum are expected because of changes in the selection rules. This principle has been used extensively to determine whether acido anions such as SO_4^{2-} and CO_3^{2-} coordinate to metals as unidentate, chelating bidentate, or bridging bidentate ligands. Although symmetry lowering is also caused by the crystalline environment, this effect is generally much smaller than the effect of coordination.

(1) Sulfato (SO_4) Complexes

The free sulfate ion belongs to the high-symmetry point group T_d. Of the four fundamentals, only ν_3 and ν_4 are infrared active. If the symmetry of the ion is lowered by complex formation, the degenerate vibrations split and

Raman-active modes appear in the infrared spectrum. The lowering of symmetry caused by coordination is different for the unidentate and bidentate complexes, as shown below:

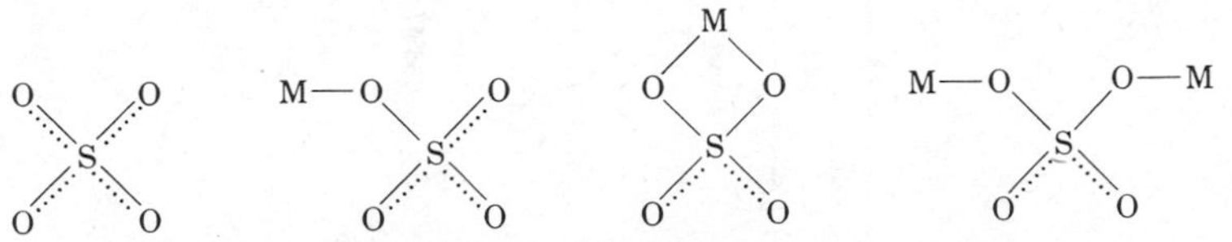

Free ion (T_d) | Unidentate complex (C_{3v}) | Bidentate complex (C_{2v}) | Bridged bidentate complex (C_{2v})

The change in the selection rules caused by the lowering of symmetry was shown in Table II-6*f*. Table III-23 and Fig. III-31 give the frequencies and the spectra of typical Co(III) sulfato complexes obtained by Nakamoto et al.[294] In $[Co(NH_3)_6]_2(SO_4)_3{\cdot}5H_2O$, ν_3 and ν_4 do not split and ν_2 does not appear, although ν_1 is observed, it is very weak. We conclude, therefore, that

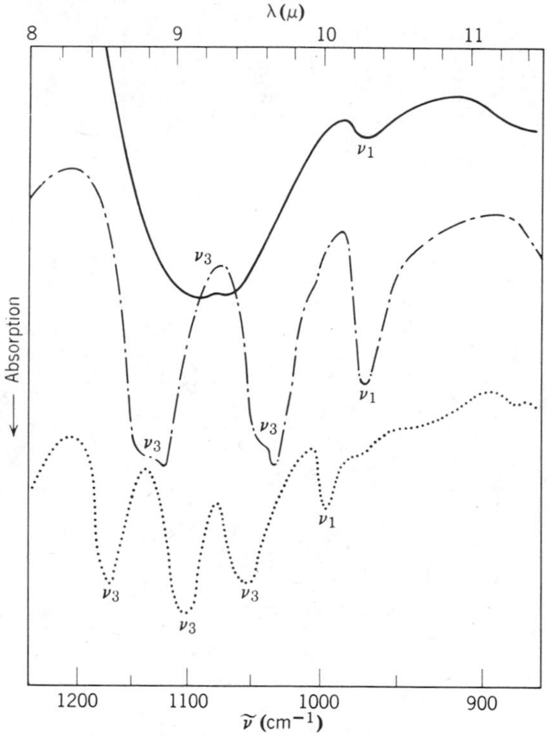

Fig. III-31. Infrared spectra of $[Co(NH_3)_6](SO_4)_3{\cdot}5H_2O$ (solid line); $[Co(NH_3)_5SO_4]Br$ (dot-dash line), and

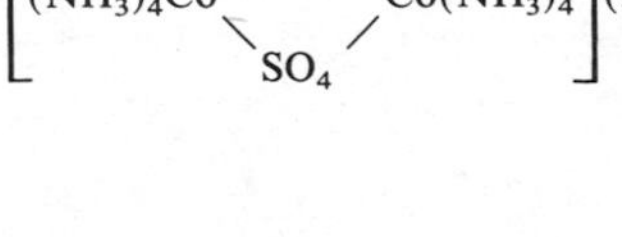

$(NO_3)_3$ (dotted line).[294]

TABLE III-23. VIBRATIONAL FREQUENCIES OF Co(III) SULFATO COMPLEXES (CM^{-1})[294]

Compound	Symmetry	ν_1	ν_2	ν_3	ν_4
Free SO_4^{2-} ion	$\mathbf{T}_d$	—	—	1104 (vs)[a]	613 (s)
$[Co(NH_3)_6]_2(SO_4)_3\cdot 5H_2O$	$\mathbf{T}_d$	973 (vw)	—	1130–1140 (vs)	617 (s)
$[Co(NH_3)_5SO_4]Br$	$\mathbf{C}_{3v}$	970 (m)	438 (m)	1032–1044 (s) 1117–1143 (s)	645 (s) 604 (s)
$\left[(NH_3)_4Co\begin{matrix} NH_2 \\ SO_4 \end{matrix}Co(NH_3)_4\right][NO_3]_3$	$\mathbf{C}_{2v}$	995 (m)	462 (m)	1050–1060 (s) 1170 (s) 1105 (s)	641 (s) 610 (s) 571 (m)

[a] vs = very strong; s = strong; m = medium; vw = very weak.

the symmetry of the SO_4^{2-} ion is approximately T_d. In $[Co(NH_3)_5SO_4]Br$, both ν_1 and ν_2 appear with medium intensity; moreover, ν_3 and ν_4 each splits into two bands. This result can be explained by assuming a lowering of symmetry from T_d to C_{3v} (unidentate coordination). In

$$\left[(NH_3)_4Co \overset{\displaystyle NH_2}{\underset{\displaystyle SO_4}{\diagup \diagdown}} Co(NH_3)_4 \right](NO_3)_3$$

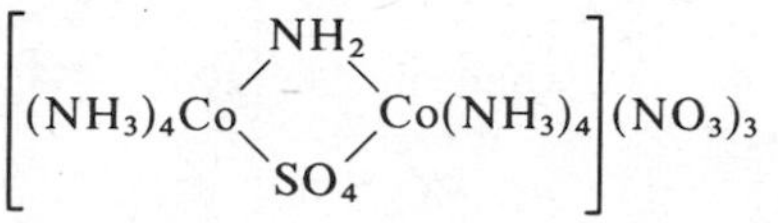

both ν_1 and ν_2 appear with medium intensity, and ν_3 and ν_4 each splits into three bands. These results suggest that the symmetry is further lowered and probably reduced to C_{2v}, as indicated in Table II-6f. Thus the SO_4^{2-} group in this complex is concluded to be a bridging bidentate as depicted in the foregoing diagram.

The chelating bidentate SO_4^{2-} group was discovered by Barraclough and Tobe,[295] who observed three bands (1211, 1176, and 1075 cm^{-1}) in the ν_3 region of $[Co(en)_2SO_4]Br$. These frequencies are higher than those of the bridging bidentate complex listed in Table III-23. Eskenazi et al.[296] also found the same trend in Pd(II) sulfato complexes. Thus the distinction between bridging and chelating sulfato complexes can be made on this basis. Table III-24 lists the observed frequencies of the sulfato groups and the modes of coordination as determined from the spectra.

The symmetries of the sulfate and nitrate ions of metal salts at various stages of hydration have been discussed on the basis of their infrared spectra.[303] Normal coordinate analyses on the sulfato, nitrato, and carbonato groups in Co(III) ammine complexes have been carried out by Tanaka et al.,[304] Goldsmith et al.[305] carried out normal coordinate analyses on the skeletons of $[Co(NH_3)_5X]^{2+}$ ($X = SO_4^{2-}$ and CO_3^{2-}) and cis-$[Co(NH_3)_4X_2]^{n+}$- ($X_2 = CO_3^{2-}$ and PO_4^{3-}) type complexes.

(2) Perchlorato (ClO_4) Complexes

In general, the perchlorate (ClO_4^-) ion coordinates to a metal when its complexes are prepared in nonaqueous solvents. The structure and bonding of metal complexes containing these weakly coordinating ligands have been reviewed briefly by Rosenthal.[306] Infrared and Raman spectroscopy has been used extensively to determine the mode of coordination of the ClO_4^- ligand. The structures listed in Table III-25 were determined on the basis of the same symmetry selection rules as are used for sulfato complexes. Chausse et al.[312] concluded from their IR and Raman study that $[Al(ClO_4)_n]^{-(n-3)}$ contain two unidentate and two bidentate for $n = 4$, four unidentate and one bidentate for $n = 5$, and six unidentate ligands for $n = 6$.

(3) Complexes of Other Tetrahedral Anions

Many tetrahedral anions coordinate to a metal as unidentate and bidentate ligands, and their modes of coordination have been determined by the same

TABLE III-24. VIBRATIONAL FREQUENCIES AND MODES OF COORDINATION OF VARIOUS SULFATO COMPLEXES (CM^{-1})

Compound	Mode of Coordination	ν_1	ν_2	ν_3	ν_4	Ref.
$[Cr(H_2O)_5SO_4]Cl \cdot \frac{1}{2}H_2O$	Unidentate	1002	—	1118 1068	—	297
$[Cu(bipy)SO_4] \cdot 2H_2O$ (polymeric)	Bridging bidentate	971	—	1163 1096 1053–1035	—	298
$Ni(morpholine)_2SO_4$ (polymeric)	Bridging bidentate	973	493	1177 1094 1042	628 612 593	299
$[Co_2\{(SO_4)_2OH\}(NH_3)_6]Cl$	Bridging bidentate	966	—	1180 1101 1048	645 598	300
$Pd(NH_3)_2SO_4$	Bridging bidentate	960	—	1195 1110 1035	—	296
$Pd(phen)SO_4$	Chelating bidentate	955	—	1240 1125 1040–1015	—	296
$Pd(PPh_3)_2SO_4$	Chelating bidentate	920	—	1265 1155 1110	—	301
$Ir(PPh_3)_2(CO)I(SO_4)$	Chelating bidentate	856	549	1296 1172 880	662 610	302

method as is used for SO_4^{2-} and ClO_4^- ions. For example, the PO_4^{3-} ion is a unidentate in $[Co(NH_3)_5PO_4]$ and a bidentate in $[Co(NH_3)_4PO_4]$.[313] Similar pairs have been made with the AsO_4^{3-},[314] CrO_4^{2-}, and MoO_4^{2-} ions.[315] The SeO_4^{2-} ion in $[Co(NH_3)_5SeO_4]Cl$ is a unidentate,[316] whereas it is a bridging bidentate in $[Co_2\{(SeO_4)_2OH\}$-$(NH_3)_6]Cl$.[300]

The $S_2O_3^{2-}$ ion can coordinate to a metal in a variety of ways. According to Freedman and Straughan,[317] $\nu_a(SO_3)$ near 1130 cm^{-1} is most useful as a structural diagnosis: >1175 (S-bridging); 1175–1130 (S-coordination); ~1130 (ionic $S_2O_3^{2-}$); <1130 cm^{-1} (O-coordination). On the basis of this criterion, they proposed polymeric structures linked by O-bridges for thiosulfates of UO_2^{2+} and ZrO_2^{2+}. The infrared (and Raman) spectra show that the SO_3F^- ion is a unidentate in $[Sn(SO_3F)_6]^{2-}$ [318] but is a unidentate as well as a bidentate in $VO(SO_3F)_3$.[319]

(4) Carbonato(CO_3) Complexes

The unidentate and bidentate(chelating) coordinations shown below are found in the majority of carbonato complexes.

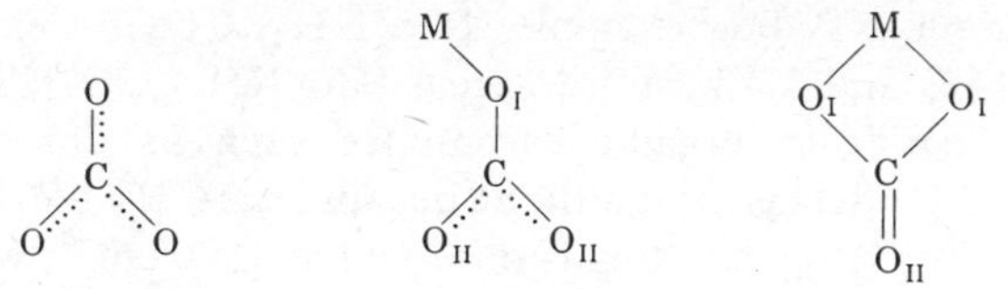

Free ion (D_{3h}) Unidentate (C_s) Bidentate (C_{2v})

The selection rule changes as shown in Table I-11. In $\mathbf{C}_{2v}$ and $\mathbf{C}_s$,* the ν_1 vibration, which is forbidden in the free ion, becomes infrared active and each of the doubly degenerate vibrations, ν_3 and ν_4, splits into two bands. Although the number of infrared-active fundamentals is the same for $\mathbf{C}_{2v}$ and $\mathbf{C}_s$, the splitting of the degenerate vibrations is larger in the bidentate than in the

TABLE III-25. ClO STRETCHING FREQUENCIES OF PERCHLORATO COMPLEXES (CM^{-1})

Complex	Structure	ν_3	ν_4	Ref.
$K[ClO_4]$	Ionic	1170–1050	(935)[a]	
$Cu(ClO_4)_2 \cdot 6H_2O$	Ionic	1160–1085	(947)[a]	307
$Cu(ClO_4)_2 \cdot 2H_2O$	Unidentate	1158 1030	920	307
$Cu(ClO_4)_2$	Bidentate	1270–1245 1130 948–920	1030	307
$Mn(ClO_4)_2 \cdot 2H_2O$	Bidentate	1210 1138 945	1030	308
$Co(ClO_4)_2 \cdot 2H_2O$	Bidentate	1208 1125 935	1025	308
$[Ni(en)_2(ClO_4)_2]$[b]	Bidentate	1130 1093 1058	962	309
$Ni(CH_3CN)_4(ClO_4)_2$	Unidentate	1135 1012	912	310
$Ni(CH_3CN)_2(ClO_4)_2$	Bidentate	1195 1106 1000	920	310
$[Ni(4\text{-}Me\text{-}py)_4](ClO_4)_2$	Ionic	1040–1130	(931)[a]	311
$Ni(3\text{-}Br\text{-}py)_4(ClO_4)_2$	Unidentate	1165–1140 1025	920	311

[a] Weak.

[b] Blue form.

* The symmetry of the unidentate carbonato group is $\mathbf{C}_{2v}$ if the metal atom is ignored.

unidentate complex.[294] For example, $[Co(NH_3)_5CO_3]Br$ exhibits two CO stretchings at 1453 and 1373 cm^{-1}, whereas $[Co(NH_3)_4CO_3]Cl$ shows them at 1593 and 1265 cm^{-1}. In organic carbonates such as dimethyl carbonate, $(CH_3O_I)_2CO_{II}$, this effect is more striking because the $CH_3—O_I$ bond is strongly covalent. Thus the CO_{II} stretching is observed at 1870 cm^{-1}, whereas the CO_I stretching is at 1260 cm^{-1}. Gatehouse and co-workers[320] showed that the separation of the CO stretching bands increases along the following series:

basic salt < carbonato complex < acid < organic carbonate

Fujita et al.[321] carried out normal coordinate analysis on unidentate and bidentate carbonato complexes of Co(III). According to their results the CO stretching force constant, which is 5.46 for the free ion, becomes 6.0 for the $C\text{-}O_{II}$ bonds and 5.0 for the $C\text{-}O_I$ bond of the unidentate complex, whereas it becomes 8.5 for the $C\text{-}O_{II}$ bond and 4.1 for the $C\text{-}O_I$ bonds of the bidentate complex (all are UBF force constants in units of mdyn/Å). The observed and calculated frequencies and theoretical band assignments are shown in Table III-26. Normal coordinate analyses on carbonato complexes have also been carried out by Hester and Grossman[322] and Goldsmith and Ross.[323]

As is shown in Table III-26, normal coordinate analysis predicts that the highest-frequency CO stretching band belongs to the B_2 species in the unidentate and the A_1 species in the bidentate complex. Elliott and Hathaway[324] studied the polarized infrared spectra of single crystals of $[Co(NH_3)_4CO_3]Br$ and confirmed these symmetry properties. As will be shown later for nitrato complexes, Raman polarization studies are also useful for this purpose.

According to X-ray analysis, the carbonato groups in $[(NH_3)_3Co(\mu\text{-}OH)_2(\mu\text{-}CO_3)Co(NH_3)_3]SO_4\cdot 5H_2O$[325] and [(teed)CuCl(CO$_3$)CuCl(teed)] (teed: N,N,N′,N′-tetraethyl-ethylenediamine)[326] take the bridging and tridentate(bridging) structures, respectively.

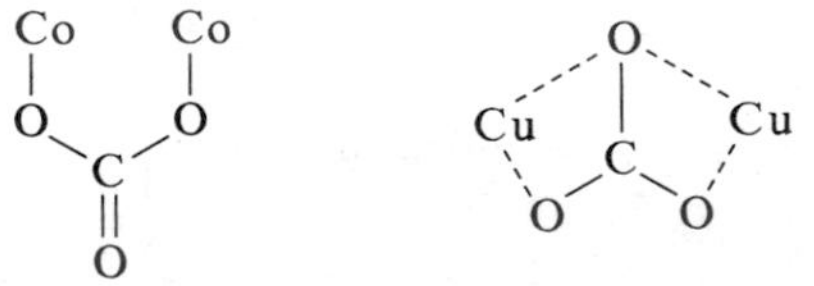

No simple criteria have been established to distinguish these structures from common unidentate and bidentate (chelating) coordination based on vibrational frequencies. Finally, the IR spectrum of K_2CO_3 in a N_2 matrix indicates that the CO_3 group coordinates in a bidentate fashion to one of the K atom and in a unidentate fashion to the other K atom.[327]

(5) Nitrato (NO_3) Complexes

The structures and vibrational spectra of a large number of nitrato complexes have been reviewed by Addison et al.[328] and Rosenthal.[306] X-ray analyses show that the NO_3^- ion coordinates to a metal as a unidentate, symmetric and asymmetric chelating bidentate, and bridging bidentate ligand of various

TABLE III-26. CALCULATED AND OBSERVED FREQUENCIES OF UNIDENTATE AND BIDENTATE CO(III) CARBONATO COMPLEXES (CM^{-1})[321]

Species ($\mathbf{C}_{2v}$)[a]	$\nu_1(A_1)$	$\nu_2(A_1)$	$\nu_3(A_1)$	$\nu_4(A_1)$	$\nu_5(B_2)$	$\nu_6(B_2)$	$\nu_7(B_2)$	$\nu_8(B_1)$
Calculated frequency	1376	1069	772	303	1482	676	92	—
Assignment	$\nu(CO_{II}) + \nu(CO_I)$	$\nu(CO_I) + \nu(CO_{II})$	$\delta(O_{II}CO_{II})$	$\nu(CoO_I)$	$\nu(CO_{II})$	$\rho_r(O_{II}CO_{II})$	$\delta(CoO_IC)$	π
$[Co(NH_3)_5CO_3]Br$	1373	1070	756	362	1453	678	—	850
$[Co(ND_3)_5CO_3]Br$	1369	1072	751	351	1471	687	—	854
$[Co(NH_3)_5CO_3]I$	1366	1065	776	360	1449	679	—	850
$[Co(ND_3)_5CO_3]I$	1360	1063	742	341	1467	687	—	853

Species ($\mathbf{C}_{2v}$)	$\nu_1(A_1)$	$\nu_2(A_1)$	$\nu_3(A_1)$	$\nu_4(A_1)$	$\nu_5(B_2)$	$\nu_6(B_2)$	$\nu_7(B_2)$	$\nu_8(B_1)$
Calculated frequency	1595	1038	771	370	1282	669	429	—
Assignment	$\nu(CO_{II})$	$\nu(CO_I)$	Ring def. $+ \nu(CoO_I)$	$\nu(CoO_I)$ + ring def.	$\nu(CO_I) + \delta(O_ICO_{II})$	$\delta(O_ICO_{II}) + \nu(CO_I) + \nu(CoO_I)$	$\nu(CoO_I)$	π
$[Co(NH_3)_4CO_3]Cl$	1593	1030	760	395	1265	673	430	834
$[Co(ND_3)_4CO_3]Cl$	1635, 1607	(1031)[b]	753	378	1268	672	418	832
$[Co(NH_3)_4CO_3]ClO_4$	1602	—[c]	762	392	1284	672	428	836
$[Co(ND_3)_4CO_3]ClO_4$	1603	—[c]	765	374	1292	676	415	835

[a] Symmetry assuming a linear Co–O–C bond (see Ref. 321).
[b] Overlapped with $\delta_s(ND_3)$.
[c] Hidden by $[ClO_4]^-$ absorption.

structures. It is rather difficult to differentiate these structures by vibrational spectroscopy since the symmetry of the nitrate ion differs very little among them ($\mathbf{C}_{2v}$ or $\mathbf{C}_s$). Even so, vibrational spectroscopy is still useful in distinguishing unidentate and bidentate ligands.

Originally, Gatehouse et al.[329] noted that the unidentate NO_3 group exhibits three NO stretching bands, as expected for its $\mathbf{C}_{2v}$ symmetry. For example, $[Ni(en)_2(NO_3)_2]$ (unidentate) exhibits three bands as follows:

$\nu_5(B_2)$	1420 cm^{-1}	$\nu_a(NO_2)$
$\nu_1(A_1)$	1305 cm^{-1}	$\nu_s(NO_2)$
$\nu_2(A_1)$	(1008) cm^{-1}	$\nu(NO)$

whereas $[Ni(en)_2NO_3]ClO_4$ (chelating bidentate) exhibits three bands at the following:

$\nu_1(A_1)$	1476 cm^{-1}	$\nu(N{=}O)$
$\nu_5(B_2)$	1290 cm^{-1}	$\nu_a(NO_2)$
$\nu_2(A_1)$	(1025) cm^{-1}	$\nu_s(NO_2)$

The separation of the two highest-frequency bands is 115 cm^{-1} for the unidentate complex, whereas it is 186 cm^{-1} for the bidentate complex. Thus Curtis and Curtis[330] concluded that $[Ni(dien)(NO_3)_2]$ contains both types, since it exhibits bands due to unidentate (1440 and 1315 cm^{-1}) and bidentate (1480 and 1300 cm^{-1}) groups. In general, the separation of the two highest-frequency bands is larger for bidentate than for unidentate coordination if the complexes are similar. However, this rule does not hold if the complexes are markedly different. This is clearly shown in Table III-27.

Lever et al.[340] proposed using the combination band, $\nu_1 + \nu_4$, of free NO_3^- which appears in the 1800–1700 cm^{-1} region for structural diagnosis. Upon coordination, ν_4 (E', in-plane bending) near 700 cm^{-1} splits into two bands, and the magnitude of this splitting is expected to be larger for bidentate than for unidentate ligands. This should be reflected in the separation of two ($\nu_1 + \nu_4$) bands in the 1800–1700 cm^{-1} region. According to Lever et al., the NO_3^- ion is bidentate if the separation is ca. 66–20 cm^{-1} and is unidentate if it is ca. 26–5 cm^{-1}.

As stated previously, the highest-frequency CO stretching band of the carbonato complexes belongs to the A_1 species in the bidentate and to the B_2 species in the unidentate complex. The same holds true for the nitrato complex. Ferraro et al.[341] showed that all the nitrato groups in $Th(NO_3)_4(TBP)_2$ coordinate to the metal as bidentate ligands since the Raman band at 1550 cm^{-1} is polarized (TBP: tributylphosphate). This rule holds very well for other compounds.[342] According to Addison et al.,[328] the intensity pattern of the three NO stretching bands in the Raman spectrum can also be used to distinguish unidentate and symmetrical bidentate NO_3 ligands. The middle band is very strong in the former, whereas it is rather weak in the latter.

TABLE III-27. NO STRETCHING FREQUENCIES OF UNIDENTATE AND BIDENTATE NITRATO COMPLEXES (cm^{-1})

Compound	Mode of Coordination	ν_5	ν_1	ν_2	$\nu_5 - \nu_1$	Ref.
$Re(CO)_5NO_3$	Unidentate	1497	1271	992	226	331
cis-$[Pt(NH_3)_2(NO_3)_2]$	Unidentate	1510	1275	997	235	332
$Sn(NO_3)_4$	Chelating bidentate	1630	1255	983	375	333
$K[UO_2(NO_3)_3]$	Chelating bidentate	1555 1521	1271	1025	284 250	334
$Co(NO_3)_3$	Chelating bidentate	1619	1162	963	457	335
$Na_2[Mn(NO_3)_4]$	Chelating bidentate	1490	1280	1041 1036	210	336
$Cu(NO_3)_2MeNO_2$	Bridging bidentate	1519	1291	1008	228	337
$Zn(bt)_2(NO_3)_2$ [a]	Chelating bidentate	1485	1300	—	185	338
$Ni(dmpy)_2(NO_3)_2$ [b]	Chelating bidentate	1513	1270	1013	243	339

[a] bt = benzothiazole.
[b] dmpy = 2,6-dimethyl-4-pyrone.

The use of far-infrared spectra to distinguish unidentate and bidentate nitrato coordination has been controversial. Nuttall and Taylor[343] suggested that unidentate and bidentate complexes exhibit one and two MO stretching bands, respectively, in the 350–250 cm^{-1} region. Bullock and Parrett[344] showed, however, that such a simple rule is not applicable to many known nitrato complexes. Ferraro and Walker[345] assigned the MO stretching bands of anhydrous metal nitrates such as $Cu(NO_3)_2$ and $Pr(NO_3)_3$.

Several workers studied the Raman spectra of metal nitrates in aqueous solution and molten states. For example, Irish and Walrafen[346] found that E' mode degeneracy is removed even in dilute solutions of $Ca(NO_3)_2$. This, combined with the appearance of the A_1' mode in the infrared, suggests $\mathbf{C}_{2v}$ symmetry of the NO_3^- ion. Hester and Krishnan[347] studied the Raman spectra of $Ca(NO_3)_2$ dissolved in molten KNO_3 and $NaNO_3$. Their results suggest an asymmetric perturbation of the NO_3^- ion by the Ca^{2+} ion through ion-pair formation.

(6) Sulfito (SO_3), Selenito (SeO_3), and Sulfinato (RSO_2) Complexes

The pyramidal sulfite (SO_3^{2-}) ion may coordinate to a metal as a unidentate, bidentate, or bridging ligand. The following two structures are probable for unidentate coordination:

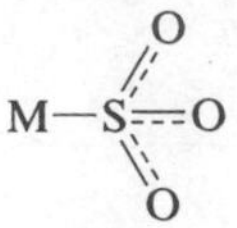

C_{3v}

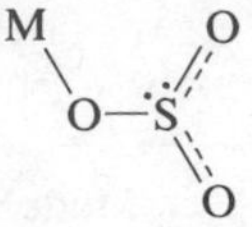

C_s

If coordination occurs through sulfur, the $\mathbf{C}_{3v}$ symmetry of the free ion will be preserved. If coordination occurs through oxygen, the symmetry may be lowered to $\mathbf{C}_s$. In this case, the doubly degenerate vibrations of the free ion will split into two bands. It is anticipated[348] that coordination through sulfur will shift the SO stretching bands to higher frequencies, whereas coordination through oxygen will shift them to lower frequencies, than those of the free ion. On the basis of these criteria, Newman and Powell[349] showed that the sulfito groups in $K_6[Pt(SO_3)_4]\cdot 2H_2O$ and $[Co(NH_3)_5(SO_3)]Cl$ are S-bonded and those in $Tl_2[Cu(SO_3)_2]$ are O-bonded. Baldwin[350] suggested that the sulfito groups in *cis*- and *trans*-$Na[Co(en)_2(SO_3)_2]$ and $[Co(en)_2(SO_3)X]$ (X=Cl or OH) are S-bonded, since they show only two SO stretchings between 1120 and 930 cm^{-1}. According to Nyberg and Larsson,[351] the appearance of a strong SO stretching band above 975 and below 960 cm^{-1} is an indication of S- and O-coordination, respectively. Table III-28 lists typical results obtained for unidentate complexes.

The structures of complexes containing bidentate sulfito groups are rather difficult to deduce from their infrared spectra. Bidentate sulfito groups may be chelating or bridging through either oxygen or sulfur or both, all resulting in $\mathbf{C}_s$ symmetry. Baldwin[350] prepared a series of complexes of the type $[Co(en)_2(SO_3)]X$ (X = Cl, I, or SCN), which are monomeric in aqueous solution. They show four strong bands in the SO stretching region (one of them may be an overtone or a combination band). She suggests a chelating structure in which two oxygens of the sulfito group coordinate to the Co(III) atom. Newman and Powell[349] obtained the infrared spectra of $K_2[Pt(SO_3)_2]\cdot 2H_2O$, $K_3[Rh(SO_3)_3]\cdot 2H_2O$, and other complexes for which

TABLE III-28. INFRARED SPECTRA OF UNIDENTATE SULFITO COMPLEXES (CM^{-1})

Complex	Structure	$\nu_3(E)$	$\nu_1(A_1)$	$\nu_2(A_1)$	$\nu_4(E)$	Ref.
Free SO_3^{2-}	—	933	967	620	469	
$K_6[Pt(SO_3)_4]\cdot 2H_2O$	S-bonded	1082–1057	964	660	540	349
$[Co(NH_3)_5(SO_3)]Cl$	S-bonded	1110	985	633	519	349
trans-$Na[Co(en)_2(SO_3)_2]$	S-bonded	1068	939	630	—	350
$[Co(en)_2(SO_3)Cl]$	S-bonded	1117–1075	984	625	—	350
$Tl_2[Cu(SO_3)_2]$	O-bonded	902, 862	989	673	506, 460	349
$(NH_4)_9[Fe(SO_3)_6]$	O-bonded	943	815	638	520	352

bidentate coordination of the sulfito group is expected. It was not possible, however, to determine their structures from infrared spectra alone.

The mode of coordination of the selenite ion(SeO_3^{2-}) is similar to that of the sulfite ion. Two types of unidentate complexes are expected. The O-coordinated complex exhibits $\nu_3(E)$ and $\nu_1(A_1)$ at 755 and 805 cm^{-1}, respectively, for $[Co(NH_3)_5(SeO_3)]Br \cdot H_2O$[353] whereas the Se-coordinated complex, $[Co(NH_3)_5(SeO_3)]ClO_4$,[354] shows them at 823 and 860 cm^{-1}, respectively.

Four types of coordination are probable for sulfinato (RSO_2^-, R = CH_3, CF_3, Ph, etc.) groups:

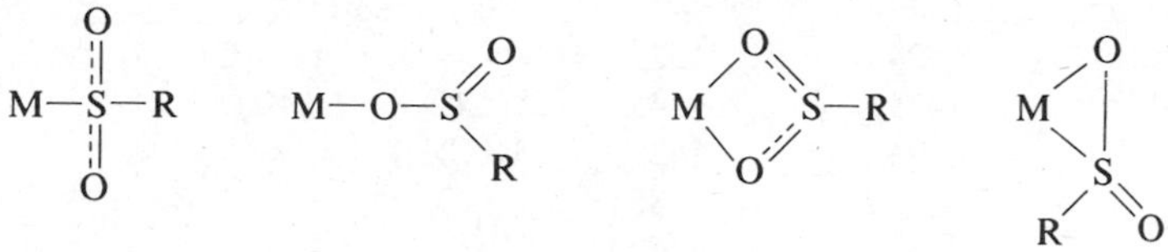

The SO stretching bands at 1200–850 cm^{-1} are useful in distinguishing these structures.[355,356]

III-12. COMPLEXES OF β-DIKETONES

(1) Complexes of Acetylacetonato Ion

A number of β-diketones form metal chelate rings of Type A:

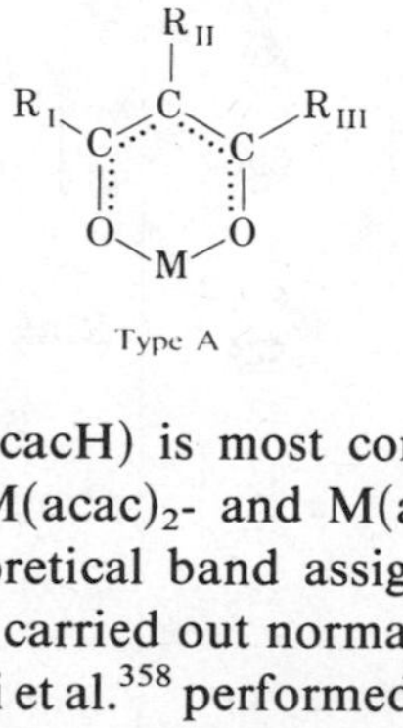

Type A

Among them, acetylacetone (acacH) is most common ($R_I = R_{III} = CH_3$ and $R_{II} = H$). Infrared spectra of $M(acac)_2$- and $M(acac)_3$-type complexes have been studied extensively. Theoretical band assignments were first made by Nakamoto and Martell,[357] who carried out normal coordinate analysis on the 1:1 model of $Cu(acac)_2$. Mikami et al.[358] performed normal coordinate analyses on the 1:2 (square-planar) and 1:3 (octahedral) models of various acac complexes. Figure III-32 shows the infrared spectra of six acac complexes, and Table III-29 lists the observed frequencies and band assignments for the Cu(II), Pd(II), and Fe(III) complexes obtained by Mikami et al. In this table, the 1577- and 1529-cm^{-1} bands of $Cu(acac)_2$ are assigned to $\nu(C \cdots C)$ coupled with $\nu(C \cdots O)$ and $\nu(C \cdots O)$ coupled with $\nu(C \cdots C)$, respectively. Junge and Musso[359] have measured the ^{13}C and ^{18}O isotope shifts of these bands and concluded that the above assignments must be reversed.

The $\nu(MO)$ of acac complexes are most interesting since they provide direct information about the M–O bond strength. Using the metal isotope technique, Nakamoto et al.[360] assigned the MO stretching bands of acetylacetonato

TABLE III-29. OBSERVED FREQUENCIES AND BAND ASSIGNMENTS OF ACETYLACETONATO COMPLEXES (cm^{-1})[358]

$Cu(acac)_2$	$Pd(acac)_2$	$Fe(acac)_3$	Predominant Mode
3072	3070	3062	$\nu(CH)$
2987, 2969, 2920	2990, 2965, 2920	2895, 2965, 2920	$\nu(CH_3)$
1577	1569	1570	$\nu(C\!\cdots\!C)+\nu(C\!\cdots\!O)$
1552	1549	—	combination
1529	1524	1525	$\nu(C\!\cdots\!O)+\nu(C\!\cdots\!C)$
1461	(1425)	1445	$\delta(CH)+\nu(C\!\cdots\!C)$
1413	1394	1425	$\delta_d(CH_3)$
1353	1358	1385, 1360	$\delta_s(CH_3)$
1274	1272	1274	$\nu(C-CH_3)+\nu(C\!\cdots\!C)$
1189	1199	1188	$\delta(CH)+\nu(C-CH_3)$
1019	1022	1022	$\rho_r(CH_3)$
936	937	930	$\nu(C\!\cdots\!C)+\nu(C\!\cdots\!O)$
780	786, 779	801, 780, 771	$\pi(CH)$
684	700	670, 664[a]	$\nu(C-CH_3)+$ring deformation $+\nu(MO)$
653	678	656	$\pi\left(CH_3-C\begin{smallmatrix}C\\O\end{smallmatrix}\right)$
612	661	559, 548[a]	Ring deformation $+\nu(MO)$
451	463	433	$\nu(MO)+\nu(C-CH_3)$
431	441	415, 408	Ring deformation
291	294	298	$\nu(MO)$
1.45	1.85	1.30	$K(M-O)$ (mdyn/Å)(UBF)

[a] Pure ring deformation.

complexes at the following frequencies (cm^{-1}):

$Cr(acac)_3$	$Fe(acac)_3$	$Pd(acac)_2$	$Cu(acac)_2$	$Ni(acac)_2(py)_2$
463.4	436.0	466.8	455.0	438.0
358.4	300.5	297.1	290.5	270.8
		265.9		

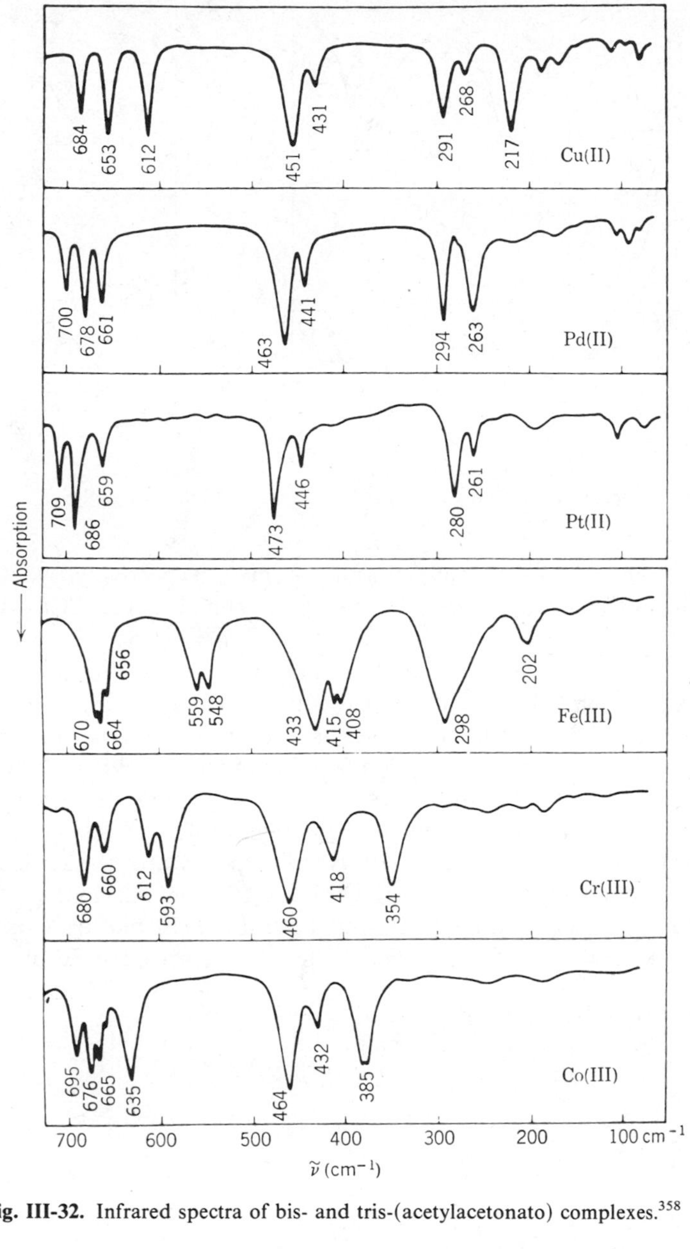

Fig. III-32. Infrared spectra of bis- and tris-(acetylacetonato) complexes.[358]

Both normal coordinate calculations and isotope shift studies show that the bands near 450 cm^{-1} are coupled with the C—CH$_3$ bending mode whereas those in the low-frequency region are relatively pure MO stretching vibrations. Figure III-33 shows the actual tracings of the infrared spectra of ^{50}Cr(acac)$_3$ and its ^{53}Cr analog. It is seen that two bands at 463.4 and 358.4 cm^{-1} of the

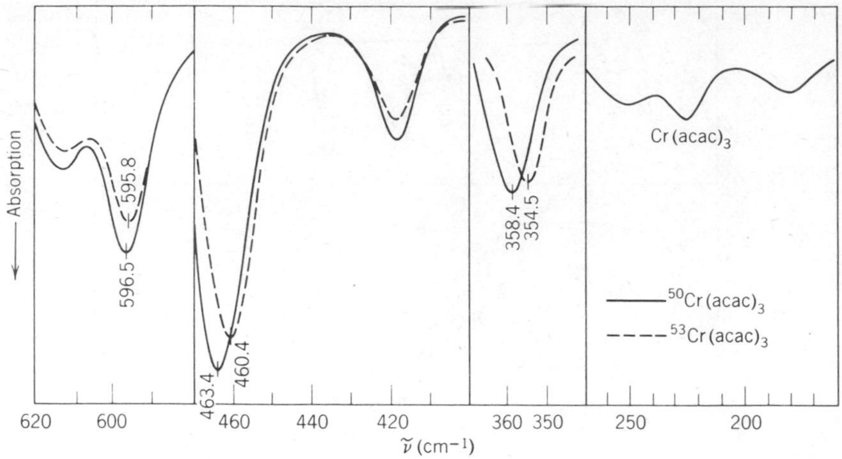

Fig. III-33. Infrared spectra of $^{50}Cr(acac)_3$ and its ^{53}Cr analog.[360]

former give negative shifts of 3.0 and 3.9 cm^{-1}, respectively, whereas other bands (ligand vibrations) produce negligible shifts by the ^{50}Cr–^{53}Cr substitution.

Complexes of the $M(acac)_2X_2$-type may take the *cis* or *trans* structure. Although steric and electrostatic considerations would favor the *trans*-isomer, the greater stability of the *cis*-isomer is expected in terms of metal–ligand π-bonding. This is the case for $Ti(acac)_2F_2$, which is "*cis*" with two ν(TiF) at 633 and 618 cm^{-1}.[361] In the case of $Re(acac)_2Cl_2$, however, both forms can be isolated; the *trans*-isomer exhibits ν(ReO) and ν(ReCl) at 464 and 309 cm^{-1}, respectively, while each of these bands splits into two in the *cis*-isomer [472 and 460 cm^{-1} for ν(ReO) and 346 and 333 cm^{-1} for ν(ReCl) in the infrared].[362] For $VO(acac)_2L$, where L is a substituted pyridine, *cis*- and *trans*-isomers are expected. According to Caira et al.,[363] these structures can be distinguished

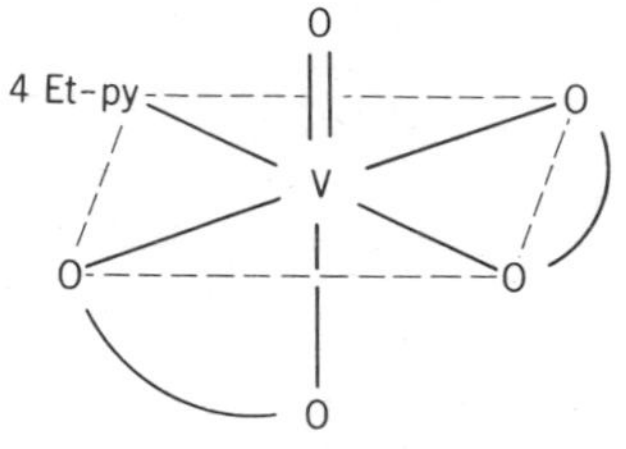

cis-isomer

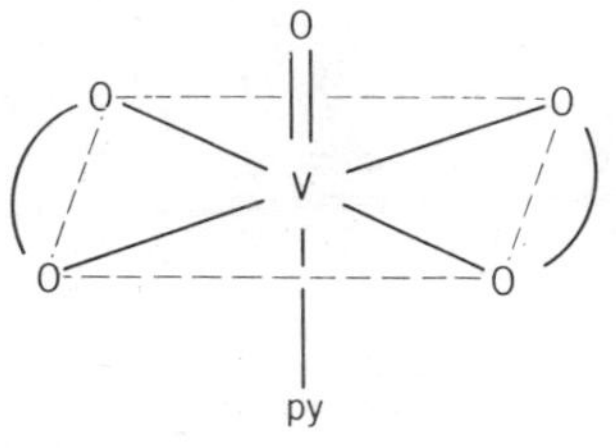

trans-isomer

by their infrared spectra. The ν(V=O) and ν(V—O) of the *cis*-isomer are lower than those of the *trans*-isomer. For example, ν(V=O) of $VO(acac)_2$ is 999 cm^{-1}, and this band shifts to 959 cm^{-1} for 4-Et-py (*cis*) and to 973 cm^{-1} for py (*trans*). Furthermore, ν(V—O) of the *cis*-isomer splits into two bands.

Vibrational spectra of acetylacetonato complexes have been studied by many other workers. Only references are cited for the following: Raman spectra of tris(acac) complexes,[288] infrared spectra of acac complexes of rare-earth metals,[364] relationships between CH and CH_3 stretching frequencies and ^{13}C—H spin-spin coupling constants,[365] and relationships between $\nu(C \cdots O)$, $\nu(C \cdots C)$, and ^{13}C NMR shifts of CO groups.[366]

According to X-ray analysis,[367] the hexafluoroacetylacetonato ion (hfa) in $[Cu(hfa)_2\{MeN\text{-}(CH_2)_2\text{-}NH_2\}_2]$ coordinates to the metal as a unidentate via one of its O atoms. This compound exhibits $\nu(C{=}O)$ at 1675 and 1615 cm^{-1}, values slightly higher than those for $Cu(hfa)_2$, in which the hfa ion is chelated to the metal (1644 and 1614 cm^{-1}).

The following dimeric bridging structure has been proposed for $[CoBr(acac)]_2$:

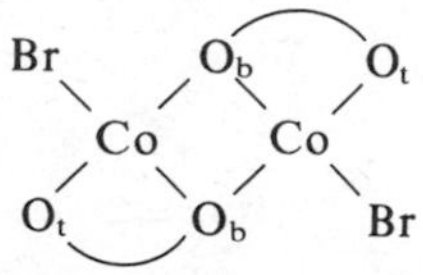

$\nu(CoO_t)$ and $\nu(CoO_b)$ were assigned at 435 and 260 cm^{-1}, respectively.[368] In $[Ni(acac)_2]_3$ and $[Co(acac)_4]_4$, the O atoms of the acac ion serve as a bridge between two metal atoms.[369] However, no band assignments are available on these polymeric species.

(2) Complexes of Neutral Acetylacetone

In some compounds, the keto form of acetylacetone forms a chelate ring of type B:

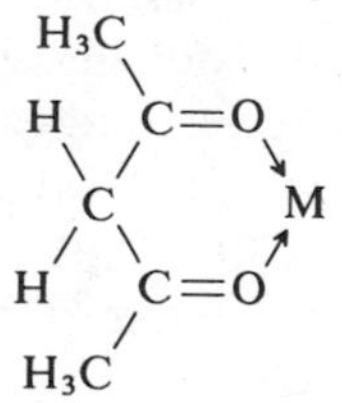

Type B

This particular type of coordination was found by van Leeuwen[370] in $[Ni(acacH)_3](ClO_4)_2$ and its derivatives, and by Nakamura and Kawaguchi[371] in $Co(acacH)Br_2$. These compounds were prepared in acidic or neutral media, and exhibit strong $\nu(C{=}O)$ bands near 1700 cm^{-1}. Similar ketonic coordination was proposed for $Ni(acacH)_2Br_2$[372] and $M(acacH)Cl_2$ (M = Co and Zn).[373].

According to X-ray analysis,[374] the acetylacetone molecule in $Mn(acacH)_2Br_2$ is in the enol form and is bonded to the metal as a unidentate via one of its O atoms:

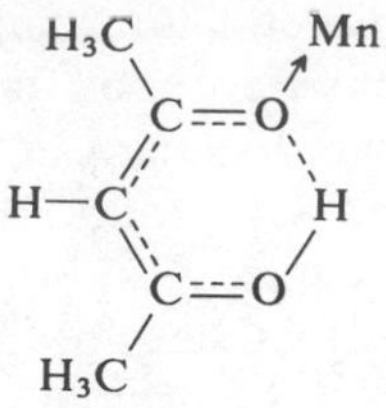

Type C

The C$\cdots$O and C$\cdots$C stretching bands of the enol ring were assigned at 1627 and 1564 cm^{-1}, respectively.

(3) C-bonded Acetylacetonato Complexes

Lewis and co-workers[375] reported the infrared and NMR spectra of a number of Pt(II) complexes in which the metal is bonded to the γ-carbon atom of the acetylacetonato ion:

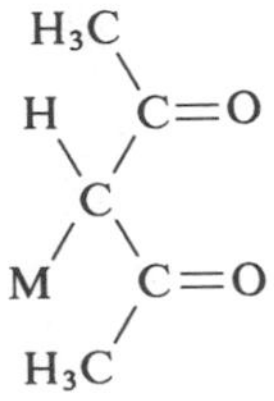

Type D

Behnke and Nakamoto carried out normal coordinate analysis on the $[Pt(acac)Cl_2]^-$ ion, in which the acac ion is chelated to the metal (Type A),[376] and on the $[Pt(acac)_2Cl_2]^{2-}$ ion, in which the acac ion is C-bonded to the metal (Type D).[377] Table III-30 lists the observed frequencies and band assignments for these two types, and Fig. III-34 shows the infrared spectra of these two compounds. The results indicate that (1) two ν(C=O) of Type D are higher than those of Type A, (2) two ν(C—C) of type D are lower than those of Type A, and (3) ν(PtC) of Type D is at 567 cm^{-1}, while ν(PtO) of Type A are at 650 and 478 cm^{-1}. Figure III-34 also shows that the structure of $K[Pt(acac)_2Cl]$ is as follows:

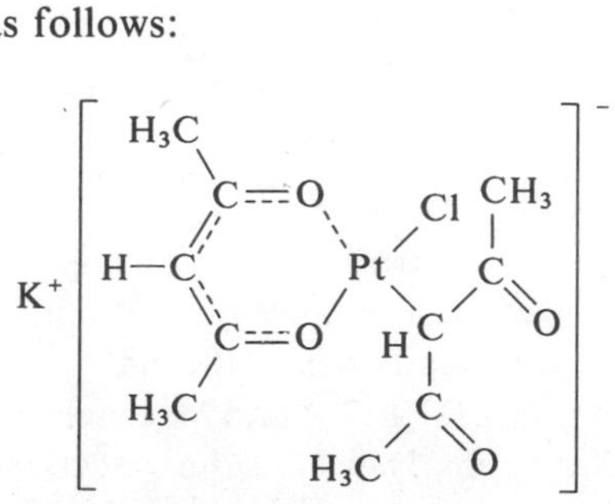

since its spectrum is roughly a superposition of those types A and D. Similarly,

TABLE III-30. OBSERVED FREQUENCIES, BAND ASSIGNMENTS, AND FORCE CONSTANTS FOR $K[Pt(acac)Cl_2]$ AND $Na_2[Pt(acac)_2Cl_2]\cdot 2H_2O$

$K[Pt(acac)Cl_2]$ (O-bonded, Type A)	$Na_2[Pt(acac)_2Cl_2]\cdot 2H_2O$ (C-bonded, Type D)	Band Assignment
—	1652, 1626	ν(C=O)
1563, 1380	—	ν(C⋯O)
1538, 1288	—	ν(C⋯C)
—	1350, 1193	ν(C—C)
1212, 817	1193, 852	δ(CH) or π(CH)
650, 478	—	ν(PtO)
—	567	ν(PtC)
K(C⋯O) = 6.50 K(C⋯C) = 5.23 K(C—CH$_3$) = 3.58 K(Pt—O) = 2.46 K(C—H) = 4.68 $\rho = 0.43^a$	K(C=O) = 8.84 K(C—C) = 2.52 K(C—CH$_3$) = 3.85 K(Pt—C) = 2.50 K(C—H) = 4.48	UBF constant (mdyn/Å)

[a] The stretching-stretching interaction constant (ρ) was used for Type A because of the presence of resonance in the chelate ring.

the infrared spectrum of $K[Pt(acac)_3)]$[375] is interpreted as a superposition of spectra of Types A, D, and D′, in which two C—O bonds are transoid.[378]

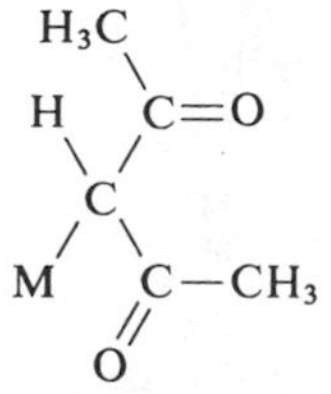

Type D′

The C-bonded acac ion was found in $Hg_2Cl_2(acac)$,[379] $Au(acac)(PPh_3)$,[380] and $Pd(acac)_2(PPh_3)$.[381] In the last compound, one acac group is Type A and the other Type D. In all these cases, the ν(C=O) of the Type-D acac groups are at 1700–1630 cm^{-1}.

As discussed above, $K[Pt(acac)_2Cl]$ contains one Type-A and one type-D acac group. If a solution of $K[Pt(acac)_2Cl]$ is acidified, its Type-D acac group is converted into Type-E:

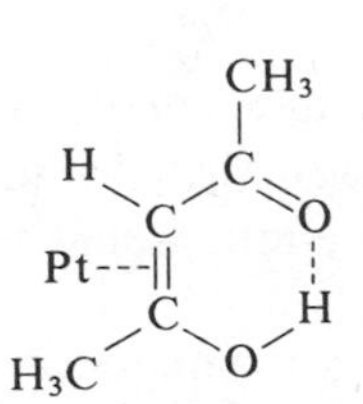

Type E

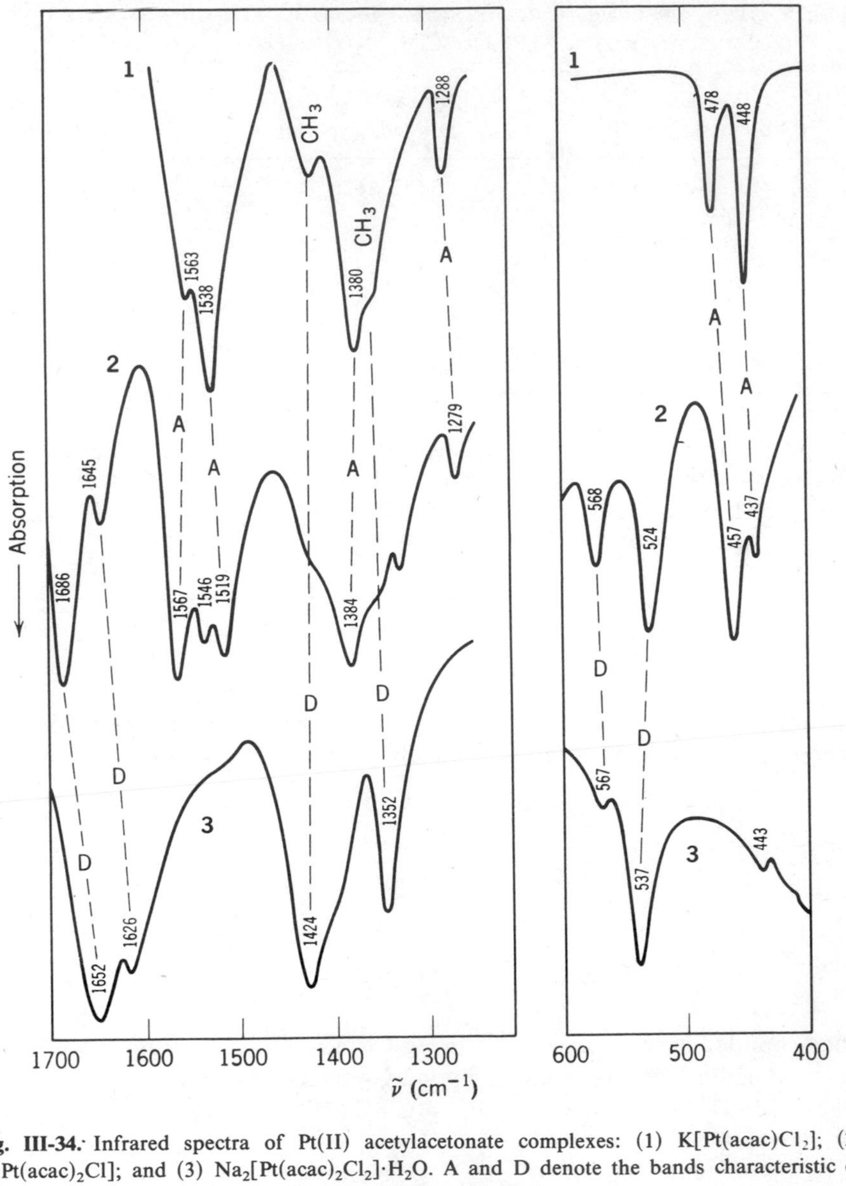

Fig. III-34. Infrared spectra of Pt(II) acetylacetonate complexes: (1) $K[Pt(acac)Cl_2]$; (2) $K[Pt(acac)_2Cl]$; and (3) $Na_2[Pt(acac)_2Cl_2]\cdot H_2O$. A and D denote the bands characteristic of Types A and D, respectively.

This structure was first suggested by Allen et al.,[382] based on NMR evidence. Behnke and Nakamoto[383] showed that the infrared spectrum of [Pt(acac)-(acacH)Cl] thus obtained can be interpreted as a superposition of spectra of Types A and E.

That the two O atoms of the C-bonded acac group (Type D) retain the ability to coordinate to a metal was first demonstrated by Lewis and Oldham,[384] who prepared neutral complexes of the following type:

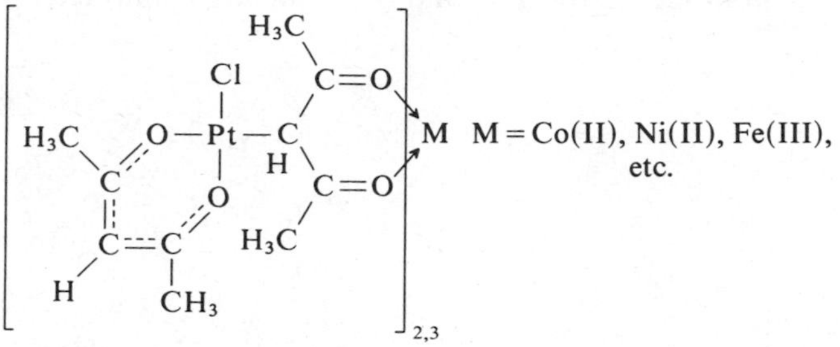

Using the metal isotope technique, Nakamura and Nakamoto[385] assigned the ν(NiO) of Ni[Pt(acac)$_2$Cl]$_2$ at 279 and 266 cm^{-1}. These values are relatively close to the ν(NiO) of Ni(acacH)$_2$Br$_2$ (264 and 239 cm^{-1}), discussed previously. Thus the newly formed Ni-acac ring retains its keto character and is close to Type B. Other types prepared by Kawaguchi and co-workers include the following:

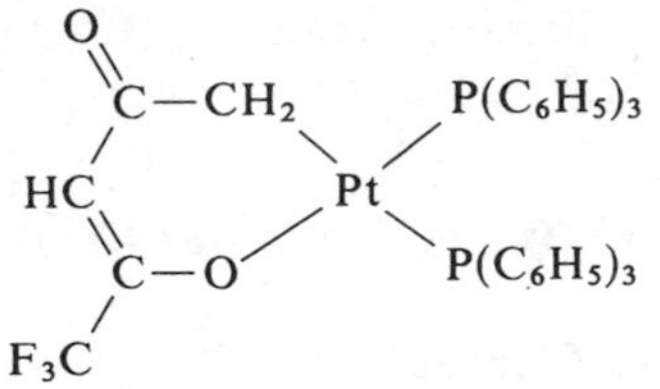

C,O-chelate[386]

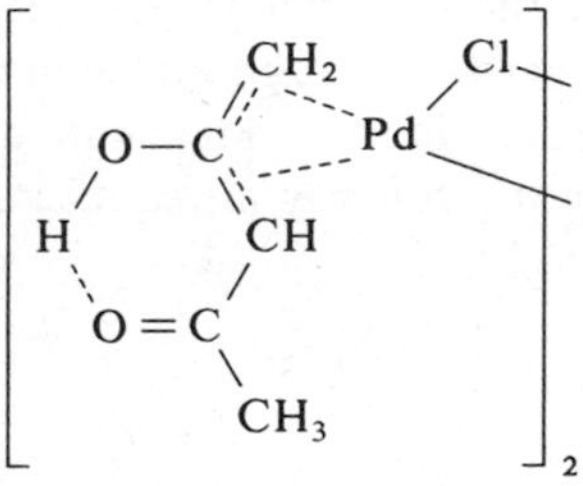

η^3-allylic[387]

(4) Complexes of Other β-Diketones

In a series of metal tropolonato complexes, Hulett and Thornton[388] noted a parallel relationship between the ν(MO) and the CFSE energy. These workers assigned the ν(MO) of trivalent metal tropolonates in the 660–580 cm^{-1} region, based on the ^{16}O–^{18}O isotope shifts observed for the Cu(II) complex.[389] Using the metal isotope technique, Hutchinson et al.[390] assigned the ν(MO) at the following frequencies (cm^{-1});

V(III)		Cr(III)		Mn(III)		Fe(III)		Co(III)
377	~	361	>	338	>	317	<	371
319	<	334	>	268	>	260	<	360

It was found that these frequencies still follow the order predicted by the CFSE.

2,4,6-Heptanetrione forms 1:1 and 1:2 (metal:ligand) complexes with Cu(II):[391]

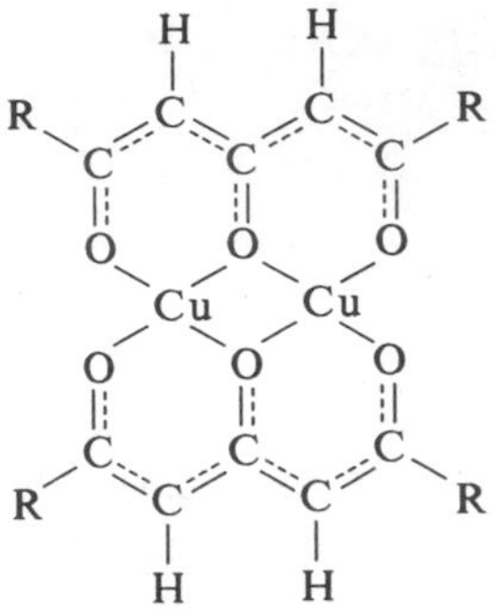

1:1 Complex   1:2 Complex

Both complexes exhibit multiple bands due to Type-A rings in the 1600–1500 cm^{-1} region. However, the 1:2 complex exhibits $\nu(C=O)$ of the uncoordinated C=O groups near 1720 cm^{-1}.

III-13 COMPLEXES OF UREA, SULFOXIDES, AND RELATED COMPOUNDS

(1) Complexes of Urea and Related Ligands

Penland et al.[392] first studied the infrared spectra of urea complexes to determine whether coordination occurs through nitrogen or oxygen. The electronic structure of urea may be represented by a resonance hybrid of structures I, II, and III, with each contributing roughly an equal amount:

$$O=C\begin{subarray}{l} NH_2 \\ \\ NH_2 \end{subarray} \qquad {}^-O-C\begin{subarray}{l} NH_2^+ \\ \\ NH_2 \end{subarray} \qquad {}^-O-C\begin{subarray}{l} NH_2 \\ \\ NH_2^+ \end{subarray}$$

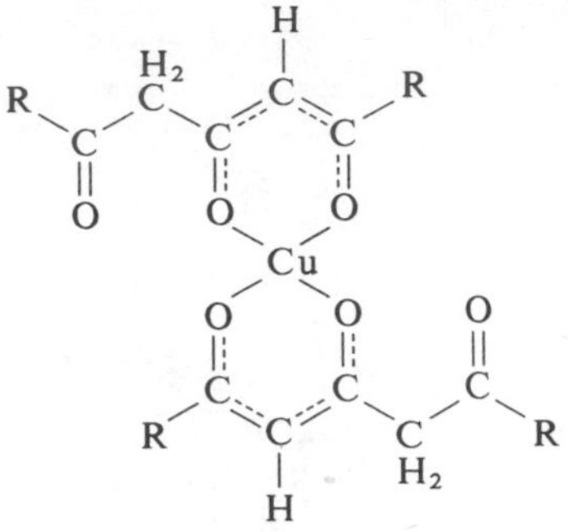

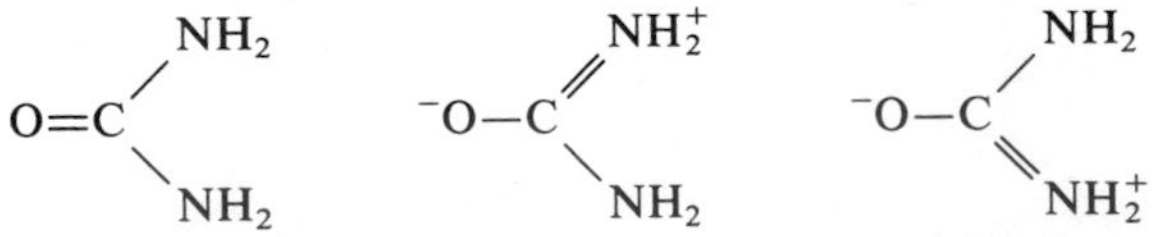

I   II   III

If coordination occurs through nitrogen, the contributions of structures II and III will decrease. This results in an increase of the CO stretching frequency with a decrease in the CN stretching frequency. The NH stretching frequency in this case may fall in the same range as the value for the amido complexes (Sec. III-1). If coordination occurs through oxygen, the contribution of structure I will decrease. This may result in a decrease of the CO stretching frequency but no appreciable change in the NH stretching frequency. Since the spectrum of urea itself has been analyzed completely,[393] band shifts caused by coordination can be checked immediately. The results shown in Table III-31 indicate

TABLE III-31. SOME VIBRATIONAL FREQUENCIES OF UREA AND ITS METAL COMPLEXES (CM^{-1})[392]

$[Pt(urea)_2Cl_2]$	Urea	$[Cr(urea)_6]Cl_3$	Predominant Mode
3390 } 3290	3500 } 3350	3440 } 3330	$\nu(NH_2)$, free
3130 } 3030		3190	$\nu(NH_2)$, bonded
1725	1683	1505[a]	$\nu(C{=}O)$
1395	1471	1505[a]	$\nu_a(CN)$

[a] $\nu(C{=}O)$ and $\nu(C{-}N)$ couple in the Cr complex.

that coordination occurs through nitrogen in the Pt(II) complex, and through oxygen in the Cr(III) complex. It was also found that Pd(II) coordinates to the nitrogen, whereas Fe(III), Zn(II), and Cu(II) coordinate to the oxygen of urea. The infrared spectra of tetramethylurea (tmu) complexes of lanthanide elements, $[Ln(tmu)_6](ClO_4)_3$, indicate the presence of O-coordination.[394]

From infrared studies on thiourea $[(NH_2)_2CS]$ complexes, Yamaguchi et al.[395] found that all the metals studied (Pt, Pd, Zn, and Ni) form M-S bonds, since the CN stretching frequency increases and the CS stretching frequency decreases upon coordination, without an appreciable change in the NH stretching frequency. On the basis of the same criterion, thiourea complexes of Fe(II),[396] Mn(II), Co(II), Cu(I), Hg(II), Cd(II), and Pb(II) were shown to be S-bonded.[397] Several investigators[398-400] studied the far-infrared spectra of thiourea complexes and assigned the MS stretching bands between 300 and 200 cm^{-1}. Thus far, the only metal reported to be N-bonded is Ti(IV).[401] Infrared spectra of alkylthiourea complexes have also been studied. Lane and colleagues[402] studied the infrared spectra of methylthiourea complexes and concluded that methylthiourea forms M-S bonds with Zn(II) and Cd(II) and M-N bonds with Pd(II), Pt(II), and Cu(I). For other alkylthiourea complexes, see Refs. 403 and 404. Infrared spectra of selenourea (su) complexes of Co(II), Zn(II), Cd(II), and Hg(II) exhibit $\nu(MSe)$ in the 245-167 cm^{-1} region.[405] The Raman spectra of $[Pd(su)_4]^{2+}$ and $[Pt(su)_4]^{2+}$ ions exhibit the A_{1g} $\nu(MSe)$ at 178 and 191 cm^{-1}, respectively.[406]

Linkage isomerism was found for the formamidopentamminecobalt (III): $[(NH_3)_5Co(-NH_2CHO)]^{3+}$ and $[(NH_3)_5Co(-OCHNH_2)]^{3+}$. Although little difference was found in the $\nu(C{=}O)$ region, the N-isomer showed the aldehyde $\nu(CH)$ at 2700 cm^{-1}, whereas such a band was not obvious in the O-isomer.[407]

(2) Complexes of Sulfoxides and Related Compounds

Cotton et al.[408] studied the infrared spectra of sulfoxide complexes to see whether coordination occurs through oxygen or sulfur. The electronic structure of sulfoxides may be represented by a resonance hybrid of these structures:

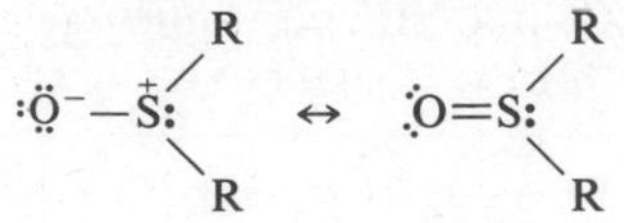

IV V

If coordination occurs through oxygen, the contribution of structure V will decrease and result in a decrease in $\nu(S{=}O)$. If coordination occurs through sulfur, contribution of structure IV will decrease and may result in an increase in $\nu(S{=}O)$. It has been concluded that coordination occurs through oxygen in the $Co(DMSO)_6^{2+}$ ion, since the $\nu(S{=}O)$ of this ion absorbs at 1100–1055 cm^{-1}. On the other hand, coordination may occur through sulfur in $PdCl_2(DMSO)_2$ and $PtCl_2(DMSO)_2$, since $\nu(S{=}O)$ of these compounds (1157–1116 cm^{-1}) are higher than the value for the free ligand. Other ions such as Mn(II), Fe(II, III), Ni(II), Cu(II), Zn(II), and Cd(II) are all coordinated through oxygen, since the DMSO complexes of these metals exhibit $\nu(S{=}O)$ between 960 and 910 cm^{-1}. Drago and Meek,[409] however, assigned $\nu(S{=}O)$ of O-bonded complexes in the 1025–985 cm^{-1} region, since they are metal sensitive. The bands between 960 and 930 cm^{-1}, which were previously assigned to $\nu(S{=}O)$ are not metal sensitive and are assigned to $\rho_r(CH_3)$. Even so, $\nu(S{=}O)$ of O-bonded complexes are lower than the value for free DMSO. To confirm $\nu(S{=}O)$ assignments, it is desirable to compare the spectra of the corresponding DMSO-d_6 complexes since $\rho_r(CD_3)$ is outside the $\nu(S{=}O)$ region. Table III-32 lists $\nu(S{=}O)$ of typical compounds.

Wayland and Schramm[410] found the first example of mixed coordination of DMSO in the $[Pd(DMSO)_4]^{2+}$ ion; it exhibits two S-bonded $\nu(S{=}O)$ at 1150 and 1140 cm^{-1}, and two O-bonded $\nu(S{=}O)$ at 920 and 905 cm^{-1}. Thus the infrared spectrum is most consistent with a configuration in which two

TABLE III-32. SO STRETCHING FREQUENCIES OF DMSO COMPLEXES (CM^{-1})

Compound	$\nu(S{=}O)$	Bonding	Ref.
$Sn(DMSO)_2Cl_4$	915	O	410
$[Cr(DMSO)_6](ClO_4)_3$	928	O	410
$[Ni(DMSO)_6](ClO_4)_2$	955	O	410
$[Ln(DMSO)_8](ClO_4)_3$, (Ln = La, Ce, Pr, Nd)	998–992	O	411
$[Al(DMSO)_6]X_3$, (X = Cl, Br, I)	1000–1008	O	412
$[Ru(NH_3)_5(DMSO)](PF_6)_2$	1045	S	413
trans-$[Pd(DMSO)_2Cl_2]$	1116	S	415
cis-$[Pt(DMSO)_2Cl_2]$	1135 1160	S	414

S-bonded and two O-bonded DMSO are in the *cis* position. The infrared and NMR spectra of $Ru(DMSO)_4Cl_2$ suggested a mixing of O- and S-coordination; $\nu(S{=}O)$ at 1120 and 1090 cm^{-1} for S-coordination and at 915 cm^{-1} for O-coordination.[416] X-ray analysis[417] has since shown that two Cl atoms are in the *cis* positions of an octahedron and the remaining positions are occupied by one O-bonded and three S-bonded DMSO ligands. Infrared spectra show that all DMSO ligands in $Ru(DMSO)_3Cl_3$ are O-bonded while O- and S-bonded DMSO ligands are mixed in $M(DMSO)_3Cl_3$ (M = Os and Rh).[418]

Complete assignments on infrared and Raman spectra of *trans*-$Pd(DMSO)_2X_2$ (X = Cl and Br) and their deuterated analogs have been made by Tranquille and Forel.[419] Berney and Weber[420] found the order of $\nu(MO)$ in the $[M(DMSO)_6]^{n+}$ ion to be as follows:

M =	Cr(III)		Ni(II)		Co(II)		Zn(II)		Fe(II)		Mn(II)
$\nu(MO)$ (cm^{-1})	529	>	444	>	436	>	431	>	438 415	>	418

Griffiths and Thornton[421] made band assignments of these DMSO complexes based on d_6 and ^{18}O substitution of DMSO.

Ligands such as DPSO(diphenylsulfoxide) and TMSO (tetramethylenesulfoxide) do not exhibit the CH_3 rocking bands near 950 cm^{-1}. Thus the SO stretching bands of metal complexes containing these ligands can be assigned without difficulty. In a series of O-bonded DMSO and TMSO complexes, the S=O stretching force constant decreases linearly as the M–O stretching force constant increases.[421a] Table III-33 lists the SO stretching frequencies and the magnitude of band shifts in DPSO complexes.[422] van Leeuwen and

TABLE III-33. SHIFTS OF SO STRETCHING BANDS IN DPSO AND DMSO COMPLEXES (CM^{-1})[422]

	DPSO Complex		DMSO Complex
Metal	$\nu(SO)$	Shift	Shift
Ca(II)	1012–1035	0–(–23)	—
Mg(II)	1012	–23	—
Mn(II)	983–991	–45	–41
Zn(II)	987–988	–47	—
Fe(II)	987	–48	—
Ni(II)	979–982	–55	–45
Co(II)	978–980	–56	–51
Cu(II)	1012, 948	–23, –87	–58
Al(III)	942	–93	—
Fe(III)	931	–104	—

Groeneveld[422] noted that the shift becomes larger as the electronegativity of the metal increases. In Table III-33, the metals are listed in the order of increasing electronegativity.

In $[M(DTHO_2)_3]^{2+}$ [M = Co(II), Ni(II), Mn(II), etc.], the metals are O-bonded since the $\nu(S=O)$ of free ligand (1055–1015 cm^{-1}) are shifted to lower frequencies by 40–22 cm^{-1}:

$$
\begin{array}{ccc}
O & & O \\
\parallel & & \parallel \\
S & & S \\
\diagup\;\diagdown & & \diagup\;\diagdown \\
CH_3 \qquad (CH_2)_2 & & CH_3
\end{array}
$$

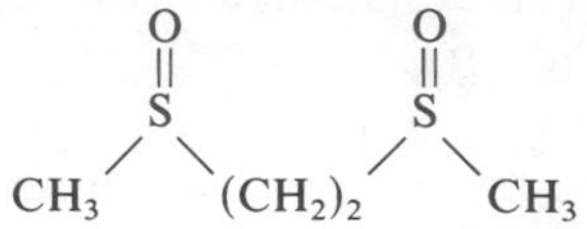

2,5-Dithiahexane-2,5-dioxide (DTHO$_2$)

On the other hand, the metals are S-bonded in $M(DTHO_2)Cl_2$ [M = Pt(II) and Pd(II)] since $\nu(S=O)$ are shifted to higher frequencies by 108–77 cm^{-1}.[423] Dimethylselenoxide, $(CH_3)_2Se=O$, forms complexes of the $MCl_2(DMSeO)_n$ type, where M is Hg(II), Cd(II), Cu(II), and so on, and n is 1, $1\frac{1}{2}$, or 2. The $\nu(Se=O)$ of the free ligand (800 cm^{-1}) is shifted to the 770–700 cm^{-1} region, indicating the O-bonding in these complexes.[424]

III-14. CYANO AND NITRILE COMPLEXES

(1) Cyano Complexes

The vibrational spectra of cyano complexes have been studied extensively and these investigations are reviewed by Sharp,[425] Griffith,[426] Rigo and Turco,[427] and Jones and Swanson.[428]

(a) CN Stretching Bands. Cyano complexes can be identified easily since they exhibit sharp $\nu(CN)$ at 2200–2000 cm^{-1}. The $\nu(CN)$ of free CN$^-$ is 2080 cm^{-1} (aqueous solution). Upon coordination to a metal, the $\nu(CN)$ shift to higher frequencies, as shown in Table III-34. The CN$^-$ ion acts as a σ-donor by donating electrons to the metal and also as a π-acceptor by accepting electrons from the metal. σ-Donation tends to raise the $\nu(CN)$ since electrons are removed from the 5σ orbital, which is weakly antibonding, while π-backbonding tends to decrease the $\nu(CN)$ because the electrons enter into the antibonding $2p\pi^*$ orbital. In general, CN$^-$ is a better σ-donor and a poorer π-acceptor than CO. Thus the $\nu(CN)$ of the complexes are generally higher than the value for free CN$^-$, whereas the opposite prevails for the CO complexes (Sec. III-16).

According to El-Sayed and Sheline,[436] the $\nu(CN)$ of cyano complexes are governed by (1) the electronegativity, (2) the oxidation state, and (3) the coordination number of the metal. The effect of electronegativity is seen in the order:

$$
\begin{array}{ccccc}
[Ni(CN)_4]^{2-} & & [Pd(CN)_4]^{2-} & & [Pt(CN)_4]^{2-} \\
2128 & < & 2143 & < & 2150 \qquad cm^{-1}
\end{array}
$$

TABLE III-34. C≡N STRETCHING FREQUENCIES OF CYANO COMPLEXES (cm^{-1})

Compound	Symmetry	ν(CN)	Ref.
$Tl[Au(CN)_2]$	$D_{\infty h}$	2164 (Σ_g^+), 2141 (Σ_u^+)	429, 430
$K[Ag(CN)_2]$	$D_{\infty h}$	2146 (Σ_g^+), 2140 (Σ_u^+)	431
$K_2[Ni(^{12}C^{14}N)_4]$	D_{4h}	2143.5 (A_{1g}), 2134.5 (B_{1g}), 2123.5 (E_u)	432
$K_2[Pd(^{12}C^{14}N)_4]$	D_{4h}	2160.5 (A_{1g}), 2146.4 (B_{1g}), 2135.8 (E_u)	432
$K_2[Pt(^{12}C^{14}N)_4]$	D_{4h}	2168.0 (A_{1g}), 2148.8 (B_{1g}), 2133.4 (E_u)	432
$Na_3[Ni(CN)_5]$	C_{4v}	2130 (A_1), 2117 (B_1), 2106 (E), 2090 (A_1)	433
$Na_3[Co(CN)_5]$	C_{4v}	2115 (A_1), 2110 (B_1), 2096 (E), 2080 (A_1)	433
$K_3[Mn(CN)_6]$	O_h	2129 (A_{1g}), 2129 (E_g), 2112 (F_{1u})	434, 435
$K_4[Mn(CN)_6]$	O_h	2082 (A_{1g}), 2066 (E_g), 2060 (F_{1u})	434
$K_3[Fe(CN)_6]$	O_h	2135 (A_{1g}), 2130 (E_g), 2118 (F_{1u})	434
$K_4[Fe(CN)_6]\cdot 3H_2O$	O_h	2098 (A_{1g}), 2062 (E_g), 2044 (F_{1u})	434
$K_3[Co(CN)_6]$	O_h	2150 (A_{1g}), 2137 (E_g), 2129 (F_{1u})	434
$K_4[Ru(CN)_6]\cdot 3H_2O$	O_h	2111 (A_{1g}), 2071 (E_g), 2048 (F_{1u})	434
$K_3[Rh(CN)_6]$	O_h	2166 (A_{1g}), 2147 (E_g), 2133 (F_{1u})	434
$K_2[Pd(CN)_6]$	O_h	2185 (F_{1u})	435a
$K_4[Os(CN)_6]\cdot 3H_2O$	O_h	2109 (A_{1g}), 2062 (E_g), 2036 (F_{1u})	434
$K_3[Ir(CN)_6]$	O_h	2167 (A_{1g}), 2143 (E_g), 2130 (F_{1u})	434

Since the electronegativity of Ni(II) is smallest, the σ-donation will be the least, and the ν(CN) is expected to be the lowest. The effect of oxidation state is seen in the frequency order.[437]

$$\begin{array}{ccccc} [V(CN)_6]^{5-} & & [V(CN)_6]^{4-} & & [V(CN)_6]^{3-} \\ 1910 & < & 2065 & < & 2077 \quad cm^{-1} \end{array}$$

The higher the oxidation state, the stronger the σ-bonding, and the higher the ν(CN). The effect of coordination number[438–440] is evident in the frequency order:

$$\begin{array}{ccccc} [Ag(CN)_4]^{3-} & & [Ag(CN)_3]^{2-} & & [Ag(CN)_2]^{-} \\ 2092 & < & 2105 & < & 2135 \quad cm^{-1} \end{array}$$

Here an increase in the coordination number results in a decrease in the positive charge on the metal, which, in turn, weakens the σ-bonding, thus decreasing the ν(CN).

Other cyano complexes which are not included in Table III-34 are $Na[Cu(CN)_2]2H_2O$ (polymeric chain),[441] $Na_2[Cu(CN)_3]\cdot 3H_2O$ (D_{3h}),[442] $Cs[Hg(CN)_3]$ (D_{3h}),[443] and $K_2[Zn(CN)_4]$ (T_d).[444] The symmetry of the $[Mo(CN)_7]^{4-}$ ion may be D_{5h}[445] or C_{2v}.[446] The pentagonal-bipyramidal structure (D_{5h}) has been proposed for $[Re(CN)_7)]^{4-}$,[447] $[Tc(CN)_7]^{4-}$,[448] and $[W(CN)_7]^{5-}$,[449] based on their IR and Raman spectra either in the solid state or in solution or both. According to X-ray analysis,[450] the $[Mo(CN)_8]^{4-}$ ion in $K_4[Mo(CN)_8]\cdot 2H_2O$ is definitely $\mathbf{D}_{2d}$ (dodecahedron). On the other hand, a Raman study[451] supported the $\mathbf{D}_{4d}$ (archimedean-antiprism) structure of the $[Mo(CN)_8]^{4-}$ ion in aqueous solution. The stereochemical conversion of the $[Mo(CN)_8]^{4-}$ ion from $\mathbf{D}_{2d}$ (solid) to $\mathbf{D}_{4d}$ (solution) symmetry was confirmed by Hartman and Miller[452] and Parish et al.[453] Similar conversions were proposed for the $[W(CN)_8]^{4-}$ [452,453] and $[Nb(CN)_8]^{4-}$ [454] ions. However, Long and Vernon[455] claim that the $\mathbf{D}_{2d}$ geometry is maintained even in aqueous solution. Both X-ray and Raman studies confirm the $\mathbf{D}_{2d}$ structure for $K_5[Nb(CN)_8]$ in the solid state although the $\mathbf{D}_{4d}$ structure prevails in solution.[456]

According to the results of X-ray analysis,[457] the unit cell of $[Cr(en)_3]$-$[Ni(CN)_5]\cdot 1\frac{1}{2}H_2O$ contains both square-pyramidal ($\mathbf{C}_{4v}$) and trigonal-bipyramidal ($\mathbf{D}_{3h}$) structures of the $[Ni(CN)_5]^{3-}$ ion. Terzis et al.[458] showed that the complicated vibrational spectrum of this crystal in the ν(CN) region is simplified dramatically when it is dehydrated. These spectral changes suggest that the $\mathbf{D}_{3h}$ (somewhat distorted) units have been converted to $\mathbf{C}_{4v}$ geometry upon dehydration. Basile et al.[459] showed that such conversion from $\mathbf{D}_{3h}$ to $\mathbf{C}_{4v}$ also occurs when the crystal is subjected to high pressure. Hellner et al.[460] observed the splitting of the degenerate ν(CN) of $K_2[Zn(CN)_4]$ and a partial reduction of the central metal in $K_3[M(CN)_6]$ [M = Fe(III) and Mn(III)] when these crystals are under high external pressure.

Jones and Penneman[438–440] made an extensive infrared study of the equilibria of cyano complexes in aqueous solution. (For aqueous infrared spectroscopy, see Sec. III-9.) Figure III-35 shows the infrared spectra of aqueous silver cyano complexes obtained by changing the ratio of Ag^+ to CN^- ions. Table III-35 lists the frequencies and extinction coefficients from which equilibrium constants can be calculated. Chantry and Plane[462] studied the same equilibria using Raman spectroscopy.

(b) Lower-Frequency Bands. In addition to ν(CN), the cyano complexes exhibit ν(MC), δ(MCN), and δ(CMC) bands in the low-frequency region. Figure III-36 shows the infrared spectra of $K_3[Co(CN)_6]$ and $K_2[Pt(CN)_4]\cdot 3H_2O$. Normal coordinate analyses have been carried out on various hexacyano complexes to assign these low-frequency bands (Table III-36). The results of these calculations indicate that the ν(MC), δ(MCN), and δ(CMC) vibrations appear in the regions 600–350, 500–350, and 130–60 cm^{-1}, respectively. The MC and C≡N stretching force constants obtained are also given in Table III-36.

Nakagawa and Shimanouchi[467] noted that the MC stretching force constant increases in the order Fe(III) < Co(III) < Fe(II) < Ru(II) < Os(II), and the C≡N stretching force constant decreases in the same order of metals. This result was interpreted as indicating that the M–C π-bonding is increasing in the above order. The degree of M–C π-bonding may be proportional to the number of d-electrons in the t_{2g} electronic level. According to Jones,[468] the

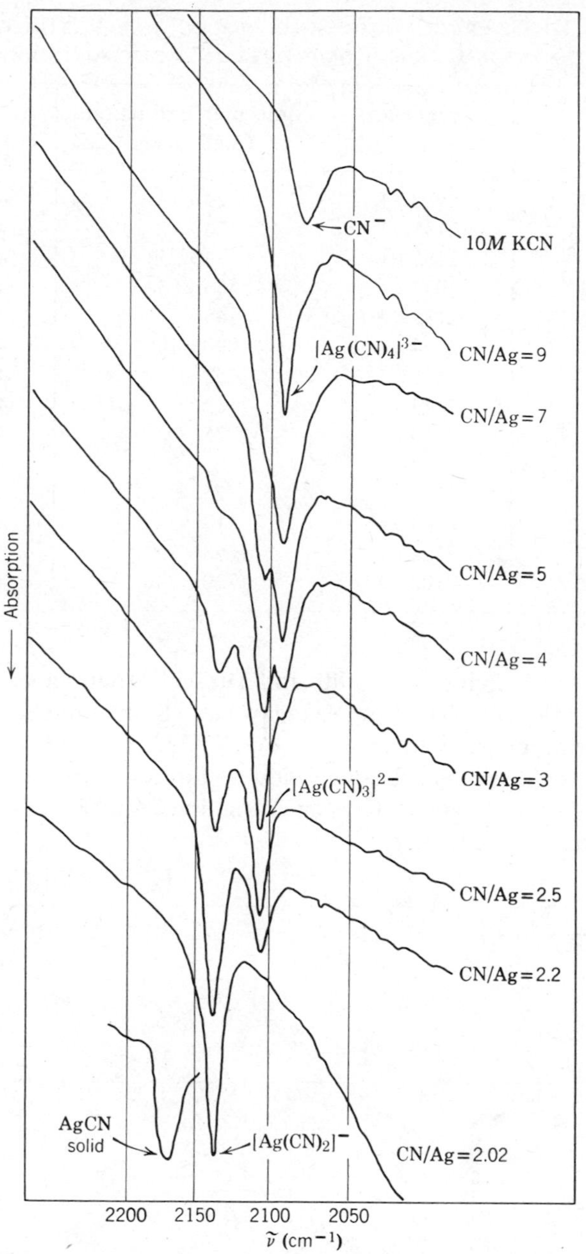

Fig. III-35. Infrared spectra of silver cyano complexes in aqueous solutions.[438]

integrated absorption coefficient of the C≡N stretching band (F_{1u}) becomes larger as the number of d-electrons in the t_{2g} level increases. Thus the results shown in Table III-37 suggest that the M–C π-bonding increases in the order Cr(III) < Mn(III) < Fe(III) < Co(III). The order of ν(MC) shown in the same

TABLE III-35. FREQUENCIES AND MOLECULAR EXTINCTION COEFFICIENTS OF CYANO COMPLEXES IN AQUEOUS SOLUTIONS

Ion	Frequency (cm^{-1})	Molecular Extinction Coefficient	Ref.
Free $[CN]^-$	2080 ± 1	29 ± 1	438
$[Ag(CN)_2]^-$	2135 ± 1	264 ± 12	438
$[Ag(CN)_3]^{2-}$	2105 ± 1	379 ± 23	438
$[Ag(CN)_4]^{3-}$	2092 ± 1	556 ± 83	438
$[Cu(CN)_2]$	2125 ± 3	165 ± 25	439
$[Cu(CN)_3]^{2-}$	2094 ± 1	1090 ± 10	439
$[Cu(CN)_4]^{3-}$	2076 ± 1	1657 ± 15	439
$[Zn(CN)_4]^{2-}$	2149	113	440
$[Cd(CN)_4]^{2-}$	2140	75	440
$Hg(CN)_2$	2194	3	440
$[Hg(CN)_3]^-$	2161	26	440
$[Hg(CN)_4]^{2-}$	2143	113	440
$[Ni(CN)_4]^{2-}$	2124 ± 1	1068 ± 95	461
$[Ni(CN)_5]^{3-}$	2102 ± 2	1730 ± 230	461

table confirms this conclusion. Griffith and Turner[434] found a similar trend in the Fe(II) < Ru(II) < Os(II) series. Nakagawa and Shimanouchi[469] carried out complete normal coordinate analyses on $K_3[M(CN)_6]$ [M = Fe(III) and Cr(III)] crystals, including all lattice modes. Jones et al.[470] also performed complete normal coordinate analyses on crystalline $Cs_2Li[Fe(CN)_6]$, including its ^{13}C, ^{15}N, and ^{6}Li analogs.

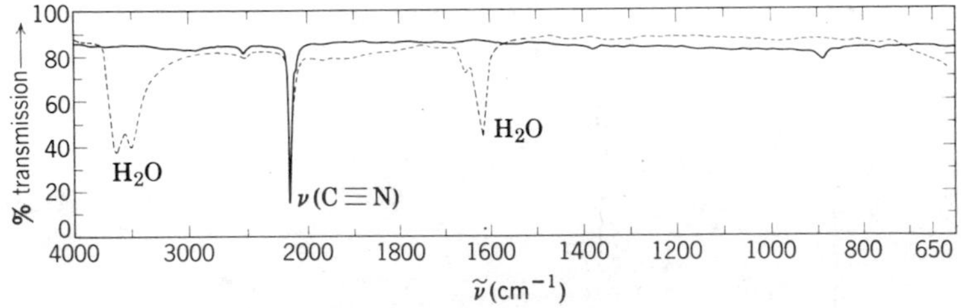

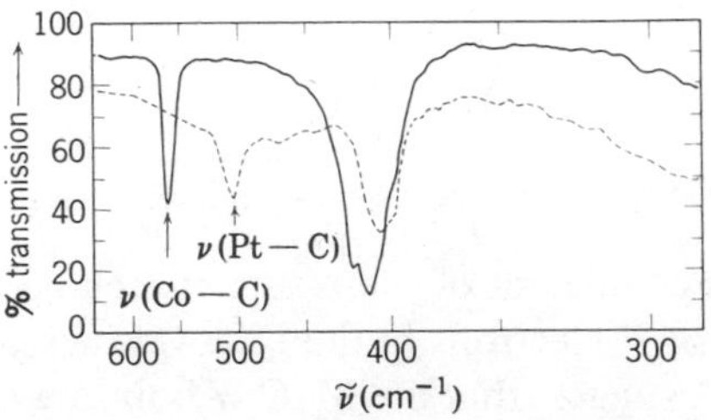

Fig. III-36. Infrared spectra of $K_3[Co(CN)_6]$ (solid line) and $K_2[Pt(CN)_4] \cdot 3H_2O$ (broken line).

Table III-36. Vibrational Frequencies and Band Assignments of Hexacyano Complexes (cm^{-1})

		$[Cr(CN)_6]^{3-}$	$[Co(CN)_6]^{3-}$	$[Ir(CN)_6]^{3-}$	$[Rh(CN)_6]^{3-}$	$[Co(CN)_6]^{3-}$	$[Fe(CN)_6]^{4-}$	$[Fe(CN)_6]^{3-}$	$[Ru(CN)_6]^{4-}$	$[Os(CN)_6]^{4-}$
A_{1g}	$\nu(MC)$	374	408	469	445	406	(410)	(390)	(460)	(480)
E_g	$\nu(MC)$	336	(391)	450	435	(375)	(390)	—	(410)	(450)
F_{1g}	$\delta(MCN)$	536	(358)	(415)	(380)	(380)	(350)	—	(340)	(360)
F_{1u}	$\nu(MC)$	457	564	520	520	565	585	511	550	554
	$\delta(MCN)$	694	416	398	386	414	414	387	376	392
	$\delta(CMC)$	124	(84)	(82)	(88)	—	—	89	—	—
F_{2g}	$\delta(MCN)$	536	(480)	483	(475)	—	(420)	—	(400)	(430)
	$\delta(CMC)$	106	98	95	94	98	—	99	—	—
F_{2u}	$\delta(MCN)$	496	(440)	445	—	(395)	—	—	(365)	(390)
	$\delta(CMC)$	95	(72)	(69)	—	—	—	70	—	—
Force field		GVF	GVF	GVF	GVF	UBF	UBF	UBF	UBF	UBF
$K(M-C)$ (mdyn/Å)		1.928	2.063	2.704	2.366	2.308	2.428	1.728	2.793	3.343
$K(C\equiv N)$ (mdyn/Å)		16.422	16.767	16.678	16.831	16.5	15.1	17.0	15.3	14.9
References		463	464 465 466	465	465	467	467	467 466	467	467

TABLE III-37. RELATION BETWEEN INFRARED SPECTRUM AND ELECTRONIC STRUCTURE IN HEXACYANO COMPLEXES[468]

Compound	Number of *d*-Electrons in t_{2g} Level	ν(CN) (cm^{-1})	ν(MC) (cm^{-1})	Integrated Absorption Coefficient ($mole^{-1}$ cm^{-2})
$K_3[Cr(CN)_6]$	3	2128	339	2,100
$K_3[Mn(CN)_6]$	4	2112	361	8,200
$K_3[Fe(CN)_6]$	5	2118	389	12,300
$K_3[Co(CN)_6]$	6	2129	416	18,300

Normal coordinate analyses have been made on tetrahedral, square-planar, and linear cyano complexes of various metals; Table III-38 gives the results of these studies. Far-infrared spectra of various cyano complexes have been measured.[482] Jones and co-workers[483] carried out an extensive study on mixed cyano–halide complexes such as $[Au(CN)_2X_2]^-$, where X is Cl^-, Br^-, or I^-. An ultraviolet and infrared study[484] showed that the $[Ni(CN)_4]^{2-}$ and $[Ni(CN)_5]^{3-}$ ions are in equilibrium in a solution containing $Na_2[Ni(CN)_4]$, KCN, and KF. The integrated absorption coefficient of the C≡N stretching band increases in the order Hg(II) < Ag(I) < Au(I) in linear dicyano complexes, indicating that the M–C π-bonding increases in the same order.[468] From the measurements of infrared dichroism, Jones determined the orientation of $[Ag(CN)_2]^-$ and $[Au(CN)_2]^-$ ions in their potassium salts.[480,481] His results are in good agreement with those of X-ray analysis.

(c) Cyano Complexes Containing NO and Halogens. The infrared spectra of nitroprusside salts have been reviewed briefly.[485] Khanna et al.[486] assigned the infrared and Raman spectra of the $Na_2[Fe(CN)_5NO]\cdot 2H_2O$ crystal and its deuterated analog. On the basis of a comparison of ν(CN), δ(MCN), and ν(MC) between Fe(II) and Fe(III) complexes of $[Fe(CN)_5X]^{n-}$-type ions, Brown suggests that the Fe–NO bonding of the $[Fe(CN)_5NO]^{2-}$ ion be formulated as Fe(III)—NO and not as Fe(II)—NO^+.[487] The infrared spectra of $K_3[Mn(CN)_5NO]$ and its ^{15}NO analog have been reported.[488,489] Tosi and Danon[490] studied the infrared spectra of $[Fe(CN)_5X]^{n-}$ ions (X = H_2O, NH_3, NO_2^-, NO^-, and SO_3^{2-}). The ν(C≡N) of the $[Fe(CN)_5NO]^{2-}$ ion (2170, 2160, and 2148 cm^{-1}) are unusually high in this series because the M–C π-bonding in this ion is much less than in other compounds due to extensive M–NO π-bonding. The infrared spectra of $K_2[Pt(CN)_4X_2]$[491] and $K_2[Pt(CN)_5X]$[492] (X = Cl^-, Br^-, I^-, etc.) have been reported.

(d) Bridged Cyano Complexes. If the M—C≡N group forms a M—C≡N—M′ type bridge, ν(C≡N) shifts to a higher, and ν(MC) to a lower, frequency. The higher-frequency shift of ν(C≡N) should be noted since the opposite trends are observed for bridging carbonyl and halogeno complexes. Shriver[493]

TABLE III-38. FREQUENCIES AND BAND ASSIGNMENTS OF THE LOWER-FREQUENCY BANDS OF CYANO COMPLEXES (cm^{-1})

Ion	Symmetry	ν(MC)	δ(MCN)	δ(CMC)	Force Constant[a] K(M—C)	K(C≡N)	Ref.
$[Cu(CN)_4]^{3-}$	$\mathbf{T}_d$	364(IR)	324(R)	(74)	1.25- }	16.10- }	471
		288(R)	306(IR)	(63)	1.30 }	16.31 }	472
$[Zn(CN)_4]^{2-}$	$\mathbf{T}_d$	359(IR)[b]	315(IR)[b]	71(R)	1.30	17.22	473
		342(R)	230(R)				444
$[Cd(CN)_4]^{2-}$	$\mathbf{T}_d$	316(IR)[b]	250(R)[b]	61(R)	1.28	17.13	473
		324(R)	194(R)				
$[Hg(CN)_4]^{2-}$	$\mathbf{T}_d$	330(IR)[b]	235(R)[b]	54(R)	1.53	17.08	473
		335(R)	180(R)				
$[Pt(CN)_4]^{2-}$	$\mathbf{D}_{4h}$	505(IR)	318(R)	95(R)	3.425	16.823	474
		465(R)	300(IR)				475
		455(R)					
$[Ni(CN)_4]^{2-}$	$\mathbf{D}_{4h}$	543(IR)	433(IR)		2.6	16.67	476
		(419)	421(IR)	(54)			
		(405)	488(IR)				
			(325)				
$[Au(CN)_4]^-$	$\mathbf{D}_{4h}$	462(IR)	415(IR)	110(R)	3.28- }	17.40- }	477
		459(R)			3.42 }	17.44 }	
		450(R)					
$[Hg(CN)_2]$	$\mathbf{D}_{\infty h}$	442(IR)	341(IR)	(100)	2.607	17.62	478
		412(R)	276(R)				479
$[Ag(CN)_2]^-$	$\mathbf{D}_{\infty h}$	390(IR)	(310)	(107)	1.826	17.04	480
		(360)	250(R)				
$[Au(CN)_2]^-$	$\mathbf{D}_{\infty h}$	427(IR)	(368)	(100)	2.745	17.17	481
		445(R)	305(R)				

[a] Force constants (mdyn/Å) were obtained by using the GVF field for all ions except the $[Pt(CN)_4]^{2-}$ ion, for which the UBF field was used.
[b] Coupled vibrations between ν(MC) and δ(MCN).

observed that ν(C≡N) of $K_2[Ni(CN)_4]$ at 2130 cm^{-1} shifts to 2250 cm^{-1} in $K_2[Ni(CN)_4]\cdot 4BF_3$ because of the formation of the Ni—C≡N—BF_3 type bridge. They[494] also found that, for $KFeCr(CN)_6$, the green isomer containing the Fe(II)—C≡N—Cr(III) bridges exhibits ν(C≡N) at 2092 cm^{-1}, while the red isomer containing the Cr(III)—C≡N—Fe(II) bridges shows ν(C≡N) at 2168 and 2114 cm^{-1}. Brown et al.[495] studied the mechanism of conversion from green to red isomer by combining infrared and Mossbauer spectroscopy with other techniques. The ν(C≡N) and ν(Fe-C) of crystalline $Cs_2Mg[Fe(CN)_6]$ are higher by 40 cm^{-1} than those of the $[Fe(CN)_6]^{4-}$ ion in aqueous solution.[496] The same trend is seen for crystalline $Mn_3[Co(CN)_6]\cdot xH_2O$ and the $[Co(CN)_6]^{3-}$ ion in aqueous solution.[497] These observations suggest the presence of strong interaction of the Fe—C≡N···Mg or Co—C≡N···Mn type in the solid state. The bridging $\nu(C{\equiv}N)$ of the $[(NC)_5Fe^{II}{-}CN{-}Co^{III}(CN)_5]^{6-}$ and $[(NC)_5Fe^{III}{-}CN{-}Co^{III}(CN)_5]^{5-}$ ions are at 2130 and 2185 cm^{-1}, respectively.[498] The infrared and Mossbauer spectra of $K_4[Fe(CN{-}SbX_3)_6]$ (X = F and Cl) and $K_4[Fe(CN{-}SbX_3)_4(CN)_2]$ (X = Cl and Br) have been studied.[499] As expected, the infrared spectrum of Prussian blue is identical to that of Turnbull's blue.[500]

Finally, partially oxidized tetracyanoplatinates such as $K_2[Pt(CN)_4]Br_{0.3}\cdot 3H_2O$ are known as one-dimensional (linear chain) conductors.[501] In these compounds, the planar $Pt(CN)_4^{2-}$ ions are stacked in one direction, and the Pt···Pt distances (2.88 Å) are much shorter than that of the parent compound, $K_2[Pt(CN)_4]\cdot 3H_2O$ (3.478 Å). The oxidation state of the Pt atom in $K_2[Pt(CN)_4]Br_{0.33}\cdot 3H_2O$ is +2.33. As a result, its $\nu(CN)$ [2182(A_{1g}), 2165(B_{1g}) cm^{-1}] are between those of $K_2[Pt^{II}(CN)_4]\cdot 3H_2O$ (2168, 2149 cm^{-1}) and $K_2[Pt^{IV}(CN)_4Cl_2]$ (2196 and 2186 cm^{-1}).[502]

(2) Nitrile and Isonitrile Complexes

Nitriles (R—C≡N, R = alkyl or phenyl) form a number of metal complexes by coordination through their N atoms. Again, $\nu(CN)$ becomes higher upon complex formation. For example, Walton[503] measured the infrared spectra of $MX_2(RCN)_2$-type compounds, where M is Pt(II) and Pd(II) and X is Cl^- and Br^-. When R is phenyl, the $\nu(CN)$ are near 2285 cm^{-1}, which is higher than the value for the $\nu(CN)$ of free benzonitrile (2231 cm^{-1}). It was noted that the $\nu(CN)$ of benzonitrile (2231 cm^{-1}) shifts to a higher frequency (2267 cm^{-1}) when it coordinates to the pentammine Ru(III) species but to a lower frequency (2188 cm^{-1}) when coordinated to the pentammine Ru(II) species. This result may indicate that the latter species has unusually strong π-back-bonding ability.[504] A strong band at 174 cm^{-1} of $ZnCl_2(CH_3CN)_2$ was suggested to be $\nu(ZnN)$.[505] The $\nu(MN)$ bands of other acetonitrile complexes have been assigned in the 450–160 cm^{-1} region.[506]

Farona and Kraus[507] observed $\nu(CN)$ of $Mn(CO)_3(NC{-}CH_2{-}CH_2{-}CN)Cl$ at 2068 cm^{-1}, although $\nu(CN)$ of free succinonitrile (sn) is at 2257 cm^{-1}. This large shift to a lower frequency was attributed to the chelating bidentate coordination through its CN triple bonds:

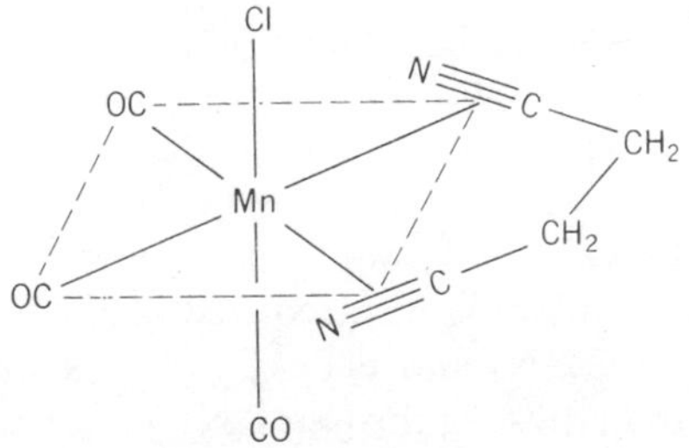

According to X-ray analysis,[508] the complex ion in $[Cu(sn)_2]NO_3$ takes a polymeric chain structure in which the ligand is in the *gauche* conformation:

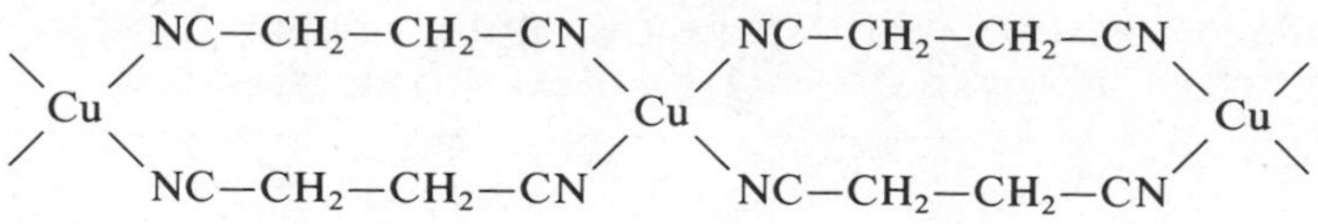

In these dinitrile complexes, ν(CN) are shifted to higher frequencies upon coordination. As in the case of ethylenediamine complexes (Sec. III-2), infrared spectroscopy has been used to determine the conformation of the ligand in metal complexes. The Cu(I) complex, which is known to contain the *gauche* conformation, exhibits two CH_2 rocking modes at 966 and 835 cm^{-1}, whereas the Ag(I) complex, $Ag(sn)_2BF_4$, shows a single CH_2 rocking mode at 770 cm^{-1}, which is characteristic of the *trans* conformation.[509]

There are four rotational isomers for glutaronitrile (gn), NC—CH_2—CH_2—CH_2—CN, which are spectroscopically distinguishable. Figure III-37 shows the conformation, the symmetry, and the number of infrared-active CH_2 rocking vibrations for each isomer. According to X-ray analysis on $Cu(gn)_2NO_3$,[510] the ligand in this complex is in the *gg* conformation. The infrared spectrum of this complex is very similar to that of solid glutaronitrile in the stable form. Matsubara[511] therefore concluded that the latter also takes the *gg* conformation. However, the spectrum of solid glutaronitrile in the metastable form (produced by rapid cooling) is different from that of the *gg* conformation and it could have been *tt*, *tg*, or *gg'*. The *tt* conformation was excluded because of the absence of the 730 cm^{-1} band characteristic of the *trans*-planar methylene chain,[512] and the *gg'* conformation was considered to be improbable because of steric repulsion between two CN groups. This left only the *tg* conformation for the metastable solid. The complicated spectrum of liquid glutaronitrile was accounted for by assuming that it is a mixture of the *tg*, *gg*, and *tt* conformations. Kubota and Johnston,[513] using these results, have been able to show that the glutaronitrile molecules in $Ag(gn)_2ClO_4$ and $Cu(gn)_2ClO_4$ are in the *gg* conformation, while those in $TiCl_4gn$ and $SnCl_4gn$ have the *tt* conformation. Table III-39 summarizes the CH_2 rocking frequencies of glutaronitrile and its metal complexes. An infrared study similar to the above has been extended to adiponitrile (NC—$(CH_2)_4$—CN) and its Cu(I) complex.[514]

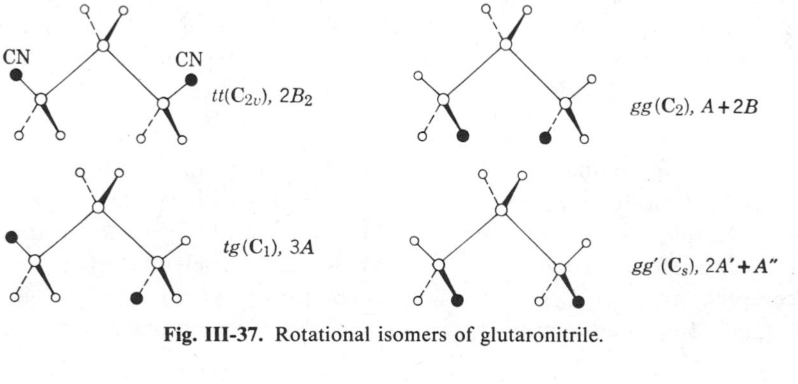

Fig. III-37. Rotational isomers of glutaronitrile.

TABLE III-39. INFRARED-ACTIVE CH_2 ROCKING FREQUENCIES OF GLUTARONITRILE AND ITS METAL COMPLEXES (CM^{-1})

Liquid[a]	945 (*tg*)	904 (*gg*)	835 (*tg*, *gg*)	757 (*tg*, *gg*)	737 (*tt*)[b]
Solid[a] (metastable)	943 (*tg*)	—	839 (*tg*)	757 (*tg*)	—
Solid[a] (stable)	—	903 (*gg*)	837 (*gg*)	768 (*gg*)	—
$Cu(gn)_2NO_3$[a]	—	913 (*gg*)	830 (*gg*)[c]	778 (*gg*)	—
$Cu(gn)_2ClO_4$[d]	—	908 (*gg*)	875 (*gg*)	767 (*gg*)	—
$Ag(gn)_2ClO_4$[d]	—	904 (*gg*)	872 (*gg*)	772 (*gg*)	—
$SnCl_4(gn)$[d]	—	—	—	—	733 (*tt*)
$TiCl_4(gn)$[d]	—	—	—	—	730 (*tt*)

[a] Reference 511.
[b] The *tt* form should exhibit two infrared-active CH_2 rocking vibrations. The other one is not known, however.
[c] Overlapped with a NO_3^- absorption.
[d] Reference 513.

Cotton and Zingales[515] studied the N≡C stretching bands of isonitrile complexes. When isonitriles are coordinated to zero-valent metals such as Cr(O), back donation of electrons from the metal to the ligand is extensive and the N≡C stretching band is shifted to a lower frequency. For monopositive and dipositive metal ions, little or no back donation occurs and the N≡C stretching band is shifted to a higher frequency as a result of the inductive effect of the metal ion. Sacco and Cotton[516] obtained the infrared spectra of $Co(CH_3NC)_4X_2$ and $[Co(CH_3NC)_4][CoX_4]$-type compounds (X = Cl, Br, etc.). Dart et al.[517] report the ν(NC) of bis(phosphine) tris(isonitrile) complexes of Co(I). Boorman et al.[518] made rather complete assignments of vibrational spectra of some isonitrile complexes of Co(I) and Co(II) in the 4000–33 cm^{-1} region.

III-15. THIOCYANATO AND OTHER PSEUDOHALOGENO COMPLEXES

The CN^-, OCN^-, SCN^-, $SeCN^-$, CNO^-, and N_3^- ions are called pseudohalide ions, since they resemble halide ions in their chemical properties. These ions may coordinate to a metal through either one of the end atoms. As a result, the following linkage isomers are possible:

M—CN, cyano complex	M—NC, isocyano complex
M—OCN, cyanato complex	M—NCO, isocyanato complex
M—SCN, thiocyanato complex	M—NCS, isothiocyanato complex
M—SeCN, selenocyanato complex	M—NCSe, isoselenocyanato complex
M—CNO, fulminato complex	M—ONC, isofulminato complex

Two compounds are called true linkage isomers if they have exactly the same composition and two of the different linkages mentioned above. A well-known example is nitro (and nitrito) pentammine Co(III) chloride, discussed in Sec. III-5. A pair of true linkage isomers is difficult to obtain since, in general, one form is much more stable than the other. As will be shown later, a number of new linkage isomers have been isolated, and infrared spectroscopy has proved to be very useful in distinguishing them. Burmeister[519] reviewed linkage isomerism in metal complexes. Bailey et al.[520] and Norbury[521] reviewed the infrared spectra of SCN, SeCN, NCO, and CNO complexes and their linkage isomers in detail.

(1) Thiocyanato (SCN) Complexes

The SCN group may coordinate to a metal through the nitrogen or the sulfur or both (M—NCS—M′). In general, Class A metals (first transition series, such as Cr, Mn, Fe, Co, Ni, Cu, and Zn) form M-N bonds, whereas Class B metals (second half of the second and third transition series, such as Rh, Pd, Ag, Cd, Ir, Pt, Au, and Hg) form M-S bonds.[522] However, other factors, such as the oxidation state of the metal, the nature of other ligands in a complex, and steric consideration, also influence the mode of coordination.

Several empirical criteria have been developed to determine the bonding type of the NCS group in metal complexes.

1. The CN stretching frequencies are generally lower in N-bonded complexes (near and below 2050 cm^{-1}) than in S-bonded complexes (near 2100 cm^{-1}).[523] The bridging (M—NCS—M′) complexes exhibit $\nu(CN)$ well above 2100 cm^{-1}. However, this rule must be applied with caution since $\nu(CN)$ are affected by many other factors.[520]

2. Several workers[524–526] considered $\nu(CS)$ as a structural diagnosis: 860–780 cm^{-1} for N-bonded, and 720–690 cm^{-1} for S-bonded, complexes. However, this band is rather weak and is often obscured by the presence of other bands in the same region.

3. It was suggested[525,526] that the N-bonded complex exhibits a single sharp $\delta(NCS)$ near 480 cm^{-1}, whereas the S-bonded complex shows several bands of low intensity near 420 cm^{-1}. However, these bands are also weak and tend to be obscured by other bands.

4. Several workers[527–530] used the integrated intensity of $\nu(CN)$ as a criterion; it is larger than $9 \times 10^4\ M^{-1}\ cm^{-2}$ per NCS^- for N-bonded complexes, and close to or smaller than $2 \times 10^4\ M^{-1}\ cm^{-2}$ for S-bonded complexes. However, this rule is also difficult to apply when the spectrum consists of multiple components or when the dissociation occurs in solution.

5. Some workers[531,532] proposed using $\nu(MN)$ and $\nu(MS)$ in the far-infrared region as a criterion; in general, $\nu(MN)$ is higher than $\nu(MS)$. However, these frequencies are very sensitive to the overall structure of the complex and the nature of the central metal. Thus extreme caution must be taken in applying this criterion.

It is clear that only a combination of these five criteria would provide reliable structural diagnosis. Table III-40 lists the vibrational frequencies of typical isothiocyanato and thiocyanato complexes. The ν(MN) and ν(MS) vibrations of some of these and other complexes were assigned by using metal isotopes[536] and ^{15}N-substituted ligands.[537] The $[Cd(SCN)_3]^-$ ion is S-bonded in DMSO but N-bonded in water.[538]

Clark and Williams[531] measured the infrared spectra of tetrahedral $M(NCS)_2L_2$, monomeric octahedral $M(NCS)_2L_4$, and polymeric octahedral $M(NCS)_2L_2$-type complexes (M = Fe, Co, Ni, etc.; L = py, α-pic, etc.), and studied the relationship between the spectra and stereochemistry. They found that ν(CS) are higher by 40–50 cm^{-1} for tetrahedral than for octahedral complexes for the same metal, although ν(CN) are very similar for both.

TABLE III-40. VIBRATIONAL FREQUENCIES OF ISOTHIOCYANATO AND THIOCYANATO COMPLEXES (CM^{-1})[a]

Compound	ν(CN)	ν(CS)	δ(NCS)	Ref.
K[NCS]	2053	748	486, 471	II-255
$(NEt_4)_2[Co(-NCS)_4]$	2062 (s)	837 (w)	481 (m)	520
$K_3[Cr(-NCS)_6]$	2098 (vs)	820 (vw)	474 (s)	533
	2058 (vs)			
$(NEt_4)_2[Cu(-NCS)_4]$	2074 (s)	835 (w)	—	534
$(NEt_4)_3[Fe(-NCS)_6]$	2098 (sh)	822 (w)	479 (m)	520
	2052 (s)			
$(NEt_4)_4[Ni(-NCS)_6]$	2109 (sh)	818 (w)	469 (m)	520
	2102 (s)			
$(NEt_4)_2[Zn(-NCS)_4]$	2074 (s)	832 (w)	480 (m)	520
$(NH_4)[Ag(-SCN)_2]$	2101 (s)	718 (w)	453 (m)	520
	2086 (s)			
$K[Au(-SCN)_4]$	2130 (s)	700 (w)	458 (w)	520
			413 (s)	
$K_2[Hg(-SCN)_4]$	2134 (m)	716 (m)	461 (m)	520
	2122 (sh)	709 (sh)	448 (m)	
	2109 (s)	703 (sh)	432 (sh)	
			419 (m)	
$(NBu_4)_3[Ir(-SCN)_6]$	2127 (m)	822 (m)	430 (w)	535
	2098 (s)	693 (w)		
$K_2[Pd(-SCN)_4]$	2125 (s)	703 (w)	474 (w)	520
	2095 (s)	697 (sh)	467 (w)	
			442 (m)	
			432 (m)	
$K_2[Pt(-SCN)_4]$	2128 (s)	696 (w)	477 (w)	520
	2099 (s)		469 (w)	
	2077 (sh)		437 (m)	
			426 (m)	

[a] vs = very strong; s = strong; m = medium; w = weak; sh = shoulder.

The *cis*- and *trans*-isomers of $[Co(en)_2(NCS)_2]Cl \cdot H_2O$, for example, can be distinguished by infrared spectra in the $\nu(CN)$ region: *trans*, 2136 cm^{-1}; *cis*, 2122 and 2110 cm^{-1}.[539] Lever et al.[540] have found, however, that no splittings of $\nu(CN)$ are observed at room temperature for *cis*-octahedral $ML_2(NCS)_2$, where M is Co(II) and Ni(II) and L is 1,2-bis-(2′-imidazolin-2′-yl)benzene. The splitting of $\nu(CN)$ of this complex was observed only at liquid-nitrogen temperature.

Turco and Pecile[524] noted that the presence of other ligands in a complex influences the mode of the N-C-S bonding. For example, in $Pt(NCS)_2L_2$, the NCS ligand is N-bonded if L is a phosphine (π-acceptor), and is S-bonded if L is an amine. In the solid state, $Ni(NCS)_2(PMePh_2)_2$ is N-bonded (*trans*) but its Pd analog is S-bonded (*trans*), and the Pt analog is N-bonded(*cis*).[541] For $[Cr(NCS)_4L_2]^{n-}$ ions, Contreras and Schmidt[542] proposed, based on the $\nu(CN)$ and $\nu(CS)$ of these ions, N-bonding for L = urea, glycinate ion, and so on, and S-bonding for L = thiourea, acetamide, and so on. These results have been explained in terms of the steric and electronic effects of L.

Since 1963, a variety of true linkage isomers involving the NCS group have been prepared. Table III-41 lists the $\nu(CN)$ and $\nu(CS)$ of typical pairs of these linkage isomers. Epps and Marzilli[550] isolated three linkage isomers of $AsPh_4[Co(DMG)_2(NCS)_2]$:

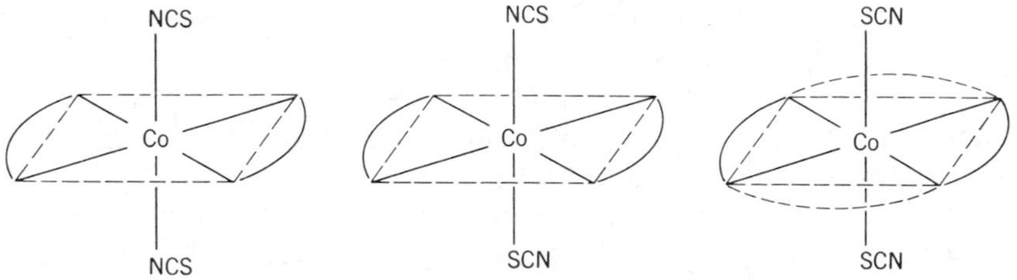

TABLE III-41. VIBRATIONAL FREQUENCIES OF TRUE LINKAGE ISOMERS INVOLVING THE NCS GROUP (CM^{-1})

Compound	Type	$\nu(CN)$	$\nu(CS)$	Refs.
trans-$[Pd(AsPh_3)_2(NCS)_2]$	N-bonded	2089	854	543, 545, 546
	S-bonded	2119	—	
$Pd(bipy)(NCS)_2$	N-bonded	2100	842	544, 545
	S-bonded	2117	700	
		2108		
$(\pi\text{-}Cp)Mo(CO)_3(NCS)$	N-bonded	2099	—	547
	S-bonded	2114	699	
$K_3[Co(CN)_5(NCS)]$	N-bonded	2065	810	548
	S-bonded	2110	718	
trans-$[Co(DMG)_2(py)(NCS)]$	N-bonded	2128	—	549
	S-bonded	2118	—	

Although all these isomers exhibit ν(CN) at 2110 cm^{-1}, they can be distinguished by the differences in the intensity of the ν(CN) band; the (NCS, NCS) isomer is the strongest, the (SCN, SCN) isomer is the weakest, and the (NCS, SCN) isomer is in between.

Both N-bonded and S-bonded NCS groups have been found in [Pd(4,4'-dimethyl-bipy)(NCS)(SCN)][551], [Pd{$Ph_2P(CH_2)_3NMe_2$}(NCS)(SCN)],[552] and similar Pd complexes.[553,554] Preetz and co-workers[555] isolated 9 out of 10 possible isomers of $[Os(NCS)_n(SCN)_{6-n}]^{3-}$ ions ($n = 1 \sim 6$) and characterized them by IR and Raman spectra. Similar mixed NCS–SCN bonding was found for [PdL(NCS)(SCN)], where L is $Ph_2P(o\text{-}C_6H_4)AsPh_2$ and $Ph_2P(CH_2)_2NMe_2$.[556] These bidentate ligands contain two different donor atoms which give different electronic effects on the NCS groups *trans* to them. Thus the *trans* effect, together with the steric effect of these ligands, may be responsible for the mixing of the N- and S-bonding. In the case of $[Ni(DPEA)(NCS)_2]_2$[DPEA: di(2-pyridyl-β-ethyl)amine], IR spectra suggest that terminal N-bonded and bridging NCS groups are mixed [ν(CN): 2094 and 2128 cm^{-1}, respectively].[557]

Burmeister et al.[558] found that in the ML_2X_2-type complexes [M = Pd(II), Pt(II); L = a neutral ligand; X = SCN, SeCN, NCO, etc.], the mode of bonding of X to the metal is determined by the nature of the solvent. For example, $Pd(AsPh_3)_2(NCS)_2$ is N-bonded in pyridine and acetone solution, whereas it is S-bonded in DMF and DMSO solution. However, the bonding of the NCO group is insensitive to the nature of the solvent.

The NCS group also forms a bridge between two metal atoms. The CN stretching frequency of a bridging group is generally higher than that of a terminal group. For example, $HgCo(NCS)_4$(Co—NCS—Hg) absorbs at 2137 cm^{-1}, whereas $(NEt_4)_2[Co(—NCS)_4]$ absorbs at 2065 cm^{-1}. According to Chatt and Duncanson,[559] the CN stretching frequencies of Pt(II) complexes are 2182–2150 cm^{-1} for the bridging and 2120–2100 cm^{-1} for the terminal NCS group. $[(P(n\text{-}Pr)_3)_2Pt_2(SCN)_2Cl_2]$ (compound I) exhibits one bridging CN stretching, whereas $[(P(n\text{-}Pr)_3)_2Pt_2(SCN)_4]$ (compound II) exhibits both bridging and terminal CN stretching bands. Thus the IR spectra suggest that the structure of each compound is as follows:

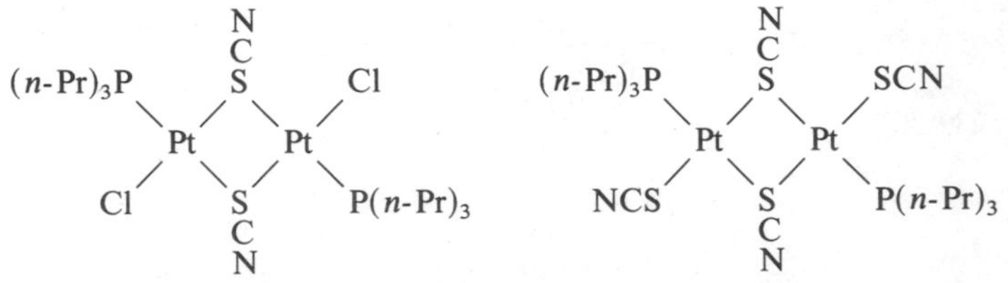

I

n-Pr: *n*-Propyl

II

Compound I, however, exists as two isomers, α and β, which absorb at 2162 and 2169 cm^{-1}, respectively. Chatt and Duncanson[559] originally suggested a geometrical isomerism in which two SCN groups were in a *cis* or *trans* position with respect to the central ring. Later,[560–562] "bridge isomerism" such as the following was demonstrated by X-ray analysis:

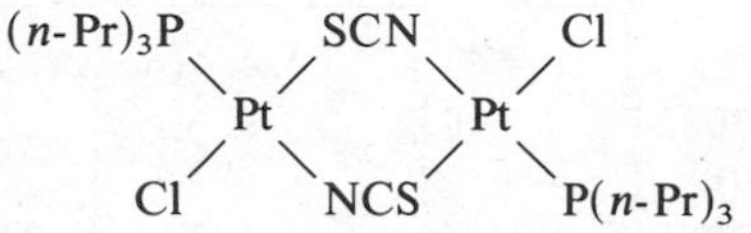

α

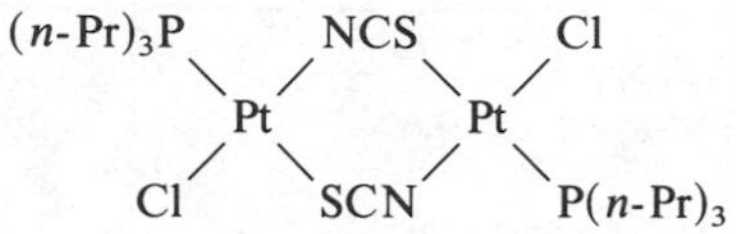

β

The IR spectra of metal complexes containing bridging NCS groups have been reported for $Sn(NCS)_2$,[563] $M(py)_2(NCS)_2$[M = Mn(II), Co(II), and Ni(II)],[564] $[Me_3Pt(NCS)]_4$,[565] and $M[Pt(SCN)_6]$ [M = Co(II). Ni(II), Fe(II), etc.].[566]

According to X-ray analysis,[567] the $[Re_2(NCS)_{10}]^{3-}$ ion contains solely N-bonded bridging thiocyanate groups which exhibit ν(CN) near 1900 cm^{-1}.

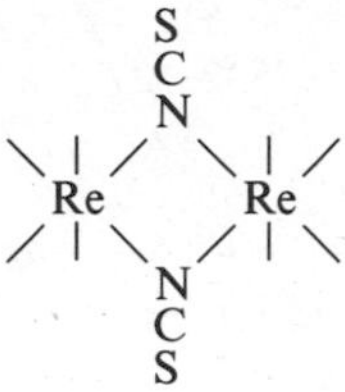

(All terminal NCS groups are also N-bonded.)

(2) Selenocyanato (SeCN) Complexes

The SeCN group also coordinates to a metal through the nitrogen (M—NCSe) or the selenium (M—SeCN) or both (M—NCSe—M′). Again, Class A metals tend to form M–N bonds, while Class B metals prefer to form M–Se bonds. Although the number of SeCN complexes studied is much smaller than that of SCN complexes, these studies suggest the following trends:

1. ν(CN) is below 2080 cm^{-1} for N-bonded, but higher for Se-bonded, complexes. The ν(CN) of a bridged complex $[HgCo(NCSe)_4]$ is at 2146 cm^{-1}.[568]
2. The ν(CSe) is at 700–620 cm^{-1} for N-bonded and 550–500 cm^{-1} for Se-bonded complexes.
3. The δ(NCSe) of N-bonded complexes are above 400 cm^{-1}, whereas Se-bonded complexes show at least one component of δ(NCSe) below 400 cm^{-1}.
4. The integrated intensity of ν(CN) is larger for the N-bonded than for the Se-bonded group.[569]

Table III-42 lists the observed frequencies of typical N-bonded and Se-bonded complexes.

Burmeister and Gysling[574] observed that in $[PdL_2(SeCN)_2]$-type compounds the effect of changing the π-bonding ability and basicity of L on the Pd–SeCN

TABLE III-42. VIBRATIONAL FREQUENCIES OF ISOSELENOCYANATO AND SELENOCYANATO COMPLEXES (CM^{-1})

Compound[a]	ν(CN)	ν(CSe)	δ(NCSe)	Ref.
K[NCSe]	2070	558	424, 416	II-256
$R_4[Mn(-NCSe)_6]$	2079, 2082 2070	640 617	424	569
$R_2[Fe(-NCSe)_4]$	2067, 2055	673, 666	432	569
$R_4[Ni(-NCSe)_6]$	2118, 2102	625	430	569
$[Ni(pn)_2(-NCSe)_2]$	2096, 2083	692	—	570
$R'_2[Co(-NCSe)_4]$	2053	672	433, 417	571
$[Co(NH_3)_5(-NCSe)](NO_3)_2$	2116	624	—	572
$R_2[Zn(-NCSe)_4]$	2087	661	429	569
$[Cu(pn)_2(-SeCN)_2]$	2053, 2028	—	—	570
$R_3[Rh(-SeCN)_6]$	2104, 2071	515	—	569
$R''_2[Pd(-SeCN)_4]$	2114, 2105	521	410, 374	573
$R_2[Pt(-SeCN)_4]$	2105, 2060	516	—	569
$[Pt(bipy)(-SeCN)_2]$	2135, 2125	532, 527	—	572
$K_2[Pt(-SeCN)_6]$	2130	519	390, 379 367	571

[a] $R=[N(n\text{-}C_4H_9)_4]^+$; $R'=[N(C_2H_5)_4]^+$; $R''=[N(CH_3)_4]^+$; pn = propylenediamine; bipy = 2,2′-bipyridine.

bonding is negligible in contrast to the analogous SCN complexes. A pair of true linkage isomers has been isolated and characterized by infrared spectra for $[(\pi\text{-Cp})Fe(CO)(PPh_3)(SeCN)]$[575] and $[Pd(Et_4dien)(SeCN)]BPh_4$,[576] where Et_4dien is 1,1,7,7-tetraethyldiethylenetriamine.

(3) Cyanato (OCN) Complexes

The OCN group may coordinate to a metal through the nitrogen (M—NCO) or the oxygen (M—OCN) or both. Thus far, the majority of complexes are reported to be N-bonded. Table III-43 lists the observed frequencies of N-bonded NCO groups in typical complexes; ν_a(NCO) and ν_s(NCO) denote vibrations consisting mainly of ν(CN) and ν(CO), respectively. For $ML_2(NCO)_2$ (M = Pd or Pt; L = NH_3, py, etc.) and $In(NCO)_3L_3$ (L = py, DMSO, etc.), see Refs. 585 and 586, respectively.

Thus far, O-bonded structures have been suggested for $[M(OCN)_6]^{n-}$ [M = Mo(III), Re(IV), and Re(V)].[583] Anderson and Norbury[587] prepared the first example of linkage isomers: yellow $Rh(PPh_3)_3(NCO)$ and orange $Rh(PPh_3)_3(OCN)$. The integrated ν(CN) intensity of the former is smaller than that of the free ion, whereas the intensity of the latter is larger than that of the free ion. Also, the latter exhibits two δ(OCN) at 607 and 590 cm^{-1}, whereas the former shows only one band at 592 cm^{-1}.

TABLE III-43. VIBRATIONAL FREQUENCIES OF ISOCYANATO COMPLEXES (cm^{-1})

Compound	ν_a(NCO)	ν_s(NCO)	δ(NCO)	Refs.
K[NCO]	2155	1282, 1202	630	II-253
$Si(NCO)_4$	2284	1482	608, 546	577
$Ge(NCO)_4$	2247	1426	608, 528	577
$[Zn(NCO)_4]^{2-}$	2208	1326	624	578
$[Mn(NCO)_4]^{2-}$	2222	1335	623	579, 580
$[Fe(NCO)_4]^{2-}$	2182	1337	619	579, 580
$[Co(NCO)_4]^{2-}$	2217, 2179	1325	620, 617	579, 580
$[Ni(NCO)_4]^{2-}$	2237, 2186	1330	619, 617	579
$[Fe(NCO)_4]^{-}$	2208, 2171	1370	626, 619	579
$[Pd(NCO)_4]^{2-}$	2200-2190	1319	613, 604, 594	581
$[Sn(NCO)_6]^{2-}$	2270, 2188	1307	667, 622	581
$[Zr(NCO)_6]^{2-}$	2205	1340	628	582
$[Mo(OCN)_6]^{3-}$	2205	1296, 1140	595	583
$[Ln(NCO)_6]^{3-}$ [a]	2190	1333	633	584

[a] Ln = Yb or Lu.

A bridging NCO group of the type:

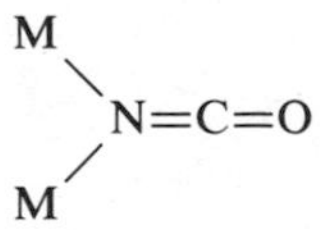

has been proposed for $ML_2(NCO)_2$ (M = Mn, Fe, Co, and Ni; L = 3- or 4-CN-py)[588] and $Re_2(CO)_8(NCO)_2$.[589] Forster and Horrocks[578] carried out normal coordinate analyses on $[Zn(NCX)_4]^{2-}$ anions, where X is O, S, or Se.

(4) Fulminato (CNO) Complexes

The fulminate (CNO^-) ion may coordinate to a metal through the carbon (M—CNO) or the oxygen (M—ONC) or both. So far, all the complexes containing the CNO group are presumed to be C-bonded. Table III-44 lists the observed CNO frequencies of typical fulminato complexes. Thus far, not much work has been done on fulminato complexes.

TABLE III-44. OBSERVED FREQUENCIES OF TYPICAL FULMINATO
COMPLEXES (CM^{-1})

Ion	$\nu(CN)$	$\nu(NO)$	$\delta(CNO)$	Ref.
$[CNO]^-$	2052	1057	471	590
$[Ag(CNO)_2]^-$	2119	1144	—	591
$[Fe(CNO)_6]^{4-}$	2187	1040	514	591
			466	
$[Hg(CNO)_4]^{2-}$	2130	1143	—	592
$[Ni(CNO)_4]^{2-}$	2184	1122	479	593
			470	
$[Zn(CNO)_4]^{2-\ a}$	2146 (A_1)	1177 (A_1)	498 (E)	594
	2130 (F_2)	1154 (F_2)	475 (F_2)	
$[Pt(CNO)_4]^{2-\ b}$	2194 (A_{1g})	1174 (A_{1g})	476 (B_{2g})	594
	2189 (B_{1g})	1140 (B_{1g})	453 (E_g)	

^{a} T_d symmetry.
^{b} D_{4h} symmetry.

(5) Azido (N_3) Complexes

Table III-45 lists the observed frequencies of typical azido complexes. The two N_3 groups around the Hg atom in $Hg_2(N_3)_2$ are in the *trans* position (C_{2h}), whereas they are in a twisted configuration (C_2) in $Hg(N_3)_2$. The former exhibits one $\nu_a(N_3)$ at 2080 cm^{-1}, whereas the latter shows two $\nu_a(N_3)$ at 2090 and 2045 cm^{-1}.[598] For Co(III) azido ammine complexes and $[M(N_3)_2(py)_2]$ (M = Cu, Zn, and Cd), see Refs. 599 and 600, respectively. The bridging azido groups are found in

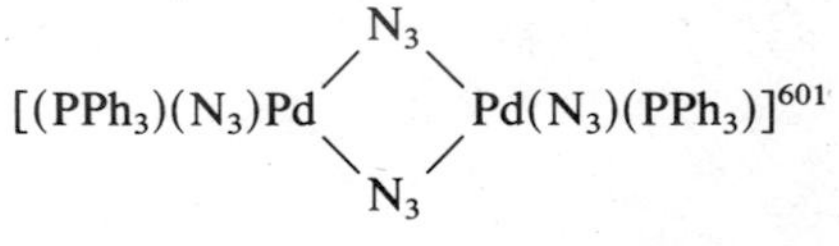

$$[(PPh_3)(N_3)Pd \diagdown^{N_3}_{N_3} \diagdown Pd(N_3)(PPh_3)]^{601}$$

and

$$[(acac)_2Co \diagdown^{N_3}_{N_3} \diagdown Co(acac)_2].^{602}$$

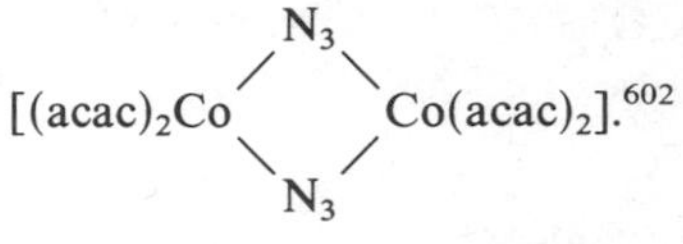

However, it is not possible to determine the structures of these bridges from infrared spectra. Forster and Horrocks[596] made complete assignments of vibrational spectra of the $[Co(N_3)_4]^{2-}$ and $[Zn(N_3)_4]^{2-}$ (D_{2d}) and $[Sn(N_3)_6]^{2-}$ (D_{3d}) ions. The spectra suggest that the M—NNN bonds in these anions are not linear.

TABLE III-45. VIBRATIONAL FREQUENCIES OF AZIDO COMPLEXES (CM^{-1})

Compound[a]	ν_a(NNN)	ν_s(NNN)	δ(NNN)	ν(MN)	Refs.
$K[N_3]$	2041	1344	645	—	II-254
$R_2[Pt(N_3)_4]$	2075, 2060 2024, 2029	1276	582	394	595
$R[Au(N_3)_4]$	2030, 2034	1261 1251	578	432	595
$R''_2[Zn(N_3)_4]$	2097, 2058	1330 1282	—	—	595
$R_2[VO(N_3)_4]$	2088, 2051 2092, 2060 2005	1340	652	442 405	595
$R_2[Pb(N_3)_6]$	2045, 2056 2037	1262 1253	640 597	327 313	595
$R_2[Pt(N_3)_6]$	2022, 2028	1275 1262 1253	578	402 397 320	595
$R'_2[Co(N_3)_4]$	2089, 2050	1338 1280	642 610	368	596
$R'_2[Zn(N_3)_4]$	2098, 2055	1342 1290	649 615	351 295	596
$R'_2[Sn(N_3)_4]$	2115, 2080	1340	659 601	390 330	596
trans-$R_2[TiCl_4(N_3)_2]$	2072, 2060	1344	610	—	597

[a] $R = [As(Ph)_4]^+$; $R' = [N(C_2H_5)_4]^+$; $R'' = [P(Ph)_4]^+$.

III-16. CARBONYL AND NITROSYL COMPLEXES

In the last few decades, a large number of carbonyl complexes have been synthesized, and their spectra and structures have been studied exhaustively. This section describes only typical results obtained from these investigations. For more comprehensive information, several review articles[603–608] should be consulted.

Most carbonyl complexes exhibit strong and sharp ν(CO) bands at ca. 2100–1800 cm^{-1}. Since ν(CO) is generally free from coupling with other modes and is not obscured by the presence of other vibrations, studies of ν(CO) alone often provide valuable information about the structure and bonding of carbonyl complexes. In the majority of compounds, ν(CO) of free CO (2155 cm^{-1}) is shifted to lower frequencies. In terms of simple MO theory, this observation has been explained as follows. First, the σ-bond is formed by donating 5σ electrons of CO to the empty orbital of the metal (see Fig. III-38). This tends to raise ν(CO), since the 5σ orbital is slightly antibonding. Second, the π-bond is formed by back-donating the $d\pi$-electrons of the metal to an empty antibonding orbital, the $2p\pi^*$ orbital of CO. This tends to lower

TABLE III-46. VIBRATIONAL FREQUENCIES AND BAND ASSIGNMENTS OF MONONUCLEAR METAL CARBONYLS (CM^{-1})[a]

Compound	Symmetry	State	$\nu(CO)$	$\nu(MC)$	$\delta(MCO)$	$\delta(CMC)$	Ref.
$Ni(CO)_4$	$\mathbf{T}_d$	Gas	2131 (A_1)	367.5 (A_1)	380 (E)	64 (E)	610
			2057.6 (F_2)	421 (F_2)	458.8 (F_2)	80 (F_2)	
					300 (F_1)		
$[Co(CO)_4]^-$	$\mathbf{T}_d$	DMF sol'n	2002 (A_1)	431 (A_1)	523 (F_2)	91 (E)	611
			1890 (F_2)	556 (F_2)			
$[Fe(CO)_4]^{2-}$	$\mathbf{T}_d$	Aqueous sol'n	1788 (A_1)	464 (A_1)	550 (F_2)	100–85	612
			1788 (F_2)	644 (F_2)	785 (E)	(E, F_2)	
$Fe(CO)_5$	$\mathbf{D}_{3h}$	Liquid	2116 (A_1')	418 (A_1')	278 (A_2')	107 (E')	613
			2030 (A_1')	381 (A_1')	653 (E')	64 (E')	614
			1989 (E')	482 (E')	559 (E')		
					491 (E'')		
					448 (E'')		
$Cr(CO)_6$	$\mathbf{O}_h$	Gas	2118.7 (A_{1g})	379.2 (A_{1g})	364.1 (F_{1g})	97.2 (F_{1u})	615
			2026.7 (E_g)	390.6 (E_g)	668.1 (F_{1u})	89.7 (F_{2g})	
			2000.4 (F_{1u})	440.5 (F_{1u})	532.1 (F_{2g})	67.9 (F_{2u})	
					510.9 (F_{2u})		

$Mo(CO)_6$	$\mathbf{O}_h$	Gas	2120.7 (A_{1g}) 2024.8 (E_g) 2000.3 (F_{1u})	391.2 (A_{1g}) 381 (E_g) 367.2 (F_{1u})	341.6 (F_{1g}) 595.6 (F_{1u}) 477.4 (F_{2g}) 507.2 (F_{2u})	81.6 (F_{1u}) 79.2 (F_{2g}) 60 (F_{2u})	615
$W(CO)_6$	$\mathbf{O}_h$	Gas	2126.2 (A_{1g}) 2021.1 (E_g) 1997.6 (F_{1u})	426 (A_{1g}) 410 (E_g) 374.4 (F_{1u})	361.6 (F_{1g}) 586.6 (F_{1u}) 482.0 (F_{2g}) 521.3 (F_{2u})	82.0 (F_{1u}) 81.4 (F_{2g}) 61.4 (F_{2u})	615
$[V(CO)_6]^-$	$\mathbf{O}_h$	CH_3CN sol'n	2020 (A_{1g}) 1894 (E_g) 1858 (F_{1u})	374 (A_{1g}) 393 (E_g) 460 (F_{1u})	356 (F_{1g}) 650 (F_{1u}) 517 (F_{2g}) 506 (F_{2u})	92 (F_{1u}) 84 (F_{2g})	616
$[Re(CO)_6]^+$	$\mathbf{O}_h$	CH_3CN sol'n	2197 (A_{1g}) 2122 (E_g) 2085 (F_{1u})	441 (A_{1g}) 426 (E_g) 356 (F_{1u})	354 (F_{1g}) 584 (F_{1u}) 486 (F_{2g}) 522 (F_{2u})	82 (F_{1u}) 82 (F_{2g})	616
$[Mn(CO)_6]^+$	$\mathbf{O}_h$	CH_3CN Sol'n (IR) Solid (R)	2192 (A_{1g}) 2125 (E_g) 2095 (F_{1u})	384 (A_{1g}) 390 (E_g) 412 (F_{1u})	347 (F_{1g}) 636 (F_{1u}) 500 (F_{2g}) 500 (F_{2u})	101 (F_{1u}) 101 (F_{2g})	617

[a] The three low-frequency vibrations may be coupled.

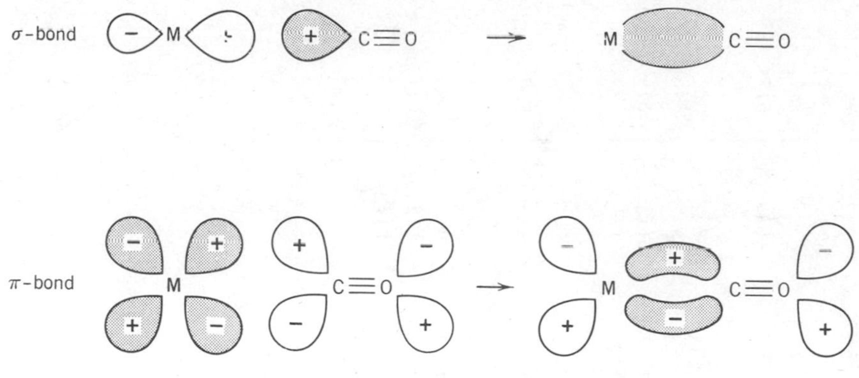

Fig. III-38. The σ- and π-bonding in metal carbonyls.

ν(CO). Although these two components of bonding are synergic, the net result is a drift of electrons from the metal to CO when the metal is in a relatively low oxidation state. Thus the ν(CO) of metal carbonyl complexes are generally lower than the value for free CO. The opposite trend is observed, however, when CO is complexed with metal halides in which the metals are in a relatively higher oxidation state [see Sec. III-16(6)].

If CO forms a bridge between two metals, its ν(CO) (1900–1800 cm^{-1}) is much lower than that of the terminal CO group (2100–2000 cm^{-1}). An extremely low ν(CO) (ca. 1300 cm^{-1}) is observed when the bridging CO group forms an adduct via its O atom [see Sec. III-16(2)].[609]

(1) Mononuclear Carbonyls

Table III-46 lists the observed frequencies and band assignments of mononuclear carbonyls of tetrahedral ($\mathbf{T}_d$), trigonal-bipyramidal ($\mathbf{D}_{3h}$), and octahedral ($\mathbf{O}_h$) structures. Complete normal coordinate analyses have been made on most of these carbonyls. Jones and co-workers[614,615] carried out extensive vibrational studies on $Fe(CO)_5$ and $M(CO)_6$ (M = Cr, Mo, and W), including their ^{13}C and ^{18}O species. They obtained the following F_{1u} force constants (mdyn/Å) from gas-phase spectra:

	$Cr(CO)_6$	$Mo(CO)_6$	$W(CO)_6$	
F(C≡O)	17.22	17.39	17.21	(GVF)
F(M—C)	1.64	1.43	1.80	

This result indicates that the M-C bond strength increases in the order Mo < Cr < W, an order also supported by a Raman intensity study of these compounds.[618] On the other hand, Hendra and Qurashi[619] related the Raman intensity ratio of two A_{1g} modes, $I(\nu_1$, CO stretching)/$I(\nu_2$, MC stretching), to the π-character of the M-C bond, and concluded that the M-C bond strength increases in the order Cr < W < Mo < Re(I).

For more recent Raman studies on hexacarbonyls, see Refs. 620 (vapor phase) and 621 (Ar matrix).

The existence of $Ru(CO)_5$ and $Os(CO)_5$ has been confirmed by their IR spectra in heptane solution. Both compounds seem to be trigonal-bipyramidal.[622] In the $M(CO)_4$ series ($\mathbf{T}_d$), $\nu(CO)$ decreases and $\nu(MC)$ increases in going from $Ni(CO)_4$ to $[Co(CO)_4]^-$ to $[Fe(CO)_4]^{2-}$. This result indicates that the $M \rightarrow CO$ π-back-donation is increasing in the order $Ni(0) < Co(-I) < Fe(-II)$. Highly reduced species such as $Na_4[M(CO)_4]$ (M = Cr, Mo, and W of formal oxidation state, −IV) exhibit very low $\nu(CO)$ (1530–1460 cm^{-1}).[623] Edgell and co-workers[624] attributed the band at 413 cm^{-1} of $Li[Co(CO)_4]$ in a THF solution to the vibrations of the alkali ions, which form ion pairs with $[Co(CO)_4]^-$: For sodium and potassium salts, the corresponding bands are observed at 192 and 142 cm^{-1}, respectively. Based on computer-aided curve analysis of IR spectra of $Na[Co(CO)_4]$ in the $\nu(CO)$ region, they[625] also demonstrated that there are three kinds of ion sites in THF solution each of which exhibits different spectra. Their structures and $\nu(CO)$ are shown in Fig. III-39.

S S S S Na$^+$ S A$^-$ S S S S	S S S Na$^+$ A$^-$ S S S	S S S S A^{-1} Na$^+$ A$^-$ S S S S
Solvent separated ion pair	Contact ion pair	Triple ion pair
1887 cm^{-1} ($\mathbf{T}_d$)	1899, 1856 cm^{-1} ($\mathbf{C}_{3v}$)	1906, 1846 cm^{-1} ($\mathbf{C}_{3v}$)

Fig. III-39. Structures and $\nu(CO)$ of three ion sites of $Na[Co(CO)_4]$ in THF solution. (A^- and S indicate the anion and solvent, respectively.)

(2) Polynuclear Carbonyls

Since polynuclear carbonyls take a variety of structures, elucidation of their structures by vibrational spectroscopy has been a subject of considerable interest in the past. The principles involved in these structure determinations were described in Sec. I-10. However, the structures of some polynuclear complexes are too complicated to allow elucidation by simple application of selection rules based on symmetry. Thus the results are often ambiguous. In these cases, one must resort to X-ray analysis to obtain definitive and accurate structural information. However, vibrational spectroscopy is still useful in elucidating the structures of metal carbonyls in solution.

According to X-ray analysis,[626] $Co_2(CO)_8$ takes structure III of Fig. III-40. For this $\mathbf{C}_{2v}$ structure, five terminal and two bridging $\nu(CO)$ are expected to be infrared active; the former are observed at 2075, 2064, 2047, 2035, and 2028 cm^{-1}, and the latter are located at 1867 and 1859 cm^{-1}.[627] In addition to the usual two isomers (III and IV of Fig. III-40), the IR spectra of $CO_2(CO)_8$

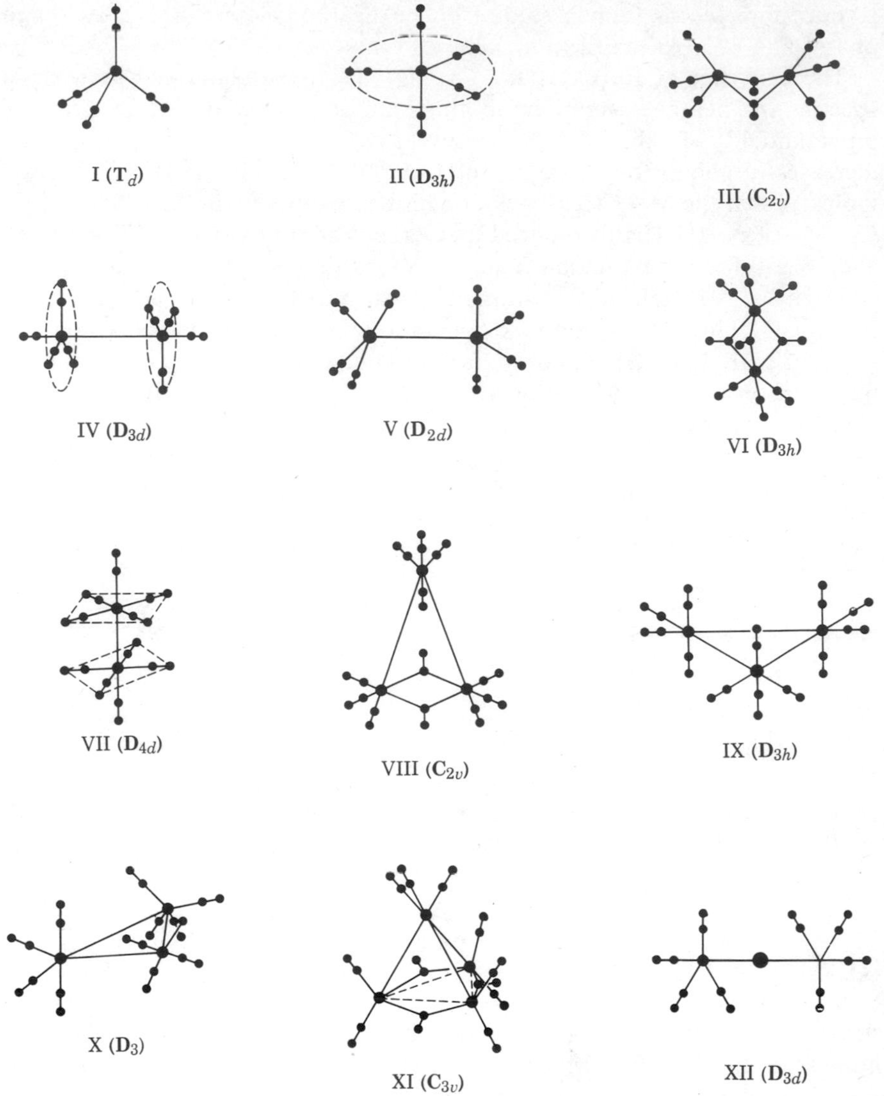

Fig. III-40. Structures of metal carbonyls.

in various matrices show the presence of the third isomer (structure V) after the matrices are photolyzed.[628]

The infrared spectrum of $Fe_2(CO)_9$ was first obtained by Sheline and Pitzer,[629] who observed two terminal (2080 and 2034 cm^{-1}) and one bridging (1828 cm^{-1}) CO stretching bands. This result agrees with that expected from structure VI, determined by X-ray analysis.[630] Again according to X-ray analysis,[631] the structure of $Mn_2(CO)_{10}$ is that shown by VII of Fig. III-40. This D_{4d} structure predicts four Raman- and three infrared-active CO stretching bands.

Adams et al.[632] observed the former at 2116 (A_1), 1997 (A_1), 2024 (E_2), and 1981 (E_3) cm^{-1}, and Bor[633] observed the latter at 2046 (B_2), 1984 (B_2), and 2015 (E_1) cm^{-1} in solution. Levenson et al.[634] confirmed these infrared assignments by studying the polarization properties of the three bands in a nematic liquid crystal. The structures of $Re_2(CO)_{10}$ and $Tc_2(CO)_{10}$ are similar to that of $Mn_2(CO)_{10}$. The vibrational spectra of $Re_2(CO)_{10}$, $Tc_2(CO)_{10}$, $MnTc(CO)_{10}$, and $TcRe(CO)_{10}$ are reported.[632,635–637]

Figure III-40 (VIII) shows the structure of $Fe_3(CO)_{12}$ determined by X-ray analysis.[638] This structure can account for Mossbauer[639] and solid-state infrared spectra. In solution, the infrared spectrum does not agree with that expected for structure VIII; the bridging ν(CO) is very weak and the terminal ν(CO) region is broad without resolution. Cotton and Hunter[640] suggest that a whole range of structures varying from $\mathbf{D}_{3h}$ (structure IX) to $\mathbf{C}_{2v}$ (structure VIII) are in equilibrium, the majority being close to $\mathbf{D}_{3h}$. Johnson suggests the presence of a new isomer of $\mathbf{D}_3$ symmetry, shown by structure X.[641] According to X-ray analysis,[642] $Os_3(CO)_{12}$ takes the $\mathbf{D}_{3h}$ structure (IX), for which four terminal ν(CO) should be infrared active. Huggins et al.[643] assigned them at 2068 (E'), 2035 (A_2''), 2014 (E'), and 2002 (E') cm^{-1}. Quicksall and Spiro[644] assigned the Raman spectra of $Os_3(CO)_{12}$ and analogous $Ru_3(CO)_{12}$, for which six ν(CO) are expected in the Raman spectrum. For $Os_3(CO)_{12}$, they are observed at 2130 (A_1'), 2028 (E''), 2019 (E'), 2006 (A_1'), 2000 (E'), and 1989 (E') cm^{-1}.

According to X-ray analysis,[645] $Co_4(CO)_{12}$ takes the $\mathbf{C}_{3v}$ structure (XI of Fig. 111-40), for which six terminal and two bridging ν(CO) are infrared active. Vibrational analyses have been made on $Co_4(CO)_{12}$ and $Rh_4(CO)_{12}$.[646] The ν(CO) frequencies are reported for $[CoRu_3(CO)_{13}]^-$.[647] Stammreich et al.[648] proposed structure XII of $\mathbf{D}_{3d}$ symmetry from a Raman study of $M[Co(CO)_4]_2$ (M = Cd or Hg). For this structure, three ν(CO) are Raman active and the other three are infrared active. The former were observed at 2107 (A_{1g}), 2030 (A_{1g}), and 1990 (E_g) cm^{-1}.[648] and the latter were located at 2072 (A_{2u}), 2022 (A_{2u}), and 2007 (E_u) cm^{-1}.[649] Ziegler et al.[650] made complete vibrational assignments of the $M[Co(CO)_4]_2$ series, where M is Zn, Cd, and Hg.

Figure III-41 shows the structures of unusual bridging carbonyl compounds found by X-ray analysis. "Semibridging" carbonyls (structure I) are present in $(Cp)_2V_2(CO)_5$, and their ν(CO) have been assigned to the bands at 1871 and 1832 cm^{-1}.[651] A "semitriple bridging" carbonyl group (structure II) was found in $[(Cp)_2Rh_3(CO)_4]^-$ [652] and the band at 1693 cm^{-1} is probably due to this carbonyl. Another "semitriple bridging" carbonyl group (structure III) in $(Cp)_3Nb_3(CO)_7$ exhibits an extremely low ν(CO) at 1330 cm^{-1}.[653] In $(Cp)_2Ti(CO)R$ (structure IV), the ν(CO) was observed at 1470 cm^{-1}.[654] The carbide carbon atom in $Co_6C(CO)_{12}S_2$ is at the center of the trigonal prism formed by six Co atoms (structure V). Based on ^{12}C–^{13}C isotope shifts, its Co–C vibrations have been assigned at 819 and 548 cm^{-1}.[655] In the $[Os_{10}C(CO)_{24}]^-$ ion, the carbide carbon atom is located at the center of the Os_{10} skeleton.[656] The ^{12}C–^{13}C isotope experiments show the presence of three Os–C (carbide) vibrations at 772.8, 760.3, and 735.4 cm^{-1}.[657]

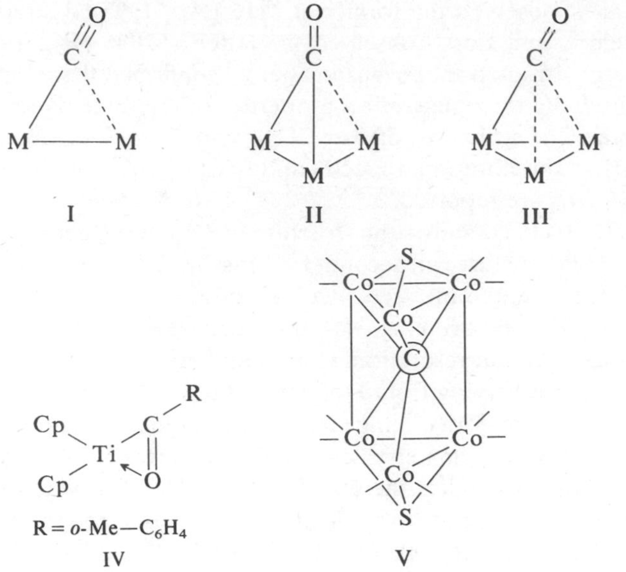

Fig. III-41. Structures of unusual metal carbonyls.

As stated before, Shriver et al.[658] discovered that the O atom of the bridging CO group can form a bond with a Lewis acid such as $AlEt_3$. Kristoff and Shriver[609] found that $Co_2(CO)_8$ forms an adduct of the following type:

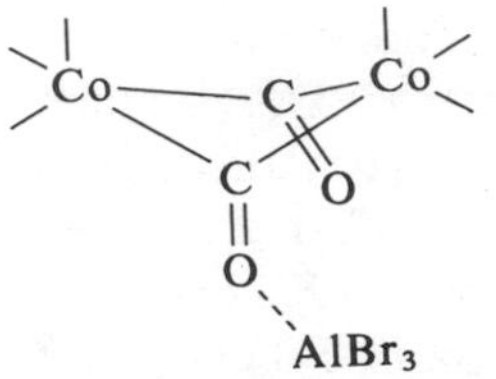

As expected from this structure, the adduct exhibits two bridging ν(CO) in the infrared: one at 1867 cm^{-1}, which is 15 cm^{-1} higher, and the other at 1600 cm^{-1}, which is 232 cm^{-1} lower, than that of the parent compound. In the case of $Fe_2(CO)_9AlBr_3$, only one bridging ν(CO) is observed at 1557 cm^{-1}. This suggests the following structure, which resulted from rearrangement of the CO groups of the parent compound.

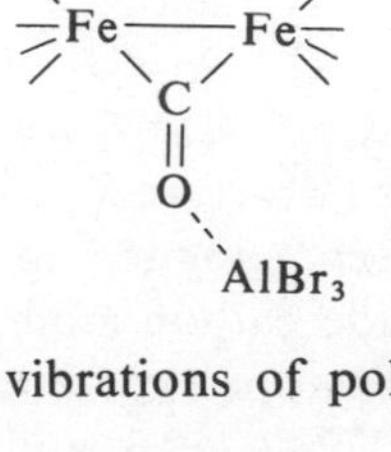

For metal–metal stretching vibrations of polynuclear carbonyls, see Sec. III-20.

(3) Metal Carbonyls Containing Other Ligands

Carbonyl halides exhibit bands characteristic of both M—CO and M—X (X: a halogen) groups. The MX vibrations will be discussed in Sec. III-19. If the CO group is substituted by a halogen, ν(CO) tends to shift to a higher frequency, since the M–CO π-back-bonding decreases as the metal becomes more electropositive by forming a M–X bond. In a series of halogenocarbonyls of the same type, ν(CO) is highest for the chloro compound and lowest for the iodo compound, the bromo compound being between the two, as expected from their electronegativities. Table III-47 lists the observed frequencies for ν(CO) and ν(MX). Complete assignments have been made for all these compounds.

El-Sayed and Kaesz[664] studied the ν(CO) of $M_2(CO)_8X_2$ (M = Mn, Tc, and Re; X = Cl, Br, and I), and proposed the halogen-bridging structure I shown in Fig. III-42. Four infrared-active ν(CO) have been observed in accordance with this structure. Garland and Wilt[665] interpreted the infrared spectrum of $Rh_2(CO)_4X_2$ (X = Cl and Br) on the basis of the $\mathbf{C}_{2v}$ structure II (Fig. III-42) found by X-ray analysis.[666] As predicted, three infrared-active ν(CO) have been observed for this compound. Johnson et al.[667] studied the exchange of $C^{18}O$ with CO groups of $Rh_2(CO)_4X_2$ (X = Cl, Br, I, etc.) with time by following the variation of infrared spectra in the ν(CO) region. Cotton and Johnson[668]

TABLE III-47. VIBRATIONAL FREQUENCIES OF METAL CARBONYL HALIDES (CM^{-1})

Compound	IR or Raman and Symmetry	ν(CO)	ν(MX)	Ref.
$Mn(CO)_5Cl$	IR ($\mathbf{C}_{4v}$)	2138 (A_1) 2056 (E) 2000 (A_1)	291 (A_1)	659
$Mn(CO)_5Br$	IR ($\mathbf{C}_{4v}$)	2138 (A_1) 2052 (E) 2007 (A_1)	222 (A_1)[a]	660
fac-$[Os(CO)_3Cl_3]^-$	Raman ($\mathbf{C}_{3v}$)	2125 (A_1) 2022 (E) 2033 (E)	321 (A_1) 287 (E)	661
cis-$[Os(CO)_2Cl_4]^{2-}$	Raman ($\mathbf{C}_{2v}$)	2016 (A_1) 1910 (B_2)	316 (A_1) 281 (A_1) 308 (B_2)	661, 662
$[Os(CO)Cl_5]^{3-}$	IR ($\mathbf{C}_{4v}$)	1968 (A_1)	332 (A_1)[a] 316 (A_1)[a] 306 (E)	661
$[Pt(CO)Cl_3]^-$	IR ($\mathbf{C}_{2v}$)	2120 (A_1)	331 (A_1) 310 (A_1)	661 663

[a] Raman frequency.

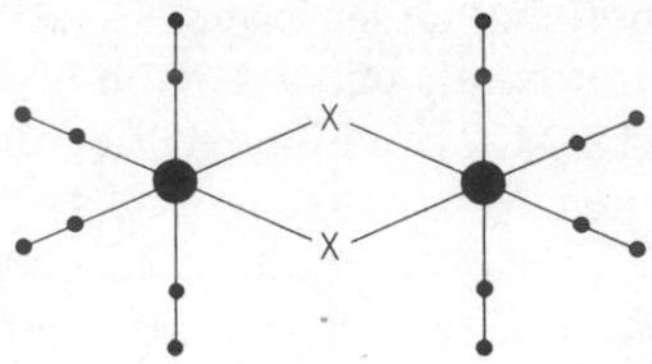

I (D_{2h})

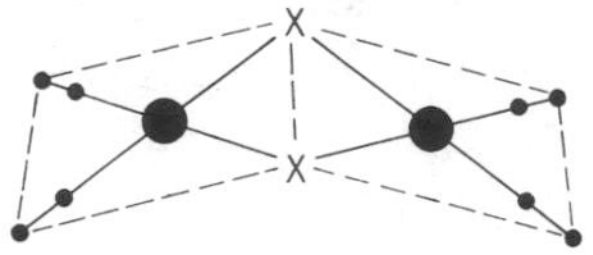

II (C_{2v})

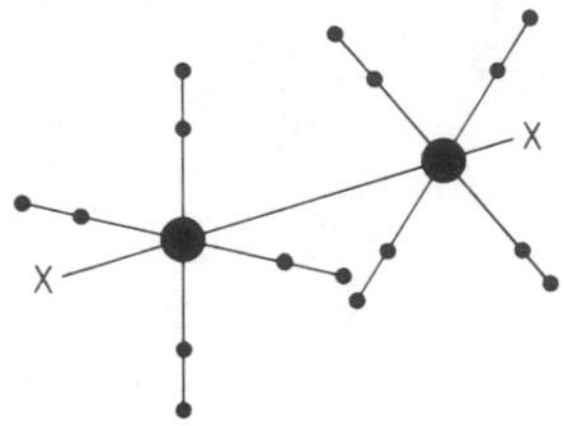

III (D_{4d})

Fig. III-42. Structures of metal carbonyl halides.

proposed the staggered structure (III) for $Fe_2(CO)_8I_2$, since only two ν(CO) were observed in the infrared.

If CO is replaced by a phosphine, ν(CO) decreases since the latter is a strong σ-donor but a weak π-acceptor. Ligands such as arsines, amines, and isonitriles give similar trends. Vibrational assignments have been made for ν(CO) of most of these carbonyl derivatives. Table III-48 lists the observed ν(CO) of typical compounds; the ν(CO) of the last compound are very low

TABLE III-48. CO STRETCHING FREQUENCIES OF METAL CARBONYLS CONTAINING OTHER LIGANDS (CM^{-1})

Compound	IR or Raman and Symmetry	ν(CO)	Ref.
$Ni(CO)_3(PMe_3)$	Raman (C_{3v})	2069 (A_1), 1980 (E)	669
$Fe(CO)_4(PMe_3)$	Raman (C_{3v})	2051 (A_1), 1967 (A_1) 1911 (E)	670
$Fe(CO)_4(AsMe_3)$	Raman (C_{3v})	2050 (A_1), 1964 (A_1) 1911 (E)	670
$Co(CO)_5(PEt_3)$	IR (C_{4v})	2060 (A_1), 1973 (B_1) 1943 (A_1), 1935 (E)	671
$W(CO)_5(NMe_3)$	IR (C_{4v})	2073 (A_1), 1932 (E) 1920 (A_1)	672
$W(CO)_4(bipy)$	IR (C_{2v})	2010 (A_1), 1900 (B_1) 1874 (A_1), 1832 (B_2)	673
$W(CO)_2(bipy)_2$	IR (C_2)	1778 (A_1), 1719 (B_2)	673

(1778 and 1719 cm^{-1}).[673] Vibrational assignments have been reported for *cis*-$M(CO)_4(L\text{-}L)$ [M = Cr, Mo, and W, and L-L = $(C_6H_5)_2P\text{-}(CH_2)_n\text{-}P(C_6H_5)_2$ ($n = 1 \sim 3$)][674] and $M(CO)_5(CS)$ (M = Cr and W) and their CSe analogs.[675]

Low-frequency infrared spectra have been reported for $M(CO)_{6-n}(PR_3)_n$ (M = Cr, Mo, and W),[676] $M(CO)_{6-n}(CH_3CN)_n$ (M = Cr and W: $n = 1$ and 2),[677] and $Fe(CO)_4L$ (L = PPh_3, $AsPh_3$, and $SbPh_3$).[678] For phosphine and arsine complexes, see Sec. III-21.

(4) Normal Coordinate Calculations

Normal coordinate analyses on metal carbonyl compounds have been carried out by many investigators. Among them, Jones and co-workers have made the most extensive study in this field. For example, they performed rigorous calculations on the $M(CO)_6$ (M = Cr, Mo, and W) series,[615] $Fe(CO)_5$,[614] and $Mn(CO)_5Br$,[660] including their ^{13}C and ^{18}O analogs. For the last compound, 5 stretching, 16 stretching-stretching interaction, and 33 bending-bending interaction constants (GVF) were used to calculate its 30 normal vibrations.

On the other hand, Cotton and Kraihanzel[679] developed an approximation (C-K) method for calculating the CO stretching and CO—CO stretching interaction constants, while neglecting all other low-frequency modes. For $Mn(CO)_5Br$, they used only the five force constants[680] shown in Fig. III-43. Since only four CO stretching bands are observed for this type of compound, it was assumed that $\frac{1}{2}k_t = k_c = k_d$ holds. This was justified on the basis of the symmetry properties of the metal $d\pi$ orbitals involved. This C-K method has since been applied to many other carbonyls in making band assignments, in interpreting intensity data, and in discussing the bonding schemes of metal carbonyls.[605] It is clear that the choice of a rigorous approach (Jones) or a simplified method (C—K) depends on the availability of observed data and the purpose of the investigation. Jones[681] and Cotton[682] discuss the merits of their respective approaches relative to the alternative.

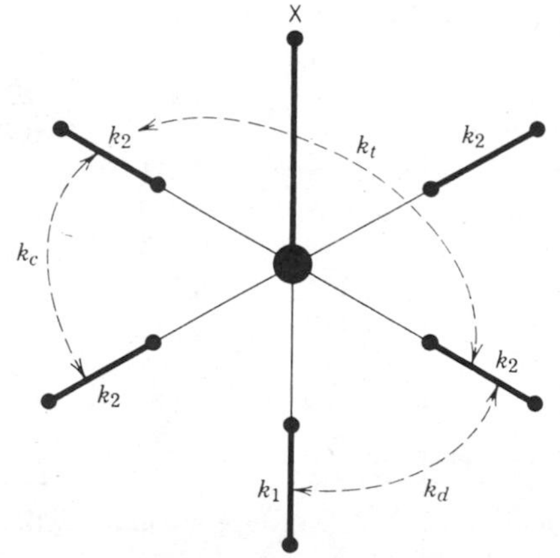

Fig. III-43. Definition of force constants for $M(CO)_5X$.

As mentioned earlier, ν(CO) of metal carbonyls are determined by two factors: (1) donation of the 5σ-electrons to the empty metal orbital tends to raise the ν(CO) since the 5σ orbital is slightly antibonding; and (2) back donation of metal $d\pi$-electrons to the $2p\pi$ orbitals of CO tends to lower ν(CO) since the $2p\pi$ orbitals are antibonding. Vibrational spectroscopy does not allow observation of these two effects separately since the observed ν(CO) and the corresponding force constant reflect only the net result of the two counteracting components. It is possible, however, to correlate the CO stretching force constants (C–K) with the occupancies of the 5σ and $2p\pi$ orbitals as calculated by MO theory. Table III-49 lists the results obtained for d^6 carbonyl halides and dihalides by Hall and Fenske.[683] It is interesting that the *trans*-CO in $Fe(CO)_4I_2$ and the *cis*-CO in $Mn(CO)_5Cl$ have almost the same force constants since the 5σ occupancy of the former is smaller by 0.102 than that of the latter, while the $2p\pi$ occupancy of the former is larger by 0.108 than that of the

TABLE III-49. CARBONYL ORBITAL OCCUPANCIES[a] AND FORCE CONSTANTS

Compound	Structure	5σ	$2\pi_x$	$2\pi_y$	k (mdyn/Å)[b]
$Cr(CO)_5Cl^-$	*trans*	1.407	0.355	0.355	14.07
$Cr(CO)_5Br^-$	*trans*	1.405	0.353	0.353	14.10
$Mn(CO)_4I_2^-$	*trans*	1.354	0.302	0.330	15.48
$Mn(CO)_4IBr^-$	*trans*	1.355	0.302	0.327	15.48
$Mn(CO)_4Br_2^-$	*trans*	1.357	0.302	0.325	15.50
$Cr(CO)_5Br^-$	*cis*	1.456	0.261	0.282	15.56
$Cr(CO)_5Cl^-$	*cis*	1.457	0.261	0.276	15.58
$Mn(CO)_5Cl$	*trans*	1.352	0.286	0.286	16.28
$Mn(CO)_5Br$	*trans*	1.350	0.286	0.286	16.32
$Mn(CO)_5I$	*trans*	1.349	0.286	0.286	16.37
$Mn(CO)_4I_2^-$	*cis*	1.402	0.251	0.251	16.75
$Mn(CO)_4IBr^-$	*cis*	1.404	0.241	0.252	16.77
$Mn(CO)_4Br_2^-$	*cis*	1.406	0.242	0.242	16.91
$Mn(CO)_5I$	*cis*	1.394	0.213	0.240	17.29
$Mn(CO)_5Br$	*cis*	1.394	0.212	0.228	17.39
$Fe(CO)_4I_2$	*trans*	1.293	0.252	0.285	17.43
$Mn(CO)_5Cl$	*cis*	1.395	0.211	0.218	17.46
$Fe(CO)_4Br_2$	*trans*	1.295	0.250	0.272	17.53
$Fe(CO)_5Br^+$	*trans*	1.287	0.233	0.233	17.93
$Fe(CO)_5Cl^+$	*trans*	1.289	0.233	0.233	17.95
$Fe(CO)_4I_2$	*cis*	1.337	0.221	0.221	17.95
$Fe(CO)_4Br_2$	*cis*	1.338	0.205	0.205	18.26
$Fe(CO)_5Cl^+$	*cis*	1.325	0.171	0.177	18.99
$Fe(CO)_5Br^+$	*cis*	1.325	0.171	0.193	19.00

[a] The *cis* and *trans* designations of the CO groups are made with respect to the position of the halogen or halogens.
[b] C–K force constants (see Ref. 683).

latter. It is also noteworthy that the *trans*-CO in $Fe(CO)_4I_2$ and the *cis*-CO in $Cr(CO)_5Cl^-$ have identical $2p\pi$ occupancies (0.537) but substantially different force constants (17.43 and 15.58 mdyn/Å, respectively). In this case, the difference in force constants originates in the difference in the 5σ occupancies (1.293 vs. 1.457). Hall and Fenske[683] found a linear relationship between the C-K CO stretching force constants and the occupancies of the 5σ and $2p\pi$ levels:

$$k = -11.73[2\pi_x + 2\pi_y + (0.810)5\sigma] + 35.81$$

A similar attempt has been made for a series of Mn carbonyls containing isocyanide groups.[684]

(5) Hydrocarbonyls

Hydrocarbonyls exhibit bands characteristic of both M—H and M—CO groups. Kaesz and Saillant[685] reviewed the vibrational spectra of metal carbonyls containing the hydrido group. Vibrational spectra of hydrido complexes containing other groups will be discussed in Sec. III-18. In general, the terminal M-H group exhibits a relatively sharp- and medium-intensity ν(MH) band in the 2200–1800 cm^{-1} region. The MH stretching band can be distinguished easily from the CO stretching band by the deuteration experiment.

Edgell and co-workers[686] assigned the infrared bands at 1934 and 704 cm^{-1} of $HCo(CO)_4$ to ν(CoH) and δ(CoH), respectively, and proposed structure I of Fig. III-44, in which the H atom is on the C_3 axis. Stammreich et al.[612] reported the Raman spectrum of $HFe(CO)_4^-$, which is expected to have a structure similar to that of $HCo(CO)_4$. According to X-ray analysis,[687] the $Mn(CO)_5$ skeleton of $HMn(CO)_5$ takes the $\mathbf{C}_{4v}$ structure shown in structure II of Fig. III-44. Kaesz and co-workers[688,689] assigned the infrared spectrum of $HMn(CO)_5$ in the ν(CO) region on the basis of this structure. The Raman spectra of $HMn(CO)_5$ and $HRe(CO)_5$ exhibit their ν(MH) at 1780 and 1824 cm^{-1}, respectively.[690] A complete vibrational assignment of gaseous $HMn(CO)_5$ has been made by Edgell et al.[691] The infrared spectrum of $H_2Fe(CO)_4$ in hexane at −78°C exhibits three or more ν(CO) above 2000 cm^{-1} and a weak, broad ν(FeH) at 1887 cm^{-1}. Thus Farmery and Kilner[692] suggested structure III of Fig. III-44. Table III-50 lists the observed frequencies of other hydrocarbonyl compounds.

It is rather difficult to locate the bridging ν(MH) in polynuclear hydrocarbonyls. These vibrations appear in the region from 1600 to 800 cm^{-1}, and are rather broad at room temperature although they are sharpened at low temperatures. Huggins et al.[697] were the first to suggest the presence of bridging hydrogens in $Re_3H_3(CO)_{12}$ (structure IV of Fig. III-44) since no terminal ν(ReH) bands were observed. Smith et al.[698] observed a very weak and broad band at 1100 cm^{-1} in the Raman spectrum of $Re_3H_3(CO)_{12}$ and assigned it to the bridging ν(ReH) since it shifted to 787 cm^{-1} upon deuteration.

Although structure V was proposed for $[M_2H(CO)_{10}]^-$ (M = Cr, Mo and W),[699] the W-H-W angle of the tungsten complex was found to be 137°.[700]

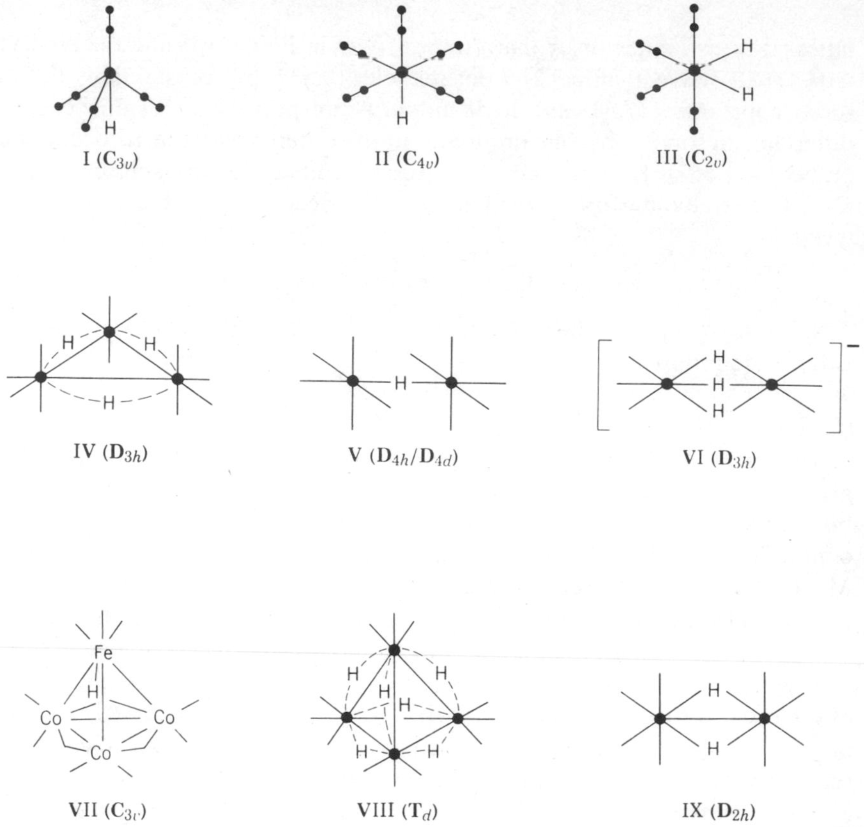

Fig. III-44. Structures of hydrocarbonyls.

TABLE III-50. VIBRATIONAL FREQUENCIES OF METAL HYDROCARBONYL COMPOUNDS (cm^{-1})[a]

Compound	ν(CO)	ν(MH)	δ(MH)	Ref.
$RhH(CO)(PPh_3)_3$	1926	2004	784	693
$IrH(CO)(PPh_3)_3$	1930	2068	822	693
$IrHCl_2(CO)(PEt_2Ph)_2$	2101	2008	—	694
$IrHBr_2(CO)(PEt_2Ph)_2$	2035	2232	—	694
$IrHCl_2(CO)(PPh_3)_2$	2027	2240	—	695
$OsHCl(CO)(PPh_3)_2$	1912	2097	—	695
$OsH_2(CO)_2(PPh_3)_2$	2014	1928	—	696
	1990	1873		

[a] For the configurations of these molecules, see the original references.

Shriver and co-workers[701] located the antisymmetric and symmetric stretching vibrations of the W-H-W bridge at 1683 and ~900 cm^{-1}, respectively. However, the latter splits into four bands at 960, 869, 832, and 702 cm^{-1}. Although the origin of this splitting is not clear, the possibility of Fermi resonance with an overtone or a combination band involving the ν(WC) or δ(WCO) was ruled out based on CO-$C^{18}O$ isotopic shifts.[701]

Ginsberg and Hawkes[702] suggested structure VI for $[Re_2H_3(CO)_6]^-$ since they could not observe any terminal ν(ReH) vibrationals. The bridging ν(FeH) band of $FeHCo_3(CO)_{12}$ in the infrared was finally located at 1114 cm^{-1} by Mays and Simpson,[703] using a highly concentrated KBr pellet. This band shifts to 813 cm^{-1} upon deuteration. On the basis of mass spectroscopic and infrared evidence, they proposed structure VII, in which the H atom is located inside the metal atom cage. From the spectra in the ν(CO) region, together with X-ray evidence, Kaesz et al.[704] proposed the $\mathbf{T}_d$ skeleton (structure VIII) for $[Re_4H_6(CO)_{12}]^{2-}$. It showed no terminal ν(ReH), but a broad bridging ν(ReH) centered at 1165 cm^{-1} was observed in its Raman spectrum. This band shifts to 832 cm^{-1} with less broadening upon deuteration. Bennett et al.[705] found no terminal ν(ReH) in the infrared spectrum of $Re_2H_2(CO)_8$. However, its Raman spectrum exhibits bands at 1382 and 1272 cm^{-1}, which shift to 974 and 924 cm^{-1}, respectively, upon deuteration. The $\mathbf{D}_{2h}$ structure (IX) was proposed for this compound.

Figure III-45 shows the Raman spectra of $Ru_4H_4(CO)_{12}$ and $Ru_4D_4(CO)_{12}$ obtained by Knox et al.[706] Two ν(RuH) bands at 1585 and 1290 cm^{-1} of the former compound are shifted to 1153 and 909 cm^{-1}, respectively, upon deuteration. Its infrared spectrum exhibits five ν(CO) instead of the two expected for $\mathbf{T}_d$ symmetry. Thus a structure of $\mathbf{D}_{2d}$ symmetry, which lacks two H atoms from structure VIII, was proposed.[707] In the $[Ru_6H(CO)_{18}]^-$ ion, the H atom is located at the center of an octahedron consisting of six Ru atoms.[708] Oxton et al.[709] located its ν(R–H) at 845 and 806 cm^{-1} (95 K) which are probably split by Fermi resonance.

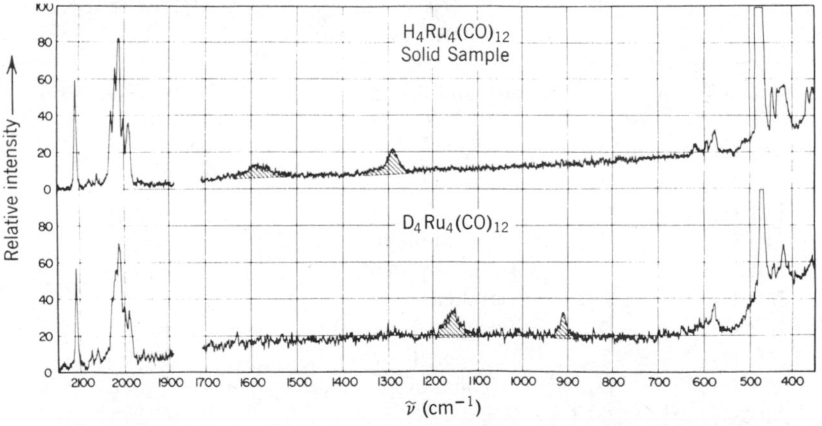

Fig. III-45. Raman spectra of $Ru_4H_4(CO)_{12}$ and its deuterated analog.[706]

As stated in Sec. III-6 (aquo complexes), the inelastic neutron scattering (INS) technique is very effective in locating hydrogen vibrations. White and Wright[710] found two hydrogen vibrations at 608 and 312 cm^{-1} in the INS spectrum of $Mn_3H_3(CO)_{12}$. However, the nature of these vibrations is not clear.

(6) Metal Carbonyls in Inert Gas Matrices

A number of unstable and transient metal carbonyls have been synthesized and their structures determined by vibrational spectroscopy in inert gas matrices. In most cases, only ν(CO) vibrations have been measured to determine the structures of these compounds since it is rather difficult to observe low-frequency modes in inert gas matrices.

These transient carbonyls can be prepared by two methods. The first involves direct reaction of metal vapor with CO diluted in inert gas matrices: $M + xCO \rightarrow M(CO)_x$. As discussed in Sec. I-22, DeKock[711] first succeeded in preparing the $Ni(CO)_x$ ($x = 1$-3) series by this method. Although the method yields a mixture of carbonyls of various stoichiometries, bands characteristic of each species can be determined in several ways: warm-up experiments, concentration-dependence studies, isotope substitutions, and so on.[712] Table III-51 lists the number of infrared-active ν(CO) predicted for possible structures. The structures of $M(CO)_2$, $M(CO)_3$, and $M(CO)_4$ (M = Ni,[711], Pd,[713,714] and Pt[715]) have been found to be linear, trigonal-planar, and tetrahedral, respectively, since all of these compounds exhibit only one ν(CO). In the case of the $M(CO)_{1-6}$ series (M = Ta,[711] U,[716] Pr, etc.[717]), it was more difficult to determine the structures of the transient species because the spectra were more complicated.

Moskovits and Ozin[712] determined the structures of a large number of metal carbonyls produced via matrix co-condensation reactions. They proposed an unusual isocarbonyl structure, O≡C—Au—O≡C for $Au(CO)_2$,[718] and the

TABLE III-51. NUMBER OF INFRARED-ACTIVE CO STRETCHING VIBRATIONS FOR $M(CO)_x$

Molecule	Symmetry	Structure	IR-Active ν(CO)
M(CO)	$\mathbf{C}_{\infty v}$	Linear	Σ^+
$M(CO)_2$	$\mathbf{D}_{\infty h}$	Linear	Σ_u^+
$M(CO)_2$	$\mathbf{C}_{2v}$	Bent	$A_1 + B_2$
$M(CO)_3$	$\mathbf{D}_{3h}$	Trigonal-planar	E'
$M(CO)_3$	$\mathbf{C}_{3v}$	Trigonal-pyramidal	$A_1 + E$
$M(CO)_4$	$\mathbf{T}_d$	Tetrahedral	F_2
$M(CO)_4$	$\mathbf{D}_{4h}$	Tetragonal-pyramidal	E_u
$M(CO)_5$	$\mathbf{C}_{4v}$	Tetragonal-pyramidal	$2A_1 + E$
$M(CO)_5$	$\mathbf{D}_{3h}$	Trigonal-bipyramidal	$A_2'' + E'$
$M(CO)_6$	$\mathbf{O}_h$	Octahedral	F_{1u}

monocarbonyl peroxyformate structure for the co-condensation product of Au with a mixture of CO and O_2.[719]

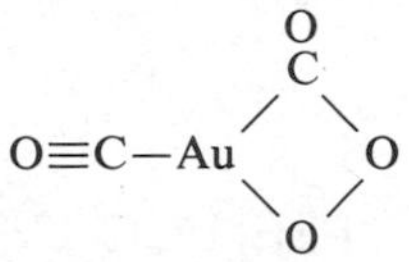

Sheline and Slater[720] reviewed the IR spectra of lanthanoid and actinoid carbonyls prepared in inert gas matrices.

The second method utilizes *in situ* photolysis of stable metal carbonyls in inert gas matrices. For example, Poliakoff and Turner[721] carried out UV photolysis of ^{13}CO-enriched $Fe(CO)_5$ in SF_6 and Ar matrices $[Fe(CO)_5 \xrightarrow{h\nu} Fe(CO)_4 + CO]$, and concluded that the structure of $Fe(CO)_4$ is $\mathbf{C}_{2v}$ since it exhibits four $\nu(CO)$ $(2A_1 + B_1 + B_2)$ in the infrared spectrum. Graham et al.[722] proposed the $\mathbf{C}_{4v}$ structure for $Cr(CO)_5$ produced by the photolysis of $Cr(CO)_6$ in inert gas matrices. On the other hand, Kündig and Ozin[723] proposed the $\mathbf{D}_{3h}$ structure for $Cr(CO)_5$ prepared by co-condensation of Cr atoms with CO in inert gas matrices. They derived a general rule that $M(CO)_5$ species take the $\mathbf{D}_{3h}$ structure when the number of valence-shell electrons is even [Cr (16), Fe (18)], and the $\mathbf{C}_{4v}$ structure when it is odd [V (15), Mn (17)]. However, the $\mathbf{D}_{3h}$ structure of $Cr(CO)_5$ has been questioned by Black and Braterman[724] and Perutz and Turner.[725] Nernst glower of an IR spectrophotometer accelerates the formation of $Fe(CO)_4(CH_4)$ via the reaction of $Fe(CO)_4$ with methane in a methane matrix.[726] Charged carbonyls such as $Cr(CO)_5^-$, $Fe(CO)_4^-$, and $Ni(CO)_3^-$ can also be prepared in inert gas matrices using several techniques.[727] Turner[728] reviewed photochemical fragmentation of hexacarbonyls of Cr, Mo, and W.

Carbonyl complexes of the type MX_2CO are formed by reacting metal halide vapor directly with CO in inert gas matrices.[729,730] In this case, $\nu(CO)$ shifts to higher frequencies by complexation, since the bonding is dominated by the donation of σ-electrons to the metal. On the other hand, $\nu(MX)$ shifts to lower frequencies because the oxidation state of the metal is lowered by accepting σ-electrons from CO. Figure III-46 shows infrared spectra of the PbF_2—L system (L = CO, NO, and N_2) in Ar matrices obtained by Tevault and Nakamoto.[730] In this series, the magnitudes of the shifts of the PbF_2 and L stretching bands (cm^{-1}) relative to the free state are as follows:

	PbF_2CO	PbF_2NO	PbF_2N_2
$\nu(PbF_2)$	−10.8	−8.8	−5.8
$\nu_a(PbF_2)$	−10.9	−8.5	−5.0
$\nu(L)$	+38.4	+16.4	—

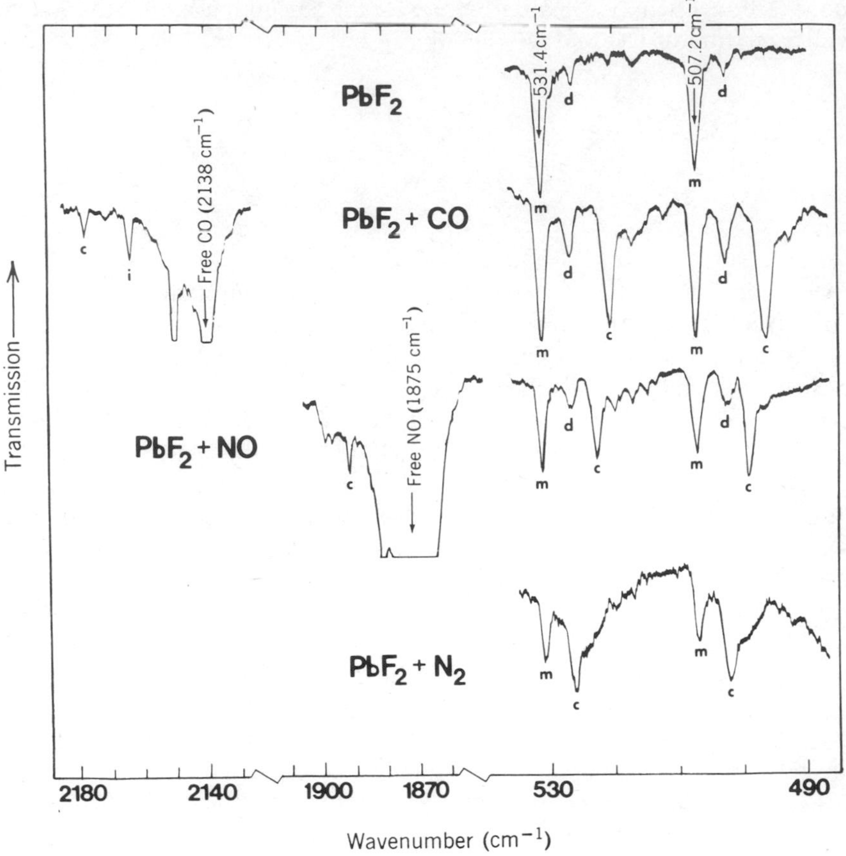

Fig. III-46. Infrared spectra of PbF_2, PbF_2CO, PbF_2NO, and PbF_2N_2 in Ar matrices: (m) monomeric PbF_2; (d) dimeric PbF_2; (c) complex; (i) impurity (HF–CO).

This result definitely indicates that CO is the best, NO is the next best, and N_2 is the poorest σ-donor.

Other work involves the direct deposition of stable carbonyls in inert gas matrices, mainly to study the effect of matrix environments on the structure. Both $Fe(CO)_5$[731] and $M_3(CO)_{12}$ (M = Ru and Os)[732] were found to be distorted from $\mathbf{D}_{3h}$ symmetry in inert gas matrices. If a thick deposit is made on a cryogenic window while maintaining a relatively high sample/inert gas dilution ratio, it is possible to observe low-frequency modes such as ν(MC) and δ(MCO). It was found that these bands show splittings due to the mixing of metal isotopes. For example, the $F_{1u}\nu$(CrC) of $Cr(CO)_6$ in a N_2 matrix exhibits four bands due to ^{50}Cr, ^{52}Cr, ^{53}Cr, and ^{54}Cr (see Fig. I-20). The magnitude of these isotope splittings may be used to estimate the degree of the ν(MC)–δ(MCO) mixing in the low-frequency vibrations.[733]

(7) Nitrosyl (NO) Complexes

Many review articles[734–739] are available for nitrosyl complexes. Like CO, NO acts as a σ-donor and a π-acceptor. NO contains one more electron than CO, and this electron is in the $2p\pi^*$ orbital. The loss of this electron gives the nitrosonium ion (NO^+), which is much more stable than NO. Thus the ν(NO) of the nitrosonium ion (ca. 2200 cm^{-1}) is much higher than that of the latter (1876 cm^{-1}). In nitrosyl complexes, ν(NO) ranges from 1900 to 1500 cm^{-1}. Recent X-ray studies on nitrosyl complexes have revealed the presence of linear and bent M—NO groups:

I II

In the valence-bond theory, the hybridizations of the N atom in (I) and (II) are *sp* and sp^2, respectively. If the pair of electrons forming the M-N bond is counted as the ligand electrons, the nitrosyl groups in (I) and (II) are regarded as NO^+ and NO^-, respectively. Thus, one is tempted to correlate ν(NO) with the charge on NO and the MNO angle. It was not possible, however, to find simple relationships between them since ν(NO) is governed by several other factors (electronic effects of other ligands, nature of the metal, structure, and charge of the whole complex etc.).[738] According to Haymore and Ibers,[740] the distinction of linear and bent geometry can be made by using properly corrected ν(NO) values; the MNO group is linear or bent, respectively, if this value is above or below 1620–1610 cm^{-1}.

According to X-ray analysis, $RuCl(NO)_2(PPh_3)_2PF_6$ contains one linear and one bent M—NO group which absorb at 1845 and 1687 cm^{-1}, respectively.[741] $CoCl_2(NO)L_2[L=P(CH_3)Ph_2]$ exists in two isomeric forms:

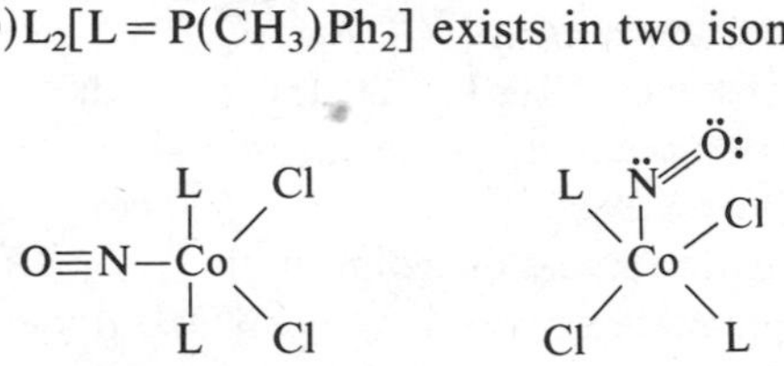

The ν(NO) of the former is at 1750 cm^{-1}, whereas that of the latter is at 1650 cm^{-1}.[742]

Table III-52 lists the vibrational frequencies of typical nitrosyl complexes. Although the M—NO group is expected to show ν(NO), ν(MN), and δ(MNO), only ν(NO) have been observed in most cases. The latter two modes are often coupled since their frequencies are close to each other. Jones et al.[745] carried out a complete analysis of the vibrational spectra of $Co(CO)_3(NO)$ and its ^{13}C, ^{18}O, and ^{15}N analogs. According to Quinby-Hunt and Feltham,[752] vibrational spectra of a wide variety of nitrosyl complexes can be accounted for on the basis of the simple three-body (M-N-O) model as long as the complex does not contain two or more NO groups attached to the metal.

TABLE II-52. VIBRATIONAL FREQUENCIES OF NITROSYL COMPLEXES (CM^{-1})

Compound	ν(NO)	ν(MN)	δ(MNO)	Ref.
$Cr(NO)_4$	1721	650	496	743
$Co(NO)_3$	1860, 1795	—	—	744
$Co(CO)_3(NO)$	1822	609	566	745
$Mn(CO)_4(NO)$	1781	524	657	746
$Mn(PF_3)(NO)_3$	1836, 1744	—	—	747
cis-$[MoCl_4(NO)_2]^{2-}$	1720, 1600	—	—	748
$NiCl_2(NO)_2$	1872, 1842	—	—	749
$[RuCl_5(NO)]^{2-}$	1904	606	588	750
$[RuBr_5(NO)]^{2-}$	1870	572	300	751

The ν(NO) of the bridging nitrosyl group is much lower than that of the terminal nitrosyl group. For example, $(C_5H_5)_2Cr_2(NO)_3(NXY)$ (X = OH and Y = *t*-Bu) shown below exhibits the terminal ν(NO) at 1683 and 1625 cm^{-1} and the bridging ν(NO) at 1499 cm^{-1}.[753]

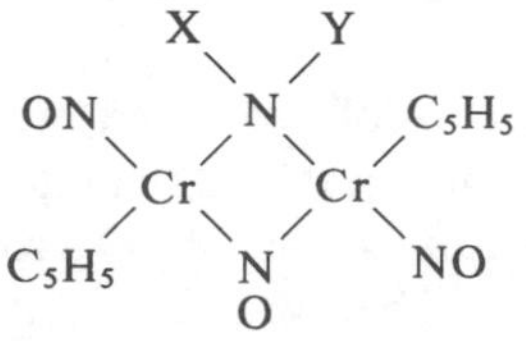

Similar frequencies are reported for an analogous compound [X = Et and Y = $B(Et)_2$].[754] The structure of $M_3(CO)_{10}(NO)_2$ (M = Ru and Os) resembles that of $Fe_3(CO)_{12}$ (structure VIII in Fig. III-40) with double nitrosyl bridges in place of the double carbonyl bridges in the latter. As expected, ν(NO) of these nitrosyl groups are very low: 1517 and 1500 cm^{-1} for the Ru compound, and 1503 and 1484 cm^{-1} for the Os compound.[755]

The ν(NO) is spin-state sensitive in Fe(NO)(salphen) [salphen: N,N'-*o*-phenylenebis(salicylideneimine)]: 1724 cm^{-1} for the high-spin (room temperature) and 1643 cm^{-1} for the low-spin (liquid-N_2 temperature) state.[756] Photolysis of $Cr(NO)_4$ in Ar matrices produces $Cr(NO)_3(NO^*)$ where NO^* denotes a bent NO group with an unusually low ν(NO)(1450 cm^{-1}).[757] A similar observation was made for $Mn(CO)(NO)_2(NO^*)$ which was produced by the photolysis of $Mn(CO)(NO)_3$ in inert gas matrices.[758]

III-17. COMPLEXES OF DIOXYGEN(O_2)

Dioxygen (molecular oxygen) adducts of metal complexes have been studied extensively because of their importance as oxygen carriers in biological systems (Sec. V-1) and as catalytic intermediates in oxidation reactions of organic compounds. A number of review articles are available on the chemistry of dioxygen adducts.[738,759–766b]

As discussed in Sec. II-1, the bond order of the O–O linkage decreases as the number of electrons in the antibonding $2p\pi^*$ orbital increases in the following order:

	$[O_2^+]AsF_6$		O_2		$K[O_2^-]$		$Na_2[O_2^{2-}]$
Bond order	2.5	>	2.0	>	1.5	>	1.0
Bond distance (Å)	1.123	<	1.207	<	1.28	<	1.49
$\nu(O_2)$ (cm^{-1})	1858	>	1555	>	1108	>	~760

The decrease in the bond order causes an increase in the O–O distance and a decrease in the $\nu(O_2)$. In fact, there is a good linear relationship between the O–O bond order and the $\nu(O_2)$ of these simple dioxygen compounds.

Dioxygen adducts of more complex molecules are generally classified into two groups; complexes which exhibit $\nu(O_2)$ in the 1200–1100 cm^{-1} region are called "superoxo" because their frequencies are close to that of KO_2, and complexes whose $\nu(O_2)$ are in the 920–750 cm^{-1} are called "peroxo" because their frequencies are close to that of Na_2O_2. As will be shown later, there are many compounds which exhibit $\nu(O_2)$ outside of these regions. Thus, this distinction of dioxygen adducts is not always clear-cut.

Structurally, the dioxygen adducts are classified into three types:

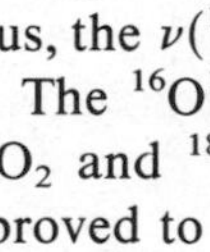
Asymmetric (end-on), I

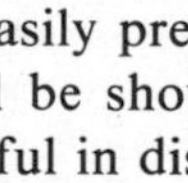
Symmetric (side-on), II

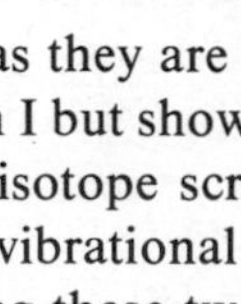
Bridging III

In I, the two oxygen atoms are not equivalent whereas they are equivalent in II and III. Thus, the $\nu(^{16}O^{18}O)$ splits into two bands in I but shows no splitting in II and III. The $^{16}O^{18}O$ gas is easily prepared by isotope scrambling of a mixture of $^{16}O_2$ and $^{18}O_2$. As will be shown later, vibrational studies with $^{16}O^{18}O$ have proved to be very useful in distinguishing these two types.

In the following, we classify dioxygen adducts according to the type of ligand involved, and review typical results obtained for each type. For dioxygen adducts of biological compounds, see Section V.

(1) Dioxygen Adducts of Metal Atoms

As stated in Sec. I-22, a number of stable and unstable complexes of the ML_n type have been synthesized via matrix co-condensation reactions of metal vapor(M) with gaseous ligands(L). Table III-53 lists typical results obtained

TABLE III-53. VIBRATIONAL FREQUENCIES OF $M(O_2)$-TYPE COMPOUNDS (CM^{-1})

Compound	$\nu(O_2)$	$\nu_s(MO)$	$\nu_a(MO)$	Ref.
$^6LiO_2$	1097.4	743.8	507.3	767
$^7LiO_2$	1096.9	698.8	492.4	767
NaO_2	1094	390.7	332.8	767
KO_2	1108	307.5	—	767
RbO_2	1111.3	255.0	282.5	767
CsO_2	1115.6	236.5	268.6	767
AgO_2	1082/1077	—	—	768
RhO_2	900	—	422	769
InO_2	1084	332	277.7	770
GaO_2	1089	380	285.5	770
AuO_2	1092	—	—	771
TlO_2	1082	296	250	772
PdO_2	1024.0	427	—	773
NiO_2	966.2	504	—	773
FeO_2	946	—	—	774
PtO_2	926.6	—	—	773

for the $M(O_2)$-type compounds. It is seen that the $\nu(O_2)$ of these dioxygen adducts scatter over a wide range from 1120 to 920 cm^{-1}. Previously, we noted that the $\nu(O_2)$ decreases as the negative charge on the O_2 increases. Thus, these results seem to suggest that the negative charge on the O_2 can be varied continuously by changing the metal. In fact, Lever et al.[775] noted that there is a linear relationship between the electron affinity of the M^{2+} ion and the $M{-}O_2$ CT transition energy in the $M(O_2)_2$ series and that the latter is linearly related to the $\nu(O_2)$.

The dioxygen ligand may coordinate to a metal in the end-on or side-on fashion. These two structures can be distinguished by using the isotope scrambling technique. Andrews[767] first applied this method to the structure determination of the ion-pair complex $Li^+O_2^-$; a mixture of $^{16}O_2$, $^{16}O^{18}O$, and $^{18}O_2$ was prepared by Tesla coil discharge of a $^{16}O_2$-$^{18}O_2$ mixture, and reacted with Li vapor in an Ar matrix. Three $\nu(O_2)$ were observed in the Raman spectrum:

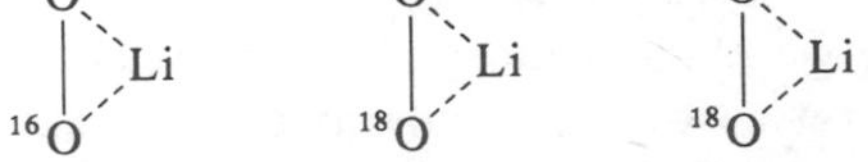

1096.1 cm^{-1} 1065.7 cm^{-1} 1034.6 cm^{-1}

This result clearly indicates side-on coordination since four bands are expected for end-on coordination (see above). Using the same technique, Ozin and co-workers[771,773] showed that, in all cases they studied, O_2 coordinates to a metal in the side-on fashion and that, in $M(O_2)_2$ (M = Ni, Pd, and Pt), the complexes take the spiro $\mathbf{D}_{2d}$ structure. Some metal superoxides and peroxides are prepared by ordinary methods and their $\nu(O_2)$ are reported by Evans[776] and Eysel and Thym.[777] For linear O–M–O type compounds, see Table II-2b.

(2) Dioxygen Adducts of Cobalt Ammine and Schiff-Base Complexes

Extensive vibrational studies have been made on dioxygen adducts of cobalt ammine and Schiff-base complexes. Table III-54 lists the $\nu(O_2)$ and $\nu(CoO)$ of representative compounds.

The $\nu(O_2)$ of dinuclear cobalt complexes such as $\{[Co(NH_3)_5]_2O_2\}^{n+}$ ($n=4$ or 5) are markedly different depending upon whether the O_2 group is of superoxo or peroxo type. The $\nu(O_2)$ of the $\{[Co(NH_3)_5]_2O_2\}^{5+}$ ion appears strongly in Raman spectra (1122 cm^{-1}) but is forbidden in IR spectra because the O–O bridge is centrosymmetric. However, the $\nu(O_2)$ of a dibridged complex ion, $[(NH_3)_4Co(NH_2)(O_2)Co(NH_3)_4]^{4+}$, is observed at 1068 cm^{-1} in IR spectra.[780]

N,N′-ethylenebis(salicylideniminato)cobalt, Co(salen), binds dioxygen

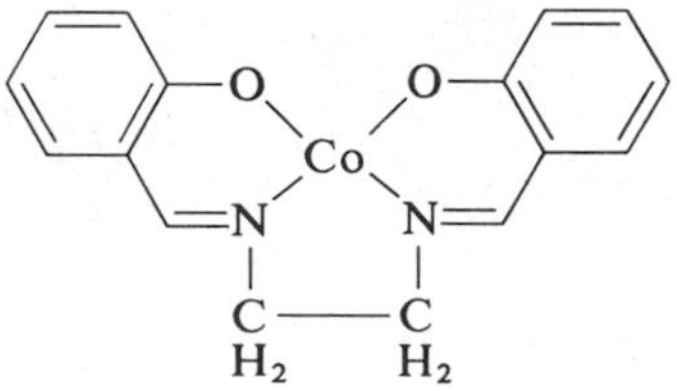

TABLE III-54. VIBRATIONAL FREQUENCIES OF DIOXYGEN ADDUCTS OF COBALT AMMINE AND SCHIFF-BASE COMPLEXES (CM^{-1})

Compound	$\nu(O_2)$	$\nu(CoO)$	Ref.
$Co(J\text{-}en)(py)O_2$ [a]	1146	—	778
$Co(salen)(py)O_2$	1144	527	779
$\{[Co(NH_3)_5]_2O_2\}Cl_5$	1122	620, 441	780
$\{[Co(NH_3)_5]_2O_2\}(NO_3)_5$	1122	—	781
$K_5\{[Co(CN)_5]_2O_2\}$	1104	493	782
$[Co(salen)]_2O_2$	1011	533	783, 784
$[Co(salen)(pyO)]_2O_2$	910	535	785
$[Co(salen)(py)]_2O_2$	884	543	784
$[Co(J\text{-}en)(py)]_2O_2$ [a]	841	562	778
$[Co(DMG)(PPh_3)]_2O_2$ [b]	818	551	786
$K_6\{[Co(CN)_5]_2O_2\}$	804	602	782
$\{[Co(NH_3)_5]_2O_2\}(NO_3)_4$	805	642, 547	780
$\{[Co(NH_3)_5]_2\}(NCS)_4$	786	—	781

[a] $H_2(J\text{-}en)$ = N,N′-ethylenebis(2,2′-diacetylethylideneamine).
[b] H(DMG) = dimethylglyoxime.

reversibility in the solid state.[787] Figure III-47 shows the resonance Raman spectra of $[Co(salen)]_2O_2$ at ~100 K.[783] The bands at 1011 and 533 cm^{-1} are shifted to 943 and 514 cm^{-1}, respectively, by $^{16}O_2$-$^{18}O_2$ substitution, and thus assigned to the $\nu(O_2)$ and $\nu(CoO)$, respectively. The former frequency is unique in that it is between those of superoxo and peroxo complexes. However, this band is shifted to the normal peroxo range when the base ligands are coordinated trans to the dioxygen. Evidently, electron donation from the base to the dioxygen is responsible for the shift of $\nu(O_2)$ to a lower frequency.

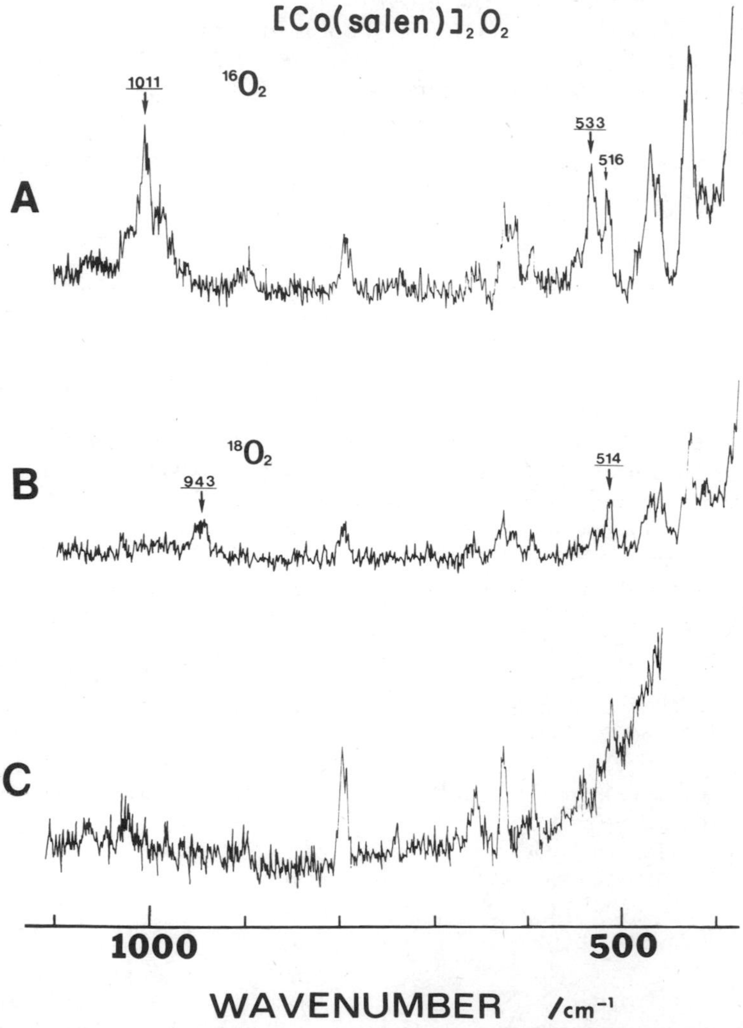

Fig. III-47. Resonance Raman spectra of (A) $[Co(salen)]_2{}^{16}O_2$, (B) $[Co(salen)]_2{}^{18}O_2$, and (C) Co(salen) by 579 nm excitation at ~100 K.

When a Co Schiff-base (SB) complex in a nonaqueous solvent absorbs oxygen in the presence of a base (B), the following equilibria are established:

$$[Co(SB)B] + O_2 \rightleftharpoons [Co(SB)B]O_2$$

$$[Co(SB)B]O_2 + [Co(SB)B] \rightleftharpoons [Co(SB)B]_2O_2$$

The $\nu(O_2)$ of the 1:1 (Co/O_2) adduct is near 1140 cm^{-1} whereas that of the 1:2 adduct is between 920 and 820 cm^{-1}. Using these bands as the markers, it is possible to examine the effects of oxygen pressure, temperature, and solvent polarity on the above equilibria. Figure III-48 shows the RR spectra of Co(J-en) in CH_2Cl_2 containing pyridine which were saturated with oxygen at various oxygen pressures and temperatures.[778] It is seen that the concentra-

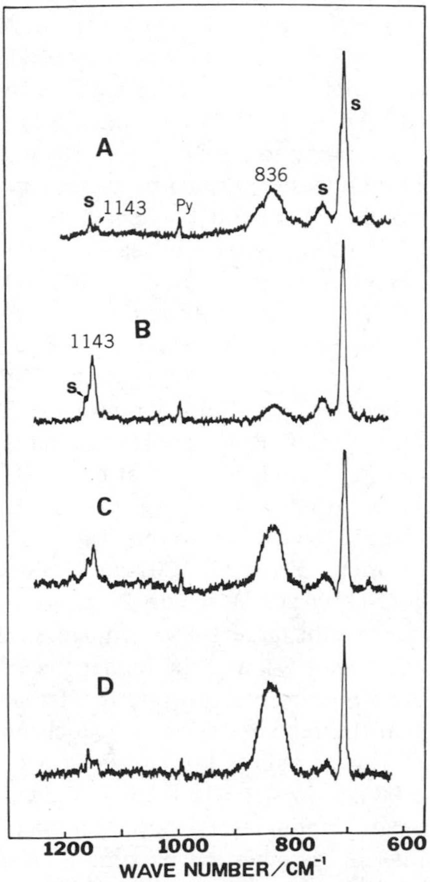

Fig. III-48. Resonance Raman spectra of Co(J-en) in CH_2Cl_2 containing 3% pyridine which were saturated with O_2 at various O_2 pressures and temperatures (580 nm excitation): (A) 1 atm, −78°C; (B) ~3 atm, −80°C; (C) ~3 atm, −30°C; (D) ~3 atm, +20°C. S and py denote the solvent and pyridine bands, respectively.

tion of the 1:1 adduct (1143 cm^{-1}) increases and that of the 1:2 adduct (836 cm^{-1}) decreases as the oxygen pressure increases (A → B) and as the temperature decreases (D → C → B). It was also noted that 1:1 adduct is favored in a polar solvent containing a relatively strong base.

(3) Dioxygen Adducts of "Base-free" Metalloporphyrins and Related Compounds

"Base-bound" dioxygen adducts are formed when iron(II) or cobalt(II) porphyrins are reacted with dioxygen in a nonaqueous solvent containing a base ligand (Sec. V-1). On the other hand, "base-free" dioxygen adducts are prepared via matrix co-condensation reactions of metalloporphyrins with dioxygen in gas matrices. Kozuka and Nakamoto[788] first observed the $\nu(O_2)$ of Co(TPP)O_2 at 1276 cm^{-1} in argon matrices. This is the highest $\nu(O_2)$ known for dioxygen adducts of metal complexes. It is higher by 133 cm^{-1} than that of "base-bound" Co(TPP)(1-MeIm)O_2 (1143 cm^{-1}). A shift of a similar magnitude (127 cm^{-1}) was observed in going from $[Co(salen)(py)]_2O_2$ to $[Co(salen)]_2O_2$. These results indicate that the negative charge on the dioxygen increases markedly and the adduct becomes much more stable as a result of electron donation from the base to the dioxygen. The end-on coordination of the dioxygen in Co(TPP)O_2 was confirmed by the observation that the $^{16}O^{18}O$ adduct exhibits two bands at 1252 and 1241 cm^{-1}.

Matrix co-condensation reactions have been employed to prepare many other "base-free" dioxygen adducts, and their $\nu(O_2)$ are Co(OEP)O_2 (1275 cm^{-1}),[789] Fe(TPP)O_2 (1195 and 1106 cm^{-1}),[790,791] Fe(OEP)O_2 (1190 and 1104 cm^{-1}),[791] Mn(TPP)O_2 (983 cm^{-1}),[792] and Mn(OEP)O_2 (~991 cm^{-1}).[793] It is interesting to note that the order of $\nu(O_2)$ is Co(TPP)O_2 > Fe(TPP)O_2 > Mn(TPP)O_2, and that Fe(TPP)O_2 exhibits two $\nu(O_2)$ as seen in Fig. III-49. As stated above, the dioxygen in Co(TPP)O_2 is end-on whereas that in Mn(TPP)O_2 is side-on. Watanabe et al.[791] have shown that co-condensation reactions of Fe(TPP) with O_2 yield two isomers, one (1195 cm^{-1}, end-on) being thermally more stable than the other (1106 cm^{-1}, side-on). They also noted that the $\nu(O_2)$ decreases in the order Fe(Pc)O_2(1207 cm^{-1}) > Fe(TPP)O_2(1195 and 1106 cm^{-1}) > Fe(salen)O_2(1106 cm^{-1}), where Pc denotes the phthalocyanato ion. This result indicates that the larger the π-conjugated chelating ring system, the less negative charge on the O_2, and the higher the $\nu(O_2)$.

Formation of ferrylporphyrin, FeO(porphyrin), from oxyironporphyrin is the most crucial step in the reaction cycle of cytochrome P-450 (Sec. V-2). Bajdor and Nakamoto[794] were able to prepare FeO(TPP) via laser photolysis of Fe(TPP)O_2 in pure O_2 matrices at ~15 K. As is shown in Fig. III-50, a new band appears at 852 cm^{-1} upon laser irradiation (406.7 nm, 1~2 mW) of Fe(TPP)O_2, and its intensity reaches the maximum after about 20 min. This band is shifted to 818 cm^{-1} by $^{16}O_2$–$^{18}O_2$ substitution. Similar experiments with a mixture of $^{16}O_2$, $^{16}O^{18}O$, and $^{18}O_2$ produce only two bands at 852 and 818 cm^{-1}. These results clearly indicate that the 852- and 818-cm^{-1} bands are due to the $\nu(Fe–^{16}O)$ and $\nu(Fe–^{18}O)$, respectively, of FeO(TPP) which is formed by the

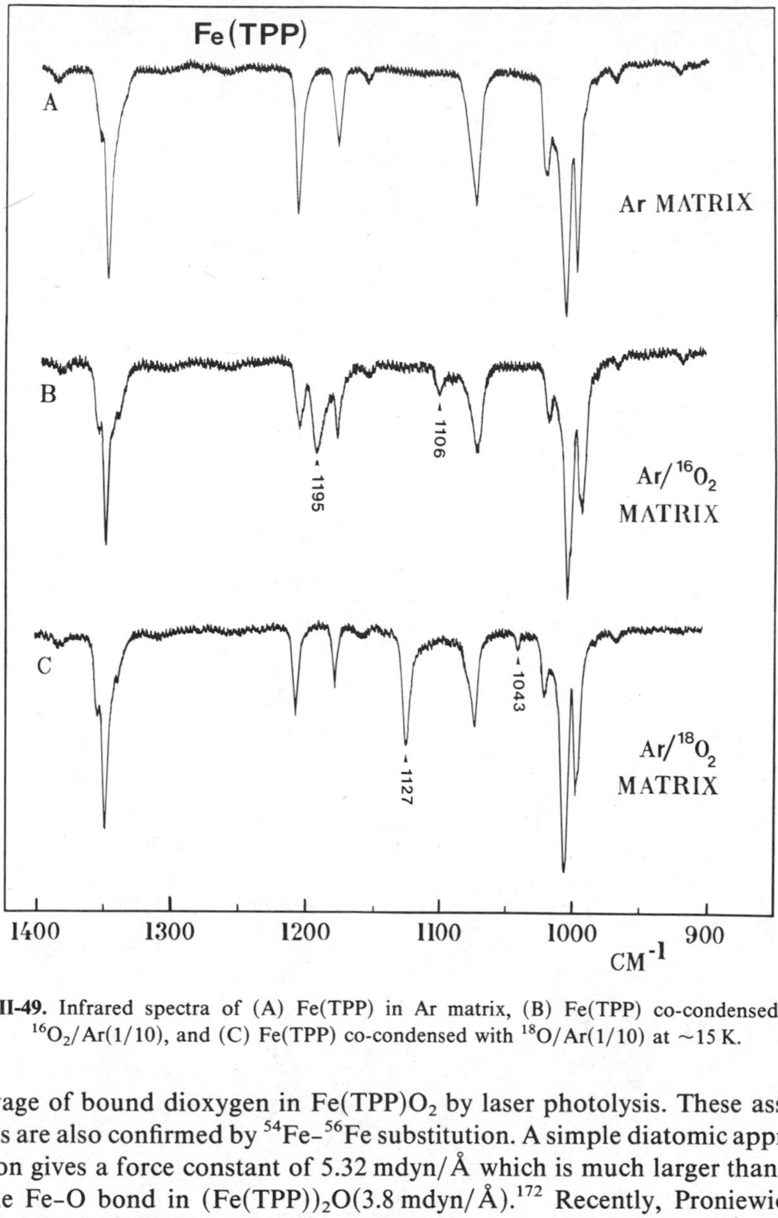

Fig. III-49. Infrared spectra of (A) Fe(TPP) in Ar matrix, (B) Fe(TPP) co-condensed with $^{16}O_2$/Ar(1/10), and (C) Fe(TPP) co-condensed with ^{18}O/Ar(1/10) at ~15 K.

cleavage of bound dioxygen in Fe(TPP)O_2 by laser photolysis. These assignments are also confirmed by ^{54}Fe–^{56}Fe substitution. A simple diatomic approximation gives a force constant of 5.32 mdyn/Å which is much larger than that of the Fe–O bond in $(Fe(TPP))_2O$(3.8 mdyn/Å).[172] Recently, Proniewicz et al.[794a] formulated the ferryl bond of FeO(TPP) as (TPP)Fe(IV) ⇚ O^{2-}. Here, the arrowed line indicates a σ-bond formed via the $d_{z^2}-p_z$ overlap, and the broken lines represent two π-bonds formed via the $d_{xz}-p_x$ and $d_{yz}-p_y$ overlaps. The latter two bonds are expected to be weak due to poor orbital overlap resulting from the doming of the porphyrin core. Similar experiments on Fe(Pc)O_2 produced no ferryl group. In contrast, the ferryl species was readily produced upon laser irradiation of Fe(salen)O_2. These observations seem to

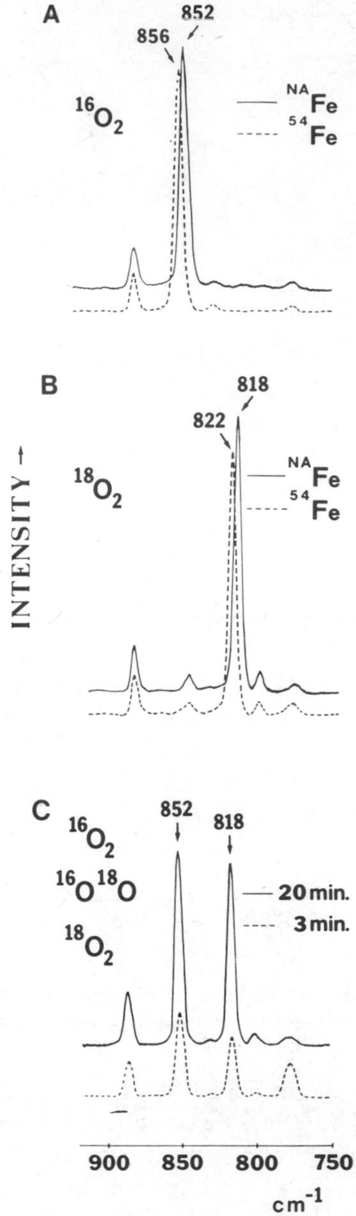

Fig. III-50. Resonance Raman spectra of Fe(TPP) co-condensed with O_2 at ~15 K (406.7 nm excitation): (A) ^{NA}Fe(TPP) with $^{16}O_2$; (B) ^{NA}Fe(TPP) with $^{18}O_2$; and (C) ^{NA}Fe(TPP) with isotopically scrambled $O_2(^{16}O_2/^{16}O^{18}O/^{18}O_2 = 1/2/1)$. The broken lines in (A) and (B) denote the spectra of ^{54}Fe(TPP) co-condensed with respective gases. All the spectra in (A), (B), and (C) (solid line) were obtained after 20-min laser irradiation. The dotted line in (C) indicates the spectrum obtained only after 3-min laser irradiation. ^{NA}Fe (Fe in natural abundance) contains 91.66% ^{56}Fe.

suggest that the O–O bond strength decreases in the order Fe(Pc)O_2 > Fe(TPP)O_2 > Fe(salen)O_2. This trend has already been confirmed by the order of the $\nu(O_2)$ of these compounds as mentioned earlier.

Recently, the ν(FeO) band of Horseradish Peroxidase Compound II was observed at 779 cm^{-1} by Terner et al.[794b] and at 787 cm^{-1} by Kitagawa et al.[794c] These frequencies are by ~70 cm^{-1} lower than that of FeO(TPP) because the FeO bond of the former is weakened by the presence of the proximal histidine at the *trans* position to the ferryl group.

(4) Dioxygen Adducts of Other Transition-Metal Complexes

Trans-planar [IrCl(CO)$(PPh_3)_2$] (Vaska's salt) binds dioxygen reversibly to form a trigonal-bipyramidal complex shown below:[795]

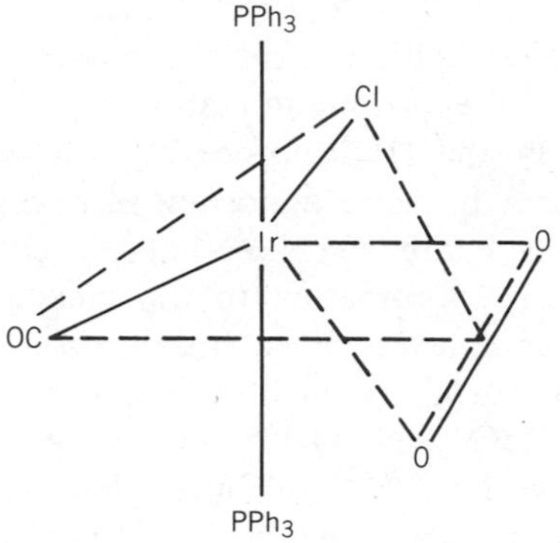

The O–O distance of this compound (1.30 Å)[796] is close to that of the O_2^- ion (1.28 Å) although its $\nu(O_2)$ is at 857 cm^{-1},[797] which is typical of peroxo adducts. Table III-55 lists the $\nu(O_2)$ and ν(MO) of representative compounds belonging to this type. All these adducts exhibit $\nu(O_2)$ in the 900–820 cm^{-1} region. Several $^{18}O_2$ isotopic studies show that $\nu(O_2)$ couples strongly with ν(MO) in the 600–400 cm^{-1} region. Nakamura et al.[798] carried out normal coordinate analyses on several dioxygen complexes of this type.

TABLE III-55. OBSERVED FREQUENCIES OF MOLECULAR OXYGEN COMPOUNDS (CM^{-1})

Compound	$\nu(O_2)$	ν(MO)	Ref.
$Pt(O_2)(PPh_3)_2$	828	472	798
$Ni(O_2)(t\text{-}BuNC)_2$	898	552	798
$Pd(O_2)(t\text{-}BuNC)_2$	893	484	798
$Rh(O_2)Cl(PPh_3)_2(t\text{-}BuNC)$	892	576	798
$[VO(O_2)(H_2O)(bipicoline)]^-$	839	610–570	799
$Ir(O_2)F(CO)(PPh_3)_2$	850	—	800

III-18. DINITROGEN(N_2), NITRIDO(N), AND HYDRIDO(H) COMPLEXES

(1) Dinitrogen Adducts of Transition-Metal Complexes

Since Allen and Senoff[801] prepared the first stable molecular nitrogen compounds, $[Ru(N_2)(NH_3)_5]X_2$ ($X = Br^-$, I^-, BF_4^-, etc.), a large number of molecular nitrogen compounds have been synthesized. The chemistry and spectroscopy of these compounds have been reviewed extensively.[738,802-805] The structures of molecular nitrogen compounds are classified into three types:

$$M—N≡N \qquad\qquad M \overset{\displaystyle N}{\underset{\displaystyle N}{\big<\ \vert\vert\vert}} \qquad\qquad M—N≡N—M$$

End-on   Side-on   Bridging
(linear)   (symmetrical)   (linear)

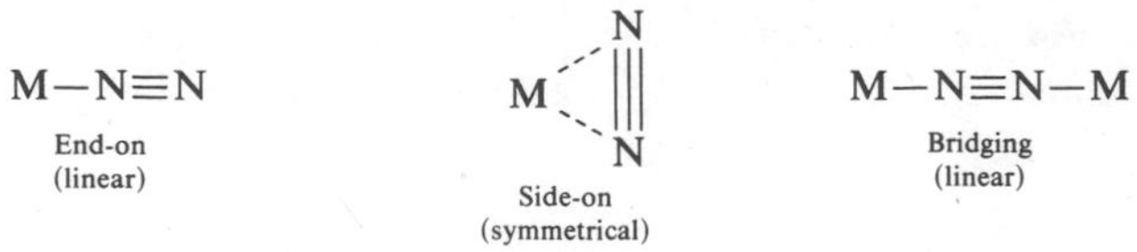

The terminal end-on coordination is most common. The $M–N_2$ bonding is interpreted in terms of the σ-donation and π-back-bonding, which were discussed in Secs. III-14 and III-16. Since N_2 is a weaker Lewis base than CO, π-back-bonding may be more important in nitrogen complexes than in CO complexes.[806] Free N_2 exhibits $\nu(N≡N)$ at 2331 cm^{-1}, and this band shifts to 2220–1850 cm^{-1} upon coordination to the metal. Table III-56 lists the $\nu(N≡N)$ of typical complexes. Very little information is available for $\nu(M—N_2)$ and $\delta(M—N≡N)$ in the low-frequency region. Allen et al.[806] assigned $\nu(Ru–N_2)$ of $[Ru(N_2)(NH_3)_5]^{2+}$-type compounds in the 508–474 cm^{-1} region, whereas other workers[807,808] attributed these bands to $\delta(Ru—N≡N)$. Figure III-51 shows the infrared spectrum of $[Ru(NH_3)_5N_2]Br_2$ obtained by Allen et al.[806]

According to Srivastava and Bigorgne,[816] $Co(N_2)H(PPh_3)_3$ exists in two forms in the solid state; one form exhibits $\nu(N≡N)$ at ca. 2087 cm^{-1}, and the

TABLE III-56. OBSERVED N≡N STRETCHING
FREQUENCIES (cm^{-1})

Complex	$\nu(N≡N)$	Ref.
$[Ru(N_2)(NH_3)_5]Br_2$	2105	807
$[Ru(N_2)(NH_3)_5]I_2$	2124	808
$[Os(N_2)(NH_3)_5]Cl_2$	2022, 2010	809
$Co(N_2)(PPh_3)_3$	2093	810
$Co(N_2)H(PPh_3)_3$	2105	811
$Ir(N_2)Cl(PPh_3)_2$	2105	812
$trans$-$Mo(N_2)_2(DPE)_2^b$	1970, (2020)	813
cis-$W(N_2)_2(PMePh)_4$	1998, 1931	814
$Co(N_2)(PR_3)(PR_2)_2^{2-\ a}$	1904 ~ 1864	815

^{a} R = n-Bu or phenyl.
^{b} DPE = $Ph_2P—CH_2—CH_2—PPh_2$.

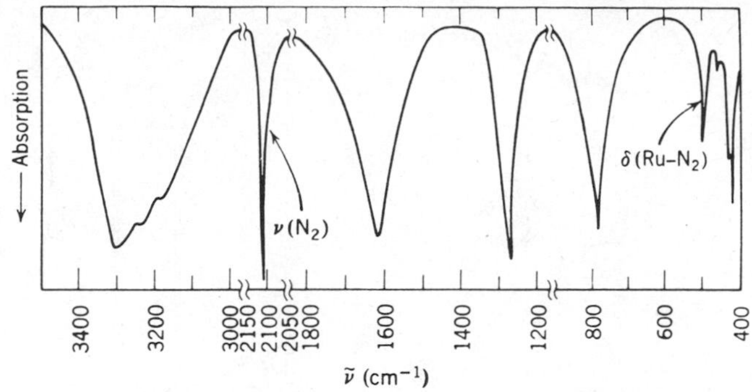

Fig. III-51. Infrared spectrum of $[Ru(NH_3)_5N_2]Br_2$.[806]

other shows two bands of equal intensity at 2101 and 2085 cm^{-1}. However, their structural differences are unknown. Darensbourg[817] obtained a linear relationship between $\nu(N{\equiv}N)$ and the absolute integrated intensity in a series of molecular nitrogen compounds.

Armor and Taube[818] postulated the occurrence of the side-on structure as a possible transition state in linkage isomerization: $[(NH_3)_5Ru{-}^{14}N{\equiv}{}^{15}N]Br_2 \leftrightarrow [(NH_3)_5Ru{-}^{15}N{\equiv}{}^{14}N]Br_2$. Krüger and Tsay[819] carried out X-ray analysis on $[\{(C_6H_5Li)_3Ni\}_2(N_2)\{(C_2H_5)_2O\}]_2$ and confirmed the presence of the side-on coordination in this compound; the N—N distance was found to be extremely long (1.35 Å).

The bridging M—N≡N—M type complex should not show $\nu(N{\equiv}N)$ in the IR spectrum. However, it may show a strong $\nu(N{\equiv}N)$ in the Raman spectrum. Thus $[\{Ru(NH_3)_5\}_2(N_2)]^{4+}$ shows no infrared bands in the 2220–1920 cm^{-1} region, whereas a strong $\nu(N{\equiv}N)$ band appears at 2100 cm^{-1} in the Raman.[820] If N_2 forms a bridge between two different metals, $\nu(N{\equiv}N)$ is observed in the infrared. For example, $\nu(N{\equiv}N)$ is at 1875 cm^{-1} in the infrared spectrum of $[(PMe_2Ph)_4ClRe{-}N_2{-}CrCl_3(THF)_2]$.[821] According to X-ray analysis,[822] an analogous compound, $[(PMe_2Ph)_4ClRe{-}N_2{-}MoCl_4(OMe)]$, has a N≡N distance of 1.21 Å, and its $\nu(N{\equiv}N)$ is at 1660 cm^{-1}. As expected, the $\nu(N{=}N)$ of $[(CO)_5Cr{-}NH{=}NH{-}Cr(CO)_5]$ is very low (1415 cm^{-1}).[823]

(2) Dinitrogen Adducts of Metal Atoms

Similar to $M(CO)_n$- and $M(O_2)_n$-type compounds discussed previously (Secs. III-16 and 17), it is possible to prepare simple $M(N_2)_n$-type adducts by reacting metal atoms with N_2 in inert gas matrices. Again the distinction of end-on and side-on geometry can be made by using the isotope scrambling techniques ($^{14}N_2 + {}^{14}N^{15}N + {}^{15}N_2$). Figure III-52 shows the IR spectra of $Ni(N_2)$(end-on)[824] and $Co(N_2)$(side-on).[825] The observed frequencies (cm^{-1}) and assignments of the four bands of the former are as follows:

$Ni{-}^{14}N{\equiv}{}^{14}N$	$Ni{-}^{14}N{\equiv}{}^{15}N$	$Ni{-}^{15}N{\equiv}{}^{14}N$	$Ni{-}^{15}N{\equiv}{}^{15}N$
2089.9	2057.4	2053.6	2020.6

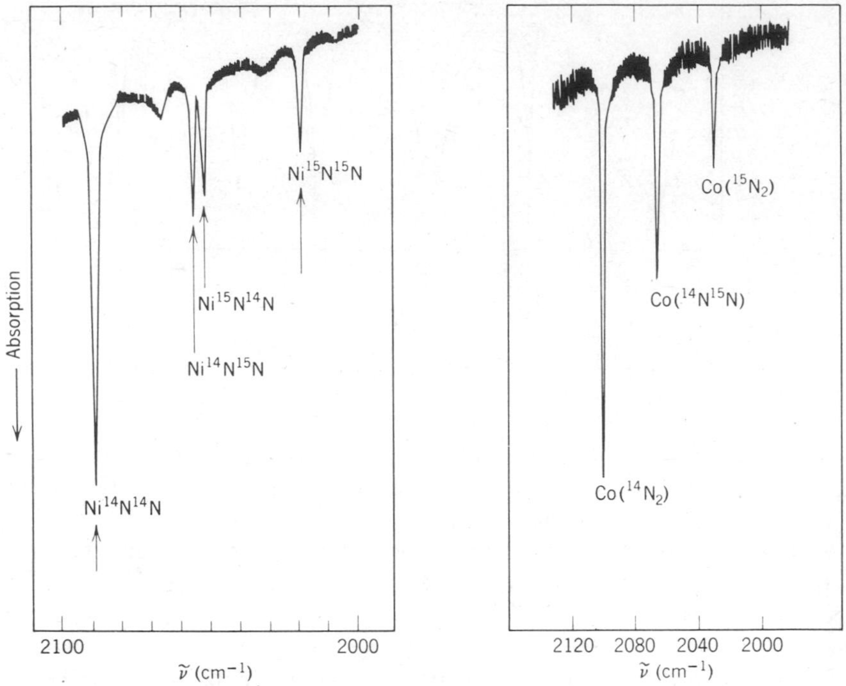

Fig. III-52. Matrix isolation infrared spectra of Ni and Co atom vapors co-condensed with $^{14}N_2/^{14}N^{15}N/^{15}N_2/Ar$ at 10 K.[824,825]

Table III-57 lists the $\nu(N_2)$ of $M(N_2)$-type complexes. All these adducts take the end-on structure except for $Co(N_2)$ and $Th(N_2)$.[830] The structures of $M(N_2)_4$, $M(N_2)_3$, and $M(N_2)_2$ are tetrahedral, trigonal-planar, and linear, respectively, although slight distortion from these ideal symmetries occur due to the matrix effect. Most of these studies have been made in the $\nu(N_2)$ region

TABLE III-57. TYPICAL 1:1 (METAL/N_2) ADDUCTS PREPARED BY MATRIX CO-CONDENSATION TECHNIQUES

Adduct	$\nu(N_2)$	Ref.
$Ni(N_2)$	2088	824
$Pd(N_2)$	2211	824
$Pt(N_2)$	2070	826
$Co(N_2)$	2101	825
$V(N_2)$	2216	827
$Nb(N_2)$	1926/1931	828
$Cr(N_2)$	2215	829
$Th(N_2)$	1829	830

since low-frequency vibrations are generally weak and difficult to measure in inert gas matrices. Finally, UN_2 and PuN_2 prepared by the spattering techniques take the linear N–M–N structure (Sec. II-2).

(3) Nitrido complexes

If the N^{3-} ion coordinates to a metal, it is called a nitrido complex. Nitrido complexes of transition metals can be prepared by several methods, and their preparations, structures, and spectra have been reviewed by Griffith.[831] The M≡N triple bonds are formed as a result of the strong π-donating property of the N^{3-} ion. The ν(M≡N) of nonbridging nitrido complexes are in the 1200–950 cm^{-1} region.[832] For example, ν(M≡N) of $[Mo(N)Cl_5]^{2-}$ is at 1023 cm^{-1},[833] and those of $[M(N)X_5]^{2-}$ [M = Ru(VI) and Os(VI); X = Cl^- and Br^-] are at 1120–1000 cm^{-1}.[834]

Dinuclear nitrido complexes of the types $[M_2(N)X_8(H_2O)_2]^{3-}$ and $[M_2(N)(NH_3)_8Y_2]^{3+}$ (M = Ru and Os; X = Cl^- and Br^-; Y = Cl^-, Br^-, NCS^-, and N_3^-) contain the linear M—N—M bridging group, which exhibits $\nu_a(NM_2)$ and $\nu_s(NM_2)$ at 1120–1050 and 350–280 cm^{-1}, respectively. In trinuclear nitrido complexes containing the trigonal-planar NIr_3 unit, $\nu_a(NIr_3)$ and $\nu_s(NIr_3)$ are at 800–700 and ca. 230 cm^{-1}, respectively. In both cases, the symmetric modes have been observed only in the Raman spectra.[832]

(4) Hydrido Complexes

As stated in Sec. III-16(5) (hydrocarbonyls), the terminal M–H group exhibits a relatively sharp band of medium intensity in the 2250–1700 cm^{-1} region.[685] In addition, it shows δ(MH) in the 800–600 cm^{-1} region. These assignments can be confirmed by deuteration experiments. Table III-58 lists the observed M–H group frequencies of typical hydrido complexes.

The ν(MH) is sensitive to other ligands, particularly those in the *trans*-position in the square-planar Pt(II) complexes. Thus Chatt et al.[844] found that

TABLE III-58. M—H FREQUENCIES OF TYPICAL HYDRIDO COMPLEXES (CM^{-1})

Complex	ν(MH)	δ(MH)	Ref.
$Co(H)_2(PPh_3)_3$	1755	—	835
mer-$Co(H)_3(PPh_3)_3$	1933, 1745	—	836
$[Co(H)(CN)_5]^{3-}$	1840	774	837
$[Ir(H)(CN)_5]^{3-}$	2043	810	838
trans-$[Fe(H)(Cl)\{C_2H_4(PEt_2)_2\}_2]$	1849	656	839
trans-$[Fe(H)_2\{o\text{-}C_6H_4(PEt_2)_2\}_2]$	1726	716	839
$Ir(H)(Cl)_2(PPh_3)_3$ (isomer I)	2197	840, 804	840
cis-$[Ir(H)_2(CO)(PPh_3)_3]$	2160, 2107	—	841
cis-$[Ir(H)_2(Ph_2P\text{-}(CH_2)_2\text{-}PPh_2)_2]^-$	2091, 2080	—	842
trans-$[Os(H)_2\{C_2H_4(PEt_2)_2\}_2]$	1721	—	843

the order of ν(PtH) in *trans*-[Pt(H)X(PEt$_3$)$_2$] is as follows:

X =	NO_3^-	<	Cl^-	<	Br^-	<	I^-	<	NO_2^-	<	SCN^-	<	CN^-
ν(PtH)(cm^{-1})	2242	>	2183	>	2178	>	2156	>	2150	>	2112	>	2041

This is the increasing order of *trans*-influence. Church and Mays[845] found that the NMR Pt—H coupling constant and ν(PtH) decrease in the same order in the *trans*-[Pt(H)L(PEt$_3$)$_2$]$^+$ series:

L =	py	<	CO	<	PPh_3	<	$P(OPh)_3$	<	$P(OMe)_3$	<	PEt_3
J(PtH)(Hz)	1106	>	967	>	890	>	872	>	846	>	790
ν(PtH)(cm^{-1})	2216	>	2167	>	2100	>	2090	>	2067	<	2090

In the above series, the σ-donor strength of L increases as the J(PtH) value decreases and ν(PtH) shifts to a lower frequency. Atkins et al.[846] found linear relationships between the chemical shift of the hydride, the Pt-H coupling constant, ν(PtH), and the pK_a value of the parent carboxylic acid in a series of *trans*-[Pt(H)L(PEt$_3$)$_2$], where L is a carboxylate ligand.

Vibrational frequencies of bridging hydrido complexes have been discussed in Sec. III-16(5).

III-19. HALOGENO COMPLEXES

Halogens (X) are the most common ligands in coordination chemistry. Several review articles[847–849] summarize the results of extensive infrared studies on halogeno complexes. Part II of this book lists the vibrational frequencies of many halogeno complexes. Here the vibrational spectra of halogeno complexes containing other ligands are discussed. In most cases ν(MX) can readily be assigned by halogen or metal (isotope) substitution.

(1) Terminal Metal–Halogen Bond

Terminal MX stretching bands appear in the regions of 750–500 cm^{-1} for MF, 400–200 cm^{-1} for MCl, 300–200 cm^{-1} for MBr, and 200–100 cm^{-1} for MI. According to Clark and Williams,[100] the ν(MBr)/ν(MCl) and ν(MI)/ν(MCl) ratios are 0.77–0.74 and 0.65, respectively. Several factors govern ν(MX).[850] If other conditions are equal, ν(MX) is higher as the oxidation state of the metal is higher. Examples have already been given for tetrahedral MX_4- and octahedral MX_6-type compounds, discussed in Part II. It is interesting to note, however, that in the [M(dias)$_2$Cl$_2$]$^{n+}$ series* ν(MCl) changes rather drastically in going from Ni(III) to Ni(IV) (Fig. III-62), while very little change is observed between Fe(III) and Fe(IV):

	d^4	d^5	d^6	d^7
	Fe(IV)	Fe(III)	Ni(IV)	Ni(III)
ν(MCl) (cm^{-1})	390	384	421	240

* dias: *o*-phenylenebis(dimethylarsine).

This was attributed to the presence of one electron in the antibonding e_g* orbital in the Ni(III) complex.[851]

If other conditions are equal, ν(MX) is higher as the coordination number of the metal is smaller. Table III-59 indicates the structure dependence of ν(NiX), obtained by Saito et al.[102] According to Wharf and Shriver,[856] the SnX stretching force constants of halogenotin compounds are approximately proportional to the oxidation number of the metal divided by the coordination number of the complex.

It is interesting to note that the ν(SnCl) of free $SnCl_3^-$ ion [289 (A_1) and 252 (E) cm^{-1}] are shifted to higher frequencies upon coordination to a metal. Thus ν(SnCl) of $[Rh_2Cl_2(SnCl_3)_4]^{2-}$ are at 339 and 323 cm^{-1}. According to Shriver and Johnson,[857] the L–X force constant of the LX_n-type ligand will increase upon coordination to a metal if X is significantly more electronegative than L. In the above example, chlorine is more electronegative than tin. In metal amine complexes (see Sec. III-1), ν(NH) shifts to lower frequencies because nitrogen is more electronegative than hydrogen. As expected, the ν(GeCl) of free $GeCl_3^-$ ion [303 (A_1) and 285 (E) cm^{-1}] are also shifted to higher frequencies in $[Pd(PhNC)(PPh_3)(GeCl_3)Cl]$ (384 and 360 cm^{-1}).[858]

The MX vibrations are very useful in determining the stereochemistry of the complex. Appendix III tabulates the number of infrared- and Raman-active vibrations of various MX_nY_m-type compounds. Using these tables, it is possible to determine the stereochemistry of a halogeno complex simply by counting

TABLE III-59. STRUCTURAL DEPENDENCE OF NiX STRETCHING FREQUENCIES (CM^{-1})[a]

Stretching Frequency	Linear Triatomic	*trans*-Planar	*cis*-Planar	Tetrahedral	*trans*-Octahedral
ν(NiCl)	$NiCl_2$[b] 521	$Ni(PEt_3)_2Cl_2$[c] 403	$Ni(DPE)Cl_2$[d] 341, 328	$Ni(PPh_3)_2Cl_2$[c] 341, 305	$Ni(py)_4Cl_2$ 207
ν(NiBr)	$NiBr_2$[b] 414	$Ni(PEt_3)_2Br_2$[c] 338	$Ni(DPE)Br_2$[d] 290, 266	$Ni(PPh_3)_2Br_2$[e] 265, 232	$Ni(py)_4Br_2$ 140
ν(NiI)			$Ni(DPE)I_2$[d] 260, 212	$Ni(PPh_3)_2I_2$[e] 215	$Ni(py)_4I_2$ 105
$\frac{\nu(NiBr)}{\nu(NiCl)}$	0.80	0.84	0.83[f]	0.77[f]	0.68
$\frac{\nu(NiI)}{\nu(NiCl)}$			0.70[f]	0.67[f]	0.51

[a] DPE = 1, 2-bis(diphenylphosphino)ethane.
[b] Reference 852.
[c] Reference 853.
[d] Reference 854.
[e] Reference 855.
[f] This value was calculated by using average frequencies of two bands.

the number of $\nu(MX)$ fundamentals observed. Examples of this method will be given in the following sections.

(a) Square-Planar complexes. Vibrational spectra of planar $M(NH_3)_2X_2$ [M = Pt(II) and Pd(II)] were discussed in Sec. III-1. The *trans*-isomer ($\mathbf{D}_{2h}$) exhibits one $\nu(MX)$ (B_{3u}), whereas the *cis*-isomer ($\mathbf{C}_{2v}$) exhibits two $\nu(MX)$ (A_1 and B_2) bands in the infrared. The infrared spectra of *cis*- and *trans*-$[Pd(NH_3)_2Cl_2]$ were shown in Fig. III-4. Similar results have been obtained for a pair of *cis*- and *trans*-$[Pt(py)_2Cl_2]$[859] and PtL_2X_2, where L is one of a variety of neutral ligands.[860]

In planar Pt(II) and Pd(II) complexes, $\nu(MX)$ is sensitive to the ligand *trans* to the M–X bond. Thus the effect of "*trans*-influence"[861] has been studied extensively by using infrared spectroscopy. In the $[PtCl_3L]^-$ series,[862] $\nu(PtCl_{trans})$ follows the order:

L =	CO		SMe_2		C_2H_4		SEt_2		$AsEt_3$		PPh_3		PMe_3		$AsMe_3$		PEt_3
$\nu(PtCl)$ (cm^{-1})	322	>	310	~	309	~	307	>	280	~	279	~	275	~	272	~	271

Their order represents an increasing degree of *trans*-influence, since $\nu(PtCl)$ becomes lower as a ligand of stronger *trans*-influence is introduced *trans* to the Pt–Cl bond. It was found that $\nu(PtCl_{cis})$ is insensitive to the change in L. An order of *trans*-influence such as

$$Cl^- < Br^- < I^- \sim CO < CH_3 < PR_3 \sim AsR_3 < H$$

was noted from the order of $\nu(M\text{–}Cl_{trans})$ in a series of octahedral Rh(III) and Os(III) complexes.[863]

Fujita et al.[864] prepared two isomers of $PtCl(C_2H_4)$(L-ala), where L-ala is L-alanino anion:

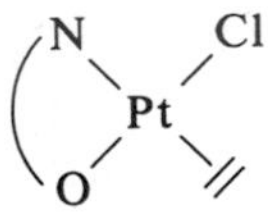

(N-isomer)

I

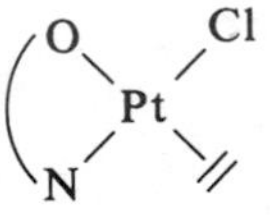

(O-isomer)

II

Isomers I and II exhibit their $\nu(PtCl)$ at 360 and 340 cm^{-1}, respectively. Since the *trans*-influence of the N-donor is expected to be stronger than that of the O-donor, the structures of these two isomers have been assigned as shown above.

Complexes of the type $Ni(PPh_2R)_2Br_2$ (R = alkyl) exist in two isomeric forms: tetrahedral (green) and *trans*-planar (brown). Distinction between these two can be made easily since the numbers and frequencies of infrared-active $\nu(NiBr)$ and $\nu(NiP)$ are different for each isomer. Wang et al.[865] studied the infrared spectra of a series of compounds of this type, and confirmed that

ν(NiBr) and ν(NiP) are at ca. 330 and 260 cm^{-1}, respectively, for the planar form and at ca. 270–230 and 200–160 cm^{-1}, respectively, for the tetrahedral form. The presence or absence of the 330-cm^{-1} band is particularly useful in distinguishing these two isomers. According to X-ray analysis,[866] the green form of $Ni(PPh_2Bz)_2Br_2$ (Bz = benzyl) is a mixture of the planar and tetrahedral molecules in a 1:2 ratio. Ferraro et al.[867] studied the effect of high pressure on the infrared spectra of this compound, and found that all the bands characteristic of the tetrahedral form disappear as the pressure is increased to ca. 20,000 atm. This result indicates that the tetrahedral molecule can be converted to the planar form under high pressure if the energy difference between the two is relatively small. This conversion is completely reversible; the original green form is recovered as the pressure is reduced. High-pressure infared spectroscopy has also been used to distinguish symmetric and antisymmetric MX stretching vibations. For example, Fig. III-53 shows the effect of pressure on ν_a(PtCl) and ν_s(PtCl) of $Pt(NBD)Cl_2$ (NBD: norbornadiene).[868] It is seen that by increasing pressure the intensity of ν_s(PtCl) is suppressed to a greater degree than that of ν_a(PtCl). For high-pressure vibrational spectroscopy, see a review by Ferraro.[869,870]

(b) Octahedral Complexes. *cis*-MX_2L_4 ($\mathbf{C}_{2v}$) should exhibit two ν(MX), while *trans*-MX_2L_4 ($\mathbf{D}_{4h}$) should give only one ν(MX) in the infrared. Thus *cis*-$[IrCl_2(py)_4]Cl$ shows two ν(IrCl) at 333 and 327 cm^{-1}, while *trans*-$[IrCl_2(py)_4]Cl$ exhibits only one ν(IrCl) at 335 cm^{-1}.[100] If MX_3L_3 is *fac* ($\mathbf{C}_{3v}$),

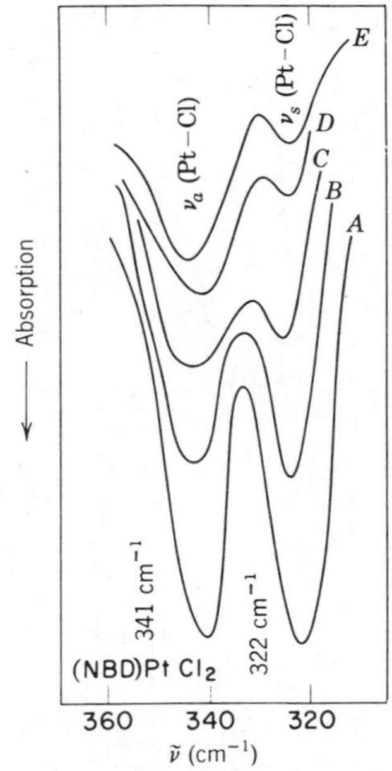

Fig. III-53. Effect of pressure on Pt–Cl stretching bands: (A) 1 atm; (B) 6,000 atm; (C) 12,000 atm; (D) 18,000 atm; (E) 24,000 atm.

two $\nu(MX)$ are expected in the infrared. If it is *mer* ($\mathbf{C}_{2v}$), three $\nu(MX)$ should be infrared active. As is shown in Fig. III-54, *fac*-[$RhCl_3(py)_3$] gives two bands at 341 and 325 cm^{-1} and *mer*-[$RhCl_3(py)_3$] shows three bands at 355, 322, and 295 cm^{-1}.

In MX_4L_2-type compounds, the number of IR-active $\nu(MX)$ is one for the *trans*-isomer ($\mathbf{D}_{4h}$) and four for the *cis*-isomer ($\mathbf{C}_{2v}$). For example, *trans*-[$PtCl_4(NH_3)_2$] exhibits one $\nu(PtCl)$ at 352 cm^{-1} (with a shoulder at 346 cm^{-1}), whereas *cis*-[$PtCl_4(NH_3)_2$] exhibits four $\nu(PtCl)$ at 353, 344, 330, and 206 cm^{-1}.[871] Using Sn isotopes, Ohkaku and Nakamoto[872] confirmed that *trans*-[$SnCl_4L_2$] (L = py, THF, etc.) exhibits one $\nu(SnCl)$ in the 342–370 cm^{-1} region, while *cis*-[$SnCl_4(L—L)$] (L—L = bipy,phen, etc.) shows four $\nu(SnCl)$ in the 340–280 cm^{-1} region. For $MX_5L(\mathbf{C}_{4v})$, one expects three $\nu(MX)$ in the infrared. The $\nu(InCl)$ of $[InCl_5(H_2O)]^{2-}$ were observed at 280, 271, and 256 cm^{-1}.[873]

(2) Bridging Metal–Halogen Bond

Halogens tend to form bridges between two metal atoms. In general, bridging MX stretching frequencies [$\nu_b(MX)$] are lower than terminal MX stretching frequencies [$\nu_t(MX)$]. Vibrational spectra of simple M_2X_6-type ions having bridging halogens were discussed in Sec. II-10. Table III-60 lists the $\nu_t(MX)$ and $\nu_b(MX)$ of bridging halogeno complexes containing other ligands.

The *trans*-planar $M_2X_4L_2$-type compounds ($\mathbf{C}_{2h}$) exhibit three infrared-active (B_u) $\nu(MX)$ modes: one $\nu(MX_t)$ and two $\nu(MX_b)$. For the latter two,

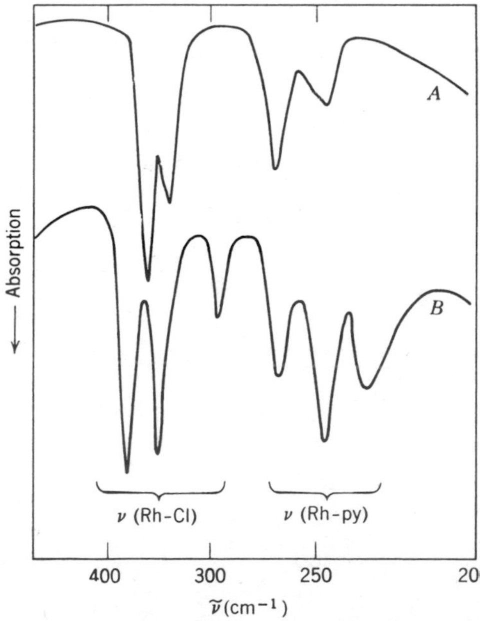

Fig. III-54. Far-infrared spectra of (A) *fac*- and (B) *mer*-$Rh(py)_3Cl_3$.

TABLE III-60. TERMINAL AND BRIDGING METAL–HALOGEN STRETCHING FREQUENCIES (CM^{-1})

Compound[a]	ν_t(MX)	ν_b(MX)	ν_b/ν_t[b]	Ref.
trans-$Pd_2Cl_4L_2$	360–339	308–294	0.86	874
		283–241	0.75	
trans-$Pt_2Cl_4L_2$	368–347	331–317	0.91	874
		301–257	0.78	
$Pd_2Br_4L_2$	285–265	220–185	0.74	875
		200–165	0.66	
$Pt_2Br_4L_2$	260–235	230–210	0.89	875
		190–175	0.74	
$Pt_2I_4L_2$	200–170	190–150	0.92	875
		150–135	0.77	
$Ni(py)_2Cl_2$	—	193, 182	—	876
$Ni(py)_2Br_2$	—	147	—	876
$Co(py)_2Cl_2$				
Monomeric	347, 306	—		109
Polymeric	—	186, 174		109

[a] L = PMe_3, PEt_3, PPh_3, and so on.
[b] These values were calculated using average frequencies.

the higher-frequency band corresponds to $\nu(MX_b)$ *trans* to *X*, whereas the lower-frequency mode is assigned to $\nu(MX_b)$ *trans* to L since it is sensitive to the nature of L.[874] Strong coupling is expected, however, among these modes since they belong to the same symmetry species.

$Co(py)_2Cl_2$ is known to exist in two forms: the monomeric tetrahedral (blue) and the polymeric octahedral (lilac). The latter is an infinite chain polymer bonded through chlorine bridges:

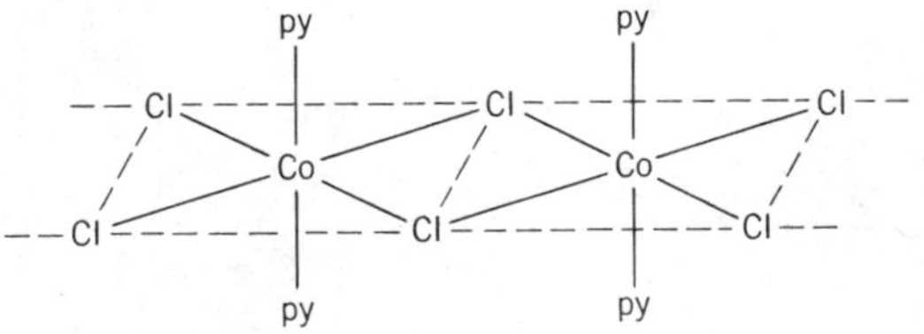

Figure III-55 shows the infrared spectra of both forms. The $\nu(CoCl_b)$ of the polymer is very low relative to $\nu(CoCl_t)$ of the monomer because of an increase in coordination number and the effect of bridging.[109] Polymeric $Ni(py)_2X_2$ also exhibits $\nu(NiX_b)$ below 200 cm^{-1}.[876]

Figure III-56*a* shows the structure of the metal cluster ion such as $[(Mo_6X_8)Y_6]^{2-}$. Because of its high symmetry ($\mathbf{O}_h$) only five F_{1u} modes are infrared active. For $[(Mo_6Cl_8)Cl_6]^{2-}$, Cotton et al.[877] proposed assigning two

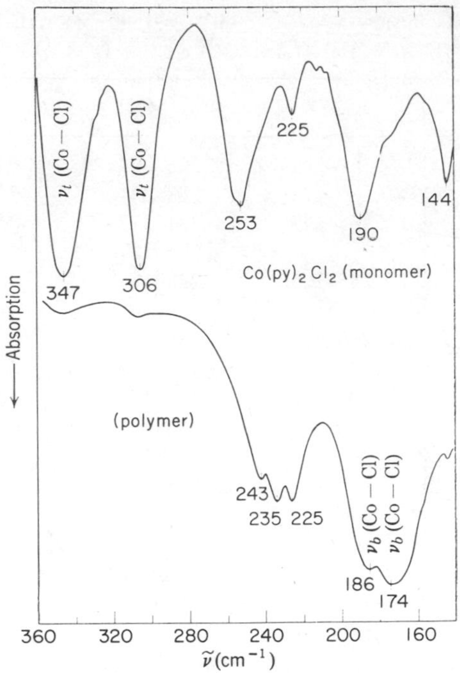

Fig. III-55. Infrared spectra of monomeric and polymeric forms of $Co(py)_2Cl_2$.

$\nu(MoCl_X)$, $\nu(MoCl_Y)$, and $\delta(Cl_XMoCl_Y)$ at 350–310, 246, and 110 cm^{-1}, respectively; $\nu(MoMo)$, which will be discussed in Sec. III-20, was assigned at 220 cm^{-1}. On the other hand, Hogue and McCarley[878] assigned five bands of $[(W_6X_8)Y_n]$-type compounds to the following five modes: three W_6X_8 unit

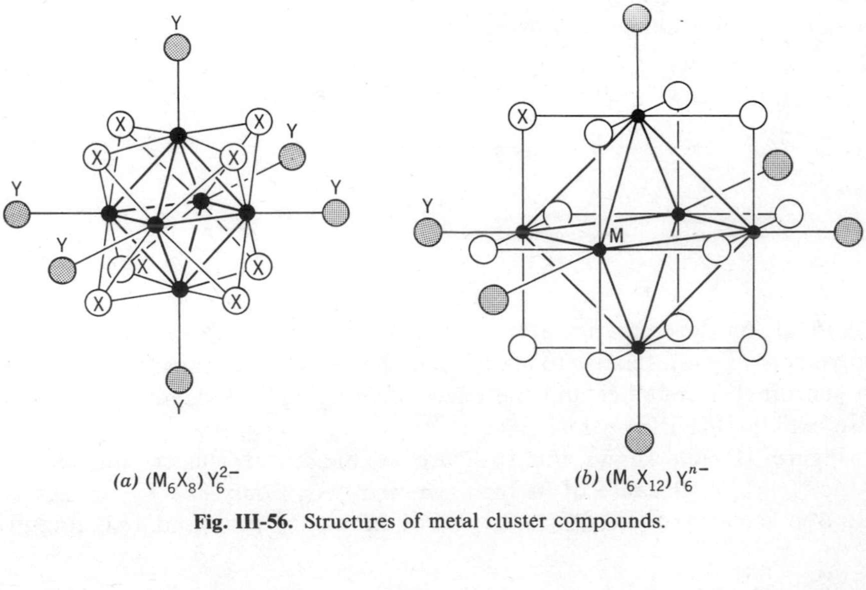

(a) $(M_6X_8)Y_6^{2-}$ *(b)* $(M_6X_{12})Y_6^{n-}$

Fig. III-56. Structures of metal cluster compounds.

vibrations, ν(WY), and δ(XWY). For X = Y = Cl, these bands are at 318, 284, 225, 305, and 105 cm^{-1}, respectively. The W_6X_8 unit vibrations were not assigned to individual internal modes because of the strong coupling between them.

Figure III-56*b* shows the structure of the $[(M_6X_{12})Y_6]^{n-}$-type metal cluster. The infrared spectra of the M_6X_{12} unit (M = Nb and Ta; X = Cl, Br)[879,880] have been reported. Under $\mathbf{O}_h$ symmetry, the M_6X_{12} unit is expected to show four infrared-active fundamentals: three ν(MX) and one ν(MM). For the Nb_6Cl_{12} unit, these were assigned at 340, 280, 232, and 145 [ν(NbNb)]cm^{-1}, respectively.[880] For the $[(Nb_6Cl_{12})Cl_6]^{3-}$ ion, two more vibrations, $\nu(NbCl_Y)$ and $\delta(Cl_XNbCl_Y)$, are expected to be infrared active. The former was assigned at 200 cm^{-1}, but the latter was not located. Fleming et al.[881] concluded that ν(MM) cannot be identified in these compounds because of strong coupling among individual modes.

III-20. COMPLEXES CONTAINING METAL–METAL BONDS

A large number of complexes containing metal-metal bonds are known, and their vibrational spectra have been reviewed extensively.[882–886] Vibrational spectra of polynuclear carbonyls were reviewed in Sec. III-16 and will also be discussed in Sec. IV-7. Vibrational spectra of metal clusters containing halogen bridges were reviewed in Sec. III-19. In this section, only metal-metal stretching, ν(MM), of these and other compounds are discussed. In general, ν(MM) appear in the low-frequency region (250–100 cm^{-1}) because the M–M bonds are relatively weak and the masses of metals are relatively large. If the complex is perfectly centrosymmetric with respect to the M–M bond at the center, ν(MM) is forbidden in the infrared. If it is not, ν(MM) may appear weakly in the infrared. The ν(M—M′) vibration of a heteronuclear complex is expected to be stronger because of the presence of a dipole moment along the M–M′ bond. In contrast, Raman spectroscopy has distinct advantages in that both ν(MM) and ν(MM′) appear strongly since large changes in polarizabilities are expected as a result of stretching long, covalent M–M(M′) bonds. Special caution must be taken, however, in measuring Raman spectra since metal-metal bonded compounds may undergo thermal and/or photochemical decomposition.

(1) Polynuclear Carbonyls

The ν(MM) of polynuclear carbonyls have been assigned for a number of compounds, and the MM stretching force constants obtained from normal coordinate analysis have been used to discuss the nature of the M–M bond. Figure III-57 shows the Raman spectra of $Mn_2(CO)_{10}$, $MnRe(CO)_{10}$, and $Re_2(CO)_{10}$ obtained by Quicksall and Spiro.[887] Risen and co-workers[888–890] carried out normal coordinate analyses on many dinuclear and trinuclear metal carbonyls. Table III-61 lists the observed ν(MM) and the corresponding force constants obtained by these and other workers. It is noted that the MM

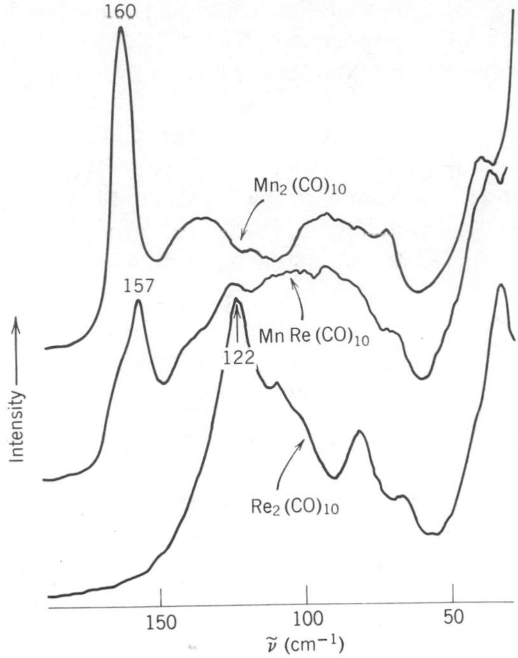

Fig. III-57. Low-frequency Raman spectra of polycrystalline $Mn_2(CO)_{10}$, $MnRe(CO)_{10}$, and $Re_2(CO)_{10}$ (632.8 nm excitation).[887]

stretching force constants obtained by rigorous calculations are surprisingly close to those obtained by approximate calculations considering only metal atoms. There is a general trend that as the MM stretching force constant increases, the $\nu(MM)$ frequency decreases in going from lighter to heavier metals in the $M_2(CO)_{10}$ (M = Mn, Tc and Re)[887] and $[M_2(CO)_{10}]^{2-}$ (M = Cr, Mo and W)[892] series.

In Sec. III-16(3), we discussed the spectra of $M_2(CO)_4X_2$-type compounds (M = Mn, Tc, Re, Rh, etc.) in which the metals are bonded through halogen (X) bridges. Goggin and Goodfellow[893] concluded, however, that the $[Pt_2(CO)_2X_4]^{2-}$ ion (X = Cl and Br) contains the direct Pt–Pt bond:

$$
\begin{array}{cc}
O & O \\
C & C \\
| & | \\
X-Pt-Pt-X \\
| & | \\
X & X
\end{array}
$$

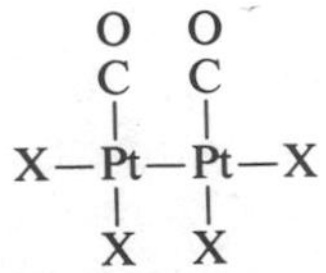

They isolated two isomers of $[N(n-Pr)_4]_2[Pt_2(CO)_2Cl_4]$ which differ only in the angle of rotation about the Pt–Pt bond. Both isomers exhibit $\nu(PtPt)$ at $\sim$170 cm^{-1}.

TABLE III-61. METAL-METAL STRETCHING FREQUENCIES (CM^{-1}) AND FORCE CONSTANTS

Compound	ν(MM)	Force Constant (mdyn/Å)		Ref.
		Rigorous Calculation	Approximate Calculation[a]	
$(CO)_5Mn—Mn(CO)_5$	160	0.59	0.41	887
$(CO)_5Tc—Tc(CO)_5$	148	0.72	0.63	887
$(CO)_5Re—Re(CO)_5$	122	0.82	0.82	887
$(CO)_5Re—Mn(CO)_5$	157	0.81	0.62	887
$(CO)_5Mn—W(CO)_5^-$	153	0.71	0.55	888
$(CO)_5Mn—Mo(CO)_5^-$	150	0.60	0.47	888
$(CO)_5Mn—Cr(CO)_5^-$	149	0.50	0.37	888
$Cl_3Sn—Co(CO)_4$	204	1.23	0.97	889
$Cl_3Ge—Co(CO)_4$	240	1.05	1.11	889
$Cl_3Si—Co(CO)_4$	309	1.32	1.07	889
$Br_3Ge—Co(CO)_4$	200	0.96	—	890
$I_3Ge—Co(CO)_4$	161	0.52	—	890
$Br_3Sn—Co(CO)_4$	182	1.05	—	890
$I_3Sn—Co(CO)_4$	156	0.64	—	890
$H_3Ge—Re(CO)_5$	209	—	1.34	891
$H_3Ge—Mn(CO)_5$	219	—	0.88	891
$H_3Ge—Co(CO)_4$	228	—	1.00	891
$(CO)_4Co—Zn—Co(CO)_4$	170, 284[b]	1.30	—	650
$(CO)_4Co—Cd—Co(CO)_4$	163, 218[b]	1.28	—	650
$(CO)_4Co—Hg—Co(CO)_4$	163, 195[b]	1.26	—	650

[a] Calculations considering only metal atom skeletons.
[b] Under $\mathbf{D}_{3d}$ symmetry, these frequencies correspond to the A_{1g} (symmetric) and A_{2u} (antisymmetric) MCo stretching modes, respectively.

$Fe_2(CO)_9$ and $Fe_3(CO)_{12}$ exhibit very strong Raman bands at 225 and 219 cm^{-1}, respectively. San Filippo and Sniadoch[894] assigned them to ν(FeFe). Later studies[895] showed, however, that these bands are due to decomposition products resulting from strong laser irradiation. Thus the appearance of strong Raman bands in the low-frequency region does not necessarily mean that they are due to ν(MM). It is also noted that $Re_2(CO)_8X_2$ (X = Cl and Br), which does not contain Re-Re bonds, shows strong Raman bands at 125 cm^{-1} where ν(ReRe) of $Re_2(CO)_{10}$ appears.[895] Cooper et al.[896] were able to locate the ν(MM) of $Fe_2(CO)_9$ and $Fe_3(CO)_{12}$ at 260 and at 240 and 176 cm^{-1}, respectively. These assignments are based on the ^{54}Fe-^{56}Fe isotopic shifts observed in Raman spectra at ~10 K. Onaka and Shriver[897] observed three ν(MM) bands at 235, 185, and 159 cm^{-1} in acetone solution of $Co_2(CO)_8$ which correspond to the three isomers discussed in Sec. III-16(2). They have shown that the ν(MM) is higher than 200 cm^{-1} for bridging carbonyls and between 190 and 140 cm^{-1} for single-bonded nonbridged complexes.

Trinuclear complexes such as $Ru_3(CO)_{12}$ and $Os_3(CO)_{12}$ contain a triangular M_3 skeleton for which two ν(MM) are expected under $\mathbf{D}_{3h}$ symmetry. Quicksall and Spiro[644] assigned the Raman bands at 185 and 149 cm^{-1} of the Ru complex to ν(RuRu) of the A_1' and E' species, respectively. The latter is coupled with other modes. The corresponding RuRu stretching force constant is 0.82 mdyn/Å. Kettle and co-workers[898,899] have assigned the ν(MM) of the $[Os_xRu_{3-x}(CO)_{12}]$- ($x = 0, 1, 2$, and 3) type complexes. Quicksall and Spiro[900] assigned the Raman spectrum of $Ir_4(CO)_{12}$, which consists of a tetrahedral Ir skeleton; three ν(IrIr) bands were assigned at 207 (A_1), 161 (F_2), and 131 (E) cm^{-1}. The ratio of these three frequencies, 2:1.56:1.27, is far from that predicted by a "simple cluster model" ($2:\sqrt{2}:1$),[901] indicating the substantial coupling between the individual stretching modes. Their rigorous calculations gave K(Ir–Ir) of 1.69 mdyn/Å, together with interaction constants of −0.13 and +0.13 mdyn/Å for the adjacent and opposite Ir–Ir bonds, respectively. The ν(MM) of $Rh_4(CO)_{12}$ [902] and $Co_4(CO)_{12}$ [903] have been assigned, and the corresponding force constants calculated.[903]

(2) Metal Cluster Compounds

As discussed in Sec. II-11(3), the metal atoms in $[M_2Cl_9]^{3-}$ (M = Cr and W) are bonded directly through the M–M bond or indirectly through the bridging chlorine atom or both. Normal coordinate analysis does not allow one to resolve the M–M interaction into these two components. Based on X-ray and magnetic data, Ziegler and Risen[904] assumed that the Cr–Cr stretching force constant is zero and that the M–Cl–M bending force constant is the same for the Cr and W compounds. These assumptions yielded the W–W stretching force constant of 1.15 mdyn/Å. The ν(WW) at 139 cm^{-1} was found to contain 44% W–W stretching character. The $[Cr_2Cl_9]^{3-}$ ion exhibits a strong Raman band at 161 cm^{-1} although it does not possess any direct M–M restoring force.

The structures of the $[(M_6X_8)Y_6]^{2-}$- and $[(M_6X_{12})Y_6]^{n-}$-type metal clusters are shown in Fig. III-56. As stated in Sec. III-19(2), Cotton et al.[877] assigned ν(MoMo) of $[(Mo_6Cl_8)Cl_6]^{2-}$ at 220 cm^{-1}. Hogue and McCarley[878] could not assign ν(MoMo) empirically since strong coupling between ν(MoCl) and ν(MoMo) of the Mo_6Cl_8 unit was expected. Although Mattes[905] carried out normal coordinate analysis on the $(Mo_6X_8)Y_4$ cluster (M = Mo, W; X, Y = Cl, Br), he did not calculate the potential energy distribution, which would indicate the degree of coupling between individual modes.

The infrared spectra of $[(M_6X_{12})Y_6]^{2-}$-type clusters have been reported by several investigators.[879] Mackay and Schneider[880] assigned the ν(NbNb) of the Nb_6Cl_{12} cluster at 140 cm^{-1}. Again, Fleming et al.[881] pointed out that ν(MM) of such a cluster cannot be assigned because of strong coupling with other modes. Although Mattes[906] carried out normal coordinate analysis on the $(M_6X_{12})Y_n$ system (M = Nb, Ta; $n = 2$–4), the potential energy distribution in individual modes was not calculated. He concluded that ν(MM) are too low (below 100 cm^{-1}) to be observed.

Figure III-58*a* shows the structure of $[M_6O_{19}]^{8-}$ ions (M = Nb, Ta). Farrell et al.[907] obtained the following set of force constants for the $[Nb_6O_{19}]^{8-}$ ion:

K(Nb—O)			K(Nb—Nb)
5.66 terminal	2.92 bridging	0.91 central	1.01 mdyn/Å

The ratio of the first three force constants is about 6:3:1. Although the NbNb stretching constant was estimated to be ca. 1 mdyn/Å, this value does not represent the strength of this bond, since such a value can be obtained without any M-M interaction.[907] Mattes[908] carried out normal coordinate analysis on the same system, and obtained a ratio of 8:4:1 for the three MO stretching force constants mentioned above.

The metal skeleton of the $[Pb_6O(OH)_6]^{4+}$ ion takes the very unusual structure shown in Fig. III-58*b*.[909] A band at 150 cm^{-1} was assigned to ν(PbPb), which corresponds to the shortest Pb-Pb bond (Pb_3—Pb_4 of Fig. III-58*b*).[910]

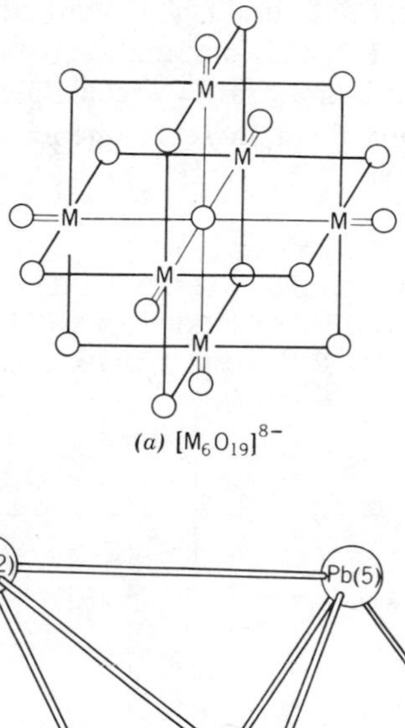

(a) $[M_6O_{19}]^{8-}$

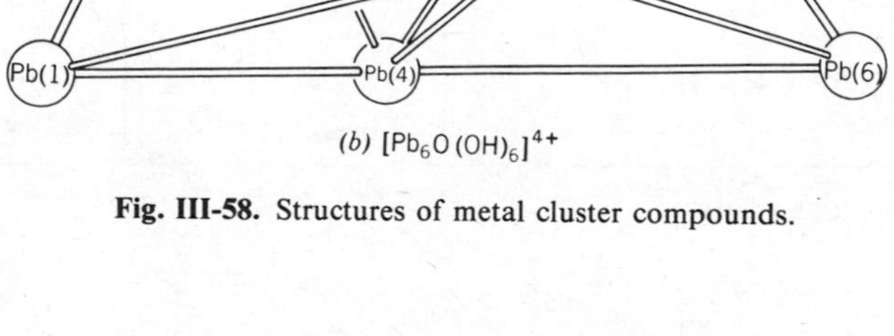

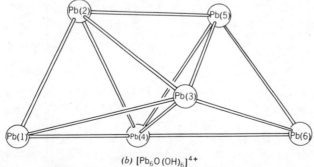

(b) $[Pb_6O(OH)_6]^{4+}$

Fig. III-58. Structures of metal cluster compounds.

(3) Compounds Containing Metal–Metal Multiple Bonds

A number of compounds containing unusually short M–M bonds exhibit unusually high ν(MM). For example, the Mo–Mo distance of $Mo_2(OAc)_4$ is only 2.09 Å and its ν(MM) is at 406 cm^{-1}. According to Cotton,[911] this Mo–Mo bond consists of one σ-bond, two π-bonds and a δ-bond (bond order 4). Such a quadruple bond is also expected for $[Re_2Cl_8]^{2-}$ which exhibits the ν(ReRe) at 272 cm^{-1} with the Re–Re distance of 2.22 Å.[912,913] Table III-62 lists ν(MM) of typical compounds. It is seen that the ν(MM) of dimolybdenum compounds of bond order 4 scatter over a wide range. In contrast, dirhenium compounds exhibit a nice ν(MM)-bond order relationship as demonstrated by Fig. III-59. Table III-62 also show that the ν(MoMo) and Mo–Mo distance are rather sensitive to the nature of the axial ligand. Normal coordinate analyses have been made on $M_2(O_2CCH_3)_4$ and $[M_2X_8]^{n-}$ (M = Mo and Re; X = Cl and Br).[917,918] The (MoMo) of $Mo_2(O_2CCH_3)_4$ is shifted by 9 cm^{-1} by ^{92}Mo–^{NA}Mo substitution (^{NA}Mo approximately represents ^{96}Mo).[919]

Most of the ν(MM) frequencies discussed above were obtained by RR spectroscopy. As an example, Fig. III-60 shows the RR spectra of the $[Mo_2Cl_8]^{4-}$ and $[Mo_2Cl_9]^{5-}$ ions.[920] The former exhibits a strong electronic band near 19,000 cm^{-1} which is due to a transition from the $(\sigma)^2(\pi)^4(\delta)^2$ to $(\sigma)^2(\pi)^4(\delta)(\delta^*)$ state. When the laser frequency approaches the energy of this transition, the ν(MoMo) at 345 cm^{-1} (ν_1) is markedly enhanced, and a series of overtones, $n\nu$ (up to $n = 11$), and combination bands, $n\nu_1 + \nu_4$(ν_4: A_{1g} Mo–Cl stretch), are observed. Clark and Franks[920] calculated the anharmonicity constants of the ν_1 vibrations from these frequencies. The resonance Raman

TABLE III-62. METAL–METAL STRETCHING FREQUENCIES (CM^{-1}), BOND ORDERS, AND BOND DISTANCES (A) OF MOLYBDENUM AND RHENIUM COMPOUNDS

Compound	Bond Order	Bond Distance	ν(MM)	Ref.
$Mo_2(O_2CCH_3)_4$	4	2.09	406	914
$Mo_2(O_2CCF_3)_4$	4	2.09	397	915
$Mo_2(O_2CCF_3)_4(py)_2$	4	2.22	367	915
$K_4[Mo_2Cl_8]\cdot 2H_2O$	4	2.14	345	915
$K_3[Mo_2(SO_4)_4]\cdot 3.5H_2O$	3.5	2.16	386	916
			373	
$Re_2(O_2CCH_3)_4Cl_2$	4	2.24	289	914
$[Bu_4N]_2[Re_2Cl_8]$	4	2.22	272	912
$Re_2Cl_5(DTH)_2$ [a]	3	2.29	267	914
$Re_2OCl_5(O_2CCH_2CH_3)(PPh_3)_2$	2	2.52	216	914
$Re_2(CO)_{10}$	1	3.02	122	914

[a] DTH = 2,5-dithiahexane.

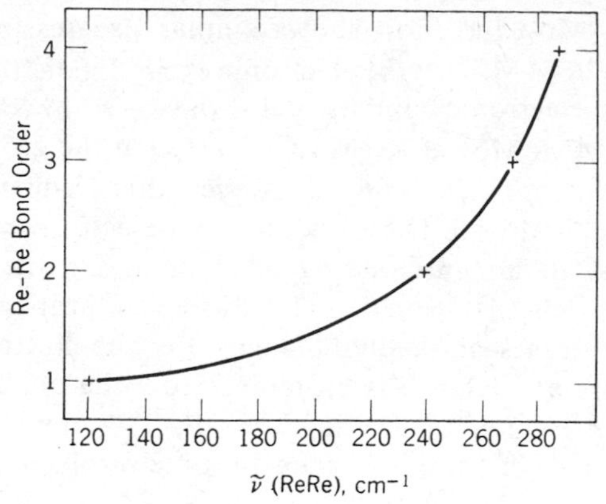

Fig. III-59. ν(Re–Re) vs Re–Re bond order.

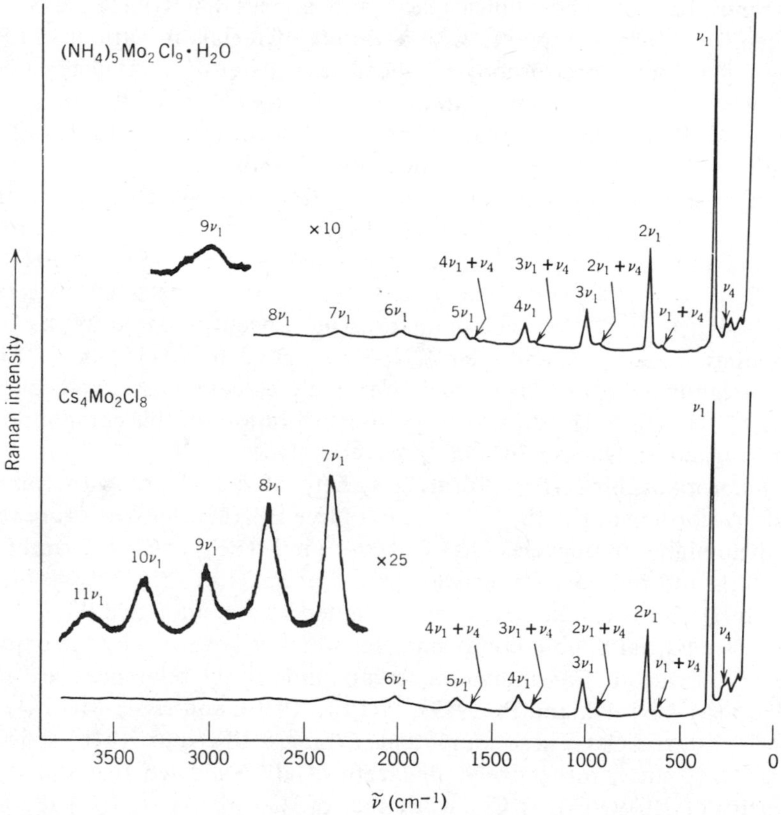

Fig. III-60. Resonance Raman spectra of $(NH_4)_5Mo_2Cl_9H_2O$ and $Cs_4Mo_2Cl_8$ (514.5 nm excitation).

spectrum of the $[Mo_2Cl_9]^{5-}$ ion shows similar progressions. Finally, it is possible to obtain the ν(MM) at the electronic excited state from the vibrational fine structure of an electronic band involving the $\delta - \delta^*$ or $\delta - \pi^*$ transition.[921] For example, the ν(MoMo) of $K_4[Mo_2Cl_8]\cdot 2H_2O$ at the ground state (^1A_g) is 345 cm^{-1} as shown in Table III-62. However, this frequency is reduced to 336 cm^{-1} at the excited state ($^1A_{2u}$) because an electron is promoted from a bonding δ orbital to an antibonding δ^* orbital. On the other hand, the ν(Rh–Rh) of the $[Rh_2L_4]^{2+}$ ion (L = 1,3-diisocyanopropane) in the ground state (79 cm^{-1}) increases markedly (144 cm^{-1}) in the electronic excited state $[E_u(^3A_{2u})]$ because an electron is promoted from the antibonding $4d_{z^2}\sigma^*$ to the bonding $5p_z\sigma$ orbital.[921a] Dallinger et al.[921b] were able to obtain the vibrational frequency of the latter state by time-resolved resonance Raman spectroscopy [see Sec. III-3(2)].

III–21. COMPLEXES OF PHOSPHINES AND ARSINES

Ligands such as phosphines (PR_3) and arsines (AsR_3) (R = alkyl, aryl, halogen, etc.) form complexes with a variety of metals in various oxidation states. Vibrational spectroscopy has been used extensively to determine the structures of these compounds and to discuss the nature of the metal-phosphorus (M–P) bonding. Verkade[922] reviewed spectroscopic studies of M–P bonding with emphasis on cyclic phosphine ligands.

The most simple phosphine ligand is PH_3. The vibrational spectra of $Ni(PH_3)_4$,[923] $Ni(PH_3)(CO)_3$,[924] and $Ni(PH_3)(PF_3)_3$[925] have been reported by Bigorgne and co-workers. All these compounds exhibit ν(PH), $\delta(PH_3)$, and ν(NiP) at 2370–2300, 1120–1000, and 340–295 cm^{-1}, respectively. A series of the $Ni(PH_3)_n$- ($n = 1 \sim 4$) type complexes have been prepared by matrix co-condensation reactions, and their ν(Ni–P) assigned at 390–350 cm^{-1}.[926] Complete assignments based on normal coordinate calculations have been made on $Ni(P(CH_3)_3)_4$.[927] The A_1 and F_2 ν(Ni–P) vibrations of this compound have been assigned at 296 and 343 cm^{-1}, respectively.

Trifluorophosphine (PF_3) forms a variety of complexes with transition metals. According to Kruck,[928] the ν(PF) of free PF_3 (892, 860 cm^{-1}) are shifted slightly to higher frequencies (960–850 cm^{-1}) in $M(PF_3)_n$ ($n = 4$, 5, and 6) and $HM(PF_3)_4$ and to lower frequencies (850–750 cm^{-1}) in $[M(PF_3)_4]^-$ (M = Co, Rh, and Ir). These results have been explained by assuming that the P–F bond possesses a partial double-bond character which is governed by the oxidation state of the metal. For individual compounds, only references are given: $M(PF_3)_4$ (M = Ni, Pd, and Pt) (929), $M(PF_3)_5$ (930), and *cis*-$H_2M(PF_3)_4$ (931) (M = Fe, Ru, and Os). These compounds exhibit $\delta(PF_3)$ and ν(MP) at 590–280 and 250–180 cm^{-1}, respectively. Bénazeth et al.[932] showed that the skeletal symmetry of $HCo(PF_3)_4$ is $\mathbf{C}_{3v}$, while that of $[Co(PF_3)_4]^-$ is $\mathbf{T}_d$. The ν(CoP) of these compounds are at 250–210 cm^{-1}. Woodward and co-workers[933] carried out complete vibrational analyses of the $M(PF_3)_4$ (M = Ni, Pd, and Pt) series.

Their results give the following:

		Ni	Pd	Pt	
$\nu(MP)(cm^{-1})$	A_1	195	204	213	
	F_2	219	222	219	
$K(M—P)(mdyn/Å)$		2.71	3.17	3.82	(GVF)

The infrared spectra of trialkyl phosphine halogeno complexes have been studied by many investigators. The main interest in these investigations has been to determine the stereochemistry and the nature of the M–P bond from the vibrational spectra. Although metal–halogen stretching vibrations can be assigned easily (see Sec. III-19), it is much more difficult to assign the metal–phosphorus (MP) stretching bands because they appear in the region where alkylphosphines exhibit many bands. To distinguish these two types of vibration, Shobatake and Nakamoto[853] utilized the metal isotope technique (Sec. I-16). Figure III-61 shows the infrared spectra of *trans*-[$^{58,62}Ni(PEt_3)_2X_2$] (X = Cl and Br), and Table III-63 lists the observed frequencies, metal isotope shifts, and band assignments. It is clear that the $\nu(NiP)$ of these complexes must be assigned near 270 cm^{-1}, in contrast to previous investigations, which placed these vibrations near 450–410 cm^{-1}.[875,934–936]

Triphenylphosphine (PPh_3) is most common among phosphine ligands. It is not simple, however, to assign the $\nu(MP)$ of PPh_3 complexes since PPh_3 exhibits a number of ligand vibrations in the low-frequency region.[938] Using the metal isotope technique, Nakamoto et al. showed that tetrahedral $Ni(PPh_3)_2Cl_2$, for example, exhibits two $\nu(NiP)$ at 189.6 and 164.0 cm^{-1}, in agreement with the result of previous workers.[939]

TABLE III-63. INFRARED FREQUENCIES, ISOTOPIC SHIFTS, AND BAND ASSIGNMENTS OF $NiX_2(PEt_3)_2$ (X = Cl AND Br) (CM^{-1})[853]

PEt_3	$^{58}NiCl_2(PEt_3)_2$		$^{58}NiBr_2(PEt_3)_2$		
ν	ν	$\Delta\nu^a$	ν	$\Delta\nu^a$	Assignment^b
408	416.7	0.0	413.6	1.2	$\delta(CCP)$
—	403.3	6.7	337.8	10.5^c	$\nu(NiX)$
365	372.5	−0.1	374.0	1.1	$\delta(CCP)$
330	329.0	−0.5	327.8	0.5^c	$\delta(CCP)$
—	273.4	5.9	265.0	4.7	$\nu(NiP)$
245	(hidden)		(hidden)		$\delta(CCP)$
	200.2	0.8	190.4	0.7	$\delta(CPC)$
	186.5	−0.2	155.1	1.5	$\delta(NiX)$
	161.5	−0.5	(hidden)		$\delta(NiP)$

^{a} $\Delta\nu$ indicates metal-isotope shift, $\nu(^{58}Ni)-\nu(^{62}Ni)$.
^{b} Ligand vibrations were assigned according to Ref. 937.
^{c} Since these two bands are overlapped (Fig. III-61), $\Delta\nu$ values are only approximate.

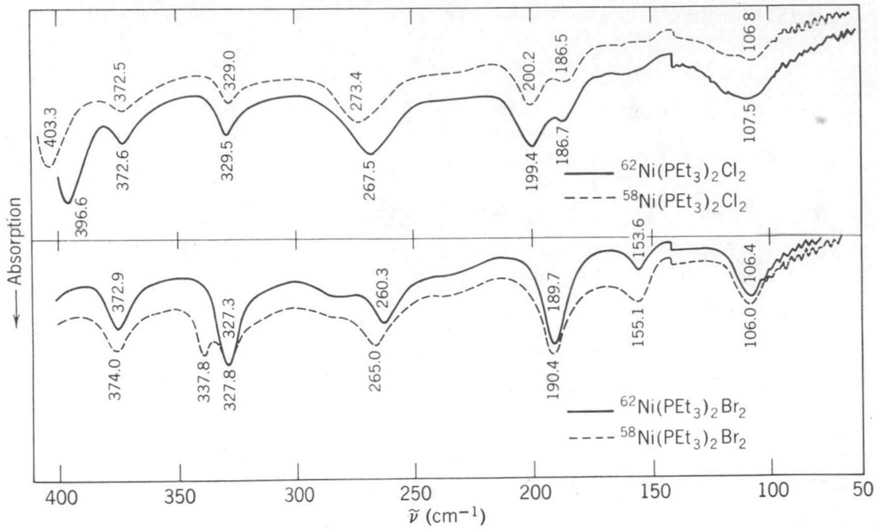

Fig. III-61. Far-infrared spectra of $^{58}NiX_2(PEt_3)_2$ and $^{62}NiX_2(PEt_3)_2$ (X = Cl and Br).

As stated in Sec. III-19, complexes of the type $Ni(PPh_2R)_2Br_2$ (R = alkyl) exist in two forms (tetrahedral and square-planar) which can be distinguished by the ν(NiBr) and ν(NiP).[865] For R = Et, the ν(NiP) of the planar complex is at 243 cm^{-1}, whereas these vibrations are at 195 and 182 cm^{-1} in the tetrahedral complex.

Udovich et al.[854] studied the infrared spectra of $Ni(DPE)X_2$, where DPE is 1,2-bis(diphenylphosphino)ethane and X is Cl, Br, and I, by using the metal isotope technique. It was found that the ν(NiX) are always lower and ν(NiP) are always higher in the *cis*-$Ni(DPE)X_2$ than in the corresponding *trans*-$Ni(PEt_3)_2X_2$. This difference has been attributed to the strong *trans*-influence of phosphine ligands.

A number of investigators have discussed the nature of the M–P bonding based on electronic, vibrational, and NMR spectra,[922] and controversy has arisen about the degree of π-back bonding in the M–P bond. For example, Park and Hendra[940] suggest the presence of a considerable degree of π-bonding in square-planar Pd(II) and Pt(II) complexes of PMe_3 and $AsMe_3$. On the other hand, Venanzi[941] claims from NMR evidence that the Pt–P π-bonding is much less than originally predicted.[942] It is rather difficult, however, to discuss the degree of π-bonding from vibrational spectra alone since the MP stretching frequency and force constant are determined by the net effect, which involves both σ- and π-bonding.

Complexes of the type $M(CO)_5L$, where L is arsine (AsH_3) and stibine (SbH_3) and M is Cr, Mo, and W, have been prepared by Fischer et al.[943] ν(AsH) and $\delta(AsH_3)$ are near 2200 and 900 cm^{-1}, respectively. Complexes of trimethylarsine ($AsMe_3$) have been studied by several investigators.

Goodfellow et al.[944] and Park and Hendra[940] measured the infrared spectra of $M(AsMe_3)_2X_2$- (M = Pt and Pd; X = Cl, Br, and I) type complexes and assigned ν(MAs) in the 300–260 cm^{-1} region. The latter workers assigned ν(MSb) of analogous alkylstibine complexes at ca. 200 cm^{-1}. Konya and Nakamoto[851] assigned ν(MAs) and ν(MX) of $[M(dias)_2]^{2+}$- and $[M(dias)_2X_2]Y_n$-type complexes by using the metal isotope technique. Figure III-62 shows the infrared spectra of $[^{58}Ni(dias)_2X_2]X$ and $[^{58}Ni(dias)_2X_2]$-$(ClO_4)_2$ (X = Cl and Br) and their ^{62}Ni analogs. Their results show that the ν(MAs) are very weak and appear at 325–295 cm^{-1} for the Ni, Co, and Fe complexes and at 270–210 cm^{-1} for the Pd and Pt complexes. For the ν(MX) of these complexes, see Sec. III-19.

Tertiary phosphine oxides and arsine oxides coordinate to a metal through their O atoms. The ν(P=O) of triphenylphosphine oxide (TPPO) at 1193 cm^{-1} is shifted by ca. 35 cm^{-1} to a lower frequency when it coordinates to Zn(II).[945] The shift is much larger in $MX_4(TPPO)_2$ (160–120 cm^{-1}), where MX_4 is a tetrahalide of Pa, Np, and Pu.[946] A similar observation has been made for ν(As=O) of arsine oxide and its complexes. Exceptions to this rule are found in $MnX_2(Ph_3AsO)_2$ (X = Cl and Br); their ν(As=O) are higher by 30–20 cm^{-1} than the frequency of the free ligand (880 cm^{-1}).[947] Rodley et al.[948] have assigned the ν(MO) of tertiary arsine oxide complexes at 440–370 cm^{-1}.

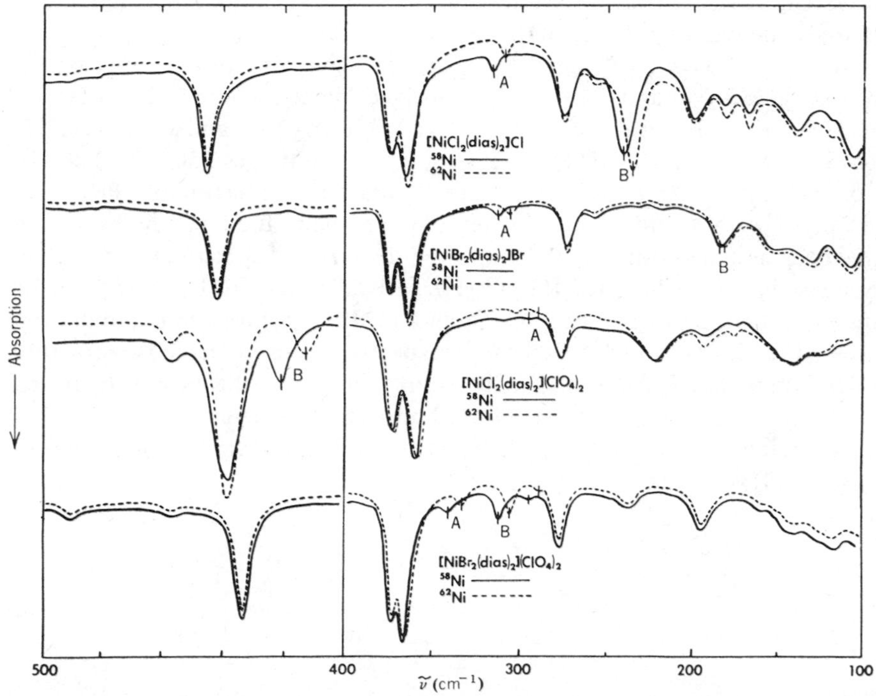

Fig. III-62. Far-infrared spectra of octahedral nickel dias complexes. Vertical lines marked by A and B indicate Ni–As and Ni–X stretching bands, respectively.

III-22. COMPLEXES OF SULFUR, SELENIUM, AND TELLURIUM-CONTAINING LIGANDS

A large number of metal complexes of ligands containing sulfur, selenium, and tellurium are known. Here the vibrational spectra of typical compounds will be reviewed briefly. For SO_3 and thiourea complexes that form metal-sulfur bonds, see Sec. III-11 and III-13, respectively.

(1) Complexes of Relatively Simple Ligands

Like molecular oxygen, S_2 and Se_2 form π-complexes with transition metals. The ν(SS) of $[Rh(S_2)(DPE)_2]Cl$* and the ν(SeSe) of $[Ir(Se_2)(DME)_2]Cl$† are at 525 and 310 cm^{-1}, respectively.[949] These values are much lower than those of free S_2 (726 cm^{-1}) and Se_2 (392 cm^{-1}). The $[Pt(S_5)_3]^{2-}$ and $[Pt(S_5)_2]^{2-}$ ions contain three and two S_5 chelate rings, respectively. The ν(PtS) of the former was assigned at 294 cm^{-1}.[950] Recently, a number of new complexes containing the S_2 groups as terminal and bridging ligands have been synthesized. Some examples are $(NH_4)_2[(S_2)_2MoS_2Mo(S_2)_2]\cdot 2H_2O$,[951] $[Mo_3S_{13}]^{2-}$,[952] $[(CH_3)_4N]_2[(S_2)MoO(S)_2MoO(S_2)]$[953] and $[(C_5H_5)Fe(SCH_3)_2(S_2)Fe(C_5H_5)]$.[954] All these compounds exhibit ν(SS) in the region from 550 to 510 cm^{-1}. The coordination chemistry of disulfur complexes has been reviewed by Müller et al.[954a]

The ν(NS) of thionitrosyl(NS) complexes such as $(C_5H_5)Cr(CO)_2(NS)$[955] and $ReCl_2(NS)(PMe_2Ph)_3$[956] are at 1180 cm^{-1}, while the ν(SO) of a thionyl(SO) complex such as $Fe_3S(CO)_9(SO)$ is at 1107 cm^{-1}.[957]

Allkins and Hendra[958] carried out an extensive vibrational study on *cis*- and *trans*-$[MX_2Y_2]$ and their halogen-bridged dimers, where M is Pd(II) and Pt(II), X is Cl, Br, and I, and Y is $(CH_3)_2S$, $(CH_3)_2Se$, and $(CH_3)_2Te$. The ν(MS), ν(MSe), and ν(MTe) were assigned in the ranges 350–300, 240–170, and 230–165 cm^{-1}, respectively. The vibrational spectra of PtX_2L_2,[959] PdX_2L_2,[960] AuX_3L and AuXL,[961] where X is a halogen and L is a dialkylsulfide, have been assigned. Aires et al.[962] reported the infrared spectra of MX_3L_3-type compounds, where M is Ru(III), Os(III), Rh(III), and Ir(III), X is Cl or Br, and L is Et_2S and Et_2Se. The ν(MS) and ν(MSe) of these compounds were assigned at 325–290 and 225–200 cm^{-1}, respectively, based on the *fac*-structure. On the other hand, Allen and Wilkinson[963] proposed the *mer*-structure for these compounds, based on far-infrared and other evidence.

According to X-ray analysis, $Pd_2Br_4(Me_2S)_2$ is bridged via Br atoms, whereas $Pt_2Br_4(Et_2S)_2$ is bridged via S atoms:[964]

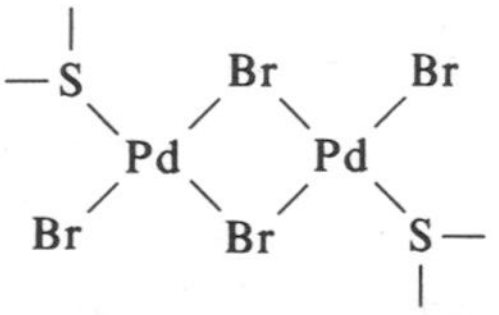

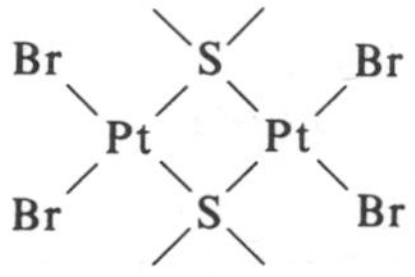

* DPE = $Ph_2P—CH_2—CH_2—PPh_2$.
† DME = $(CH_3)_2P—CH_2—CH_2—P(CH_3)_2$.

Adams and Chandler[965] have shown that halogen-bridged $Pd_2Cl_4(SEt_2)_2$ exhibits $\nu(PdCl_t)$, $\nu(PdCl_b)$, and $\nu(PdS)$ at 366, 266, and 358 cm^{-1}, respectively, whereas sulfur-bridged $Pt_2Cl_4(SEt_2)_2$ exhibits $\nu(PtCl_t)$ and $\nu(PtS)$ at 365–325 and 422–401 cm^{-1}, respectively.

Like CO, thiocarbonyl (CS) forms complexes with transition metals. Thus far, the following three structures have been found or suggested:

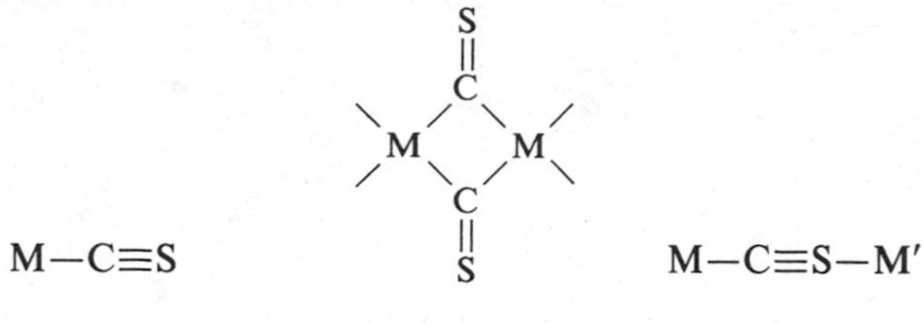

I II III

The $\nu(CS)$ of free CS is at 1275 cm^{-1}. The $\nu(CS)$ of the C-bonded terminal CS group (Structure I) is higher than that of free CS (1360–1290 cm^{-1}).[966,967] The bridging C-bonded structure (II) was suggested for $Mn_2(\pi\text{-}Cp)_2(NO)_2(CS)_2$.[968] The bridging C- and S-bonded structure (III) was suggested for $(DPE)_2(CO)W(CS)W(CO)_5$ since its $\nu(CS)$ is very low (1161 cm^{-1}).[969] Butler and co-workers made normal coordinate analyses on $M(CO)_5(CX)$ (M = Cr and W; X = S and Se)[970] and $(C_6H_6)Cr(CO)_2(CX)$ (X = 0, S, and Se).[971]

Vibrational spectra of transition-metal complexes of CS_2 and CS have been reviewed by Butler and Fenster.[972] According to Wilkinson et al.,[966,973] CS_2 coordinates to the metal in four ways:

I II III IV

Free CS_2 exhibits $\nu(CS_2)$ at 1533 cm^{-1}. The $\nu(CS_2)$ of structure I (π-bonded) and structure II (σ-bonded) are at 1100 ~ 1150 and 1510 cm^{-1}, respectively. The $\nu(CS)$ of the bridging CS groups are at ca. 980–840 cm^{-1} for structure III,[966] and 1120 cm^{-1} for structure IV.[973] For the infrared spectra of other CS_2 complexes, see Refs. 966–974. From infrared and other evidence, $[Ir(CS_2)$-

$(CO)(PPh_3)_3]BPh_4$ was originally thought to be a six-coordinate complex with a π-bonded CS_2.[974] However, X-ray analysis[975] revealed an unexpected structure; a five-coordinate complex of the type $[Ir(CO)(PPh_3)_2(S_2C{-}PPh_3)]BPh_4$. Matrix co-condensation reactions of Ni atoms with CS_2/Ar produce a mixture of $Ni(CS_2)_n$ where n is 1, 2, and 3. It was not possible, however, to determine the mode of coordination from their IR spectra.[976] A mixed carbon dichalcogenide such as SCSe can form a pair of geometrical isomers:

where L_4 represents $(CO)(CNR)(PPh_3)_2$ (R = p-tolyl). The ν(CSe) of the former and the ν(CS) of the latter have been observed at 1015 and 1066 cm^{-1}, respectively.[977]

Sulfur dioxide (SO_2) may take one of the following structures when it coordinates to a metal:

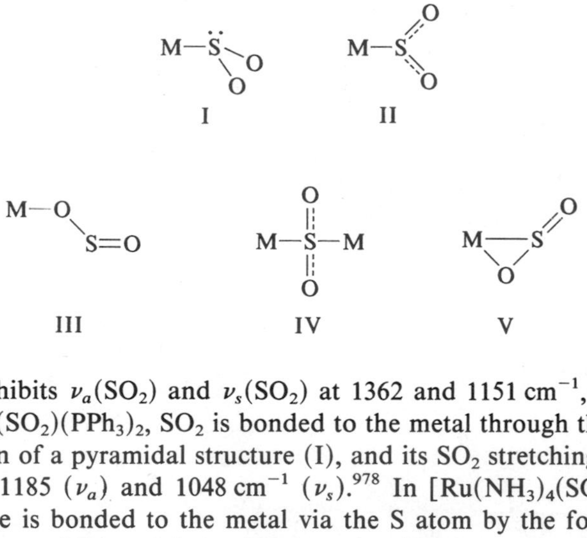

Free SO_2 exhibits $\nu_a(SO_2)$ and $\nu_s(SO_2)$ at 1362 and 1151 cm^{-1}, respectively. In $IrCl(CO)(SO_2)(PPh_3)_2$, SO_2 is bonded to the metal through the S atom by the formation of a pyramidal structure (I), and its SO_2 stretching frequencies are at 1198–1185 (ν_a) and 1048 cm^{-1} (ν_s).[978] In $[Ru(NH_3)_4(SO_2)Cl]Cl$, the SO_2 molecule is bonded to the metal via the S atom by the formation of a planar structure (II), and its $\nu_a(SO_2)$ and $\nu_s(SO_2)$ are at 1301–1278 and 1100 cm^{-1}, respectively.[979] Although the SO_2 stretching frequencies of the latter are higher than those of the former, they are lower than those of free SO_2 in both cases. In $[SbF_5(SO_2)]$, SO_2 is bonded to the metal via the O atom, as shown in structure III,[980] and its $\nu(SO_2)$ are observed at 1327 (ν_a) and 1102 cm^{-1} (ν_s).[981] According to Byler and Shriver,[981] SO_2 in a complex is O-bonded if ($\nu_a - \nu_s$) is larger than 190 cm^{-1}, and is S-bonded if it is smaller than 190 cm^{-1}. According to X-ray analysis,[982] SO_2 in $[(\pi\text{-}Cp)Fe(CO)_2]_2(SO_2)$ takes the bridging structure IV. This compound exhibits the ν_a and ν_s at 1135 and 993 cm^{-1}, respectively.[983] The SO_2 in $Fe_2(CO)_8SO_2$ also takes structure IV. Complete vibrational analysis[984] shows that the ν_a, ν_s, and $\delta(SO_2)$ are at 1203, 1048, and 530 cm^{-1}, respectively. The chelate structure (V) is found in $Ru(NO)Cl(SO_2)(PPh_3)_2$.[985] This compound exhibits a band at 895 cm^{-1} which originates in the SO_2 group.[985] Johnson and Dew[986] observed a linkage isomerization (structure II → V) in $[Ru(NH_3)_2(SO_2)Cl]Cl$ which was photochemically induced (365 nm) in the solid state. The ν_a and ν_s of structure II are at 1255 and 1110 cm^{-1}, respectively, whereas those of structure V are assigned at 1165 and 940 cm^{-1}, respectively. According to Kubas,[987] the coordination geometry of the SO_2 group can be deduced only when combinations of spectroscopic and chemical properties are considered.

Müller and co-workers[988,989] carried out an extensive vibrational study on MS_4^{n-}, MS_3O^{n-}-, and $MS_2O_2^{n-}$-type ions (Part II) and their metal complexes. Complete normal coordinate analyses have been performed on $[^{58}Ni(^{92}MoS_4)_2]^{2-}$ and its ^{62}Ni and ^{100}Mo analogs,[990] and empirical assignments have been made for $[Ni(WS_3O)_2]^{2-}$,[991] and $[Ni(WS_2O_2)_2]^{2-}$.[992] Structures such as the following have been deduced from the infrared spectra:

C_{2h} or C_{2v} D_{2h}

The infrared spectra of $M(S_2PR_2)_2$, where M is Ni(II), Pd(II), and Pt(II), and R is CH_3, C_6H_5, and so on, have been reported.[993]

The infrared spectra of thiocarbonato complexes of the type $[M(CS_3)_2]^{2-}$ [M = Ni(II), Pd(II), and Pt(II)] have been studied by Burke and Fackler[994] and Cormier et al.[995] The latter workers carried out normal coordinate analyses on the $[^{58}Ni(CS_3)_2]^{2-}$ ion and its ^{62}Ni analog. The $\nu(NiS)$ were assigned at 385 and 366 cm^{-1}, with a corresponding force constant of 1.41 mdyn/Å (UBF). Bis(dithioacetato)palladium, $Pd(CH_3CS_2)_2$, exists in three different crystalline forms. Piovesana et al.[996] characterized each of these forms by IR spectroscopy.

(2) Complexes of Relatively Large Ligands

2,5-Dithiahexane(dth) forms metal complexes such as $[ReCl_3(dth)]_n$ and $Re_3Cl_9(dth)_{1.5}$. Cotton et al.[997] showed from infrared spectra that dth of the former forms a chelate ring in the *gauche* conformation, whereas that of the latter forms a bridge between two metals by taking the *trans* conformation (see Sec. III-2). Infrared spectra have been used to show that ethanedithiol forms a chelate ring of the *gauche* conformation in $Bi(S_2C_2H_4)X$, where X is Cl and Br.[998] Schläpfer et al.[999] assigned the $\nu(NiS)$ and $\nu(NiN)$ of dth, ete [2-(ethylthio)ethylamine], and mea [mercaptoethylamine] complexes with the metal isotope technique:

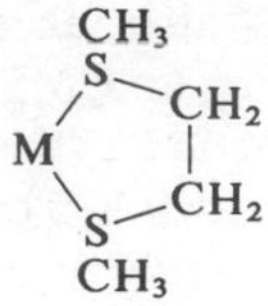

dth

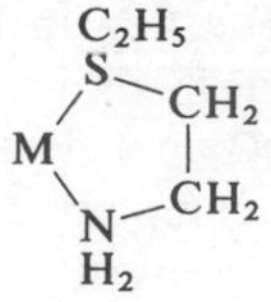

ete

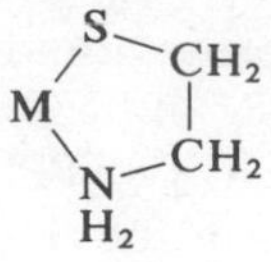

mea

The infrared spectra of N,N-dialkyldithiocarbamato complexes have been studied extensively. All these compounds exhibit strong $\nu(C{=}N)$ bands in the 1600–1450 cm^{-1} region. These compounds are roughly classified into two types:

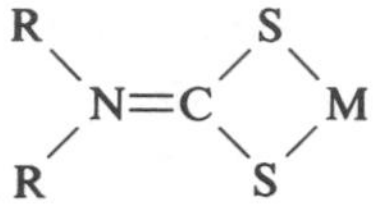

Bidentate coordination

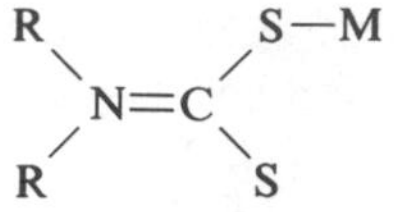

Unidentate coordination

The former exhibits $\nu(CS)$ near 1000 cm^{-1} as a single band, whereas the latter shows a doublet in the same region.[1000] Also, the $\nu(C{=}N)$ of the former (above 1485 cm^{-1}) is higher than that of the latter (below 1485 cm^{-1}).[1001] The $\nu(MS)$ of the bidentate complexes are observed at 400–300 cm^{-1}.[1002] Dithiocarbamato complexes of Fe(III) undergo the high-spin($^6A_1$) to low-spin($^2T_2$) crossover. This change can be induced by applying high pressure[1003] or by lowering temperature.[1004] Sorai[1004] assigned the $\nu(FeS)$ of $Fe(S_2CN(Et)_2)_3$ of the high- and low-spin states to the IR bands at 355 and 552 cm^{-1}, respectively. Hutchinson et al.[1004a] have shown by ^{54}Fe–^{57}Fe isotope shifts and variable-temperature studies that the $\nu(FeS)$ of Fe(III) dialkyldithiocarbamates appear at 250–205 and 350–305 cm^{-1}, respectively, for high- and low-spin states, and that intermediate-spin complexes show $\nu(FeS)$ in both regions.

The infrared spectra of diselenocarbamato complexes have been reported,[1005] and assigned on the basis of normal coordinate analysis.[1006] In the Ni(II) complex, $\nu(NiSe)$ is assigned at 298 cm^{-1}, which is lower by 85 cm^{-1} than the $\nu(NiS)$ of the corresponding dithiocarbamato complex. The infrared spectra of xanthato complexes:

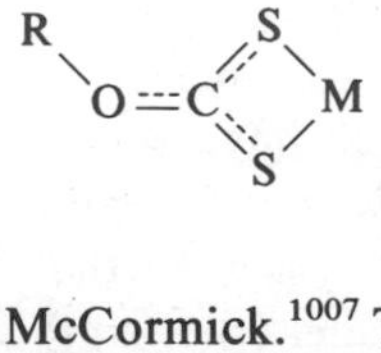

have been studied by Watt and McCormick.[1007] The $\nu(CO)$, $\nu(CS)$, and $\nu(MS)$ were assigned at 1325–1250, 760–540, and 380–340 cm^{-1}, respectively.

Savant et al.[1008] roughly classified monothiobenzoato complexes into three categories:

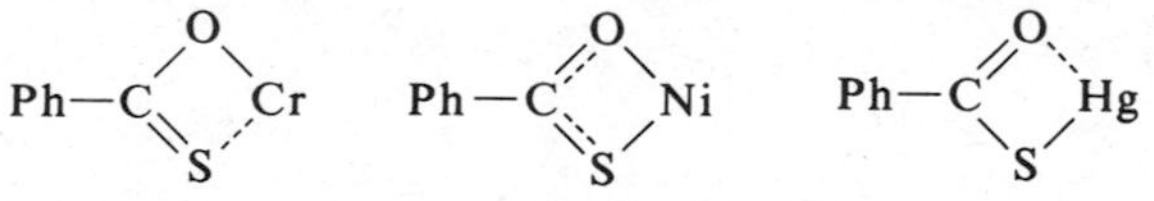

$\nu(CO)(cm^{-1})$	1465	1508	1630
$\nu(CS)(cm^{-1})$	982	958	912

In the Hg(II), Cu(I), and Ag(I) complexes, coordination occurs mainly via sulfur. In the Cr(III) complex, however, the Cr-O bond is stronger than the Cr-S bond. The Ni(II) complex is between these two cases and is close to symmetrical coordination. This is reflected in the frequency trends shown above.

In thiocarboxylato complexes of the type $Ni(R{-}COS)_2\cdot\frac{1}{2}(EtOH)$ ($R = CH_3$, C_2H_5, and Ph), the ligand serves as a bridge between two metals as shown:

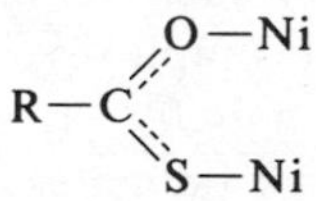

and their $\nu(CO)$ are reported to be at 1580–1520 cm^{-1}.[1009]

The infrared spectra of metal complexes of thiosemicarbazides

$$NH_2{-}\underset{\underset{S}{\|}}{C}{-}NH{-}NH_2$$

and thiosemicarbazones

$$NH_2{-}\underset{\underset{S}{\|}}{C}{-}NH{-}N{=}CR_1R_2 \quad (R_1 \text{ and } R_2 \text{ are alkyl, phenyl, etc.})$$

have been reviewed by Campbell.[1010] For metal complexes of dithiocarbazic acid, infrared spectra support the N,S-chelated structure (shown below) rather than the S,S-chelated structure normally found for dithiocarbamato complexes.[1011] In 1:1 complexes of N,N-monosubstituted dithiooxamides, Desseyn et al.[1012] concluded from infrared spectra that metals such as Ni(II) and Cu(II) are primarily bonded to the N, whereas metals such as Hg(II), Pb(II), and Pd(II) are bonded to the S atom.

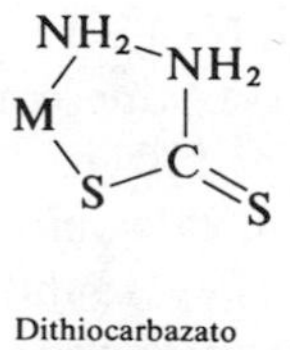

Dithiocarbazato complex

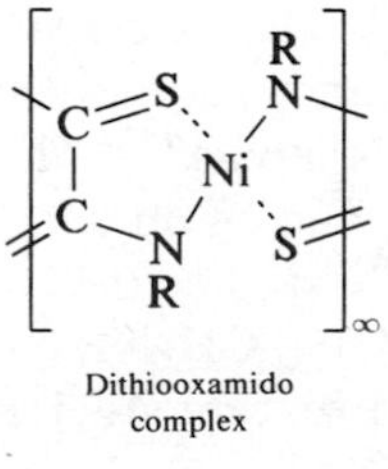

Dithiooxamido complex

Coucouvanis et al.[1013] synthesized novel tin halide adducts of Ni(II) and Pd(II) dithiooxalto (DTO) complexes:

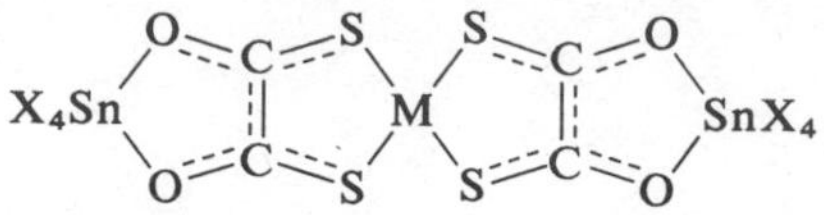

The ν(NiS) of the $[Ni(DTO)_2]^{2-}$ ion is at 349 cm^{-1}. In SnX_4 (X = Cl, Br, and I) adducts, this band shifts to 385–375 cm^{-1}, indicating a strengthening of the Ni–S bond by complexation. It was found that $Cr(DTO)_3[Cu(PPh_3)_2]_3$ exists in two isomeric forms;[1014] one in which the Cr atom is bonded to sulfur, and another in which it is bonded to oxygen of the DTO ion. As expected, ν(C=O), ν(CS), and ν[CrO(S)] are markedly different between these two isomers.

Metal complexes of 1,2-dithiolates (or dithienes) have been of great interest to inorganic chemists because of their redox properties.[1015] Schläpfer and Nakamoto[1016] prepared a series of complexes of the type $[Ni(S_2C_2R_2)_2]^n$, where R is H, Ph, CF_3, and CN and n is 0, −1, or −2, and carried out normal coordinate analysis to obtain rough estimates of the charge distribution based on the calculated force constants.

Infrared spectra of metal complexes with thio-β-diketones have been reviewed briefly.[1017,1018] Siimann and Fresco[1019] and Martin et al.[1020] carried out normal coordinate analysis on metal complexes of dithioacetylacetone, monothioacetylacetone, and related ligands. For dithioacetylacetonato complexes, two ν(MS) have been assigned at 390–340 and 300–260 cm^{-1}.

L-Cysteine has three potential coordination sites (S, N, and O), and infrared spectra have been used to determine the structures of its metal complexes. For example, the Zn(II) complex shows no ν(SH), and its carboxylate frequency indicates the presence of a free COO^- group. Thus the following structure was proposed:[1021]

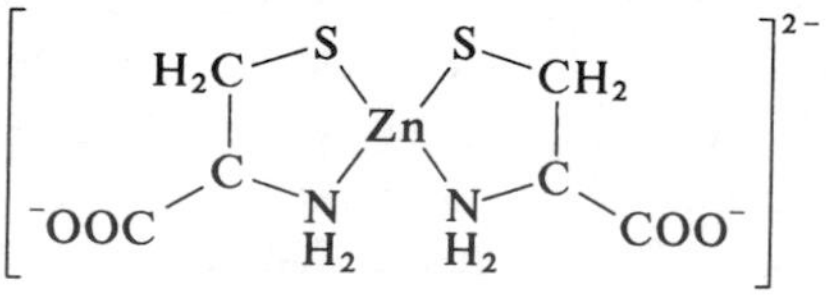

On the other hand, the (S, O) chelation has been suggested for the Pt(II) and Pd(II) complexes.[1022] IR spectra show that the Fe atom in $Fe(cyst)(H_2O)_{1.5}$ is bonded to the S, N, and O atoms whereas that in $Na_2[Fe(cyst)_2]H_2O$ is bonded only to the S and N atoms.[1023] The ν(SS) of L-cystine complexes of Cu(II), Ni(II), and Zn(II), and so on are observed near 500 cm^{-1}.[1024]

McAuliffe et al.[1025] studied the IR spectra of metal complexes of methionine [CH_3—S—CH_2—CH_2—CH(NH_2)—COOH]. They found that most of the metals they studied [except Ag(I)] coordinate through the NH_2 and COO^-

groups, and that the CH_3S groups of these complexes are available for further coordination to other metals. McAuliffe[1026] suggested that complexes of the type M(methionine)Cl_2 [M = Pd(II) and Pt(II)] take the polymeric structure:

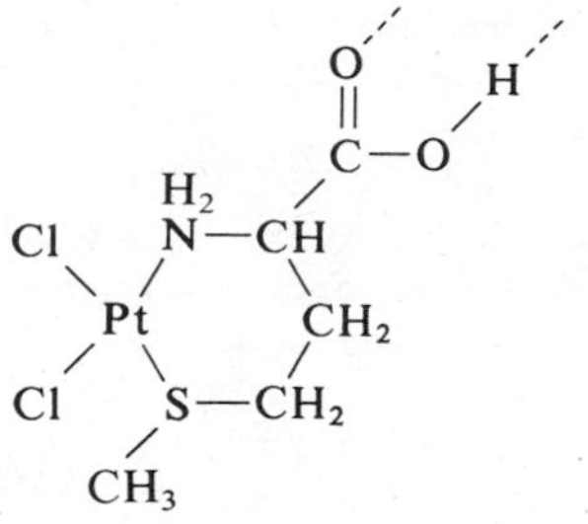

Infrared spectra of metal complexes of sulfur-containing ligands are reported for 2-methylthioaniline,[1027] 8-mercaptoquinoline,[1028] cyclic thioethers,[1029] and N-alkylthiopicolinamides.[1030]

References

1. K. H. Schmidt and A. Müller, *Coord. Chem. Rev.*, **19**, 41 (1976).
2. R. Plus, *J. Raman Spectrosc.*, **1**, 551 (1973); *Spectrochim. Acta*, **32A**, 263 (1976).
3. N. Tanaka, M. Kamada, J. Fujita, and E. Kyuno, *Bull. Chem. Soc. Jpn.*, **37**, 222 (1964).
4. T. V. Long, II and D. J. B. Penrose, *J. Am. Chem. Soc.*, **93**, 632 (1971).
5. K. H. Schmidt and A. Müller, *J. Mol. Struct.*, **22**, 343 (1974).
6. K. H. Schmidt and A. Müller, *Inorg. Chem.*, **14**, 2183 (1975).
7. A. Deak and J. L. Templeton, *Inorg. Chem.*, **19**, 1075 (1980).
8. W. P. Griffith, *J. Chem. Soc. A*, 899 (1966).
9. T. Shimanouchi and I. Nakagawa, *Inorg. Chem.*, **3**, 1805 (1964).
10. L. Sacconi, A. Sabatini, and P. Gans, *Inorg. Chem.*, **3**, 1772 (1964).
11. K. H. Schmidt, W. Hauswirth, and A. Müller, *J. Chem. Soc., Dalton Trans.*, 2199 (1975).
12. H. Siebert and H. H. Eysel, *J. Mol. Struct.*, **4**, 29 (1969).
13. T. V. Long, II, A. W. Herlinger, E. F. Epstein, and I. Bernal, *Inorg. Chem.*, **9**, 459 (1970).
14. J. M. Terrasse, H. Poulet, and J. P. Mathieu, *Spectrochim. Acta*, **20**, 305 (1964).
15. J. M. Janik, J. A. Janik, G. Pytasz, and J. Sokolowski, *J. Raman Spectrosc.*, **4**, 13 (1975).
16. M. J. Nolan and D. W. James, *J. Raman Spectrosc.*, **1**, 259, 271 (1973).
17. J. Hiraishi, I. Nakagawa, and T. Shimanouchi, *Spectrochim. Acta*, **24A**, 819 (1968).
18. J. Fujita, K. Nakamoto, and M. Kobayashi, *J. Am. Chem. Soc.*, **78**, 3295 (1956).
19. B. S. Ault, *J. Am. Chem. Soc.*, **100**, 5773 (1978).
20. K. Nakamoto, Y. Morimoto, and J. Fujita, Proceedings of the Seventh ICCC, Stockholm, 1962, p. 15.
21. I. Nakagawa and T. Shimanouchi, *Spectrochim. Acta*, **22**, 759 (1966).
22. A. Müller, P. Christophliemk, and I. Tossidis, *J. Mol. Struct.*, **15**, 289 (1973).
23. K. Nakamoto, J. Takemoto, and T. L. Chow, *Appl. Spectrosc.*, **25**, 352 (1971).
24. M. Manfait, A. J. P. Alix, and J. Delaunay-Zeches, *Inorg. Chim. Acta*, **44**, L261 (1980).
25. M. Manfait, A. J. P. Alix, L. Bernard, and T. Theophanides, *J. Raman Spectrosc.*, **7**, 143 (1978).
26. S. J. Cyvin, B. N. Cyvin, K. H. Schmidt, W. Wiegeler, A. Müller, and J. Brunvoll, *J. Mol. Struct.*, **30**, 315 (1976).
27. M. Manfait, A. J. P. Alix, and C. Kappenstein, *Inorg. Chim. Acta*, **50**, 147 (1981).

28. A. L. Geddes and G. L. Bottger, *Inorg. Chem.*, **8,** 802 (1969).
29. M. G. Miles, J. H. Patterson, C. W. Hobbs, M. J. Hopper, J. Overend, and R. S. Tobias, *Inorg. Chem.* **7,** 1721 (1968).
30. E. P. Bertin, I. Nakagawa, S. Mizushima, T. L. Lane , and J. V. Quagliano, *J. Am. Chem. Soc.*, **80,** 525 (1958).
31. I. Nakagawa and T. Shimanouchi, *Spectrochim. Acta*, **22,** 1707 (1966).
32. T. Grzybek, J. M. Janik, A. Kulczycki, G. Pytasz, J. A. Janik, J. Ściesiński, and E. Ściesińska. *J. Raman Spectrosc.*, **1,** 185 (1973).
33. I. Nakagawa, *Bull. Chem. Soc. Jpn.*, **46,** 3690 (1973).
34. H. Poulet, P. Delorme, and J. P. Mathieu, *Spectrochim. Acta*, **20,** 1855 (1964).
35. D. M. Adams and J. R. Hall, *J. Chem. Soc., Dalton Trans.*, 1450 (1973).
36. J. R. Hall and D. A. Hirons, *Inorg. Chim. Acta*, **34,** L277 (1979).
37. J. A. Janik, W. Jacob, and J. M. Janik, *Acta Phys. Pol.*, Pt. A, **38,** 467 (1970).
38. J. M. Janik, A. Magdal-Mikuli, and J. A. Janik, *Acta Phys. Pol.*, Pt. A, **40,** 741 (1971).
39. A. Adamson and T. M. Dunn, *J. Mol. Spectrosc.*, **18,** 83 (1965).
40. H. H. Eysel, *Z. Phys. Chem.* (*Frankfurt am. Main*), **72,** 82 (1970).
41. C. D. Flint and P. Greenough, *J. Chem. Soc., Faraday Trans. 2*, **68,** 897 (1972).
42. T. M. Loehr, J. Zinich, and T. V. Long, II, *Chem. Phys. Lett.*, **7,** 183 (1970).
43. A. F. Schreiner and J. A. McLean, *J. Inorg. Nucl. Chem.*, **27,** 253 (1965).
44. A. D. Allen and J. R. Stevens, *Can. J. Chem.*, **51,** 92 (1973).
45. M. W. Bee, S. F. A. Kettle, and D. B. Powell, *Spectrochim. Acta*, **30A,** 139 (1974).
46. C. H. Perry, D. P. Athans, E. F. Young, J. R. Durig, and B. R. Mitchell, *Spectrochim. Acta*, **23A,** 1137 (1967).
47. P. J. Hendra, *Spectrochim. Acta*, **23A,** 1275 (1967).
47a. K. Nakamoto, P. J. McCarthy, J. Fujita, R. A. Condrate, and G. T. Behnke, *Inorg. Chem.*, **4,** 36 (1965).
48. P. J. Hendra and N. Sadasivan, *Spectrochim. Acta*, **21,** 1271 (1965).
49. R. Layton, D. W. Sink, and J. R. Durig, *J. Inorg. Nucl. Chem.*, **28,** 1965 (1966).
50. J. R. Durig, R. Layton, D. W. Sink, and B. R. Mitchell, *Spectrochim. Acta*, **21,** 1367 (1965).
51. D. B. Powell, *J. Chem. Soc.*, 4495 (1956).
52. J. R. Durig and B. R. Mitchell, *Appl. Spectrosc.*, **21,** 221 (1967).
53. R. J. H. Clark and W. R. Trumble, *Inorg. Chem.*, **15,** 1030 (1976).
54. S. Mizushima, I. Nakagawa, and D. M. Sweeny, *J. Chem. Phys.*, **25,** 1006 (1956); I. Nakagawa, R. B. Penland, S. Mizushima, T. J. Lane, and J. V. Quagliano, *Spectrochim. Acta*, **9,** 199 (1957).
55. K. Niwa, H. Takahashi, and K. Higashi, *Bull. Chem. Soc. Jpn.*, **44,** 3010 (1971).
56. K. Brodersen and H. J. Becher, *Chem. Ber.*, **89,** 1487 (1956).
57. A. Novak, J. Portier, and P. Bouvlier, *Compt. Rend.*, **261,** 455 (1965).
58. D. C. Bradley and M. H. Gitlitz, *J. Chem. Soc. A*, 980 (1969).
59. G. W. Watt, B. B. Hutchinson, and D. S. Klett, *J. Am. Chem. Soc.*, **89,** 2007 (1967).
60. Y. Y. Kharitonov, I. K. Dymina, and T. Leonova, *Izv. Akad. Nauk SSSR, Ser. Khim.*, 2057 (1966).
61. M. Goldstein and E. F. Mooney, *J. Inorg. Nucl. Chem.*, **27,** 1601 (1965).
62. J. Chatt, L. A. Duncanson, and L. M. Venanzi, *J. Chem. Soc.*, 4456, 4461 (1955); 2712 (1956); *J. Inorg. Nucl. Chem.*, **8,** 67 (1958).
63. L. A. Duncanson and L. M. Venanzi, *J. Chem. Soc.*, 3841 (1960).
64. Y. Y. Kharitonov, M. A. Sarukhanov, I. B. Baranovskii, and K. U. Ikramov, *Opt., Spektrosk.*, **19,** 460 (1965).
65. L. Sacconi and A. Sabatini, *J. Inorg. Nucl. Chem.*, **25,** 1389 (1963).
66. K. Brodersen, *Z. Anorg. Allg. Chem.*, **290,** 24 (1957).
67. K. H. Linke, F. Dürholz, and P. Hädicke, *Z. Anorg. Allg. Chem.*, **356,** 113 (1968).
68. D. Nicholls and R. Swindells, *J. Inorg. Nucl. Chem.*, **30,** 2211 (1968).
69. S. Mizushima, I. Ichishima, I. Nakagawa, and J. V. Quagliano, *J. Phys. Chem.*, **59,** 293 (1955).
70. D. M. Sweeny, S. Mizushima, and J. V. Quagliano, *J. Am. Chem. Soc.*, **77,** 6521 (1955).
71. A. Nakahara, Y. Saito, and H. Kuroya, *Bull. Chem. Soc. Jpn.*, **25,** 331 (1952).

72. R. E. Cramer and J. T. Huneke, *Inorg. Chem.*, **14,** 2565 (1975).
73. J. Gouteron, *J. Inorg. Nucl. Chem.*, **38,** 63 (1976).
74. G. Borch, P. H. Nielsen, and P. Klaboe, *Acta Chem. Scand.*, **A31,** 109 (1979); *Spectrochim Acta*, **34A,** 87 and 93 (1978).
75. C. D. Flint and A. P. Matthews, *Inorg. Chem.*, **14,** 1219 (1975).
76. Z. Gabelica, *Spectrochim. Acta*, **32A,** 327 (1976).
77. Y. Omura, I. Nakagawa, and T. Shimanouchi, *Spectrochim. Acta*, **27A,** 2227 (1971).
78. A. B. P. Lever and E. Mantovani, *Can. J. Chem.*, **51,** 1567 (1973).
79. G. W. Rayner-Canham and A. B. P. Lever, *Can. J. Chem.*, **50,** 3866 (1972).
80. A. B. P. Lever and E. Mantovani, *Inorg. Chem.*, **10,** 817 (1971).
81. J. M. Rigg and E. Sherwin, *J. Inorg. Nucl. Chem.*, **27,** 653 (1965).
82. M. N. Hughes and W. R. McWhinnie, *J. Inorg. Nucl. Chem.*, **28,** 1659 (1966).
83. S. Kida, *Bull. Chem. Soc. Jpn.*, **39,** 2415 (1966).
84. E. B. Kipp and R. A. Haines, *Can. J. Chem.*, **47,** 1073 (1969).
85. Y. Omura, I. Nakagawa, and T. Shimanouchi, *Spectrochim. Acta*, **27A,** 1153 (1971).
86. R. W. Berg and K. Rasmussen, *Spectrochim. Acta*, **29A**, 37 (1973).
87. G. C. Papavassiliou, T. Theophanides, and R. Rapsomanikis, *J. Raman Spectrosc.*, **8,** 227 (1979).
88. D. B. Powell and N. Sheppard, *J. Chem. Soc.*, 3089 (1959).
89. G. Newman and D. B. Powell, *J. Chem. Soc.*, 447 (1961); 3447 (1962).
90. K. Brodersen and T. Kahlert, *Z. Anorg. Allg. Chem.*, **348,** 273 (1966).
91. K. Brodersen, *Z. Anorg. Allg. Chem.*, **298,** 142 (1959).
92. T. Iwamoto and D. F. Shriver, *Inorg. Chem.*, **10,** 2428 (1971).
93. G. W. Watt and D. S. Klett, *spectrochim. Acta*, **20,** 1053 (1964).
94. A. R. Gainsford and D. A. House, *Inorg. Chim. Acta*, **3,** 367 (1969).
95. H. H. Schmidtke and D. Garthoff, *Inorg. Chim. Acta*, **2,** 357 (1968).
96. K W. Kuo and S. K. Madan, *Inorg. Chem.*, **8,** 1580 (1969).
97. J. H. Forsberg, T. M. Kubik, T. Moeller, and K. Gucwa, *Inorg. Chem.*, **10,** 2656 (1971).
98. D. A. Buckingham and D. Jones, *Inorg. Chem.*, **4,** 1387 (1965).
99. K. W. Bowker, E. R. Gardner, and J. Burgess, *Inorg. Chim. Acta*, **4,** 626 (1970).
100. R. J. H. Clark and C. S. Williams, *Inorg. Chem.*, **4,** 350 (1965).
101. Y. Saito, M. Cordes, and K. Nakamoto, *Spectrochim. Acta*, **28A,** 1459 (1972).
102. Y. Saito, C. W. Schläpfer, M. Cordes, and K. Nakamoto, *Appl. Spectrosc.*, **27,** 213 (1973).
103. M. Choca, J. R. Ferraro, and K. Nakamoto, *J. Chem. Soc., Dalton Trans.*, 2297 (1972).
104. R. H. Nuttall, A. F. Cameron, and D. W. Taylor, *J. Chem. Soc. A*, 3103 (1971).
105. M. Fleischmann, P. J. Hendra, and A. McQuillan, *Chem. Phys. Lett.*, **26,** 163 (1974); *J. Electroanal. Chem.*, **65,** 933 (1975).
106. D. J. Jeanmaire and R. P. Van Duyne, *J. Electroanal. Chem.*, **66,** 235 (1975).
107. R. P. Van Duyne, in C. B. Moore, ed., *Chemical and Biological Applications of Lasers*, Vol. 4, Academic Press, New York, 1979, p. 101.
108. H. Yamada, *Appl. Spectrosc. Rev.*, **17,** 227 (1981).
109. C. Postmus, J. R. Ferraro, A. Quattrochi, K. Shobatake, and K. Nakamoto, *Inorg. Chem.*, **8,** 1851 (1969).
110. J. R. Allan, D. H. Brown, R. H. Nuttall, and D. W. A. Sharp, *J. Inorg. Nucl. Chem.*, **27,** 1305 (1965).
111. N. S. Gill and H. J. Kingdon, *Aust. J. Chem.*, **19,** 2197 (1966).
112. L. Cattalini, R. J. H. Clark, A. Orio, and C. K. Poon, *Inorg. Chim. Acta*, **2,** 62 (1968).
113. J. Burgess, *Spectrochim. Acta*, **24A,** 277 (1968).
114. M. Goldstein, E. F. Mooney, A. Anderson, and H. A. Gebbie, *Spectrochim. Acta*, **21,** 105 (1965).
115. A. B. P. Lever and B. S. Ramaswamy, *Can. J. Chem.*, **51,** 1582 (1973).
116. W. R. McWhinnie, *J. Inorg. Nucl. Chem.*, **27,** 2573 (1965).
117. D. E. Billing and A. E. Underhill, *J. Inorg. Nucl. Chem.*, **30,** 2147 (1968).
118. F. Farha and R. T. Iwamoto, *Inorg. Chem.*, **4,** 844 (1965).

119. D. G. Brewer, and P. T. T. Wong, *Can. J. Chem.*, **44,** 1407 (1966).
120. J. Burgess, *Spectrochim. Acta,* **24A,** 1645 (1968).
121. L. El-Sayed and R. O. Ragsdale, *J. Inorg. Nucl. Chem.*, **30,** 651 (1968).
122. R. G. Garvey, J. H. Nelson, and R. O. Ragsdale, *Coord. Chem. Rev.*, **3,** 375 (1968).
123. C. P. Prabhakaran and C. C. Patel, *J. Inorg. Nucl. Chem.*, **34,** 3485 (1972).
124. I. S. Ahuja and P. Rastogi, *J. Chem. Soc. A,* 378 (1970).
125. F. A. Cotton and J. F. Gibson, *J. Chem. Soc. A,* 2105 (1970).
126. D. M. L. Goodgame, M. Goodgame, P. J. Hayward, and G. W. Rayner-Canham, *Inorg. Chem.*, **7,** 2447 (1968).
127. C. E. Taylor and A. E. Underhill, *J. Chem. Soc. A,* 368 (1969).
128. B. Cornilsen and K. Nakamoto, *J. Inorg. Nucl. Chem.*, **36,** 2467 (1974).
129. W. J. Eilbeck, F. Holmes, C. E. Taylor, and A. E. Underhill, *J. Chem. Soc. A,* 128 (1968).
130. D. M. L. Goodgame, M. Goodgame, and G. W. Rayner-Canham, *Inorg. Chim. Acta,* **3,** 399 (1969).
131. D. M. L. Goodgame, M. Goodgame, and G. W. Rayner-Canham, *Inorg. Chim. Acta,* **3,** 406 (1969).
132. W. J. Eilbeck, F. Holmes, C. E. Taylor, and A. E. Underhill, *J. Chem. Soc. A,* 1189 (1968).
133. G. A. Melson and R. H. Nuttall, *J. Mol. Struct.*, **1,** 405 (1968).
134. M. M. Cordes and J. L. Walter, *Spectrochim. Acta,* **24A,** 1421 (1968).
135. J. B. Hodgson, G. C. Percy, and D. A. Thornton, *J. Mol. Struct.*, **66,** 81 (1980).
136. S. Salama and T. G. Spiro, *J. Am. Chem. Soc.*, **100,** 1105 (1978).
137. B. Hutchinson, J. Takemoto, and K. Nakamoto, *J. Am. Chem. Soc.*, **92,** 3335 (1970).
138. Y. Saito, J. Takemoto, B. Hutchinson, and K. Nakamoto, *Inorg. Chem.*, **11,** 2003 (1972); J. Takemoto, B. Hutchinson, and K. Nakamoto, *Chem. Commun.* 1007 (1971).
139. E. König and E. Lindner, *Spectrochim. Acta,* **28A,** 1393 (1972).
140. R. Wilde, T. K. K. Srinivasan, and N. Ghosh, *J. Inorg. Nucl. Chem.*, **35,** 1017 (1973).
141. J. S. Strukl and J. L. Walter, *Spectrochim. Acta,* **27A,** 223 (1971).
142. K. Krishnan and R. A. Plane, *Spectrochim. Acta,* **25A,** 831 (1969).
143. R. J. H. Clark, P. C. Turtle, D. P. Strommen, B. Streusand, J. Kincaid, and K. Nakamoto, *Inorg. Chem.*, **16,** 84 (1977).
144. R. F. Dallinger and W. H. Woodruff, *J. Am. Chem. Soc.*, **101,** 1355 (1979).
144a. P. G. Bradley, N. Kress, B. A. Hornberger, R. F. Dallinger, and W. H. Woodruff, *J. Am. Chem. Soc.*, **103,** 7441 (1981).
144b. M. Forster and R. E. Hester, *Chem. Phys. Lett.*, **81,** 42 (1981).
144c. S. McClanahan, T. Hayes, and J. Kincaid, *J. Am. Chem. Soc.*, **105,** 4486 (1983).
144d. W. K. Smothers and M. S. Wrighton, *J. Am. Chem. Soc.*, **105,** 1067 (1983).
145. J. Takemoto and B. Hutchinson, *Inorg. Nucl. Chem. Lett.*, **8,** 769 (1972).
146. E. König and K. J. Watson, *Chem. Phys. Lett.*, **6,** 457 (1970).
147. J. Takemoto and B. Hutchinson, *Inorg. Chem.*, **12,** 705 (1973).
148. J. R. Ferraro and J. Takemoto, *Appl. Spectrosc.*, **28,** 66 (1974).
149. P. F. B. Barnard, A. T. Chamberlain, G. C. Kulasingam, R. J. Dosser, and W. R. McWhinnie, *Chem. Commun.*, 520 (1970).
150. H. van der Poel, G. van Koten, and K. Vrieze, *Inorg. Chem.*, **19,** 1145 (1980).
151. J. Takemoto, *Inorg. Chem.*, **12,** 949 (1973).
152. B. Hutchinson and A. Sunderland, *Inorg. Chem.*, **11,** 1948 (1972).
153. A. Bigotto, G. Costa, V. Galasso, and G. DeAlti, *Spectrochim. Acta,* **26A,** 1939 (1970).
154. A. Bigotto, V. Galasso, and G. DeAlti, *Spectrochim. Acta,* **27A,** 1659 (1971).
155. P. E. Rutherford and D. A. Thornton, *Spectrochim. Acta,* **35A,** 711 (1979).
156. N. Ohkaku and K. Nakamoto, *Inorg. Chem.*, **10,** 798 (1971).
157. B. Hutchinson, A. Sunderland, M. Neal, and S. Olbricht, *Spectrochim. Acta* **29A,** 2001 (1973).
158. H. Ogoshi, Y. Saito, and K. Nakamoto, *J. Chem. Phys.*, **57,** 4194 (1972).
159. H. Susi and J. S. Ard, *Spectrochim. Acta,* **33A,** 561 (1977).
160. S. Sunder and H. J. Bernstein, *J. Raman Spectrosc.*, **5,** 351 (1976).

161. L. L. Gladkov, A. T. Gradyushko, A. M. Shulga, K. N. Solovyov, and A. S. Starukhin, *J. Mol. Struct.*, **47,** 463 (1978); **45,** 267 (1978).
162. H. Ogoshi, N. Masai, Z. Yoshida, J. Takemoto, and K. Nakamoto, *Bull. Chem. Soc. Jpn.*, **44,** 49 (1971).
163. J. R. Kincaid, M. W. Urban, T. Watanabe, and K. Nakamoto, *J. Phys. Chem.*, **87,** 3096 (1983).
164. T. Kitagawa, M. Abe, and H. Ogoshi, *J. Chem. Phys.*, **69,** 4516 (1978); M. Abe, T. Kitagawa, and Y. Kyogoku, *J. Chem. Phys.*, **69,** 4526 (1978).
165. T. Kitagawa, H. Ogoshi, E. Watanabe, and Z. Yoshida, *J. Phys. Chem.*, **79,** 2629 (1975).
165a. J. Teraoka and T. Kitagawa, *J. Phys. Chem.*, **84,** 1928 (1980).
166. T. Kitagawa, M. Abe, Y. Kyogoku, H. Ogoshi, E. Watanabe, and Z. Yoshida, *J. Phys. Chem.*, **80,** 1181 (1976).
167. H. Ogoshi, E. Watanabe, Z. Yoshida, J. Kincaid, and K. Nakamoto, *J. Am. Chem. Soc.*, **95,** 2845 (1973).
168. J. Kincaid and K. Nakamoto, *Spectrosc. Lett.*, **9,** 19 (1976).
169. T. G. Spiro, "The Resonance Raman Spectroscopy of Metalloporphyrins and Heme Proteins," in A. B. P. Lever and H. B. Gray, ed., *Iron Porphyrins*, Addison-Wesley, Reading, Mass., 1983.
170. H. Oshio, T. Ama, T. Watanabe, J. Kincaid, and K. Nakamoto, *Spectrochim. Acta*, **40A,** 863 (1984).
171. J. Kincaid and K. Nakamoto, *J. Inorg. Nucl. Chem.*, **37,** 85 (1975).
172. J. M. Burke, J. R. Kincaid, and T. G. Spiro, *J. Am. Chem. Soc.*, **100,** 6077 (1978).
173. J. M. Burke, J. R. Kincaid, S. Peters, R. R. Gagne, J. P. Collman, and T. G. Spiro, *J. Am. Chem. Soc.*, **100,** 6083 (1978).
174. J. D. Stong, T. G. Spiro, R. J. Kubaska, and S. I. Shupack, *J. Raman Spectrosc.*, **9,** 312 (1980).
175. G. Chottard, P. Battioni, J. P. Battioni, M. Lange, and D. Mansuy, *Inorg. Chem.*, **20,** 1718 (1981).
176. S. Asher and K. Sauer, *J. Chem. Phys.*, **64,** 4115 (1976).
177. P. G. Wright, P. Stein, J. M. Burke, and T. G. Spiro, *J. Am. Chem. Soc.*, **101,** 3531 (1979).
178. A. L. Verma, R. Mendelsohn, and H. J. Bernstein, *J. Chem. Phys.*, **61,** 383 (1974).
179. L. J. Boucher and J. J. Katz, *J. Am. Chem. Soc.*, **89,** 1340 (1967).
180. S. Choi, T. G. Spiro, K. C. Langry, and K. M. Smith, *J. Am. Chem. Soc.*, **104,** 4337 (1982).
181. S. Choi, T. G. Spiro, K. C. Langry, K. M. Smith, D. L. Budd, and G. N. La Mar, *J. Am. Chem. Soc.*, **104,** 4345 (1982).
182. H. Ogoshi, E. Watanabe, Z. Yoshida, J. Kincaid, and K. Nakamoto, *Inorg. Chem.*, **14,** 1344 (1975).
183. Y. Ozaki, T. Kitagawa, and H. Ogoshi, *Inorg. Chem.*, **18,** 1772 (1979).
184. Y. Murakami and K. Sakata, *Inorg. Chim. Acta*, **2,** 273 (1968); Y. Murakami, Y. Matsuda, K. Sakata, and K. Harada, *Bull. Chem. Soc. Jpn.*, **47,** 458 and 3021 (1974).
185. Y. Murakami, K. Sakata, Y. Tanaka, and T. Matsuo, *Bull. Chem. Soc., Jpn.*, **48,** 3622 (1975).
186. B. Stymne, F. X. Sauvage, and G. Wettermark, *Spectrochim. Acta*, **35A,** 1195 (1979); **36A,** 397 (1980).
187. I. V. Aleksandrov, Y. S. Bobovich, V. G. Maslov, and A. N. Sidorov, *Opt. Spectrosc.*, **37,** 265 (1974).
188. I. Nakagawa, T. Shimanouchi, and K. Yamasaki, *Inorg. Chem.*, **3,** 772 (1964); **7,** 1332 (1968).
189. M. Le Postolloe, J. P. Mathieu, and H. Poulet, *J. Chim. Phys.*, **60,** 1319 (1963).
190. M. J. Cleare and W. P. Griffith, *J. Chem. Soc.* **A,** 1144 (1967).
191. M. J. Nolan and D. W. James, *Aust. J. Chem.*, **23,** 1043 (1970).
192. J. T. Huneke, B. Meisner, L. Walford, and R. L. Bain, *Spectrosc. Lett.*, **7,** 91 (1974).
193. D. W. James and M. J. Nolan, *Aust. J. Chem.*, **26,** 1433 (1973).
194. P. E. Merritt and S. E. Wiberley, *J. Phys. Chem.*, **59,** 55 (1955).
195. R. B. Hagel and L. F. Druding, *Inorg. Chem.*, **9,** 1496 (1970).
196. I. Nakagawa and T. Shimanouchi, *Spectrochim. Acta*, **23A,** 2099 (1967).
197. M. J. Nolan and D. W. James, *Aust. J. Chem.*, **26,** 1413 (1973).
198. K. Nakamoto, J. Fujita, and H. Murata, *J. Am. Chem. Soc.*, **80,** 4817 (1958).
199. D. M. L. Goodgame and M. A. Hitchman, *Inorg. Chem.*, **3,** 1389 (1964).

200. W. W. Fee, C. S. Garner, and J. N. M. Harrowfield, *Inorg. Chem.*, **6,** 87 (1967).
201. D. M. L. Goodgame, M. A. Hitchman, D. F. Marsham, and C. E. Souter, *J. Chem. Soc. A,* 2464 (1969).
202. D. M. L. Goodgame and M. A. Hitchman, *Inorg. Chem.*, **6,** 813 (1967).
203. L. El-Sayed and R. O. Ragsdale, *Inorg. Chem.*, **6,** 1640 (1967).
204. R. B. Penland, T. J. Lane, and J. V. Quagliano, *J. Am. Chem. Soc.*, **78,** 887 (1956).
205. I. R. Beattie and D. P. N. Satchell, *Trans. Faraday Soc.*, **52,** 1590 (1956).
206. J. L. Burmeister, *Coord. Chem. Rev.*, **3,** 225 (1968).
207. D. M. L. Goodgame and M. A. Hitchman, *Inorg. Chem.*, **4,** 721 (1965).
208. D. M. L. Goodgame and M. A. Hitchman, *J. Chem. Soc. A,* 612 (1967).
209. D. M. L. Goodgame, M. A. Hitchman, and D. F. Marsham, *J. Chem. Soc. A,* 1933 (1970).
210. U. Thewalt and R. E. Marsh, *Inorg. Chem.*, **9,** 1604 (1970).
211. S. Kubo, T. Shibahara, and M. Mori, *Bull. Chem. Soc. Jpn.*, **52,** 101 (1979).
212. K. Wieghardt and H. Siebert, *Z. Anorg. Allg. Chem.*, **374,** 186 (1970).
213. D. M. L. Goodgame, M. A. Hitchman, D. F. Marsham, P. Phavanantha, and D. Rogers, *Chem. Commun.*, 1383 (1969).
214. D. M. L. Goodgame, M. A. Hitchman, and D. F. Marsham, *J. Chem. Soc. A,* 259 (1971).
215. M. Hass and G. B. B. M. Sutherland, *Proc. R. Soc. London, Ser. A,* **236,** 427 (1956).
216. V. P. Tayal, B. K. Srivastava, D. P. Khandelwal, and H. D. Bist, *Appl. Spectrosc. Rev.*, **16,** 43 (1980).
217. J. O. Lundgren and I. Olovsson, *Acta Crystallogr.*, **23,** 966 (1967).
218. A. C. Pavia and P. A. Giguère, *J. Chem. Phys.*, **52,** 3551 (1970).
219. J. M. Williams, *Inorg. Nucl. Chem. Lett.*, **3,** 297 (1967).
220. J. Roziere and J. Potier, *J. Inorg. Nucl. Chem.*, **35,** 1179 (1973).
221. I. Nakagawa and T. Shimanouchi, *Spectrochim. Acta,* **20,** 429 (1964).
222. D. M. Adams and P. J. Lock, *J. Chem. Soc. A,* 2801 (1971).
223. H. L. Schlafer and H. P. Fritz, *Spectrochim. Acta,* **23A,** 1409 (1967).
224. T. G. Chang and D. E. Irish, *Can. J. Chem.*, **51,** 118 (1973).
225. R. E. Hester and R. A. Plane, *Inorg. Chem.*, **3,** 768 (1964).
226. H. Boutin, G. J. Safford, and H. R. Danner, *J. Chem. Phys.*, **40,** 2670 (1964); H. J. Prask and H. Boutin, *J. Chem. Phys.*, **45,** 699, 3284 (1966).
227. J. J. Rush, J. R. Ferraro, and A. Walker, *Inorg. Chem.*, **6,** 346 (1967).
228. T. Dupuis, C. Duval, and J. Lecomte, *Compt. Rend.*, **257,** 3080 (1963).
229. M. Maltese and W. J. Orville-Thomas, *J. Inorg. Nucl. Chem.*, **29,** 2533 (1967).
230. J. R. Ferraro and W. R. Walker, *Inorg. Chem.*, **4,** 1382 (1965).
231. W. R. McWhinnie, *J. Inorg. Nucl. Chem.*, **27,** 1063 (1965).
232. J. R. Ferraro, R. Driver, W. R. Walker, and W. Wozniak, *Inorg. Chem.*, **6,** 1586 (1967).
233. G. Blyholder and N. Ford, *J. Phys. Chem.*, **68,** 1496 (1964).
234. V. A. Maroni and T. G. Spiro, *J. Am. Chem. Soc.*, **89,** 45 (1967).
235. R. W. Adams, R. L. Martin, and G. Winter, *Aust. J. Chem.*, **20,** 773 (1967).
236. R. C. Mehrotra and J. M. Batwara, *Inorg. Chem.*, **9,** 2505 (1970).
237. L. M. Brown and K. S. Mazdiyasni, *Inorg. Chem.*, **9,** 2783 (1970).
238. P. W. N. M. Van Leeuwen, *Rec. Trav. Chim.*, **86,** 247 (1967).
239. D. Knetsch and W. L. Groeneveld, *Inorg. Chim. Acta,* **7,** 81 (1973).
240. H. Wieser and P. J. Krueger, *Spectrochim. Acta,* **26A,** 1349 (1970).
241. G. W. A. Fowles, D. A. Rice, and R. A. Walton, *Spectrochim. Acta,* **26A,** 143 (1970).
242. D. G. Parsons, M. R. Truter, and J. N. Wingfield, *Inorg. Chim. Acta,* **14,** 45 (1975).
243. W. L. Driessen and W. L. Groeneveld, *Rec. Trav. Chim.*, **88,** 977 (1969).
244. W. L. Driessen and W. L. Groeneveld, *Rec. Trav. Chim.*, **90,** 258 (1971).
245. W. L. Driessen and W. L. Groeneveld, *Rec. Trav. Chim.*, **90,** 87 (1971).
246. W. L. Driessen, W. L. Groeneveld, and F. W. Van der Wey, *Rec. Trav. Chim.*, **89,** 353 (1970).
247. K. Itoh and H. J. Bernstein, *Can. J. Chem.*, **34,** 170 (1956).
248. S. D. Robinson and M. F. Uttley, *J. Chem. Soc.*, 1912 (1973).
248a. G. B. Deacon and R. J. Phillips, *Coord. Chem. Rev.*, **33,** 227 (1980).

249. T. A. Stephenson and G. Wilkinson, *J. Inorg. Nucl. Chem.*, **29,** 2122 (1967).
250. G. Csontos, B. Heil and C. Markó, *J. Organometal. Chem.*, **37,** 183 (1972).
251. D. Stoilova, G. Nikolov, and K. Balarev, *Izv. Akad. Nauk SSSR, Ser. Khim.*, **9,** 371 (1976).
252. G. Busca and V. Lorenzelli, *Minerals Chem.*, **7,** 89 (1982).
253. S. Baba and S. Kawaguchi, *Inorg. Nucl. Chem. Lett.*, **9,** 1287 (1973).
254. K. Nakamoto, P. J. McCarthy, and B. Miniatus, *Spectrochim. Acta*, **21,** 379 (1965).
255. M. Tsuboi, T. Onishi, I. Nakagawa, T. Shimanouchi, and S. Mizuschima, *Spectrochim. Acta*, **12,** 253 (1958).
256. K. Fukushima, T. Onishi, T. Shimanouchi, and S. Mizushima, *Spectrochim. Acta*, **14,** 236 (1959).
257. A. J. Stosick, *J. Am. Chem. Soc.*, **67,** 365 (1945).
258. K. Nakamoto, Y. Morimoto, and A. E. Martell, *J. Am. Chem. Soc.*, **83,** 4528 (1961).
259. R. A. Condrate and K. Nakamoto, *J. Chem. Phys.*, **42,** 2590 (1965).
260. J. Kincaid and K. Nakamoto, *Spectrochim. Acta*, **32A,** 277 (1976).
261. M. L. Niven and D. A. Thornton, *Inorg. Chim. Acta*, **32,** 205 (1979).
262. J. B. Hodgson, G. C. Percy, and D. A. Thornton, *Spectrochim. Acta*, **35A, 949** (1979).
263. G. C. Percy and H. S. Stenton, *J. Chem. Soc., Dalton Trans.*, 2429 (1976).
264. A. W. Herlinger, S. L. Wenhold, and T. V. Long, II, *J. Am. Chem. Soc.*, **92,** 6474 (1970).
265. A. W. Herlinger and T. V. Long, II, *J. Am. Chem. Soc.*, **92,** 6481 (1970).
266. J. A. Kieft and K. Nakamoto, to be published.
267. J. A. Kieft and K. Nakamoto, *J. Inorg. Nucl. Chem.*, **29,** 2561 (1967).
268. J. R. Kincaid, J. A. Larrabee, and T. G. Spiro, *J. Am. Chem. Soc.*, **100,** 334 (1978).
269. D. H. Busch and J. C. Bailar, Jr., *J. Am. Chem. Soc.*, **75,** 4574 (1953); **78,** 716 (1956); M. L. Morris and D. H. Busch, *J. Am. Chem. Soc.*, **78,** 5178 (1956); K. Swaminathan and D. H. Busch, *J. Inorg. Nucl. Chem.*, **20,** 159 (1961); R. E. Sievers and J. C. Bailar, Jr., *Inorg. Chem.*, **1,** 174 (1962).
270. D. Chapman, *J. Chem. Soc.*, 1766 (1955).
271. Y. Tomita and K. Ueno, *Bull. Chem. Soc. Jpn.*, **36,** 1069 (1963).
272. K. Krishnan and R. A. Plane, *J. Am. Chem. Soc.*, **90,** 3195 (1968).
273. A. A. McConnell and R. H. Nuttall, *Spectrochim. Acta*, **33A,** 459 (1977).
273a. E. G. Bartick and R. G. Messerschmidt, *Am. Lab.*, November 1984, p. 56.
274. S. Fronaeus and R. Larsson, *Acta Chem. Scand.*, **16,** 1433, 1447 (1962).
275. S. Fronaeus and R. Larsson, *Acta Chem. Scand.*, **14,** 1364 (1960).
276. R. Larsson, *Acta Chem. Scand.*, **19,** 783 (1965).
277. K. Nakamoto, Y. Morimoto, and A. E. Martell, *J. Am. Chem. Soc.*, **84,** 2081 (1962); **85,** 309 (1963).
278. Y. Tomita, T. Ando, and K. Ueno, *J. Phys. Chem.*, **69,** 404 (1965).
279. Y. Tomita and K. Ueno, *Bull. Chem. Soc. Jpn.*, **36,** 1069 (1963).
280. A. E. Martell and M. K. Kim, *J. Coord. Chem.*, **4,** 9 (1974).
281. M. K. Kim and A. E. Martell, *J. Am. Chem. Soc.*, **85,** 3080 (1963).
282. M. K. Kim and A. E. Martell, *Biochemistry*, **3,** 1169 (1964).
283. M. K. Kim and A. E. Martell, *J. Am. Chem. Soc.*, **88,** 914 (1966).
284. J. Fujita, A. E. Martell, and K. Nakamoto, *J. Chem. Phys.*, **36,** 324, 331 (1962).
285. R. D. Hancock and D. A. Thornton, *J. Mol. Struct.*, **6,** 441 (1970).
286. J. Gouteron, *J. Inorg. Nucl. Chem.*, **38,** 55 (1976).
287. K. L. Scott, K. Wieghardt, and A. G. Sykes, *Inorg. Chem.*, **12,** 655 (1973); K. Wieghardt, *Z. Anorg. Allg. Chem.*, **391,** 142 (1972).
288. R. E. Hester and R. A. Plane, *Inorg. Chem.*, **3,** 513 (1964); E. C. Gruen and R. A. Plane, *Inorg. Chem.*, **6,** 1123 (1967).
289. P. X. Armendarez and K. Nakamoto, *Inorg. Chem.*, **5,** 796 (1966).
290. P. Fischer, R. Graf, and J. Weidlein, *J. Organometal. Chem.*, **144,** 95 (1978).
291. B. B. Kedzia, P. X. Armendarez, and K. Nakamoto, *J. Inorg. Nucl. Chem.*, **30,** 849 (1968).
292. L. Cavalca, M. Nardelli, and G. Fava, *Acta Crystallogr.*, **13,** 594 (1960).
293. Y. Saito, K. Machida, and T. Uno, *Spectrochim. Acta*, **26A,** 2089 (1970).

294. K. Nakamoto, J. Fujuta, S. Tanaka, and M. Kobayashi, *J. Am. Chem. Soc.*, **79**, 4904 (1957).
295. C. G. Barraclough and M. L. Tobe, *J. Chem. Soc.*, 1993 (1961).
296. R. Eskenazi, J. Rasovan, and R. Levitus, *J. Inorg. Nucl. Chem.*, **28**, 521 (1966).
297. J. E. Finholt, R. W. Anderson, J. A. Fyfe, and K. G. Caulton, *Inorg. Chem.*, **4**, 43 (1965).
298. W. R. McWhinnie, *J. Inorg. Nucl. Chem.*, **26**, 21 (1964).
299. I. S. Ahuja, *Inorg. Chim. Acta*, **3**, 110 (1969).
300. K. Wieghardt and J. Eckert, *Z. Anorg. Allg. Chem.*, **383**, 240 (1971).
301. R. Ugo, F. Conti, S. Cenini, R. Mason, and G. B. Robertson, *Chem. Commun.*, 1498 (1968).
302. R. W. Horn, E. Weissberger, and J. P. Collman, *Inorg. Chem.*, **9**, 2367 (1970).
303. J. R. Ferraro and A. Walker, *J. Chem. Phys.*, **42**, 1278 (1965).
304. N. Tanaka, H. Sugi, and J. Fujita, *Bull. Chem. Soc. Jpn.*, **37**, 640 (1964).
305. J. A. Goldsmith, A. Hezel, and S. D. Ross, *Spectrochim. Acta*, **24A**, 1139 (1968).
306. M. R. Rosenthal, *J. Chem. Ed.*, **50**, 331 (1973).
307. B. J. Hathaway and A. E. Underhill, *J. Chem. Soc.*, 3091 (1961).
308. B. J. Hathaway, D. G. Holah, and M. Hudson, *J. Chem. Soc.*, 4586 (1963).
309. M. E. Farago, J. M. James, and V. C. G. Trew, *J. Chem. Soc. A*, 820 (1967).
310. A. E. Wickenden and R. A. Krause, *Inorg. Chem.*, **4**, 404 (1965).
311. L. E. Moore, R. B. Gayhart, and W. E. Bull, *J. Inorg. Nucl. Chem.*, **26**, 896 (1964).
312. T. Chausse, A. Potier, and J. Potier, *J. Chem. Res. (S)*, 316 (1980).
313. S. F. Lincoln and D. R. Stranks, *Aust. J. Chem.*, **21**, 37 (1968).
314. T. A. Beech and S. F. Lincoln, *Aust. J. Chem.*, **24**, 1065 (1971).
315. R. Coomber and W. P. Griffith, *J. Chem. Soc.*, 1128 (1968).
316. S. D. Ross and N. A. Thomas, *Spectrochim. Acta*, **26A**, 971 (1970).
317. A. N. Freedman and B. P. Straughan, *Spectrochim. Acta*, **27A**, 1455 (1971).
318. P. A. Yeats, J. R. Sams, and F. Aubke, *Inorg. Chem.*, **12**, 328 (1973).
319. S. D. Brown and G. L. Gard, *Inorg. Chem.*, **17**, 1363 (1978).
320. B. M. Gatehouse, S. E. Livingstone, and R. S. Nyholm, *J. Chem. Soc.*, 3137 (1958).
321. J. Fujita, A. E. Martell, and K. Nakamoto, *J. Chem. Phys.*, **36**, 339 (1962).
322. R. E. Hester and W. E. L. Grossman, *Inorg. Chem.*, **5**, 1308 (1966).
323. J. A. Goldsmith and S. D. Ross, *Spectrochim. Acta*, **24A**, 993 (1968).
324. H. Elliott and B. J. Hathaway, *Spectrochim. Acta*, **21**, 1047 (1965).
325. M. R. Churchill, R. A. Lashewycz, K. Koshy, and T. P. Dasgupta, *Inorg. Chem.*, **20**, 376 (1981).
326. M. R. Churchill, G. Davies, M. A. El-Sayed, M. F. El-Shazly, J. P. Hutchinson, and M. W. Rupich, *Inorg. Chem.*, **19**, 201 (1980).
327. J. S. Ogden and S. J. Williams, *J. Chem. Soc., Dalton Trans.*, 456 (1981).
328. C. C. Addison, N. Logan, S. C. Wallwork, and C. D. Barner, *Quart. Rev.*, **25**, 289 (1971).
329. B. M. Gatehouse, S. E. Livinstone, and R. S. Nyholm, *J. Chem. Soc.*, 4222 (1957); *J. Inorg. Nucl. Chem.*, **8**, 75 (1958).
330. N. F. Curtis and Y. M. Curtis, *Inorg. Chem.*, **4**, 804 (1965).
331. C. C. Addison, R. Davis, and N. Logan, *J. Chem. Soc. A*, 3333 (1970).
332. B. Lippert, C. J. L. Lock, B. Rosenberg, and M. Zvagulis, *Inorg. Chem.*, **16**, 1525 (1977).
333. C. C. Addison and W. B. Simpson, *J. Chem. Soc.*, 598 (1965).
334. J. G. Allpress and A. N. Hambly, *Aust. J. Chem.*, **12**, 569 (1959).
335. R. J. Fereday, N. Logan, and D. Sutton, *J. Chem. Soc. A*, 2699 (1969).
336. D. W. Johnson and D. Sutton, *Can. J. Chem.*, **50**, 3326 (1972).
337. N. Logan and W. B. Simpson, *Spectrochim. Acta*, **21**, 857 (1965).
338. E. J. Duff, M. N. Hughes, and K. J. Rutt, *J. Chem. Soc. A*, 2126 (1969).
339. E. M. Briggs and A. E. Hill, *J. Chem. Soc. A*, 2008 (1970).
340. A. B. P. Lever, E. Mantovani, B. S. Ramaswamy, *Can. J. Chem.*, **49**, 1957 (1971).
341. J. R. Ferraro, A. Walker, and C. Cristallini, *Inorg. Nucl. Chem. Lett.*, **1**, 25 (1965).
342. C. C. Addison, D. W. Amos, D. Sutton, and W. H. H. Hoyle, *J. Chem. Soc. A*, 808 (1967).
343. R. H. Nuttall and D. W. Taylor, *Chem. Commun.*, 1417 (1968).
344. J. I. Bullock and F. W. Parrett, *Chem. Commun.*, 157 (1969).
345. J. R. Ferraro and A. Walker, *J. Chem. Phys.*, **42**, 1273 (1965); **43**, 2689 (1965); **45**, 550 (1966).

346. D. E. Irish and G. E. Walrafen, *J. Chem. Phys.*, **46,** 378 (1967).
347. R. E. Hester and K. Krishnan, *J. Chem. Phys.*, **46,** 3405 (1967); **47,** 1747 (1967).
348. F. A. Cotton and R. Francis, *J. Am. Chem. Soc.*, **82,** 2986 (1960).
349. G. Newman and D. B. Powell, *Spectrochim. Acta,* **19,** 213 (1963).
350. M. E. Baldwin, *J. Chem. Soc.*, 3123 (1961).
351. B. Nyberg and R. Larsson, *Acta Chem. Scand.*, **27,** 63 (1973).
352. J. P. Hall and W. P. Griffith, *Inorg. Chim. Acta,* **48,** 65 (1981).
353. A. D. Fowless and D. R. Stranks, *Inorg. Chem.*, **16,** 1271 (1977).
354. R. C. Elder and P. E. Ellis, Jr., *Inorg. Chem.*, **17,** 870 (1978).
355. E. Lindner and G. Vitzthum, *Chem. Ber.*, **102,** 4062 (1969).
356. G. Vitzthum and E. Lindner, *Angew. Chem. Int. Ed.*, **10,** 315 (1971).
357. K. Nakamoto and A. E. Martell, *J. Chem. Phys.*, **32,** 588 (1960).
358. M. Mikami, I. Nakagawa, and T. Shimanouchi, *Spectrochim. Acta,* **23A,** 1037 (1967).
359. H. Junge and H. Musso, *Spectrochim. Acta,* **24A,** 1219 (1968).
360. K. Nakamoto, C. Udovich, and J. Takemoto, *J. Am. Chem. Soc.*, **92,** 3973 (1970).
361. R. C. Fay and R. N. Lowry, *Inorg. Nucl. Chem. Lett.*, **3,** 117 (1967).
362. W. D. Courrier, C. J. L. Lock, and G. Turner, *Can. J. Chem.*, **50,** 1797 (1972).
363. M. R. Caira, J. M. Haigh, and L. R. Nassimbeni, *J. Inorg. Nucl. Chem.*, **34,** 3171 (1972).
364. M. F. Richardson, W. F. Wagner, and D. E. Sands, *Inorg. Chem.*, **7,** 2495 (1968).
365. J. C. Hammel, J. A. S. Smith, and E. J. Wilkins, *J. Chem. Soc. A,* 1461 (1969).
366. J. C. Hammel and J. A. S. Smith, *J. Chem. Soc. A,* 2883 (1969).
367. M. A. Bush, D. E. Fenton, R. S. Nyholm, and M. R. Truter, *Chem. Commun.*, 1335 (1970).
368. Y. Nakamura, N. Kanehisa, and S. Kawaguchi, *Bull. Chem. Soc. Jpn.*, **45,** 485 (1972).
369. F. A. Cotton and R. C. Elder, *J. Am. Chem. Soc.*, **86,** 2294 (1964); *Inorg. Chem.*, **4,** 1145 (1965).
370. P. W. N. M. van Leeuwen, *Rec. Trav. Chim.*, **87,** 396 (1968).
371. Y. Nakamura and S. Kawaguchi, *Chem. Commun.*, 716 (1968).
372. S. Koda, S. Ooi, H. Kuroya, K. Isobe, Y. Nakamura, and S. Kawaguchi, *Chem. Commun.*, 1321 (1971).
373. Y. Nakamura, K. Isobe, H. Morita, S. Yamazaki, and S. Kawaguchi, *Inorg. Chem.*, **11,** 1573 (1972).
374. S. Koda, S. Ooi, H. Kuroya, Y. Nakamura, and S. Kawaguchi, *Chem. Commun.*, 280 (1971).
375. J. Lewis, R. F. Long, and C. Oldham, *J. Chem. Soc.*, 6740 (1965); D. Gibson, J. Lewis, and C. Oldham, *J. Chem. Soc. A,* 1453 (1966).
376. G. T. Behnke and K. Nakamoto, *Inorg. Chem.*, **6,** 433 (1967).
377. G. T. Behnke and K. Nakamoto, *Inorg. Chem.*, **6,** 440 (1967).
378. G. T. Behnke and K. Nakamoto, *Inorg. Chem.*, **7,** 330 (1968).
379. F. Bonati and G. Minghetti, *Angew. Chem. Int. Ed.*, **7,** 629 (1968).
380. D. Gibson, B. F. G. Johnson, and J. Lewis, *J. Chem. Soc. A,* 367 (1970).
381. S. Baba, T. Ogura, and S. Kawaguchi, *Inorg. Nucl. Chem. Lett.*, **7,** 1195 (1971).
382. G. Allen, J. Lewis, R. F. Long, and C. Oldham, *Nature,* **202,** 589 (1964).
383. G. T. Behnke and K. Nakamoto, *Inorg. Chem.*, **7,** 2030 (1968).
384. J. Lewis and C. Oldham, *J. Chem. Soc. A,* 1456 (1966).
385. Y. Nakamura and K. Nakamoto, *Inorg. Chem.*, **14,** 63 (1975).
386. S. Okeya, T. Nakamura, S. Kawaguchi, and T. Hinomoto, *Inorg. Chem.*, **20,** 1576 (1981).
387. Z. Kanda, Y. Nakamura, and S. Kawaguchi, *Inorg. Chem.*, **17,** 910 (1978).
388. L. G. Hulett and D. A. Thornton, *Spectrochim. Acta,* **27A,** 2089 (1971).
389. H. Junge, *Spectrochim. Acta,* **24A,** 1957 (1968).
390. B. Hutchinson, D. Eversdyk, and S. Olbricht, *Spectrochim. Acta,* **30A,** 1605 (1974).
391. F. Sagara, H. Kobayashi, and K. Ueno, *Bull. Chem. Soc. Jpn.*, **45,** 794 (1972).
392. R. B. Penland, S. Mizushima, C. Curran, and J. V. Quagliano, *J. Am. Chem. Soc.*, **79,** 1575 (1957).
393. A. Yamaguchi, T. Miyazawa, T. Shimanouchi, and S. Mizushima, *Spectrochim. Acta,* **10,** 170 (1957).
394. E. Giesbrecht and M. Kawashita, *J. Inorg. Nucl. Chem.*, **32,** 2461 (1970).

395. A. Yamaguchi, R. B. Penland, S. Mizushima, T. J. Lane, C. Curran, and J. V. Quagliano, *J. Am. Chem. Soc.*, **80**, 527 (1958).
396. R. A. Bailey and T. R. Peterson, *Can. J. Chem.*, **45**, 1135 (1967).
397. K. Swaminathan and H. M. N. H. Irving, *J. Inorg. Nucl. Chem.*, **26**, 1291 (1964).
398. C. D. Flint and M. Goodgame, *J. Chem. Soc. A*, 744 (1966).
399. P. J. Hendra and Z. Jović, *J. Chem. Soc. A*, 735 (1967).
400. D. M. Adams and J. B. Cornell, *J. Chem. Soc. A*, 884 (1967).
401. R. Rivest, *Can. J. Chem.*, **40**, 2234 (1962).
402. T. J. Lane, A. Yamaguchi, J. V. Quagliano, J. A. Ryan, and S. Mizushima, *J. Am. Chem. Soc.*, **81**, 3824 (1959).
403. M. Schafer and C. Curran, *Inorg. Chem.*, **5**, 265 (1966).
404. R. K. Gosavi and C. N. R. Rao, *J. Inorg. Nucl. Chem.*, **29**, 1937 (1967).
405. G. B. Aitken, J. L. Duncan, and G. P. McQuillan, *J. Chem. Soc., Dalton Trans.*, 2103 (1972).
406. P. J. Hendra and Z. Jović, *Spectrochim. Acta*, **24A**, 1713 (1968).
407. R. J. Balahura and R. B. Jordan, *J. Am. Chem. Soc.*, **92**, 1533 (1970).
408. F. A. Cotton, R. Francis, and W. D. Horrocks, *J. Phys. Chem.*, **64**, 1534 (1960).
409. R. S. Drago and D. W. Meek, *J. Phys. Chem.*, **65**, 1446 (1961): D. W. Meek, D. K. Straub, and R. S. Drago, *J. Am. Chem. Soc.*, **82**, 6013 (1960).
410. B. B. Wayland and R. F. Schramm, *Inorg. Chem.*, **8**, 971 (1969); *Chem. Commun.*, 1465 (1968).
411. V. N. Krishnamarthy and S. Soundararajan, *J. Inorg. Nucl. Chem.*, **29**, 517 (1967); S. K. Ramalingam and S. Soundararajan, *Z. Anorg. Allg. Chem.*, **353**, 216 (1967).
412. C. G. Fuentes and S. J. Patel, *J. Inorg. Nucl. Chem.*, **32**, 1575 (1970).
413. C. V. Senoff, E. Maslowsky, Jr., and R. G. Goel, *Can. J. Chem.*, **49**, 3585 (1971).
414. W. Kitchings, C. J. Moore, and D. Doddrell, *Inorg. Chem.*, **9**, 541 (1970).
415. D. A. Langs, C. R. Hare, and R. G. Little, *Chem. Commun.*, 1080 (1967).
416. I. P. Evans, A. Spencer, and G. Wilkinson, *J. Chem. Soc., Dalton Trans.*, 204 (1973).
417. A. Mercer and J. Trotter, *J. Chem. Soc., Dalton Trans.*, 2480 (1975).
418. P. G. Antonov, Y. N. Kukushkin, V. I. Konnov, and B. I. Ionin, *Russ. J. Inorg. Chem.*, **23**, 245 (1978).
419. M. Tranquille and M. T. Forel, *Spectrochim. Acta*, **28A**, 1305 (1972).
420. C. V. Berney and J. H. Weber, *Inorg. Chem.*, **7**, 283 (1968).
421. G. Griffiths and D. A. Thornton, *J. Mol. Struct.*, **52**, 39 (1979).
421a. B. R. James and R. H. Morris, *Spectrochim. Acta*, **34A**, 577 (1978).
422. P. W. N. M. van Leeuwen, *Rec. Trav. Chim.*, **86**, 201 (1967); P. W. N. M. van Leeuwen and W. L. Groeneveld, *Rec. Trav. Chim.*, **86**, 721 (1967).
423. S. K. Madan, C. M. Hull, and L. J. Herman, *Inorg. Chem.*, **7**, 491 (1968).
424. K. A. Jensen and K. Krishnan, *Scand. Chim. Acta*, **21**, 1988 (1967).
425. A. G. Sharp, *The Chemistry of Cyano Complexes of the Transition Metals*, Academic Press, New York, 1976.
426. W. P. Griffith, *Coord. Chem. Rev.*, **17**, 177 (1975).
427. P. Rigo and A. Turco, *Coord. Chem. Rev.*, **13**, 133 (1974).
428. L. H. Jones and B. I. Swanson, *Accounts Chem. Res.*, **9**, 128 (1976).
429. H. Stammreich, B. M. Chadwick, and S. G. Frankiss, *J. Mol. Struct.*, **1**, 191 (1968).
430. B. M. Chadwick and S. G. Frankiss, *J. Mol. Struct.*, **31**, 1 (1976).
431. B. M. Chadwick and S. G. Frankiss, *J. Mol. Struct.*, **2**, 281 (1968).
432. G. J. Kubas and L. H. Jones, *Inorg. Chem.*, **13**, 2816 (1974).
433. W. P. Griffiths and J. R. Lane, *J. Chem. Soc., Dalton Trans.*, 158 (1972).
434. W. P. Griffith and G. T. Turner, *J. Chem. Soc. A*, 858 (1970).
435. P. W. Jensen, *J. Raman Spectrosc.*, **4**, 75 (1975).
435a. H. Siebert and A. Siebert, *Angew. Chem. Int. Ed.*, **8**, 6009 (1969); *Z. Anorg. Allg. Chem.*, **378**, 160 (1970).
436. M. F. A. El-Sayed and R. K. Sheline, *J. Inorg. Nucl. Chem.*, **6**, 187 (1958).
437. R. Nast and D. Rehder, *Chem. Ber.*, **104**, 1709 (1971).
438. L. H. Jones and R. A. Penneman, *J. Chem. Phys.*, **22**, 965 (1954).

439. R. A. Penneman and L. H. Jones, *J. Chem. Phys.*, **24,** 293 (1956).
440. R. A. Penneman and L. H. Jones, *J. Inorg. Nucl. Chem.*, **20,** 19 (1961).
441. C. Kappenstein and R. P. Hugel, *Inorg. Chem.*, **16,** 250 (1977).
442. C. Kappenstein and R. P. Hugel, *Inorg. Chem.*, **17,** 1945 (1978).
443. B. M. Chadwick, D. A. Long, and S. U. Qureshi, *J. Mol. Struct.*, **63,** 167 (1980).
444. L. H. Jones and B. I. Swanson, *J. Chem. Phys.*, **63,** 5401 (1975).
445. G. R. Rossman, F.-D. Tsay, and H. B. Gray, *Inorg. Chem.*, **12,** 824 (1973).
446. P. M. Kiernan and W. P. Griffith, *Inorg. Nucl. Chem. Lett.*, **12,** 377 (1976).
447. W. P. Griffith, P. M. Kiernan, and J.-M. Brégeault, *J. Chem. Soc., Dalton Trans.*, 1411 (1978).
448. H. S. Trop, A. G. Jones, and A. Davison, *Inorg. Chem.*, **19,** 1993 (1980).
449. A. M. Soares, P. M. Kiernan, D. J. Cole-Hamilton, and W. P. Griffith, *Chem. Commun.*, 84 (1981).
450. J. L. Hoard, T. A. Hamor, and M. D. Glick, *J. Am. Chem. Soc.*, **90,** 3177 (1968).
451. H. Stammreich and O. Sala, *Z. Elektrochem.*, **64,** 741 (1960); **65,** 149 (1961).
452. K. O. Hartman and F. A. Miller, *Spectrochim. Acta*, **24A,** 669 (1968).
453. B. V. Parish, P. G. Simms, M. A. Wells, and L. A. Woodward, *J. Chem. Soc.*, 2882 (1968).
454. P. M. Kiernan and W. P. Griffith, *J. Chem. Soc., Dalton Trans.*, 2489 (1975).
455. T. V. Long, II and G. A. Vernon, *J. Am. Chem. Soc.*, **93,** 1919 (1971).
456. M. B. Hursthouse and A. M. Galas, *Chem. Commun.*, 1167 (1980).
457. K. N. Raymond, P. W. R. Corfield, and J. A. Ibers, *Inorg. Chem.*, **7,** 1362 (1968).
458. A. Terzis, K. N. Raymond, and T. G. Spiro, *Inorg. Chem.*, **9,** 2415 (1970).
459. L. J. Basile, J. R. Ferraro, M. Choca, and K. Nakamoto, *Inorg. Chem.*, **13,** 496 (1974).
460. E. Hellner, H. Ahsbahs, G. Dehnicke, and K. Dehnicke, *Ber. Bunsensenged. Phys. Chem.*, **77,** 277 (1973).
461. R. L. McCullough, L. H. Jones, and R. A. Penneman, *J. Inorg. Nucl. Chem.*, **13,** 286 (1960).
462. G. W. Chantry and R. A. Plane, *J. Chem. Phys.*, **33,** 736 (1960); **34,** 1268 (1961); **35,** 1027 (1961).
463. V. Caglioti, G. Sartori, and C. Furlani, *J. Inorg. Nucl. Chem.*, **13,** 22 (1960); **8,** 87 (1958).
464. L. H. Jones, *J. Mol. Spectrosc.*, **8,** 105 (1962); *J. Chem. Phys.*, **36,** 1209 (1962).
465. L. H. Jones, *J. Chem. Phys.*, **41,** 856 (1964).
466. D. Bloor, *J. Chem. Phys.*, **41,** 2573 (1964).
467. I. Nakagawa and T. Shimanouchi, *Spectrochim. Acta*, **18,** 101 (1962).
468. L. H. Jones, *Inorg. Chem.*, **2,** 777 (1963).
469. I. Nakagawa and T. Shimanouchi, *Spectrochim. Acta*, **26A,** 131 (1970).
470. L. H. Jones, B. I. Swanson, and G. J. Kubas, *J. Chem. Phys.*, **61,** 4650 (1974); B. I. Swanson and L. H. Jones, *Inorg. Chem.*, **13,** 313 (1974).
471. L. H. Jones, *J. Chem. Phys.*, **29,** 463 (1958).
472. H. Poulet and J. P. Mathieu, *Spectrochim. Acta*, **15,** 932 (1959).
473. L. H. Jones, *Spectrochim. Acta*, **17,** 188 (1961).
474. D. M. Sweeny, I. Nakagawa, S. Mizushima, and J. V. Quagliano, *J. Am. Chem. Soc.*, **78,** 889 (1956).
475. C. W. F. T. Pistorius, *Z. Phys. Chem.*, **23,** 197 (1960).
476. R. L. McCullough, L. H. Jones, and G. A. Crosby, *Spectrochim. Acta*, **16,** 929 (1960).
477. L. H. Jones and J. M. Smith, *J. Chem. Phys.*, **41,** 2507 (1964).
478. L. H. Jones, *J. Chem. Phys.*, **27,** 665 (1957).
479. L. H. Jones, *Spectrochim. Acta*, **19,** 1675 (1963).
480. L. H. Jones, *J. Chem. Phys.*, **26,** 1578 (1957); **25,** 379 (1956).
481. L. H. Jones, *J. Chem. Phys.*, **27,** 468 (1957); **21,** 1891 (1953); **22,** 1135 (1954).
482. V. Lorenzelli and P. Delorme, *Spectrochim. Acta*, **19,** 2033 (1963).
483. L. H. Jones, *Inorg. Chem.* **3,** 1581 (1964); **4,** 1472 (1965); J. M. Smith, L. H. Jones, I. K. Kressin, and R. A. Penneman, *Inorg. Chem.*, **4,** 369 (1965).
484. J. C. Coleman, H. Peterson, and R. A. Penneman, *Inorg. Chem.*, **4,** 135 (1965).
485. J. H. Swinebart, *Coord, Chem. Rev.*, **2,** 385 (1967).
486. R. K. Khanna, C. W. Brown, and L. H. Jones, *Inorg. Chem.*, **8,** 2195 (1969); J. B. Bates and R. K. Khanna, *Inorg. Chem.*, **9,** 1376 (1970).

487. D. B. Brown, *Inorg. Chim. Acta*, **5,** 314 (1971).
488. A. Poletti, A. Santucci, and G. Paliani, *Spectrochim. Acta*, **27A,** 2061 (1971).
489. E. Miki, S. Kubo, K. Mizumachi, T. Ishimori, and H. Okuno, *Bull. Chem. Soc. Jpn.*, **44,** 1024 (1971).
490. L. Tosi and J. Danon, *Inorg. Chem.*, **3,** 150 (1964).
491. L. H. Jones and J. M. Smith, *Inorg. Chem.*, **4,** 1677 (1965).
492. M. N. Memering, L. H. Jones, and J. C. Bailar, Jr., *Inorg. Chem.*, **12,** 2793 (1973).
493. D. F. Shriver, *J. Am. Chem. Soc.*, **84,** 4610 (1962); **85,** 1405 (1963); D. F. Shriver and J. Posner, *J. Am. Chem. Soc.*, **88,** 1672 (1966).
494. D. F. Shriver, S. A. Shriver, and S. E. Anderson, *Inorg. Chem.*, **4,** 725 (1965).
495. D. B. Brown, D. F. Shriver, and L. H. Schwartz, *Inorg. Chem.*, **7,** 77 (1968).
496. B. I. Swanson and J. J. Rafalko, *Inorg. Chem.*, **15,** 249 (1976).
497. B. I. Swanson, *Inorg. Chem.*, **15,** 253 (1976).
498. R. E. Hester and E. M. Nour, *J. Chem. Soc., Dalton Trans.*, 939 (1981).
499. H. G. Nadler, J. Pebler, and K. Dehnicke, *Z. Anorg. Allg. Chem.*, **404,** 230 (1974).
500. R. E. Wilde, S. N. Ghosh, and B. J. Marshall, *Inorg. Chem.*, **9,** 2513 (1970).
501. J. R. Ferraro, *Coord. Chem. Rev.*, **43,** 205 (1982).
502. J. R. Ferraro, L. J. Basile, J. M. Williams, J. I. McOmber, D. F. Shriver, and D. R. Greig, *J. Chem. Phys.*, **69,** 3871 (1978).
503. R. A. Walton, *Spectrochim. Acta*, **21,** 1795 (1965); *Can. J. Chem.*, **44,** 1480 (1966).
504. R. E. Clarke and P. C. Ford, *Inorg. Chem.*, **9,** 227 (1970).
505. J. C. Evans and G. Y.-S. Lo, *Spectrochim. Acta*, **21,** 1033 (1965).
506. J. Reedijk and W. L. Groeneveld, *Rec. Trav. Chim.*, **86,** 1127 (1967).
507. M. F. Farona and K. F. Kraus, *Inorg. Chem.*, **9,** 1700 (1970).
508. Y. Kinoshita, I. Matsubara, and Y. Saito, *Bull. Chem. Soc. Jpn.*, **32,** 741 (1959).
509. M. Kubota, D. L. Johnston, and L. Matsubara, *Inorg. Chem.*, **5,** 386 (1966).
510. Y. Kinoshita, I. Matsubara, and Y. Saito, *Bull. Chem. Soc. Jpn.*, **32,** 1216 (1959).
511. I. Matsubara, *Bull. Chem. Soc. Jpn.*, **34,** 1719 (1961); *J. Chem. Phys.*, **35,** 373 (1961).
512. J. K. Brown, N. Sheppard, and D. M. Simpson, *Philos. Trans. R. Soc.*, **A247,** 35 (1954).
513. M. Kubota and D. L. Johnston, *J. Am. Chem. Soc.*, **88,** 2451 (1966).
514. I. Matsubara, *Bull. Chem. Soc. Jpn.*, **35,** 27 (1962).
515. F. A. Cotton and F. Zingales, *J. Am. Chem. Soc.*, **83,** 351 (1961).
516. A. Sacco and F. A. Cotton, *J. Am. Chem. Soc.*, **84,** 2043 (1962).
517. J. W. Dart, M. K. Lloyd, R. Mason, J. A. McCleverty, and J. Williams, *J. Chem. Soc., Dalton Trans.*, 1747 (1973).
518. P. M. Boorman, P. J. Craig, and T. W. Swaddle, *Can. J. Chem.*, **48,** 838 (1970).
519. J. L. Burmeister, *Coord. Chem. Rev.*, **3,** 225 (1968); **1,** 205 (1966).
520. R. A. Bailey, S. L. Kozak, T. W. Michelsen, and W. N. Mills, *Coord. Chem. Rev.*, **6,** 407 (1971).
521. A. H. Norbury, *Adv. Inorg. Chem. Radiochem.*, **17,** 231 (1975).
522. S. Ahrland, J. Chatt, and N. R. Davies, *Quart. Rev.*, **12,** 265 (1958).
523. P. C. H. Mitchell and R. J. P. Williams, *J. Chem. Soc.*, 1912 (1960).
524. A. Turco and C. Pecile, *Nature*, **191,** 66 (1961).
525. J. Lewis, R. S. Nyholm, and P. W. Smith, *J. Chem. Soc.*, 4590 (1961).
526. A. Sabatini and I. Bertini, *Inorg. Chem.*, **4,** 959 (1965).
527. C. Pecile, *Inorg. Chem.*, **5,** 210 (1966).
528. S. Fronaeus and R. Larsson, *Acta Chem. Scand.*, **16,** 1447 (1962).
529. R. A. Bailey, T. W. Michelsen, and W. N. Mills, *J. Inorg. Nucl. Chem.*, **33,** 3206 (1971).
530. R. Larsson and A. Miezis, *Acta Chem. Scand.*, **23,** 37 (1969).
531. R. J. H. Clark and C. S. Williams, *Spectrochim. Acta*, **22,** 1081 (1966).
532. D. Forster and D. M. L. Goodgame, *Inorg. Chem.*, **4,** 715 (1965).
533. M. A. Bennett, R. J. H. Clark, and A. D. J. Goodwin, *Inorg. Chem.*, **6,** 1625 (1967).
534. D. Forster and D. M. L. Goodgame, *Inorg. Chem.*, **4,** 823 (1965).
535. H. H. Schmidtke and D. Garthoff, *Helv. Chim. Acta*, **50,** 1631 (1967).
536. K. H. Schmidt, A. Müller, and M. Chakravorti, *Spectrochim. Acta*, **32A,** 907 (1976).

537. C. Engelter and D. A. Thornton, *J. Mol. Struct.*, **33,** 119 (1976).
538. I. Persson, Å. Iverfeldt, and S. Ahrland, *Acta Chem. Scand.*, **A35,** 295 (1981).
539. M. M. Chamberlain and J. C. Bailar, Jr., *J. Am. Chem. Soc.*, **81,** 6412 (1959).
540. A. B. P. Lever, B. S. Ramaswamy, S. H. Simonsen, and L. K. Thompson, *Can. J. Chem.*, **48,** 3076 (1970).
541. J. J. MacDougall, J. H. Nelson, M. W. Babich, C. C. Fuller, and R. A. Jacobson, *Inorg. Chim. Acta*, **27,** 201 (1978).
542. G. Contreras and R. Schmidt, *J. Inorg. Nucl. Chem.*, **32,** 1295, 127 (1970).
543. F. Basolo, J. L. Burmeister, and A. J. Poe, *J. Am. Chem. Soc.*, **85,** 1700 (1963).
544. J. L. Burmeister and F. Basolo, *Inorg. Chem.*, **3,** 1587 (1964).
545. A. Sabatini and I. Bertini, *Inorg. Chem.*, **4,** 1665 (1965).
546. D. M. L. Goodgame and B. W. Malerbi, *Spectrochim. Acta*, **24A,** 1254 (1968).
547. T. E. Sloan and A. Wojcicki, *Inorg. Chem.*, **7,** 1268 (1968).
548. I. Stotz, W. K. Wilmarth, and A. Haim, *Inorg. Chem.*, **7,** 1250 (1968).
549. R. L. Hassel and J. L. Burmeister, *Chem. Commun.*, 568 (1971).
550. L. A. Epps and L. G. Marzilli, *Inorg. Chem.*, **12,** 1514 (1973).
551. I. Bertini and A. Sabatini, *Inorg. Chem.*, **5,** 1025 (1966).
552. G. R. Clark, G. J. Palenik, and D. W. Meek, *J. Am. Chem. Soc.*, **92,** 1077 (1970).
553. G. P. McQuillan and I. A. Oxton, *J. Chem. Soc., Dalton Trans.*, 1460 (1978).
554. K. K. Chow, W. Levason, and C. A. McAuliffe, *Inorg. Chim. Acta*, **15,** 79 (1975).
555. G. Peters and W. Preetz, *Z. Naturforsch.*, **B35,** 994 (1980); W. Preetz and G. Peters, *Z. Naturforsch.*, **B34,** 1243 (1979); H.-J. Schwerdtfeger and W. Preetz, *Angew. Chem. Int. Ed.*, **16,** 108 (1977).
556. D. W. Meek, P. E. Nicpon, and V. I. Meek, *J. Am. Chem. Soc.*, **92,** 5351 (1970).
557. S. M. Nelson and J. Rodgers, *Inorg. Chem.*, **6,** 1390 (1967).
558. J. L. Burmeister, R. L. Hassel, and R. J. Phelan, *Inorg. Chem.*, **10,** 2032 (1971).
559. J. Chatt and L. A. Duncanson, *Nature*, **178,** 997 (1956).
560. J. Chatt, L. A. Duncanson, F. A. Hart, and P. G. Owston, *Nature*, **181,** 43 (1958).
561. P. G. Owston and J. M. Rowe, *Acta Crystallogr.*, **13,** 253 (1960).
562. J. Chatt and F. A. Hart, *J. Chem. Soc.*, 1416 (1961).
563. B. R. Chamberlain and W. Moser, *J. Chem. Soc. A*, 354 (1969).
564. G. Liptay, K. Burger, E. Papp-Molnár, and Sz. Szebeni, *J. Inorg. Nucl. Chem.*, **31,** 2359 (1969).
565. J. M. Homan, J. M. Kawamoto, and G. L. Morgan, *Inorg. Chem.*, **9,** 2533 (1970).
566. R. A. Bailey and T. W. Michelsen, *J. Inorg. Nucl. Chem.*, **34,** 2671 (1972).
567. F. A. Cotton, A. Davison, W. H. Ilsley, and H. S. Trop, *Inorg. Chem.*, **18,** 2719 (1979).
568. F. A. Cotton, D. M. L. Goodgame, M. Goodgame, and T. E. Hass, *Inorg. Chem.*, **1,** 565 (1962).
569. J. Chatt and L. A. Duncanson, *Nature*, **178,** 997 (1956).
570. M. E. Farago and J. M. James, *Inorg. Chem.*, **4,** 1706 (1965).
571. A. Turco, C. Pecile, and M. Nicolini, *J. Chem. Soc.*, 3008 (1962).
572. J. L. Burmeister and Y. Al-Janabi, *Inorg. Chem.*, **4,** 962 (1965).
573. D. Forster and D. M. L. Goodgame, *Inorg. Chem.*, **4,** 1712 (1965).
574. J. L. Burmeister and H. J. Gysling, *Inorg. Chim. Acta*, **1,** 100 (1967).
575. M. A. Jennings and A. Wojcicki, *Inorg. Chim. Acta*, **3,** 335 (1969).
576. J. L. Burmeister, H. J. Gysling, and J. C. Lim, *J. Am. Chem. Soc.*, **91,** 44 (1969).
577. F. A. Miller and G. L. Carlson, *Spectrochim. Acta*, **17,** 977 (1961).
578. D. Forster and W. D. Horrocks, *Inorg. Chem.*, **6,** 339 (1967).
579. D. Forster and D. M. L. Goodgame, *J. Chem. Soc.*, 262 (1965).
580. A. R. Chugtai and R. N. Keller, *J. Inorg. Nucl. Chem.*, **31,** 633 (1969).
581. D. Forster and D. M. L. Goodgame, *J. Chem. Soc.*, 1286 (1965).
582. E. J. Peterson, A. Galliart, and J. M. Brown, *Inorg. Nucl. Chem. Lett.*, **9,** 241 (1973).
583. R. A. Bailey and S. L. Kozak, *J. Inorg. Nucl. Chem.*, **31,** 689 (1969).
584. H.-D. Amberger, R. D. Fischer, and G. G. Rosenbauer, *Z. Naturforsch.*, **31b,** 1 (1976).
585. A. H. Norbury and A. I. P. Sinha, *J. Chem. Soc. A*, 1598 (1968).
586. S. J. Patel and D. G. Tuck, *J. Chem. Soc. A*, 1870 (1968).

587. S. J. Anderson and A. H. Norbury, *Chem. Commun.*, 37 (1974).
588. J. Nelson and S. M. Nelson, *J. Chem. Soc. A*, 1597 (1969).
589. R. B. Saillant, *J. Organometal. Chem.*, **39,** C71 (1972).
590. W. Beck, *Chem. Ber.*, **95,** 341 (1962).
591. W. Beck, P. Swoboda, K. Feldl, and E. Schuierer, *Chem. Ber.*, **103,** 3591 (1970).
592. W. Beck and E. Schuierer, *Z. Anorg. Allg. Chem.*, **347,** 304 (1966).
593. W. Beck and E. Schuierer, *Chem. Ber.*, **98,** 298 (1965).
594. W. Beck, C. Oetker, and P. Swoboda, *Z. Naturforsch*, **28b,** 229 (1973).
595. W. Beck, W. P. Fehlhammer, P. Pöllmann, E. Schuierer, and K. Feldl, *Chem. Ber.*, **100,** 2335 (1967); *Angew. Chem.*, **77,** 458 (1965).
596. D. Forster and W. D. Horrocks, *Inorg. Chem.*, **5,** 1510 (1966).
597. W.-M. Dyck, K. Dehnicke, F. Weller, and U. Müller, *Z. Anorg. Allg. Chem.*, **470,** 89 (1980).
598. D. Seybold and K. Dehnicke, *Z. Anorg. Allg. Chem.*, **361,** 277 (1968).
599. L. F. Druding, H. C. Wang, R. E. Lohen, and F. D. Sancilio, *J. Coord. Chem.*, **3,** 105 (1973).
600. I. Agrell, *Acta Chem. Scand.*, **25,** 2965 (1971).
601. W. Beck, W. P. Felhammer, P. Pöllman, and R. S. Tobias, *Inorg. Chim. Acta*, **2,** 467 (1968).
602. D. R. Herrington and L. J. Boucher, *Inorg. Nucl. Chem. Lett.*, **7,** 1091 (1971).
603. E. W. Abel and F. G. A. Stone, *Quart. Rev.*, **23,** 325 (1969).
604. L. M. Haines and M. H. B. Stiddard, *Adv. Inorg. Chem. Radiochem.*, **12,** 53 (1969).
605. P. S. Braterman, *Metal Carbonyl Spectra*, Academic Press, New York, 1974.
606. M. Bigorgne, *J. Organometal Chem.*, **94,** 161 (1975).
607. S. F. A. Kettle, *Topics Curr. Chem.*, **71,** 111 (1977).
608. C. P. Horwitz and D. F. Shriver, *Adv. Organometal. Chem.*, **23,** 219 (1984).
609. J. S. Kristoff and D. F. Shriver, *Inorg. Chem.*, **13,** 499 (1974).
610. G. Bouquet and M. Bigorgne, *Spectrochim. Acta*, **27A,** 139 (1971).
611. W. F. Edgell and J. Lyford, IV, *J. Chem. Phys.*, **52,** 4329 (1970).
612. H. Stammreich, K. Kawai, Y. Tavares, P. Krumholz, J. Behmoiras, and S. Bril, *J. Chem. Phys.*, **32,** 1482 (1960).
613. M. Bigorgne, *J. Organometal. Chem.*, **24,** 211 (1970).
614. L. H. Jones, R. S. McDowell, M. Goldblatt, and B. I. Swanson, *J. Chem. Phys.*, **57,** 2050 (1972).
615. L. H. Jones, R. S. McDowell, and M. Goldblatt, *Inorg. Chem.*, **8,** 2349 (1969).
616. E. W. Abel, R. A. N. McLean, S. P. Tyfield, P. S. Braterman, A. P. Walker, and P. J. Hendra, *J. Mol. Spectrosc.*, **30,** 29 (1969).
617. R. A. N. McLean, *Can. J. Chem.*, **52,** 213 (1974).
618. A. Terzis and T. G. Spiro, *Inorg. Chem.*, **10,** 643 (1971).
619. P. J. Hendra and M. M. Qurashi, *J. Chem. Soc. A*, 2963 (1968).
620. M. R. Afiz, R. J. H. Clark, and N. R. D'Urso, *J. Chem. Soc., Dalton Trans.*, 250 (1977).
621. W. Scheuermann and K. Nakamoto, *J. Raman Spectrosc.*, **7,** 341 (1978).
622. F. Calderazzo and F. L'Eplattenier, *Inorg. Chem.*, **6,** 1220 (1967).
623. J. E. Ellis, C. P. Parnell, and G. P. Hagen, *J. Am. Chem. Soc.*, **100,** 3605 (1978).
624. W. F. Edgell, J. Lyford, R. Wright, W. M. Risen, Jr., and A. T. Watts, *J. Am. Chem. Soc.*, **92,** 2240 (1970); W. F. Edgell and J. Lyford, *J. Am. Chem. Soc.*, **93,** 6407 (1971).
625. W. F. Edgell, S. Hegde, and A. Barbetta, *J. Am. Chem. Soc.*, **100,** 1406 (1978).
626. G. G. Summer, H. P. Klug, and L. E. Alexander, *Acta Crystallogr.*, **17,** 732 (1964).
627. F. A. Cotton and R. R. Monchamp, *J. Chem. Soc.*, 1882 (1960).
628. R. L. Sweany and T. L. Brown, *Inorg. Chem.*, **16,** 415 (1977).
629. R. K. Sheline and K. S. Pitzer, *J. Am. Chem. Soc.*, **72,** 1107 (1950).
630. H. M. Powell and R. V. G. Ewens, *J. Chem. Soc.*, 286 (1939).
631. L. F. Dahl and R. E. Rundle, *Acta Crystallogr.*, **16,** 419 (1963).
632. D. M. Adams, M. A. Hooper, and A. Squire, *J. Chem. Soc. A*, 71 (1971).
633. G. Bor, *Chem. Commun.*, 641 (1969).
634. R. A. Levenson, H. B. Gray, and G. P. Ceasar, *J. Am. Chem. Soc.*, **92,** 3653 (1970).
635. I. J. Hyams, D. Jones, and E. R. Lippincott, *J. Chem. Soc. A*, 1987 (1967).
636. N. Flitcroft, D. K. Huggins, and H. D. Kaesz, *Inorg. Chem.*, **3,** 1123 (1964).

637. G. D. Michels and H. J. Svec, *Inorg. Chem.*, **20,** 3445 (1981).
638. L. F. Dahl and J. F. Blount, *Inorg. Chem.*, **4,** 1373 (1965); C. H. Wei and L. F. Dahl, *J. Am. Chem. Soc.*, **91,** 1351 (1969).
639. N. E. Erickson and A. W. Fairhall, *Inorg. Chem.*, **4,** 1320 (1965).
640. F. A. Cotton and D. L. Hunter, *Inorg. Chim. Acta*, **11,** L9 (1974).
641. B. F. G. Johnson, *Chem. Commun.*, 703 (1976).
642. E. R. Corey and L. F. Dahl, *Inorg. Chem.*, **1,** 521 (1962).
643. D. K. Huggins, N. Flitcroft, and H. D. Kaesz, *Inorg. Chem.*, **4,** 166 (1965).
644. C. O. Quicksall and T. G. Spiro, *Inorg. Chem.*, **7,** 2365 (1968).
645. P. Corradini, *J. Chem. Phys.*, **31,** 1676 (1959).
646. G. Bor, G. Sbrignadello, and K. Noack, *Helv. Chim. Acta*, **58,** 815 (1975).
647. P. C. Steinhardt, W. L. Gladfelter, A. D. Harley, J. R. Fox, and G. L. Geoffroy, *Inorg. Chem.*, **19,** 332 (1980).
648. H. Stammreich, K. Kawai, O. Sala, and P. Krumholz, *J. Chem. Phys.*, **35,** 2175 (1961).
649. G. Bor, *Inorg. Chim. Acta*, **3,** 196 (1969).
650. R. J. Ziegler, J. M. Burlitch, S. E. Hayes, and W. M. Risen, Jr., *Inorg. Chem.*, **11,** 702 (1972).
651. F. A. Cotton, L. Kruczynski, and B. A. Frenz, *J. Organometal Chem.*, **160,** 93 (1978).
652. W. D. Jones, M. A. White, and R. G. Bergman, *J. Am. Chem. Soc.*, **100,** 6770 (1978).
653. W. A. Herrmann, M. L. Ziegler, K. Weidenhammer, and H. Biersack, *Angew. Chem. Int. Ed.*, **18,** 960 (1979).
654. E. J. M. De Boer, L. C. Ten Cate, A. G. J. Staring, and J. H. Teuben, *J. Organometal Chem.*, **181,** 61 (1979).
655. G. Bor and P. L. Stanghellini, *Chem. Commun.*, 886 (1979).
656. P. F. Jackson, B. F. G. Johnson, J. Lewis, M. McPartlin, and W. J. H. Nelson, *Chem. Commun.*, 224 (1980).
657. I. A. Oxton, S. G. A. Kettle, P. F. Jackson, B. F. G. Johnson, and J. Lewis, *J. Mol. Struct.*, **71,** 117 (1981).
658. N. J. Nelson, N. E. Kime, and D. F. Shriver, *J. Am. Chem. Soc.*, **91,** 5173 (1969).
659. I. J. Hyams and E. R. Lippincott, *Spectrochim. Acta*, **25A,** 1845 (1969).
660. D. K. Ottesen, H. B. Gray, L. H. Jones, and M. Goldblatt, *Inorg. Chem.*, **12,** 1051 (1973).
661. M. J. Cleare and W. P. Griffith, *J. Chem. Soc. A*, 372 (1969).
662. F. H. Johannsen, W. Preetz, and A. Scheffler, *J. Organometal. Chem.*, **102,** 527 (1975).
663. J. Browning, P. L. Goggin, R. J. Goodfellow, M. G. Norton, A. J. M. Rattray, B. F. Taylor, and J. Mink, *J. Chem. Soc., Dalton Trans.*, 2061 (1977).
664. M. A. El-Sayed and H. D. Kaesz, *Inorg. Chem.*, **2,** 158 (1963).
665. C. W. Garland and J. R. Wilt, *J. Chem. Phys.*, **36,** 1094 (1962).
666. L. F. Dahl, C. Martell, and D. L. Wampler, *J. Am. Chem. Soc.*, **83,** 1761 (1961).
667. B. F. G. Johnson, J. Lewis, P. W. Robinson, and J. R. Miller, *J. Chem. Soc. A*, 2693 (1969).
668. F. A. Cotton and B. F. G. Johnson, *Inorg. Chem.*, **6,** 2113 (1967).
669. A. Loutellier and M. Bigorgne, *J. Chim. Phys.*, **67,** 78, 99, 107 (1970).
670. M. Bigorgne, *J. Organometal. Chem.*, **24,** 211 (1970).
671. J. Dalton, I. Paul, J. G. Smith, and F. G. A. Stone, *J. Chem. Soc. A*, 1195 (1968).
672. R. J. Angelici and M. D. Malone, *Inorg. Chem.*, **6,** 1731 (1967).
673. B. Hutchinson and K. Nakamoto, *Inorg. Chim. Acta*, **3,** 591 (1969).
674. H. Gäbelein and J. Ellermann, *J. Organometal. Chem.*, **156,** 389 (1978).
675. A. M. English, K. R. Plowman, and I. S. Butler, *Inorg. Chem.*, **20,** 2553 (1981).
676. A. A. Chalmers, J. Lewis, and R. Whyman, *J. Chem. Soc. A*, 1817 (1967).
677. M. F. Farona, J. G. Grasselli, and B. L. Ross, *Spectrochim. Acta*, **23A,** 1875 (1967).
678. S. Singh, P. P. Singh, and R. Rivest, *Inorg. Chem.*, **7,** 1236 (1968).
679. F. A. Cotton and C. S. Kraihanzel, *J. Am. Chem. Soc.*, **84,** 4432 (1962); *Inorg. Chem.*, **2,** 533 (1963); **3,** 702 (1964).
680. F. A. Cotton, M. Musco, and G. Yagupsky, *Inorg. Chem.*, **6,** 1357 (1967).
681. L. H. Jones, *Inorg. Chem.*, **7,** 1681 (1968); **6,** 1269 (1967).
682. F. A. Cotton, *Inorg. Chem.*, **7,** 1683 (1968).

683. M. B. Hall and R. F. Fenske, *Inorg. Chem.*, **11**, 1619 (1972).
684. A. C. Sarapu and R. F. Fenske, *Inorg. Chem.*, **14**, 247 (1975).
685. H. D. Kaesz and R. B. Saillant, *Chem. Rev.*, **72**, 231 (1972).
686. W. F. Edgell, C. Magee, and G. Gallup, *J. Am. Chem. Soc.*, **78**, 4185, 4188 (1956); W. F. Edgell and R. Summitt, *J. Am. Chem. Soc.*, **83**, 1772 (1961).
687. S. J. LaPlaca, W. C. Hamilton, and J. A. Ibers, *Inorg. Chem.*, **3**, 1491 (1964); *J. Am. Chem. Soc.*, **86**, 2288 (1964).
688. D. K. Huggins and H. D. Kaesz, *J. Am. Chem. Soc.*, **86**, 2734 (1964).
689. P. S. Braterman, R. W. Harrill, and H. D. Kaesz, *J. Am. Chem. Soc.*, **89**, 2851 (1967).
690. A. Davison and J. W. Faller, *Inorg. Chem.*, **6**, 845 (1967).
691. W. F. Edgell, J. W. Fisher, G. Asato, and W. M. Risen, Jr., *Inorg. Chem.*, **8**, 1103 (1969).
692. K. Farmery and M. Kilner, *J. Chem. Soc. A*, 634 (1970).
693. S. S. Bath and L. Vaska, *J. Am. Chem. Soc.*, **85**, 3500 (1963).
694. J. Chatt, N. P. Johnson, and B. L. Shaw, *J. Chem. Soc.*, 1625 (1964).
695. L. Vaska, *J. Am. Chem. Soc.*, **88**, 4100 (1966).
696. F. L'Eplattenier and F. Calderazzo, *Inorg. Chem.*, **7**, 1290 (1968).
697. D. K. Huggins, W. Fellman, J. M. Smith, and H. D. Kaesz, *J. Am. Chem. Soc.*, **86**, 4841 (1964).
698. J. M. Smith, W. Fellmann, and L. H. Jones, *Inorg. Chem.*, **4**, 1361 (1965).
699. R. G. Hayter, *J. Am. Chem. Soc.*, **88**, 4376 (1966).
700. R. Bau and T. F. Koetzle, *Pure Appl. Chem.*, **50**, 55 (1978).
701. C. B. Cooper, III, D. F. Shriver, D. J. Darensbourg, and J. A. Froelich, *Inorg. Chem.*, **18**, 1407 (1979); C. B. Cooper, III, D. F. Shriver, and S. Onaka, *Adv. Chem. Ser.*, **167**, 232 (1978).
702. A. P. Ginsberg and M. J. Hawkes, *J. Am. Chem. Soc.*, **90**, 5931 (1968).
703. M. J. Mays and R. N. F. Simpson, *J. Chem. Soc. A*, 1444 (1968); *Chem. Commun.*, 1024 (1967).
704. H. D. Kaesz, F. Fontal, R. Bau, S. W. Kirtley, and M. R. Churchill, *J. Am. Chem. Soc.*, **91**, 1021 (1969).
705. M. J. Bennett, W. A. G. Graham, J. K. Hoyano, and W. L. Hutcheon, *J. Am. Chem. Soc.*, **94**, 6232 (1972).
706. S. A. R. Knox, J. W. Koepke, M. A. Andrews, and H. D. Kaesz, *J. Am. Chem. Soc.*, **97**, 3942 (1975).
707. S. A. R. Knox and H. D. Kaesz, *J. Am. Chem. Soc.*, **93**, 4594 (1971).
708. C. R. Eady, B. F. G. Johnson, J. Lewis, M. C. Malatesta, P. Machin, and M. McPartlin, *Chem. Commun.*, 945 (1976).
709. I. A. Oxton, S. F. A. Kettle, P. F. Jackson, B. F. G. Johnson, and J. Lewis, *Chem. Commun.*, 687 (1979).
710. J. W. White and C. J. Wright, *Chem. Commun.*, 971 (1970).
711. R. L. DeKock, *Inorg. Chem.*, **10**, 1205 (1971).
712. M. Moskovits and G. A. Ozin, "Characterization of the Products of Metal Atom-Molecule Condensation Reactions by Matrix Infrared and Raman Spectroscopy," in J. R. Durig, ed., *Vibrational Spectra and Structure*, Vol. 4, Elsevier, Amsterdam, 1975.
713. J. H. Darling and J. S. Ogden, *Inorg. Chem.*, **11**, 666 (1972); *J. Chem. Soc., Dalton Trans.*, 1079 (1973).
714. H. Huber, E. P. Kündig, M. Moskovits, and G. A. Ozin, *Nature, Phys. Sci.*, **235**, 98 (1972); E. P. Kündig, M. Moskovits, and G. A. Ozin, *Can. J. Chem.*, **50**, 3587 (1972).
715. E. P. Kündig, D. McIntosh, M. Moskovits, and G. A. Ozin, *J. Am. Chem. Soc.*, **95**, 7234 (1973).
716. J. L. Slater, R. K. Sheline, K. C. Lin, and W. Weltner, *J. Chem. Phys.*, **55**, 5129 (1971).
717. J. L. Slater, T. C. DeVore, and V. Calder, *Inorg. Chem.*, **12**, 1918 (1973); **13**, 1808 (1974).
718. D. McIntosh and G. A. Ozin, *Inorg. Chem.*, **16**, 51 (1977).
719. H. Huber, D. McIntosh, and G. A. Ozin, *Inorg. Chem.*, **16**, 975 (1977).
720. R. K. Sheline and J. L. Slater, *Angew. Chem. Int. Ed.*, **14**, 309 (1975).
721. M. Poliakoff and J. J. Turner, *J. Chem. Soc., Dalton Trans.*, 1351 (1973); 2276 (1974).
722. M. A. Graham, M. Poliakoff, and J. J. Turner, *J. Chem. Soc. A*, 2939 (1971).
723. E. P. Kündig and G. A. Ozin, *J. Am. Chem. Soc.*, **96**, 3820 (1974).
724. J. D. Black and P. S. Braterman, *J. Am. Chem. Soc.*, **97**, 2908 (1975).

725. R. N. Perutz and J. J. Turner, *Inorg. Chem.*, **14,** 262 (1975).
726. A. McNeish, M. Poliakoff, K. P. Smith, and J. J. Turner, *Chem. Commun.*, 859 (1976).
727. P. A. Breeze, J. K. Burdett, and J. J. Turner, *Inorg. Chem.*, **20,** 3369 (1981).
728. J. J. Turner, *Angew. Chem. Int. Ed.*, **14,** 304 (1975).
729. D. A. Van Leirsburg and C. W. DeKock, *J. Phys. Chem.*, **78,** 134 (1974).
730. D. Tevault and K. Nakamoto, *Inorg. Chem.*, **15,** 1282 (1976).
731. B. I. Swanson, L. H. Jones, and R. R. Ryan, *J. Mol. Spectrosc.*, **45,** 324 (1973).
732. M. Poliakoff and J. J. Turner, *J. Chem. Soc. A*, 654 (1971).
733. D. Tevault and K. Nakamoto, *Inorg. Chem.*, **14,** 2371 (1975); A. Cormier, J. D. Brown, and K. Nakamoto, *Inorg. Chem.*, **12,** 3011 (1973).
734. P. Gans, A. Sabatini, and L. Sacconi, *Coord. Chem. Rev.*, **1,** 187 (1966).
735. J. Masek, *Inorg. Chim. Acta Rev.*, **3,** 99 (1969).
736. B. F. G. Johnson and J. A. McCleverty, *Progr. Inorg. Chem.*, **7,** 277 (1966).
737. W. P. Griffith, *Adv. Organometal. Chem.*, **7,** 211 (1968).
738. J. A. McGinnety, *MTP, Int. Rev. Sci. Inorg. Chem.*, **5,** 229 (1972).
739. J. H. Enemark and R. D. Feltham, *Coord. Chem. Rev.*, **13,** 339 (1974).
740. B. L. Haymore and J. A. Ibers, *Inorg. Chem.*, **14,** 3060 (1975).
741. C. G. Pierpont, D. G. Van Derveer, W. Durland, and R. Eisenberg, *J. Am. Chem. Soc.*, **92,** 4760 (1970).
742. C. P. Brock, J. P. Collman, G. Dolcetti, P. H. Farnham, J. A. Ibers, J. E. Lester, and C. A. Reed, *Inorg. Chem.*, **12,** 1304 (1973).
743. M. Herberhold and A. Razavi, *Angew, Chem. Int. Ed.*, **11,** 1092 (1972).
744. I. H. Sabberwal and A. B. Burg, *Chem. Commun.*, 1001 (1970).
745. L. H. Jones, R. S. McDowell, and B. I. Swanson, *J. Chem. Phys.*, **58,** 3757 (1973).
746. G. Barna and I. S. Butler, *Can. J. Spectrosc.*, **17,** 2 (1972).
747. O. Crichton and A. J. Rest, *Inorg. Nucl. Chem. Lett.*, **9,** 391 (1973).
748. B. F. G. Johnson, *J. Chem. Soc. A*, 475 (1967).
749. Z. Iqbal and T. C. Waddington, *J. Chem. Soc. A*, 1092 (1969).
750. E. Miki, T. Ishimori, H. Yamatera, and H. Okuno, *J. Chem. Soc. Jpn.*, **87,** 703 (1966).
751. J. R. Durig, W. A. McAllister, J. N. Willis, Jr., and E. E. Mercer, *Spectrochim. Acta*, **22,** 1091 (1966).
752. M. Quinby-Hunt and R. D. Feltham, *Inorg. Chem.*, **17,** 2515 (1978).
753. J. Müller and S. Schmitt, *J. Organometal. Chem.*, **160,** 109 (1978).
754. R. G. Ball, B. W. Hames, P. Legzdins, and J. Trotter, *Inorg. Chem.*, **19,** 3626 (1980).
755. J. R. Norton, J. P. Collman, G. Dolcetti, and W. T. Robinson, *Inorg. Chem.*, **11,** 382 (1972).
756. B. W. Fitzsimmons, L. F. Larkworthy, and K. A. Rogers, *Inorg. Chim. Acta*, **44,** L53 (1980).
757. S. K. Satija, B. I. Swanson, O. Crichton, and A. J. Rest, *Inorg. Chem.*, **17,** 1737 (1978).
758. O. Crichton and A. J. Rest, *J. Chem. Soc., Dalton Trans.*, 202 and 208 (1978).
759. V. J. Choy and C. H. O'Connor, *Coord. Chem. Rev.*, **9,** 145 (1972/73).
760. F. Basolo, B. M. Hoffman, and J. A. Ibers, *Acc. Chem. Res.*, **8,** 384 (1975).
761. L. Vaska, *Acc. Chem. Res.*, **9,** 175 (1976).
762. J. P. Collman, *Acc. Chem. Res.*, **10,** 265 (1977).
763. G. McLendon and A. E. Martell, *Coord. Chem. Rev.*, **19,** 1 (1976).
764. R. W. Erskine and B. O. Field, *Structure and Bonding*, **28,** 1 (1976).
765. R. D. Jones, D. A. Summerville, and F. Basolo, *Chem. Rev.*, **79,** 139 (1979).
766. T. D. Smith and J. R. Pilbrow, *Coord. Chem. Rev.*, **39,** 295 (1981).
766a. M. H. Gubelmann and A. F. Williams, *Structure and Bonding*, **55,** 1 (1983).
766b. E. C. Niederhoffer, J. H. Timmons and A. E. Martell, *Chem. Rev.*, **84,** 137 (1984).
767. L. Andrews, "Infrared and Raman Spectroscopic Studies of Alkali-Metal-Atom Matrix-Reaction Products," in M. Moskovits and G. A. Ozin, eds, *Cryochemistry*, Wiley-Interscience, New York, 1976, p. 211.
768. D. McIntosh and G. A. Ozin, *Inorg. Chem.*, **16,** 59 (1977).
769. A. J. L. Hanlan and G. A. Ozin, *Inorg. Chem.*, **16,** 2848 (1977).
770. M. J. Zehe, D. A. Lynch, Jr., B. J. Kelsall, and K. D. Carlson, *J. Phys. Chem.*, **83,** 656 (1979).

771. D. McIntosh and G. A. Ozin, *Inorg. Chem.*, **15**, 2869 (1976).
772. B. J. Kelsall and K. D. Carlson, *J. Phys. Chem.*, **84**, 951 (1980).
773. H. Huber, W. Klotzbücher, G. A. Ozin, and A. Vander Voet, *Can. J. Chem.*, **51**, 2722 (1973).
774. S. Chang, G. Blyholder, and J. Fernandez, *Inorg. Chem.*, **20**, 2813 (1981).
775. A. B. P. Lever, G. A. Ozin, and H. B. Gray, *Inorg. Chem.*, **19**, 1823 (1980).
776. J. C. Evans, *Chem. Commun.*, 682 (1969).
777. H. H. Eysel and S. Thym, *Z. Anorg. Allg. Chem.*, **411**, 97 (1975).
778. K. Nakamoto, Y. Nonaka, T. Ishiguro, M. W. Urban, M. Suzuki, M. Kozuka, Y. Nishida, and S. Kida, *J. Am. Chem. Soc.*, **104**, 3386 (1982).
779. K. Bajdor, K. Nakamoto, H. Kanatomi, and I. Murase, *Inorg. Chim. Acta*, **82**, 207 (1984).
780. T. Shibahara and M. Mori, *Bull. Chem. Soc. Jpn.*, **51**, 1374 (1978).
781. C. G. Barraclough, G. A. Lawrence, and P. A. Lay, *Inorg. Chem.*, **17**, 3317 (1978).
782. R. E. Hester and E. M. Nour, *J. Raman Spectrosc.*, **11**, 43 (1981).
783. M. Suzuki, T. Ishiguro, M. Kozuka, and K. Nakamoto, *Inorg. Chem.*, **20**, 1993 (1981).
784. E. M. Nour and R. E. Hester, *J. Mol. Struct.*, **62**, 77 (1980).
785. K. Nakamoto, M. Suzuki, T. Ishiguro, M. Kozuka, Y. Nishida, and S. Kida, *Inorg. Chem.*, **19**, 2822 (1980).
786. R. E. Hester and E. M. Nour, *J. Raman Spectrosc.*, **11**, 59 (1981).
787. T. Tsumaki, *Bull. Chem. Soc. Jpn.*, **13**, 252 (1938).
788. M. Kozuka and K. Nakamoto, *J. Am. Chem. Soc.*, **103**, 2162 (1981).
789. M. W. Urban, K. Nakamoto, and J. Kincaid, *Inorg. Chim. Acta*, **61**, 77 (1982).
790. K. Nakamoto, T. Watanabe, T. Ama, and M. W. Urban, *J. Am. Chem. Soc.*, **104**, 3744 (1982).
791. T. Watanabe, T. Ama, and K. Nakamoto, *J. Phys. Chem.*, **88**, 440 (1984).
792. M. W. Urban, K. Nakamoto, and F. Basolo, *Inorg. Chem.*, **21**, 3406 (1982).
793. T. Watanabe, T. Ama, and K. Nakamoto, *Inorg. Chem.*, **22**, 2470 (1983).
794. K. Bajdor and K. Nakamoto, *J. Am. Chem. Soc.*, **106**, 3045 (1984).
794a. L. M. Proniewicz, K. Bajdor, and K. Nakamoto, *J. Phys. Chem.*, **90**, 1760 (1986).
794b. J. Terner, A. J. Sitter, and C. M. Reczek, *Biochem. Biophys. Acta*, **828**, 73 (1985).
794c. S. Hashimoto, T. Tatsuno, and T. Kitagawa, *Proc. Japan. Acad.*, **60B**, 345 (1984).
795. L. Vaska, *Science*, **140**, 809 (1963).
796. S. J. La Placa and J. A. Ibers, *J. Am. Chem. Soc.*, **87**, 2581 (1965).
797. P. B. Chock and J. Halpern, *J. Am. Chem. Soc.*, **88**, 3511 (1966).
798. A. Nakamura, Y. Tatsuno, M. Yamamoto, and S. Otsuka, *J. Am. Chem. Soc.*, **93**, 6052 (1971).
799. F. Offner and J. Dehand, *Compt. Rend.*, **273**, C50 (1971).
800. C. A. Reed and W. R. Roper, *J. Chem. Soc., Dalton Trans.*, 1370 (1973).
801. A. D. Allen and C. V. Senoff, *Chem. Commun.*, 621 (1965).
802. A. D. Allen and F. Bottomley, *Acc. Chem. Res.*, **1**, 360 (1968).
803. P. C. Ford, *Coord. Chem. Rev.*, **5**, 75 (1970).
804. R. Murray and D. C. Smith, *Coord. Chem. Rev.*, **3**, 429 (1968).
805. G. Henrici-Olive and S. Olive, *Angew. Chem. Int. Ed.*, **8**, 650 (1969).
806. A. D. Allen, F. Bottomley, R. O. Harris, V. P. Reinsalu, and C. V. Senoff, *J. Am. Chem. Soc.*, **89**, 5595 (1967).
807. S. Pell, R. H. Mann, H. Taube, and J. N. Armor, *Inorg. Chem.*, **13**, 479 (1974).
808. M. W. Bee, S. F. A. Kettle, and D. B. Powell, *Spectrochim. Acta*, **30A**, 585 (1974).
809. A. D. Allen and J. R. Stevens, *Chem. Commun.*, 1147 (1967).
810. G. Speier and L. Markó, *Inorg. Chim. Acta*, **3**, 126 (1969).
811. J. H. Enemark, B. R. Davis, J. A. McGinnety, and J. A. Ibers, *Chem. Commun.*, 96 (1968).
812. J. P. Collman, M. Kubota, F. D. Vastine, J. Y. Sun, and J. W. Kang, *J. Am. Chem. Soc.*, **90**, 5430 (1968).
813. M. Hidai, K. Tominari, Y. Uchida, and A. Misono, *Chem. Commun.*, 1392 (1969).
814. B. Bell, J. Chatt, and G. J. Leigh, *Chem. Commun.*, 842 (1970).
815. G. Speier and L. Markó, *J. Organometal. Chem.*, **21**, P46 (1970).
816. S. C. Srivastava and M. Bigorgne, *J. Organometal. Chem.*, **19**, 241 (1969).
817. D. J. Darensbourg, *Inorg. Chem.*, **11**, 1436 (1972).

818. J. N. Armor and H. Taube, *J. Am. Chem. Soc.*, **92,** 2560 (1970).
819. C. Krüger and Y.-H. Tsay, *Angew. Chem. Int. Ed.*, **12,** 998 (1973).
820. J. Chatt, A. B. Nikolsky, R. L. Richards, and J. R. Sanders, *Chem. Commun.*, 154 (1969).
821. J. Chatt, R. C. Fay, and R. L. Richards, *J. Chem. Soc. A*, 702 (1971).
822. M. Mercer, R. H. Crabtree, and R. L. Richards, *Chem. Commun.*, 808 (1973).
823. D. Sellman, A. Brandl, and R. Endell, *J. Organometal. Chem.*, **49,** C22 (1973).
824. H. Huber, E. P. Kündig, M. Moskovits, and G. A. Ozin, *J. Am. Chem. Soc.*, **95,** 332 (1973).
825. G. A. Ozin and A. Vander Voet, *Can. J. Chem.*, **51,** 637 (1973).
826. D. W. Green, J. Thomas, and D. M. Gruen, *J. Chem. Phys.*, **58,** 5453 (1973).
827. H. Huber, T. A. Ford, W. Klotzbücher, and G. A. Ozin, *J. Am. Chem. Soc.*, **98,** 3176 (1976).
828. D. W. Green, R. V. Hodges, and D. M. Gruen, *Inorg. Chem.*, **15,** 970 (1976).
829. T. C. DeVore, *Inorg. Chem.*, **15,** 1315 (1976).
830. D. W. Green and G. T. Reedy, *J. Mol. Spectrosc.*, **74,** 423 (1979).
831. W. P. Griffith, *Coord. Chem. Rev.*, **8,** 369 (1972).
832. M. J. Cleare and W. P. Griffith, *J. Chem. Soc. A*, 1117 (1970).
833. W. Kolitsch and K. Dehnicke, *Z. Natursforsch.*, **25b,** 1080 (1970).
834. W. P. Griffith and D. Pawson, *J. Chem. Soc., Dalton Trans.*, 1315 (1973).
835. A. Misono, Y. Uchida, T. Saito, and K. M. Song, *Chem. Commun.*, 419 (1967).
836. A. Sacco and M. Rossi, *Chem. Commun.*, 316 (1967).
837. R. G. S. Banks and J. M. Pratt, *J. Chem. Soc. A*, 854 (1968).
838. K. Krogmann and W. Binder, *Angew. Chem. Int. Ed.*, **6,** 881 (1967).
839. J. Chatt and R. G. Hayter, *J. Chem. Soc.*, 5507 (1961).
840. L. Vaska and J. W. DiLizio, *J. Am. Chem. Soc.*, **84,** 4989 (1962).
841. L. Vaska, *Chem. Commun.*, 614 (1966).
842. L. Vaska and D. L. Catone, *J. Am. Chem. Soc.*, **88,** 5324 (1966).
843. J. Chatt and R. G. Hayter, *J. Chem. Soc.*, 2605 (1961).
844. J. Chatt, L. A. Duncanson, and B. L. Shaw, *Chem. Ind.* (*London*), 859 (1958).
845. M. J. Church and M. J. Mays, *J. Chem. Soc.*, 3074 (1968); 1938 (1970).
846. P. W. Atkins, J. C. Green, and M. L. H. Green, *J. Chem. Soc. A*, 2275 (1968).
847. R. J. H. Clark, in V. Gutmann, ed., *Halogen Chemistry*, Vol. 3, Academic Press, New York, 1967, p. 85.
848. R. H. Nuttall, *Talanta*, **15,** 157 (1968).
849. A. J. Carty, *Coord. Chem. Rev.*, **4,** 29 (1969).
850. R. J. H. Clark, *Spectrochim. Acta*, **21,** 955 (1965).
851. K. Konya and K. Nakamoto, *Spectrochim. Acta*, **29A,** 1965 (1973).
852. K. Thompson and K. Carlson, *J. Chem. Phys.*, **49,** 4379 (1968).
853. K. Shobatake and K. Nakamoto, *J. Am. Chem. Soc.*, **92,** 3332 (1970).
854. C. Udovich, J. Takemoto, and K. Nakamoto, *J. Coord. Chem.*, **1,** 89 (1971).
855. P. M. Boorman and A. J. Carty, *Inorg. Nucl. Chem. Lett.*, **4,** 101 (1968).
856. I. Wharf and D. F. Shriver, *Inorg. Chem.*, **8,** 914 (1969).
857. D. F. Shriver and M. P. Johnson, *Inorg. Chem.*, **6,** 1265 (1967).
858. B. Crociani, T. Boschi, and M. Nicolini, *Inorg. Chim. Acta*, **4,** 577 (1970).
859. F. H. Herbelin, J. D. Herbelin, J. P. Mathieu, and H. Poulet, *Spectrochim. Acta*, **22,** 1515 (1966).
860. D. M. Adams, J. Chatt, J. Gerratt, and A. D. Westland, *J. Chem. Soc.*, 734 (1964).
861. T. G. Appleton, H. C. Clark, and L. E. Manzer, *Coord. Chem. Rev.*, **10,** 335 (1973).
862. R. J. Goodfellow, P. L. Goggin and D. A. Duddell, *J. Chem. Soc. A*, 504 (1968).
863. M. A. Bennett, R. J. H. Clark, and D. L. Milner, *Inorg. Chem.*, **6,** 1647 (1967).
864. J. Fujita, K. Konya, and K. Nakamoto, *Inorg. Chem.*, **9,** 2794 (1970).
865. J. T. Wang, C. Udovich, K. Nakamoto, A. Quattrochi, and J. R. Ferraro, *Inorg. Chem.*, **9,** 2675 (1970).
866. B. T. Kilbourn and H. M. Powell, *J. Chem. Soc. A*, 1688 (1970).
867. J. R. Ferraro, K. Nakamoto, J. T. Wang, and L. Lauer, *Chem. Commun.*, 266 (1973).
868. C. Postmus, K. Nakamoto, and J. R. Ferraro, *Inorg. Chem.*, **6,** 2194 (1967).
869. J. R. Ferraro and G. J. Long, *Acc. Chem. Res.*, **8,** 171 (1975).

870. J. R. Ferraro, *Coord. Chem. Rev.*, **29,** 1 (1979).
871. D. M. Adams and P. J. Chandler, *J. Chem. Soc. A*, 1009 (1967).
872. N. Ohkaku and K. Nakamoto, *Inorg. Chem.*, **12,** 2440, 2446 (1973).
873. D. M. Adams and D. C. Newton, *J. Chem. Soc., Dalton Trans.*, 681 (1972).
874. R. J. Goodfellow, P. L. Goggin, and L. M. Venanzi, *J. Chem. Soc. A*, 1897 (1967).
875. D. M. Adams and P. J. Chandler, *Chem. Commun.*, 69 (1966).
876. M. Goldstein and W. D. Unsworth, *Inorg. Chim. Acta*, **4,** 342 (1970).
877. F. A. Cotton, R. M. Wing, and R. A. Zimmerman, *Inorg. Chem.*, **6,** 11 (1967).
878. R. D. Hogue and R. E. McCarley, *Inorg. Chem.*, **9,** 1354 (1970).
879. P. M. Boorman and B. P. Straughan, *J. Chem. Soc.*, 1514 (1966).
880. R. A. Mackay and R. F. Schneider, *Inorg. Chem.*, **7,** 455 (1968).
881. P. B. Fleming, J. L. Meyer, W. K. Grindstaff, and R. E. McCarley, *Inorg. Chem.*, **9,** 1769 (1970).
882. T. G. Spiro, *Progr. Inorg. Chem.*, **11,** 1 (1970).
883. K. L. Watters and W. M. Risen, Jr., *Inorg. Chim. Acta Rev.*, **3,** 129 (1969).
884. E. Maslowsky, Jr., *Chem. Rev.*, **71,** 507 (1971).
885. B. J. Bulkin and C. A. Rundell, *Coord. Chem. Rev.*, **2,** 371 (1967).
886. D. F. Shriver and C. B. Cooper, III, *Adv. Infrared Raman Spectrosc.*, **6,** 127 (1980).
887. C. O. Quicksall and T. G. Spiro, *Inorg. Chem.*, **8,** 2363 (1969).
888. J. R. Johnson, R. J. Ziegler, and W. M. Risen, Jr., *Inorg. Chem.*, **12,** 2349 (1973).
889. K. L. Watters, J. N. Britain, and W. M. Risen, Jr., *Inorg. Chem.*, **8,** 1347 (1969).
890. K. L. Watters, W. M. Butler, and W. M. Risen, Jr., *Inorg. Chem.*, **10,** 1970 (1971).
891. K. M. Mackay and S. R. Stobart, *J. Chem. Soc., Dalton Trans.*, 214 (1973).
892. S. Onaka, C. B. Cooper, III, and D. F. Shriver, *Inorg. Chim. Acta*, **37,** L467 (1979).
893. P. L. Goggin and R. J. Goodfellow, *J. Chem. Soc., Dalton Trans.*, 2355 (1973).
894. J. San Filippo, Jr., and H. J. Sniadoch, *Inorg. Chem.*, **12,** 2326 (1973).
895. B. I. Swanson, J. J. Rafalko, D. F. Shriver, J. San Filippo, Jr., and T. G. Spiro, *Inorg. Chem.*, **14,** 1737 (1975).
896. C. B. Cooper, III, S. Onaka, D. F. Shriver, L. Daniels, R. L. Hance, B. Hutchinson and R. Shipley, *Inorg. Chim. Acta*, **24,** L92 (1977).
897. S. Onaka and D. F. Shriver, *Inorg. Chem.*, **15,** 915 (1976).
898. S. F. A. Kettle and P. L. Stanghellini, *Inorg. Chem.*, **18,** 2749 (1979).
899. G. A. Battiston, G. Bor, U. K. Dietler, S. F. A. Kettle, R. Rossetti, G. Sbrignadello, and P. L. Stanghellini, *Inorg. Chem.*, **19,** 1961 (1980).
900. C. O. Quicksall and T. G. Spiro, *Inorg. Chem.*, **8,** 2011 (1969).
901. C. O. Quicksall and T. G. Spiro, *Chem. Commun.*, 839 (1967).
902. J. A. Creighton and B. T. Heaton, *J. Chem. Soc., Dalton Trans.*, 1498 (1981).
903. C. Sourisseau, *J. Raman Spectrosc.*, **6,** 303 (1977).
904. R. J. Ziegler and W. M. Risen, Jr., *Inorg. Chem.*, **11,** 2796 (1972).
905. R. Mattes, *Z. Anorg. Allg. Chem.*, **357,** 30 (1968).
906. R. Mattes, *Z. Anorg. Allg. Chem.*, **364,** 279 (1969).
907. F. J. Farrell, V. A. Maroni, and T. G. Spiro, *Inorg. Chem.*, **8,** 2638 (1969).
908. R. Mattes, H. Bierbüsse, and J. Fuchs, *Z. Anorg. Allg. Chem.*, **385,** 230 (1971).
909. T. G. Spiro, D. H. Templeton, and A. Zalkin, *Inorg. Chem.*, **8,** 856 (1969).
910. T. G. Spiro, V. A. Maroni, and C. O. Quicksall, *Inorg. Chem.*, **8,** 2524 (1969).
911. F. A. Cotton, *Chem. Soc. Rev.*, 27 (1975); *Acc. Chem. Res.*, **11,** 225 (1978).
912. R. J. H. Clark and M. L. Franks, *J. Am. Chem. Soc.*, **98,** 2763 (1976).
913. F. A. Cotton, B. A. Frenz, B. R. Stults, and T. R. Webb, *J. Am. Chem. Soc.*, **98,** 2768 (1976).
914. J. San Filippo, Jr. and H. J. Sniadoch, *Inorg. Chem.*, **12,** 2326 (1973).
915. C. L. Angell, F. A. Cotton, B. A. Frenz, and T. R. Webb, *Chem. Commun.*, 399 (1973).
916. A. Loewenschuss, J. Shamir, and M. Ardon, *Inorg. Chem.*, **15,** 238 (1976).
917. W. K. Bratton, F. A. Cotton, M. Debeau, and R. A. Walton, *J. Coord. Chem.*, **1,** 121 (1971).
918. A. P. Ketteringham, C. Oldham, and C. J. Peacock, *J. Chem. Soc., Dalton Trans.*, 1640 (1976).
919. B. Hutchinson, J. Morgan, C. B. Cooper, III, Y. Mathey, and D. F. Shriver, *Inorg. Chem.*, **18,** 2048 (1979).

920. R. J. H. Clark and M. L. Franks, *J. Am. Chem. Soc.*, **97,** 2691 (1975).
921. W. C. Trogler and H. B. Gray, *Acc. Chem. Res.*, **11,** 232 (1978).
921a. K. R. Mann and H. B. Gray, *Adv. Chem. Ser.*, **173,** 225 (1979).
921b. R. F. Dallinger, V. M. Miskowski, H. B. Gray, and W. H. Woodruff, *J. Am. Chem. Soc.*, **103,** 1595 (1981).
922. J. G. Verkade, *Coord. Chem. Rev.*, **9,** 1 (1972).
923. M. Trabelsi, A. Loutellier, and M. Bigorgne, *J. Organometal. Chem.*, **40,** C45 (1972).
924. M. Bigorgne, A. Loutellier, and M. Pańkowski, *J. Organometal. Chem.*, **23,** 201 (1970).
925. M. Trabelsi, A. Loutellier, and M. Bigorgne, *J. Organometal. Chem.*, **56,** 369 (1973).
926. M. Trabelsi and A. Loutellier, *J. Mol. Struct.*, **43,** 151 (1978).
927. A. Loutellier, M. Trabelsi, and M. Bigorgne, *J. Organometal. Chem.*, **133,** 201 (1977).
928. Th. Kruck, *Angew. Chem. Int. Ed.*, **6,** 53 (1967).
929. Th. Kruck and K. Bauer, *Z. Anorg. Allg. Chem.*, **364,** 192 (1969).
930. Th. Kruck and A. Prasch, *Z. Anorg. Allg. Chem.*, **356,** 118 (1968).
931. Th. Kruck and A. Prasch, *Z. Anorg. Allg. Chem.*, **371,** 1 (1969).
932. S. Bénazeth, A. Loutellier, and M. Bigorgne, *J. Organometal. Chem.*, **24,** 479 (1970).
933. L. A. Woodward and J. R. Hall, *Spectrochim. Acta*, **16,** 654 (1960); H. G. M. Edwards and L. A. Woodward, *Spectrochim. Acta*, **26A,** 897 (1970).
934. P. L. Goggin and R. J. Goodfellow, *J. Chem. Soc. A*, 1462 (1966).
935. G. D. Coates and C. Parkin, *J. Chem. Soc.*, 421 (1963).
936. M. A. Bennett, R. J. H. Clark, and A. D. J. Goodwin, *Inorg. Chem.*, **6,** 1625 (1967).
937. J. H. S. Green, *Spectrochim. Acta*, **24A,** 137 (1968).
938. K. Shobatake, C. Postmus, J. R. Ferraro, and K. Nakamoto, *Appl. Spectrosc.*, **23,** 12 (1969).
939. J. Bradbury, K. P. Forest, R. H. Nuttall, and D. W. A. Sharp, *Spectrochim. Acta*, **23A,** 2701 (1967).
940. P. J. D. Park and P. J. Hendra, *Spectrochim. Acta*, **25A,** 227, 909 (1969).
941. L. M. Venanzi, *Chem. Brit.* **4,** 162 (1968).
942. J. Chatt, G. A. Gamlen, and L. E. Orgel, *J. Chem. Soc.*, 486 (1958).
943. E. O. Fischer, W. Bathelt, and J. Müller, *Chem. Ber.*, **103,** 1815 (1970).
944. R. J. Goodfellow, J. G. Evans, P. L. Goggin, and D. A. Duddell, *J. Chem. Soc. A*, 1604 (1968).
945. G. B. Deacon and J. H. S. Green, *Spectrochim. Acta*, **24A,** 845 (1968).
946. D. Brown, J. Hill, and C. E. F. Richard, *J. Chem. Soc. A*, 497 (1970).
947. D. M. L. Goodgame and F. A. Cotton, *J. Chem. Soc.*, 2298, 3735 (1961).
948. G. A. Rodley, D. M. L. Goodgame, and F. A. Cotton, *J. Chem. Soc.*, 1499 (1965).
949. A. P. Ginsberg and W. E. Lindsell, *Chem. Commun.*, 232 (1971).
950. A. E. Wickenden and R. A. Krause, *Inorg. Chem.*, **8,** 779 (1969).
951. A. Müller, W.-O. Nolte, and B. Krebs, *Angew. Chem. Int. Ed.*, **17,** 279 (1978).
952. A. Müller, S. Sarkar, R. G. Bhattacharyya, S. Pohl, and M. Dartmann, *Angew. Chem. Int. Ed.*, **17,** 535 (1978).
953. W. Rittner, A. Müller, A. Neumann, W. Bäther, and R. C. Sharma, *Angew. Chem. Int. Ed.*, **18,** 530 (1979).
954. G. J. Kubas and P. J. Vergamini, *Inorg. Chem.*, **20,** 2667 (1981).
954a. A. Müller, W. Jaegermann and J. H. Enemark, *Coord. Chem. Rev.*, **46,** 245 (1982).
955. T. J. Greenhough, B. W. S. Kolthammer, P. Legzdins, and J. Trotter, *Inorg. Chem.*, **18,** 3548 (1979).
956. M. W. Bishop, J. Chatt, and J. R. Dilworth, *Chem. Commun..*, 780 (1975).
957. L. Markó, B. Markó-Monostory, T. Madach, and H. Vahrenkamp, *Angew. Chem. Int. Ed.*, **19,** 226 (1980).
958. J. R. Allkins and P. J. Hendra, *J. Chem. Soc. A*, 1325 (1967); *Spectrochim. Acta*, **22,** 2075 (1966); **23A,** 1671 (1967); **24A,** 1305 (1968).
959. E. A. Allen and W. Wilkinson, *Spectrochim. Acta*, **28A,** 725 (1972).
960. R. J. H. Clark, G. Natile, U. Belluco, L. Cattalini, and C. Filippin, *J. Chem. Soc. A*, 659 (1970).
961. E. A. Allen and W. Wilkinson, *Spectrochim. Acta*, **28A,** 2257 (1972).
962. B. E. Aires, J. E. Fergusson, D. T. Howarth, and J. M. Miller, *J. Chem. Soc. A*, 1144 (1971).

963. E. A. Allen and W. Wilkinson, *J. Chem. Soc., Dalton Trans.*, 613 (1972).
964. P. L. Goggin, R. J. Goodfellow, D. L. Sales, J. Stokes, and P. Woodward, *Chem. Commun.*, 31 (1968).
965. D. M. Adams and P. J. Chandler, *J. Chem. Soc. A*, 588 (1969).
966. M. C. Baird, G. Hartwell, and G. Wilkinson, *J. Chem. Soc. A*, 2037 (1967).
967. J. D. Gilbert, M. C. Baird, and G. Wilkinson, *J. Chem. Soc. A*, 2198 (1968).
968. A. Efraty, R. Arneri, and M. H. A. Huang, *J. Am. Chem. Soc.*, **98**, 639 (1976).
969. B. D. Dombek and R. J. Angelici, *J. Am. Chem. Soc.*, **96**, 7568 (1974).
970. A. M. English, K. R. Plowman, and I. S. Butler, *Inorg. Chem.*, **20**, 2553 (1981).
971. A. M. English, K. R. Plowman, and I. S. Butler, *Inorg. Chem.*, **21**, 338 (1982).
972. I. S. Butler and A. E. Fenster, *J. Organometal. Chem.*, **66**, 161 (1974).
973. M. C. Baird and G. Wilkinson, *Chem. Commun.*, 514 (1966); *J. Chem. Soc. A*, 865 (1967).
974. M. P. Yagupsky and G. Wilkinson, *J. Chem. Soc. A*, 2813 (1968).
975. G. R. Clark, T. J. Collins, S. M. James, W. R. Roper, and K. G. Town, *Chem. Commun.*, 475 (1976).
976. H. Huber, G. A. Ozin, and W. J. Power, *Inorg. Chem.*, **16**, 2234 (1977).
977. P. J. Brothers, C. E. L. Headford, and W. R. Roper, *J. Organometal. Chem.*, **195**, C29 (1980).
978. L. Vaska and S. S. Bath, *J. Am. Chem. Soc.*, **88**, 1333 (1966).
979. L. H. Vogt, J. L. Katz, and S. E. Wiberley, *Inorg. Chem.*, **4**, 1157 (1965).
980. J. W. Moore, H. W. Baird, and H. B. Miller, *J. Am. Chem. Soc.*, **90**, 1358 (1968).
981. D. M. Byler and D. F. Shriver, *Inorg. Chem.*, **15**, 32 (1976).
982. M. R. Churchill, B. G. DeBoer, K. L. Kalra, P. Reich-Rohrwig, and A. Wojcicki, *Chem. Commun.*, 981 (1972).
983. P. Reich-Rohrwig, A. C. Clark, R. L. Downs, and A. Wojcicki, *J. Organometal. Chem.*, **145**, 57 (1978).
984. C. Sourisseau and J. Corset, *Inorg. Chim. Acta*, **39**, 153 (1980).
985. R. D. Wilson and J. A. Ibers, *Inorg. Chem.*, **17**, 2134 (1978).
986. D. A. Johnson and V. C. Dew, *Inorg. Chem.*, **18**, 3273 (1979).
987. G. J. Kubas, *Inorg. Chem.*, **18**, 182 (1979).
988. K. H. Schmidt and A. Müller, *Coord. Chem. Rev.*, **14**, 115 (1974).
989. A. Müller, E. Diemann, R. Jostes, and H. Bögge, *Angew. Chem. Int. Ed.*, **20**, 934 (1981).
990. E. Königer-Ahlborn, A. Müller, A. D. Cormier, J. D. Brown, and K. Nakamoto, *Inorg. Chem.*, **14**, 2009 (1975).
991. A. Müller and H.-H. Heinsen, *Chem. Ber.*, **105**, 1730 (1972).
992. A. Müller, H.-H. Heinsen, and G. Vandrish, *Inorg. Chem.*, **13**, 1001 (1974).
993. R. G. Cavell, W. Byers, E. D. Day, and P. M. Watkins, *Inorg. Chem.*, **11**, 1598 (1972).
994. J. M. Burke and J. P. Fackler, *Inorg. Chem.*, **11**, 2744 (1972).
995. A. Cormier, K. Nakamoto, P. Christophliemk, and A. Müller, *Spectrochim. Acta*, **30A**, 1059 (1974).
996. O. Piovesana, C. Bellitto, A. Flamini, and P. F. Zanazzi, *Inorg. Chem.*, **18**, 2258 (1979).
997. F. A. Cotton, C. Oldham, and R. A. Walton, *Inorg. Chem.*, **6**, 214 (1967).
998. M. Ikram and D. B. Powell, *Spectrochim. Acta*, **28A**, 59 (1972).
999. C. W. Schläpfer, Y. Saito, and K. Nakamoto, *Inorg. Chim. Acta*, **6**, 284 (1972); C. W. Schläpfer and K. Nakamoto, *Inorg. Chim. Acta*, **6**, 177 (1972).
1000. F. Bonati and R. Ugo, *J. Organometal. Chem.*, **10**, 257 (1967).
1001. C. O'Connor, J. D. Gilbert, and G. Wilkinson, *J. Chem. Soc. A*, 84 (1969).
1002. D. C. Bradley and M. H. Gitlitz, *J. Chem. Soc. A*, 1152 (1969).
1003. R. J. Butcher, J. R. Ferraro, and E. Sinn, *Inorg. Chem.*, **15**, 2077 (1976).
1004. M. Sorai, *J. Inorg. Nucl. Chem.*, **40**, 1031 (1978).
1004a. B. Hutchinson, P. Neill, A. Finkelstein, and J. Takemoto, *Inorg. Chem.*, **20**, 2000 (1981).
1005. K. A. Jensen and V. Krishnan, *Acta Chem. Scand.*, **24**, 1088 (1970).
1006. K. Jensen, B. M. Dahl, P. Nielsen, and G. Borch, *Acta Chem. Scand.*, **26**, 2241 (1972).
1007. G. W. Watt and B. J. McCormick, *Spectrochim. Acta*, **21**, 753 (1965).
1008. V. V. Savant, J. Gopalakrishnan, and C. C. Patel, *Inorg. Chem.*, **9**, 748 (1970).

1009. G. A. Melson, N. P. Crawford, and B. J. Geddes, *Inorg. Chem.*, **9,** 1123 (1970).
1010. M. J. M. Campbell, *Coord. Chem. Rev.*, **15,** 279 (1975).
1011. M. A. Ali, S. E. Linvingstone, and D. J. Phillips, *Inorg. Chim. Acta*, **5,** 119 (1971).
1012. H. O. Desseyn, W. A. Jacob, and M. A. Herman, *Spectrochim. Acta*, **25A,** 1685 (1969).
1013. D. Coucouvanis, N. C. Baenziger, and S. M. Johnson, *J. Am. Chem. Soc.*, **95,** 3875 (1973).
1014. D. Coucouvanis and D. Piltingsrud, *J. Am. Chem. Soc.*, **95,** 5556 (1973).
1015. J. A. McCleverty, *Progr. Inorg. Chem.*, **10,** 49 (1968).
1016. C. W. Schläpfer and K. Nakamoto, *Inorg. Chem.*, **14,** 1338 (1975).
1017. M. Cox and J. Darken, *Coord. Chem. Rev.*, **7,** 29 (1971).
1018. S. E. Linvingstone, *Coord. Chem. Rev.*, **7,** 59 (1971).
1019. O. Siimann and J. Fresco, *Inorg. Chem.*, **8,** 1846 (1969); *J. Chem. Phys.*, **54,** 734 (1971); *J. Chem. Phys.*, **54,** 740 (1971).
1020. C. G. Barraclough, R. L. Martin, and I. M. Stewart, *Aust. J. Chem.*, **22,** 891 (1969); G. A. Heath and R. L. Martin, *Aust. J. Chem.*, **23,** 1721 (1970).
1021. H. Shindo and T. L. Brown, *J. Am. Chem. Soc.*, **87,** 1904 (1965).
1022. M. Chandrasekharan, M. R. Udupa, and G. Aravamudan, *Inorg. Chim. Acta*, **7,** 88 (1973).
1023. R. Panossian, G. Terzian, and M. Guiliano, *Spectrosc. Lett.*, **12,** 715 (1979).
1024. R. J. Gale and C. A. Winkler, *Inorg. Chim. Acta*, **21,** 151 (1977).
1025. C. A. McAuliffe, J. V. Quagliano, and L. M. Vallarino, *Inorg. Chem.*, **5,** 1996 (1966).
1026. C. A. McAuliffe, *J. Chem. Soc. A*, 641 (1967).
1027. M. Ikram and D. B. Powell, *Spectrochim. Acta*, **27A,** 1845 (1971).
1028. Y. Mido and E. Sekido, *Bull. Chem. Soc. Jpn.*, **44,** 2130 (1971).
1029. J. A. W. Dalziel, M. J. Hitch, and S. D. Ross, *Spectrochim. Acta*, **25A,** 1055 (1969).
1030. W. W. Fee and J. D. Pulsford, *Inorg. Nucl. Chem. Lett.*, **4,** 227 (1968).

Organometallic Compounds

Part IV

IV-1. METAL ALKANES

The methyl group bonded to a metal ($M—CH_3$) exhibits six normal vibrations such as those shown in Fig. III-2. In addition, CMC bending and CH_3 torsional modes are expected for $M(CH_3)_n$- ($n \geq 2$) type compounds. Table IV-1 lists the vibrational frequencies of typical $M(CH_3)_4$-type compounds. It is seen that the CH_3 rocking, MC stretching, and CMC bending frequencies are most sensitive to the change in metals. Tables IV-2 and IV-3 list the MC stretching and CMC bending frequencies of various $M(CH_3)_n$-type molecules and ions. As shown in Appendix III, the number of these skeletal modes which are infrared or Raman active provides direct information about the structure of the MC_n skeleton.

Some metal alkyls are polymerized in condensed phases. $Li(CH_3)$ forms a tetramer containing $Li—CH_3—Li$ bridges in the solid state,[33] and its CH_3 frequencies are lower than those of nonbridging compounds [$\nu_a(CH_3)$ and $\nu_s(CH_3)$ are 2840 and 2780 cm^{-1}, respectively].[34]

Solid $Be(CH_3)_2$ and $Mg(CH_3)_2$ also form long chain polymers through CH_3 bridges,[35] while $Al(CH_3)_3$ is dimeric in the solid state.[36,37] The infrared spectra of $Li[Al(CH_3)_4]$ and $Li_2[Zn(CH_3)_4]$ have been interpreted on the basis of linear polymeric chains in which the Al (or Zn) atom and the Li atom are bonded alternately through two CH_3 groups.[38] Normal coordinate analyses have been carried out on $M(CH_3)_2$- (M = Zn, Cd, and Hg),[4,39] dimeric $Al(CH_3)_3$-,[36,37] and linear $[M(CH_3)_2]^{n+}$-type cations.[23,26]

The ethyl group bonded to a metal ($M—CH_2—CH_3$) exhibits bands characteristic of both the CH_3 and CH_2 groups. It is difficult, however, to give complete assignments of the $M—C_2H_5$ group vibrations because of band overlapping and vibrational coupling. Table IV-4 lists the MC_n skeletal

TABLE IV-1. RAMAN FREQUENCIES[a] OF $M(CH_3)_4$-TYPE MOLECULES (CM^{-1})[1,2]

Compound	$\nu_a(CH_3)$	$\nu_s(CH_3)$	$\delta_d(CH_3)$	$\delta_s(CH_3)$	$\rho_r(CH_3)$	$\nu(MC)$	$\delta(CMC)$
$C(CH_3)_4$	2959	2922	(1475)	—	926	733	418
	2963		1457		(926)	1260	332
$Si(CH_3)_4$	(2959)	2913	(1430)	1271	870	593	239
	2964		1421		(870)	698	190
	(2910)						
$Ge(CH_3)_4$	(2981)	2920	(1430)	1259	—	561	196
	2982		1420		(828)	599	188
$Sn(CH_3)_4$	(2984)	2920	(1447)	1211	—	509	137
	2988		—		(768)	527	133
$Pb(CH_3)_4$	2996	2924	1450	1170	767	478	145
	2924		1400	1154	700	459	130

[a] ( ) = IR frequency.

TABLE IV-2. METAL–CARBON SKELETAL FREQUENCIES OF $M(CH_3)_n$-TYPE COMPOUNDS (cm^{-1})[1]

Compound	Structure	ν_a(MC)	ν_s(MC)	δ(CMC)	Refs.
$Be(CH_3)_2$	Linear	1081	—	—	3
$Zn(CH_3)_2$	Linear	604	503	157	4, 4a, 5
$Cd(CH_3)_2$	Linear	525	460	140	4, 5
$Hg(CH_3)_2$	Linear	538	515	160	4, 5
$Se(CH_3)_2$ [a]	Bent	604	589	233	6, 7
$Te(CH_3)_2$ [a]	Bent	528	528	198	6, 7
$B(CH_3)_3$	Planar	1177	680	341, 321	8, 9, 9a
$Al(CH_3)_3$	Planar	760	530	170	10
$Ga(CH_3)_3$	Planar	577	521.5	162.5	11, 12, 12a
$In(CH_3)_3$	Planar	500	467	132	12
$P(CH_3)_3$	Pyramidal	703	653	305, 263	13, 14, 14a
$As(CH_3)_3$	Pyramidal	583	568	238, 223	13, 14, 14b
$Sb(CH_3)_3$	Pyramidal	513	513	188	15
$Bi(CH_3)_3$	Pyramidal	460	460	171	15
$Si(CH_3)_4$	Tetrahedral	696	598	239, 202	16, 16a
$Ge(CH_3)_4$	Tetrahedral	595	558	195, 175	2, 16a
$Sn(CH_3)_4$	Tetrahedral	529	508	157	16a, 17
$Pb(CH_3)_4$	Tetrahedral	476	459	120	16a, 18
$Ti(CH_3)_4$	Tetrahedral	577	489	180	19
$Sb(CH_3)_5$	Trigonal-bipyramidal	514[b] 456[c]	493[b] 414[c]	213[b] 199[b] 104[c]	20
$W(CH_3)_6$	Octahedral	482	—	—	21

[a] New assignments have been proposed based on the $\mathbf{D}_{3d}$ model containing a linear C–M–C skeleton (Ref. 7a).
[b] Equatorial.
[c] Axial.

frequencies of typical $M(C_2H_5)_n$-type compounds. The MC stretching frequencies of the ethyl compounds are lower than those of the corresponding methyl compounds (Table IV-2) due to the larger mass of the ethyl, relative to the methyl group.

$Li(C_2H_5)$ is hexameric in hydrocarbon solvents[54] and is polymeric in the solid state.[55] The LiC stretching bands of these polymers are at 530–300 cm^{-1}.[56] The vibrational spectra of other polymeric ethyl compounds such as $Be(C_2H_5)_2$ (dimer),[57] $Mg(C_2H_5)_2$,[57a] $Al(C_2H_5)_3$ (dimer),[58,59] and $Li[Al(C_2H_5)_4]$ (polymer)[60] have been reported. There are many other compounds containing higher alkyl groups. References for some typical compounds are as follows: $[Tl(n\text{-}C_3H_7)_2]Cl$ (25), $Al(n\text{-}C_3H_7)_3$ (58), $Ge(n\text{-}C_4H_9)_4$ (61), and $[Li(t\text{-}C_4H_9)]_4$ (62). Vibrational spectra have also been reported for cycloalkyl compounds such as $Zn(c\text{-}C_3H_5)_2$,[63] $M(c\text{-}C_3H_5)_4$ (M = Si, Ge, and Sn),[64] $Pb(c\text{-}C_3H_5)_4$,[64a] and $Sb(c\text{-}C_3H_5)_5$.[65]

TABLE IV-3. METAL–CARBON SKELETAL FREQUENCIES OF $[M(CH_3)_n]^{m+}$-TYPE COMPOUNDS (cm^{-1})

Compound	Structure	ν_a(MC)	ν_s(MC)	δ(CMC)	Refs.
$[Zn(CH_3)]^+$	—	—	557	—	22
$[In(CH_3)_2]^+$	Linear	566	502	—	23
$[Tl(CH_3)_2]^+$	Linear	559	498	114	24, 25
$[Sn(CH_3)_2]^{2+}$	Linear	582	529	180	26
$[Sn(CH_3)_3]^+$	Planar	557	521	152	27
$[Sb(CH_3)_3]^{2+}$	Planar	582	536	166	28
$[Se(CH_3)_3]^+$	Nonplanar	602	580	272	29
$[Te(CH_3)_3]^+$	Nonplanar	534	—	—	30
$[P(CH_3)_4]^+$	Tetrahedral	783	649	285	31
				170	32
$[As(CH_3)_4]^+$	Tetrahedral	652	590	217	15
$[Sb(CH_3)_4]^+$	Tetrahedral	574	535	178	15

TABLE IV-4. METAL–CARBON SKELETAL FREQUENCIES OF $M(C_2H_5)_n$-TYPE COMPOUNDS (cm^{-1})

Compound	Structure	ν_a(MC)	ν_s(MC)	δ(MCC)	δ(CMC)	Refs.
$Zn(C_2H_5)_2$	Linear	563	474	261	205	40, 40a
$Hg(C_2H_5)_2$	Linear	515	488	267	140	41, 42
				262	85	
$^{10}B(C_2H_5)_3$	Planar	1135	—	—	287	8, 43
$Ga(C_2H_5)_3$	Planar	496	—	—	—	44
$P(C_2H_5)_3$	Pyramidal	697	619	410–249	—	45, 45a
		669				
$As(C_2H_5)_3$	Pyramidal	540	570	—	—	45a, 46
			563			
$Sb(C_2H_5)_3$	Pyramidal	505	505	—	—	45, 47
$Bi(C_2H_5)_3$	Pyramidal	~450	~450	253	124	48
				213		
$Si(C_2H_5)_4$	Tetrahedral	731	549	392	170	31, 49
				233		
$Ge(C_2H_5)_4$	Tetrahedral	572	532	332	152	50, 51
$Sn(C_2H_5)_4$	Tetrahedral	508	490	272	132	52, 53
					86	
$Pb(C_2H_5)_4$	Tetrahedral	461	443	243	107	48, 52
				213		

IV-2. METAL ALKENES, ALKYNES, AND PHENYLS

A vinyl group σ-bonded to a metal ($M—CH=CH_2$) exhibits CH stretching, C=C stretching, CH_2 scissoring, rocking, twisting and wagging, and CH in-plane and out-of-plane bending in addition to the MC stretching and CMC bending modes. Table IV-5 lists the C=C and MC stretching frequencies for typical vinyl compounds. The C=C stretching vibrations are generally strong in the Raman. Their intensity in the infrared, however, depends on the metal. Vibrational spectra of halovinyl compounds have been reported for $Hg(CH=CHCl)_2$ [73] and $M(CF=CF_2)_n$ (M = Hg, As, and Sn).[74] Complete vibrational assignments are available for σ-bonded allyl compounds such as $M(CH_2—CH=CH_2)_4$ (M = Si and Sn),[75] $M(CH_2—CH=CH_2)_3$ (M = P and As)[75a] and $Hg(CH_2—CH=CH_2)_2$.[76,77] Table IV-5 also lists the C≡C and MC stretching frequencies of acetylenic compounds. Again, the C≡C stretching vibrations are strong in the Raman, but vary from strong to weak in the infrared, depending on the metal involved. Recently, IR spectra of several compounds containing double-bonded silicon have been measured in inert gas matrices: $(CH_3)_2Si=CH_2$ [80a] and $(CH_3)HSi=CH_2$.[80b]

The phenyl group when σ-bonded to a metal exhibits bands characteristic of monosubstituted benzenes.[81] The $M—C_6H_5$-type molecule exhibits 30 fundamentals, only six of which, shown in Fig. IV-1, are sensitive to the change in metals. Table IV-6 lists the observed frequencies of these six modes for typical

TABLE IV-5. CARBON-CARBON AND METAL-CARBON STRETCHING FREQUENCIES OF VINYL AND ACETYLENIC COMPOUNDS (cm^{-1})

Compound	ν(C=C) or ν(C≡C)	ν(MC)	Refs.
$Zn(CH=CH_2)_2$	1565	—	66
$Hg(CH=CH_2)_2$	1603	541, 513	67
$^{10}B(CH=CH_2)_3$	1604	1186	68, 69
$P(CH=CH_2)_3$	1590	715, 667	69a
$Si(CH=CH_2)_4$	1592	732, 583	70, 71
$Ge(CH=CH_2)_4$	1595	600, 561	71
$Sn(CH=CH_2)_4$	1583	527, 513	71, 72
$Pb(CH=CH_2)_4$	1580	495, 481	71
$Si(C≡CH)_4$	2053	534, 708, or 687	78, 79
$Ge(C≡CH)_4$	2057	507, 523	78, 79
$Sn(C≡CH)_4$	2043	504 or 447	78, 79
$As(C≡CH)_3$	2053	517, 526	79a
$Sb(C≡CH)_3$	—	474	79a
$(CH_3)_2Si(C≡CH)_2$	2041	595[a]	79, 80
$(CH_3)_2Ge(C≡CH)_2$	2041	538,[a] 521,[a]	79, 80
$(CH_3)_2Sn(C≡CH)_2$	2016	454,[a] 445[a]	79, 80

[a] ν(MC) indicates ν(M—C≡C).

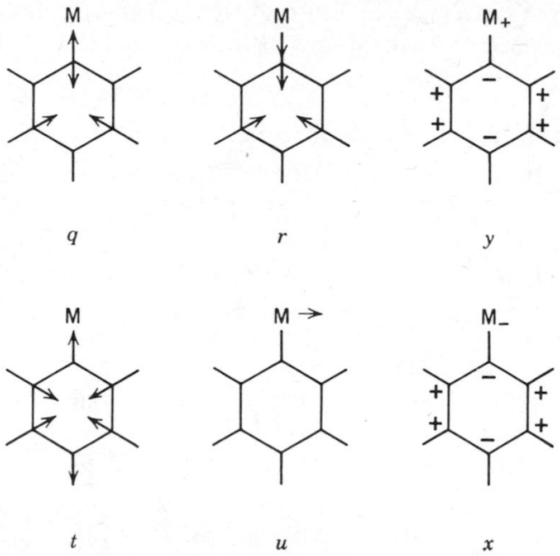

Fig. IV-1. Metal-sensitive modes of monosubstituted benzenes.[81]

phenyl compounds. It is seen that the *t*, *x*, and *u* modes are most sensitive to the change in metals. There are many metal-phenyl compounds containing other functional groups. The spectra of these compounds can be interpreted roughly as the overlap of the $M—C_6H_5$ group vibrations and the vibrations of other functional groups discussed later.

IV-3. HALOGENO AND PSEUDOHALOGENO COMPOUNDS

As discussed in Sec. III-19, terminal $\nu(MX)$ bands appear in the following regions:

$$\begin{array}{cccc} \nu(\mathrm{MF}) & \nu(\mathrm{MCl}) & \nu(\mathrm{MBr}) & \nu(\mathrm{MI}) \\ 750\text{–}500 > & 400\text{–}200 > & 300\text{–}200 > & 200\text{–}100 \end{array} \ \mathrm{cm}^{-1}$$

For a series with a fixed halogen, $\nu(MX)$ is higher, as M is lighter in the same family of the periodic table. Thus the $\nu(MCl)$ of the $M(CH_3)_3Cl$ series follow this order:

$$\begin{array}{rccc} \mathrm{M} = & \mathrm{Si} & \mathrm{Ge} & \mathrm{Sn} \\ & 487 > & 398 > & 325\ \mathrm{cm}^{-1} \end{array}$$

In general, the infrared intensity of $\nu(MX)$ decreases in the order $\nu(MF) > \nu(MCl) > \nu(MBr) > \nu(MI)$, whereas the opposite order prevails for the Raman intensity. Table IV-7 lists $\nu(MC)$ and $\nu(MX)$ of typical compounds.

In condensed phases, halogeno compounds tend to polymerize by forming halogeno bridges between two metals. As discussed in Sec. III-19, the bridging

TABLE IV-6. VIBRATIONAL FREQUENCIES OF METAL-SENSITIVE MODES OF METAL PHENYLS (CM^{-1})

Compound	*q*	*r*	*y*	*t*	*x*	*u*	Refs.
$Hg(C_6H_5)_2$	1067	661	456	258	—	207	82, 83, 83a
				252			
				248			
$^{10}B(C_6H_5)_3$	1285	893	600	650	245	408	84, 85
	1248						
$Al(C_6H_5)_3$	1085	670	460	420	207	332	85, 86
		643					
$Ga(C_6H_5)_3$	1085	665	453	315	180	245	85
			445			225	
$In(C_6H_5)_3$	1070	673	465	270	180	248	85, 87
						195	
$Si(C_6H_5)_4$	1108	709	519	435	185	261	88, 89
			511	239	171	223	
$Ge(C_6H_5)_4$	1091	—	481	332	187	232	89, 90
			465		168	214	
$Sn(C_6H_5)_4$	1075	616	459	268	193	225	89, 91, 91a
			448	212	152		
$Pb(C_6H_5)_4$	1061	645	450	223	147	181	89
			440	201			
$P(C_6H_5)_3$	1089	—	501	428	248	209	91, 92
				398		190	
$As(C_6H_5)_3$	1082	667	474	313	237	192	91, 92
	1074						
$Sb(C_6H_5)_3$	1065	651	457	270	216	166	91, 92
$Bi(C_6H_5)_3$	1055	—	448	237	207	157	91, 92
				220			

frequencies are much lower than the terminal frequencies. Thus it is possible to distinguish monomeric and polymeric (halogen-bridged) structures by vibrational spectroscopy. It was found that $(CH_3)_2BX$ (X = F, Cl, and Br) is monomeric,[100] whereas $(CH_3)_2AlF$, $(CH_3)_2GaF$, and $(CH_3)_2InCl$ are tetrameric,[101] trimeric,[102] and dimeric,[103] respectively, in benzene solution. Alkyl silicon and germanium halides tend to be monomeric, whereas alkyl tin and lead halides tend to be polymeric, in the liquid and solid phases. For example, $(CH_3)_2SnF_2$ and $(CH_3)_3SnF$ are polymerized through the fluorine bridges:[104]

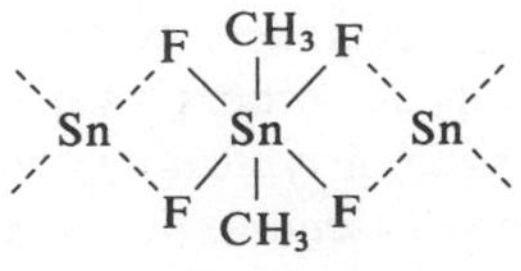

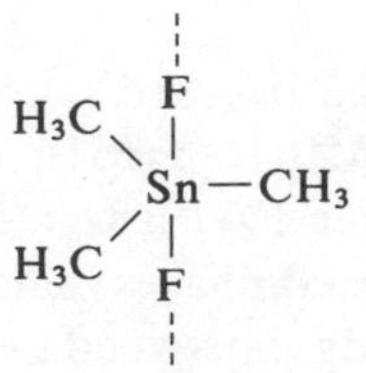

TABLE IV-7. METAL-CARBON AND METAL-HALOGEN STRETCHING FREQUENCIES OF TYPICAL METHYLHALOGENO COMPOUNDS (CM^{-1})

Compound	ν(MC)	ν(MX)	Refs.
CH_3CdCl	476	247	93
CH_3CdBr	475	206	93
CH_3CdI	482	167	93
$(CH_3)_3SiF$	704, 635	898	94
$(CH_3)_3SiCl$	704, 635	472	95
$(CH_3)_3GeF$	623, 576	623	96
$(CH_3)_3GeCl$	612, 569	378	97
$(CH_3)_3SbF_2$	591, 546	484, 465	98
$(CH_3)_3SbCl_2$	577, 538	282, 272	98
$(CH_3)_3SbBr_2$	569, 526	215, 168	98
$(CH_3)_3SbI_2$	559, 508	144, 122	98
$[(CH_3)_2AuCl]_2$	571, 561	273	99
$[(CH_3)_2AuBr]_2$	561, 550	181	99
$[(CH_3)_2AuI]_2$	550, 545	141, 131	99

The ν(SnF) for terminal bonds are at 650–625 cm^{-1}, whereas those for bridging bonds are at 425–335 cm^{-1}. In the solid state, the coordination number of tin is five or six. Dialkyl compounds prefer six-coordinate structures, while trialkyl compounds tend to form five-coordinate structures. In both cases, the favored positions of the alkyl groups are those shown in the above diagrams. Normal coordinate calculations have been made on the *trans*-$[(CH_3)_2SnX_4]^{2-}$ (X = F, Cl, and Br) series.[105]

$(CH_3)_3PbX$ (X = F, Cl, Br, and I) are monomeric in benzene but polymeric in the solid state; ν(PbCl) of the monomer and polymer are at 281 and 191 cm^{-1}, respectively.[106] In the $[(CH_3)_2AuX]_2$ series,[99] the Au atom takes a square-planar arrangement with two methyl groups in the *cis* position:

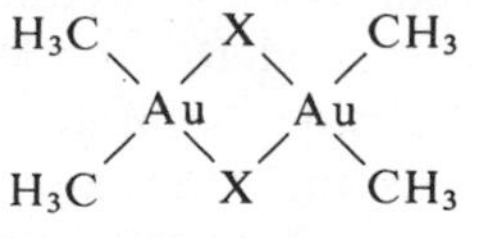

In this case, the bridging ν(AuX) are surprisingly similar to those observed for the corresponding *cis*-$[(CH_3)_2AuX_2]^-$.[99] Figure IV-2 shows the tetrameric structure of $[(CH_3)_3PtX]_4$-type compounds. Vibrational spectra have been reported for $[(CH_3)_3PtX]_4$ (X = Cl, Br, I[107] and Cl, I[108], N_3[108a] and SCN,[108b]) and for $[(CH_3)_2PtX_2]_n$ (X = Cl, Br, and I; *n* is probably 4).[109]

The vibrational spectra of coordination compounds containing pseudohalogeno groups were discussed in Sec. III-15. The vibrational spectra of azido complexes have been reported for CH_3ZnN_3,[110] CH_3HgN_3,[111]

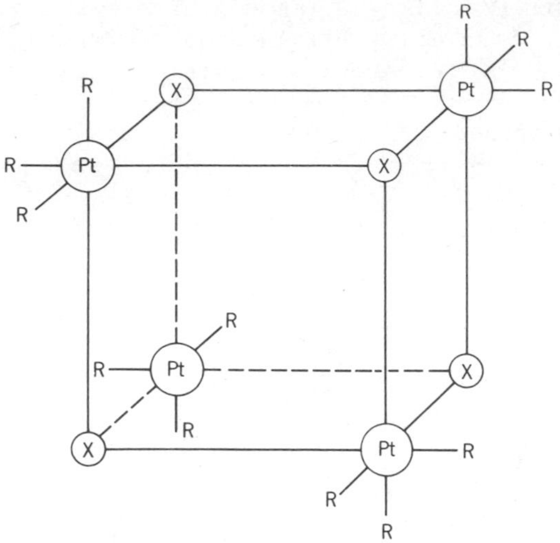

Fig. IV-2. Structure of $[Pt(CH_3)_3X]_4$. R denotes CH_3.

$(CH_3)_2AlN_3$,[112] $(CH_3)_3SiN_3$,[112a] and $(CH_3)_3SnN_3$.[113] The MN stretching frequencies of these compounds are in the 600-400 cm^{-1} region. The NCO group is always bonded to the metal through the N atom (isocyanato complex). The vibrational spectra of monomeric $CH_3Hg—NCO$,[114] $(CH_3)_3Si—NCO$,[112a,115] $(CH_3)_3Ge—NCO$,[115a] and $(CH_3)_3Sb(NCO)_2$ [114] have been reported.

The NCS group may be bonded to a metal through the N or S atom or may form a bridge between two metals by using the N, the S, or both atoms. It is not easy to distinguish all these possible structures by vibrational spectra. Table IV-8 lists the modes of coordination and the vibrational frequencies of the M—NCS group. The NCSe group is N-bonded in $(CH_3)_3Si—NCSe$[121] and $(CH_3)_3Ge—NCSe$[121a] but is Se-bonded in $(CH_3)_3Pb—SeCN$.[122] Only a very few compounds containing the M-CNO (fulminato) group are known. The spectrum of $(CH_3)_2Tl(CNO)$ is similar to that of K[CNO], and the Tl-CNO bond may be ionic.[123] No isofulminato complexes are reported. The CN group is usually bonded to the metal through the C atom. In the case of $(CH_3)_3M(CN)$ (M = Si and Ge), however, the cyano and isocyano complexes are in equilibrium in the liquid phase, although the mole fraction of the latter is rather small. The CN stretching frequencies (cm^{-1}) of these isomers are as follows:

	$(CH_3)_3M—CN$	$(CH_3)_3M—NC$	
M = Si	2198	2095	(gas phase)[124]
M = Ge	2182	2090	($CHCl_3$ solution)[125]

TABLE IV-8. VIBRATIONAL FREQUENCIES OF THIOCYANATO AND ISOTHIOCYANATO COMPOUNDS (CM^{-1})

Compound	Mode of Coordination	ν(CN)	ν(CS)	δ(NCS)	ν(MS) or ν(MN)	Ref.
$[CH_3Zn(SCN)]_\infty$	Zn\SCN—Zn (S bridging to two Zn: Zn/ , Zn\ )	2190, 2140	685	455	—	116
$[CH_3Hg(SCN)_3]^{2-}$	Hg—SCN	2119	—	—	276	117
$[(CH_3)_3Al(SCN)]^-$	Al—SCN	2097	845	485	335	118
$[(CH_3)_2Al(SCN)]_3$	Al\SCN/Al	2075	627	501, 438	—	119
$(CH_3)_3Sn(NCS)$, solid (polymer)	Sn—NCS—Sn	2098, 2079, 2046	779	474, 467	—	120
$(CH_3)_3Sn(NCS)$, CS_2 solution (monomer)	Sn—NCS	2050	781	485	478	120
$[(CH_3)_2Au(NCS)]_2$	Au(NCS)(SCN)Au (bridged ring)	2163	775	444, 430 [a]	—	99

[a] These bands may be assigned to ν(AuN).

Complete band assignments for $(CH_3)_3GeCN$ and its deuterated analog have been made based on normal coordinate analysis.[125a]

IV-4. COMPOUNDS CONTAINING OTHER FUNCTIONAL GROUPS

The ν(MN) of $[CH_3HgNH_3]^+$ [126] and $[(CH_3)_3SnNH_3]^+$ [127] have been assigned at 585 and 503 cm^{-1}, respectively. The infrared and Raman spectra of $[(CH_3)_3Pt(NH_3)_3]^+$ have been interpreted on the basis of the *fac* structure ($\mathbf{C}_{3v}$). Its ν(PtN) are at 390 cm^{-1} (A_1, Raman) and at 410 and 377 cm^{-1} (E, infrared).[128] The ν(SnN) of alkyl tin halide adducts with bipy and phen were suggested to be close to 200 cm^{-1}.[129]

Compounds containing the hydroxo group exhibit ν(OH), δ(MOH), and ν(MO) at 3760–3000, 1200–700, and 900–300 cm^{-1}, respectively. As expected, these frequencies depend heavily on the strength of the hydrogen bond involved. References for typical hydroxo compounds are as follows: $[(CH_3)_2GaOH]_4$ (130), $(CH_3)_3SiOH$ (131), $(CH_3)_3SnOH$ (132), $(CH_3)_4SbOH$ (133), and $[(CH_3)_3PtOH]_4$ (108, 134). The vibrational spectra of aquo compounds are characterized by the bands discussed in Sec. III-6; there are some pertinent references: $[CH_3Hg(OH_2)]^+$ (135) and $[(CH_3)_3Pt(OH_2)_3]^+$ (136). As stated in Sec. III-7, the alkoxides exhibit ν(CO) in the 1200–950 cm^{-1} region: $(CH_3)_2Al(OCH_3)$,[137] $(CH_3)_2Si(OCH_3)_2$,[138] and $(CH_3)_2Sn(OCH_3)_2$.[139] The carboxylates, such as $(CH_3)_nSi(OCOCH_3)_{4-n}$, exhibit ν(CO) at 1580 cm^{-1}.[140] The band assignments of the O-bonded (chelated) acac complexes were discussed in Sec. III-12. References for acac complexes of metal alkyls are as follows: $(CH_3)_2Ga(acac)$ (130), $(CH_3)_2Sn(acac)_2$ (141), $(CH_3)_2Pb(acac)_2$ (142), $(CH_3)_2SbCl_2(acac)$ (143), and $(CH_3)_2Au(acac)$ (144). The structure of $(CH_3)_2Sn(acac)_2$ has been controversial. Originally, it was suggested, on the basis of NMR and vibrational spectra, that its structure in the solid state and in solution was *trans*.[145] Later, the *cis* structure was proposed because of the large dipole moment (2.95 D) in benzene solution.[146] X-ray analysis shows that the compound is *trans* in the solid state.[147] Ramos and Tobias[141] suggest that the structure remains *trans* in solution and that the large dipole moment may originate in the nonplanarity of the SnO_4 plane with the remainder of the acac ring. The ν(MS) of $(CH_3)_3Si(SC_2H_5)$,[148] $(CH_3)_3Si(SC_6H_5)$,[149] $(CH_3)_3Ge(SCH_3)$,[149a] and $(CH_3)_2Sn(SCH_3)_2$[150] are assigned at 486, 459, 394, and 347 cm^{-1}, respectively.

A large number of metal alkyls containing acido groups are known. As discussed in Sec. III-11, vibrational spectra are very useful in elucidating the mode of coordination. References are given for the following acido groups: $(CH_3)_3SnNO_3$ (151), $[(CH_3)_3Sn]_2SO_4$ (152), $(C_2H_5)_2SnSO_3$ (153), $(CH_3)_2SnCO_3$ (154), and $(CH_3)_3SnClO_4$ (155).

As discussed in Sec. III-18, metal hydrido complexes exhibit sharp ν(MH) in the 2200–1700 cm^{-1} region. Table IV-9 lists ν(MH) and ν(MC) of typical

TABLE IV-9. METAL–HYDROGEN AND METAL–CARBON STRETCHING FREQUENCIES OF TYPICAL HYDRIDO COMPOUNDS (CM^{-1})

Compound	ν(MH)	ν(MC)	Ref.
CH_3SiH_3	2166, 2169	701	156
$(CH_3)_2SiH_2$	2145, 2142	728, 659	157
$(CH_3)_3SiH$	2123	711, 624	157
CH_3GeH_3	2086, 2085	604	158
$(CH_3)_2GeH_2$	2080, 2062	604, 590	159
$(CH_3)_3GeH$	2049	601, 573	159
CH_3SnH_3	1875	527	160
$(CH_3)_2SnH_2$	1869	536, 514	161
$(CH_3)_3SnH$	1837	521, 516	161
$(CH_3)_3PbH$	1709	—	162
$(CH_3)_2PH$	2288	703, 660	162a
$(CH_3)_2AsH$	2080	580, 565	162b

compounds. It is seen that ν(MH) decreases as the mass of M increases in the same family of the periodic table and as more halogens are substituted by alkyl groups. The MH_3 torsional modes of CH_3MH_3-type compounds have been assigned at 142 and 113 cm^{-1} for M = Si and Ge, respectively.[163] Vibrational spectra of B_2H_6-type molecules were discussed in Sec. II-11. The compounds $(CH_3)_nB_2H_{6-n}$ (n = 1-4) exhibit the terminal and bridging ν(BH) at 2600-2500 and 2150-1525 cm^{-1}, respectively.[164] Dialkylaluminum hydride exists as a trimer in solution and pure liquid, and its ν(AlH) is at ca. 1800 cm^{-1}.[165]

As discussed in Sec. III-20, ν(MM′) are generally strong in the Raman and weak in the infrared. Figure IV-3 shows the SnMn stretching bands in the far-infrared spectra of $(CH_3)_{3-n}Cl_nSn{-}Mn(CO)_5$-type compounds.[173] The vibrational spectra of $(CH_3)_3M{-}M(CH_3)_3$ (M = Si, Ge, Sn, and Pb) and $M[Si(CH_3)_3]_4$ (M = Si, Ge, and Sn) are reported in Refs. 166, 167, and 168, respectively. Table IV-10 lists the MM′ stretching frequencies of metal alkylcarbonyl compounds.

The band at 284 cm^{-1} in the resonance Raman spectrum of (t-Bu—C≡C—t-Bu)$_2$Fe$_2$(CO)$_4$:

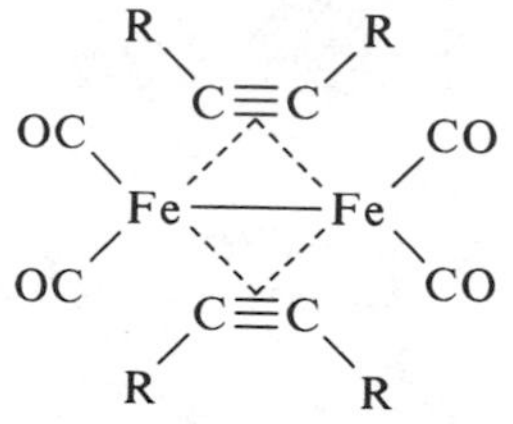

was assigned to ν(FeFe) by Kubas and Spiro.[173a] Its FeFe stretching force constant (3.0 mdyn/Å) is about twice that of the Fe–Fe single bond (1.3 mdyn/Å), found in $Fe_2S_2(CO)_6$.[173b] Thus the Fe–Fe bond of the former compound must be close to a double bond.

TABLE IV-10. METAL-METAL STRETCHING FREQUENCIES OF METAL ALKYL COMPOUNDS (CM^{-1})

Compound	ν(MM′)	Ref.
$(CH_3)_3Si{-}Mn(CO)_5$	297 (R)	169
$(CH_3)_3Ge{-}Cr(CO)_3(\pi\text{-}Cp)$	119 (R)	170
$(CH_3)_3Ge{-}Mn(CO)_5$	191 (R)	171
$(CH_3)_3Ge{-}Co(CO)_4$	192 (R)	172
$(CH_3)_3Sn{-}Mo(CO)_3(\pi\text{-}Cp)$	172 (IR)	173
$(CH_3)_3Sn{-}Mn(CO)_5$	182 (IR, R)	173
$(CH_3)_3Sn{-}Re(CO)_5$	147 (R)	171
$(CH_3)_3Sn{-}Co(CO)_4$	176 (IR, R)	172

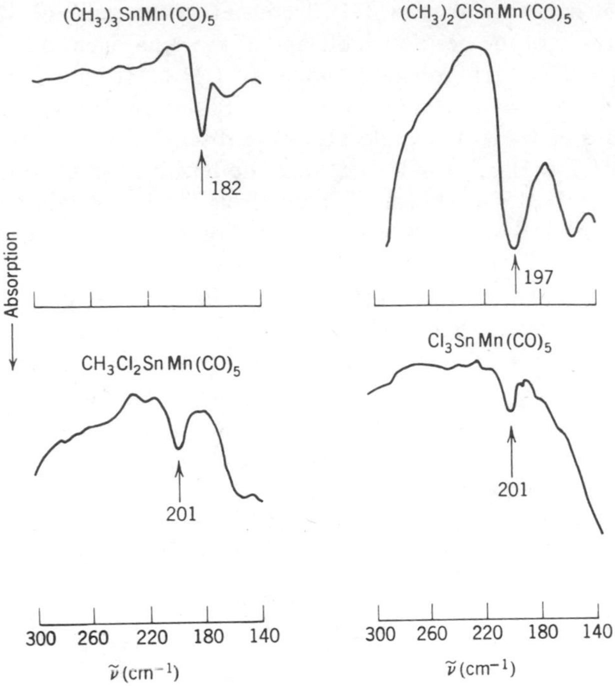

Fig. IV-3. Far-infrared spectra of $(CH_3)_{3-n}Cl_nSnMn(CO)_5$ where $n = 0$, 1, 2 or 3.[173] The arrow indicates the SnMn stretching band.

IV-5. π-COMPLEXES OF ALKENES, ALKYNES, AND RELATED LIGANDS

Vibrational spectra of π-bonded complexes of alkenes and alkynes with transition metals have been reviewed by Davidson.[174] In contrast to σ-bonded complexes (Sec. IV-2), the C=C and C≡C stretching bands of π-bonded complexes show marked shifts to lower frequencies relative to those of free ligands.

(1) Complexes of Monoolefins

Ethylene and other olefins form π-complexes with transition metals. The simplest and best-studied complex is Zeise's salt, $K[Pt(C_2H_4)Cl_3]\cdot H_2O$. Several investigators[175-177] have given complete assignments of the infrared spectra of Zeise's salt and its deuterated analog. Table IV-11 lists the observed frequencies and band assignments reported by Grogan and Nakamoto.[175] Normal coordinate analysis shows that the C=C stretching mode couples strongly with the CH_2 bending mode.[175-177] Thus the negative shift of the C=C stretching mode upon complexation cannot be used as a quantitative measure of the strength of the metal–olefin bond.

The assignments of the low-frequency modes of Zeise's salt have been controversial. According to Chatt et al.,[178] two types of bonding are involved

TABLE IV-11. OBSERVED FREQUENCIES OF FREE ETHYLENE, ZEISE'S SALT, AND THEIR DEUTERATED ANALOGS (CM^{-1})[175]

	C_2H_4	Zeise's Salt	C_2D_4	Zeise's Salt-d_4	Assignment[a]
Coordinated ethylene	3019	2920	2251	2115	$\nu(CH)$
	1623	1526	1515	1428	$\nu(C{=}C)+\delta_s(CH_2)$[b]
	1342	1418	981	978	$\delta_s(CH_2)+\nu(C{=}C)$[b]
	3108	2975	2304	2219	$\nu(CH)$
	1236	1251	1009	1021	$\rho_r(CH_2)$
	3106	3098	2345	2335	$\nu(CH)$
	810	844	586	536	$\rho_r(CH_2)$
	2990	3010	2200	2185	$\nu(CH)$
	1444	1428	1078	1067	$\delta_a(CH_2)$
	1007	730	726	450	$\rho_t(CH_2)$
	943	1023	780	811	$\rho_w(CH_2)$
	949	1023	721	818	$\rho_w(CH_2)$
Square-planar skeleton		331		329	$\nu_s(PtCl)$
		407		387	$\nu(Pt{-}C_2H_4)$
		310		305	$\nu(PtCl_t)$[c]
		183		185	$\delta(ClPtCl)$
		339		339	$\nu_a(PtCl)$
		210		198	$\delta(ClPtC_2H_4)+\delta(ClPtCl)$
		161		160	$\delta(ClPtCl_t)+\delta(ClPtC_2H_4)$
		121		117	$\pi(C_2H_4PtCl_t)$
		92		92	$\pi(ClPtCl)$

[a] Band assignments are for nondeuterated Zeise's salt.
[b] This coupling does not exist for Zeise's salt-d_4.
[c] Cl_t denotes the Cl atom *trans* to C_2H_4.

in the Pt–ethylene bond: (1) a σ-type bond is formed by the overlap of the filled $2p\pi$ bonding orbital of the olefin with the vacant dsp^2 bonding orbital of the metal, and (2) a π-type bond is formed by the overlap of the $2p\pi^*$ antibonding orbital of the olefin with a filled dp hybrid orbital of the metal.

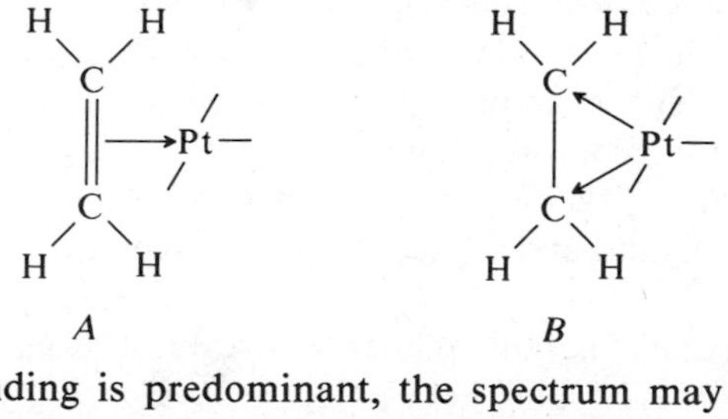

A *B*

If the σ-type bonding is predominant, the spectrum may be interpreted in terms of bonding scheme *A*, which predicts one Pt–olefin stretching mode.

Grogan and Nakamoto[175] preferred this interpretation and assigned the 407-cm^{-1} band of Zeise's salt to this mode. On the other hand, other workers[176,177] preferred bonding scheme *B*, which involves a five coordinate Pt atom and assigned two bands at 491 and 403 cm^{-1} to the symmetric and antisymmetric PtC stretching modes, respectively. The real bonding is somewhere between *A* and *B*, and the latter may become more predominant as the oxidation state of the metal becomes lower. In the case of Zeise's salt involving the Pt(II) atom, X-ray analysis[179] and MO calculations,[180] together with infrared studies,[175] seem to favor bonding scheme *A*.

The vibrational spectra of ethylene complexes with other metals have been reported: $Fe(C_2H_4)(CO)_4$ (356 cm^{-1}),[181] $Fe(C_2X_4)(CO)_4$ (X = Cl, Br),[181a] $[Ag(C_2H_4)]^+$,[182] $[Rh(C_2H_4)_2Cl]_2$ (399 cm^{-1}),[183] $Ni(C_2H_4)_3$,[184] and $Ir(C_2H_4)_4Cl$ (502 and 308 cm^{-1}, IR).[184a] Here, the numbers in parentheses indicate the metal–ethylene stretching frequencies. Moskovits and Ozin[184b] measured the IR spectra of a number of complexes of the type $M(C_2H_4)_n$ (M = Ni, Co, Cu, Ag, Pd, Au, etc.) which were prepared via matrix co-condensation reactions of ethylene with the respective metal vapor. Tevault et al.[184c] measured the IR spectra and proposed the structures of $(HgCl_2)L$ (L = ethylene and other olefins) which were prepared by matrix co-condensation reactions.

(2) Allyl Complexes

If the allyl group is π-bonded to a metal, the C=C stretching band characteristic of the σ-bonded allyl group (Sec. IV-2) does not appear. Instead, three bands of medium or strong intensity are observed at 1510–1375 cm^{-1}. In the low-frequency region, the metal–olefin vibrations appear in the range from 570 to 320 cm^{-1}. Complete band assignments have been reported for $M(\pi\text{-}C_3H_5)_2$ (M = Ni, Pd), $M(\pi\text{-}C_3H_5)_3$ (M = Rh, Ir),[185] $[Pd(\pi\text{-}C_3H_5)X]_2$ (X = Cl and Br),[186,187] $Fe(\pi\text{-}C_3H_5)(CO)_3X$ (X = Br, NO_3),[186a] and $Mn(\pi\text{-}C_3H_5)(CO)_4$.[187a] Chenskaya et al.[187b] assigned the metal–olefin and metal–halogen vibrations of π-allyl complexes of transition metals.

(3) Complexes of Diolefins and Oligoolefins

Nonconjugated diolefins such as norbornadiene (NBD, C_7H_8) and 1,5-hexadiene (C_6H_{10}) form metal complexes via their C=C double bonds (Figs. IV-4*a*, IV-4*b*). Complete vibrational assignments have been made for $M(NBD)(CO)_4$ (M = Cr, Mo, and W),[188] $Cr(NBD)(CO)_4$,[189] and $Pd(NBD)X_2$ (X = Cl and Br).[189] The metal–olefin vibrations are assigned in the region from 305 to 200 cm^{-1}. The spectrum of $K_2[(PtCl_3)_2(C_6H_{10})]$ is similar to that of the free ligand in the *trans* conformation.[190] Thus its structure may be shown as in Fig. IV-4*b*. However, the spectrum of $Pt(C_6H_{10})Cl_2$ is more complicated than that of the free ligand and suggests a chelate structure such as that shown in Fig. IV-4*a*.

Free butadiene (C_4H_6) is *trans*-planar. However, it takes a *cis*-planar structure in $Fe(C_4H_6)(CO)_3$ [191] and $Fe(C_4H_6)_2CO$.[192] For $K_2[C_4H_6(PtCl_3)_2]$, the infrared spectrum indicates the *trans*-planar structure of the olefin.[193] The

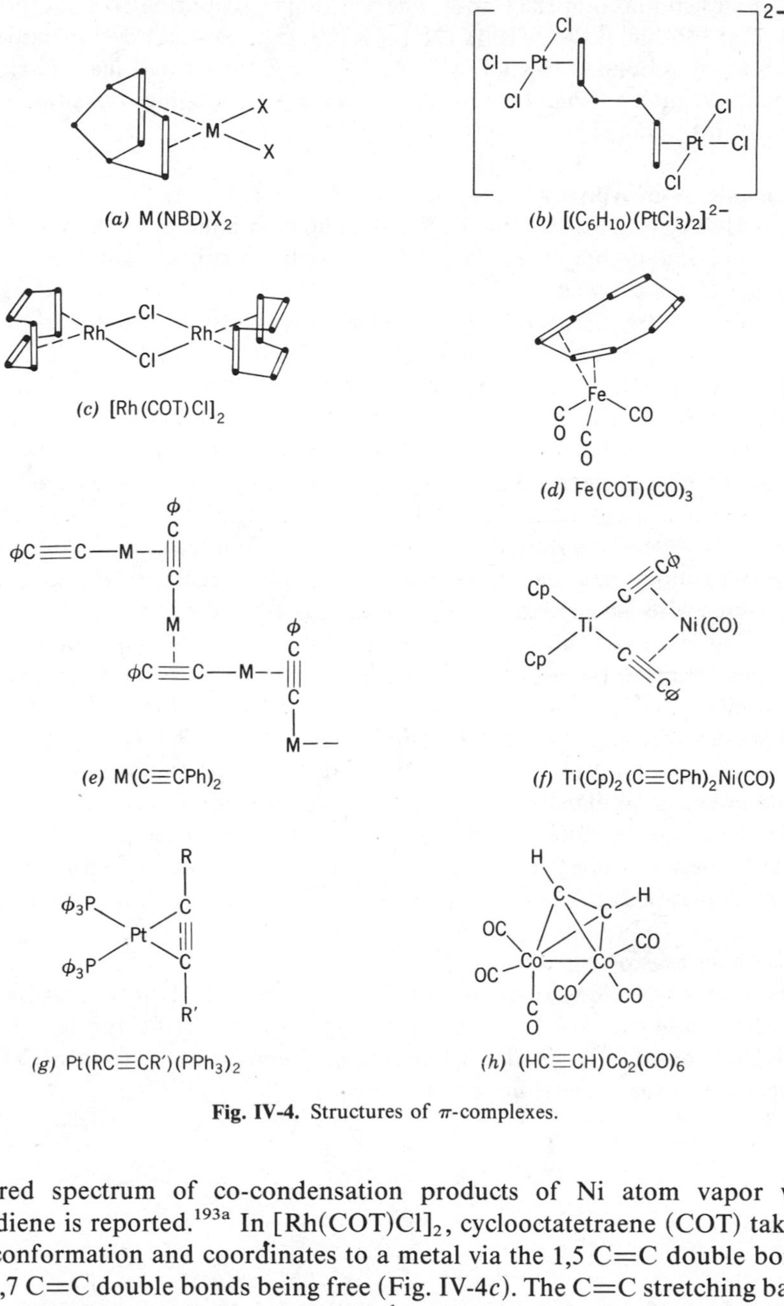

Fig. IV-4. Structures of π-complexes.

infrared spectrum of co-condensation products of Ni atom vapor with butadiene is reported.[193a] In $[Rh(COT)Cl]_2$, cyclooctatetraene (COT) takes a tub conformation and coordinates to a metal via the 1,5 C=C double bonds, the 3,7 C=C double bonds being free (Fig. IV-4*c*). The C=C stretching bands of free COT are at 1630 and 1605 cm^{-1}, whereas those of the complex are at 1630 (free) and 1410 (bonded) cm^{-1}.[194] According to X-ray analysis,[195] only two of the four C=C double bonds of COT are bonded to the metal in $Fe(COT)(CO)_3$ (Fig. IV-4*d*). In this case, the C=C stretching band for free

C=C double bonds is at 1562 cm^{-1}, whereas that for coordinated C=C double bonds is at 1460 cm^{-1}.[196] In $[Rh(COD)Cl]_2$ (COD, C_8H_{12}, 1,5-cyclooctadiene), the Rh atom is bonded to COD via the 1,5 C=C bonds in a manner similar to its COT analog (Fig. IV-4*c*). The Rh-olefin stretching vibrations were assigned in the range from 490 to 385 cm^{-1}.[196a]

(4) Complexes of Alkynes

Free $HC{\equiv}C(C_6H_5)$ exhibits the C≡C stretching band at 2111 cm^{-1}. In the case of σ-bonded complexes (Sec. IV-2), this band shifts slightly to a lower frequency (2036–2017 cm^{-1}).[197] In $M[-C{\equiv}C(C_6H_5)]_2$ [M = Cu(I) and Ag(I)], it shifts to 1926 cm^{-1}. This relatively large shift was attributed to the formation of both σ- and π-type bonding, shown in Fig. IV-4*e*.[198,199] $Ti[C{\equiv}C(C_6H_5)]_2(\pi\text{-}Cp)_2$ reacts with $Ni(CO)_4$ to form the complex shown in Fig. IV-4*f*. The C≡C stretching band of the parent compound at 2070 cm^{-1} is shifted to 1850 cm^{-1} by complex formation.[200] According to Chatt and co-workers,[201] the C≡C stretching bands of disubstituted alkynes (2260–2190 cm^{-1}) are lowered to ca. 2000 cm^{-1} in $Na[Pt(RC{\equiv}CR')Cl_3]$ and $[Pt(RC{\equiv}CR')Cl_2]_2$, and to ca. 1700 cm^{-1} in $Pt(RC{\equiv}CR')(PPh_3)_2$. Here R and R′ denote various alkyl groups. The former represents a relatively weak π-bonding similar to that found for Zeise's salt, whereas the latter indicates strong π-bonding in which the C≡C triple bond is almost reduced to the double bond (Fig. IV-4*g*). Similar results were found for $(RC{\equiv}CR')Co_2(CO)_6$, which exhibits the C≡C stretching bands near 1600 cm^{-1}.[202] In the case of $(HC{\equiv}CH)Co_2(CO)_6$, the C≡C stretching band was observed at 1402 cm^{-1}, which is ca. 570 cm^{-1} lower than the value for free acetylene (1974 cm^{-1}). The spectrum of the coordinated acetylene in this complex is similar to that of free acetylene in its first excited state, at which the molecule takes a *trans*-bent structure. Considering possible steric repulsion between the hydrogens and the $Co(CO)_3$ moiety, a structure such as that shown in Fig. IV-4*h* was proposed.[202a]

(5) π-Complexes of Nitriles

The C≡N stretching frequency of $CF_3{-}C{\equiv}N$ is 2271 cm^{-1}. This band is shifted to 1734 cm^{-1} in $Pt(CF_3CN)(PPh_3)_2$ because of the formation of a Pt-nitrile π-bond.[202b] A similar π-bonding has been proposed for $Mn(CO)_3IL$, where L is *o*-cyanophenyldiphenylphosphine:

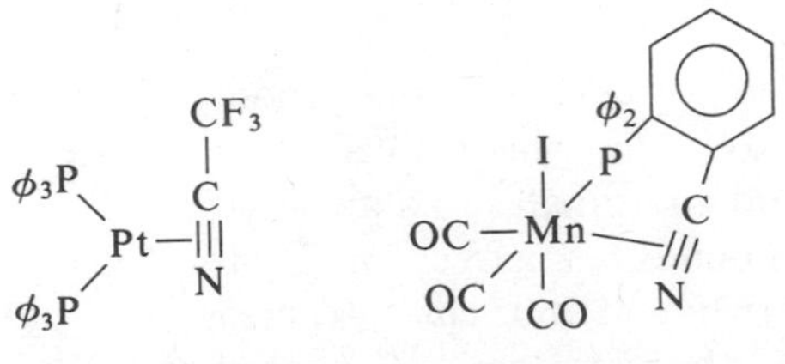

In the latter case, the C≡N stretching band of the free ligand at 2225 cm^{-1} is shifted to 1973 cm^{-1} in the complex.[202c]

IV-6. CYCLOPENTADIENYL COMPLEXES

The infrared spectra of cyclopentadienyl (C_5H_5 or Cp) complexes have been reviewed extensively by Fritz.[203] According to Fritz, they are roughly classified into four groups, each of which exhibits its own characteristic spectrum.

(1) Ionic Complexes

These are complexes such as MCp ($M = K^+$, Rb^+, and Cs^+) and MCp_2 ($M = Sr^{2+}$, Ba^{2+}, Mn^{2+}, and Eu^{2+}),[204–207] in which M^{n+} and Cp^- are ionically bonded. The spectra of these compounds are essentially the same as the spectrum of the $C_5H_5^-$ ion ($\mathbf{D}_{5h}$ symmetry) and consist of the following bands; ν(CH), 3100–3000 cm^{-1}; ν(CC), 1500–1400 cm^{-1}; δ(CH), 1010–1000 cm^{-1}; π(CH), 750–650 cm^{-1}.

(2) Centrally σ-Bonded Complexes

These are complexes such as MCp (M = Li and Na) and MCp_2 (M = Be, Mg, and Ca)[204,207–209] in which the metal is bonded to the center of the ring through a σ-bond. In this case, the local symmetry of the ring is regarded as $\mathbf{C}_{5v}$, and the following bands are IR active; ν(CH), 3100–3000 cm^{-1}; ν(CH), 2950–2900 cm^{-1}; ν(CC), 1450–1400 cm^{-1}; ν(CC), 1150–1100 cm^{-1}; δ(CH), 1010–990 cm^{-1}; and two π(CH), 890–700 cm^{-1}. In addition, these compounds exhibit metal-ring (MR) stretching and ring-tilt vibrations below 550 cm^{-1}. As will be discussed in the next subsection, MCp_2-type compounds exhibit one MR stretching and one ring-tilt vibration in the infrared if two rings are parallel to each other. According to electron diffraction studies,[210] the two rings of $SnCp_2$ and $PbCp_2$ form angles of 45° and 55°, respectively, in the vapor state. On the assumption of angular structure in the solid state, two bands at 240 and 170 cm^{-1} of $SnCp_2$ have been assigned to the antisymmetric and symmetric MR stretching modes, respectively.[211] The Raman spectrum of $BeCp_2$ in the solid state suggests the presence of one ring of $\mathbf{C}_{5v}$ symmetry and another ring of much lower symmetry.[208]

(3) Centrally π-Bonded Complexes*

These are complexes such as $FeCp_2$ and $RuCp_2$, in which the transition metals are bonded to the center of the ring via the d-π-bond. Figure IV-5 shows the infrared spectra of $FeCp_2$ and $NiCp_2$, and Table IV-12 lists the observed frequencies and band assignments for the compounds belonging to this group. In the high-frequency region, these compounds exhibit seven bands, listed previously for group (2). In addition, MCp_2 compounds with parallel rings exhibit the six skeletal modes shown in Fig. IV-6, three of which (ν_3, ν_5, and ν_6) are infrared active.

---

* Or pentahapto (h^5) complexes.

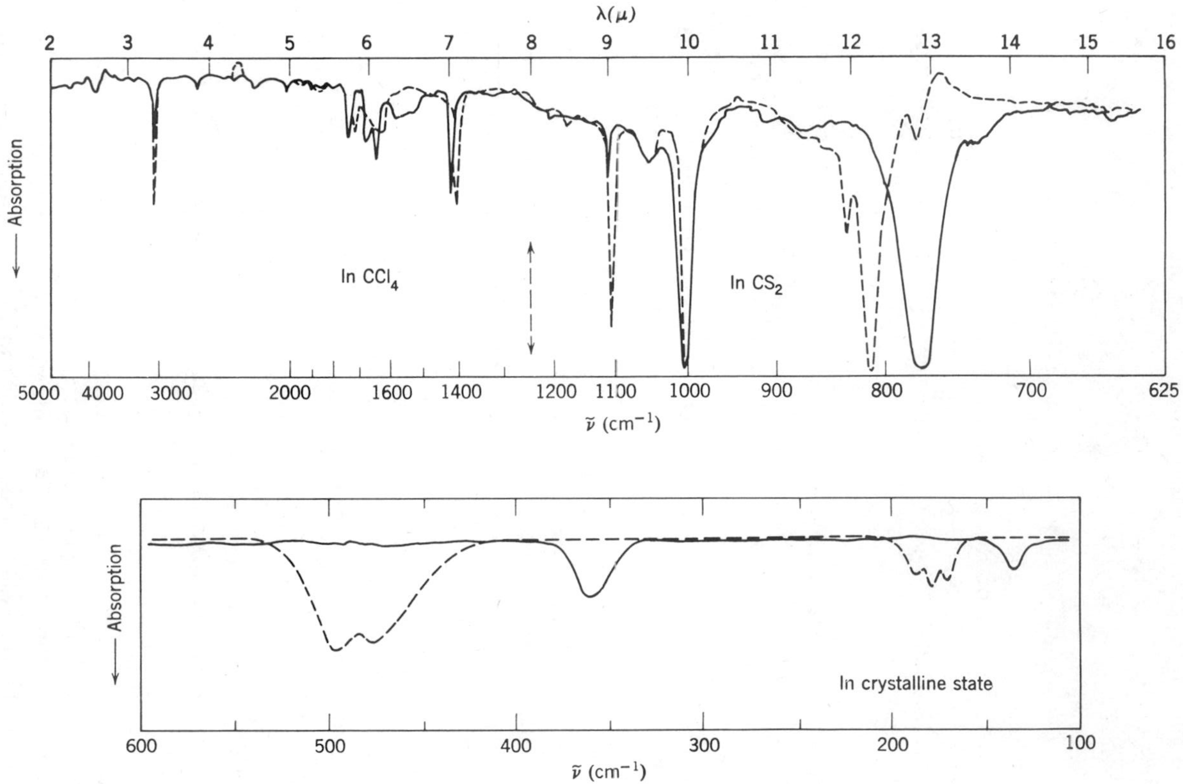

Fig. IV-5. Infrared spectra of $Ni(C_5H_5)_2$ (solid line) and $Fe(C_5H_5)_2$ (dashed line).[211a]

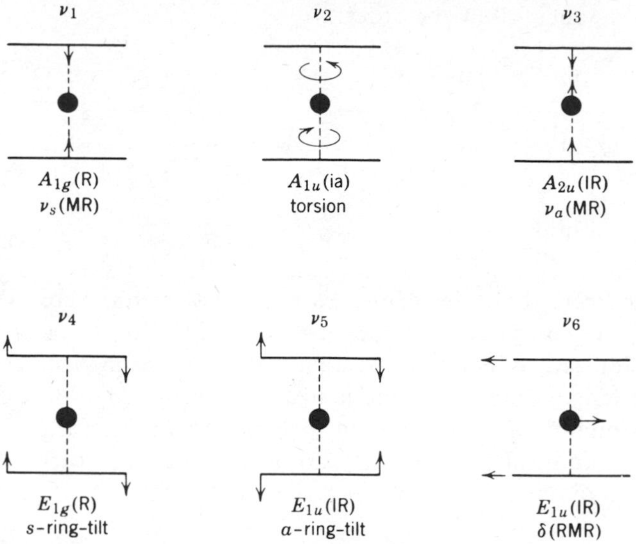

Fig. IV-6. Skeletal vibrations of dicyclopentadienyl metal complexes ($\mathbf{D}_{5d}$ symmetry).

It is interesting to note that two rings in solid ferrocene take the staggered configuration ($\mathbf{D}_{5d}$), while those in ruthenocene take the eclipsed configuration ($\mathbf{D}_{5h}$):

$\mathbf{D}_{5d}$ $\mathbf{D}_{5h}$

Since the number of infrared- or Raman-active fundamentals is the same for both conformations, they cannot be distinguished on the basis of the number of fundamentals observed. According to Fritz,[203] the MR stretching frequencies

TABLE IV-12. OBSERVED INFRARED FREQUENCIES AND BAND ASSIGNMENTS OF CENTRALLY π-BONDED MCp_2-TYPE COMPOUNDS (cm^{-1})

Compound	ν(CH)		ν(CC)		δ(CH)	π(CH)		Ring Tilt	ν(MR)[a]	δ(RMR)[a]	Refs.
$FeCp_2$	—	3077	1110	1410	1005	820	855	492	478	179	211a, 213
$RuCp_2$	—	3076	1095	1410	1005	808	834	450	380	170	213
$OsCp_2$	3061	3061	1098	1400	998	823	831	428	353	160	203, 214
$CoCp_2$	3041	3041	1101	1412	995	778	828	464	355	—	203, 214
$NiCp_2$	3075	3075	1110	1430	1000	773	839	355	355	—	211a, 215
$FeCp_2^+$	3108	3108	1116	1421	1017	805	860	501	423	—	216
	3100	3100	1110	1412	1001	779	841	490	405		
$CoCp_2^+$	3094	3094	1113	1419	1010	860	895	495	455	172	217
$IrCp_2^+$	3077	3077	1106	1409	1009	818	862	—	—	—	217

[a] R denotes the Cp ring. For the Raman spectra of MCp_2 (M = Mn, Cr, V, Ru, and Os), see Ref. 217a.

and force constants follow the order:

	Os	Fe	Ru	Cr	Co	V	Ni	Zn
k(MR)(mdyn/Å)	2.8	>2.7	>2.4	≫1.6	~1.5	~1.5	~1.5	~1.5
ν(MR)(cm^{-1})	353	478	379	408	355	379	355	345

Here, R denotes the Cp ring.

This may indicate the order of the M–R bond strength. Table IV-12 shows that the MR stretching band of $FeCp_2$ at 478 cm^{-1} is shifted to a lower frequency when it is ionized to $FeCp_2^+$. Apparently, the deviation from the inert gas electronic configuration due to the ionization weakens the M–R bond.

Many references are available on the vibrational spectra of $FeCp_2$ (212, 213, 218–221) and $RuCp_2$ (213, 222–223). Brunvoll et al.[224] carried out normal coordinate analysis on the whole ferrocene molecule.

(4) Diene-Type (σ-Bonded) Complexes*

These are complexes such as $HgCp_2$ and CH_3HgCp,[225,226] in which the metal is σ-bonded to one of the C atoms of the Cp ring:

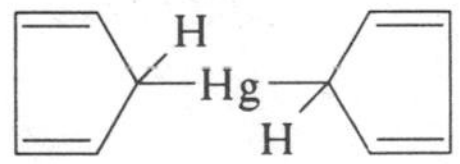

The spectra of these compounds are similar to that of C_5H_6 (cyclopentadiene), and are markedly different from those of the other groups discussed previously. Figure IV-7 shows the infrared spectrum of $HgCp_2$.[227] Band assignments of these compounds can be made based on those obtained for C_5H_6.[228] Infrared

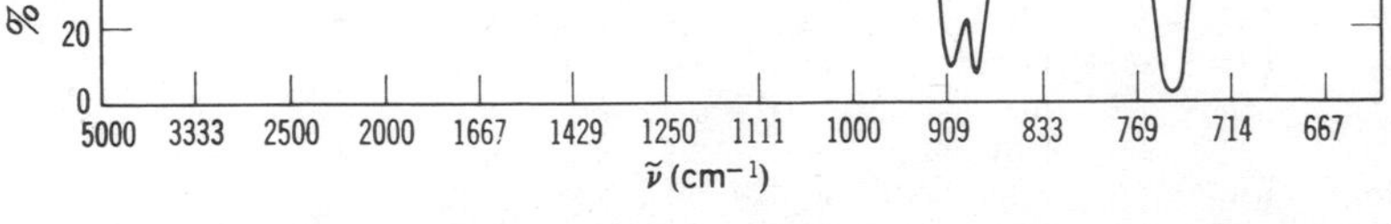

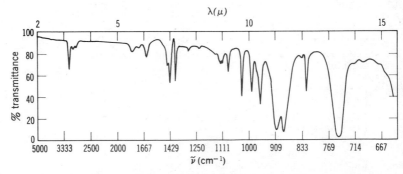

Fig. IV-7. Infrared spectrum of $Hg(C_5H_5)_2$ in CS_2(2–6μ and 7.1–15.5μ), $CHCl_3$(6–6.6μ), and CCl_4(6.6–7.1μ).[227]

* Or monohapto (h^1) complexes.

and NMR evidence suggests the presence of diene-type bonding for $(Cp)M(CH_3)_3$ (M = Si, Ge, and Sn).[229]

There are many other complexes in which the π-bonded (type 3) and the σ-bonded (type 4) cyclopentadienyl groups are mixed. As expected, these compounds exhibit bands characteristic of both groups. Typical examples are as follows: VCp_3 (two π and one σ),[230] $NbCp_4$ (two π and two σ),[231] $ZrCp_4$ (three π and one σ),[232,233,233a] and $MoCp_4$ (three π and one σ).[234] Infrared,[231] X-ray,[233] and NMR[235] evidence indicates the presence of two π- and two σ-bonded Cp rings in $TiCp_4$.

(5) Complexes of Other Types

In addition to the complexes discussed above, recent X-ray studies have revealed the existence of other types. For example, an allylic (or a trihapto, h^3) bonding was found in $[Ni(h^3-Cp)(C_3H_4)]_2$, whose structure is shown in Fig. IV-8*a*.[236] In $TiCp_3$, two rings are π-bonded while the third is bonded to the metal through only two adjacent C atoms.[237] It is rather difficult, however, to distinguish these structures from other types by vibrational spectroscopy. In the case of $NbCp_2$, the very unusual structure shown in Fig. IV-8*b* was suggested from NMR and other evidence.[238] Here the Cp ring serves as a bridge between two metal atoms by forming one π- and one σ-bond.

IV-7. CYCLOPENTADIENYL COMPOUNDS CONTAINING OTHER GROUPS

(1) Carbonyl Compounds

The vibrational spectra of carbonyl compounds were discussed in Sec. III-16. Here we discuss only those containing cyclopentadienyl rings.* It has been well established that the number of CO stretching bands observed in the

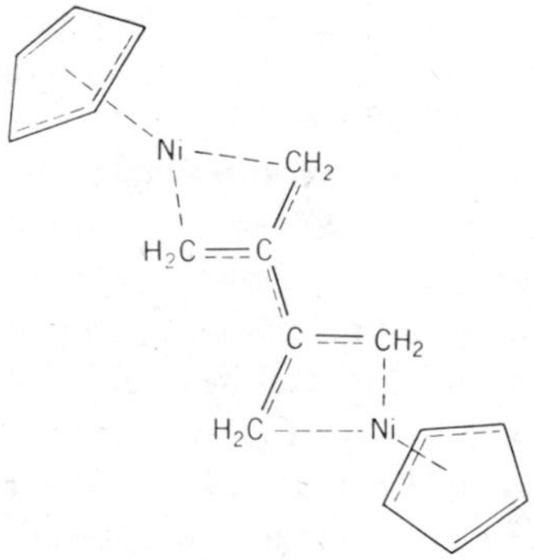

(*a*)

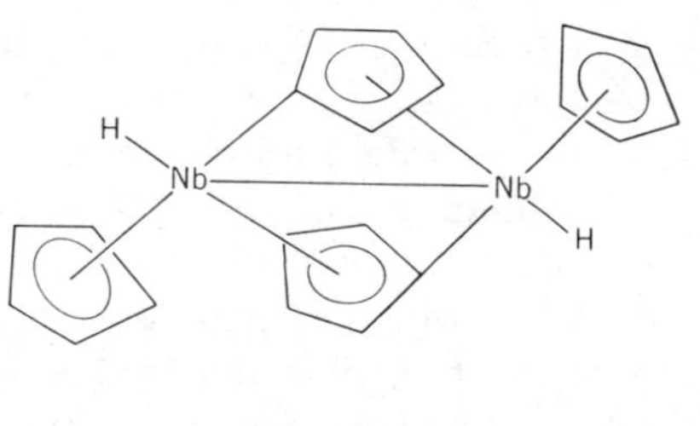

(*b*)

Fig. IV-8. Structures of some cyclopentadienyl compounds.

* Hereafter, Cp denotes a π-bonded C_5H_5 group.

infrared depends on the local symmetry of the $M(CO)_n$ group in $M(Cp)_m(CO)_n$-type compounds.[203,239] For example, only two CO stretching bands have been observed for the following compounds, in accordance with the prediction from local symmetry:

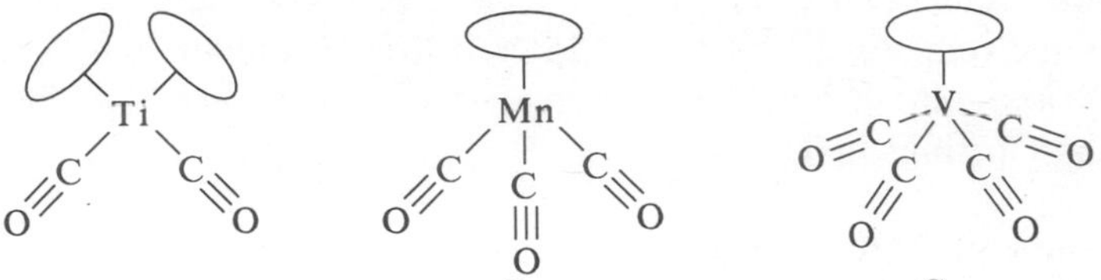

C_{2v}	C_{3v}	C_{4v}
1890 (B_2)	1938 (E)	1916 (E)
1969 (A_1)	2025 (A_1)	2016 (A_1)

In the case of $MCp(CO)_3$ (M = Mn and Re), the breakdown of the C_{5v} selection rule for the Cp vibrations was noted in solution IR spectra.[240] Typical references for carbonyl compounds are $[M(Cp)(CO)_3]^-$ (M = Cr, Mo, and W),[241] $Mn(Cp)(CO)_3$,[242] $Re(Cp)(CO)_3$,[243] and $V(Cp)(CO)_4$.[244] In $M(Cp)(CO)_3$-type compounds,[245] the CO stretching frequencies increase in the order $V^{-1} < Cr^0 < Mn^{+1} < Fe^{2+}$. This indicates that the higher the oxidation state of the metal, the less the M–C π-back bonding and the higher the CO stretching frequency.

Originally, $Fe(Cp)_2(CO)_2$ was thought to contain two π-bonded Cp rings.[246] However, an infrared and NMR study[247] showed that one ring is π-bonded and the other σ-bonded to the metal. Later, X-ray analysis confirmed this structure.[248] The structure of $Fe_2(Cp)_2(CO)_4$ has been studied extensively. In the solid state, it takes a *trans*-bridged structure (Fig. IV-9*a*),[249] or a *cis*-bridged structure (Fig. IV-9*b*) if crystallized in polar solvents at lower temperatures.[250] The *cis*-isomer exhibits two terminal (1975 and 1933 cm^{-1}) and two bridging (1801 and 1766 cm^{-1}) bands. Although the *trans*-isomer also exhibits two terminal (1956 and 1935 cm^{-1}) and two bridging (1769 and 1755 cm^{-1}) bands, these splittings are probably due to the crystal-field effect.

The structure of $Fe_2(Cp)_2(CO)_4$ in solution has been controversial. Early infrared studies[251,252] suggested the presence of the *cis*-bridged structure (Fig. IV-9*b*) mixed with a trace of noncentrosymmetric, nonbridging isomer (Fig. IV-9*c*). Manning[253] proposed, however, an equilibrium involving the three isomers, *a*, *b*, and *c* of Fig. IV-9. This was confirmed by Bullitt et al.,[254] who gave the following assignments for the spectrum in a $CS_2—C_6D_5CD_3$ solution: *trans*-isomer (*a*), 1954 and 1781 cm^{-1}; *cis*-isomer (*b*), 1998, 1954, 1810, and 1777 cm^{-1}. The frequencies of nonbridged species could not be determined because of their very low concentration. In the case of $Ru_2(Cp)_2(CO)_4$, Bullitt et al.[254] proposed an equilibrium containing four isomers: *a*, *b*, *d*, and *e* of Fig. IV-9.

It is interesting to note that the bridging CO groups of $Fe_2(Cp)_2(CO)_4$ form an adduct with trialkylaluminum[255] (see Sec. III-16):

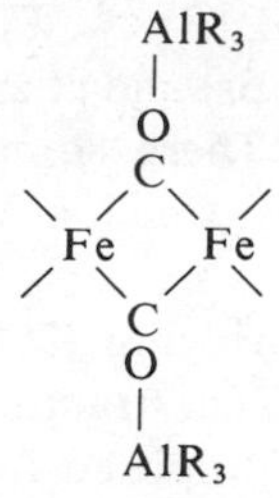

This indicates that the basicity of the bridging CO group is greater than that of the terminal CO group. The CO stretching bands of the parent compound (R: isobutyl) are at 2005 and 1962 (terminal) and 1794 (bridging) cm^{-1} in heptane solution. These bands are shifted to 2041 and 2003 (terminal) and 1680 (bridging) cm^{-1} by adduct formation. X-ray analysis has been carried

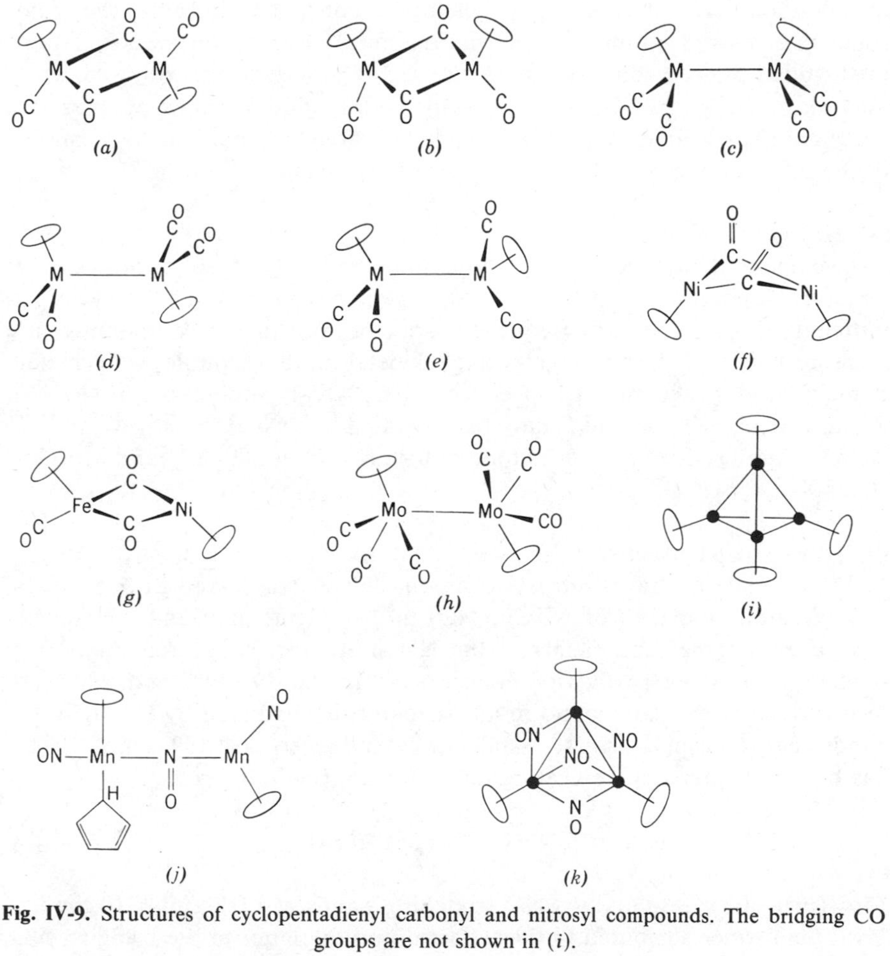

Fig. IV-9. Structures of cyclopentadienyl carbonyl and nitrosyl compounds. The bridging CO groups are not shown in (*i*).

$[Fe_2(Cp)_2(CO)_4][Al(C_2H_5)_3]_2$.[256] Formation of adducts such as $)_2(CO)_4]BX_3$ (X = Cl and Br) and $[Fe(Cp)(CO)]_4 \cdot BX_3$ (X = F, Cl, and also been confirmed.[257] These compounds exhibit bands at 1470–1290 cm^{-1} for bridging CO groups, which are bonded to a Lewis acid via the O atom.

$Ni_2(Cp)_2(CO)_2$ exhibits two bridging CO stretching bands at 1854 and 1896 cm^{-1} in heptane solution. The structure shown in Fig. IV-9*f* with a puckered $Ni(CO)_2Ni$ bridge was proposed for this compound.[258] In heptane solution $FeNi(Cp)_2(CO)_3$ shows a strong terminal CO stretching at 2004 cm^{-1} and two bridging CO stretching bands at 1855 and 1825 cm^{-1}. Since the 1855 cm^{-1} band (symmetric type) is very weak, the $Ni(CO)_2Fe$ bridge in this compound was thought to be virtually planar, as shown in Fig. IV-9*g*.[258]

According to X-ray analysis,[259] the structure of $Mo_2(Cp)_2(CO)_6$ is *trans*-centrosymmetric, as shown in Fig. IV-9*h*. The infrared spectrum in the CO stretching region is consistent with this structure, both in the solid state and in solution.[260] In solvents of high dielectric constants, however, the *trans*-rotamer is rearranged into the *gauche*-rotamer.[261] For the infrared spectra of analogous tungsten compounds, see Refs. 241 and 262. According to X-ray analysis,[263] $Fe_4(Cp)_4(CO)_4$ takes a regular tetrahedral structure such as that shown in Fig. IV-9*i*. It exhibits a bridging CO stretching band at 1649 cm^{-1} [255] in the infrared and a FeFe stretching band at 214 cm^{-1} [264] in the Raman.

(2) Halogeno Compounds

Cyclopentadienyl complexes containing metal–halogen bonds exhibit metal–halogen vibrations (Sec. III-19), together with those of the cyclopentadienyl rings. The low-frequency spectra of these compounds are complicated[265,266] because metal-ring skeletal modes couple with metal-halogen modes. The infrared spectra of $M(Cp)_2X_2$-type compounds (M = Ti, Zr, and Hf; X = Cl, Br, and I) have been studied by several investigators.[265–268] Also, infrared spectra have been reported for $Mo(Cp)(CO)_3X$[240] and $Mo(Cp)(CO)_2X_3$ (X = Cl, Br, and I).[269]

(3) Nitrosyl Compounds

Vibrational spectra of nitrosyl compounds were discussed in Sec. III-16. The vibrational spectra of Ni(Cp)(NO) and its deuterated and ^{15}N species have been assigned completely:[270] the NO stretching, NiN stretching, NiCp stretching, and NiCp tilt vibrations are at 1809, 649, 322, and 290 cm^{-1}, respectively. If this compound in an Ar matrix is irradiated by UV light, the bands near 1830 cm^{-1} disappear and a new band emerges at 1390 cm^{-1}.[271] This has been interpreted as indicating the following photoionization:

$$(Cp)NiNO \xrightarrow{h\nu} (Cp)Ni^{+}NO^{-}.$$

$Mn_2(Cp)_3(NO)_3$ exhibits two NO stretching bands at 1732 and 1510 cm^{-1}.[272] With the former attributed to the terminal and the latter to the bridging NO, the structure which is shown in Fig. IV-9*j* was proposed. The infrared spectrum of $Mo(Cp)(CO)_2(NO)$ has been reported.[273,240] Figure IV-9*k* shows the structure of $Mn_3(Cp)_3(NO)_4$, containing doubly and triply bridging NO groups. The bands at 1530 and 1480 cm^{-1} were assigned to the doubly bridged NO groups, whereas the 1320-cm^{-1} band was attributed to the triply bridged NO group.[274]

(4) Hydrido Complexes

Vibrational spectra of hydrido complexes were reviewed in Sec. III-18. The metal-hydrogen stretching bands for $Mo(Cp)_2H_2$,[275] $Re(Cp)_2H_2$, and $W(Cp)_2H_2$ [276,277] have been observed in the 2100–1800 cm^{-1} region. X-ray analysis on $Mo(Cp)_2H_2$ [278] suggests that the coordination around the Mo atom is approximately tetrahedral. In polymeric $[Zr(Cp)_2H_2]_n$,[279] the bridging ZnH stretching vibration is observed as a strong, broad band at 1540 cm^{-1}.[280] A similar bridging TiH vibration is found at 1450 cm^{-1} for $[Ti(Cp)_2H]_2$.[281] In $[\{Rh(Cp')\}_2HCl_3]$, where Cp′ denotes the pentamethyl-Cp group, the bridging RhH vibration was assigned at 1151 cm^{-1}.[282]

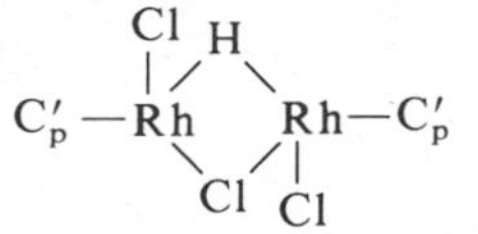

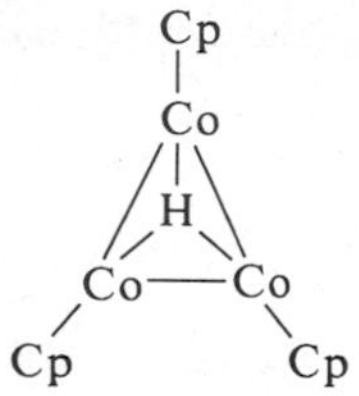

An extremely low CoH stretching frequency (950 cm^{-1}), together with an unusually high-field proton chemical shift observed for $[Co(Cp)H]_4$, was attributed to the triply bridged structure shown above (only one face of the tetrahedron is shown).[283]

(5) Complexes Containing Other Groups

As discussed in Sec. III-15, the mode of coordination of the pseudohalide ion can be determined by vibrational spectroscopy. Burmeister et al.[284] found that all NCS and NCSe groups are N-bonded in $M(Cp)_2X_2$-type compounds, where M is Ti, Zr, Hf, or V, and X is NCS or NCSe. In the case of analogous NCO complexes, Ti, Zr, and Hf form O-bonded complexes, whereas V forms an N-bonded complex. Later, Jensen et al.[285] suggested the N-bonded structure for the titanium complex.

A strong N≡N stretching band is observed at 1910 cm^{-1} in the Raman spectrum of $L_2(Cp)Mo-N\equiv N-Mo(Cp)L_2$ (L: PPh_3).[286] Thiocarbonyl complexes of the $(Cp)Mn(CO)_{3-n}(CS)_n$- ($n = 1, 2, 3$) type exhibit the C≡S stretching bands at 1340–1235 cm^{-1}.[287] In $(Cp)Nb(S_2)X$-type compounds (X = Cl, Br, I, and SCN), the S_2 is probably coordinated to the metal in a side-on fashion, and its SS stretching band may be assigned at 540 cm^{-1}.[288]

IV-8. COMPLEXES OF OTHER CYCLIC UNSATURATED LIGANDS

The vibrational spectra of a cyclobutadiene complex, $Fe(C_4H_4)(CO)_3$, have been measured,[289] and the FeR stretching and tilt vibrations have been assigned at 406 and ~500 cm^{-1}, respectively. The infrared spectra of $Ni(C_4(CH_3)_4)Cl_2$ and $M(C_4(C_6H_5)_4)X_2$ (M = Ni and Pd; X = Cl, Br, and I) have been reported.[290]

Dibenzene chromium, $Cr(C_6H_6)_2$, takes a ferrocenelike sandwich structure, and its vibrational spectra have been assigned by Fritz and Fischer.[291] Table IV-13 lists the vibrational frequencies of $Cr(C_6H_6)_2$ and its analogs. Figure IV-10 shows the infrared spectrum of $Cr(C_6H_6)_2$.[291a] On the basis of the three-body approximation, the Cr—R (ring) stretching force constant of $Cr(C_6H_6)_2$ was estimated to be 2.39 mdyn/Å, which is smaller than that of ferrocene (2.7 mdyn/Å). For complete normal coordinate analysis of $Cr(C_6H_6)_2$, see Ref. 292. Complete assignments have been made for the vibrational spectra of $Cr(C_6H_6)(C_6F_6)$.[292a] As expected, the IR spectra of $M(C_6H_6)(C_5H_5)$-type complexes show bands characteristic of both C_6H_6 and C_5H_5 rings.[293] Complete band assignments are available for $Cr(C_6H_6)(CO)_3$.[294,295]

Originally, two CH stretching bands of $M(C_5H_6)(C_5H_5)$ (M = Co and Rh) near 2750 and 2945 cm^{-1} were assigned to $\nu(CH_{endo})$ and $\nu(CH_{exo})$, respectively.[296]

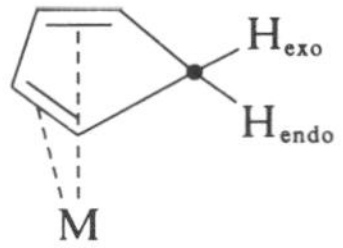

A later study[297] shows, however, that the lower-frequency band near 2750 cm^{-1} must be assigned to $\nu(CH_{exo})$, since replacement of the exo hydrogen by the phenyl or perfluorophenyl group results in the disappearance of this band. In the case of $Mn(C_6H_7)(CO)_3$, the bands at 2970 and 2830 cm^{-1} were assigned to $\nu(CH_{endo})$ and $\nu(CH_{exo})$, respectively.[298] Vibrational frequencies have

TABLE IV-13. INFRARED FREQUENCIES OF DIBENZENE-METAL COMPLEXES (CM^{-1})[203]

Complex	ν(CH)		ν(CC)	δ(CH)	ν(CC)	π(CH)		Ring Tilt	$\nu(MR)^a$	$\delta(RMR)^a$
$Cr(C_6H_6)_2$	3037	—	1426	999	971	833	794	490	459	(140)
$Cr(C_6H_6)_2^+$	3040	—	1430	1000	972	857	795	466	415	(144)
$Mo(C_6H_6)_2$	3030	2916	1425	995	966	811	773	424	362	—
$W(C_6H_6)_2$	3012	2898	1412	985	963	882	798	386	331	—
$V(C_6H_6)_2$	3058	—	1416	985	959	818	739	470	424	—

^{a} R denotes the C_6H_6 ring.

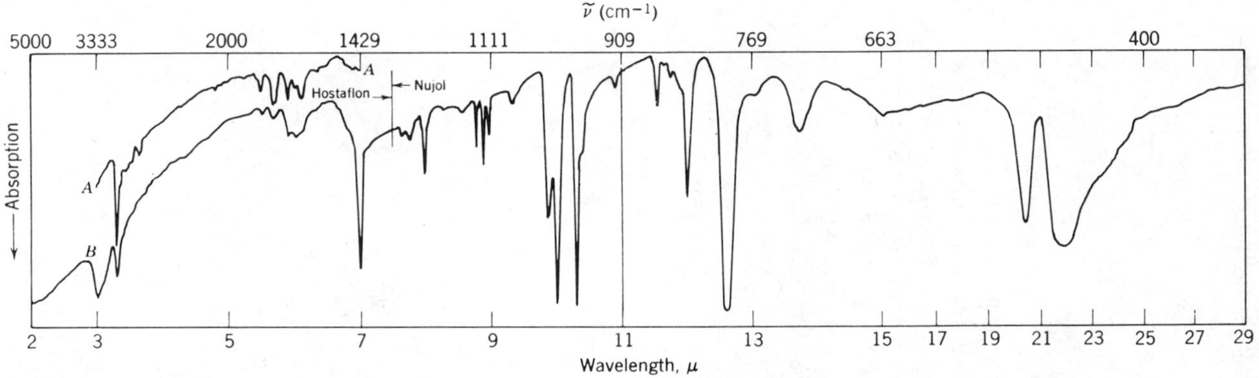

Fig. IV-10. Infrared spectrum of crystalline $Cr(C_6H_6)_2$: (A) KBr pellet, (B) Hostaflon-oil suspension (2–7.5μ), and Nujol mull suspension (7.5–29μ).[291a]

been reported for $MCl_2(C_7H_7)_2$,[299] $MCl_2(C_8H_7)_2$ (M = Ti and Zr),[300] $Ti(C_8H_8)(C_5H_5)$,[300a] and $M(C_8H_8)_2$ (M = Th[300b] and U).[301] The indenyl group may coordinate to the metal through a σ- or a π-bond:

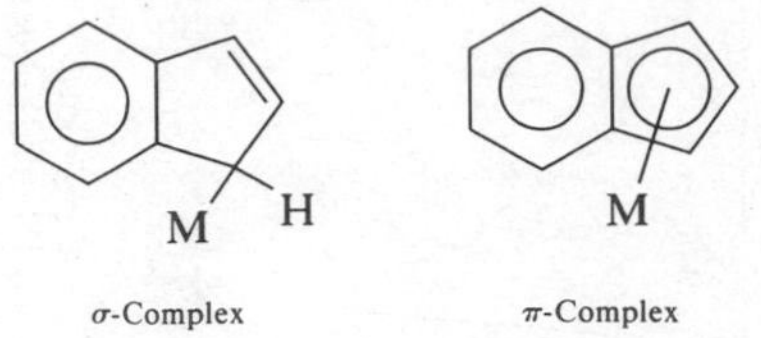

An example of the former is seen in $Hg(C_9H_7)Cl$, which exhibits an aromatic CH stretching at 3060–3050 and an aliphatic CH stretching band at 2920–2850 cm^{-1}. The latter band should be absent in the π-bonded complex.[302]

Infrared spectra are reported for a mixed-valence-state complex, biferrocene (Fe^{2+}, Fe^{3+}) picrate[303] and bis(pentalenyl)Ni,[304] whose structures are shown in Fig. IV-11*a* and IV-11*b*, respectively. The spectrum of a triple-decker compound, $[Ni_2(Cp')_3]BF_4$ (Cp′: CH_3-Cp) (Fig. IV-11*c*), is similar to that of $Ni(Cp')_2$.[305]

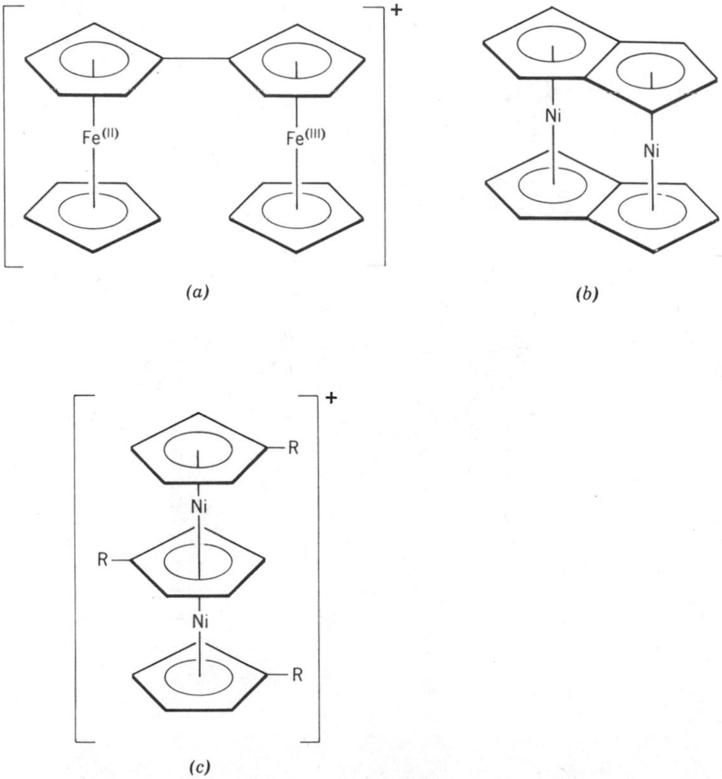

Fig. IV-11. Structures of some metal sandwich compounds.

IV-9. MISCELLANEOUS COMPOUNDS

There are many other organometallic compounds which have not been covered in the preceding sections. For these, the reader should consult general references such as the following:

1. N. N. Greenwood, ed., *Spectroscopic Properties of Inorganic and Organometallic Compounds*, Vol. 1 (1968) to present, The Chemical Society, London.
2. M. Dub, ed., *Organometallic Compounds: Methods of Synthesis, Physical Constants and Chemical Reactions*, Vols. 1–3, Springer-Verlag, New York, 1966.
3. E. Maslowsky, Jr., *Vibrational Spectra of Organometallic Compounds*, Wiley, New York, 1977.
4. K. Nakamoto, "Characterization of Organometallic Compounds by Infrared Spectroscopy", in M. Tsutsui ed., *Characterization of Organometallic Compounds*, Part I, Wiley-Interscience, New York, 1969.

Other review articles on specific groups of compounds are as follows:

Alkyldiboranes: W. J. Lehmann and I. Shapiro, *Spectrochim. Acta*, **17**, 396 (1961).
Organoaluminum Compounds: E. G. Hoffman, *Z. Elektrochem.*, **64**, 616 (1960).
Organosilicon Compounds: A. L. Smith, *Spectrochim. Acta*, **16**, 87 (1960).
Organogermanes: R. J. Cross and R. Glockling, *J. Organometal. Chem.*, **3**, 146 (1965).
Organotin Compounds: R. Okawara and W. Wada, *Adv. Organometal. Chem.*, **5**, 137 (1967).
Organophosphorus Compounds: D. E. C. Corbridge, *The Structural Chemistry of Phosphorus*, Elsevier, Amsterdam, 1974; L. C. Thomas, *Interpretation of the Infrared Spectra of Organophosphorus Compounds*, Heyden, London, 1974.
Organometallic Compounds of P, As, Sb, and Bi: E. Maslowsky, Jr., *J. Organometal. Chem.*, **70**, 153 (1974).

References

1. E. Maslowsky, Jr., *Chem. Soc. Rev.*, **9**, 25 (1980).
2. A. M. Pyndyk, M. R. Aliev, and V. T. Aleksanyan, *Opt. Spectrosc.* (English), **36**, 393 (1974).
3. R. A. Kovar and G. L. Morgan, *Inorg. Chem.*, **8**, 1099 (1969).
4. I. S. Butler and M. L. Newbury, *Spectrochim. Acta*, **33A**, 669 (1977).
4a. B. Nagel and W. Brüser, *Z. Anorg. Allg. Chem.*, **468**, 148 (1980).
5. J. R. Durig and S. C. Brown, *J. Mol. Spectrosc.*, **45**, 338 (1973).
6. J. R. Allkins and P. J. Hendra, *Spectrochim. Acta*, **22**, 2075 (1966).
7. J. R. Durig, C. M. Plater, Jr., J. Bragin, and Y. S. Li, *J. Chem. Phys.*, **55**, 2895 (1971).
7a. K. Hamada and H. Morishita, *J. Mol. Struct.*, **44**, 119 (1978).
8. W. J. Lehmann, C. O. Wilson, and I. Shapiro, *J. Chem. Phys.*, **31**, 1071 (1959).
9. L. A. Woodward, J. R. Hall, R. N. Nixon, and N. Sheppard, *Spectrochim. Acta*, **15**, 249 (1959).
9a. H. J. Becher and F. Bramsiepe, *Spectrochim. Acta*, **35A**, 53 (1979).
10. R. J. O'Brien and G. A. Ozin, *J. Chem. Soc. A*, 1136 (1971).
11. G. E. Coates and A. J. Downs, *J. Chem. Soc.*, 3353 (1964).
12. J. R. Hall, L. A. Woodward, and E. A. V. Ebsworth, *Spectrochim. Acta*, **20**, 1249 (1964).
12a. J. R. Durig and K. K. Chatterjee, *J. Raman Spectrosc.*, **11**, 168 (1981).
13. G. Bouquet and M. Bigorgne, *Spectrochim. Acta*, **23A**, 1231 (1967).
14. P. J. D. Park and P. J. Hendra, *Spectrochim. Acta*, **24A**, 2081 (1968).
14a. H. Rojhantalab, J. W. Nibler, and C. J. Wilkins, *Spectrochim. Acta*, **32A**, 519 (1976).
14b. H. Rojhantalab and J. W. Nibler, *Spectrochim. Acta*, **32A**, 947 (1976).
15. H. Siebert, *Z. Anorg. Allg. Chem.*, **273**, 161 (1953).
16. S. Sportouch, C. Lacoste, and R. Gaufrès, *J. Mol. Struct.*, **9**, 119 (1971).

16a. F. Watari, *Spectrochim. Acta*, **34A**, 1239 (1978).
17. W. F. Edgell and C. H. Ward, *J. Am. Chem. Soc.*, **77**, 6486 (1955).
18. G. A. Crowder, G. Gorin, F. H. Kruse, and D. W. Scott, *J. Mol. Spectrosc.*, **16**, 115 (1965).
19. H. H. Eysel, H. Siebert, G. Groh, and H. J. Berthold, *Spectrochim. Acta*, **26A**, 1595 (1970).
20. A. J. Downs, R. Schmutzler, and I. A. Steer, *Chem. Commun.*, 221 (1966).
21. A. J. Shortland and G. Wilkinson, *J. Chem. Soc., Dalton Trans.* 872 (1973).
22. J. W. Nibler and T. H. Cook, *J. Chem. Phys.*, **58**, 1596 (1973).
23. C. W. Hobbs and R. S. Tobias, *Inorg. Chem.*, **9**, 1998 (1970).
24. P. L. Goggin and L. A. Woodward, *Trans. Faraday Soc.*, **56**, 1591 (1960).
25. G. B. Deacon and J. H. S. Green, *Spectrochim. Acta*, **24A**, 885 (1968).
26. M. G. Miles, J. H. Patterson, C. W. Hobbs, M. J. Hopper, J. Overend, and R. S. Tobias, *Inorg. Chem.*, **7**, 1721 (1968).
27. H. Kriegsmann and S. Pischtschan, *Z. Anorg. Allg. Chem.*, **308**, 212 (1961).
28. A. J. Downs and I. A. Steer, *J. Organometal. Chem.*, **8**, P21 (1967).
29. K. J. Wynne and J. W. George, *J. Am. Chem. Soc.*, **91**, 1649 (1969).
30. M. T. Chen and J. W. George, *J. Am. Chem. Soc.*, **90**, 4580 (1968).
31. J. A. Creighton, G. B. Deacon, and J. H. S. Green, *Aust. J. Chem.*, **20**, 583 (1967).
32. R. Baumgärtner, W. Sawodny, and J. Goubeau, *Z. Anorg. Allg. Chem.*, **333**, 171 (1964).
33. E. Weiss and E. A. Lucken, *J. Organometal. Chem.*, **2**, 197 (1964).
34. R. West and W. Glaze, *J. Am. Chem. Soc.*, **83**, 3580 (1961).
35. P. Krohmer and J. Goubeau, *Z. Anorg. Allg. Chem.*, **369**, 238 (1969).
36. T. Ogawa, K. Hirota, and T. Miyazawa, *Bull. Chem. Soc. Jpn.*, **38**, 1105 (1965).
37. T. Ogawa, *Spectrochim. Acta*, **23A**, 15 (1968).
38. J. Yamamoto and C. A. Wilkie, *Inorg. Chem.*, **10**, 1129 (1971).
39. J. Mink and B. Gellai, *J. Organometal. Chem.*, **66**, 1 (1974).
40. S. Inoue and T. Yamada, *J. Organometal. Chem.*, **25**, 1 (1970).
40a. B. Nagel and W. Brüser, *Z. Anorg. Allg. Chem.*, **468**, 148 (1980).
41. J. Mink and Y. A. Pentin, *J. Organometal. Chem.*, **23**, 293 (1970).
42. J. L. Bribes and R. Gaufrès, *Spectrochim. Acta*, **27A**, 2133 (1971).
43. W. J. Lehmann, C. O. Wilson, and I. Shapiro, *J. Chem. Phys.*, **28**, 781 (1958).
44. J. Chouteau, G. Davidovics, F. D'Amato, and L. Savidan, *Compt. Rend.*, **260**, 2759 (1965).
45. J. H. S. Green, *Spectrochim. Acta*, **24A**, 137 (1968).
45a. C. Crocker and P. L. Goggin, *J. Chem. Soc., Dalton Trans.*, 388 (1978).
46. A. E. Borisov, N. V. Novikova, N. A. Chumaevskii, and E. B. Shkirtil, *Dokl. Akad. Nauk SSSR*, **173**, 855 (1967).
47. R. L. McKenney and H. H. Sisler, *Inorg. Chem.*, **6**, 1178 (1967).
48. J. A. Jackson and R. J. Nielson, *J. Mol. Spectrosc.*, **14**, 320 (1964).
49. M. I. Batuev, A. D. Petrov, V. A. Ponomarenko, and A. D. Mateeva, *Izv. Akad. Nauk SSSR, Otd. Kim. Nauk*, 1070 (1956).
50. L. A. Leites, Y. P. Egorov, J. Y. Zueva, and V. A. Ponomarenko, *Izv. Akad. Nauk SSSR, Otd. Kim. Nauk*, 2132 (1961).
51. W. R. Cullen, G. B. Deacon, and J. H. S. Green, *Can. J. Chem.*, **43**, 3193 (1965).
52. P. Taimasalu and J. L. Wood, *Trans. Faraday Soc.*, **59**, 1754 (1963).
53. C. R. Dillard and J. R. Lawson, *J. Opt. Soc. Am.*, **50**, 1271 (1960).
54. T. L. Brown, R. L. Gerteis, D. A. Bafus, and J. A. Ladd, *J. Am. Chem. Soc.*, **86**, 2134 (1964); T. L. Brown, J. A. Ladd, and C. N. Newmann, *J. Organometal. Chem.*, **3**, 1 (1965).
55. H. Dietrich, *Acta Crystallogr.*, **16**, 681 (1963).
56. T. L. Brown, in F. G. A. Stone and R. West, eds., *Advances in Organometallic Chemistry*, Vol. III, Academic Press, New York, 1965, p. 374.
57. C. N. Atam, H. Müller, and K. Dehnicke, *J. Organometal. Chem.*, **37**, 15 (1972).
57a. J. Kress and A. Novak, *J. Organometal. Chem.*, **121**, 7 (1976).
58. E. G. Hoffmann, *Z. Elektrochem.*, **64**, 616 (1960).
59. O. Yamamoto, *Bull. Chem. Soc. Jpn.*, **35**, 619 (1962).
60. C. A. Wilkie, *J. Organometal. Chem.*, **32**, 161 (1971).

61. R. J. Cross and F. Glocking, *J. Organometal. Chem.*, **3,** 146 (1965).
62. W. M. Scovell, B. Y. Kimura, and T. G. Spiro, *J. Coord. Chem.*, **1,** 107 (1971).
63. K. H. Thiele, S. Wilcke, and M. Ehrhardt, *J. Organometal. Chem.*, **14,** 13 (1968).
64. B. Busch and K. Dehnicke, *J. Organometal. Chem.*, **67,** 237 (1974).
64a. L. Czuchajowski, J. Habdas, S. A. Kucharski, and K. Rogosz, *J. Organometal. Chem.*, **155,** 185 (1978).
65. A. H. Cowley, J. L. Mills, T. M. Leohr, and T. V. Long, *J. Am. Chem. Soc.*, **93,** 2150 (1971).
66. H. D. Kaesz and F. G. A. Stone, *Spectrochim. Acta*, **15,** 360 (1959).
67. J. Mink and Y. A. Pentin, *Acta Chim. Acad. Sci. Hung.*, **66,** 277 (1970).
68. A. K. Holliday, W. Reade, K. R. Seddon, and I. A. Steer, *J. Organometal. Chem.*, **67,** 1 (1974).
69. J. D. Odom, L. W. Hall, S. Riethmiller, and J. R. Durig, *Inorg. Chem.*, **13,** 170 (1974).
69a. G. Davidson and S. Phillips, *Spectrochim. Acta*, **34A,** 949 (1978).
70. G. Davidson, *Spectrochim. Acta*, **27A,** 1161 (1971).
71. G. Masetti and G. Zerbi, *Spectrochim. Acta*, **26A,** 1891 (1970).
72. U. Kunze, E. Lindner, and J. Koola, *J. Organometal. Chem.*, **57,** 319 (1973).
73. A. N. Nesmeyanov, A. E. Borisov, N. V. Novikova, and E. I. Fedin, *J. Organometal. Chem.*, **15,** 279 (1968).
74. S. L. Stafford and F. G. A. Stone, *Spectrochim. Acta*, **17,** 412 (1961).
75. G. Davidson, P. G. Harrison, and E. M. Riley, *Spectrochim. Acta*, **29A,** 1265 (1973).
75a. G. Davidson and S. Phillips, *Spectrochim. Acta*, **35A,** 83 (1979).
76. J. Mink and Y. A. Pentin, *J. Organometal. Chem.*, **23,** 293 (1970).
77. C. Souresseau and B. Pasquier, *J. Organometal. Chem.*, **39,** 51 (1972).
78. R. E. Sacher, D. H. Lemmon, and F. A. Miller, *Spectrochim. Acta*, **32A,** 1169 (1967).
79. D. I. Maclean and R. E. Sacher, *J. Organometal. Chem.*, **74,** 197 (1974).
79a. W. M. A. Smit and G. Dijkstra, *J. Mol. Struct.*, **8,** 263 (1971).
80. R. E. Sacher, W. Davidsohn, and F. A. Miller, *Spectrochim. Acta*, **26A,** 1011 (1970).
80a. L. E. Gusel'nikov, V. V. Volkova, V. G. Avakyan, and N. S. Nametkin, *J. Organometal. Chem.*, **201,** 137 (1980).
80b. T. J. Drahnak, J. Michl, and R. West, *J. Am. Chem. Soc.*, **103,** 1845 (1981).
81. D. H. Whiffen, *J. Chem. Soc.*, 1350 (1956).
82. J. H. S. Green, *Spectrochim. Acta*, **24A,** 863 (1968).
83. J. Mink, G. Végh, and Y. A. Pentin, *J. Organometal. Chem.*, **35,** 225 (1972).
83a. C. G. Barraclough, G. E. Berkovic, and G. B. Deacon, *Aust. J. Chem.*, **30,** 1905 (1977).
84. G. Costa, A. Camus, N. Marsich, and L. Gatti, *J. Organometal. Chem.*, **8,** 339 (1967).
85. A. N. Rodionov, N. I. Rucheva, I. M. Viktorova, D. N. Shigorin, N. I. Sheverdina, and K. A. Kocheshkov, *Izv. Akad. Nauk SSSR, Ser. Khim.*, 1047 (1969).
86. H. F. Shurvell, *Spectrochim. Acta*, **23A,** 2925 (1967).
87. N. Kumar, B. L. Kalsotra, and R. K. Multani, *J. Inorg. Nucl. Chem.*, **35,** 3019 (1973).
88. A. L. Smith, *Spectrochim. Acta*, **23A,** 1075 (1967).
89. A. L. Smith, *Spectrochim. Acta*, **24A,** 695 (1968).
90. J. R. Durig, C. W. Sink, and J. B. Turner, *Spectrochim. Acta*, **26A,** 557 (1970).
91. D. H. Brown, A. Mohammed, and D. W. A. Sharp, *Spectrochim. Acta*, **21,** 663 (1965).
91a. N. S. Dance, W. R. McWhinnie, and R. C. Poller, *J. Chem. Soc., Dalton Trans.*, 2349 (1976).
92. K. Shobatake, C. Postmus, J. R. Ferraro, and K. Nakamoto, *Appl. Spectrosc.*, **23,** 12 (1969).
93. K. Cavanaugh and D. F. Evans, *J. Chem. Soc. A*, 2890 (1969).
94. H. Kriegsmann, *Z. Anorg. Allg. Chem.*, **294,** 113 (1958).
95. H. Bürger, *Spectrochim. Acta*, **24A,** 2015 (1968).
96. K. Licht and P. Koehler, *Z. Anorg. Allg. Chem.*, **383,** 174 (1971).
97. J. R. Durig, K. K. Lau, J. B. Turner, and J. Bragin, *J. Mol. Spectrosc.*, **31,** 419 (1971).
98. B. A. Nevett and A. Perry, *J. Organometal. Chem.*, **71,** 399 (1974).
99. W. M. Scovell, G. C. Stocco, and R. S. Tobias, *Inorg. Chem.*, **9,** 2682 (1970); W. M. Scovell and R. S. Tobias, *Inorg. Chem.*, **9,** 945 (1970).
100. H. J. Becher, *Z. Anorg. Allg. Chem.*, **294,** 183 (1958).
101. J. Weidlein and V. Krieg, *J. Organometal. Chem.*, **11,** 9 (1968).

102. H. Schmidbaur, J. Weidlein, H. F. Klein, and K. Eiglmeier, *Chem. Ber.*, **101**, 2268 (1968).
103. H. C. Clark and A. L. Pichard, *J. Organometal. Chem.*, **8**, 427 (1967).
104. L. E. Levchuk, J. R. Sams, and F. Aubke, *Inorg. Chem.*, **11**, 43 (1972).
105. C. W. Hobbs and R. S. Tobias, *Inorg. Chem.*, **9**, 1037 (1970).
106. E. Amberger and R. Honigschmid-Grossich, *Chem. Ber.*, **98**, 3795 (1965).
107. D. E. Clegg and J. R. Hall, *J. Organometal. Chem.*, **22**, 491 (1970).
108. P. A. Bulliner, V. A. Maroni, and T. G. Spiro, *Inorg. Chem.*, **9**, 1887 (1970).
108a. K.-H. von Dahlen and J. Lorberth, *J. Organometal. Chem.*, **65**, 267 (1974).
108b. G. C. Stocco and R. S. Tobias, *J. Coord. Chem.*, **1**, 133 (1971).
109. J. R. Hall and G. A. Swile, *J. Organometal. Chem.*, **56**, 419 (1973).
110. J. Müller and K. Dehnicke, *J. Organometal. Chem.*, **10**, P1 (1967).
111. K. Dehnicke and D. Seybold, *J. Organometal. Chem.*, **11**, 227 (1968).
112. J. Müller and K. Dehnicke, *J. Organometal. Chem.*, **12**, 37 (1968).
112a. J. R. Durig, J. F. Sullivan, A. W. Cox, Jr., and B. J. Streusand, *Spectrochim. Acta*, **34A**, 719 (1978).
113. J. Müller, *Z. Naturforsch.*, **34b**, 536 (1979).
114. H. Leimeister and K. Dehnicke, *J. Organometal. Chem.*, **31**, C3 (1971).
115. R. G. Goel and D. R. Ridley, *Inorg. Chem.*, **13**, 1252 (1974).
115a. J. R. Durig, J. F. Sullivan, and A. W. Cox, Jr., *J. Mol. Struct.*, **44**, 31 (1978).
116. J. E. Förster, M. Vargas, and H. Müller, *J. Organometal. Chem.*, **59**, 97 (1973).
117. J. Relf, R. P. Cooney, and H. F. Henneike, *J. Organometal. Chem.*, **39**, 75 (1972).
118. F. Weller, I. L. Wilson, and K. Dehnicke, *J. Organometal. Chem.*, **30**, C1 (1971).
119. K. Dehnicke, *Angew. Chem.*, **79**, 942 (1967).
120. M. Wada and R. Okawara, *J. Organometal. Chem.*, **8**, 261 (1967).
121. H. Bürger and U. Goetze, *J. Organometal. Chem.*, **10**, 380 (1967).
121a. J. S. Thayer, *Inorg. Chem.*, **7**, 2599 (1968).
122. E. E. Aynsley, N. N. Greenwood, G. Hunter, and M. J. Sprague, *J. Chem. Soc. A*, 1344 (1966).
123. W. Beck and E. Schuierer, *J. Organometal. Chem.*, **3**, 55 (1965).
124. M. R. Booth and S. G. Frankiss, *Spectrochim. Acta*, **26A**, 859 (1970).
125. J. R. Durig, Y. S. Li, and J. B. Turner, *Inorg. Chem.*, **13**, 1495 (1974).
125a. F. Watari, *J. Mol. Struct.* **32**, 285 (1976).
126. N. Q. Dao and D. Breitinger, *Spectrochim. Acta* **27A**, 905 (1971).
127. H. C. Clark, R. J. O'Brien, and A. L. Pickard, *J. Organometal. Chem.*, **4**, 43 (1965).
128. D. E. Clegg and J. R. Hall, *Spectrochim. Acta*, **23A**, 263 (1967).
129. R. J. H. Clark, A. G. Davies, and R. J. Puddenphatt, *J. Chem. Soc. A*, 1828 (1968).
130. R. S. Tobias, M. J. Sprague, and G. E. Glass, *Inorg. Chem.*, **7**, 1714 (1968).
131. J. Rouviere, V. Tabacik, and G. Fleury, *Spectrochim. Acta*, **29A**, 229 (1973).
132. J. M. Brown, A. C. Chapman, R. Harper, D. J. Mowthorpe, A. G. Davies, and P. J. Smith, *J. Chem. Soc., Dalton Trans*, 338 (1972).
133. H. Schmidbauer, J. Weidlein, and K. H. Mitschke, *Chem. Ber.*, **102**, 4136 (1969).
134. P. A. Bulliner and T. G. Spiro, *Inorg. Chem.*, **8**, 1023 (1969).
135. P. L. Goggin and L. A. Woodward, *Trans. Faraday Soc.*, **58**, 1495 (1962).
136. D. E. Clegg and R. J. Hall, *J. Organometal. Chem.*, **17**, 175 (1969); *Spectrochim. Acta*, **21**, 357 (1965).
137. G. Mann, A. Haaland, and J. Weidlein, *Z. Anorg. Allg. Chem.*, **398**, 231 (1973).
138. T. Tanaka, *Bull. Chem. Soc. Jpn.*, **33**, 446 (1960).
139. J. Lorberth and M. R. Kula, *Chem. Ber.*, **97**, 3444 (1964).
140. R. Okawara, *J. Am. Chem. Soc.*, **82**, 3287 (1960).
141. V. B. Ramos and R. S. Tobias, *Spectrochim. Acta*, **29A**, 953 (1973).
142. Y. Kawasaki, T. Tanaka, and R. Okawara, *Spectrochim. Acta*, **22**, 1571 (1966).
143. H. A. Meinema, A. Mackor, and J. G. Noltes, *J. Organometal. Chem.*, **37**, 285 (1972).
144. M. G. Miles, G. E. Glass, and R. S. Tobias, *J. Am. Chem. Soc.*, **88**, 5738 (1966).
145. M. M. McGrady and R. S. Tobias, *Inorg. Chem.*, **3**, 1161 (1964); *J. Am. Chem. Soc.*, **87**, 1909 (1965).

146. C. Z. Moore and W. H. Nelson, *Inorg. Chem.*, **8,** 138 (1969).
147. G. A. Miller and E. O. Schlemper, *Inorg. Chem.*, **12,** 677 (1973).
148. E. W. Abel, *J. Chem. Soc.*, 4406 (1960).
149. K. A. Hooton and A. L. Allred, *Inorg. Chem.*, **4,** 671 (1965).
149a. J. E. Drake, H. E. Henderson, and L. N. Khasrou, *Spectrochim. Acta*, **38A,** 31 (1982).
150. P. G. Harrison and S. R. Stobart, *J. Organometal. Chem.*, **47,** 89 (1973).
151. D. Potts, H. D. Sharma, A. J. Carty, and A. Walker, *Inorg. Chem.*, **13,** 1205 (1974).
152. H. C. Clark and R. G. Goel, *Inorg. Chem.*, **4,** 1428 (1965).
153. U. Kunze, E. Lindner, and J. Koola, *J. Organometal. Chem.*, **38,** 1 (1972).
154. H. C. Clark and R. G. Goel, *J. Organometal. Chem.*, **7,** 263 (1967).
155. B. J. Hathaway and A. E. Underhill, *J. Chem. Soc.*, 3091 (1961).
156. D. F. Ball, T. Carter, D. C. McKean, and L. A. Woodward, *Spectrochim. Acta*, **20,** 1721 (1964).
157. D. F. Ball, P. L. Goggin, D. C. McKean, and L. A. Woodward, *Spectrochim. Acta*, **16,** 1358 (1960).
158. M. W. Mackenzie, *Spectrochim. Acta*, **38A,** 1083 (1982).
159. D. F. Van de Vondel and G. P. Van der Kelen, *Bull. Soc. Chim. Belg.*, **74,** 467 (1965).
160. H. Kimmel and C. R. Dillard, *Spectrochim. Acta*, **24A,** 909 (1968).
161. C. R. Dillard and L. May, *J. Mol. Spectrosc.*, **14,** 250 (1964).
162. E. Amberger, *Angew. Chem.*, **72,** 494 (1960).
162a. A. J. F. Clark and J. E. Drake, *Spectrochim. Acta*, **34A,** 307 (1978).
162b. A. J. F. Clark, J. E. Drake, and Q. Shen, *Spectrochim. Acta*, **34A,** 311 (1978).
163. J. R. Durig and C. W. Hawley, *J. Phys. Chem.*, **75,** 3993 (1971).
164. W. J. Lehmann, C. O. Wilson, and I. Shapiro, *J. Chem. Phys.*, **32,** 1088, 1786 (1960); **33,** 590 (1960); **34,** 476, 783 (1961).
165. E. G. Hoffmann and G. Schomburg, *Z. Elektrochem.*, **61,** 1101 (1957).
166. B. Fontal and T. G. Spiro, *Inorg. Chem.*, **10,** 9 (1971).
167. R. J. H. Clark, A. G. Davies, R. J. Puddenphatt, and W. McFarlane, *J. Am. Chem. Soc.*, **91,** 1334 (1969).
168. H. Bürger and U. Goetze, *Spectrochim. Acta*, **26A,** 685 (1970).
169. R. A. Burnham and S. R. Stobart, *J. Chem. Soc., Dalton Trans.*, 1269 (1973).
170. D. J. Cardin, S. A. Keppie, and M. F. Lappert, *Inorg. Nucl. Chem. Lett.*, **4,** 365 (1968).
171. A. Terzis, T. C. Strekas, and T. G. Spiro, *Inorg. Chem.*, **13,** 1346 (1974).
172. G. F. Bradley and S. R. Stobart, *J. Chem. Soc., Dalton Trans.*, 264 (1974).
173. N. A. D. Carey and H. C. Clark, *Chem. Commun.*, 292 (1967); *Inorg. Chem.*, **7,** 94 (1968).
173a. G. J. Kubas and T. G. Spiro, *Inorg. Chem.*, **12,** 1797 (1973).
173b. W. M. Scovell and T. G. Spiro, *Inorg. Chem.*, **13,** 304 (1974).
174. G. Davidson, *Organometal. Chem. Rev. A*, 303 (1972).
175. M. J. Grogan and K. Nakamoto, *J. Am. Chem. Soc.*, **88,** 5454 (1966); **90,** 918 (1968).
176. J. P. Sorzano and J. P. Fackler, *J. Mol. Spectrosc.*, **22,** 80 (1967).
177. J. Hiraishi, *Spectrochim. Acta*, **25A,** 749 (1969).
178. J. Chatt, L. A. Duncanson, and R. G. Guy, *Nature*, **184,** 526 (1959).
179. J. A. J. Jarvis, B. T. Kilbourn, and P. G. Owston, *Acta Crystallogr.*, **B27,** 366 (1971).
180. N. Rösch, R. P. Messmer, and K. H. Johnson, *J. Am. Chem. Soc.*, **96,** 3855 (1974).
181. D. C. Andrews and G. Davidson, *J. Organometal. Chem.*, **35,** 161 (1972).
181a. M. Bigorgne, *J. Organometal. Chem.*, **127,** 55 (1977).
182. D. B. Powell, J. G. V. Scott, and N. Sheppard, *Spectrochim. Acta*, **28A,** 327 (1972).
183. M. A. Bennett, R. J. H. Clark, and D. L. Miller, *Inorg. Chem.*, **6,** 1647 (1967).
184. K. Fischer, K. Jonas, and G. Wilke, *Angew. Chem. Int. Ed.*, **12,** 565 (1973).
184a. J. Howard and T. C. Waddington, *J. Chem. Soc., Faraday Trans. 2*, **74,** 1275 (1978).
184b. M. Moskovits and G. A. Ozin, *Cryochemistry*, Wiley, New York, 1976, p. 263.
184c. D. Tevault, D. P. Strommen, and K. Nakamoto, *J. Am. Chem. Soc.*, **99,** 2997 (1977).
185. D. C. Andrews and G. Davidson, *J. Organometal. Chem.*, **55,** 383 (1973).
186. K. Shobatake and K. Nakamoto, *J. Am. Chem. Soc.*, **92,** 3339 (1970).
186a. D. C. Andrews and G. Davidson, *J. Organometal. Chem.*, **124,** 181 (1977).

187. D. M. Adams and A. Squire, *J. Chem. Soc. A*, 1808 (1970).
187a. G. Davidson and D. C. Andrews, *J. Chem. Soc., Dalton Trans.*, 126 (1972).
187b. T. B. Chenskaya, L. A. Leites, and V. T. Aleksanyan, *J. Organometal. Chem.*, **148,** 85 (1978).
188. I. S. Butler and G. G. Barna, *J. Raman Spectrosc.*, **1,** 141 (1973).
189. J. Howard and T. C. Waddington, *Spectrochim. Acta*, **34A,** 807 (1978).
190. P. J. Hendra and D. B. Powell, *Spectrochim. Acta*, **17,** 909 (1961).
191. G. Davidson, *Inorg. Chim. Acta*, **3,** 596 (1969).
192. G. Davidson and D. A. Duce, *J. Organometal. Chem.*, **44,** 365 (1972).
193. M. J. Grogan and K. Nakamoto, *Inorg. Chim. Acta*, **1,** 228 (1967).
193a. G. A. Ozin and W. J. Power, *Inorg. Chem.*, **19,** 3860 (1980).
194. M. A. Bennett and J. D. Saxby, *Inorg. Chem.*, **7,** 321 (1968).
195. B. Dickens and W. N. Lipscomb, *J. Am. Chem. Soc.*, **83,** 4062 (1961); *J. Chem. Phys.*, **37,** 2084 (1962).
196. R. T. Bailey, E. R. Lippincott, and D. Steele, *J. Am. Chem. Soc.*, **87,** 5346 (1965).
196a. G. G. Barna and I. S. Butler, *J. Raman Spectrosc.*, **7,** 168 (1978).
196b. D. W. Wertz and M. A. Moseley, *Inorg. Chem.*, **19,** 705 (1980).
197. M. A. Coles and F. A. Hart, *J. Organometal. Chem.*, **32,** 279 (1971).
198. R. Nast and H. Schindel, *Z. Anorg. Allg. Chem.*, **326,** 201 (1963).
199. I. A. Garbusova, V. T. Alexanjan, L. A. Leites, I. R. Golding, and A. M. Sladkov, *J. Organometal. Chem.*, **54,** 341 (1973).
200. K. Yasufuku and H. Yamazaki, *Bull. Chem. Soc. Jpn.*, **45,** 2664 (1972).
201. J. Chatt, G. A. Rowe, and A. A. Williams, *Proc. Chem. Soc.*, 208 (1957); J. Chatt, R. Guy, and L. A. Duncanson, *J. Chem. Soc.*, 827 (1961).
202. Y. Iwashita, A. Ishikawa, and M. Kainosho, *Spectrochim. Acta*, **27A,** 271 (1971).
202a. Y. Iwashita, F. Tamura, and A. Nakamura, *Inorg. Chem.*, **8,** 1179 (1969).
202b. W. J. Bland, R. D. Kemmitt, and R. D. Moore, *J. Chem. Soc., Dalton Trans.*, 1292 (1973).
202c. D. H. Payne, Z. A. Payne, R. Rohmer, and H. Frye, *Inorg. Chem.*, **12,** 2540 (1973).
203. H. P. Fritz, *Adv. Organometal. Chem.*, **1,** 239 (1964).
204. H. P. Fritz and L. Schäfer, *Chem. Ber.*, **97,** 1827 (1964).
205. E. O. Fischer and H. Fischer, *J. Organometal. Chem.*, **3,** 181 (1965).
206. E. O. Fischer and G. Stölzle, *Chem. Ber.*, **94,** 2187 (1961).
207. E. R. Lippincott, J. Xavier, and D. Steele, *J. Am. Chem. Soc.*, **83,** 2262 (1961).
208. J. Lusztyk and K. B. Starowieyski, *J. Organometal. Chem.*, **170,** 293 (1979).
209. K. A. Allan, B. G. Gowenlock, and W. E. Lindsell, *J. Organometal. Chem.*, **55,** 229 (1973).
210. A. Almenningen, A. Haaland, and T. Motzfeldt, *J. Organometal. Chem.*, **7,** 97 (1967).
211. P. G. Harrison and M. A. Healy, *J. Organometal. Chem.*, **51,** 153 (1973).
211a. G. Wilkinson, P. L. Pauson, and F. A. Cotton, *J. Am. Chem. Soc.*, **76,** 1970 (1954).
212. R. T. Bailey, *Spectrochim. Acta*, **27A,** 199 (1971).
213. J. S. Bodenheimer and W. Low, *Spectrochim. Acta*, **29A,** 1733 (1973).
214. B. V. Lokshin, V. T. Aleksanian, and E. B. Rusach, *J. Organometal. Chem.*, **86,** 253 (1975).
215. E. R. Lippincott and R. D. Nelson, *Spectrochim. Acta*, **10,** 307 (1958).
216. I. Pavlik and J. Klilorka, *Collect. Czech. Chem. Commun.*, **30,** 664 (1965).
217. D. Hartley and M. J. Ware, *J. Chem. Soc. A*, 138 (1969).
217a. V. T. Aleksanyan, B. V. Lokshin, G. K. Borisov, G. G. Devyatykh, A. S. Smirnov, R. V. Nazarova, J. A. Koningstein, and B. F. Gachter, *J. Organometal. Chem.*, **124,** 293 (1977).
218. F. Rocquet, L. Berreby, and J. P. Marsault, *Spectrochim. Acta*, **29A,** 1101 (1973).
219. I. J. Hyams, *Spectrochim. Acta*, **29A,** 839 (1973).
220. K. Nakamoto, C. Udovich, J. R. Ferraro, and A. Quattrochi, *Appl. Spectrosc.*, **24,** 606 (1970).
221. L. Schäfer, J. Brunvoll, and S. J. Cyvin, *J. Mol. Struct.*, **11,** 459 (1972).
222. D. M. Adams and W. S. Fernado, *J. Chem. Soc., Dalton Trans.*, 2507 (1972).
223. J. Brunvoll, S. J. Cyvin, and L. Schäfer, *Chem. Phys. Lett.*, **13,** 286 (1972).
224. J. Brunvoll, S. J. Cyvin, and L. Schäfer, *J. Organometal. Chem.*, **27,** 107 (1971).
225. E. Maslowsky, Jr., and K. Nakamoto, *Inorg. Chem.*, **8,** 1108 (1969).
226. F. A. Cotton and T. J. Marks, *J. Am. Chem. Soc.*, **91,** 7281 (1969).

227. G. Wilkinson and T. S. Piper, *J. Inorg. Nucl. Chem.*, **2,** 32 (1956).
228. E. Gallinella, B. Fortunato, and P. Mirone, *J. Mol. Spectrosc.*, **24,** 345 (1967).
229. A. Davison and P. E. Rakita, *Inorg. Chem.*, **9,** 289 (1970).
230. F. W. Siegert and H. J. de Liefde Meijer, *J. Organometal. Chem.*, **15,** 131 (1968).
231. F. W. Siegert and H. J. de Liefde Meijer, *J. Organometal. Chem.*, **20,** 141 (1969).
232. V. I. Kulishov, E. M. Brainina, N. G. Bokiy, and Yu. T. Struchkov, *Chem. Commun.* 475 (1970).
233. J. L. Calderon, F. A. Cotton, B. G. DeBoer, and J. Takats, *J. Am. Chem. Soc.*, **93,** 3592 (1971).
233a. R. D. Rogers, R. V. Bynum, and J. L. Atwood, *J. Am. Chem. Soc.*, **100,** 5238 (1978).
234. E. O. Fischer and Y. Hristidu, *Chem. Ber.*, **95,** 253 (1962).
235. J. L. Calderon, F. A. Cotton, and J. Takats, *J. Am. Chem. Soc.*, **93,** 3587 (1971).
236. A. E. Smith, *Inorg. Chem.*, **11,** 165 (1972).
237. R. A. Forder and K. Prout, *Acta Crystallogr.*, **B30,** 491 (1974).
238. F. N. Tebbe and G. W. Parshall, *J. Am. Chem. Soc.*, **93,** 3793 (1971).
239. H. P. Fritz and E. F. Paulus, *Z. Naturforsch.*, **18b,** 435 (1963).
240. P. J. Fitzpatrick, Y. Le Page, J. Sedman, and I. S. Butler, *Inorg. Chem.*, **20,** 2852 (1981).
241. R. Feld, E. Hellner, A. Klopsch, and K. Dehnicke, *Z. Anorg. Allg. Chem.*, **442,** 173 (1978).
242. D. M. Adams and A. Squire, *J. Organometal. Chem.*, **63,** 381 (1973).
243. B. V. Lokshin, Z. S. Klemmenkova, and Yu. V. Makarov, *Spectrochim. Acta*, **28A,** 2209 (1972).
244. J. R. Durig, R. B. King, L. W. Houk, and A. L. Marston, *J. Organometal. Chem.*, **16,** 425 (1969).
245. A. Davison, M. L. H. Green, and G. Wilkinson, *J. Chem. Soc.*, 3172 (1961).
246. B. F. Hallam and P. L. Pauson, *Chem. Ind. (London)*, **23,** 653 (1955).
247. T. S. Piper and G. Wilkinson, *Chem. Ind. (London)*, **23,** 1296 (1955); *J. Inorg. Nucl. Chem.*, **3,** 104 (1956).
248. M. J. Bennett, F. A. Cotton, A. Davison, J. W. Faller, S. J. Lippard, and S. M. Morehouse, *J. Am. Chem. Soc.*, **88,** 4371 (1966).
249. O. S. Mills, *Acta Crystallogr.*, **11,** 620 (1958); R. F. Bryan and P. T. Greene, *J. Chem. Soc. A*, 3064 (1970).
250. R. F. Bryan, P. T. Greene, M. J. Newlands, and D. S. Field, *J. Chem. Soc. A*, 3068 (1970).
251. F. A. Cotton and G. Yagupsky, *Inorg. Chem.*, **6,** 15 (1967).
252. R. D. Fischer, A. Vogler, and K. Noack, *J. Organometal. Chem.*, **7,** 135 (1967).
253. A. R. Manning, *J. Chem. Soc. A*, 1319 (1968).
254. J. G. Bullitt, F. A. Cotton, and T. J. Marks, *Inorg. Chem.*, **11,** 671 (1972).
255. A. Alich, N. J. Nelson, D. Strope, and D. F. Shriver, *Inorg. Chem.*, **11,** 2976 (1972); N. J. Nelson, N. E. Kime, and D. F. Shriver, *J. Am. Chem. Soc.*, **91,** 5173 (1969).
256. N. E. Kim, N. J. Nelson, and D. F. Shriver, *Inorg. Chim. Acta*, **7,** 393 (1973).
257. J. S. Kristoff and D. F. Shriver, *Inorg. Chem.*, **13,** 499 (1974).
258. P. McArdle and A. R. Manning, *J. Chem. Soc. A*, 717 (1971).
259. F. C. Wilson and D. P. Shoemaker, *J. Chem. Phys.*, **27,** 809 (1957).
260. G. Davidson and E. M. Riley, *J. Organometal. Chem.*, **51,** 297 (1973).
261. R. D. Adams and F. A. Cotton, *Inorg. Chim. Acta*, **7,** 153 (1973).
262. A. Davison, W. McFarlane, E. Pratt, and G. Wilkinson, *J. Chem. Soc.*, 3653 (1962).
263. M. A. Neuman, Trinh-Toan, and L. F. Dahl, *J. Am. Chem. Soc.*, **94,** 3382 (1972).
264. A. Terzis and T. G. Spiro, *Chem. Commun.*, 1160 (1970).
265. E. Maslowsky, Jr., and K. Nakamoto, *Appl. Spectrosc.*, **25,** 187 (1971).
266. E. Samuel, R. Ferner, and M. Bigorgne, *Inorg. Chem.*, **12,** 881 (1973).
267. P. M. Druce, B. M. Kingston, M. F. Lappert, and R. C. Srivastava, *J. Chem. Soc. A*, 2106 (1969).
268. G. Balducci, L. Bencivenni, G. DeRosa, R. Gigli, B. Martini, and S. Nunziante, *J. Mol. Struct.*, **64,** 163 (1980).
269. R. J. Haines, R. S. Nyholm, and M. H. B. Stiddard, *J. Chem. Soc. A*, 1606 (1966).
270. G. Paliani, R. Cataliotti, A. Poletti, and A. Foffani, *J. Chem. Soc., Dalton Trans.*, 1741 (1972).
271. O. Crichton and A. J. Rest, *Chem. Commun.*, 407 (1973).
272. T. S. Piper and G. Wilkinson, *J. Inorg. Nucl. Chem.*, **2,** 38 (1956).
273. H. Brunner, *J. Organometal. Chem.*, **16,** 119 (1969).
274. R. C. Elder, F. A. Cotton, and R. A. Schunn, *J. Am. Chem. Soc.*, **89,** 3645 (1967).

275. M. J. D'Aniello, Jr., and E. K. Barefield, *J. Organometal. Chem.*, **76,** C50 (1974).
276. R. L. Cooper, M. L. H. Green, and J. T. Moelwyn-Hughes, *J. Organometal. Chem.*, **3,** 261 (1965).
277. M. P. Johnson and D. F. Shriver, *J. Am. Chem. Soc.*, **88,** 301 (1966).
278. M. Gerloch and R. Mason, *J. Chem. Soc.*, 296 (1965).
279. B. D. James, R. K. Nanda, and M. G. H. Wallbridge, *Inorg. Chem.*, **6,** 1979 (1967).
280. L. Banford and G. E. Coates, *J. Chem. Soc.*, 5591 (1964).
281. J. E. Bercaw and H. H. Brintzinger, *J. Am. Chem. Soc.*, **91,** 7301 (1969).
282. C. White, D. S. Gill, J. W. Kang, H. B. Lee, and P. M. Maitlis, *Chem. Commun.*, 734 (1971).
283. J. Müller and H. Dorner, *Angew. Chem. Int. Ed.*, **12,** 843 (1973).
284. J. L. Burmeister, E. A. Deardorff, A. Jensen, and V. H. Christiansen, *Inorg. Chem.*, **9,** 58 (1970).
285. A. Jensen, V. H. Christiansen, J. F. Hansen, T. Likowski, and J. L. Burmeister, *Acta Chem. Scand.*, **26,** 2898 (1972).
286. M. L. H. Green and W. E. Silverthorn, *Chem. Commun.*, 557 (1971).
287. A. E. Fenster and I. S. Butler, *Can. J. Chem.*, **50,** 598 (1972).
288. P. M. Treichel and G. P. Werber, *J. Am. Chem. Soc.*, **90,** 1753 (1968).
289. J. Howard and T. C. Waddington, *Spectrochim. Acta*, **34A,** 445 (1978).
290. H. P. Fritz, *Z. Naturforsch.*, **16b,** 415 (1961).
291. H. P. Fritz and E. O. Fischer, *J. Organometal. Chem.*, **7,** 121 (1967).
291a. H. P. Fritz, W. Lüttke, H. Stammreich, and R. Forneris, *Spectrochim. Acta*, **17,** 1068 (1961).
292. S. J. Cyvin, J. Brunvoll, and L. S. Schäfer, *J. Chem. Phys.*, **54,** 1517 (1971).
292a. J. D. Laposa, N. Hao, B. G. Sayer, and M. J. McGlinchey, *J. Organometal. Chem.*, **195,** 193 (1980).
293. H. P. Fritz and J. Manchot, *J. Organometal. Chem.*, **2,** 8 (1964).
294. D. M. Adams, R. E. Christopher, and D. C. Stevens, *Inorg. Chem.*, **14,** 1562 (1975).
295. E. M. Bisby, G. Davidson, and D. A. Duce, *J. Mol. Struct.*, **48,** 93 (1978).
296. M. L. H. Green, L. Pratt, and G. Wilkinson, *J. Chem. Soc.*, 3753 (1959).
297. P. M. Treichel and R. L. Shubkin, *Inorg. Chem.*, **6,** 1328 (1967).
298. G. Winkhaus, L. Pratt, and G. Wilkinson, *J. Chem. Soc.*, 3807 (1961).
299. K. M. Sharma, S. K. Anand, R. K. Multani, and B. D. Jain, *J. Organometal. Chem.*, **23,** 173 (1970).
300. K. M. Sharma, S. K. Anand, R. K. Multani, and B. D. Jain, *J. Organometal. Chem.*, **25,** 447 (1970).
300a. J. Goffart and L. Hocks, *Spectrochim. Acta*, **37A,** 609 (1981).
300b. V. T. Aleksanyan, I. A. Garbusova, T. M. Chernyshova, Z. V. Todres, M. R. Leonov, and N. I. Gramateeva, *J. Organometal. Chem.*, **217,** 169 (1981).
301. J. Goffart, J. Fuger, B. Gilbert, B. Kanellakopulos, and G. Duyckaerts, *Inorg. Nucl. Chem. Lett.*, **8,** 403 (1972).
302. E. Samuel and M. Bigorgne, *J. Organometal. Chem.*, **19,** 9 (1969); **30,** 235 (1971).
303. F. Kaufman and D. O. Cowan, *J. Am. Chem. Soc.*, **92,** 6198 (1970).
304. T. J. Katz and N. Acton, *J. Am. Chem. Soc.*, **94,** 3281 (1972).
305. A. Salzer and H. Werner, *Angew. Chem. Int. Ed.*, **11,** 930 (1972).

Bioinorganic Compounds

Part V

Metal ions in biological systems are divided into two classes. The first class consists of ions such as K^+, Na^+, Mg^{2+}, and Ca^{2+} which are found in relatively high concentrations. These ions are important in maintaining the structure of proteins by neutralizing negative charges of peptide chains and in controlling the function of cell membranes which selectively pass certain molecules. In the second class, ionic forms of Mn, Fe, Co, Cu, Zn, Mo, and so on exist in small to trace quantities, and are often incorporated into proteins (metalloproteins). The latter class is divided into two categories: (A) transport and storage proteins and (B) enzymes. Type A includes oxygen transport proteins such as hemoglobin (Fe), myoglobin (Fe), hemerythrin (Fe), and hemocyanin (Cu), and electron transfer proteins such as cytochromes (Fe), iron–sulfur proteins (Fe), and blue copper proteins (Cu), and metal storage proteins such as ferritin (Fe) and ceruloplasmin (Cu). Type B includes hydrolases such as carboxypeptidase (Zn) and aminopeptidase (Zn, Mg), oxidoreductases such as oxidase (Fe, Cu, Mo) and nitrogenase (Mo, Fe), and isomerase such as vitamin B_{12} coenzyme (Co).

In order to understand the roles of these metal ions in metalloproteins, it is first necessary to know the coordination chemistry (structure and bonding) of metal ions in their active sites. Such information is difficult to obtain since these active sites are buried in a large and complex protein backbone. Although X-ray crystallography would be ideal for this purpose, its application is hampered by the difficulties in growing single crystals of large protein molecules and in analyzing diffraction data with high resolution. As will be discussed later, these difficulties have been overcome in some cases, and knowledge of precise geometries has made great contribution to our understandings of their biological functions in terms of molecular structure. In other cases where X-ray structural information is not available or definitive, a variety of physicochemical techniques have been employed to gain structural and bonding information about the metal and its environment. These include electronic, infrared, resonance Raman, ESR, NMR, ORD, CD, Mössbauer spectroscopy, EXAFS, and electrochemical, thermodynamic, and kinetic measurements.

Recently, resonance Raman(RR) spectroscopy (Sec.I-21) has been used extensively for the study of active sites of metalloproteins. The reason for this is twofold:

1. Most metalloproteins have strong electronic absorptions in the uv-visible region which originate in a chromophore containing a metal center. By tuning the laser wavelength into these bands, it is possible to selectively enhance the vibrations localized in this chromophore without interference from the rest of the protein.

2. Due to strong resonance enhancement of these vibrations, only a dilute solution is needed to observe their RR spectra. This enables one to obtain spectra from a small volume of dilute aqueous solution under biological conditions.

In some cases, however, the vibrations of interest may not be enhanced with sufficient intensity. A typical example is the $\nu(O_2)$ of oxyhemoglobin. Then, one must resort to IR spectroscopy which exhibits all vibrations allowed by IR selection rules. It should be noted, however, that IR measurements in aqueous media are generally limited to the regions where water does not absorb strongly (Sec. III-9). Furthermore, it is often necessary to use difference techniques to cancel out interfering bands due to the solvent and some solute bands.

In the following, we will review typical results to demonstrate the utility of vibrational spectroscopy in deducing structural and bonding information about large and complex biological molecules. For a more complete coverage of the subject, the reader should consult excellent review articles quoted in each chapter. Recently, marked progress has been made in biomimetic chemistry where the active site is modeled by relatively simple coordination compounds. Vibrational studies on these model systems will also be reviewed whenever available.

V-1. MYOGLOBIN AND HEMOGLOBIN

Myoglobin (Mb, MW ~ 16,000) is an oxygen storage protein found in animal muscles. Figure V-1 shows the structure of sperm-whale myoglobin as determined by X-ray analysis. It is a monomer consisting of 153 amino acids, and its active site is an iron protoporphyrin (see Fig. III-13) which is linked axially to the proximal histidine (F8). In the deoxy state, the iron is divalent and high spin, and the Fe atom is out of the porphyrin-core plane by ~0.6 Å as shown in Fig. V-2. Upon oxygenation, the dioxygen molecule coordinates to the vacant axial position, and the heme core becomes planar. The Fe atom in oxy-Mb is low spin, and its oxidation state is close to Fe(III) (vide infra).

Hemoglobin (Hb, MW ~ 64,000) is an oxygen transport protein found in animal blood. It consists of four subunits (α_1, α_2, β_1, and β_2) each of which takes a structure similar to that of Mb. However, these four subunits are not completely independent of each other. Oxygen-uptake studies show that the oxygen affinity of each subunit depends upon the number of other subunits that are already oxygenated (cooperativity). This phenomenon has been explained in terms of two quaternary structures called the T and R states. Deoxy-Hb is in the T_0 (most tense) state. As it gradually absorbs dioxygen, the R state becomes more stable than the corresponding T state. Finally, oxy-Hb assumes the R_4 (most relaxed) state.[1]

Felton and Yu,[1a] Spiro,[2] and Asher[3] have reviewed RR spectra of heme proteins. Alben[4] has reviewed IR spectra of metalloporphyrins and heme proteins.

(1) Porphyrin-Core Vibrations

In Sec. III-4, we have discussed the correlations between certain core vibrations of simple iron porphyrins in RR spectra and structure parameters

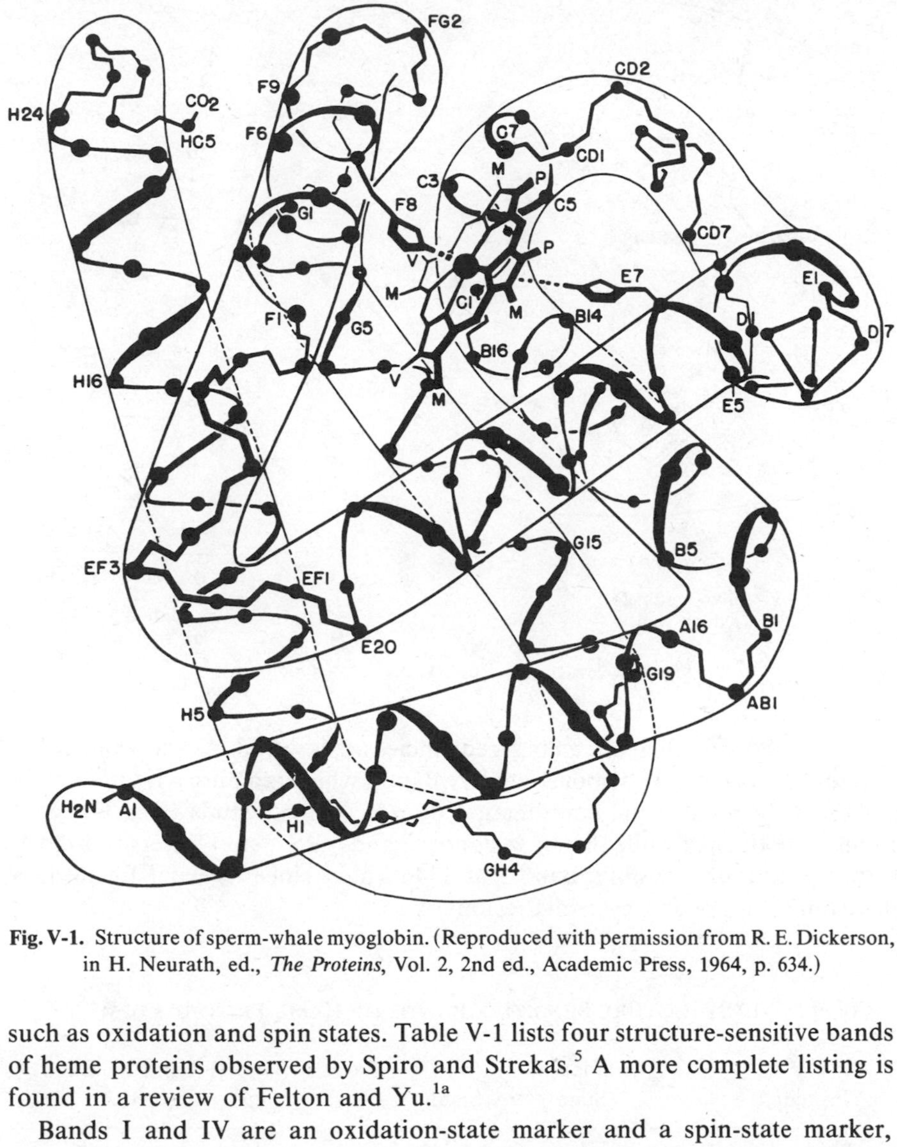

Fig. V-1. Structure of sperm-whale myoglobin. (Reproduced with permission from R. E. Dickerson, in H. Neurath, ed., *The Proteins*, Vol. 2, 2nd ed., Academic Press, 1964, p. 634.)

such as oxidation and spin states. Table V-1 lists four structure-sensitive bands of heme proteins observed by Spiro and Strekas.[5] A more complete listing is found in a review of Felton and Yu.[1a]

Bands I and IV are an oxidation-state marker and a spin-state marker, respectively, while Bands II and V are sensitive to both oxidation and spin states. Based on these results, Spiro and Strekas[5] proposed that the $Fe–O_2$ bond in oxy-Hb should be formulated as $Fe(III)–O_2^-$. Bands I, II, IV, and V correspond to the ν_4, ν_3, ν_{19}, and ν_{10}, respectively, of Ni(OEP) shown in Fig. III-16. According to normal coordinate calculations by Kitagawa et al.[6] (Table III-10), the oxidation-state-sensitive bands (ν_3, ν_4, and ν_{10}) contain $\nu(C_\alpha C_\beta)$ and $\nu(C_\alpha C_m)$ or $\nu(C_\alpha N)$ as the major contributors in their potential energy distribution. By lowering the oxidation state, back-donation of *d*-electrons to the porphyrin π^* orbital increases. Thus, the porphyrin π-bonds

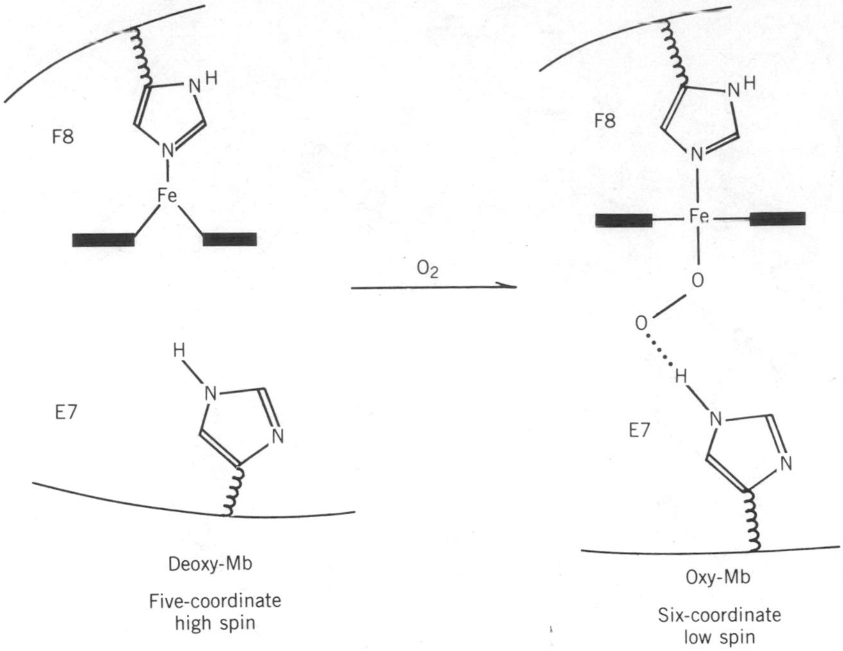

Fig. V-2. Structures of deoxy- and oxy-myoglobins.

are weakened, and their stretching frequencies are lowered. As shown in Table V-1, this is most clearly demonstrated by Band I which is a pure oxidation-state marker. In general, axial coordination of π-acceptor ligands (CO, O_2, etc.) raises its frequency while that of π-donor ligands (RS^-, etc.) lowers it. In fact, cytochrome P-450 exhibits Band I at 1346 cm^{-1}[1,7] since its axial ligand is a mercaptide sulfur of a cysteinyl residue.

TABLE V-I. STRUCTURE-SENSITIVE BANDS OF HEME PROTEINS (CM^{-1})[a,4]

Protein	Oxidation State	Spin State	Band I (*p*)	Band II (*p*)	Band IV (*ap*)	Band V (*dp*)
Ferricytochrome *c*	Fe(III)	Low spin	1374	1502	1582	1636
CN-Met-Hb	Fe(III)	Low spin	1374	1508	1588	1642
F-Met-Hb	Fe(III)	High spin	1374	1482	1555	1608
deoxy-Hb	Fe(II)	High spin	1358	1473	1552	1607
Ferrocytochrome *c*[b]	Fe(II)	Low spin	1362	1493	1584	1620
oxy-Hb	Fe(III)	Low spin	1377	1506	1586	1640

[a] The bands are numbered following the convention given by T. G. Spiro and J. M. Burke (*J. Am. Chem. Soc.*, **98,** 5482 (1976)).
[b] The RR spectrum is shown in Fig. I-19.

As discussed in Sec. III-6, the sensitivity of RR bands to spin state is attributed to expansion or out-of-plane deformation of the porphyrin core. In high-spin iron, electrons populate the antibonding $d_{x^2-y^2}$ orbital, and the lengthened Fe-N bonds are accommodated by expansion of the porphyrin core or displacement of the Fe atom from the porphyrin-core plane. This results in weakening of the methine bridge bonds in high-spin complexes. Thus, the frequencies of spin-state-sensitive bands (ν_3, ν_{19}, and ν_{10}) are lower in high-spin than in low-spin complexes since all these vibrations contain $\nu(C_\alpha C_m)$ as the major contributor in their normal modes (see Table III-10). The spin-state-sensitive bands are also metal sensitive since electron occupation in the antibonding $d_{x^2-y^2}$ orbital is varied in a series of transition metals.[8]

(2) Axial Ligand Vibrations

Asher et al.[9] observed the ν(Fe-OH), ν(Fe-N_3), and ν(Fe-F) vibrations of the OH^-, N_3^-, and F^- complexes of Met-Hb at 497, 413, and 471/443 (doublet) cm^{-1}, respectively. These vibrations are resonance-enhanced either via the direct[9] or indirect mechanism[2] (Sec. III-4).

The ν(Fe-N(Im)) vibration is of particular interest since the histidyl imidazole occupies one of the axial sites in many heme proteins. Both Nagai et al.[10] and Kincaid et al.[11] have assigned the band near 220 cm^{-1} of deoxy-Hb to the ν(Fe-N(Im)) vibration. According to Nagai et al.,[10] this band shifts from 215-218 cm^{-1} to 220-221 cm^{-1} in going from the T to R state. They interpret this result as indicating that the Fe-N bond is stretched in the T state due to strain exerted by globin. Stein et al.[12] point out, however, that such a shift may also be attributed to the effect of hydrogen bonding to the N_1-H proton of the coordinated imidazole ligand. In contrast to all these workers, Desbois and Lutz[12a] prefer to assign the same band to the ν(Fe-N(pyrrole)) mixed with pyrrole deformation.

The ν(Fe-O_2) of oxy-Hb was first observed by Brunner[13] at 567 cm^{-1} with 488 nm excitation. Oxy-Hb with $^{16}O^{18}O$ exhibits two ν(Fe-O_2) at 567 and 540 cm^{-1} which are close to that of the $^{16}O_2$ and $^{18}O_2$ adducts, respectively. By combining this observation with normal coordinate calculations, Duff et al.[14] confirmed that the O_2 in oxy-Hb takes the end-on geometry (Sec. III-17). Recently, Benko and Yu[15] suggested that the 567-cm^{-1} band of oxy-Hb should be assigned to the δ(FeOO) and not to ν(Fe-O_2). However, the work by Bajdor et al.[16] seems to support the original assignment by Brunner.

Thus far, the $\nu(O_2)$ of oxy-Hb has not been observed by RR spectroscopy either because the Fe-O_2 CT band cannot be reached by available laser lines or because its oscillator strength is too small for resonance excitation. Using IR difference techniques, Caughey et al.[17] observed two bands at 1155 and 1107 cm^{-1} which are shifted to 1096 and 1064 cm^{-1}, respectively, by $^{16}O_2$-$^{18}O_2$ substitution. These workers attributed the splitting of the $\nu(O_2)$ to Fermi resonance between the unperturbed $\nu(O_2)$ ($\sim$1135 cm^{-1}) and the first overtone of the ν(Fe-O_2) (2×567 cm^{-1}). Tsubaki and Yu[18] observed three oxygen-isotope-sensitive bands at 1153, 1137, and 1103 cm^{-1} in the RR spectrum of

Co(II)-reconstituted oxy-Hb. They present evidence that the $\nu(O_2)$ of oxy-Fe(II)-Hb are similar to its Co(II) analog, and assign the 1153-cm^{-1} band to the bound O_2 which is free from interaction with the surrounding peptide chain and ascribe the remaining pair of bands (1137 and 1103 cm^{-1}) to Fermi resonance between a porphyrin ring mode and the $\nu(O_2)$ of the bound O_2 (~1120 cm^{-1}) which is hydrogen bonded to the distal imidazole (Fig. V-2). These two isomers are different in the direction of the Fe-O-O moiety relative to the distal imidazole. Kitagawa et al.[19] observed that the 1133-cm^{-1} band (1137 cm^{-1} by previous workers[18]) is shifted to 1138 cm^{-1} in D_2O solution. This shift is consistent with the previous assignment which attributes the 1133-cm^{-1} band to the hydrogen-bonded species.

Alben and Caughey[20] first observed the $\nu(CO)$ of Hb-CO at 1950 cm^{-1} using IR difference techniques. Based on the $^{12}C^{18}O$ and $^{13}C^{16}O$ isotopic shifts, they proposed that CO is bonded to the Fe via the O atom. Tsubaki et al.[21] observed the $\nu(CO)$, $\delta(FeCO)$, and $\nu(FeC)$ at 1951, 578, and 507 cm^{-1}, respectively, in the RR spectrum of Hb-CO (406.7 nm excitation), and concluded from the isotopic shifts of these bands that the CO is bonded to the Fe via the C atom.

Stong et al.[22] observed the ν(Fe-NO) of ferrous Hb-NO at 553 cm^{-1} in RR spectra (454.5 nm excitation). When IHP (inositol hexaphosphate) was added, an additional band appeared at 592 cm^{-1} which was attributed to the ν(Fe-NO) of the five-coordinate Hb-NO resulting from the Fe-N(Im) bond breaking. However, Tsubaki and Yu[23] could not observe this band with 406.7 or 413.3 nm excitation. Recently, Benko and Yu[15] assigned the 554-cm^{-1} band of ferrous Hb-NO to the δ(FeNO) rather than the ν(Fe-NO). This new assignment is based on the zigzag isotopic shift pattern in the order of NO(554 $^{-1}$) $\rightarrow$ ^{15}NO(545 cm^{-1}) $\rightarrow$ N^{18}O(554 cm^{-1}). However, no bands corresponding to the ν(Fe-NO) have been observed by these workers. For ferric Mb-NO, Benko and Yu[15] observed two isotope-sensitive bands at 595 and 573 cm^{-1} which were assigned to the ν(Fe-NO) and δ(FeNO), respectively.

In the abnormal α subunit of Hb M Boston, the heme iron is bonded to the phenolate oxygen of tyrosine(E7)[24] instead of the proximal histidine(F8). Nagai et al.[25] assigned the 630-cm^{-1} band of this compound to the ν(Fe-O-(tyrosine)).

(3) Model Compounds

A number of model compounds have been prepared to mimic reversible oxygenation of Hb and Mb,[26] Among them, the picket-fence porphyrin ($T_{piv}PP$) (Fig. V-3a) prepared by Collman et al.[27] and Capped porphyrin (Fig. V-3*b*) prepared by Baldwin et al.[28] are studied most extensively. The $\nu(O_2)$ of Fe($T_{piv}PP$)(1-MeIm) and its Co(II) analog were observed at 1159 and 1150 cm^{-1}, respectively, by using IR difference techniques.[29] These frequencies are close to that of Co-oxy-Hb at 1153 cm^{-1} which was attributed to the bound dioxygen not hydrogen bonded to the distal imidazole.[18] Using the 457.9 nm excitation, Burke et al.[30] observed the $\nu(Fe\text{-}O_2)$ of Fe($T_{piv}PP$)(1-MeIm) at

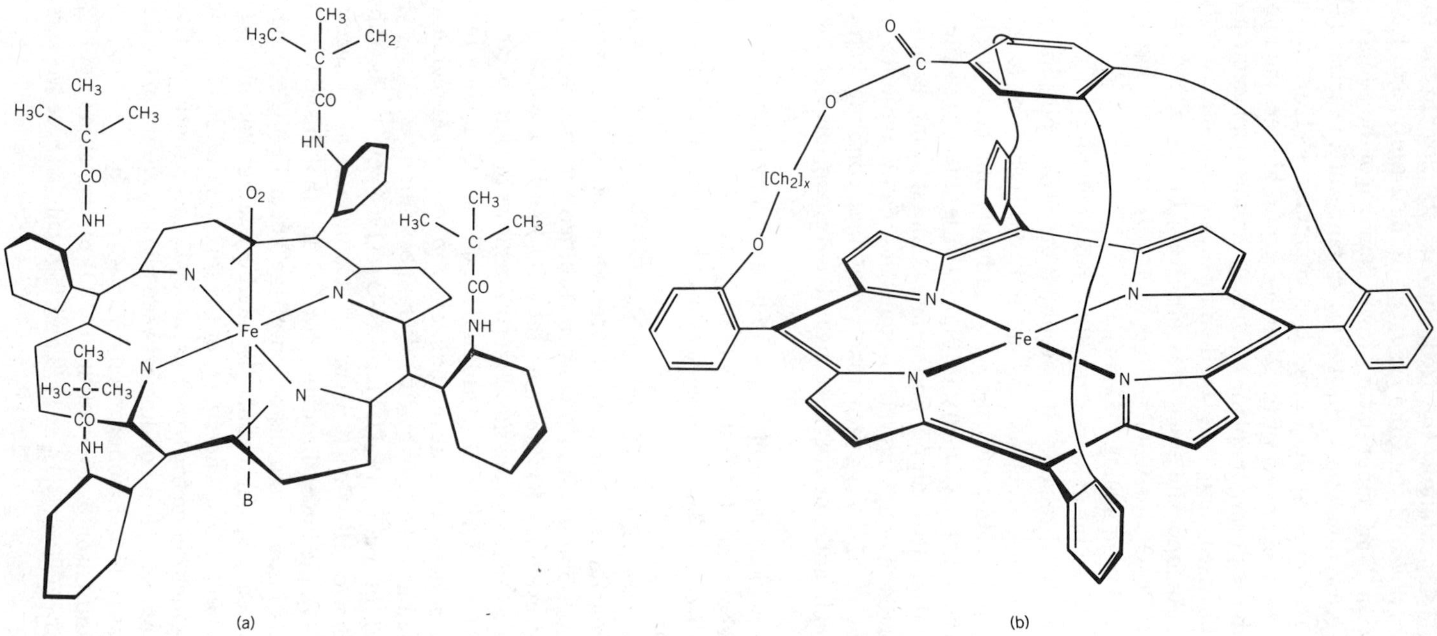

Fig. V-3. Structures of (*a*) $Fe(T_{piv}PP)(base)O_2$ and (*b*) iron capped porphyrin.

568 cm^{-1} in CH_2Cl_2 solution. This frequency is very close to that of oxy-Hb (567 cm^{-1}).[13] In the solid state, the ν(Fe-O_2) of the 2-MeIm and 1,2-DiMeIm analogs (~557 cm^{-1}) are about 10 cm^{-1} lower than that of the 1-MeIm complex. According to X-ray analysis,[32] the Fe-O_2 distance of the latter (1.75 Å) is much shorter than that of the former two compounds (1.90 Å). This result indicates that the steric effect of the 2-methyl group hinders the approach of these ligands to the porphyrin plane and results in the displacement of the Fe atom toward the base ligand, thus lengthening the Fe-O_2 distance and lowering the ν(Fe-O_2). An opposite trend was observed, however, for their Co(II) analogs; the ν(Co-O_2) of Co(T_{piv}PP)(1-MeIm)O_2 (516 cm^{-1}) is lower than that of its 1,2-DiMeIm analog (527 cm^{-1}).[33]

Figure V-4 shows the RR spectra (406.7 nm excitation) of Co(T_{piv}PP)(1-MeIm)O_2 in the $\nu(O_2)$ region.[34] It is seen that the $\nu(O_2)$ splits into two bands at 1157 and 1143 cm^{-1}. Mackin et al.[33] attributed this splitting to the presence of two conformers. However, the $^{18}O_2$ adduct exhibits a single peak at 1086 cm^{-1}. (The slight asymmetry in Fig. V-4B is due to the presence of a porphyrin band at 1078 cm^{-1}.) Furthermore, the splitting is observed only for base ligands having internal frequencies comparable to the $\nu(O_2)$. Hence, Bajdor et al.[34] suggested that vibrational coupling between $\nu(O_2)$ and an internal mode of the base ligand is responsible for the observed splitting.

The $\nu(O_2)$ of Capped porphyrins were measured by Jones et al.[35] using IR difference techniques: Fe(Cap)(1-MeIm)O_2 (1172 cm^{-1}) and Co(Cap)(1-MeIm)O_2 (1176 cm^{-1}). These frequencies are much higher than those of their picket-fence analogs.

Plain ("unprotected") porphyrins such as Fe(TPP) and Co(OEP) can also form dioxygen adducts at low temperatures. Oshio et al.[36] observed the $\nu(O_2)$ of Fe(TPP)(piperidine)O_2 at 1157 cm^{-1} in IR difference spectra. Bajdor et al.[34] carried out an extensive RR study on the $\nu(O_2)$ and ν(Co-O_2) of dioxygen adducts of plain Co(II) porphyrins. Figure V-5 shows the RR spectra (457.9 nm excitation) of Co(PPIXDME) (pyridine)O_2 and its $^{18}O_2$ analog in the $\nu(O_2)$ and ν(Co-O_2) regions. In spite of complexity in the former region, the $\nu(O_2)$ is easily located at 1143 cm^{-1} by $^{16}O_2$-$^{18}O_2$ substitution.

The ν(CO) of Fe(T_{piv}PP)(1-MeIm)(CO) (1964 cm^{-1})[37] is substantially higher than that of Hb-Co (1950 cm^{-1}).[20] Collman et al.[29] suggest that the lower the ν(CO), the lower the affinity for CO and that this lowering in heme proteins results in part from distortion of the linear Fe—C≡O bond by the heme environment. Wayland et al.[38] isolated crystalline complexes, Fe(TPP)-(CO) and Fe(TPP)$(CO)_2$. The ν(CO) of the former is 1973 cm^{-1} whereas that of the latter (antisymmetric type) is at 2042 cm^{-1} (both in IR spectra). Apparently, competitive π-bonding between two CO groups in the latter raised its ν(CO) relative to the monocarbonyl adduct.

Wayland and Olson[39] observed the ν(NO) of crystalline Fe(TPP)(NO) at 1700 cm^{-1} and Fe(TPP)$(NO)_2$ at 1870 and 1690 cm^{-1}. These workers assigned

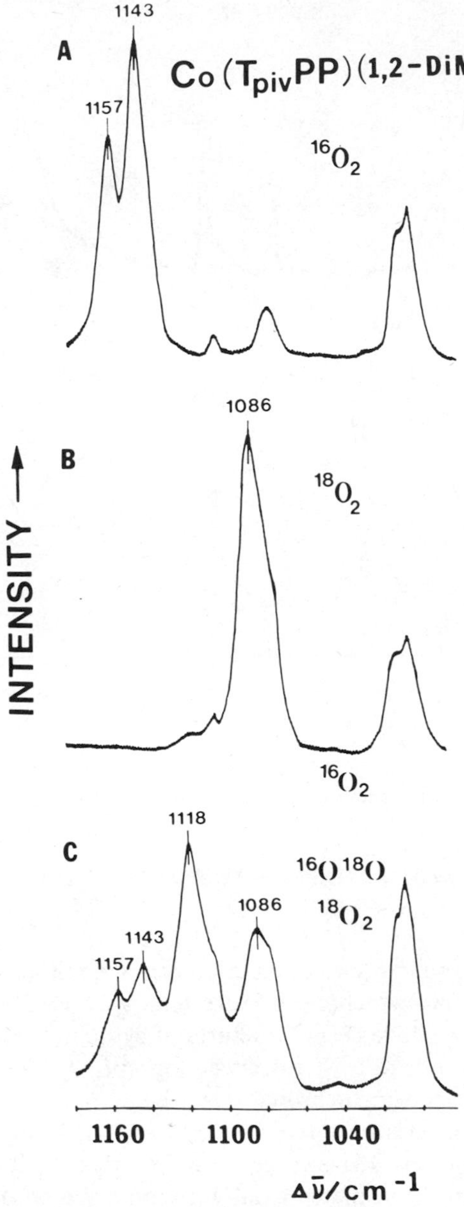

Fig. V-4. RR spectra of Co(T_{piv}PP)(1,2-DiMeIm)O_2 in CH_2Cl_2 at $-90°C$ (406.7 nm excitation).[34]

the latter two bands to the linear Fe(II)-NO^+ and bent Fe(II)-NO^- units, respectively. The ν(NO) of Hb-NO (1623 cm^{-1})[23] may suggest that the Fe-N-O bond is not linear in the heme crevice. In fact, recent X-ray analysis of Hb-NO estimates the Fe-N-O angle to be 145°.[40]

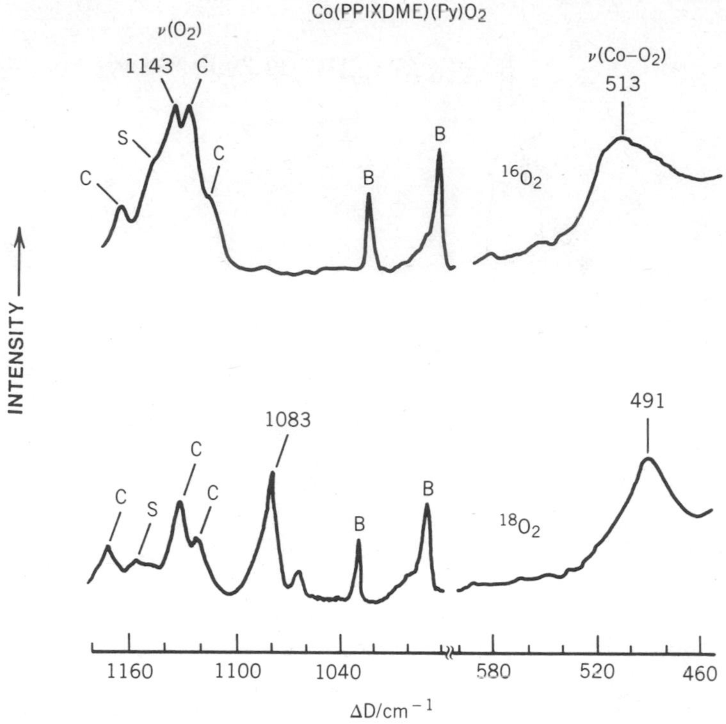

Fig. V-5. RR spectra of Co(PPIXDME)(pyridine)O_2 in CH_2Cl_2 at −90°C (457.9 nm excitation).[34] S, C, and B denote solvent, complex (porphyrin), and base (pyridine), respectively.

V-2. CYTOCHROMES AND OTHER HEME PROTEINS

(1) Cytochrome *c*

Cytochromes (*a*, *b*, and *c*) are electron carriers in the mitochondrial respiratory chain. Among them, cytochromes *c* are relatively small (MW ~ 13,000) and relatively easily crystallized. The structures of cytochromes *c* from various sources have been determined by X-ray crystallography.[41] These studies show that the prosthetic group of cytochrome *c* is a heme in which the vinyl side chains of iron protoporphyrin are replaced by cysteinyl thioether bonds and to which the imidazole(His 18) nitrogen and the methionine(Met 80) sulfur(thioether) atoms are coordinated axially. In most cytochromes *c*, the iron atoms are in the low-spin state, and the basic structure of the heme is unchanged by changing the oxidation state of the iron. As has already been shown in Table V-1, some porphyrin-core vibrations are shifted by changing the oxidation state. However, it has not been possible to relate these shifts to specific structural changes.

According to Dickinson and Chien,[42] cytochrome *c* undergoes the following structural changes by changing the pH.

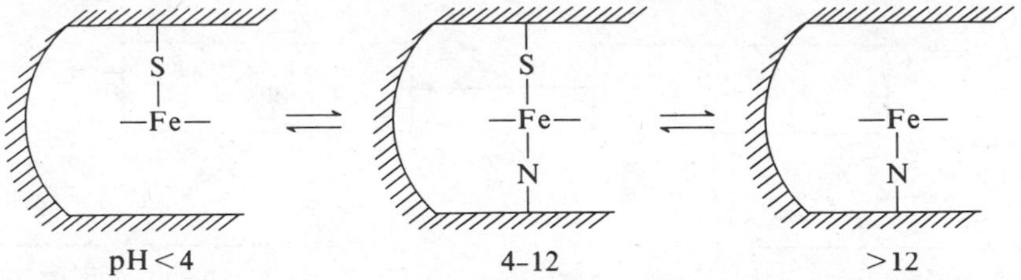

Kitagawa et al.[43] noted that the intensity ratio, I (1636 cm^{-1})/I (1582 cm^{-1}), and the frequency of the 1562-cm^{-1} band (dp) of horse heart ferri-cytochrome *c* change by increasing the pH from 7.8 to 12.4, and suggested that the axial methionine is replaced by some other strong field ligand in basic solution. Lanir et al.,[44] on the other hand, studied the RR spectra of the same cytochrome in acidic solution. They interpret their data as indicating that, at pH = 2, both axial ligands are replaced by two water molecules (the iron in high spin). The RR spectra of horse ferro-cytochrome *c* at pH = 7 ~ 11.2 exhibit many bands below 500 cm^{-1}. At pH = 13.6, however, this is replaced by a much simpler spectrum in this region. Valance and Strekas[45] interpret this in the following fashion: In neutral to alkaline solution, the heme is rigidly held by the peptide backbone, and the resulting asymmetric heme activates many Raman bands. At pH over 13, however, the protein structure is relaxed and the symmetry of the heme becomes effectively higher, resulting in fewer Raman bands.

Thus far, X-ray analysis has been reported on several model compounds of cytochrome *c* such as $[Fe(TPP)(THT)_2]ClO_4$ and $[Fe(TPP)(PMS)_2]ClO_4$ where THT and PMS denote tetrahydrothiophene and pentamethylene sulfide, respectively.[46] Oshio et al.[47] assigned the antisymmetric S–Fe–S stretching vibrations of these compounds at 328 and 323.5 cm^{-1}, respectively, based on ^{54}Fe–^{56}Fe isotope shifts in IR spectra.

(2) Cytochrome P-450

Cytochromes P-450 (MW ~ 50,000) are monooxygenase enzymes which catalyze hydroxylation reactions of substrates such as drugs, steroids, pesticides, and carcinogens. One of the microbial species in which cytochrome P-450 is found is *Pseudomonas Putida.* When this bacteria is grown in air with camphor as the substrate, cytochrome P-450_{cam} can be isolated in a crystalline form. Thus far, most spectroscopic studies have been made on this compound.

The active site of cytochrome P-450 is an iron protoporphyrin with the iron center axially bound to the mercaptide sulfur of a cysteinyl residue. The axial Fe–S linkage is retained throughout its reaction cycle shown in Fig. V-6.[48] Recent X-ray analysis[48a] on cytochrome P-450_{cam} (*B* state) confirmed this structure. The term P-450 was derived from the position of the Soret band at 450 nm of its CO adduct.

As stated in Sec. V-1, Ozaki et al.[7] observed the oxidation-state marker band of cytochrome P-450_{cam} in the C state at 1346 cm^{-1} which is much lower than those of other Fe(II) porphyrins. Similar observations have been made

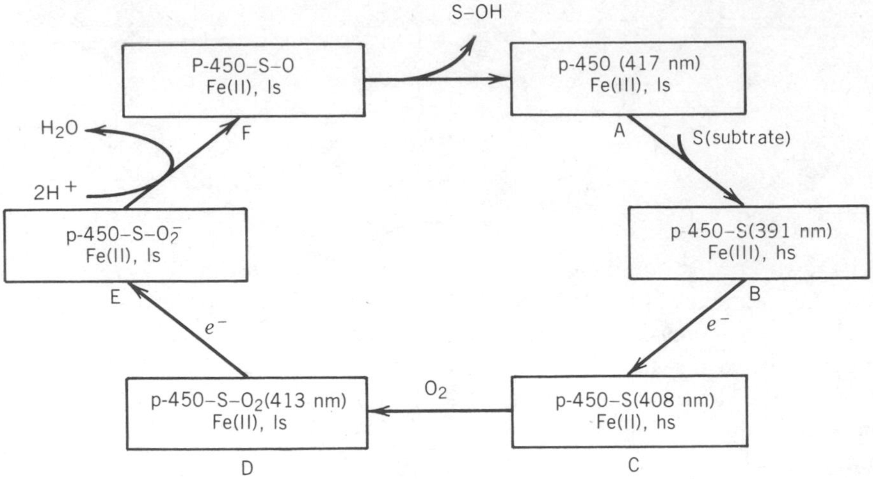

Fig. V-6. Reaction cycle of cytochrome P-450.

for cytochromes P-450 from other sources.[49] This anomaly was attributed to the extra negative charge transmitted to the porphyrin $\pi^*(e_g)$ orbital from the mercaptide sulfur (RS^-) which has two lone-pair electrons. Champion et al.[50] first observed the ν(Fe–S) of cytochrome P-450$_{cam}$ (*B* state) at 351 cm^{-1} in the RR spectrum (364 nm excitation). This band is shifted by $^{54}Fe/^{56}Fe$ and $^{32}S/^{34}S$ substitutions by 2.5 ± 0.2 and 4.9 ± 0.3 cm^{-1}, respectively. The ν(Fe–S) of model compounds such as Fe(III)(porphyrin)(thiolate) are observed at 345–335 cm^{-1} in IR spectra.[47]

O'Keefe et al.[51] observed the ν(CO) of the CO adduct of cytochrome P-450$_{cam}$ using IR spectroscopy. According to these workers, the camphor-bound compound exhibits ν(CO) at 1940 cm^{-1} (nonlinear FeCO bond) whereas the camphor-free compound exhibits two ν(CO) at 1963 (linear) and 1942 (nonlinear) cm^{-1}. Collman and Sorrell[37] observed the ν(CO) of picket-fence complexes, Fe(T_{piv}PP)(B)(CO) where B is n-C_3H_7SH or CH_3S^-; the ν(CO)(IR) of the mercaptide (1945 cm^{-1}) was much lower than that of the mercaptan (1970 cm^{-1}), indicating that the former ligand is a better π-donor than the latter. Schappacher et al.[52] observed the $\nu(O_2)$ of [Fe(T_{piv}PP)-(SC_6HF_4)O_2]$^-$ at 1139 cm^{-1} in IR spectra. This frequency is much lower than that of Fe(T_{piv}PP)(1-MeIm)O_2 (1159 cm^{-1}). Figure V-7 compares the RR spectra of Co(TPP)(B)O_2 where B is pyridine or SC_6H_5.[53] It is seen that the $\nu(O_2)$ of the thiolate complex is lower by 22 cm^{-1} than that of the pyridine adduct, indicating the stronger π-donating ability of the former relative to the latter ligand.

According to Fig. V-6, hydroxylation of the substrate molecule is accomplished by the activated oxygen released from the ferrylporphyrin (*F*-state). Although such a state has not been charcterized spectroscopically, ferryl stretching vibrations have already been located for FeO(TPP) at 852 cm^{-1} and for Horseradish Peroxidase Compound II at ~780 cm^{-1} (see Sec. III-17(3)).

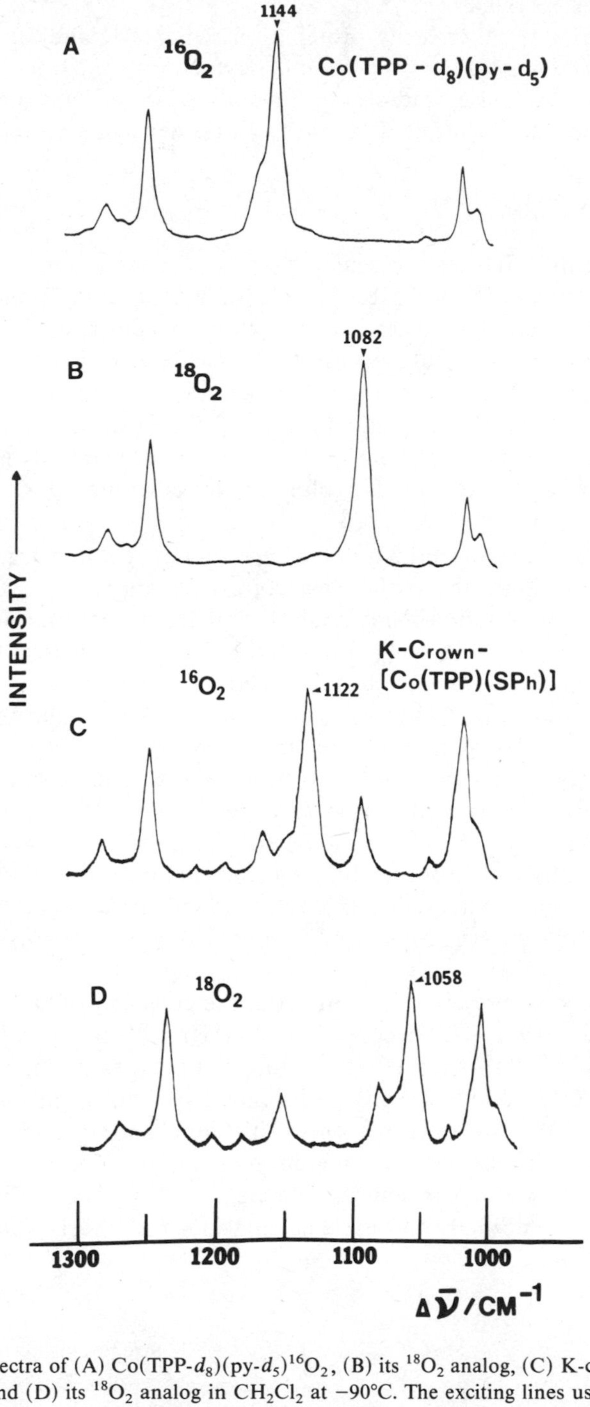

Fig. V-7. RR spectra of (A) Co(TPP-d_8)(py-d_5)$^{16}O_2$, (B) its $^{18}O_2$ analog, (C) K-crown-Co(TPP)-(SC_6H_5)$^{16}O_2$, and (D) its $^{18}O_2$ analog in CH_2Cl_2 at −90°C. The exciting lines used are 406.7 for (A) and (B), and 457.9 nm for (C) and (D).

(3) Other Heme Proteins

There are many other heme proteins on which RR and IR studies have been made. These include cytochrome b_5, cytochrome *c* oxidase, horseradish peroxidase and catalase, and so on. Several review articles mentioned previously[1a–3] should be consulted for these and other heme proteins.

V-3. HEMERYTHRINS[54,55]

Hemerythrins (Hr) are molecular oxygen carriers found in invertebrate plyla. Different from Hb and Mb, Hr have no heme groups. Thus far, spectroscopic investigations have been concentrated on hemerythrin isolated from *Golfingia gouldii*, a sipunculan worm (MW 108,000) which consists of eight identical subunits. Each unit contains 113 amino acids and two Fe atoms, and each pair of Fe atoms binds one molecule of dioxygen. However, its oxygen affinity is slightly lower than hemoglobin, and no cooperativity is found in its oxygenation reaction. Deoxy-Hr (colorless) turns to pink upon oxygenation ("pink blood").

Figure V-8 shows the primary structure of Hr obtained from *G. gouldii*, while Fig. V-9 shows the tertiary structure of monomeric myohemerythrin obtained by low-resolution X-ray analysis;[56] it consists of four nearly parallel helical segments, 30–40 Å long, connected by sharp nonhelical turns.

Figure V-10 shows the electronic spectra of deoxy-, oxy-, and Met-Hr obtained by Dunn et al.[57] The oxy form exhibits a band at 500 nm which does not exist in the deoxy form. When the laser wavelength falls under this electronic absorption, two bands are resonance enhanced at 844 and 500 cm^{-1} which are shifted to 798 and 478 cm^{-1}, respectively, by $^{16}O_2$–$^{18}O_2$ substitution (Fig. V-11). These two bands are assigned to the $\nu(O_2)$ and $\nu(Fe–O_2)$ of the oxy form, respectively. Apparently, the electronic transition at 500 nm is due to $Fe \rightarrow O_2$ charge transfer. Also, the observed frequency of $\nu(O_2)$ (844 cm^{-1}) suggests that the dioxygen is not of "superoxo" but of "peroxo" type (Sec. III-17).

In order to gain more information about the geometry of O_2 binding, Kurtz et al.[58] measured the RR spectra of the oxy-Hr with isotopically scrambled oxygen ($^{16}O_2/^{16}O^{18}O/^{18}O_2 \approx 1/2/1$). Figure V-12 shows that the central band due to the $^{16}O^{18}O$ adduct clearly splits into two peaks, indicating the nonequivalence of the two oxygen atoms. This conclusion is also supported by the RR spectrum in the $\nu(Fe–O_2)$ region. As is seen in Fig. V-13, the spectrum consists of two composite bands, one near 502 cm^{-1} and the other near 483 cm^{-1}. Simple normal coordinate calculations on models I and II indicate

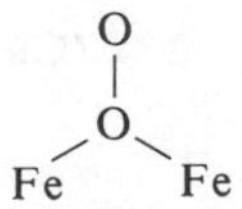

I

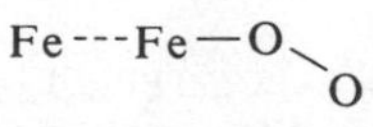

II

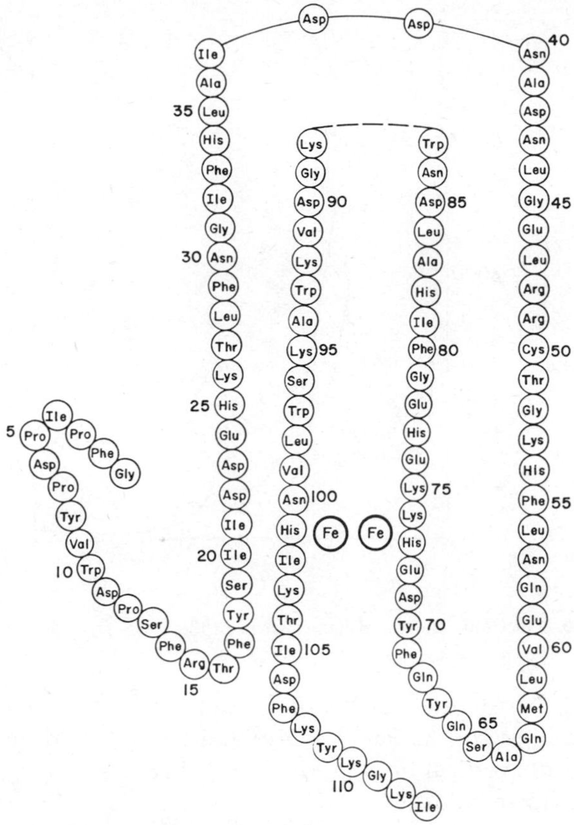

Fig. V-8. Primary structure of hemerythrin from erythrocytes of *G. gouldii.*

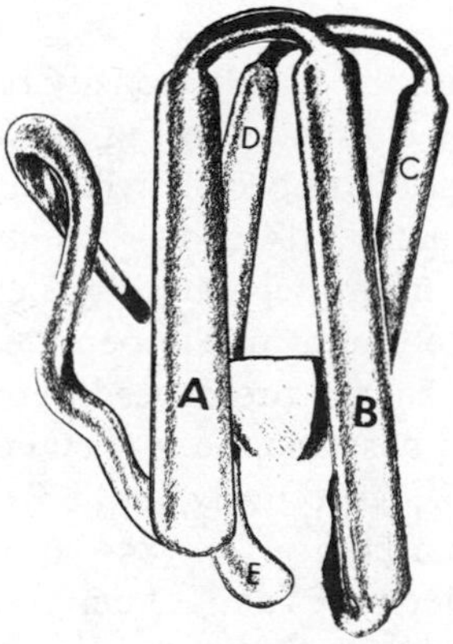

Fig. V-9. Tertiary structure of monomeric myohemerythrin (reproduced with permission).

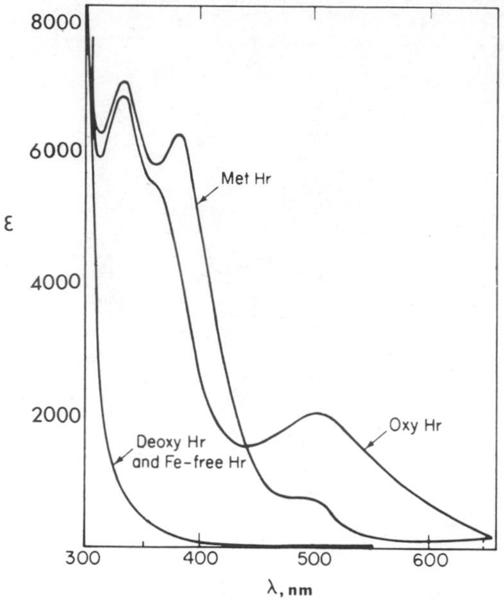

Fig. V-10. Electronic spectra of hemerythrin in the deoxy, oxy, and Met forms.

that the $\nu(Fe–O_2)$ of the $Fe–^{16}O^{16}O$(a) and $Fe–^{16}O^{18}O$(b) adducts nearly overlap, so are those of the $Fe–^{18}O^{16}O$(c) and $Fe–^{18}O^{18}O$(d) adducts (a, b, c, and d refer to the vertical lines in Fig. V-13). If the two oxygen atoms were equivalent as shown below,

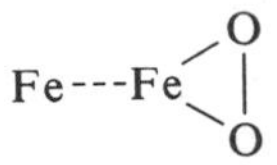

III

a three-peak spectrum with the 1:2:1 intensity ratio would have appeared in the positions indicated by e, f, and g in Fig. V-13.

Recently, X-ray analyses have been carried out on metazidohemerythrin[59] and oxyhemerythrin.[60] Figure V-14 shows the structure of the active site of the latter; the two Fe atoms are separated by 3.25 Å, and bridged by an oxo atom and two carboxylate groups of the peptide chain. The structure of the former is similar except that the protonated peroxide ion is replaced by the azide ion. Shiemke et al.[61] observed the ν_a(FeOFe) and ν_s(FeOFe) of the oxo bridge at 753 and 486 cm^{-1}, respectively, in the RR spectrum (363.8 nm excitation). They also noted that both $\nu(O_2)$ (844 cm^{-1}) and $\nu(Fe–O_2)$ (503 cm^{-1}) of oxy-Hr in H_2O are shifted by +4 and −3 cm^{-1}, respectively, in D_2O solution. These shifts are consistent with the protonated peroxide structure shown in Fig. V-14.

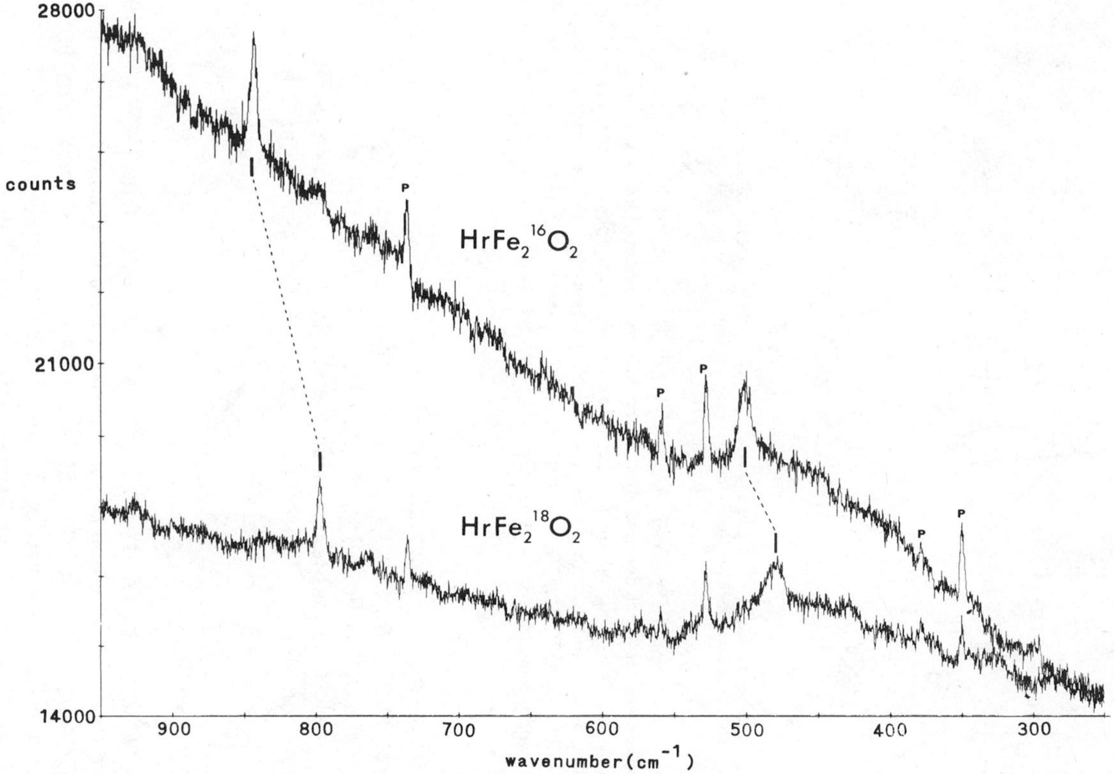

Fig. V-11. RR spectra of oxy-hemerythrin ($^{16}O_2$ and $^{18}O_2$) by 488.0 nm excitation. p denotes laser plasma lines (reproduced with permission).

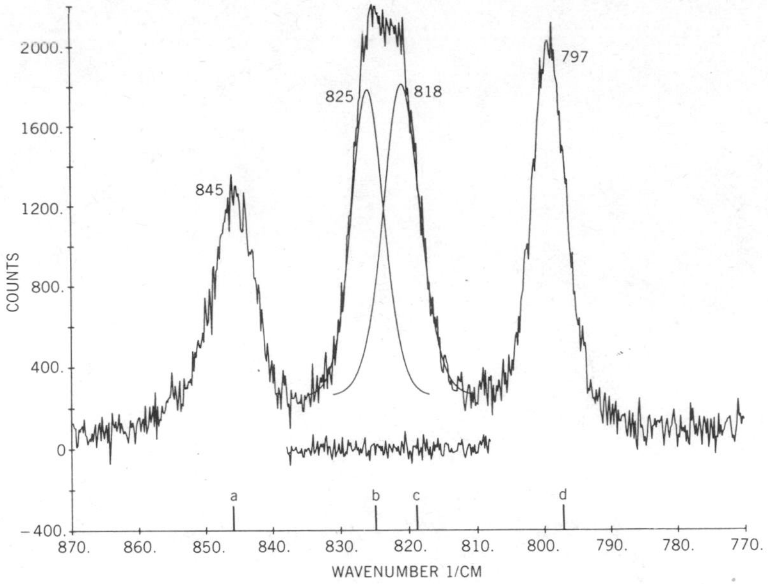

Fig. V-12. RR spectrum of oxy-hemerythrin ($^{16}O_2/^{16}O^{18}O/^{18}O_2 \approx 1/2/1$) by 514.5 nm excitation. The smooth curves represent deconvolution of the 822-cm^{-1} feature into two components. The difference between observed and fitted curves is shown below the spectrum near 822 cm^{-1}. The vertical lines, a, b, c, and d, show the calculated peak positions for models I and II of Fe–$^{16}O_2$ (845 cm^{-1}), Fe–$^{16}O^{18}O$ (825 cm^{-1}), Fe–$^{18}O^{16}O$ (818 cm^{-1}), and Fe–$^{18}O_2$ (797 cm^{-1}), respectively. [Reprinted with permission from *J. Am. Chem. Soc.*, **98**, 5034 (1976). © 1976 American Chemical Society.]

Armstrong et al.[62] prepared model compounds of the active site of Hr. According to X-ray analysis, the complexes of the type $[(HBpz_3)FeO(O_2CR)_2Fe(NBpz_3)]$ (R = H, CH_3, C_6H_5, and $HBpz_3$ = tri-1-pyrazolylborate ion) take a structure similar to that shown in Fig. V-14. The acetate complex (R = CH_3) exhibits the ν_a(FeOFe) at 751 cm^{-1} in IR and ν_s(FeOFe) at 528 cm^{-1} in RR spectra.

V-4. HEMOCYANINS[63]

Hemocyanins (Hc) are oxygen-transport nonheme proteins (MW $10^5 \sim 10^7$) which are found in the blood of some insects, crustaceans, and other invertebrates. One of the smallest Hc (MW 450,000) extracted from spiny lobster *Panulirus interruptus* consists of six subunits each containing two Cu atoms. Upon oxygenation, the deoxy form (Cu(I), colorless) turns to blue (Cu(II), "blue blood") by binding one O_2 molecule per two Cu atoms.

Oxy-Hc extracted from *C. magister* (Pacific crab) and *B. canaliculatum* (channeled whelk) exhibit absorption bands near 570 and 490 nm. Freedman et al.[64] measured the RR spectra of these compounds with 530.9 and 457.9 nm excitations. The results shown in Fig. V-15 clearly indicate that the bands near

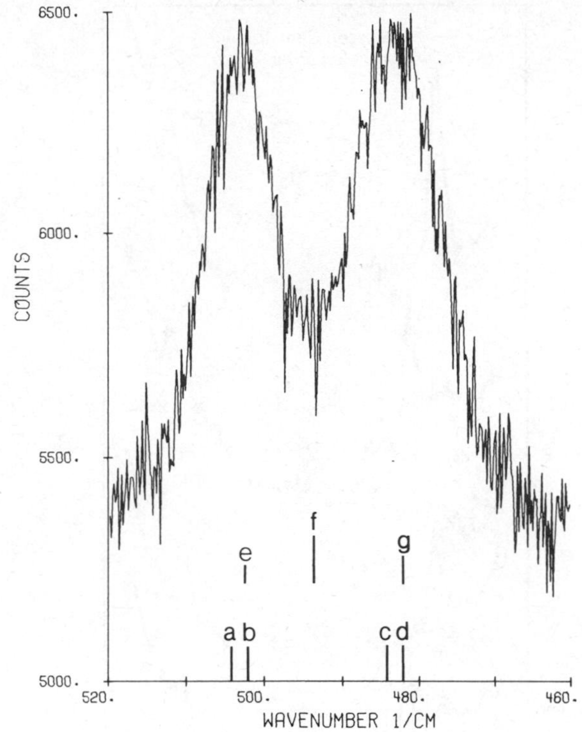

Fig. V-13. RR spectrum of oxy-hemerythrin ($^{16}O_2/^{16}O^{18}O/^{18}O_2 \approx 1/2/1$) by 514.5 nm excitation. Vertical lines a, b, c, and d show the calculated peak positions for models I and II of Fe–$^{16}O_2$ (504 cm^{-1}), Fe–$^{16}O^{18}O$ (501 cm^{-1}), Fe–$^{18}O^{16}O$ (485 cm^{-1}), and Fe–$^{18}O_2$ (482 cm^{-1}), respectively. Vertical lines e, f, and g show for model III the calculated peak positions and estimated relative intensities of the $^{16}O_2$ (502 cm^{-1}), $^{16}O^{18}O$ (495 cm^{-1}), and $^{18}O_2$ (489 cm^{-1}), respectively. [Reprinted with permission from *J. Am. Chem. Soc.*, **98**, 5034 (1976). © 1976 American Chemical Society.]

747 cm^{-1} are sensitive to $^{16}O_2$–$^{18}O_2$ substitution and must be assigned to the $\nu(O_2)$ characteristic of the peroxo(O_2^{2-}) type. Excitation profiles of the $\nu(O_2)$ consist of two components and indicate that the absorption bands near 570 and 490 nm are due to $O_2^{2-} \rightarrow Cu(II)$ charge transfer. These workers proposed

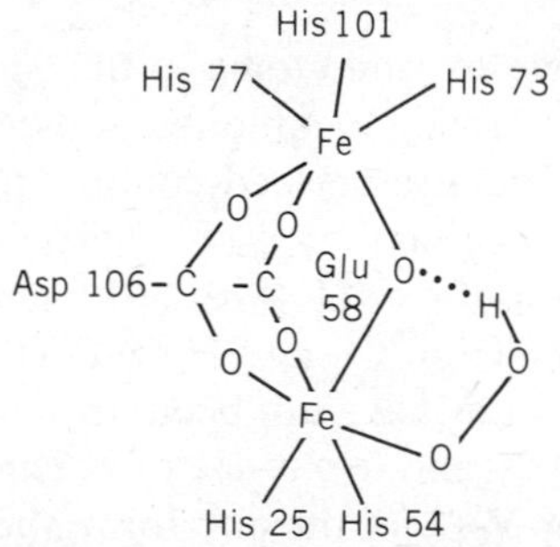

Fig. V-14. Structure of active site of hemerythrin in the oxy form.

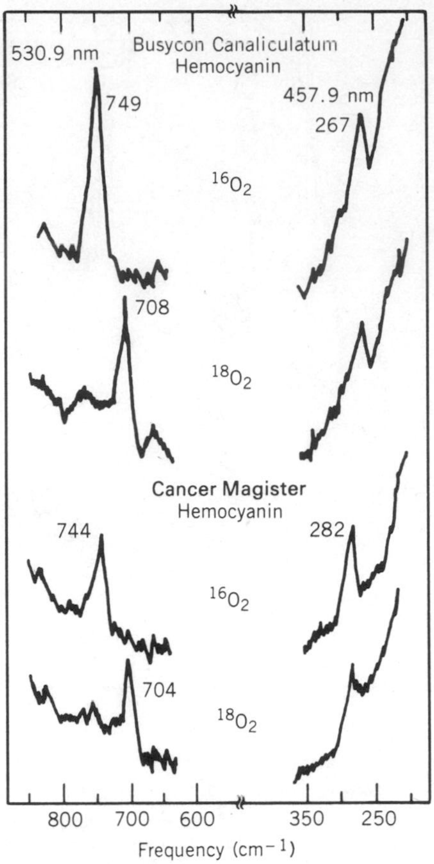

Fig. V-15. RR spectra of oxy-hemocyanins ($^{16}O_2$ and $^{18}O_2$) with 530.9 and 457.9 nm excitation.

a nonplanar (C_2) structure to account for the appearance of the two CT bands:

$$Cu\!-\!O$$
$$\diagdown$$
$$O\!-\!Cu$$

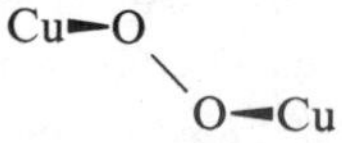

The equivalence of the two oxygen atoms in this structure was confirmed by the RR spectrum of oxy-Hc which exhibits á single $\nu(O_2)$ band at 728 cm^{-1} for the $^{16}O^{18}O$ adduct.[65] This is a marked contrast to oxy-Hr discussed in the preceding section. Larrabee and Spiro[66] observed $\nu(Cu-N(Im))$ below 350 cm^{-1} in the RR spectra of oxy-Hc with 363.8 nm excitation. Their assignments were confirmed by $^{63}Cu-^{65}Cu$ and H_2O-D_2O frequency shifts. Surprisingly, the $\nu(Cu-O_2)$ vibrations have not been observed. Brown et al.[67] carried out an EXAFS study on oxy- and deoxy-Hc of *B. canaliculatum,* and proposed the structure shown in Fig. V-16 for the oxy-form; the two Cu atoms are bound to the protein via three histidine ligands each, and bridged by the O_2^{2-} and an

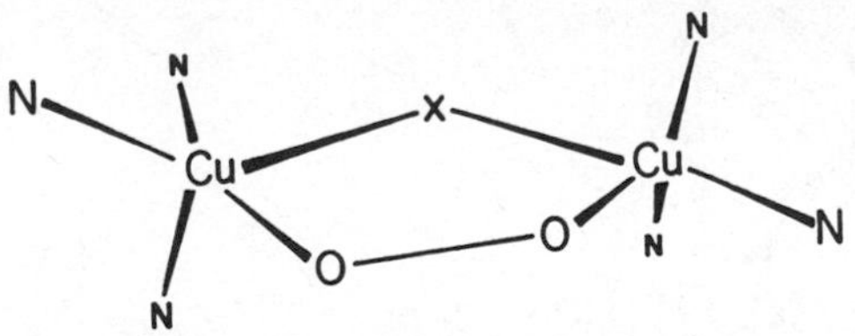

Fig. V-16. Proposed structure of the active site of oxy-hemocyanin.

X atom from a protein, possibly tyrosine. Recently, Gaykema et al.[68] carried out X-ray analysis (3.2 Å resolution) on colorless single crystals of Hc extracted from *P. interruptus.* This molecule consists of six subunits (MW 75,000), and each subunit is folded into three domains. The structure of the second domain in which two Cu atoms are located is shown in Fig. V-17. The Cu–Cu distance is 3.8 Å, and each Cu atom is coordinated by three histidyl residues as suggested by Brown et al.[67] for the deoxy form. No evidence for a bridging protein ligand was found although it is not possible to rule out such a possibility from low-resolution X-ray analysis.

Several groups of investigators prepared model compounds of Hc and demonstrated partial reversibility of oxygenation reactions of their compounds by electronic spectroscopy.[69,70] Karlin et al.[71] are the first to observe the $\nu(O_2)$ (803 cm^{-1}) of their model compound shown in Fig. V-18 using RR spectroscopy.

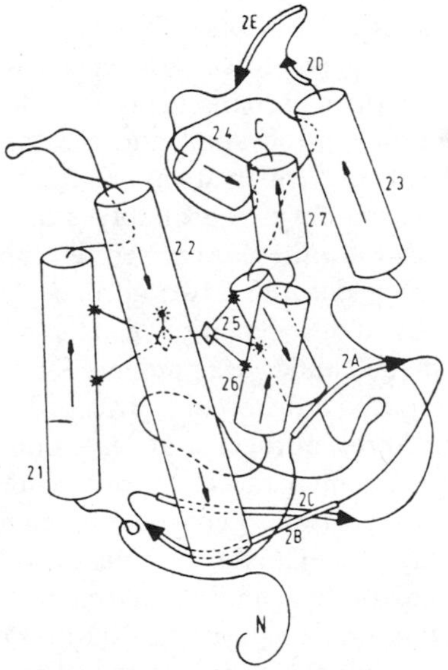

Fig. V-17. Structure of the second domain of Hc extracted from *P. interruptus.* The Cu atoms are indicated by diamonds. The cylinders (2.1 to 2.7) indicate α-helical structure while the strips (2A to 2E) represent β-structure of the peptide chain.[68] (Reprinted by permission from *Nature,* Vol. 309, p. 28, © 1984 Macmillan Journals Limited.)

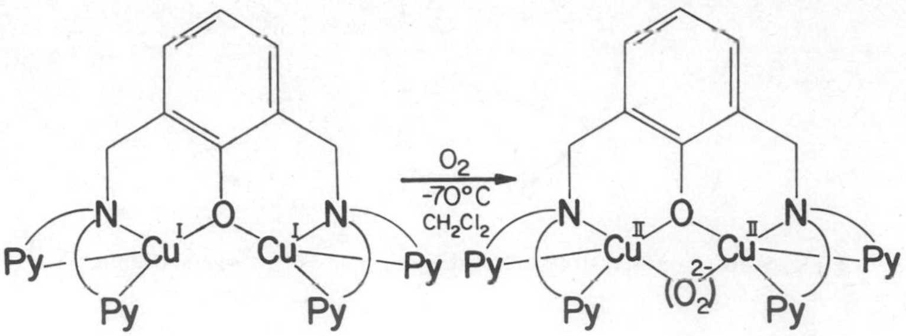

Fig. V-18. A model compound of hemocyanin.[71] Py denotes the 2-pyridyl group.

V-5. BLUE COPPER PROTEINS[72]

Blue (Type I) copper proteins are found widely in nature. Typical examples are plastocyanin (MW 10,500) and ascorbate oxidase (MW 150,000) which contain one and eight Cu atoms per protein, respectively. The former serves as a component of the electron transfer chain in plant photosynthesis while the latter is an enzyme involved in the oxidation of ascorbic acid. The oxidized form is characterized by intense blue color due to electronic absorption near 600 nm. In addition, blue copper proteins exhibit unusual properties such as extremely small hyperfine splitting constants ($0.003 \sim 0.009\ cm^{-1}$) in ESR spectra and rather high redox potential ($+0.2 \sim 0.8$ V) compared to the Cu(II)/Cu(I) couple in aqueous solution.

The structure of the active site of blue copper proteins has been a subject of many physicochemical studies.[72] Miskowski et al.[73] were the first to measure the RR spectra of three copper proteins using laser lines near 600 nm. As is shown in Fig. V-19, azurin, plastocyanin, and ceruloplasmin exhibit several bands between 470 and 350 cm^{-1} which are assignable to ν(Cu–N) or ν(Cu–O) and a weak band near 270 cm^{-1} which probably is due to ν(Cu–S). Based on these and other information, they proposed an approximately trigonal-bipyramidal structure with a sulfur and two nitrogen ligands in the equatorial plane and less strongly bound nitrogen or oxygen ligands at the axial positions.

Siiman et al.[74] also made similar assignments for RR spectra of five blue copper proteins, and proposed a distorted four-coordinate structure involving one cysteinyl sulfur and three nitrogens, at least one of which is an amide nitrogen. These workers assigned the electronic bands near 600 as well as 450 nm to the S→Cu CT transition. Ferris et al.[75] compared the ν(Cu–S) of natural proteins with those of Cu(II) complexes of macrocyclic thiaether ($280 \sim 247\ cm^{-1}$) and mercaptide ligands ($\sim 300\ cm^{-1}$), and concluded that the electrically neutral methionine or a similar sulfur rather than the ionized cysteinyl sulfur is involved in the coordination. About this time, the X-ray crystal structure of plastocyanin was determined with 2.7 Å resolution.[76] Figure V-20 shows the location of the Cu atom in the peptide chain. It was found that the Cu atom is coordinated by two histidyl nitrogens (His 37 and 87) and

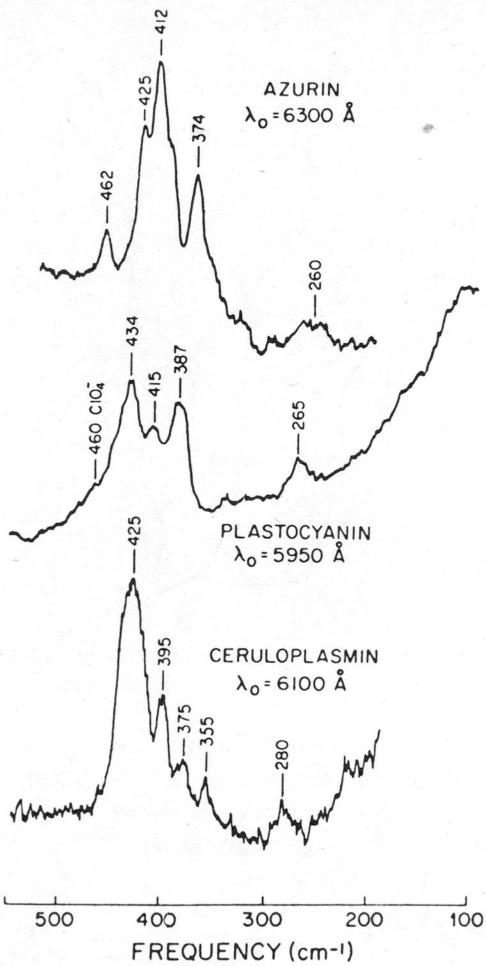

Fig. V-19. RR spectra of blue copper proteins. Excitation lines used are indicated for each compound.

one cysteinyl sulfur (Cys 84) and one methionyl sulfur (Met 92) in a distorted tetrahedral environment. Recently, Nestor et al.[77] assigned the RR spectra of stellacyanin and laccase based on $^{63}Cu/^{65}Cu$ and H_2O/D_2O shifts and normal coordinate calculations. In contrast to the previous assignments, these workers attribute two strong RR bands (647.1 nm excitation) of stellacyanin at 385 and 347 cm^{-1} to the ν(Cu–S(Cys)) and the weak band at 273 cm^{-1} to the ν(Cu–N(His)).

Thompson et al.[78] prepared a model compound of blue copper proteins: the cuprous complex, K[Cu(HB(3,5-Me_2pz)$_3$(SR)] [SR = *p*-nitrobenzenethiolate, HB(3,5-Me_2pz)$_3$ = hydrotris(3,5-dimethyl-1-pyrazolyl)borate]. According to X-ray analysis, the Cu(I) atom is coordinated to one sulfur and

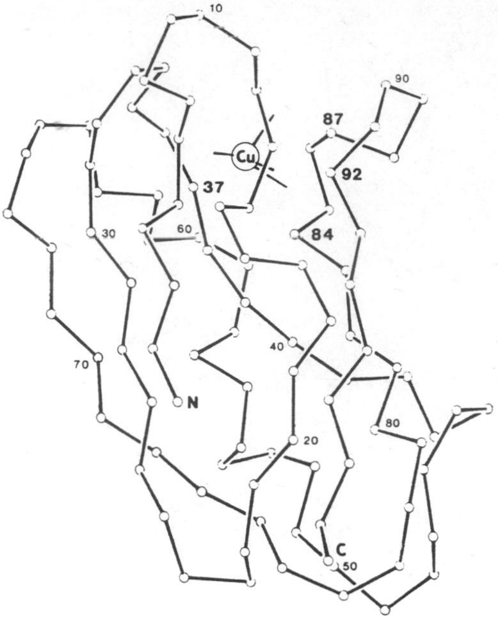

Fig. V-20. Crystal structure of plastocyanin.[76] (Reprinted by permission from *Nature*, Vol. 272, p. 320, © 1978 Macmillan Journals Limited.)

three nitrogen atoms in a trigonally distorted tetrahedron. The RR spectrum (647.1 nm excitation) of this complex shows a weak band at 276 cm^{-1} which was assigned to the ν(Cu–S), and several strong-to-weak bands in the 385–339 cm^{-1} region which were assigned to the ν(Cu–N).

V-6. IRON–SULFUR PROTEINS[79]

Iron–sulfur proteins are found in a variety of organisms, bacteria, plants, and animals, and serve as electron transfer agents via one-electron oxidation–reduction step [redox potential (E_m), −0.43 V in chloroplasts to +0.35 V in photosynthetic bacteria]. For example, ferredoxin in green plants (chloroplasts) is involved in the electron transfer system of photosynthesis. The molecular weights of iron–sulfur proteins range from 5600 (rubredoxin from *Clostridium Pasteurianum*) to 83,000 (beef heart aconitase). All these compounds show strong absorptions in the visible and near-uv regions which are due to Fe ← S CT transitions. Thus laser excitation in these regions is expected to resonance-enhance ν(Fe–S) vibrations of iron–sulfur proteins.

The most simple iron–sulfur protein is rubredoxin (Rd) which contains one Fe atom per protein. The Fe atom is coordinated by four sulfur atoms of cysteinyl residues in a tetrahedral environment. Figure V-21 shows the crystal structure of a model compound, $[Fe(S_2\text{-}o\text{-}xyl)_2]^-$ (S_2-*o*-xyl = *o*-xylene-α,α'-dithiolate).[80] Long et al.[81] first obtained the RR spectrum of

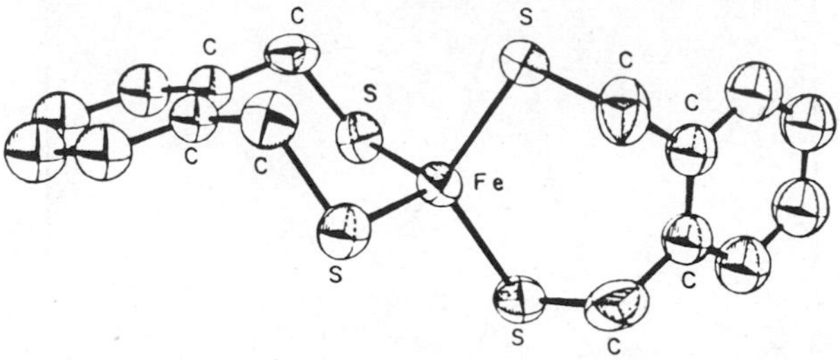

Fig. V-21. ORTEP drawing of the molecular structure of $[Fe(S_2\text{-}o\text{-xyl})_2]^-$ viewed down the C_2 axis. [Reprinted with permission from *J. Am. Chem. Soc.*, **99**, 84 (1977).]

oxidized rubredoxin, and assigned two bands at 368(ν_3) and 314(ν_1) cm^{-1} to the ν(Fe–S) and those at 150(ν_4) and 126(ν_2) cm^{-1} to the $\delta(FeS_4)$ of the FeS_4 tetrahedron. Recently, Yachandra et al.[82] observed four ν(Fe–S) bands at 371, 359, 325 and 312 cm^{-1} for oxidized rubredoxins as shown in Fig. V-22. The RR spectrum of the above model compound which has a distorted tetrahedral FeS_4 core also shows four bands at 374, 350, 321 and 297 cm^{-1}. According to normal coordinate calculations, the observed splitting of the $\nu_3(F_2)$ vibration is not caused by the distortion of tetrahedral symmetry but by vibrational coupling between ν(Fe–S) and S-ligand modes.

Two-iron proteins are found in ferredoxin from chloroplast (MW ~ 10,000) and in adrenodoxin from adrena cortex of mammals (MW ~ 13,000), and so on. These proteins contain the Fe_2S_2(cysteinyl)$_4$ cluster in which two Fe atoms are bridged by two "labile" (inorganic) sulfur atoms and each Fe atom is tetrahedrally coordinated by two bridging and two cysteinyl sulfur atoms (2Fe–2S cluster). Recently, this structure was confirmed by X-ray analysis of the ferredoxin from *Spirulina platensis* (*Sp* Fd).[83] The $Fe_2S_2S'_4$ core ($\mathbf{D}_{2h}$ symmetry) is modeled by the $[Fe_2S_2(S_2\text{-}o\text{-xyl})_2]^{2-}$ ion whose structure is shown in Fig. V-23.[84] Yachandra et al.[85] measured the RR spectra of oxidized spinach ferredoxin and its ^{34}S-enriched analog containing such a 2Fe–2S cluster, and made band assignments shown in Table V-2.

One of the most common Fe–S clusters in iron–sulfur proteins is the 4Fe–4S cube containing interpenetrating Fe_4 and S_4 tetrahedra, the Fe corners of which are bound to cysteinyl sulfur atoms. Figure V-24 shows the X-ray crystal structure of a bacterial ferredoxin from *Peptococcus aerogenes* (MW ~ 6,000) containing two such clusters.[86] The geometry of this 4Fe–4S cluster is in good agreement with that of the synthetic analog, $[Fe_4S_4(SR)_4]^{2-}$ ($R = CH_2C_6H_5$, C_6H_5, etc.) prepared by Beng and Holm.[87] In both cases, the 4Fe-4S cube is slightly squashed with four short and eight long Fe–S bonds (approximately $\mathbf{D}_{2d}$ symmetry).

Johnson et al.[88] and Czernuszewicz et al.[89] measured the RR spectra of these proteins and their model compound as shown in Fig. V-25. The bands near 385, 335, 298, 270, and 250 cm^{-1} are assigned to the bridging whereas those near 395 and 362 cm^{-1} are attributed to the terminal ν(Fe–S) vibrations.

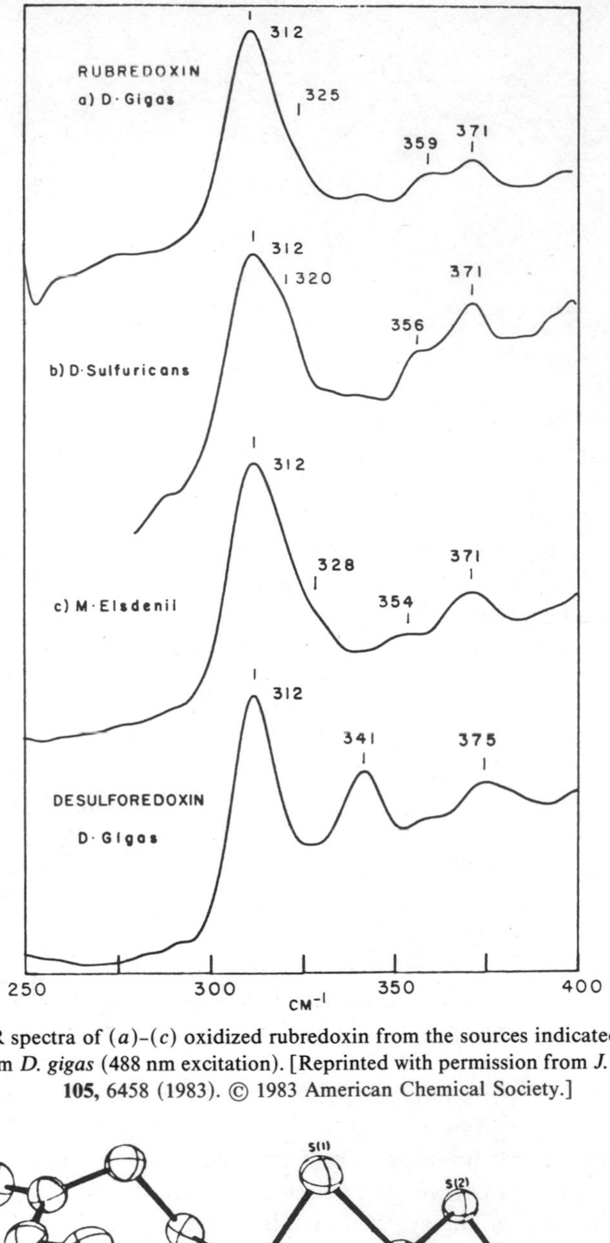

Fig. V-22. RR spectra of (*a*)–(*c*) oxidized rubredoxin from the sources indicated and (*d*) desulfuredoxin from *D. gigas* (488 nm excitation). [Reprinted with permission from *J. Am. Chem. Soc.*, **105,** 6458 (1983). © 1983 American Chemical Society.]

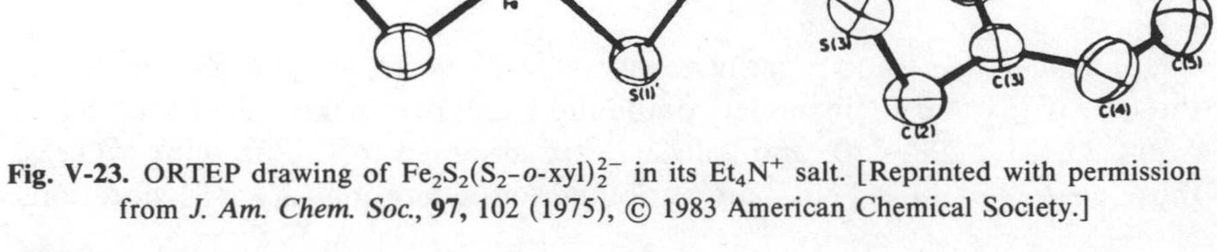

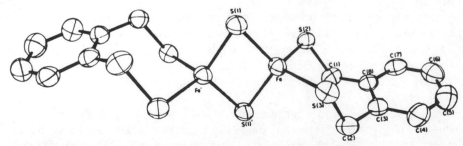

Fig. V-23. ORTEP drawing of $Fe_2S_2(S_2\text{-}o\text{-xyl})_2^{2-}$ in its Et_4N^+ salt. [Reprinted with permission from *J. Am. Chem. Soc.*, **97,** 102 (1975), © 1983 American Chemical Society.]

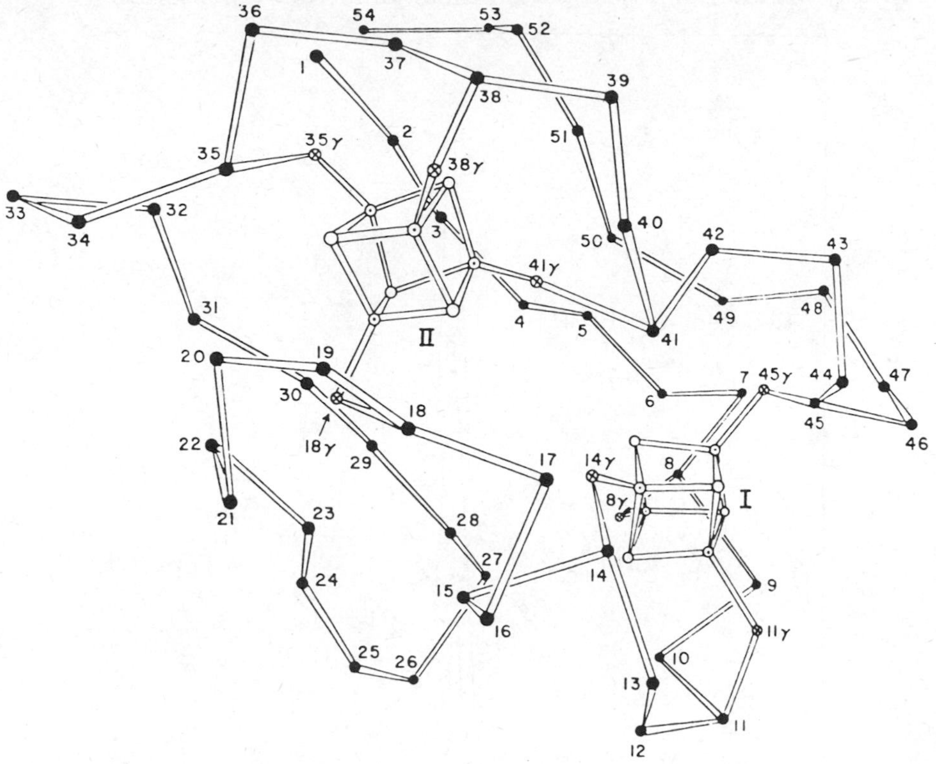

Fig. V-24. Structure of bacterial ferredoxin.[86] ⊙, ○, ⊗, and ● indicate Fe, S (inorg), S (cysteinyl), and C atoms, respectively (reproduced with permission).

TABLE V-2. VIBRATIONAL ASSIGNMENTS OF THE $Fe_2S_2(SR)_4$ CORE (CM^{-1})[85]

Vibrational Mode	Symmetry[a]	*Sp*Fd_{ox}(R)	$[Fe_2S_2(S_2\text{-}o\text{-xyl})_2]^{2-}$
Fe-S_b	B_{3u}	424 (5)[b]	417 (R)
Fe-S_b	A_{1g}	393 (6)	392 (R)
Fe-SR	B_{2g}	369 (8)	347 (R)
Fe-SR	B_{1u}	336 (2)	338 (IR)
Fe-SR	B_{2u}	—	326 (IR)
Fe-SR	B_{3g}	326 (0)	324 (R)
Fe-SR	A_{1g}	~310	~310 (R)
Fe-S_b	B_{1u}	284 (5)	279 (R)

[a] $\mathbf{D}_{2h}$ symmetry.
[b] The number in parentheses indicates the ^{32}S-^{34}S isotope shift of a band primarily due to $\nu(Fe\text{-}S_b)$ (S_b = bridging sulfur).

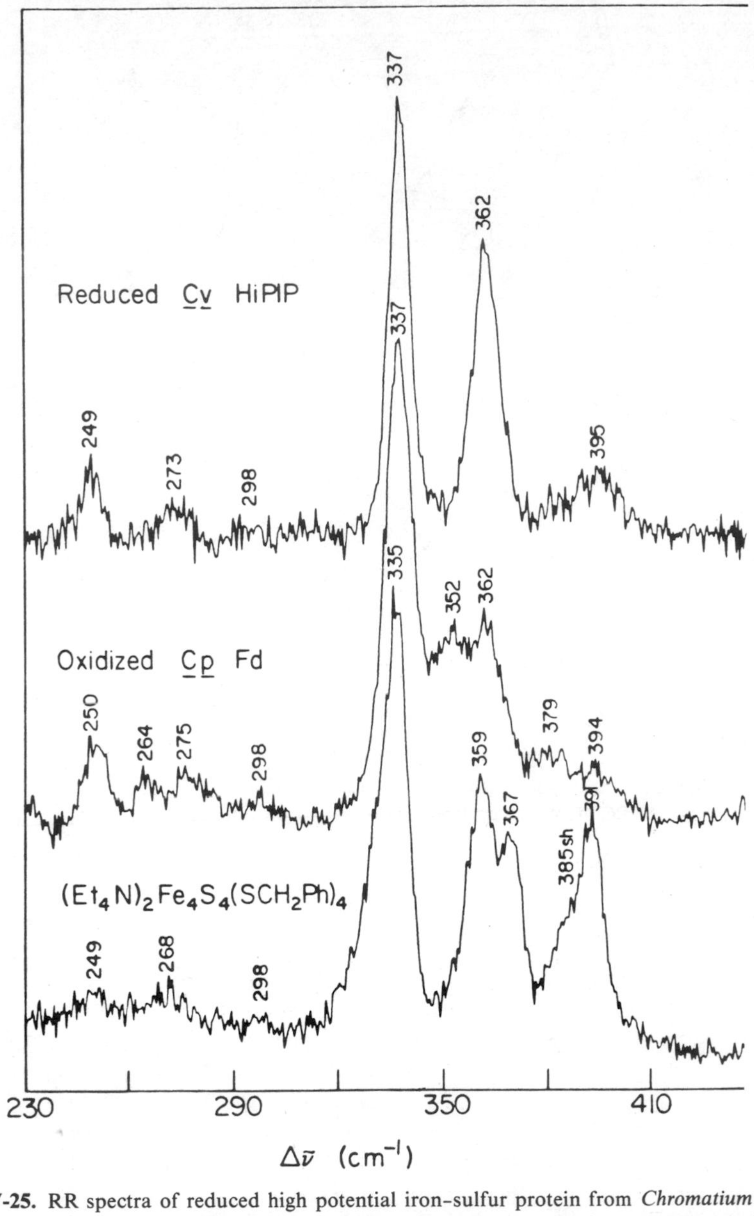

Fig. V-25. RR spectra of reduced high potential iron–sulfur protein from *Chromatium* (HiPIP) and oxidized ferredoxin (Fd) from *Clostridium Pasteurianum* (*Cp*) as frozen solution, and $(Et_4N)_2Fe_4S_4(SCH_2Ph)_4$ as crystalline state (457.9 nm excitation).

Czernuszewicz et al.[89] also carried out normal coordinate analysis on the model compound mentioned above, and proposed the band assignments shown in Table V-3. It is seen that oxidized *Clostridium Pasteurianum* (*Cp* Fd) and the model compound exhibit marked splittings of the triply degenerate modes

TABLE V-3. OBSERVED FREQUENCIES (CM^{-1}) AND BAND ASSIGNMENTS FOR THE $[Fe_4S_4(SCH_2Ph)_4]^{2-}$ ION, REDUCED *Cv* HiPIP, AND OXIDIZED *Cp Fd* [a]

$[Fe_4S_4(SCH_2Ph)_4]^{2-}$ T_d D_{2d}	IR[b]	RR[b]	Reduced *Cv* HiPIP RR[c]	Oxidized *Cp* Fd RR[c]
Terminal Fe-S stretches				
A_1-A_1	—	391 (1)	395	394 (0)
F_2 B_2	366 (0)	367 (1)	362	362 (1)
F_2 E	359 (0)	359 (2)		352 (0)
Bridging Fe-S stretches				
A_1-A_1	—	335 (8)	337	337 (6)
F_2 B_2	ob	ob		379 (4)
F_2 E	386 (6)	385 (5)	ob	
E A_1	—	298 (5)	298	298 (3)
E B_1	—	283 (4)		275 (4)
F_1 A_2	—	270 (3)	273	264 (3)
F_1 E	—	283 (4)		275 (4)
F_2 B_2	250 (4)	249 (6)	249	250 (4)
F_2 E	243 (4)	243 (5)		

[a] Numbers in parentheses indicate observed $^{34}S_b$ isotopic shifts.
[b] IR and resonance Raman (RR) in crystalline state.
[c] RR in frozen solution; ob = obscured.

(under $\mathbf{T}_d$ symmetry) at 270 (F_1, bridging), 250 (F_2, bridging), and 360 cm^{-1} (F_2, terminal) whereas reduced *Chromatium* (*Cv* HiPIP) shows no such splittings. These observations indicate significant symmetry lowering ($\mathbf{D}_{2d}$) in the former two compounds. Moulis et al.[89a] measured the RR spectrum of selenium-substituted *Cp* Fd, and assigned the ν(Fe-X) (X = S and Se) vibrations.

Recently, the existence of a 3Fe–3S cluster was confirmed by X-ray analysis of ferredoxin I extracted from *Azotobacter vinelandii* (*Av* Fd I, MW 14,000).[90] As is shown in Fig. V-26A, this protein contains a 3Fe–3S cluster (E_m, −0.42 V) in addition to a cubanelike 4Fe–4S cluster (E_m, +0.32 V). Figure V-26B shows that the three Fe and three S atoms of the 3Fe–3S cluster lie alternately on a somewhat puckered plane with a pair of cysteinyl sulfur atoms bound to each Fe atom above and below the plane although one of these six positions may be occupied by a glutamate. On the other hand, quite different structures of 3Fe clusters were suggested by EXAFS analyses of beef heart aconitase[91] and ferredoxin II from *Desulfovibrio gigas* (*Dg* Fd II).[92] Figure V-27 shows two

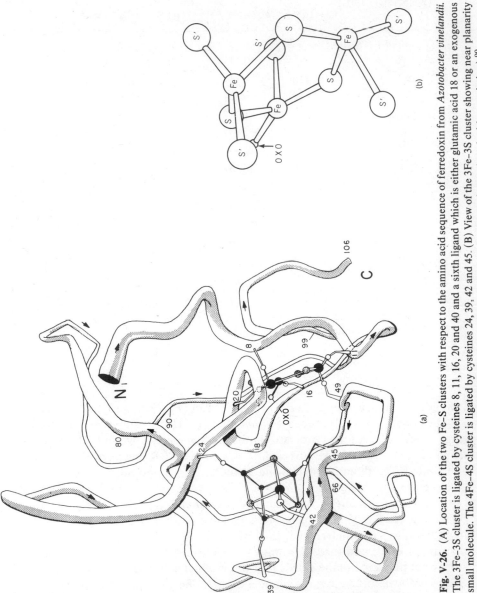

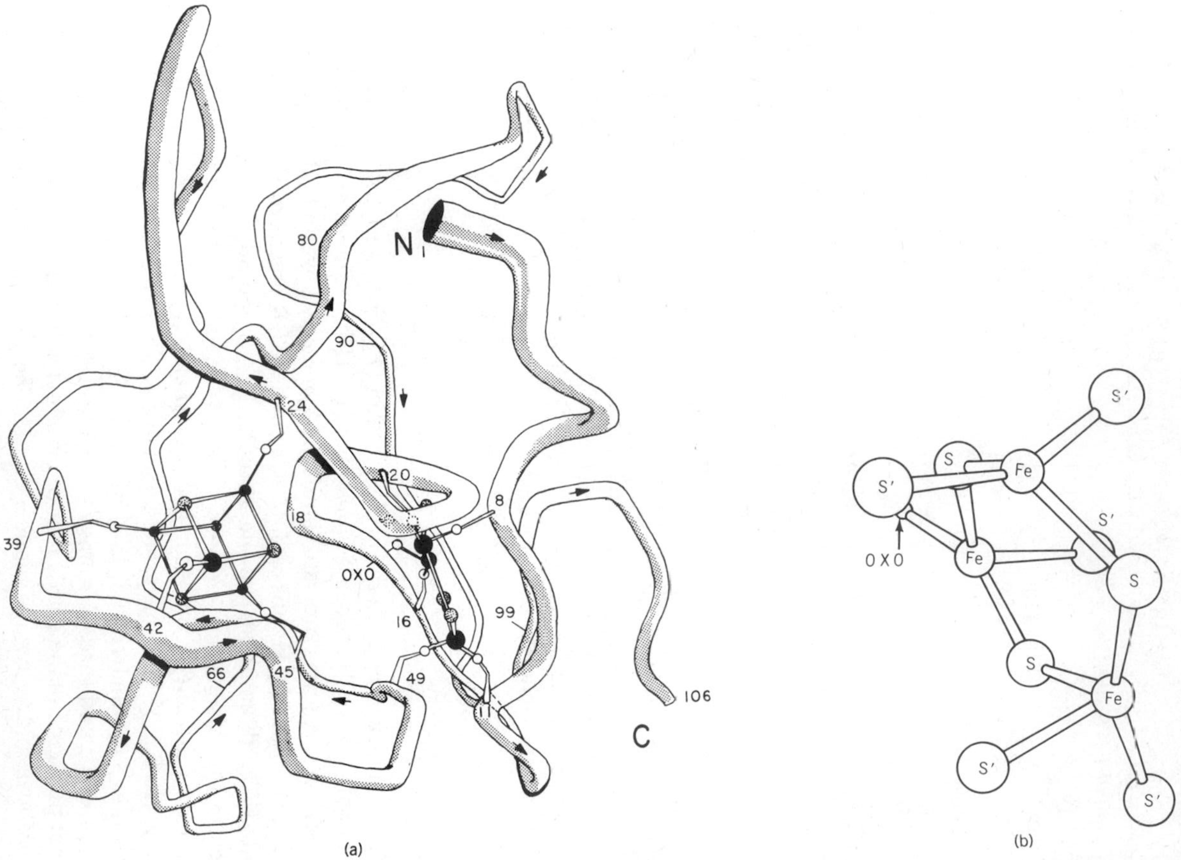

Fig. V-26. (A) Location of the two Fe–S clusters with respect to the amino acid sequence of ferredoxin from *Azotobacter vinelandii.* The 3Fe–3S cluster is ligated by cysteines 8, 11, 16, 20 and 40 and a sixth ligand which is either glutamic acid 18 or an exogenous small molecule. The 4Fe–4S cluster is ligated by cysteines 24, 39, 42 and 45. (B) View of the 3Fe–3S cluster showing near planarity of the Fe–S core and distorted tetrahedral geometry about each Fe center (reproduced with permission).[90]

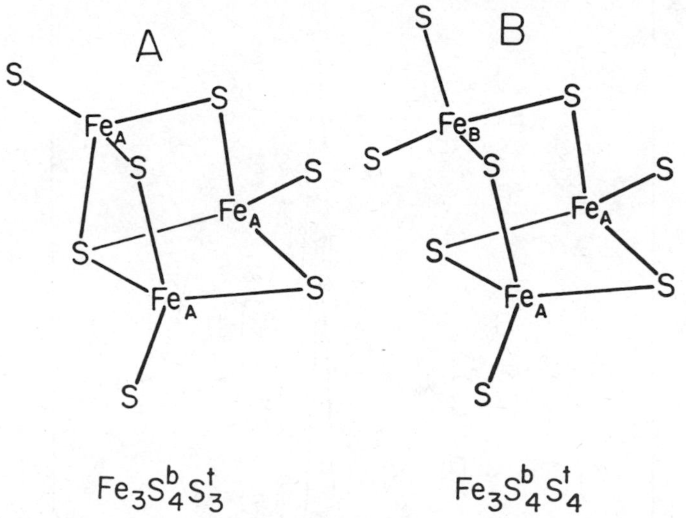

Fig. V-27. Alternative structures proposed for 3Fe–4S clusters.

cubanelike 3Fe–4S structures proposed by Beinert et al.[91] EXAFS studies on these proteins[91,92] estimate the Fe–Fe distances to be 2.7 Å which is close to those found in the 2Fe–2S and 4Fe–4S clusters.

Johnson et al.[88] measured the RR spectra of *Av* Fd I and *Thermus thermophilus* (*Tt* Fd),[93] both of which contain 3Fe as well as 4Fe clusters. As is seen in Fig. V-28, their RR spectra are dominated by the 3Fe spectra which exhibit bands at 390, 368, 347, 285, and 266 cm^{-1}. The weak band at 334 cm^{-1} is attributed to the 4Fe–4S cluster. Oxidized *Dg* Fd II[94] and ferricyanide-treated *Cp* Fd[95] which are known to contain only 3Fe clusters show no such bands. The ^{34}S sulfide substitution in *Tt* Fd and ferricyanide-treated *Cp* Fd produced downshifts of the bands near 266, 285, and 347 cm^{-1}, indicating that these bands are due to bridging ν(Fe–S). The strong band at 347 cm^{-1} is due to the totally symmetric breathing cluster mode while the remaining bands near 390 and 368 cm^{-1} are assigned to the terminal ν(Fe–S).

Normal coordinate calculations by Johnson et al.[88] have shown that the RR spectra of *Av* Fd I crystals and 3Fe bacterial ferredoxins (*Cp* Fd and *Tt* Fd) are compatible with cubanelike 3Fe–4S structures proposed by Beinert et al.[91] but not with the 3Fe–3S ring structure found for *Av* Fd I.[90] The RR spectra of aconitase and *Desulfovibrio desulfuricans*[96] are also very similar to those mentioned above, indicating the possibility of the cubanelike 3Fe–4S structures in these proteins.

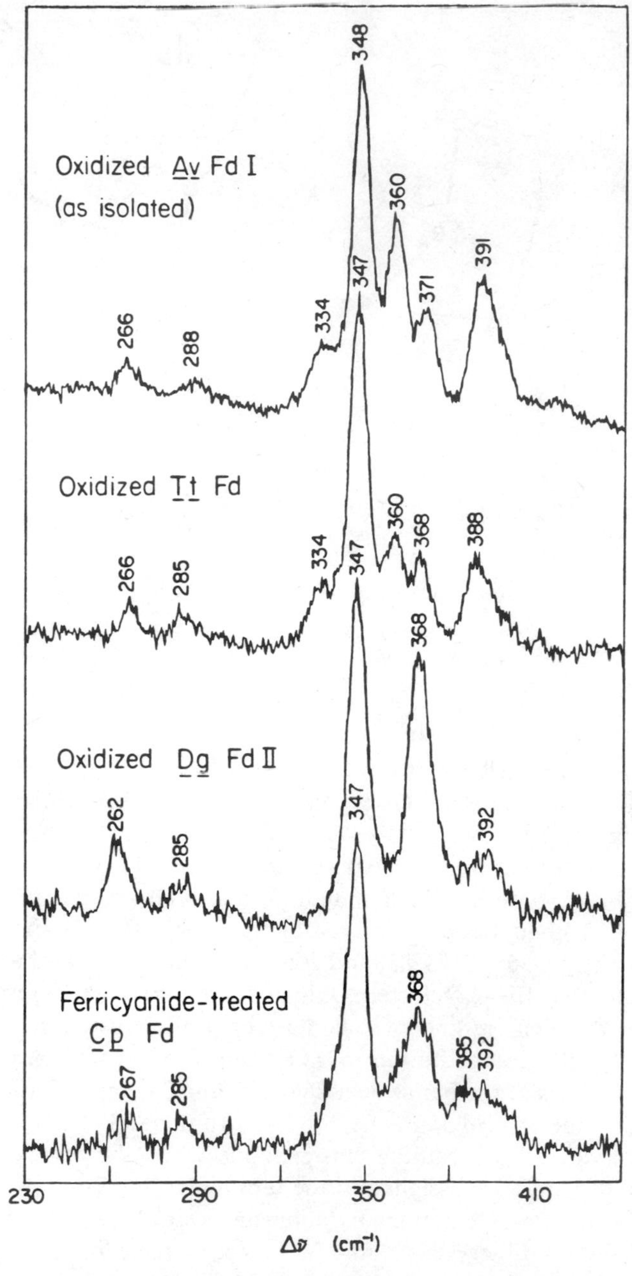

Fig. V-28. Low-temperature RR spectra of *Azotobacter vinelandii* (*Av*) ferredoxin I (Fd I), *Thermus thermophilus* (*Tt*) Fd, *Desulfovibrio gigas* (*Dg*) Fd II, and ferricyanide-treated *Cp* Fd (488 nm excitation).

References

1. R. G. Shulman, J. J. Hopfield, and S. Ogawa, *Quart. Rev. Biophys.*, **8,** 325 (1975).
1a. R. H. Felton and N.-T. Yu, "Resonance Raman Scattering from Metalloporphyrins and Hemoproteins," in D. Dolphin, ed., *The Porphyrins*, Physical Chemistry, Part A, Academic Press, New York, 1978, p. 347.
2. T. G. Spiro, "The Resonance Raman Spectroscopy of Metalloporphyrins and Heme Proteins," in A. B. P. Lever and H. B. Gray, eds., *Iron Porphyrins*, Addison-Wesley, Reading, Massachusetts, 1983.
3. S. A. Asher, "Resonance Raman Spectroscopy of Hemoglobin," in S. P. Colowick and N. O. Kaplan, eds., *Methods in Enzymology*, Vol. 76, Academic Press, New York, 1981, p. 371.
4. J. O. Alben, "Infrared Spectroscopy of Porphyrins," in D. Dolphin, ed., *The Porphyrins*, Physical Chemistry, Part A, Academic Press, New York, 1978, p. 323.
5. T. G. Spiro and T. C. Strekas, *J. Am. Chem. Soc.*, **96,** 338 (1974).
6. T. Kitagawa, N. Abe, and H. Ogoshi, *J. Chem. Phys.*, **69,** 4516 (1978); M. Abe, T. Kitagawa, and Y. Kyogoku, *J. Chem. Phys.*, **69,** 4526 (1978).
7. Y. Ozaki, T. Kitagawa, Y. Kyogoku, H. Shimada, T. Iizuka, and Y. Ishimura, *J. Biochem.*, **80,** 1447 (1976).
8. H. Oshio, T. Ama, T. Watanabe, J. Kincaid, and K. Nakamoto, *Spectrochim. Acta*, **40A,** 863 (1984).
9. S. A. Asher, L. E. Vickery, T. M. Schuster, and K. Sauer, *Biochemistry*, **16,** 5849 (1977).
10. K. Nagai, T. Kitagawa, and H. Morimoto, *J. Mol. Biol.*, **136,** 271 (1980).
11. J. Kincaid, P. Stein, and T. G. Spiro, *Proc. Natl. Acad. Sci. U.S.A.*, **76,** 549 and 4156 (1979).
12. P. Stein, M. Mitchell, and T. G. Spiro, *J. Am. Chem. Soc.*, **102,** 7795 (1980).
12a. A. Desbois and M. Lutz, *Biochim. Biophys. Acta*, **671,** 168 (1981); Brussels Hemoglobin Symposium Abstracts, 1983, p. 285.
13. H. Brunner, *Naturwissenshaften*, **61,** 129 (1974).
14. L. L. Duff, E. H. Appelman, D. F. Shriver, and I. M. Klotz, *Biochem. Biophys. Res. Commun.*, **90,** 1098 (1979).
15. B. Benko and N.-T. Yu, *Proc. Natl. Acad. Sci. U.S.A.*, **80,** 7042 (1983).
16. K. Bajdor, H. Oshio, and K. Nakamoto, *J. Am. Chem. Soc.*, **106,** 7273 (1984).
17. W. S. Caughey, M. G. Choc, and R. A. Houtchens, *Biochemical and Clinical Aspects of Oxygen*, Academic Press, New York, 1979, p. 4.
18. M. Tsubaki and N.-T. Yu, *Proc. Natl. Acad. Sci. U.S.A.*, **78,** 3581 (1981).
19. T. Kitagawa, M. R. Ondrias, D. L. Rousseau, M. Ikeda-Saito, and T. Yonetani, *Nature*, **298,** 869 (1982).
20. J. O. Alben and W. S. Caughey, *Biochemistry*, **7,** 175 (1968).
21. M. Tsubaki, R. B. Srivastava, and N.-T. Yu, *Biochemistry*, **21,** 1132 (1982).
22. J. D. Stong, J. B. Burke, P. Daly, P. Wright, and T. G. Spiro, *J. Am. Chem. Soc.*, **102,** 5815 (1980).
23. M. Tsubaki and N.-T. Yu, *Biochemistry*, **21,** 1140 (1982).
24. D. D. Pulsinelli, M. F. Perutz, and R. L. Nagel, *Proc. Natl. Acad. Sci. U.S.A.*, **70,** 3870 (1973).
25. K. Nagai, T. Kagimoto, A. Hayashi, F. Taketa, and T. Kitagawa, *Biochemistry*, **22,** 1305 (1983).
26. J. P. Collman, T. R. Halbert, and K. S. Suslick, "O_2 Binding to Heme Proteins and Their Synthetic Analogs," in T. G. Spiro, ed., *Metal Ions in Biology*, Vol. 2, Wiley, New York, 1980, p. 1.
27. J. P. Collman, R. R. Gagne, C. A. Reed, T. R. Halbert, G. Lang, and W. T. Robinson, *J. Am. Chem. Soc.*, **97,** 1427 (1975); J. P. Collman, R. R. Gagne, T. R. Halbert, J. C. Marchon, and C. A. Reed, *J. Am. Chem. Soc.*, **95,** 7868 (1973).
28. J. Almog, J. E. Baldwin, R. L. Dyer, and M. Peters, *J. Am. Chem. Soc.*, **97,** 226 (1975).
29. J. P. Collman, J. I. Brauman, T. R. Halbert, and K. S. Suslick, *Proc. Natl. Acad. Sci. U.S.A.*, **73,** 3333 (1976).
30. J. M. Burke, J. R. Kincaid, S. Peters, R. R. Gagne, J. P. Collman, and T. G. Spiro, *J. Am. Chem. Soc.*, **100,** 6083 (1978).
31. M. A. Walters, T. G. Spiro, K. S. Suslick, and J. P. Collman, *J. Am. Chem. Soc.*, **102,** 6857 (1980).

32. G. B. Jameson, F. S. Mollinaro, J. A. Ibers, J. P. Collman, J. I. Brauman, E. Rose, and K. S. Suslick, *J. Am. Chem. Soc.*, **102,** 3224 (1980).
33. H. C. Mackin, M. Tsubaki, and N.-T. Yu, *Biophys. J.*, **41,** 349 (1983).
34. K. Bajdor, K. Nakamoto, and J. R. Kincaid, *J. Am. Chem. Soc.*, **106,** 7741 (1984).
35. R. D. Jones, J. R. Budge, P. E. Ellis, Jr., J. G. Linard, D. A. Summerville, and F. Basolo, *J. Organometal. Chem.*, **181,** 151 (1979).
36. H. Oshio, T. Kuroi, and K. Nakamoto, to be published.
37. J. P. Collman and T. N. Sorrell, *J. Am. Chem. Soc.*, **97,** 4133 (1975).
38. B. B. Wayland, L. F. Mehne, and J. Swartz, *J. Am. Chem. Soc.*, **100,** 2379 (1978).
39. B. B. Wayland and L. W. Olson, *J. Am. Chem. Soc.*, **96,** 6037 (1974).
40. J. F. Deatherage and K. Moffat, *J. Mol. Biol.*, **134,** 401 (1979).
41. R. Timkovich, "Cytochrome c," in D. Dolphin, ed., *The Porphyrins*, Vol. VII, Biochemistry, Part B, Academic Press, New York, 1979.
42. L. C. Dickinson and J. C. W. Chien, *Biochemistry*, **14,** 3534 (1975).
43. T. Kitagawa, Y. Ozaki, J. Teraoka, Y. Kyogoku, and T. Yamanaka, *Biochem. Biophys. Acta*, **494,** 100 (1977).
44. A. Lanir, N.-T. Yu, and R. H. Felton, *Biochemistry*, **18,** 1656 (1979).
45. W. G. Valance and T. C. Strekas, *J. Phys. Chem.*, **86,** 1804 (1982).
46. T. Mashiko, J.-C. Marchon, D. T. Musser, C. A. Reed, M. E. Kastner, and W. R. Scheidt, *J. Am. Chem. Soc.*, **101,** 3653 (1979); T. Mashiko, C. A. Reed, K. J. Haller, M. E. Kastner, and W. R. Scheidt, *J. Am. Chem. Soc.*, **103,** 5758 (1981).
47. H. Oshio, T. Ama, T. Watanabe, and K. Nakamoto, *Inorg. Chim. Acta*, **96,** 61 (1985).
48. For example, L. S. Alexander and H. M. Goff, *J. Chem. Ed.*, **59,** 179 (1982).
48a. T. L. Poulos, B. C. Finzel, I. C. Gunsalus, G. C. Wagner and J. Kraut, *J. Biol. Chem.*, **260,** 16122 (1985).
49. Y. Ozaki, T. Kitagawa, Y. Kyogoku, Y. Imai, C. Hashimoto-Yutsudo, and R. Sato, *Biochemistry*, **17,** 5826 (1978).
50. P. M. Champion, B. R. Stallard, G. C. Wagner, and I. C. Gunsalus, *J. Am. Chem. Soc.*, **104,** 5469 (1982).
51. D. H. O'Keefe, R. E. Ebel, J. A. Peterson, J. C. Maxwell, and W. S. Caughey, *Biochemistry*, **17,** 5845 (1978).
52. M. Schappacher, L. Ricard, R. Weiss, R. Montiel-Montoya, E. Bill, U. Gonser, and A. Trautwein, *J. Am. Chem. Soc.*, **103,** 7646 (1981).
53. K. Nakamoto and H. Oshio, *J. Am. Chem. Soc.*, **107,** 6518 (1985).
54. D. M. Kurtz, Jr., D. F. Shriver, and I. M. Klotz, *Coord. Chem. Rev.*, **24,** 145 (1977).
55. I. M. Klotz and D. M. Kurtz, Jr., *Acc. Chem. Res.*, **17,** 16 (1984).
56. W. A. Henderickson, G. L. Klippenstein, and K. B. Ward, *Proc. Natl. Acad. Sci. U.S.A.*, **72,** 2160 (1975).
57. J. B. R. Dunn, D. F. Shriver, and I. M. Klotz, *Proc. Natl. Acad. Sci. U.S.A.*, **70,** 2582 (1973).
58. D. M. Kurtz, Jr., D. F. Shriver, and I. M. Klotz, *J. Am. Chem. Soc.*, **98,** 5033 (1976).
59. R. E. Stenkamp, L. C. Sieker, and L. H. Jensen, *J. Am. Chem. Soc.*, **106,** 618 (1984).
60. R. E. Stenkamp, L. C. Sieker, L. H. Jensen, J. D. McCallum, and J. Sanders-Loehr, *Proc. Natl. Acad. Sci. U.S.A.*, **82,** 713 (1985).
61. A. K. Shiemke, T. M. Loehr, and J. Sanders-Loehr, *J. Am. Chem. Soc.*, **106,** 4951 (1984).
62. W. H. Armstrong, A. Spool, G. C. Papaefthymiou, R. B. Frankel, and S. J. Lippard, *J. Am. Chem. Soc.*, **106,** 3653 (1984).
63. "Copper Proteins," in T. G. Spiro, ed., *Metal Ions in Biology*, Vol. 3, Wiley, New York, 1984.
64. T. B. Freedman, J. S. Loehr, and T. M. Loehr, *J. Am. Chem. Soc.*, **98,** 2809 (1976); J. S. Loehr, T. B. Freedman, and T. M. Loehr, *Biochem. Biophys. Res. Commun.*, **56,** 510 (1974).
65. T. J. Thamann, J. S. Loehr, and T. M. Loehr, *J. Am. Chem. Soc.*, **99,** 4187 (1977).
66. J. A. Larrabee and T. G. Spiro, *J. Am. Chem. Soc.*, **102,** 4217 (1980).
67. J. M. Brown, L. Powers, B. Kincaid, J. A. Larrabee, and T. G. Spiro, *J. Am. Chem. Soc.*, **102,** 4210 (1980).
68. W. P. J. Gaykema, W. G. J. Hol, J. M. Vereijken, N. M. Soeter, H. J. Bak, and J. J. Beintema, *Nature*, **309,** 23 (1984).

69. M. G. Simmons, C. L. Merrill, L. J. Wilson, L. A. Bottomley, and K. M. Kadish, *J. Chem. Soc., Dalton Trans.*, 1827 (1980).
70. Y. Nishida, K. Takahashi, H. Kuramoto, and S. Kida, *Inorg. Chim. Acta*, **54**, L103 (1981).
71. K. D. Karlin, R. W. Cruse, Y. Gultneh, J. C. Hayes, and J. Zubieta, *J. Am. Chem. Soc.*, **106**, 3372 (1984).
72. "Copper Proteins," in T. G. Spiro, ed., *Metal Ions in Biology*, Vol. 3, Wiley, New York, 1981.
73. V. Miskowski, S.-P. Tang, T. G. Spiro, E. Shapiro, and T. H. Moss, *Biochemistry*, **14**, 1244 (1975).
74. O. Siiman, N. M. Young, and P. R. Carey, *J. Am. Chem. Soc.*, **98**, 744 (1976).
75. N. S. Ferris, W. H. Woodruff, D. L. Tennent, and D. R. McMillin, *Biochem. Biophys. Res. Commun.*, **88**, 288 (1979).
76. P. M. Colman, H. C. Freeman, J. M. Guss, M. Murata, V. A. Norris, J. A. M. Ramshaw, and M. P. Venkatappa, *Nature*, **272**, 319 (1978).
77. L. Nestor, J. A. Larrabee, G. Woolery, B. Reinhammer, and T. G. Spiro, *Biochemistry*, **23**, 1084 (1984).
78. J. S. Thompson, T. J. Marks, and J. A. Ibers, *J. Am. Chem. Soc.*, **101**, 4180 (1979).
79. "Iron-Sulfur Proteins," in T. G. Spiro, ed., *Metal Ions in Biology*, Vol. 4, Wiley, New York, 1982.
80. R. W. Lane, J. A. Ibers, R. B. Frankel, R. H. Holm, and G. C. Papaefthymiou, *J. Am. Chem. Soc.*, **99**, 84 (1977).
81. T. V. Long and T. M. Loehr, *J. Am. Chem. Soc.*, **92**, 6384 (1970); T. V. Long, T. M. Loehr, J. R. Alkins, and W. Lovenberg, *J. Am. Chem. Soc.*, **93**, 1809 (1971).
82. V. K. Yachandra, J. Hare, I. Moura, and T. G. Spiro, *J. Am. Chem. Soc.*, **105**, 6455 (1983).
83. K. Fukuyama, T. Hase, S. Matsumoto, T. Tsukihara, Y. Katsube, N. Tanaka, M. Kakudo, K. Wada, and H. Matsubara, *Nature*, **286**, 522 (1980).
84. J. J. Mayerle, S. E. Denmark, B. V. DePamphilis, J. A. Ibers, and R. H. Holm, *J. Am. Chem. Soc.*, **97**, 1032 (1975).
85. V. K. Yachandra, J. Hare, A. Gewirth, R. S. Czernuszewicz, T. Kimura, R. H. Holm, and T. G. Spiro, *J. Am. Chem. Soc.*, **105**, 6462 (1983).
86. E. T. Adman, L. C. Sieker, and L. H. Jensen, *J. Biol. Chem.*, **248**, 3987 (1973).
87. T. M. Berg and R. H. Holm, "Iron-Sulfur Proteins," in T. G. Spiro, ed., *Metal Ions in Biology*, Vol. 4, Wiley, New York, 1982, Chapter I.
88. M. K. Johnson, R. S. Czernuszewicz, T. G. Spiro, J. A. Fee, and W. V. Sweeney, *J. Am. Chem. Soc.*, **105**, 6671 (1983).
89. R. S. Czernuszewicz, M. K. Johnson, and T. G. Spiro, private communication.
89a. J. M. Moulis, J. Meyer, and M. Lutz, *Biochem. J.*, **219**, 829 (1984).
90. C. D. Stout, D. Ghosh, V. Pattabhi, and A. H. Robbins, *J. Biol. Chem.*, **255**, 1797 (1980); D. Ghosh, W. Furey, Jr., S. O'Donnell, and C. D. Stout, *J. Biol. Chem.*, **256**, 4185 (1981); D. Ghosh, S. O'Donnell, W. Furey, Jr., A. H. Robinson, and C. D. Stout, *J. Mol. Biol.*, **158**, 73 (1982).
91. H. Beinert, M. H. Emptage, J. L. Dryer, R. A. Scott, J. E. Kahn, K. O. Hodgson, and A. Y. Thompson, *Proc. Natl. Acad. Sci. U.S.A.*, **80**, 393 (1983).
92. M. R. Antonio, B. A. Arerill, I. Moura, J. J. G. Moura, W. H. Orme-Johnson, B. K. Teo, and A. V. Xavier, *J. Biol. Chem.*, **257**, 6646 (1982).
93. J. A. Fee, M. L. Ludwig, T. Yoshida, B. H. Vuynh, T. A. Kent, and E. Munck, *Fed. Proc., Fed. Am. Soc. Exp. Biol.*, **40**, 1665 (1981).
94. M. K. Johnson, J. W. Hare, T. G. Spiro, J. J. G. Moura, A. V. Xavier, and J. Legall, *J. Biol. Chem.*, **256**, 9006 (1981).
95. M. K. Johnson, T. G. Spiro, and L. E. Mortenson, *J. Biol. Chem.*, **257**, 2447 (1982).
96. M. K. Johnson, R. S. Czernuszewicz, T. G. Spiro, R. R. Ramsay, and T. P. Singer, *J. Biol. Chem.*, **258**, 12771 (1983).

Appendixes

APPENDIX I

POINT GROUPS AND THEIR CHARACTER TABLES

The following are the character tables of the point groups that appear frequently in this book. The species (or the irreducible representations) of the point group are labeled according to the following rules: A and B denote nondegenerate species (one-dimensional representation). A represents the symmetric species (character $= +1$) with respect to rotation about the principal axis (chosen as z axis), whereas B represents the antisymmetric species (character $= -1$) with respect to rotation about the principal axis; E and F denote doubly degenerate (two-dimensional representation) and triply degenerate species (three-dimensional representation), respectively. If two species in the same point group differ in the character of C (other than the principal axis), they are distinguished by subscripts 1, 2, 3, ... If two species differ in the character of σ (other than σ_v), they are distinguished by ′ and ″. If two species differ in the character of i, they are distinguished by subscripts g and u. If these rules allow several different labels, g and u take precedence over 1, 2, 3, ..., which in turn take precedence over ′ and ″. The labels of species of point groups $\mathbf{C}_{\infty v}$ and $\mathbf{D}_{\infty h}$ (linear molecules) are exceptional and are taken from the notation for the component of the electronic orbital angular momentum along the molecular axis.

$\mathbf{C}_s$	I	$\sigma(xy)$		
A'	+1	+1	T_x, T_y, R_z	$\alpha_{xx}, \alpha_{yy}, \alpha_{zz}, \alpha_{xy}$
A''	+1	−1	T_z, R_x, R_y	α_{yz}, α_{xz}

$\mathbf{C}_2$	I	$C_2(z)$		
A	+1	+1	T_z, R_z	$\alpha_{xx}, \alpha_{yy}, \alpha_{zz}, \alpha_{xy}$
B	+1	−1	T_x, T_y, R_x, R_y	α_{yz}, α_{xz}

$\mathbf{C}_i$	i	I		
A_g	+1	+1	R_x, R_y, R_z	All components of α
A_u	+1	−1	T_x, T_y, T_x	

$\mathbf{C}_{2v}$	I	$C_2(z)$	$\sigma_v(xz)$	$\sigma_v(yz)$		
A_1	+1	+1	+1	+1	T_z	$\alpha_{xx}, \alpha_{yy}, \alpha_{zz}$
A_2	+1	+1	−1	−1	R_z	α_{xy}
B_1	+1	−1	+1	−1	T_x, R_y	α_{xz}
B_2	+1	−1	−1	+1	T_y, R_x	α_{yz}

$\mathbf{C}_{3v}$	I	$2C_3(z)$	$3\sigma_v$		
A_1	+1	+1	+1	T_z	$\alpha_{xx}+\alpha_{yy}, \alpha_{zz}$
A_2	+1	+1	−1	R_z	
E	+2	−1	0	$(T_x, T_y), (R_x, R_y)$	$(\alpha_{xx}-\alpha_{yy}, \alpha_{xy}), (\alpha_{yz}, \alpha_{xz})$

$\mathbf{C}_{4v}$	I	$2C_4(z)$	$C_4^2 \equiv C_2''$	$2\sigma_v$	$2\sigma_d$		
A_1	+1	+1	+1	+1	+1	T_z	$\alpha_{xx}+\alpha_{yy}, \alpha_{zz}$
A_2	+1	+1	+1	−1	−1	R_z	
B_1	+1	−1	+1	+1	−1		$\alpha_{xx}-\alpha_{yy}$
B_2	+1	−1	+1	−1	+1		α_{xy}
E	+2	0	−2	0	0	$(T_x, T_y), (R_x, R_y)$	$(\alpha_{yz}, \alpha_{xx})$

C_p^n (or S_p^n) denotes that C_p (or S_p) operation is carried out successively n times.

$\mathbf{C}_{\infty v}$	I	$2C_\infty^{\phi}$	$2C_\infty^{2\phi}$	$2C_\infty^{3\phi}$	...	$\infty\sigma_v$		
Σ^+	+1	+1	+1	+1	...	+1	T_z	$\alpha_{xx}+\alpha_{yy}, \alpha_{zz}$
Σ^-	+1	+1	+1	+1	...	−1	R_z	
Π	+2	$2\cos\phi$	$2\cos 2\phi$	$2\cos 3\phi$	...	0	$(T_x, T_y), (R_x, R_y)$	$(\alpha_{yz}, \alpha_{xz})$
Δ	+2	$2\cos 2\phi$	$2\cos 2\cdot 2\phi$	$2\cos 3\cdot 2\phi$	...	0		$(\alpha_{xx}-\alpha_{yy}, \alpha_{xy})$
Φ	+2	$2\cos 3\phi$	$2\cos 2\cdot 3\phi$	$2\cos 3\cdot 3\phi$	...	0		
...	...	...	...	...	...	...		

$\mathbf{C}_{2h}$	I	$C_2(z)$	$\sigma_h(xy)$	i		
A_g	+1	+1	+1	+1	R_z	$\alpha_{xx}, \alpha_{yy}, \alpha_{zz}, \alpha_{xy}$
A_u	+1	+1	−1	−1	T_z	
B_g	+1	−1	−1	+1	R_x, R_y	α_{yz}, α_{xz}
B_u	+1	−1	+1	−1	T_x, T_y	

$\mathbf{D}_3$	I	$2C_3(z)$	$3C_2$		
A_1	+1	+1	+1		$\alpha_{xx}+\alpha_{yy}, \alpha_{zz}$
A_2	+1	+1	−1	T_z, R_z	
E	+2	−1	0	$(T_x, T_y), (R_x, R_y)$	$(\alpha_{xx}-\alpha_{yy}, \alpha_{xy}), (\alpha_{yz}, \alpha_{xz})$

$\mathbf{D}_{2d} \equiv \mathbf{V}_d$	I	$2S_4(z)$	$S_4^2 \equiv C_2''$	$2C_2$	$2\sigma_d$		
A_1	+1	+1	+1	+1	+1		$\alpha_{xx}+\alpha_{yy}, \alpha_{zz}$
A_2	+1	+1	+1	−1	−1	R_z	
B_1	+1	−1	+1	+1	−1		$\alpha_{xx}-\alpha_{yy}$
B_2	+1	−1	+1	−1	+1	T_z	α_{xy}
E	+2	0	−2	0	0	$(T_x, T_y), (R_x, R_y)$	$(\alpha_{yz}, \alpha_{xz})$

$\mathbf{D}_{3d}$	I	$2S_6(z)$	$2S_6^2 \equiv 2C_3$	$S_6^3 \equiv S_2 \equiv i$	$3C_2$	$3\sigma_d$		
A_{1g}	+1	+1	+1	+1	+1	+1		$\alpha_{xx}+\alpha_{yy}, \alpha_{zz}$
A_{1u}	+1	−1	+1	−1	+1	−1		
A_{2g}	+1	+1	+1	+1	−1	−1	R_g	
A_{2u}	+1	−1	+1	−1	−1	+1	T_z	
E_g	+2	−1	−1	+2	0	0	(R_x, R_y)	$(\alpha_{xx}-\alpha_{yy}, \alpha_{xy}), (\alpha_{yz}, \alpha_{xz})$
E_u	+2	+1	−1	−2	0	0	(T_x, T_y)	

$\mathbf{D}_{4d}$	I	$2S_8(z)$	$2S_8^2 \equiv 2C_4$	$2S_8^3$	$S_8^4 \equiv C_2''$	$4C_2$	$4\sigma_d$		
A_1	+1	+1	+1	+1	+1	+1	+1		$\alpha_{xx}+\alpha_{yy}, \alpha_{zz}$
A_2	+1	+1	+1	+1	+1	−1	−1	R_z	
B_1	+1	−1	+1	−1	+1	+1	−1		
B_2	+1	−1	+1	−1	+1	−1	+1	T_z	
E_1	+2	$+\sqrt{2}$	0	$-\sqrt{2}$	−2	0	0	(T_x, T_y)	
E_2	+2	0	−2	0	+2	0	0		$(\alpha_{xx}-\alpha_{yy}, \alpha_{xy})$
E_3	+2	$-\sqrt{2}$	0	$+\sqrt{2}$	−2	0	0	(R_x, R_y)	$(\alpha_{yz}, \alpha_{xz})$

$\mathbf{D}_{2h} \equiv \mathbf{V}_h$	I	$\sigma(xy)$	$\sigma(xz)$	$\sigma(yz)$	i	$C_2(z)$	$C_2(y)$	$C_2(x)$		
A_g	+1	+1	+1	+1	+1	+1	+1	+1		$\alpha_{xx}, \alpha_{yy}, \alpha_{zz}$
A_u	+1	−1	−1	−1	−1	+1	+1	+1		
B_{1g}	+1	+1	−1	−1	+1	+1	−1	−1	R_z	α_{xy}
B_{1u}	+1	−1	+1	+1	−1	+1	−1	−1	T_z	
B_{2g}	+1	−1	+1	−1	+1	−1	+1	−1	R_y	α_{xz}
B_{2u}	+1	+1	−1	+1	−1	−1	+1	−1	T_y	
B_{3g}	+1	−1	−1	+1	+1	−1	−1	+1	R_x	α_{yz}
B_{3u}	+1	+1	+1	−1	−1	−1	−1	+1	T_x	

$\mathbf{D}_{3h}$	I	$2C_3(z)$	$3C_2$	σ_h	$2S_3$	$3\sigma_v$		
A_1'	+1	+1	+1	+1	+1	+1		$\alpha_{xx}+\alpha_{yy}, \alpha_{zz}$
A_1''	+1	+1	+1	−1	−1	−1		
A_2'	+1	+1	−1	+1	+1	−1	R_z	
A_2''	+1	+1	−1	−1	−1	+1	T_z	
E'	+2	−1	0	+2	−1	0	(T_x, T_y)	$(\alpha_{xx}-\alpha_{yy}, \alpha_{xy})$
E''	+2	−1	0	−2	+1	0	(R_x, R_y)	$(\alpha_{yz}, \alpha_{xz})$

$\mathbf{D}_{4h}$	I	$2C_4(z)$	$C_4^2 \equiv C_2''$	$2C_2$	$2C_2'$	σ_h	$2\sigma_v$	$2\sigma_d$	$2S_4$	$S_2 \equiv i$		
A_{1g}	+1	+1	+1	+1	+1	+1	+1	+1	+1	+1		$\alpha_{xx}+\alpha_{yy}, \alpha_{zz}$
A_{1u}	+1	+1	+1	+1	+1	−1	−1	−1	−1	−1		
A_{2g}	+1	+1	+1	−1	−1	+1	−1	−1	+1	+1	R_z	
A_{2u}	+1	+1	+1	−1	−1	−1	+1	+1	−1	−1	T_z	
B_{1g}	+1	−1	+1	+1	−1	+1	+1	−1	−1	+1		$\alpha_{xx}-\alpha_{yy}$
B_{1u}	+1	−1	+1	+1	−1	−1	−1	+1	+1	−1		
B_{2g}	+1	−1	+1	−1	+1	+1	−1	+1	−1	+1		α_{xy}
B_{2u}	+1	−1	+1	−1	+1	−1	+1	−1	+1	−1		
E_g	+2	0	−2	0	0	−2	0	0	0	+2	(R_x, R_y)	$(\alpha_{yz}, \alpha_{xz})$
E_u	+2	0	−2	0	0	+2	0	0	0	−2	(T_x, T_y)	

$\mathbf{D}_{5h}$	I	$2C_5(z)$	$2C_5^2$	σ_h	$5C_2$	$5\sigma_v$	$2S_5$	$2S_5^3$		
A_1'	+1	+1	+1	+1	+1	+1	+1	+1		$\alpha_{xx}+\alpha_{yy}, \alpha_{zz}$
A_1''	+1	+1	+1	−1	+1	−1	−1	−1		
A_2'	+1	+1	+1	+1	−1	−1	+1	+1	R_z	
A_2''	+1	+1	+1	−1	−1	+1	−1	−1	T_z	
E_1'	+2	2 cos 72°	2 cos 144°	+2	0	0	+2 cos 72°	+2 cos 144°	(T_x, T_y)	
E_1''	+2	2 cos 72°	2 cos 144°	−2	0	0	−2 cos 72°	−2 cos 144°	(R_x, R_y)	$(\alpha_{yz}, \alpha_{xz})$
E_2'	+2	2 cos 144°	2 cos 72°	+2	0	0	+2 cos 144°	+2 cos 72°		$(\alpha_{xx}-\alpha_{yy}, \alpha_{xy})$
E_2''	+2	2 cos 144°	2 cos 72°	−2	0	0	−2 cos 144°	−2 cos 72°		

$\mathbf{D}_{6h}$	I	$2C_6(z)$	$2C_6^2 \equiv 2C_3$	$C_6^3 \equiv C_2''$	$3C_2$	$3C_2'$	σ_h	$3\sigma_v$	$3\sigma_d$	$2S_6$	$2S_3$	$S_6^3 \equiv S_2 \equiv i$		
A_{1g}	+1	+1	+1	+1	+1	+1	+1	+1	+1	+1	+1	+1		$\alpha_{xx}+\alpha_{yy}, \alpha_{zz}$
A_{1u}	+1	+1	+1	+1	+1	+1	−1	−1	−1	−1	−1	−1		
A_{2g}	+1	+1	+1	+1	−1	−1	+1	−1	−1	+1	+1	+1	R_z	
A_{2u}	+1	+1	+1	+1	−1	−1	−1	+1	+1	−1	−1	−1	T_z	
B_{1g}	+1	−1	+1	−1	+1	−1	−1	−1	+1	+1	−1	+1		
B_{1u}	+1	−1	+1	−1	+1	−1	+1	+1	−1	−1	+1	−1		
B_{2g}	+1	−1	+1	−1	−1	+1	−1	+1	−1	+1	−1	+1		
B_{2u}	+1	−1	+1	−1	−1	+1	+1	−1	+1	−1	+1	−1		
E_{1g}	+2	+1	−1	−2	0	0	−2	0	0	−1	+1	+2	(R_x, R_y)	$(\alpha_{yz}, \alpha_{xz})$
E_{1u}	+2	+1	−1	−2	0	0	+2	0	0	+1	−1	−2	(T_x, T_y)	
E_{2g}	+2	−1	−1	+2	0	0	+2	0	0	−1	−1	+2		$(\alpha_{xx}-\alpha_{yy}, \alpha_{xy})$
E_{2u}	+2	−1	−1	+2	0	0	−2	0	0	+1	+1	−2		

$\mathbf{D}_{\infty h}$	I	$2C_\infty^\phi$	$2C_\infty^{2\phi}$	$2C_\infty^{3\phi}$	...	σ_h	∞C_2	$\infty\sigma_v$	$2S_\infty^\phi$	$2S_\infty^{2\phi}$	...	$S_2 \equiv i$		
Σ_g^+	+1	+1	+1	+1	...	+1	+1	+1	+1	+1	...	+1		$\alpha_{xx}+\alpha_{yy}, \alpha_{zz}$
Σ_u^+	+1	+1	+1	+1	...	−1	−1	+1	−1	−1	...	−1	T_z	
Σ_g^-	+1	+1	+1	+1	...	+1	−1	−1	+1	+1	...	+1	R_z	
Σ_u^-	+1	+1	+1	+1	...	−1	+1	−1	−1	−1	...	−1		
Π_g	+2	$2\cos\phi$	$2\cos 2\phi$	$2\cos 3\phi$	...	−2	0	0	$-2\cos\phi$	$-2\cos 2\phi$	...	+2	(R_x, R_y)	$(\alpha_{yz}, \alpha_{xz})$
Π_u	+2	$2\cos\phi$	$2\cos 2\phi$	$2\cos 3\phi$	...	+2	0	0	$+2\cos\phi$	$+2\cos 2\phi$	...	−2	(T_x, T_y)	
Δ_g	+2	$2\cos 2\phi$	$2\cos 4\phi$	$2\cos 6\phi$	...	+2	0	0	$+2\cos 2\phi$	$+2\cos 4\phi$	...	+2		$(\alpha_{xx}-\alpha_{yy}, \alpha_{xy})$
Δ_u	+2	$2\cos 2\phi$	$2\cos 4\phi$	$2\cos 6\phi$	...	−2	0	0	$-2\cos 2\phi$	$-2\cos 4\phi$	...	−2		
Φ_g	+2	$2\cos 3\phi$	$2\cos 6\phi$	$2\cos 9\phi$	...	−2	0	0	$-2\cos 3\phi$	$-2\cos 4\phi$	...	+2		
Φ_u	+2	$2\cos 3\phi$	$2\cos 6\phi$	$2\cos 9\phi$	...	+2	0	0	$+2\cos 3\phi$	$+2\cos 4\phi$	...	−2		
...	...	...	...	...	...	...	...	...	...	...	...	...		

$\mathbf{T}_d$	I	$8C_3$	$6\sigma_d$	$6S_4$	$3S_4^2\equiv 3C_2$		
A_1	+1	+1	+1	+1	+1		$\alpha_{xx}+\alpha_{yy}+\alpha_{zz}$
A_2	+1	+1	−1	−1	+1		
E	+2	−1	0	0	+2		$(\alpha_{xx}+\alpha_{yy}-2\alpha_{zz}, \alpha_{xx}-\alpha_{yy})$
F_1	+3	0	−1	+1	−1	(R_x, R_y, R_z)	
F_2	+3	0	+1	−1	−1	(T_x, T_y, T_z)	$(\alpha_{xy}, \alpha_{yz}, \alpha_{xz})$

$\mathbf{O}_h$	I	$8C_3$	$6C_2$	$6C_4$	$3C_4^2\equiv 3C_2''$	$S_2\equiv i$	$6S_4$	$8S_6$	$3\sigma_h$	$6\sigma_d$		
A_{1g}	+1	+1	+1	+1	+1	+1	+1	+1	+1	+1		$\alpha_{xx}+\alpha_{yy}+\alpha_{zz}$
A_{1u}	+1	+1	+1	+1	+1	−1	−1	−1	−1	−1		
A_{2g}	+1	+1	−1	−1	+1	+1	−1	+1	+1	−1		
A_{2u}	+1	+1	−1	−1	+1	−1	+1	−1	−1	+1		
E_g	+2	−1	0	0	+2	+2	0	−1	+2	0		$(\alpha_{xx}+\alpha_{yy}-2\alpha_{zz}, \alpha_{xx}-\alpha_{yy})$
E_u	+2	−1	0	0	+2	−2	0	+1	−2	0		
F_{1g}	+3	0	−1	+1	−1	+3	+1	0	−1	−1	(R_x, R_y, R_z)	
F_{1u}	+3	0	−1	+1	−1	−3	−1	0	+1	+1	(T_x, T_y, T_z)	
F_{2g}	+3	0	+1	−1	−1	+3	−1	0	−1	+1		$(\alpha_{xy}, \alpha_{yz}, \alpha_{xz})$
F_{2u}	+3	0	+1	−1	−1	−3	+1	0	+1	−1		

APPENDIX II

GENERAL FORMULAS FOR CALCULATING THE NUMBER OF NORMAL VIBRATIONS IN EACH SPECIES

These tables were quoted from G. Herzberg, *Molecular Spectra and Molecular Structure*, Vol. II (Ref. 1 of Part I).

Table A. Point Groups Including Only Nondegenerate Vibrations

Point Group	Total Number of Atoms	Species	Number of Vibrations[a]
$\mathbf{C}_2$	$2m+m_0$	A B	$3m+m_0-2$ $3m+2m_0-4$
$\mathbf{C}_s$	$2m+m_0$	A' A''	$3m+2m_0-3$ $3m+m_0-3$
$\mathbf{C}_i \equiv \mathbf{S}_2$	$2m+m_0$	A_g A_u	$3m-3$ $3m+3m_0-3$
$\mathbf{C}_{2v}$	$4m+2m_{xz}+2m_{yz}+m_0$	A_1 A_2 B_1 B_2	$3m+2m_{xz}+2m_{yz}+m_0-1$ $3m+m_{xz}+m_{yz}-1$ $3m+2m_{xz}+m_{yz}+m_0-2$ $3m+m_{xz}+2m_{yz}+m_0-2$

$\mathbf{C}_{2h}$	$4m+2m_h+2m_2+m_0$	A_g	$3m+2m_h+m_2-1$
		A_u	$3m+m_h+m_2+m_0-1$
		B_g	$3m+m_h+2m_2-2$
		B_u	$3m+2m_h+2m_2+2m_0-2$
$D_{2h}\equiv\mathbf{V}_h$	$8m+4m_{xy}+4m_{xz}+4m_{yz}+2m_{2x}+2m_{2y}+2m_{2z}+m_0$	A_g	$3m+2m_{xy}+2m_{xz}+2m_{yz}+m_{2x}+m_{2y}+m_{2z}$
		A_u	$3m+m_{xy}+m_{xz}+m_{yz}$
		B_{1g}	$3m+2m_{xy}+m_{xz}+m_{yz}+m_{2x}+m_{2y}-1$
		B_{1u}	$3m+m_{xy}+2m_{xz}+2m_{yz}+m_{2x}+m_{2y}+m_{2z}+m_0-1$
		B_{2g}	$3m+m_{xy}+2m_{xz}+m_{yz}+m_{2x}+m_{2z}-1$
		B_{2u}	$3m+2m_{xy}+m_{xz}+2m_{yz}+m_{2x}+m_{2y}+m_{2z}+m_0-1$
		B_{3g}	$3m+m_{xy}+m_{xz}+2m_{yz}+m_{2y}+m_{2x}-1$
		B_{3u}	$3m+2m_{xy}+2m_{xz}+m_{yz}+m_{2x}+m_{2y}+m_{2z}+m_0-1$

[a] Note that m is always the number of sets of equivalent nuclei not on any element of symmetry; m_0 is the number of nuclei lying on all symmetry elements present; m_{xy}, m_{xz}, m_{yz} are the numbers of sets of nuclei lying on the xy, xz, yz plane, respectively, but not on any axes going through these planes; m_2 is the number of sets of nuclei on a twofold axis but not at the point of intersection with another element of symmetry; m_{2x}, m_{2y}, m_{2z} are the numbers of sets of nuclei lying on the x, y, or z axis if they are twofold axes, but not on all of them; m_h is the number of sets of nuclei on a plane σ_h but not on the axis perpendicular to this plane.

Table B. Point Groups Including Degenerate Vibrations

Point Group	Total Number of Atoms	Species	Number of Vibrations[a]
$\mathbf{D}_3$	$6m+3m_2+2m_3+m_0$	A_1	$3m+m_2+m_3$
		A_2	$3m+2m_2+m_3+m_0-2$
		E	$6m+3m_2+2m_3+m_0-2$
$\mathbf{C}_{3v}$	$6m+3m_v+m_0$	A_1	$3m+2m_v+m_0+1$
		A_2	$3m+m_v-1$
		E	$6m+3m_v+m_0-2$
$\mathbf{C}_{4v}$	$8m+4m_v+4m_d+m_0$	A_1	$3m+2m_v+2m_d+m_0-1$
		A_2	$3m+m_v+m_d-1$
		B_1	$3m+2m_v+m_d$
		B_2	$3m+m_v+2m_d$
		E	$6m+3m_v+3m_d+m_0-2$
$\mathbf{C}_{\infty v}$	m_0	Σ^+	m_0-1
		Σ^-	0
		Π	m_0-2
		$\Delta, \Phi, \ldots$	0
$\mathbf{D}_{2d} \equiv \mathbf{V}_d$	$8m+4m_d+4m_2+2m_4+m_0$	A_1	$3m+2m_d+m_2+m_4$
		A_2	$3m+m_d+2m_2-1$
		B_1	$3m+m_d+m_2$
		B_2	$3m+2m_d+2m_2+m_4+m_0-1$
		E	$6m+3m_d+3m_2+2m_4+m_0-2$

$\mathbf{D}_{3d}$	$12m+6m_d$ $+6m_2+2m_6+m_0$	A_{1g}	$3m+2m_d+m_2+m_6$
		A_{1u}	$3m+m_d+m_2$
		A_{2g}	$3m+m_d+2m_2-1$
		A_{2u}	$3m+2m_d+2m_2+m_6+m_0-1$
		E_g	$6m+3m_d+3m_2+m_6-1$
		E_u	$6m+3m_d+3m_2+m_6+m_0-1$
$\mathbf{D}_{4d}$	$16m+8m_d$ $+8m_2+2m_8+m_0$	A_1	$3m+2m_d+m_2+m_8$
		A_2	$3m+m_d+2m_2-1$
		B_1	$3m+m_d+m_2$
		B_2	$3m+2m_d+2m_2+m_8+m_0-1$
		E_1	$6m+3m_d+3m_2+m_8+m_0-1$
		E_2	$6m+3m_d+3m_2$
		E_3	$6m+3m_d+3m_2+m_8-1$
$\mathbf{D}_{3h}$	$12m+6m_v+6m_h$ $+3m_2+2m_3+m_0$	A_1'	$3m+2m_v+2m_h+m_2+m_3$
		A_1''	$3m+m_v+m_h$
		A_2'	$3m+m_v+2m_h+m_2-1$
		A_2''	$3m+2m_v+m_h+m_2+m_3+m_0-1$
		E'	$6m+3m_v+4m_h+2m_2+m_3+m_0-1$
		E''	$6m+3m_v+2m_h+m_2+m_3-1$
$\mathbf{D}_{4h}$	$16m+8m_v+8m_d$ $+8m_h+4m_2+4m_2'$ $+2m_4+m_0$	A_{1g}	$3m+2m_v+2m_d+2m_h+m_2+m_2'+m_4$
		A_{1u}	$3m+m_v+m_d+m_h$
		A_{2g}	$3m+m_v+m_d+2m_h+m_2+m_2'-1$
		A_{2u}	$3m+2m_v+2m_d+m_h+m_2+m_2'+m_4+m_0-1$
		B_{1g}	$3m+2m_v+m_d+2m_h+m_2+m_2'$
		B_{1u}	$3m+m_v+2m_d+m_h+m_2'$
		B_{2g}	$3m+m_v+2m_d+2m_h+m_2+m_2'$
		B_{2u}	$3m+2m_v+m_d+m_h+m_2$
		E_g	$6m+3m_v+3m_d+2m_h+m_2+m_2'+m_4-1$
		E_u	$6m+3m_v+3m_d+4m_h+2m_2+2m_2'+m_4+m_0-1$

Point Group	Total Number of Atoms	Species	Number of Vibrations[a]
$\mathbf{D}_{5h}$	$20m+10m_v+10m_h+5m_2+2m_5+m_0$	A_1'	$3m+2m_v+2m_h+m_2+m_5$
		A_1''	$3m+m_v+m_h$
		A_2'	$3m+m_v+2m_h+m_2-1$
		A_2''	$3m+2m_v+m_h+m_2+m_5+m_0-1$
		E_1'	$6m+3m_v+4m_h+2m_2+m_5+m_0-1$
		E_1''	$6m+3m_v+2m_h+m_2+m_5-1$
		E_2'	$6m+3m_v+4m_h+2m_2$
		E_2''	$6m+3m_v+2m_h+m_2$
$\mathbf{D}_{6h}$	$24m+12m_v+12m_d+12m_h+6m_2+6m_2'+2m_6+m_0$	A_{1g}	$3m+2m_v+2m_d+2m_h+m_2+m_2'+m_6$
		A_{1u}	$3m+m_v+m_d+m_h$
		A_{2g}	$3m+m_v+m_d+2m_h+m_2+m_2'-1$
		A_{2u}	$3m+2m_v+2m_d+m_h+m_2+m_2'+m_6+m_0-1$
		B_{1g}	$3m+m_v+2m_d+m_h+m_2'$
		B_{1u}	$3m+2m_v+m_d+2m_h+m_2+m_2'$
		B_{2g}	$3m+2m_v+m_d+m_h+m_2$
		B_{2u}	$3m+m_v+2m_d+2m_h+m_2+m_2'$
		E_{1g}	$6m+3m_v+3m_d+2m_h+m_2+m_2'+m_6-1$
		E_{1u}	$6m+3m_v+3m_d+4m_h+2m_2+2m_2'+m_6+m_0-1$
		E_{2g}	$6m+3m_v+3m_d+4m_h+2m_2+2m_2'$
		E_{2u}	$6m+3m_v+3m_d+2m_h+m_2+m_2'$

$\mathbf{D}_{\infty h}$	$2m_\infty + m_0$	Σ_g^+	m_∞
		Σ_u^+	$m_\infty + m_0 - 1$
		Σ_g^-, Σ_u^-	0
		Π_g	$m_\infty - 1$
		Π_u	$m_\infty + m_0 - 1$
		$\Delta_g, \Delta_u,$	0
		$\Phi_g, \Phi_u, \ldots$	0
$\mathbf{T}_d$	$24m + 12m_d + 6m_2 + 4m_3 + m_0$	A_1	$3m + 2m_d + m_2 + m_3$
		A_2	$3m + m_d$
		E	$6m + 3m_d + m_2 + m_3$
		F_1	$9m + 4m_d + 2m_2 + m_3 - 1$
		F_2	$9m + 5m_d + 3m_2 + 2m_3 + m_0 - 1$
$\mathbf{O}_h$	$48m + 24m_h + 24m_d + 12m_2 + 8m_3 + 6m_4 + m_0$	A_{1g}	$3m + 2m_h + 2m_d + m_2 + m_3 + m_4$
		A_{1u}	$3m + m_h + m_d$
		A_{2g}	$3m + 2m_h + m_d + m_2$
		A_{2u}	$3m + m_h + 2m_d + m_2 + m_3$
		E_g	$6m + 4m_h + 3m_d + 2m_2 + m_3 + m_4$
		E_u	$6m + 2m_h + 3m_d + m_2 + m_3$
		F_{1g}	$9m + 4m_h + 4m_d + 2m_2 + m_3 + m_4 - 1$
		F_{1u}	$9m + 5m_h + 5m_d + 3m_2 + 2m_3 + 2m_4 + m_0 - 1$
		F_{2g}	$9m + 4m_h + 5m_d + 2m_2 + 2m_3 + m_4$
		F_{2u}	$9m + 5m_h + 4m_d + 2m_2 + m_3 + m_4$

[a] Note that m is the number of sets of nuclei not any element of symmetry; m_0 is the number of nuclei on all elements of symmetry; $m_2, m_3, m_4, \ldots$ are the numbers of sets of nuclei on a twofold, threefold, fourfold, and so on, axis but not on any other element of symmetry that does not wholly coincide with that axis; m_2' is the number of sets of nuclei on a twofold axis called C_2' in the previous character tables; m_v, m_d, m_h are the numbers of sets of nuclei on planes $\sigma_v, \sigma_d, \sigma_h$, respectively, but not on any other element of symmetry.

APPENDIX III

NUMBER OF INFRARED- AND RAMAN-ACTIVE STRETCHING VIBRATIONS FOR MX_nY_m-TYPE MOLECULES

Compound	Structure	Point Group	IR or Raman	M–X Stretching	M–Y Stretching
MX_6	Octahedral	$\mathbf{O}_h$	IR	F_{1u}	
			R	A_{1g}, E_g	
MX_5Y	Octahedral	$\mathbf{C}_{4v}$	IR	$2A_1, E$	A_1
			R	$2A_1, B_1, E$	A_1
trans-MX_4Y_2	Octahedral	$\mathbf{D}_{4h}$	IR	E_u	A_{2u}
			R	A_{1g}, B_{1g}	A_{1g}
cis-MX_4Y_2	Octahedral	$\mathbf{C}_{2v}$	IR	$2A_1, B_1, B_2$	A_1, B_1
			R	$2A_1, B_1, B_2$	A_1, B_1
mer-MX_3Y_3	Octahedral	$\mathbf{C}_{2v}$	IR	$2A_1, B_2$	$2A_1, B_1$
			R	$2A_1, B_2$	$2A_1, B_1$
fac-MX_3Y_3	Octahedral	$\mathbf{C}_{3v}$	IR	A_1, E	A_1, E
			R	A_1, E	A_1, E
MX_5	Trigonal-bipyramidal	$\mathbf{D}_{3h}$	IR	A_2'', E'	
			R	$2A_1', E'$	
MX_5	Tetragonal-pyramidal	$\mathbf{C}_{4v}$	IR	$2A_1, E$	
			R	$2A_1, B_1, E$	
MX_4	Tetrahedral	$\mathbf{T}_d$	IR	F_2	
			R	A_1, F_2	
MX_3Y	Tetrahedral	$\mathbf{C}_{3v}$	IR	A_1, E	A_1
			R	A_1, E	A_1
MX_2Y_2	Tetrahedral	$\mathbf{C}_{2v}$	IR	A_1, B_1	A_1, B_2
			R	A_1, B_1	A_1, B_2
Polymeric MX_2Y_2 [a]	Octahedral	$\mathbf{C}_i$	IR	$2A_u$	A_u
			R	$2A_g$	A_g
MX_4	Square-planar	$\mathbf{D}_{4h}$	IR	E_u	
			R	A_{1g}, B_{1g}	
MX_3Y	Planar	$\mathbf{C}_{2v}$	IR	$2A_1, B_1$	A_1
			R	$2A_1, B_1$	A_1
trans-MX_2Y_2	Planar	$\mathbf{D}_{2h}$	IR	B_{3u}	B_{2u}
			R	A_g	A_g
cis-MX_2Y_2	Planar	$\mathbf{C}_{2v}$	IR	A_1, B_2	A_1, B_2
			R	A_1, B_2	A_1, B_2
MX_3	Pyramidal	$\mathbf{C}_{3v}$	IR	A_1, E	
			R	A_1, E	
MX_3	Planar	$\mathbf{D}_{3h}$	IR	E'	
			R	A_1'', E'	

[a] Bridging through X atoms.

APPENDIX IV

DERIVATION OF EQ. 11.3 (PART I)

Using the rectangular coordinates, we write the kinetic energy as

$$2T = \tilde{\dot{\mathbf{X}}}\mathbf{M}\dot{\mathbf{X}} \tag{1}$$

where

$$\mathbf{X} = \begin{bmatrix} x_1 \\ y_1 \\ z_1 \\ x_2 \\ \vdots \\ z_N \end{bmatrix} \quad \text{and} \quad \mathbf{M} = \begin{bmatrix} m_1 & & & & & \\ & m_1 & & & & \\ & & m_1 & & & \\ & & & m_2 & & \\ & & & & \ddots & \\ & & & & & m_N \end{bmatrix}$$

By definition, the momentum p_{x_1} conjugated with x_1 is given by

$$p_{x_1} = \frac{\partial T}{\partial \dot{x}_1} = m_1 \dot{x}_1$$

$p_{y_1} \cdots p_{z_N}$ take similar forms. Using the conjugate momenta, we write T as

$$\begin{aligned} 2T &= \frac{1}{m_1} p_{x_1}^2 + \frac{1}{m_1} p_{y_1}^2 + \cdots + \frac{1}{m_N} p_{z_N}^2 \\ &= \tilde{\mathbf{P}}_x \mathbf{M}^{-1} \mathbf{P}_x \end{aligned} \tag{2}$$

where

$$\mathbf{P}_x = \begin{bmatrix} p_{x_1} \\ p_{y_1} \\ \vdots \\ p_{z_N} \end{bmatrix} \quad \text{and} \quad \mathbf{M}^{-1} = \begin{bmatrix} \mu_1 & & & \\ & \mu_1 & & \\ & & \ddots & \\ & & & \mu_N \end{bmatrix}$$

The column matrix $\mathbf{P}_x$ can be expressed as

$$\mathbf{P}_x = \mathbf{M}\dot{\mathbf{X}} \tag{3}$$

Define a set of conjugate momenta $\mathbf{P}$ associated with internal coordinates, $\mathbf{R}$. As is shown at the end of this Appendix, we have

$$\mathbf{P}_x = \tilde{\mathbf{B}}\mathbf{P} \tag{4}$$

Equations 3 and 4 give

$$\mathbf{M}\dot{\mathbf{X}} = \tilde{\mathbf{B}}\mathbf{P} \tag{5}$$

Equation 11.8 in the text gives

$$\mathbf{R} = \mathbf{B}\mathbf{X} \quad \text{and} \quad \dot{\mathbf{R}} = \mathbf{B}\dot{\mathbf{X}} \tag{6}$$

By inserting Eq. 5 into Eq. 6, we obtain

$$\dot{\mathbf{R}} = \mathbf{B}\mathbf{M}^{-1}\tilde{\mathbf{B}}\mathbf{P} \tag{7}$$

Using Eq. 4, we write Eq. 2 as

$$2T = \tilde{\mathbf{P}}\mathbf{B}\mathbf{M}^{-1}\tilde{\mathbf{B}}\mathbf{P} \tag{8}$$

If we define

$$\mathbf{G} = \mathbf{B}\mathbf{M}^{-1}\tilde{\mathbf{B}} \quad \text{(11.7 in Part I)}$$

Eq. 8 is written as

$$2T = \tilde{\mathbf{P}}\mathbf{G}\mathbf{P} \tag{9}$$

If Eq. 11.7 is combined with Eq. 7, we obtain

$$\dot{\mathbf{R}} = \mathbf{G}\mathbf{P}$$

or

$$\mathbf{G}^{-1}\dot{\mathbf{R}} = \mathbf{G}^{-1}\mathbf{G}\mathbf{P} = \mathbf{P} \tag{10}$$

Using Eq. 10, Eq. 9 can be written

$$\begin{aligned} 2T &= \tilde{\dot{\mathbf{R}}}\tilde{\mathbf{G}}^{-1}\mathbf{G}\mathbf{G}^{-1}\dot{\mathbf{R}} \\ &= \tilde{\dot{\mathbf{R}}}\mathbf{G}^{-1}\dot{\mathbf{R}} \quad \text{(11.3 in Part I)} \end{aligned}$$

Derivation of Eq. 4

The momentum p_{R_k} conjugated with the internal coordinate R_k is given by

$$p_{R_k} = \frac{\partial T}{\partial \dot{R}_k}, \qquad k = 1, 2, \ldots, s$$

If we denote the coordinates corresponding to the translational and rotational motions of the molecule by R_j^0 and its conjugate momentum by $p_{R_j}^0$

$$p_{R_j}^0 = \frac{\partial T}{\partial \dot{R}_j^0}, \qquad j = 1, 2, \ldots, 6$$

Then the momentum p_{x_1} in terms of rectangular coordinates is written as

$$\begin{aligned} p_{x_1} &= \frac{\partial T}{\partial \dot{x}_1} = \sum_k^s \frac{\partial T}{\partial \dot{R}_k}\frac{\partial R_k}{\partial x_1} + \sum_j^6 \frac{\partial T}{\partial \dot{R}_j^0}\frac{\partial R_j^0}{\partial x_1} \\ &= \sum_k^s p_{R_k} B_{k,x_1} + \sum_j^6 p_{R_j}^0 \frac{\partial R_j^0}{\partial x_1} \end{aligned}$$

The second term becomes zero since the momenta corresponding to the translational and rotational motions are zero. Thus, we have

$$p_{x_1}=\sum_k^s p_{R_k}B_{k,x_1}$$

$$p_{y_1}=\sum p_{R_k}B_{k,y_1}$$

$$\vdots \qquad \vdots$$

$$p_{z_N}=\sum p_{R_k}B_{k,z_N}$$

In a matrix form, this is written as

$$\mathbf{P}_x=\tilde{\mathbf{B}}\mathbf{P} \tag{4}$$

APPENDIX V

THE G AND F MATRIX ELEMENTS OF TYPICAL MOLECULES

In the following tables, F represents F matrix elements in the GVF field, whereas F^* denotes those in the UBF field. In the latter, $F'=-\frac{1}{10}F$ was assumed for all cases, and the *molecular tension* (Refs. I-64–66) was ignored.

(1) Bent XY_2 Molecules (C_{2v})

A_1 species—infrared and Raman active:

$$G_{11}=\mu_y+\mu_x(1+\cos\alpha)$$

$$G_{12}=-\frac{\sqrt{2}}{r}\mu_x\sin\alpha$$

$$G_{22}=\frac{2}{r^2}[\mu_y+\mu_x(1-\cos\alpha)]$$

$$F_{11}=f_r+f_{rr}$$

$$F_{12}=(\sqrt{2})rf_{r\alpha}$$

$$F_{22}=r^2f_\alpha$$

$$F_{11}^*=K+2F\sin^2\frac{\alpha}{2}$$

$$F_{12}^*=(0.9)(\sqrt{2})rF\sin\frac{\alpha}{2}\cos\frac{\alpha}{2}$$

$$F_{22}^*=r^2\left[H+F\left\{\cos^2\frac{\alpha}{2}+(0.1)\sin^2\frac{\alpha}{2}\right\}\right]$$

B_2 species—infrared and Raman active:

$$G = \mu_y + \mu_x(1 - \cos\alpha)$$

$$F = f_r - f_{rr}$$

$$F^* = K - (0.2)F\cos^2\frac{\alpha}{2}$$

(2) Pyramidal XY_3 Molecules (C_{3v})

A_1 species—infrared and Raman active:

$$G_{11} = \mu_y + \mu_x(1 + 2\cos\alpha)$$

$$G_{12} = -\frac{2}{r}\frac{(1+2\cos\alpha)(1-\cos\alpha)}{\sin\alpha}\mu_x$$

$$G_{22} = \frac{2}{r^2}\left(\frac{1+2\cos\alpha}{1+\cos\alpha}\right)[\mu_y + 2\mu_x(1-\cos\alpha)]$$

$$F_{11} = f_r + 2f_{rr}$$

$$F_{12} = r(2f_{r\alpha} + f'_{r\alpha})$$

$$F_{22} = r^2(f_\alpha + 2f_{\alpha\alpha})$$

$$F^*_{11} = K + 4F\sin^2\frac{\alpha}{2}$$

$$F^*_{12} = (1.8)rF\sin\frac{\alpha}{2}\cos\frac{\alpha}{2}$$

$$F^*_{22} = r^2\left[H + F\left(\cos^2\frac{\alpha}{2} + (0.1)\sin^2\frac{\alpha}{2}\right)\right]$$

E species—infrared and Raman active:

$$G_{11} = \mu_y + \mu_x(1 - \cos\alpha)$$

$$G_{12} = \frac{1}{r}\frac{(1-\cos\alpha)^2}{\sin\alpha}\mu_x$$

$$G_{22} = \frac{1}{r^2(1+\cos\alpha)}[(2+\cos\alpha)\mu_y + (1-\cos\alpha)^2\mu_x]$$

$$F_{11} = f_r - f_{rr}$$

$$F_{12} = r(-f_{r\alpha} + f'_{r\alpha})$$

$$F_{22} = r^2(f_\alpha - f_{\alpha\alpha})$$

$$F^*_{11} = K + \left(\sin^2\frac{\alpha}{2} - (0.3)\cos^2\frac{\alpha}{2}\right)F$$

$$F_{12}^* = -(0.9)rF \sin\frac{\alpha}{2}\cos\frac{\alpha}{2}$$

$$F_{22}^* = r^2\left[H + F\left(\cos^2\frac{\alpha}{2} + (0.1)\sin^2\frac{\alpha}{2}\right)\right]$$

Here $f_{r\alpha}$ denotes interaction between Δr and $\Delta\alpha$ having a common bond (e.g., Δr_1 and $\Delta\alpha_{12}$ or $\Delta\alpha_{31}$); $f'_{r\alpha}$ denotes interaction between Δr and $\Delta\alpha$ having no common bonds (e.g., Δr_1 and $\Delta\alpha_{23}$); see Fig. I-11*c*.

(3) Planar XY_3 Molecules (D_{3h})

A_1' species—Raman active:

$$G = \mu_y$$

$$F = f_r + 2f_{rr}$$

$$F^* = K + 3F$$

A_2'' species—infrared active:

$$G = \frac{9}{4r^2}(\mu_y + 3\mu_x)$$

$$F = F^* = r^2 f_\theta$$

E' species—infrared and Raman active:

$$G_{11} = \mu_y + \frac{3}{2}\mu_x$$

$$G_{12} = \frac{3\sqrt{3}}{2r}\mu_x$$

$$G_{22} = \frac{3}{2r^2}(2\mu_y + 3\mu_x)$$

$$F_{11} = f_r - f_{rr}$$

$$F_{12} = r(f'_{r\alpha} - f_{r\alpha})$$

$$F_{22} = r^2(f_\alpha - f_{\alpha\alpha})$$

$$F_{11}^* = K + 0.675F$$

$$F_{12}^* = -(0.9)\frac{\sqrt{3}}{4}rF$$

$$F_{22}^* = r^2(H + 0.325F)$$

The symbols $f_{r\alpha}$ and $f'_{r\alpha}$ are defined in subsection (2); f_θ denotes the force constant for the out-of-plane mode (see Fig. I-11*f*).

(4) Tetrahedral XY_4 Molecules (T_d)

A_1 species—Raman active:

$$G = \mu_y$$

$$F = f_r + 3f_{rr}$$

$$F^* = K + 4F$$

E species—Raman active:

$$G = \frac{3\mu_y}{r^2}$$

$$F = r^2(f_\alpha - 2f_{\alpha\alpha} + f'_{\alpha\alpha})$$

$$F^* = r^2(H + 0.37F)$$

F_2 species—infrared and Raman active:

$$G_{11} = \mu_y + \tfrac{4}{3}\mu_x$$

$$G_{12} = -\frac{8}{3r}\mu_x$$

$$G_{22} = \frac{1}{r^2}\left(\frac{16}{3}\mu_x + 2\mu_y\right)$$

$$F_{11} = f_r - f_{rr}$$

$$F_{12} = (\sqrt{2})r(f_{r\alpha} - f'_{r\alpha})$$

$$F_{22} = r^2(f_\alpha - f'_{\alpha\alpha})$$

$$F^*_{11} = K + \tfrac{6}{5}F$$

$$F^*_{12} = \tfrac{3}{5}rF$$

$$F^*_{22} = r^2(H + \tfrac{1}{2}F)$$

where $f_{\alpha\alpha}$ denotes interaction between two $\Delta\alpha$ having a common bond; $f'_{\alpha\alpha}$ denotes interaction between two $\Delta\alpha$ having no common bond.

(5) Square-Planar XY_4 Molecules (D_{4h})

A_{1g} species—Raman active:

$$G = \mu_y$$

$$F = f_r + 2f_{rr} + f'_{rr}$$

$$F^* = K + 2F$$

B_{1g} species—Raman active:

$$G = \mu_y$$

$$F = f_r - 2f_{rr} + f'_{rr}$$

$$F^* = K - 0.2F$$

B_{2g} species—Raman active:

$$G = \frac{4\mu_y}{r^2}$$

$$F = r^2(f_\alpha - 2f_{\alpha\alpha} + f'_{\alpha\alpha})$$

$$F^* = r^2(H + 0.55F)$$

E_u species—infrared active:

$$G_{11} = 2\mu_x + \mu_y$$

$$G_{12} = -\frac{2\sqrt{2}}{r}\mu_x$$

$$G_{22} = \frac{2}{r^2}(\mu_y + 2\mu_x)$$

$$F_{11} = f_r - f'_{rr}$$

$$F_{12} = (\sqrt{2})r(f_{r\alpha} - f'_{r\alpha})$$

$$F_{22} = r^2(f_\alpha - f'_{\alpha\alpha})$$

$$F^*_{11} = K + 0.9F$$

$$F^*_{12} = -(\sqrt{2})r(0.45)F$$

$$F^*_{22} = r^2(H + 0.55F)$$

The symbol f_{rr} denotes interaction between two Δr perpendicular to each other; f'_{rr} denotes interaction between two Δr on the same straight line. In addition, a square-planar XY_4 molecule has two out-of-plane vibrations in the A_{2u} and B_{2u} species.

(6) Octahedral XY_6 Molecules (O_h)

A_{1g} species—Raman active:

$$G = \mu_y$$

$$F = f_r + 4f_{rr} + f'_{rr}$$

$$F^* = K + 4F$$

E_g species—Raman active:

$$G = \mu_y$$

$$F = f_r - 2f_{rr} + f'_{rr}$$

$$F^* = K + 0.7F$$

F_{1u} species—infrared active:

$$G_{11} = \mu_y + 2\mu_x$$

$$G_{12} = -\frac{4}{r}\mu_x$$

$$G_{22} = \frac{2}{r^2}(\mu_y + 4\mu_x)$$

$$F_{11} = f_r - f'_{rr}$$

$$F_{12} = 2rf_{r\alpha}$$

$$F_{22} = r^2(f_\alpha + 2f_{\alpha\alpha})$$

$$F^*_{11} = K + 1.8F$$

$$F^*_{12} = 0.9rF$$

$$F^*_{22} = r^2(H + 0.55F)$$

F_{2g} species—Raman active:

$$G = \frac{4\mu_y}{r^2}$$

$$F = r^2(f_\alpha - 2f'_{\alpha\alpha})$$

$$F^* = r^2(H + 0.55F)$$

F_{2u} species—inactive:

$$G = \frac{2\mu_y}{r^2}$$

$$F = r^2(f_\alpha - 2f_{\alpha\alpha})$$

$$F^* = r^2(H + 0.55F)$$

The symbol f_{rr} denotes interaction between two Δr perpendicular to each other, whereas f'_{rr} denotes those between two Δr on the same straight line; $f_{\alpha\alpha}$ denotes interaction between two $\Delta\alpha$ perpendicular to each other, whereas $f'_{\alpha\alpha}$ denotes those between two $\Delta\alpha$ on the same plane. Only the interaction between two $\Delta\alpha$ having a common bond was considered.

APPENDIX VI

GROUP FREQUENCY CHARTS

The data cited in this book were used in the preparation of the following group frequency charts. Each section of Part III gives a number of group frequencies that are not included here. For the physical meaning of group frequency, see Sec. I-17.

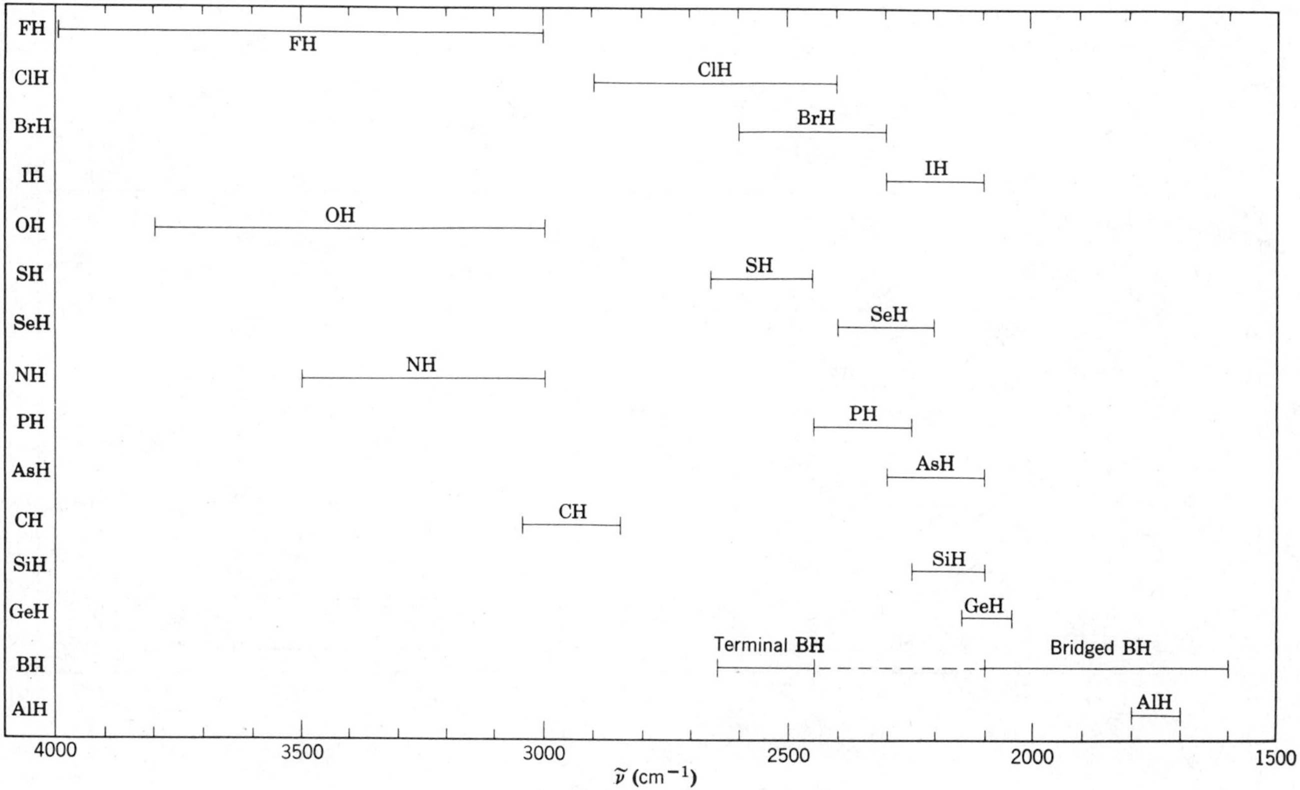

(a) Hydrogen stretching frequencies

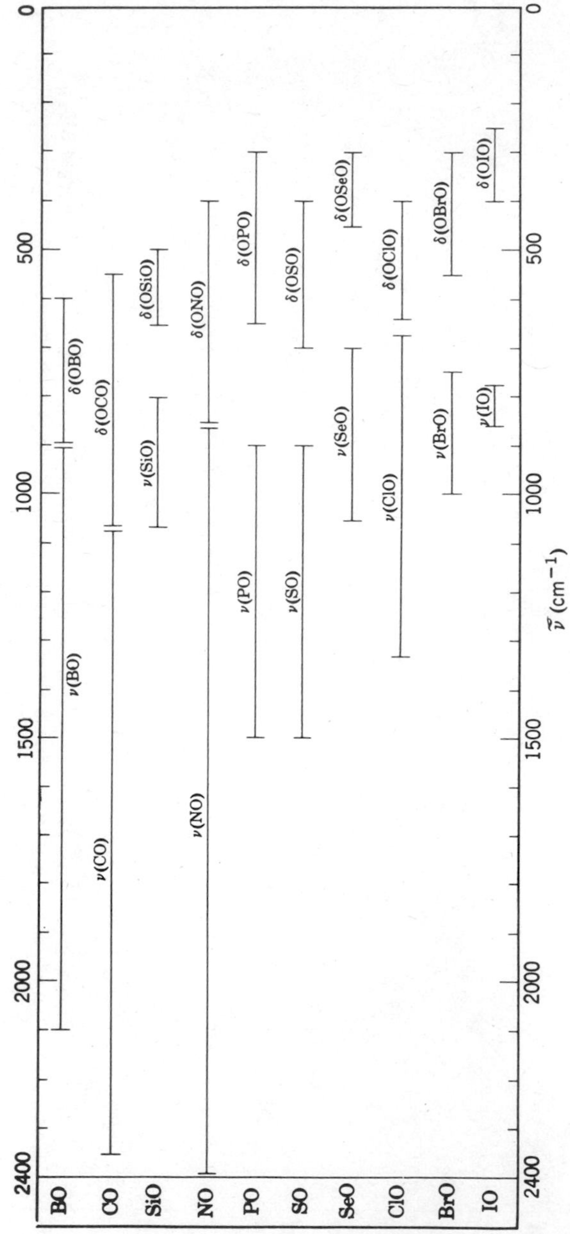

(*b*) Oxygen stretching and bending frequencies

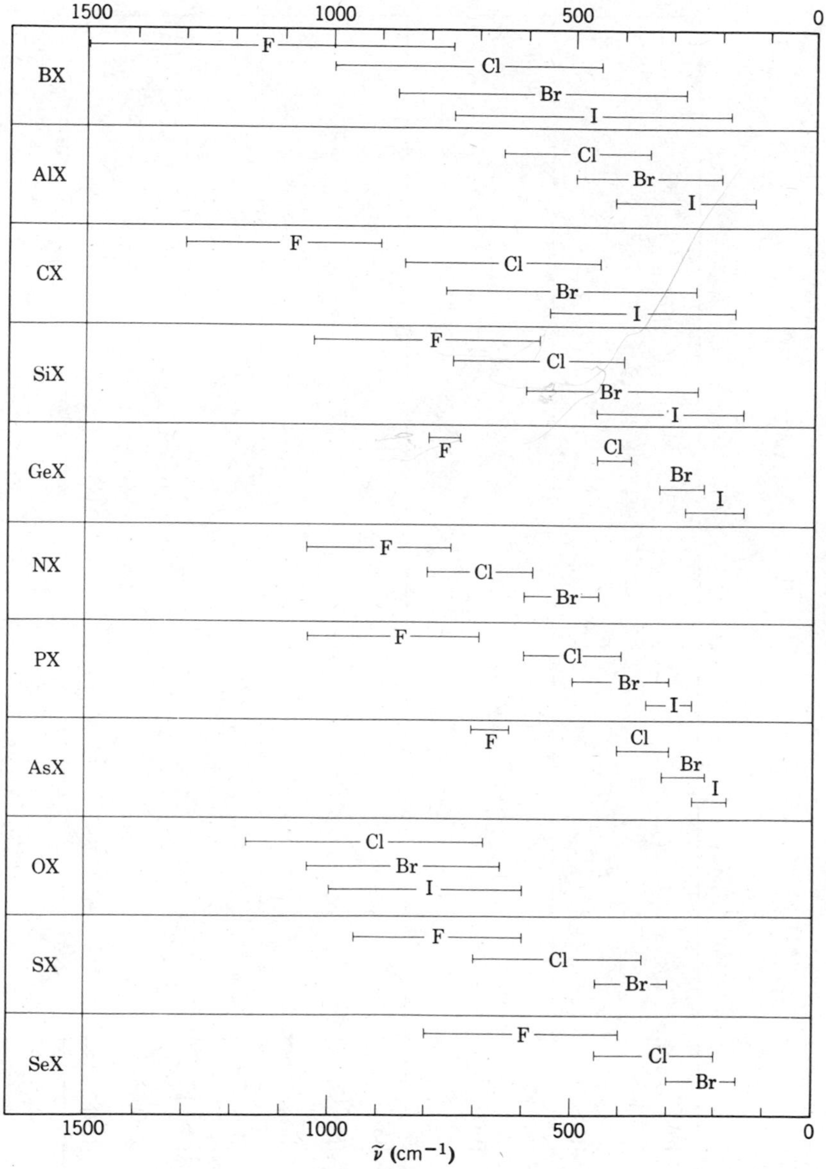

(c) Halogen (X) stretching frequencies

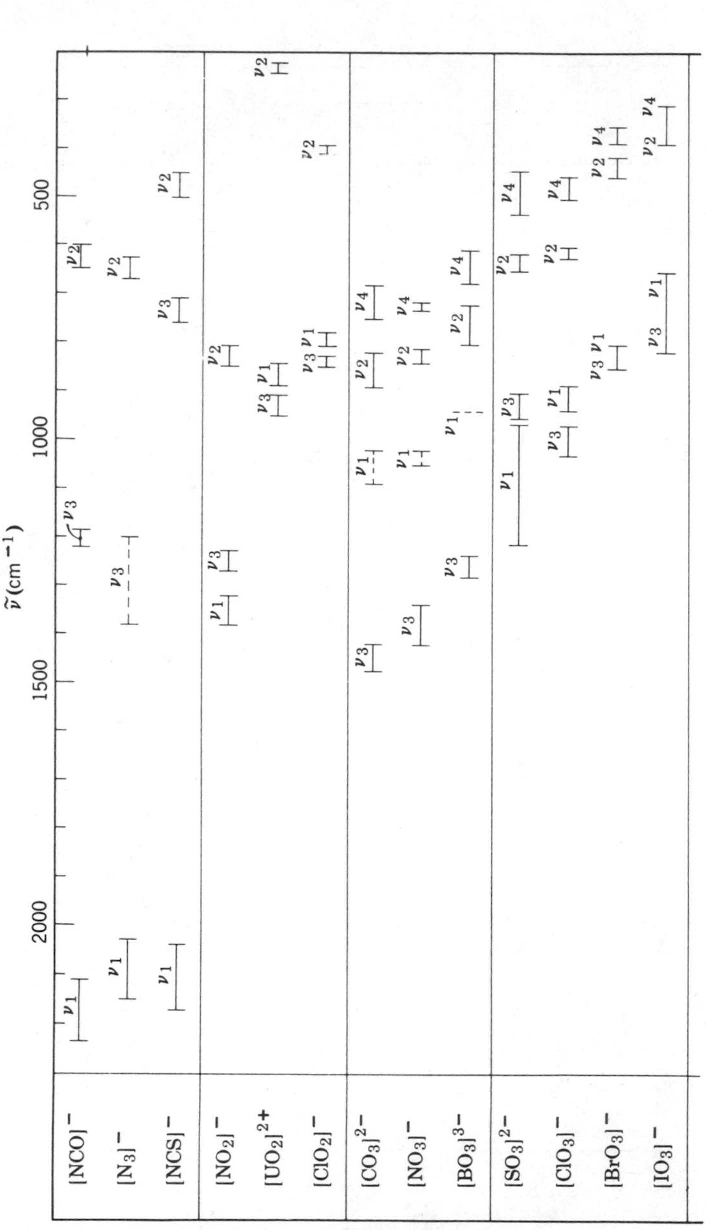
ν̃ (cm⁻¹)
500
1000
1500
2000
[NCO]⁻
[N₃]⁻
[NCS]⁻
[NO₂]⁻
[UO₂]²⁺
[ClO₂]⁻
[CO₃]²⁻
[NO₃]⁻
[BO₃]³⁻
[SO₃]²⁻
[ClO₃]⁻
[BrO₃]⁻
[IO₃]⁻
ν₁
ν₂
ν₃
ν₄

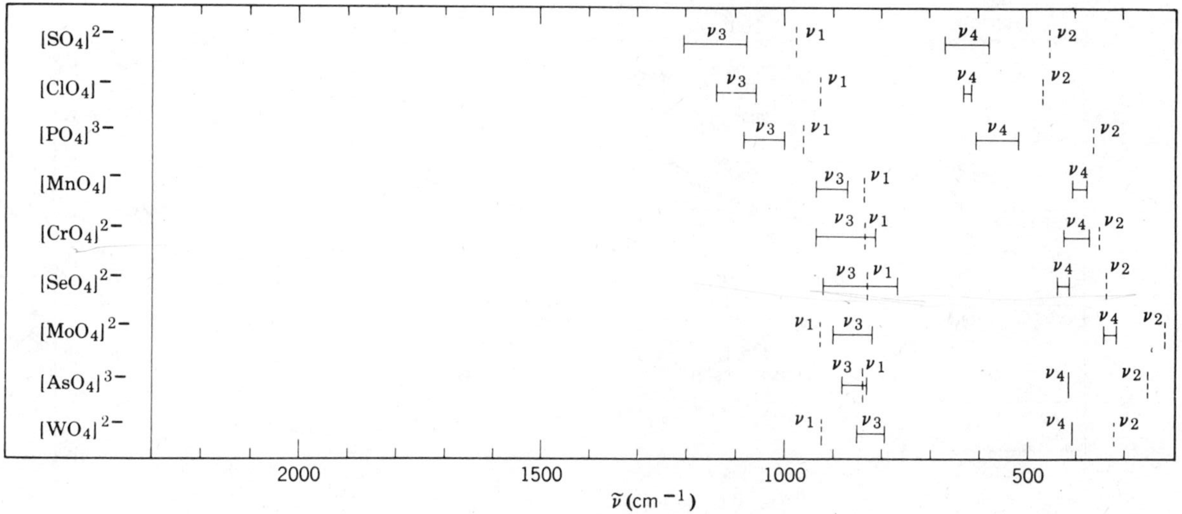

(*d*) Characteristic frequencies of some inorganic ions
(Broken lines indicate Raman active vibrations)

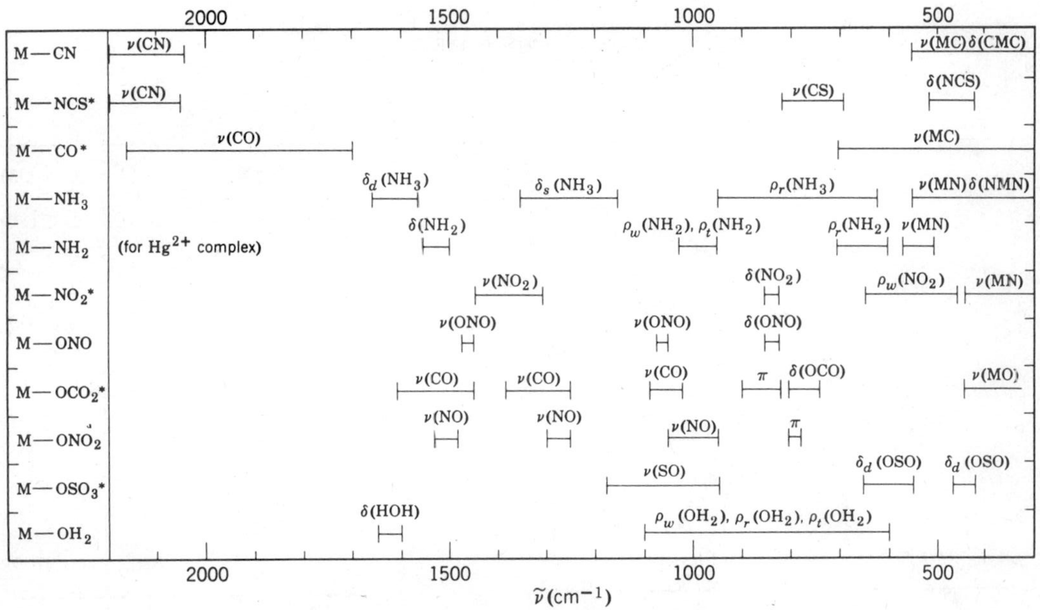

(*e*) Characteristic frequencies of metal complexes having simple ligands

(Frequency ranges include bidentate and bridged complexes for the ligands marked by an asterisk)

Index

Since the number of compounds included in this book is numerous, entries are given only for representative compounds. The majority of other compounds may be reached through general entries such as diatomic molecules, XY_4 molecules and ammine complexes. Boldface page numbers refer to figures and tables.